HEAT TRANSFER AND FLUID FLOW IN ROTATING MACHINERY

13656443

HEAT TRANSFER AND FLUID FLOW IN ROTATING MACHINERY

Edited by

Wen-Jei Yang

University of Michigan

HEMISPHERE PUBLISHING CORPORATION

A subsidiary of Harper & Row, Publishers, Inc.

Washington New York London

DISTRIBUTION OUTSIDE NORTH AMERICA

SPRINGER-VERLAG

Berlin Heidelberg Paris New York Tokyo London

HEAT TRANSFER AND FLUID FLOW IN ROTATING MACHINERY

Copyright © 1987 by Hemisphere Publishing Corporation. All rights reserved. Printed in the United States of America. Except as permitted under the United States Copyright Act of 1976, no part of this publication may be reproduced or distributed in any form or by any means, or stored in a data base or retrieval system, without the prior written permission of the publisher.

1 2 3 4 5 6 7 8 9 0 BRBR 8 9 8 7 6

Library of Congress Cataloging-in-Publication Data

Heat transfer and fluid flow in rotating machinery.

"First International Symposium on Transport Phenomena was held April 28–May 3, 1985 in Honolulu, Hawaii"—CIP preface.
Bibliography: p.
Includes index.
1. Heat—Transmission—Congresses. 2. Fluid dynamics—Congresses. 3. Turbomachines—Congresses, I. Yang, Wen-Jei, date. II. International Symposium on Transport Phenomena (1st : 1985 : Honolulu, Hawaii)
TJ260.H389 1986 621.406 86-9775
ISBN 0-89116-572-X Hemisphere Publishing Corporation

DISTRIBUTION OUTSIDE NORTH AMERICA:
ISBN 3-540-17088-X Springer-Verlag Berlin

D
621·406
HEA

Contents

GENERAL TOPICS

TURBINES AND COMPRESSORS

Preface

The First International Symposium on Transport Phenomena was held April 28–May 3, 1985, in Honolulu, Hawaii, and emphasized rotating machinery. In view of the urgent need to comprehend the complicated phenomena of heat, mass, and momentum transfer in gas turbines, compressors, hydraulic turbines and pumps, wind mills, and other rotating machinery, researchers in the Pacific Basin gathered together to present their research accomplishments and to exchange their ideas for future goals on this subject. A total of 44 papers were presented in a single session, which provided the participants with ample opportunity to expose themselves to all subjects of interest and to have direct contact with the authors. An open forum on film cooling encountered a high level of active participation and an enthusiastic exchange of participants' viewpoints over a broad range of aspects concerning this subject. Summarizing the overall picture, the symposium experienced unusually active response from the participants. Another unique feature of this symposium was the friendships established among the participants. As a result, the organizer has been urged to coordinate both the Symposium series on transport phenomena for other topics and the Advances series on other disciplines in rotating machinery.

The second symposium in this series, to be held in Tokyo, will deal with turbulent flows; the third, to be held in Taiwan, will cover environmental control equipment cooling. Since rotating machinery, such as steam and gas turbines, is essential in vehicular transportation and electric power plants, research efforts in this area will continue to be very intense. Therefore, the series on "Advances in Rotating Machinery" is needed to disseminate research results and to promote and broaden research interest. The Advances series also got its start at the meeting on "Transport Phenomena." The second meeting on "Transport Phenomena, Dynamics, and Control" will be held in Honolulu (spring 1988).

This book consists of six Parts, with 43 Chapters. Blade cooling, in Part 1, is one subject that interests thermal science researchers and includes both external surface cooling and internal cooling. The state-of-the-art review on turbine blade cooling is thorough and impressive. Major efforts are also directed toward various aspects of film cooling and heat transfer enhancement in internal convection. Part 2 covers the experimental studies of transport phenomena by flow visualization techniques and flow measurements. Flow visualization results are presented for secondary flow in curved, heated, and rotating channels, as well as for internal flow in rotating machinery. Measurement methods unique to flow in rotating devices are also introduced. Rotating tubes, channels, and heat pipes are the subjects of Part 3, which treats phase changes, turbulence, and hysteresis in flow transition. Flow and heat transfer from rotating surfaces and enclosures are grouped together in Part 4. General topics in Part 5 concern items of current interest, such as electronic generators, impinging jets, and numerical methods. Part 6 discusses the ultimate rotating devices of industrial applications, including steam and gas turbines, compressors, pumps, and wind turbines. The subjects are widespread, including lifting lines, water film behavior, thermal stress, the Wilson point, wave mechanics, stall, and work performance.

The symposium owes its success to the cooperation of the Organizing Committee, whose members actively engaged in soliciting the excellent articles. The editorial assistance of Mr. Paul P. T. Yang, a student at The University of Michigan, should be acknowledged with gratitude. This book is dedicated to all the authors whose efforts and talents have made its publication possible.

Wen-Jei Yang

BLADE COOLING

Turbine Blade Cooling

R. J. MOFFAT
Thermosciences Division
Department of Mechanical Engineering
Stanford University, Stanford, California 94305

Modern gas turbines operate with gas temperatures far above their allowable metal temperatures, and many of the components require thermal protection. In particular, the blades and vanes in the turbine stages must be cooled several hundred degrees below gas temperatures. To further complicate the problem, metal temperatures must be kept reasonably uniform to avoid thermal stresses, and the cooling must be done with the smallest possible penalty to the engine cycle. Extraction of potential working fluid for cooling exacts an immediate penalty, since the coolant has already been compressed, but re-insertion of the spent coolant into the mainstream carries with it further penalties--the irreversibilities associated with mixing the cooler, low-momentum fluid into the hotter, high-velocity mainstream.

The metal temperature is determined by a balance between the rates of heat transfer to the blade from the surrounding gas and from the blade to its coolant. The designer has a three-part problem: predicting the heat transfer from the gas stream to the outside surface of the blade, predicting the heat transfer from the coolant to the inside surface, and calculating the resultant temperature distribution in the blade material. The "bottom line" is the accurate prediction of metal temperature.

Figures 1 and 2 are from recent publications discussing the question: how well are we doing? Figure 1 compares the measured and predicted metal temperatures for one particular geometry tested in a hot cascade at NASA-Lewis Laboratories [1]. The test vanes were J-75 size, impingement cooled on the front 2/3 and pin-fin-convection cooled on the aft 1/3 with one row of injection holes (30° slant) on the pressure surface. Three different transition criteria were tried for evaluating the exterior heat load. All three worked well on the suction surface at high Reynolds number, but at low Reynolds number, none did. In fact, the best calculation for low Reynolds number was one in which the computations were forced to stay laminar over the entire suction surface.

Figure 2 compares measured and predicted heat-transfer coefficients from a nozzle-guide vane of the Garrett TFE 731-2 engine [2]. The data are from test runs in a shock-tube tunnel, using full-scale hardware. Temperatures and pressures were appropriate for engine conditions, but the vanes remained essentially at room temperature throughout the test, since the run time was typically only 20 milliseconds. Turbulence intensity was estimated at 2.6 to 6.4%. Measured values are compared with predictions from STAN5 and from the AFAPL Turbine Design System [3] using the turbulent option. STAN5 was run with a simple transition criterion (momentum-thickness Reynolds number of 200) and significantly missed the data. The TDS somewhat overpredicted near the leading edge, but otherwise did rather well. Winstanley, Booth, and Dunn [4], however, reported generally good results using STAN5 in predicting heat transfer on a full-scale vane test, at least better than flat-plate correlations or integral methods.

It seems clear from the above comparisons that there are some serious problems associated with gas-turbine blade and vane heat transfer. In a recent review paper, Rudey and Graham said, "...there is substantial evidence to indicate that ... uncertainty (in estimating metal temperatures) ... has cost much in terms of money and development time for newer engines ..." [5]. Stepka [6] estimated the uncertainties in calculated heat-transfer coefficient to be ± 35%, which, for typical engine conditions, would result in metal temperature uncertainties of ± 98°C.

No single effect is responsible for all these difficulties; hence it is unlikely that any simple correlation will ever cover a wide range of cases with good accuracy. The most common approach, today, is to establish a framework computer code and augment it by putting in new models as new effects are understood. Each time new evidence of trouble arises, "the next most likely culprit" is identified, and another parameter added to the code. Finite difference codes, which became popular in about 1968, have been developed now to the point of accounting for pressure gradients, roughness, curvature, variable fluid properties, compressibility, rotation, turbulence, secondary flows, shock/boundary layer interactions, and other factors. The end is not yet in sight, nor will it be until accurate predictions can be made at all proposed engine conditions.

The present paper treats three aspects of the turbine-blade-cooling problem: external heat transfer, with and without injection cooling, and internal heat transfer. In each section, there is a brief review of the principal works, followed by a discussion of some of the most recent issues.

External Heat Transfer: Impermeable Surfaces

In this section are collected comments on leading-edge heat transfer and boundary-layer flows on convex and concave surfaces with and without pressure gradients, without transpiration or discrete-hole injection.

Traci and Wilcox [7] analyzed turbulent flow approaching a laminar stagnation region and concluded that both intensity and scale were significant, and that reported experimental values more than 50% above those for quiescent flow were explainable. Flow around a cylinder is more complex than in a stagnation region, since it includes the laminar boundary layer and the separated region, but cylinder flows give some insight into the stagnation-line problem. Lowery and Vachon [8] reported augmentation by as much as 60% for turbulence intensities approaching 15%. In their tests, the ratio of the integral length scale of the turbulence to the cylinder diameter was 0.015-0.095. Dils and Follanbee [9] investigated the heat transfer to the stagnation line of a cylinder in a high-turbulence stream. Their data, Figure 3, show large effects of turbulence--nearly a 45% increase for turbulence intensities (RMS u'/U) of about 16%. The larger scales of turbulence in these experiments were about equal to the cylinder diameter (0.76 cm compred to 0.636). Consigny and Richards, whose data on transition are discussed below, found a 20-25% increase in leading-edge heat transfer for 5.2% turbulence in the approach flow at Reynolds numbers between 20,000 and 80,000.

Turbulence has long been known to advance the transition upstream, i.e., to smaller Reynolds numbers. Figure 4, from Blair [10], summarizes the results of nine experimental studies and one prediction describing the effect of free-stream turbulence on the transition Reynolds number on a flat plate. Blair comments on the difficulty of identifying "the" onset of transition, since it may not occur simultaneously across the entire span of a test plate. Transition starts from spots and grows into streaks, which then merge to form a turbulent boundary layer. Both the onset and the streamwise extent of transition are affected by turbulence intensity. In another work, Blair [11] showed that high turbulence can produce heat-transfer behavior which appears turbulent, even at very low Reynolds numbers. His data are shown in Figure 5. Others have noted that free-stream turbulence causes a significant increase in heat transfer for laminar layers.

Looking at cascade data, hot-rig data, and engine data, one might be tempted to say that there is no clear-cut differentiation between "laminar" and "turbulent" flow when high free-stream turbulence is involved. Studies on blades and vanes frequently show heat-transfer coefficients which lie in between "laminar" and "turbulent" values over the entire surface.

Figure 6, from Reference 10, shows transition on a flat surface subject to acceleration and moderately high turbulence. The transition behavior is considerably modified by both effects. Acceleration depresses boundary-layer turbulence on a flat plate and reduces the local Stanton number [12,13,14,15], an effect attributed to stabilization of the sublayer, and suppression of turbulence production. Free-stream turbulence has an opposite effect; hence the effects of free-stream turbulence in an accelerating velocity field are partially compensating.

Aside from the established effect of turbulence in relocating the transition zone, there is the question of whether or not the heat-transfer process in an already turbulent layer is affected by free-stream turbulence. In the years preceding Blair's work (1983), approximately a dozen studies of this effect were reported; their opinions were about equally divided. To answer this question requires one to examine the behavior in purely "local" terms, since changes in thickness would result in changes in heat transfer, but would not represent anything new to think about. Kearney, Kays, and Moffat [16] concluded that 4% turbulence did not affect the local turbulent heat transfer, with or without free-stream acceleration. Their data are shown in Figure 7, Stanton number vs. enthalpy-thickness Reynolds number. Blair [11], on the other hand, shows clear evidence of a significant effect within the turbulent region. His data, in x-Reynolds coordinates, are shown in Figure 6. These studies clearly differ in their opinion, but both experiments agreed with the same baseline data (i.e., the case of no turbulence). It is possible, but unlikely, that the difference in opinion is due to an error in one or the other experiment. A more likely explanation is that the structure of the turbulence was different in the two cases. Little is known about what aspect of turbulence affects heat transfer. Most studies report only "intensity", yet surely the scale and structural content are also important. There have been many suggestions of this sort, but no definitive studies. This problem is still worth further study.

Convex curvature on an impermeable wall depresses boundary-layer turbulence and heat transfer, and suppresses transition [17,18,19]. Figure 8 shows Stanton number vs. enthalpy-thickness Reynolds number for curvature alone and for curvature with acceleration. These coordinates were chosen to make clear the local effects of curvature and acceleration, as distinct from the effects of the upstream history of the boundary layer. Acceleration and curvature are seen to be reinforcing in that the slope of the line for the combined case is -2.0, while the slopes of the lines for each effect acting alone would have been -1.0. Figure 9 shows transitions on a convex surface with "strong" curvature (i.e., δ/R = 0.10) but no turbulence and no streamwise pressure gradient.

Consigny and Richards [20] tested the combined effects of convex curvature, acceleration, and turbulence by measuring the heat-transfer coefficient on the suction surface of a blade in a 2-D cascade. Their data are shown in Figure 10 for chordal Reynolds number of 7.23×10^5, one of four Reynolds numbers tested. Figure 11 shows the same data, along with predictions using the k-ε two-equation turbulence model and values of Jones and Launder [21]. The transition region was not well predicted on the suction side with or without turbulence. Figure 12, from the work of Graziani et al. [22] compares measured and predicted Stanton numbers for chordal Reynolds number of 5.5×10^5 with turbulence intensity of 1%. Predictions were with STAN5, using the curvature modeling of Johnston and Eide [19], with transition initiated when the momentum-thickness Reynolds number reached 200. Bayley and Priddy [23] showed a more than fourfold increase in heat-transfer coefficient on the pressure side as a consequence of 32% turbulence generated by a rotating "cage" of rods but little or no effect on the suction side. This selective response might reflect amplification, by the concave curvature of the blade of streamwise vortices coming off the rods.

The concave surface has attracted much less attention than the convex, for reasons partly related to the results in the section above. As blades and vanes of higher curvature have been developed, however, the concave side has begun to be more earnestly studied. At present, it seems clear that concave curvature increases heat transfer from a turbulent boundary layer and destabilizes a laminar layer, possibly forming streamwise vortices. Martin and Brown [24] coupled the theory for laminar boundary layers on concave surfaces with the effects of free-stream turbulence and concluded that Taylor-Görtler vortices might arise under engine conditions in the laminar boundary layers on the pressure surfaces. That would lead to early transition and higher than usual heat-transfer coefficients, as well as spanwise nonuniformities in the boundary layer. Some of those features have been seen in hydrodynamic experiments but not yet in heat transfer, except for the increase in level. The Görtler instability could cause an increase in the general level of turbulence without generating any coherent streamwise structures within the boundary layer. The increases in heat transfer which have been observed to date can be accounted for just as convincingly by an increase in turbulence level as by the direct effect of streamwise vortices.

There is hydrodynamic evidence that streamwise vortices can be produced on blades. Han and Cox [25] used kerosene vapor to visualize the flow on the concave surface of a large scale model of a blade and reported streamwise vortices at Reynolds numbers of about 1×10^6. The vortices were not steady, but appeared and disappeared, in spurts, in the regions of sharpest curvature. Water-channel studies by Jeans and Johnston [26] found augmented turbulence, with structure clearly different from flat-wall turbulence, but no clear evidence of streamwise vortices. Simonich and Moffat [27] used a liquid-crystal, heat-transfer visualization scheme in the same water channel to look for the "footprints" of streamwise vortices in the heat-transfer distribution--none was found. The heat-transfer distribution on the concave surface displayed the same spanwise streaky structure found in water-channel studies of flat-plate heat transfer, even though the level of heat transfer was about 35% above the usual flat-plate value.

External Heat Transfer: Permeable Wall

The operating temperature of a surface can be reduced by injecting cooling fluid through it. Three mechanisms are involved: (i) the coolant takes energy from the surface, thereby cooling it directly; (ii) the injected fluid reduces the average temperature in the boundary layer over the surface; and (iii) the injected fluid may alter the boundary-layer structure and change the heat-transfer coefficient. If the fluid is injected through fine pores in the surface, the process is called transpiration. In this case, the injected fluid can be assumed to be at surface temperature when it leaves the surface. Transpiration always reduces the heat-transfer coefficient. If the fluid is injected through a sparse set of holes or slots, the process is called film cooling or slot cooling. With film cooling, the fluid may or may not be at surface temperature. Injection through holes or slots does not always reduce the heat-transfer coefficient but may, in fact, increase it by increasing the turbulence in the boundary layer.

The increase in heat-transfer coefficient may more than counterbalance the thermal effect of the injection and cause a net increase in heat load.

There have been many experiments on heat transfer with transpiration, beginning in the early 1960s and covering many situations. The earliest studies were on smooth, flat plates with uniform transpiration and uniform free-stream velocity [28,29,30]. Next came studies of accelerating flows [31,32] and decelerating flows [33]. Rough surfaces with transpiration were covered by Healzer [34] and Pimenta [35] and rough walls with transpiration and acceleration by Coleman [36].

Figure 13 shows a summary of data from Moffat and Kays [30] showing Stanton number versus x-Reynolds number for uniform transpiration on a smooth, flat plate under a uniform, low-turbulence stream. Positive transpiration (i.e., blowing) thickens the boundary layer; this, by itself, would decrease the heat-transfer coefficient even if the boundary layer were otherwise unaffected. In order to determine whether or not the local behavior of the boundary layer has been affected, one must compare the transpired and untranspired behavior at the same boundary-layer thickness (using the enthalpy-thickness Reynolds number, for example), rather than at the same location on the surface (the same x-Reynolds number). Figure 14 shows the data from Figure 13 in enthalpy-thickness Reynolds-number coordinates. The heat-transfer coefficient has been reduced by transpiration, even when compared on the basis of constant boundary-layer thickness. These same data are shown in Figure 15 in relative terms: Stanton number plotted against the blowing parameter v_o/U_∞. Stanton number can be taken as close to zero as desired, by increasing the blowing. On the other hand, suction increases the Stanton number. For low values of suction, the boundary layer can remain turbulent (note x-dependence in Figure 13, for low suction); but, as suction is increased, the boundary layer will relaminarize and Stanton number will approach -F.

Various early works suggested that the Stanton number would go essentially to zero when $b = F/St(o)$ reached a value between 4.0 and 5.5. St(o) is the starting number without blowing, at the same x-location. More recent works by Mamonov [37] and by Mironov et al. [38] and work referenced by them have shown that the "blow-off" limit is affected by free-stream turbulence and suggest that the critical value of b is about 9.0 when the free-stream turbulence is 11%.

Georgieu and Louis [39] have suggested that the critical value may be as high as 16.5, even without significant turbulence. Their study also showed that the cooling effectiveness (St/St(o)) of transpiration is not significantly affected by pressure gradients (favorable or adverse) and that the effects are strongly local--upstream transpiration history has little effect on local behavior. This latter conclusion concerning local equilibrium of the heat-transfer boundary layer supports findings reported earlier by Whitten et al. [40].

Although transpiration cooling has not been widely used, it is still attractive because of the good thermal protection it provides at low coolant consumption. Studies concerning applications in coal-burning gas-turbines have recently been completed. Wolf and Moskowitz [41] report 900 hours of successful operation of transpiration-cooled blades and vanes in hot rigs and engines at temperatures up to 1650°C, using fuel doped with fly-ash particles and aluminum-oxide particles. There was no evidence of fouling beyond normal oxidation.

When the coolant is injected through discrete holes, a new degree of freedom enters the problem--the coolant need not be at surface temperature when it enters the boundary layer. The thermal effects of injection then depend on three temperatures, not two, and one new additional piece of information is needed. There are two options: an "adiabatic wall effectiveness" approach and a "superposition" approach. In experiments following the effectiveness approach, the adiabatic surface temperature is reported as a function of position and injection rate, and heat transfer is calculated using the difference between actual surface temperature and adiabatic surface temperature. The heat-transfer coefficient must be available from a second source. In the superposition approach, heat transfer is calculated using the actual surface temperature and the actual gas temperature, and the effects of injection are all accounted for by the value of the heat-transfer coefficient. Experiments following this approach record values of the heat-transfer coefficient for two reference values of injection temperature, usually $T_{inj} = t_{gas}$ (i.e., $\theta = 0.0$) and $T_{inj} = T_{surface}$ (i.e., $\theta = 1.0$), as functions of position and injection. These two approaches are equally well founded in theory and are conceptually equivalent. One can move freely between them, converting data from one system to the other.

The film-cooling literature is extensive for both approaches, and references 42-52 provide a reasonable sampling of the early reported works. Many of these studies looked at only the adiabatic effectiveness and the factors which affected it, relying on the assumption that the heat-transfer coefficient could be taken from unblown data. Typical results from an effectiveness experiment are given in Figure 16, showing adiabatic effectiveness downstream of single holes, on a pressure surface and on a suction surface. In addition to illustrating typical results, this figure shows the marked differences between the pressure and suction surface behavior on a vane. The differences in level of effectiveness and its decay downstream are believed to be due to curvature effects on the turbulence intensity in the boundary layers--high turbulence dissipating the injected plume rapidly on the concave side.

The effectiveness approach is well suited to studies in the "far field", i.e., far downstream of the injection site, and the method has been widely used. With the prospect of full-coverage film cooling, however, with higher-density hole patterns covering more and more of the surface, it became necessary to document both the effectiveness and the heat-transfer coefficient to have a complete description of the situation, or to use the superposition approach. This was first suggested by Metzger et al. [53] and later adopted by the Stanford group for all of their work on discrete-hole injection.

The heat-transfer coefficient for a given application depends on the injection temperature as well as on the injection rate, and the appropriate value is found by a weighted sum of the two reference values.

$$St(\theta) = St(0) - \theta\,[St(0) - St(1)]$$

If desired, the adiabatic effectiveness can also be found from the same data:

$$Eff(ad) = [St(0) - St(1)]/St(0)$$

To be consistent, the heat-transfer coefficient $h(0)$ from the superposition approach should be used with the adiabatic wall-temperature formulation.

Kasagi et al. [54] and Kumada et al. [55] looked at both the cooling effectiveness and the heat-tranfer properties on a film-cooled surface and also the heat-transfer properties of the impingement-cooled back face of that surface. The combination of impingement with film cooling employing the spent coolant has been used on several designs, and will be discussed later.

Early work on film cooling showed the advantages of slanting the holes (frequently about 30° to the surface), and compound-angle injection (slanted and skewed) provides even better protection. Figure 17 shows typical results for full-coverage film cooling on a flat surface with 30° slanted injection, from some of the early Stanford work [56]. Free-stream turbulence was low, and there was no axial pressure gradient. The two reference data sets are shown, one for $\theta = 1$ and the other for $\theta = 0$. Each set includes seven different M values from 0.0 to $M = 1.25$. The end of injection is marked by a vertical arrow in the $\theta = 1.0$ case. Several important features of full-coverage film cooling are shown on this figure. Injection at $\theta = 1.0$ lowers the Stanton number within the blown field for all but the highest injection rates and far downstream for all injection rates. Injection at $\theta = 0.0$ increases Stanton everywhere, except that for low values of M, there is a small decrease in the recovery region. Injection at too high a rate (above $M = 1.00$) can increase the heat-transfer coefficient above the unblown value, probably by increasing the turbulence level in the boundary layer, even for $\theta = 1.0$.

Recent work on film cooling has examined the interaction of curvature with other effects (convex curvature [57], roughness [58], secondary flow [59], and pressure gradients [60]) and the structure of turbulence in the full-coverage, film-cooled boundary layer on a convex wall [61].

Reference data from Furuhama and Moffat [57] are shown in Figure 18 as Stanton number vs. streamwise distance for full-coverage film cooling with strong convex curvature ($\delta/R = 0.1$). Both cases are shown ($\theta = 1.0$ and $\theta = 0.0$) at $M = 0.4$. This figure shows clearly the downstream benefits of injection: blowing at $M = 0.4$ in the first 70 cm protected more than 200 centimeters of surface from the start of the convex curvature. Figure 19, from the same source, shows the effects of two, four, and six rows of injection within a strong, convex curve, in enthalpy-thickness Reynolds-number coordinates. With no injection, Stanton number inside the curved region would have dropped along a line of slope -1, as shown earlier. With injection up to $M = 0.4$, the -1 slope persists. This is taken as evidence that the structure of the curved boundary layer with blowing is dominated by the curvature effects, for blowing less than $M = 0.4$. The data stay on the line of -1 slope until the end of injection, and then rise abruptly. The growth of the enthalpy thickness is dominated by the effects of injection, since the Stanton numbers are so low; hence these coordinates are not very useful for describing the region after the end of injection.

Figure 20 shows adiabatic effectiveness in the convex and recovery regions for various numbers of rows injected, compared with a dashed line representing full-coverage blowing on a flat surface (other conditions being the same).

Goldstein et al. [58] showed that surface roughness affects the spanwise-averaged effectiveness for single and double rows of holes. The effect is different at low flows and high. At $M = 0.5$, the effectiveness was reduced by 10-20%, with the greatest reduction for the highest roughness. For $M > 0.7$, there was a significant increase for single rows and a small increase for double rows. The roughness elements were stub cylinders whose diameter was half that of the injection hole and whose heights were either 1/4 or 1/2 D.

In another recent work, Goldstein and Chen [59] showed that secondary flows can strip the coolant away from the suction surface of a vane over a significntly large region near the hub. In a linear cascade, tests of effectiveness on the pressure and suction sides showed great differences near the endwall. The coolant was apparently removed entirely from the suction surface in a large, triangular region extending from about the point of maximum curvature to the trailing edge and reaching out 4.3 cm on a chord of 16.9 cm. The extent of the passage vortex and the horseshoe vortex will vary with design and load, but secondary flow interference appears to be a significant problem.

Hay, Lampard, and Saluja [60] report that the cooling effectiveness is not affected by the condition of the approach boundary layer (i.e., turbulent or transitional) or mild adverse pressure gradients, but is reduced by a favorable pressure gradient, at least at low blowing.

Most of the literature surveyed defines the cooling effectiveness in terms of the temperature of the coolant as it leaves the plane of the injection holes. There is an alternative definition in which the coolant temperature is evaluated as it approaches the back face of the injection surface. With this definition, the cooling effectiveness is credited with all of the back-face cooling. This is useful in discussing the total cooling benefit but makes difficult the separation of internal and external effects. For example, Kasagi et al., in an earlier paper [54], reported on the effect of wall conduction on film-cooling effectiveness, using the augmented definition. Tests were conducted using an acrylic wall and a brass wall, and great effects on the spanwise-averaged cooling effectiveness were reported: near the first row, the effectiveness was 15% for the acrylic plate and 50% for the brass. The authors developed an analysis relating the augmented to the conventional effectiveness, showing that impingement cooling on the back face of a full-coverage film-cooling plate offered substantial benefits in overall performance. Since only the high-conductivity wall could "share the benefit" of the back-face cooling, its apparent effectiveness was much higher than that of the acrylic plate.

Remarkably little has reached the open literature about the losses incurred with film cooling. Favorskii and Kopelov [62] in 1981 estimated that each 1% cooling air discharged through the blades dropped turbine efficiency, possibly by as much as 1%. They also remarked on increased vibration resulting from leading-edge injection. Louis [63] pointed out that profile losses for film cooling are linearly proportional to the injected flow, even though the effectiveness is not. For the same flow, suction-side and pressure-side injection yielded similar aerodynamic losses, but the pressure surface-average effectiveness was considerably less than that of the suction surface. Most recently, Kawaike et al. [64] concluded that there are significant performance advantages to be gained by passing cooling air completely through the vanes, thence to an intercooler, and using the same air to cool the associated rotor blades. This treatment avoids more than half of the thermodynamic penalty associated with the mixing of low-temperature coolant with the mainstream.

A somewhat different approach to minimizing losses was taken by Nicholson et al. [65]. They selected blade profiles to accomplish the desired aerodynamic task with minimum heat transfer to the blade surface, without film cooling, by tailoring the blade so the curvature and pressure gradient suppressed transition. Their study also showed that the aerodynamic profile losses were significantly increased by a 4% level of turbulence in the free stream.

Boundary-layer theory describes the governing equations as parabolic, leading to the expectation that downstream events should not affect upstream processes. There is evidence, however, that stator-vane heat transfer is affected by the presence of an active rotor.

Dunn and Hause [66], in 1981, reported an increase in stator-vane heat transfer associated with the presence of an active rotor, and cautioned against the use of "stator only" data to characterize full-stage behavior. They stopped short of claiming definitely to have established the fact of interaction, however. Their results differed from later findings in that they reported increases on both pressure and suction surfaces, and everywhere downstream of 10% chord, whereas more recent findings show effects only on the suction side, and then mainly downstream of transition.

Dring et al. [67] measured heat transfer on rotor blades and stator vanes in a large-scale cold rig for two spacings of rotor and stator: 15% and 65% chord.

The system was operated at scale Reynolds number and speed. Heat transfer on the suction side of the stator vane went up by about 25%, downstream of transition, when the axial clearance was 15% of nominal chord. The effect was much smaller upstream of transition. The effect was also small on the pressure surface, where the boundary layer appeared to have already been transitional everywhere. During tests with the narrow axial gap, the static pressures on both the pressure and suction surfaces of a vane fluctuated as the rotor blades passed its trailing edge. The pressure effects were pronounced near the trailing edge, and extended beyond the throat on the suction side.

Further evidence of the influence of the rotor on heat transfer to the stator was reported by Dunn [68]. He found an increase beginning at about 60% chord (on the suction side only) which raised the Stanton number by 25% at the trailing edge.

There is another problem which has not received much attention: the temperature distribution in the free stream. Migration of the burner-outlet pattern may pose serious problems for the heat-transfer designer, mainly in the first stator. Burner-outlet patterns are designed to yield a hot core surrounded by slightly cooler flow, as protection for the engine surfaces. Secondary flows may evert the pattern, bringing the hot core against the surface of a vane in the first stator ring. A pattern factor of 1.20, $(T_{max} - T_{avg})/T_{avg}$, would meen a 400°F increase in gas temperature at the surface, assuming a 2000°F burner-outlet temperature. With 1000°F cooling air, this represents a 40% increase in local heat load. Indeed, some of the ambiguity surrounding the interpretation of engine data may arise from uncertainty about what to use for the gas temperature.

There are several other issues pertaining to heat transfer on blades and vanes which have not been addressed here, because not enough has appeared in the literature to support comparisons and evaluatins: heat transfer at the tips of rotor blades, the effects of the horseshoe vortex on blade-root and end-wall heat transfer, the effects of rotation on transition, the nature of the interaction between free-stream turbulence and heat transfer, and the characteristics of turbulence in representative engines.

Internal-Flow Heat Transfer

Most of the recent activity in internal-flow heat transfer related to blade cooling has been centered around three topics: rotation effects, impingement cooling, and pin fins.

Abe et al. [69] describe a state-of-the-art turbine blade which incorporates convection cooling with different pin fins in the mid-chord and trailing edge regions, impingement cooling on the inside of the leading edge, and film cooling on the leading edge (showerhead) and pressure surface. The combination of several modes in one design is typical of modern practice.

Work on pin fins and impingement cooling has been going on for years, but rotation is a relative newcomer to the heat-transfer scene.

Mori et al. [70] reported the effects of rotation on heat transfer from a rotating circular tube, and Johnston et al. [71] showed, in 1972, that the effect of rotation was to change the structure of the turbulence near the leading and trailing walls of a rectangular passage in fully developed channel flow. Turbulence was augmented on the "pressure" surface and suppressed on the "suction" surface of a channel rotating as a spoke in a wheel, with radial flow. The changes in structure were attributed to the Coriolis forces due to the radial motion coupled with rotation.

Radial passages in turbine blades closely resemble the situation studied by Johnston and his colleagues, and applications studies soon followed.

Mityakov and colleagues [72] measured the local and average effects of rotation on turbulent heat transfer in a circular tube, again rotating as a spoke in a wheel. The wall boundary condition was approximately that of constant heat flux. The ratio of Grashof to Reynolds number squared was never larger than 0.04, so free-convection effects should have been negligible. Their results are summarized in Figure 21, which contains two parts. In the left frame are local Nusselt numbers measured along the length of the tube. Two cases are considered and, for each, two sets of data are shown: along the leading (suction) side and along the trailing (pressure) side. The upper two sets of data are pressure-side data for the two different values of Rossby number, while the lower two sets are for the suction side. The dashed lines are previous work in the Soviet literature. The dashed line labeled Re = 10,000 happens to coincide, within 10%, with the Dittus-Boelter prediction for a stationary tube at that Reynolds number and also represents, reasonably accurately, the average heat transfer in the rotating tube of this experiment. The circumferential distribution of local Nusselt number is shown in the right-hand frame. The 180° line is the center of the pressure side of the tube. At 90°, the local Nusselt number agrees reasonably well with the Dittus-Boelter equation. From this, one would conclude that rotation has only a small impact on the average heat transfer.

During the same period, Pochuev et al. [73] were investigating heat transfer to the outer surfaces of turbine blades under rotation. Only small effects were found on the convex side, but on the concave side the rotation seemed to trigger an early transition, which then increased heat transfer over a portion of the blade. Even at high rotational speeds, the Gr/Re^2 ratio is too low to have much effect on the external boundary layer.

Experiments with water-cooled turbines led to a study of heat transfer to supercritical water in a rotating round tube by Johnson [74]. The experimental values agreed reasonably well with turbulent forced-convection correlations, but were below the expected free-convection correlations based on stationary tube experience. The rotation induced a centrifugal acceleration of about 10,000 g, spinning at about 4000 rpm on a diameter of 1.83 meters.

Morris and Ayhan [75], however, presented results of an experimental study of rotation effects showing changes in the mean Nusselt number by $\pm$ 30% due to rotation. Their results are summarized in Figure 22, which shows the data plotted against a rotation and heating parameter given in terms of Raleigh number (Ra), Rossby number (S), and Reynolds number (Re). The authors predict that typical aircraft gas-turbine conditions will result in a decrease in Nusselt number by about 46%. This prediction naturally resulted in a great deal of activity in the area of rotation effects on heat transfer, and the next year or two should see some results coming to print.

The more recent studies of impingement cooling have been concerned with the effects of the crossflow of spent fluid and the geometry of the jet array [76, 77], and the influence of heat transfer around holes in the target plate [78]. The subject continues to receive considerable attention and will probably grow in importance as systems are developed which yield higher heat-trasnfer coefficients.

Hollworth et al. [78] did a detailed study of heat transfer to a target plate equipped with vent holes. The authors concluded that the high performance came, not from the increased heat transfer around the vent holes, as had been surmised in an earlier work, but because of reduced entrainment of heated air into the impinging jets. Their data are summarized in Figure 23 in terms of

$Na/Pr^{1/3}$ vs. $\bar{r}$, the distance from the jet stagnation point. All lengths are made dimensionless by jet diameter, d, used as the characteristic length of the problem. The parameters $\bar{x}$ and $\bar{z}$ are jet spacing and plate spacing. The hole arrays were square in both plates, with the vents spaced apart half as far as the jet holes. The plates were staggered so that each jet impinged at the center of a pattern of four vents. The performance advantage over the in-line vented system is about 20%, for moderate distances away from the jet centerlines.

Kreatsoulas, in a report from the MIT Gas-Turbine Lab [79] discussed the effects of rotation on impingement cooling. Tests were conducted in a large-scale, rotating, hot rig. Two principal effects were noted: a reduction in the average heat-transfer coefficient by as much as 30%, and the emergence of hot spots. The hot spots are attributed to alteration of the jet-flow field by the rotational effects. The impingement jets near the hub were deflected radially inward, while those near the tip moved outward. These movements were attributed to a combination of effects from crossflow, Coriolis forces, and buoyancy forces. This is the first work from the new rotating rig, and more can be expected in the future.

Convection cooling in almost-capillary passages is used in several proprietary structures. In general, these are laminated, diffusion-bonded, multilayer assemblies with high heat-transfer area per unit volume inside a sheet of material which appears to be solid (or, at most, perforated). Typical of these is Lamilloy (a product of Allison Gas-Turbine Products, a division of General Motors). Similar products have been announced by other major engine developers on both sides of both oceans. In general, the laminated structure is used as a non-load-bearing aerodynamic shell attached to a load-bearing spar, which is thus well protected from the gas stream. Effusion of coolant through the holes can be described either as transpiration or as discrete-hole injection, depending on the spacing and size of the holes and on the ejection velocity. High effectiveness is attainable with these designs, in the heat-exchanger sense, and the coolant frequently is very nearly at metal temperature when it leaves the surface.

Larger passages, used in cast blades, are frequently equipped with pin fins or ribs to augment the internal heat transfer and to provide more area. Early work in this area can be found in the heat-exchanger literature of the 50s and 60s, but more recently, as the gas-turbine industry became interested, there has been a resurgence of interest [80-85]. For cases in which the pins extend from wall to wall and are regularly spaced, the flow passages can be characterized by their hydraulic diameter, D_h, and the results correlated using Nusselt number vs. Reynolds number with D_h as the characteristic length. Van Fossen [80] tested arrays with four rows, which Metzger [81] subsequently showed was probably still in the entrance region.

Figure 24, cited by Rudy and Graham [5] shows Nusselt number vs. Reynolds number for short pin fins, based on channel hydraulic diameter. The solid line represents a smooth-walled channel. Pressure drop has probably gone up at least as much as the heat transfer, though no data were cited. Han et al. [84,85] have investigated the effects of height, pitch, and angle of attack of approximately transverse ribs on opposite walls of square passage. Figure 25 shows friction factor and Nusselt number for ribs with $P/\varepsilon = 10$ as a function of angle of attack ($\alpha = 90$ means the ribs are perpendicular to the channel axis). Reynolds number is based on channel width, measured to the walls.

Metzger's work on pin-fin heat transfer indicated a rise in Nusselt number through the first three to five rows, followed by a gradual decline. For the Reynolds numbers involved, this probably represented the growth of turbulence in the passage, which concealed the usual thermal-entry-length behavior. Peng [82] looked at pins which spanned only half the channel height, as well as full-height pins and found that the two data sets could not be reconciled by use of a hydraulic-diameter length scale. Both of these last two effects (the increase then decrease in Nu, and the failure to collect on D_h) were found by Arvizu and Moffat [86] in a study of heat transfer to electronic components. The two geometries, pin fins and electronic packages, have much in common. Arvizu and Moffat's hydrodynamic measurements showed that two parallel flow paths existed inside the channel. A large fraction of the total flow passed over the crests of the elements, while only a small portion stayed in the tortuous path beneath the crests. The two flow paths share a common pressure drop and apportion the flow between them according to their respective drag coefficients. As a consequence, the velocity around the elements was far smaller than the average calculated for the channel. A partition function was developed which, for the geometries tested, predicted the array heat transfer accurately as the channel height was changed.

Metzger et al. [83] described some effects of pin shape on heat transfer and pressure drop and reported one particularly favorable array geometry, one which yielded increased heat transfer and decreased pressure drop for a given flow. Pin geometry was also investigated. Pins with a generally rectangular cross section yielded high pressure drop for a given heat-transfer performance and were deemed unsatisfactory, even if they had generous radii on their corners.

There are several issues concerning internal heat transfer which seem not to have been addressed yet in the literature: sink-flow heat transfer, including crossflow; heat transfer in sharp bends in small passages, with and without rotation; optimum pin-fin geometries, including array positioning to minimize pressure drop; and rotation effects in round and rectangular passages with turbulence augmentation.

Concluding Remarks

It seems abundantly clear that our ability to predict blade metal temperatures depends on our ability to make accurate predictions of the internal and external convective heat-transfer rates for realistic engine conditions. Those heat-transfer rates are established by complex interactions among many environmental factors, each of which must be modeled by careful laboratory tests. It is not sufficient, however, for the individual factors to be investigated--their interactions must also be studied. Models describing the response of the boundary layer to one individual factor must be capable of interacting with models describing the other factor effects, and a reliable data base must be assembled against which to test the integrated models. More data are needed on the flow conditions in operating engines, and on the accuracy of predictions by the various "public domain" programs.

There remain many important issues to be dealt with before we can confidently predict the heat-transfer behavior of a new engine series, using existing programs without missing the mark by more than 10.

References

1. C. H. Liebert, R. E. Gaugler, and H. J. Gladden, "Measured and Calculated Wall Temperatures on Air-Cooled Turbine Vanes with Boundary-Layer Transition," ASME 83-GT-33.

2. M. G. Dunn and W. J. Rae, "Measurement and Analysis of Heat-Flux Data in a Turbine Stage. Part II: Discussion of Results and Comparison with Predictions," 28th Int'l. Gas-Turbine Conference, Phoenix, AZ, March, 1983.

3. R. R. Wysong et al., "Turbine Design System," AFAPL-TR-78-92, Nov. 1978.

4. D. K. Winstanley, T. C. Booth, and M. G. Dunn, "The Predictability of Turbine-Vane Heat Transfer," AIAA 81-1435.

5. R. A. Rudey and R. W. Graham, "A Review of NASA Combustor and Turbine Heat-Transfer Research," ASME 84-GT-113.

6. F. S. Stepka, "Uncertainties in Predicting Turbine-Blade Metal Temperatures," ASME 80-HT-25.

7. R. M. Traci and D. C. Wilcox, "Freestream Turbulence Effects on Stagnation Point Heat Transfer," AIAA Journal, **13**, No. 7, July, 1975, pp. 890-896.

8. G. W. Lowery and R. I. Vachon, "The Effect of Turbulence on Heat Transfer from Heated Cylinders," Int. Jn. Heat and Mass Transfer, **18**: 1229-1242, 1975.

9. R. R. Dils and P. S. Follansbee, "Heat-Transfer Coefficients around Cylinders in Crossflow in Combustor Exhaust Gases," Jn. Eng. for Power, Oct. 1977, pp. 497-508.

10. M. F. Blair, "Influence of Free-Stream Turbulence on Boundary-Layer Transition in Favorable Pressure Gradients," Jn. Eng. for Power, **104**: 743-750, October, 1982.

11. M. F. Blair, "Influence of Free-Stream Turbulence on Turbulent Boundary-Layer Heat Transfer and Mean Profile Development. Part I. Experimental Data," Jn. of Heat Trans., **105**: 33-41, Feb. 1983.

12. W. H. Thielbahr, W. M. Kays, and R. J. Moffat, "The Turbulent Boundary Layer on a Porous Plate: Experimental Heat Transfer with Moderately Strong Acceleration," Trans. ASME, Jn. of Heat Trans., **94** Series C, 1972, p. 111.

13. W. H. Thielbahr, W. M. Kays, and R. J. Moffat, "Heat Transfer to the Highly Accelerated Turbulent Boundary Layer with and without Mass Addition," Jn. of Heat Trans., Series C, No. 3, Aug. 1970.

14. D. W. Kearney, W. M. Kays, and R. J. Moffat, "Heat Transfer to a Strongly Accelerated Turbulent Boundary Layer: Some Experimental Results, Including Transpiration," Int. Jn. of Heat and Mass Transfer, **16**: 1289, June 1973.

15. H. L. Julien, W. M. Kays, and R. J. Moffat, "Experimental Hydrodynamics of the Accelerated Turbulent Boundary Layer with and without Mass Injection," ASME Jn. of Heat Transfer, **93**, Series C, No. 4, Nov. 1971, p. 373.

16. D. W. Kearney, W. M. Kays, and R. J. Moffat," The Effect of Free-Stream Turbulence on Heat Transfer to a Strongly Accelerated Turbulent Boundary

Layer," Proc. of the 1970 Heat Transfer and FLuid Mechanics Institute, Naval Postgraduate School, Monterey, CA, 1970, p. 499.

17. J. Gillis and J. P. Johnston, "Experiments on the Turbulent Boundary Layer over Convex Walls and Its Recovery to Flat-Wall Conditions," in Turbulent Shear Flows, Springer-Verlag, 1980.

18. T. Simon and R. J. Moffat, "Turbulent Boundary-Layer Heat-Transfer Experiments: A Separate Effects Study on a Convexly Curved Wall," ASME 81-HT-78.

19. S. A. Eide and J. P. Johnston, "Prediction of the Effects of Longitudinal Wall Curvature and System Rotation on Turbulent Boundary Layers," Stanford University Thermosciences Division Report PD-19, 1974.

20. H. Consigny and B. E. Richards, "Short-Duration Measurements of Heat-Transfer Rate to a Gas Turbine Rotor Blade," Jn. Eng. for Power, **104** 542-551, July 1982.

21. W. P. Jones and B. E. Launder, "The Prediction of Laminarization with a Two-Equation Model of Turbulence," Eng. Jn. of Heat and Mass Transfer, **15**: 301-314, 1972.

22. R. A. Graziani, M. F. Blair, J. R. Taylor, and R. E. Mayle, "An Experimental Study of Endwall and Airfoil Surface Heat Transfer in a Large-Scale Turbine Blade Cascade," Jn. Eng. for Power, **102**: 257-267.

23. F. J. Bayley and W. J. Priddy, "Effects of Free Stream Turbulence Intensity and Frequency on Heat Transfer to Turbine Blading," ASME 80-GT-79.

24. B. W. Martin and A. Brown, "Factors Influencing Heat Transfer to the Pressure Surfaces of Gas Turbine Blades," Int. Jn. of Heat and Fluid Flow, **1** No. 3, 1979.

25. L. S. Han and W. R. Cox, "A Visual Study of Turbine Blade Pressure-Side Boundary Layers," Jn. of Eng. for Power, **105**: 47-52, Jan. 1983.

26. A. H. Jeans and J. P. Johnston, "The Effects of Concave Curvature on Turbulent Boundary-Layer Structure," Structure of Complex Turbulent Shear Flo (R. Dumas and L. Fulachier, eds.), Springer-Verlag, 1982.

27. J. C. Simonich and R. J. Moffat, "Liquid-Crystal Visualization of Surface Heat Transfer on a Concavely Curved Turbulent Boundary Layer," Proc. Joint ASME/JSME 1983 Tokyo International Gas-Turbine Conference, 1983.

28. H. S. Mickley and R. S. Davis, "Momentum Transfer for Flow over a Flat Plate with Blowing," NACA TN-4017, 1957.

29. K. Torii, N. Nishiwaki, and M. Hirata, "Heat Transfer and Skin Friction in Turbulent Boundary Layer with Mass Addition," Proc. 3rd Int. Ht. Conf. AIChE/ASME, 1966.

30. R. J. Moffat and W. M. Kays, "The Turbulent Boundary Layer on a Porous Plate: Experimental Heat Transfer with Uniform Blowing and Sucking," Int Jn. Ht and Mass Transfer, **11**: 1547-1566, 1968.

31. see No. 12

32. see No. 13

33. A. F. Orlando, W. M. Kays, and R. J. Moffat, "Heat Transfer in Turbulent Flows under Mild and Strong Adverse Pressure-Gradient Conditions for an Arbitrary Variation of the Wall Temperature," Proc. of the 24th Heat Transfer and Fluid Mechanics Institute, Corvallis, Ore, June 1974.

34. J. M. Healzer, W. M. Kays, and R. J. Moffat, "Experimental Heat-Transfer Behavior of a Turbulent Boundary Layer on a Rough Surface with Blowing," Jn. of Ht. Trans., **100**, No. 1, pp. 134-142, Feb. 1978.

35. M. M. Pimenta, W. M. Kays, and R. J. Moffat, "The Structure of a Boundary Layer on a Rough Wall with Blowing and Heat Transfer," ASME Jn. of Ht. Trans., **101**, No. 2, pp. 193-198, May 1979.

36. H. C. Coleman, R. J. Moffat, and W. M. Kays, "The Accelerated, Fully Rough Turbulent Boundary Layer", Jn. of Fluid Mechanics, **82**, Part 3, pp. 507-528, 1977.

37. V. N. Mamonov, "Heat Transfer on a Porous Plate at High Free-Stream Turbulence," Heat Transfer, Soviet Research, **12**, No. 5, Sept.-Oct. 1980.

38. B. P. Mironov, V. N. Vasechkin, V. M. Mamonov, and N. I. Yarygina, "Heat and Mass Transfer at High Free-Stream Turbulence as a Function of Injection Rate," Heat Transfer, Soviet Research, **13**, No. 5, Sept.-Oct., 1981.

39. D. P. Georgiou and J. F. Louis, "The Transpired Turbulent Boundary Layer in Various Pressure Gradients and the Blow-Off Condition," ASME 84-WA/HT-71.

40. D. G. Whitten, W. M. Kays, and R. J. Moffat, "Heat Transfer to a Turbulent Boundary Layer with Non-Uniform Blowing and Surface Temperature," Paper FC8.8, Proc. 4th Int'l. Heat Transfer Conf., Versailles, France, Sept. 1970.

41. J. Wolf and S. Moskowitz, "Development of the Transpirations Air-Cooled Turbine for High-Temperature Dirty Gas Streams," Jn. of Eng. for Power, **105**: 821-825, Oct. 1983.

42. R. J. Goldstein, "Film Cooling," Advances in Heat Transfer, Academic Press, New York and London, **7**: 321, 1971.

43. D. E. Metzger, D. I. Takeuchi, and P. A. Kuentsler, "Effectiveness and Heat Transfer with Full-Coverage Film Cooling," Jn. Eng. for Power, **95**: 180, 1973.

44. R. E. Mayle and F. J. Camarata, "Multihole-Cooling Film Effectiveness and Heat Transfer," Jn. of Heat Transfer, **97**: 534, 1975.

45. J. F. Muska, R. W. Fish, and M. Suo, "The Additive Nature of Film Cooling from Rows of Holes," ASME 75-WA/GT-17, 1975.

46. K. Sakata, H. Usui, and K. Takahara, "Cooling Characteristics of a Film-Cooled Turbine Vane Having Multi-Rows of Ejection Holes," ASME 78-TT-21.

47. G. Bergeles, A. D. Gosman, and B. E. Launder, "Double-Row Discrete Hole Cooling: An Experimental and Numerical Study," in ASME Gas Turbine Heat Transfer, p. 1, 1978.

48. M. Sasaki, K. Takahara, T. Kumagai, and M. Hamano, "Film Cooling Effectiveness for Injection from Multirow Holes," Jn. of Eng. for Power, **101**: 101, 1979.

49. M. E. Crawford, W. M. Kays, and R. J. Moffat, "Full-Coverage Film Cooling. Part I: Comparison of Heat Transfer Data for Three Injection Angles," ASME 80-GT-43, 1980.

50. M. E. Crawford, W. M. Kays, and R. J. Moffat, "Full-Coverage Film Cooling. Part II: Heat-Transfer Data and Numerical Simulation," ASME 80-GT-44, 1980.

51. S. Ito, R. J. Goldstein, and E. R. G. Eckert, "Film Cooling of a Gas-Turbine Blade," Jn. Eng. for Power, **100**: 476-481, July 1978.

52. R. P. Dring, M. F. Blair, and H. D. Joslyn, "An Experimental Investigation of Film Cooling on a Turbine Rotor," Jn. of Eng. for Power, **102**: 81-87, Jan. 1980.

53. D. E. Metzger, H. J. Carper, and L. R. Swank, "Heat Transfer with Film Cooling near Non-Tangential Injection Slots," ASME Transactions, Jn. of Eng. for Power, 1968, pp. 157-163.

54. N. Kasagi, M. Hirata, and M. Kumada, "Studies of Full-Coverage Film Cooling. Part 1: Cooling Effectiveness of Thermally Conductive Wall," ASME 81-GT-37.

55. M. Kumada, M. Hirata, and N. Kasagi, "Studies of Full-Coverage Film Cooling. Part 2: Measurement of Local Heat Transfer Coefficient," ASME 81-GT-38.

56. M. E. Crawford, W. M. Kays, and R. J. Moffat, "Heat Transfer with Full-Coverage Film Cooling Using 30° Slant-Angle Injection," ASME 75-WA/HT-11.

57. K. Furuhama and R. J. Moffat, "Turbulent Boundary-Layer Heat Transfer on a Convex Wall with Discrete-Hole Injection," ASME/JSME Thermal Engineering Joint Conference, Hawaii, 1983.

58. R. J. Goldstein, E. R. G. Eckert, H. D. Chiang, and E. Elovic, "Effect of Surface Roughness on Film-Cooling Performance," ASME 84-GT-41.

59. R. J. Goldstein and H. P. Chen, "Film Cooling on a Gas-Turbine Blade near the End Wall," ASME 84-GT-42.

60. N. Hay, D. Lampard, and C. L. Saluja, "Effects of the Condition of the Approach Boundary Layer and of Mainstream Pressure Gradients on the Heat-Transfer Coefficient on Film-cooled Surfaces," ASME 84-GT-47.

61. K. Furuhama and R. J. Moffat, "Heat Transfer and Turbulence Measurements of a Film-Cooled Flow over a Convexly Curved Surface," Joint ASME/JSME 1983 Tokyo International Gas-Turbine Congress, Tokyo, 1983.

62. O. N. Favorskii and S. Z. Kopelev, "Air Cooling of Gas-Turbine Blades," Thermal Engineering, August, 1981, pp. 435-438 (from Teploenergetika, **28**, Part 8, pp. 7-11, 1981).

63. J. F. Louis, "Heat Transfer in Turbines," AFWAL-TR-81-2099; also ASTIA AD-A111584 (1981).

64. K. Kawaike, N. Kobayashi, and T. Ikeguchi, "Effect of New Blade Cooling System with Minimized Gas-Temperature Dilution on Gas-Turbine Performance," ASME 84-GT-89.

65. J. H. Nicholson, A. E. Forest, M. L. G. Oldfield, and D. L. Schultz, "Heat-Transfer-Optimized Turbine Rotor Blades--An Experimental Study Using Transient Techniques," ASME 82-GT-304.

66. M. G. Dunn and A. D. Hause, "Measurement of Heat Flux and Pressure in a Turbine Stage," ASME 81-GT-88.

67. R. P. Dring, H. D. Joslyn, L. W. Hardin, and J. H. Wagner, "Turbine Rotor-Stator Interaction," Jn. Eng. for Power, **104**: 729-742, Oct. 1982.

68. M. G. Dunn, "Turbine Heat-Flux Measurements. I: Influence of Slot Injection on Vane Trailing-Edge Heat Transfer. II: Influence of Rotor on Vane Heat Transfer," 29th ASME Gas Turbine Conf., Amsterdam, 1984.

69. T. Abe, N. Doi, T. Kawaguchi, and T. Yamane, "Cooling Characteristics of Film-Cooled Turbine Blades," ASME 84-GT-73.

70. Y. Mori, T. Fukada, and W. Nakayama, "Convective Heat Transfer in a Rotating Radial Circular Pipe" (2nd rpt.), Int. Jn. of Heat and Mass Transfer, **14**: 1807-1824, 1971.

71. J. P. Johnston, R. M. Halleen, and D. K. Lezius, "Effects of Spanwise Rotation on the Structure of Two-Dimensional, Fully Developed, Turbulent Channel Flow," Jn. Fluid Mech., **56**: 533-557, 1972.

72. V. Y. Mityakov, R. R. Petropavlovsky, V. V. Ris, E. M. Smirnov, and S. A. Smirnov, "Turbulent Flow and Heat Transfer in Rotating Channels and Tubes," Leningrad Polytechnical Institute, ca. 1978.

73. V. P. Pochuev, M. E. Tsaplin, and V. F. Shcherbakov, "The Effect of Mass Forces on Heat Transfer in Turbine Rotor Blades and on Hydraulic Resistance in Cooling Channels," Cent. Aeroeng. Inst., Moscow, ca. 1978.

74. B. V. Johnson, "Heat Transfer Experiments in Rotating Radial Passages with Supercritical Water," in ASME Gas Turbine Heat Transfer, 1978.

75. W. D. Morris and T. Ayhan, "Observations on the Influence of Rotation on Heat Transfer in the Coolant Channels of Gas Turbine Rotor Blades," Proc. Inst. of Mech. Engrg., **193**: 303-311, 1979.

76. L. W. Florschuetz, R. A. Berry, and D. Metzger, "Periodic Streamwise Variations of Heat Transfer Coefficients for Inline and Staggered Arrays of Circular Jets with Crossflow of Spent Air," Jn. of Heat Transfer, **102**: 132-137, 1980.

77. L. W. Florschuetz, C. R. Truman, and D. Metzger, "Streamwise Flow and Heat Transfer Distributions for Jet Array Impingement with Crossflow," ASME 81-GT-77.

78. B. R. Hollworth, G. Lehman, and J. Rosiczkowski, "Arrays of Impinging Jets with Spent Fluid Removal through Vent Holes on the Target Surface. Part 2: Local Heat Transfer," Jn. Eng. for Power, **105**: 393-402, April 1983.

79. J. C. Kreatsoulas, "Experimental Study of Impingement Cooling in Rotating Turbine Blades," MIT Gas Turbine Lab, Report #178, Sept. 1983.

80. G. J. Van Fossen, "Heat Transfer Coefficients for Staggered Arrays of Short Pin Fins," Jn. Eng. for Power, **104**: 268-274, April 1982.

81. D. E. Metzger, R. A. Berry, and J. P. Bronson, "Developing Heat Transfer in Rectangular Ducts with Staggred Arrays of Short Pin Fins," Jn. Eng. for Power, **104**: 700-706, Nov. 1982.

82. Y. Peng, "Heat Transfer and Friction Loss Characteristics of Pin-Fin Cooling Configuration," ASME 83-GT-123.

83. D. E. Metzger, C. S. Fan, and S. W. Haley, "Effects of Pin Shape and Array Orientation on Heat Transfer and Pressure Loss in Pin Fin Arrays," ASME Transactions, Jn. of Eng. for Gas Turbines and Power, **106**: 252-257, 1984.

84. J. C. Han, J. S. Park, and C. K. Lei, "Heat Transfer Enhancement in Channels with Turbulence Promotors," ASME 84-WA/HT-72.

85. J. C. Han and C. K. Lei, "Heat Transfer and Friction in Square Ducts with Two Opposite Rib-Roughened Walls," ASME 83-HT-26, 1983.

86. D. Arvizu and R. J. Moffat, "The Use of Superposition in Calculating Cooling Requirements," Proc. of the Electronic Components Conference (IEEE), 1982 pp. 133-144.

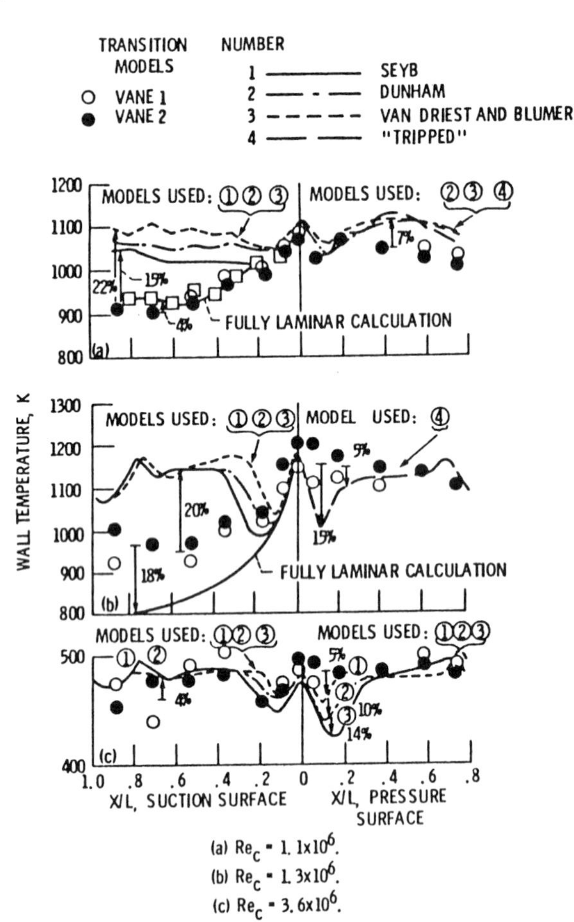

Fig. 1. Comparison of measured and predicted metal temperatures [1].

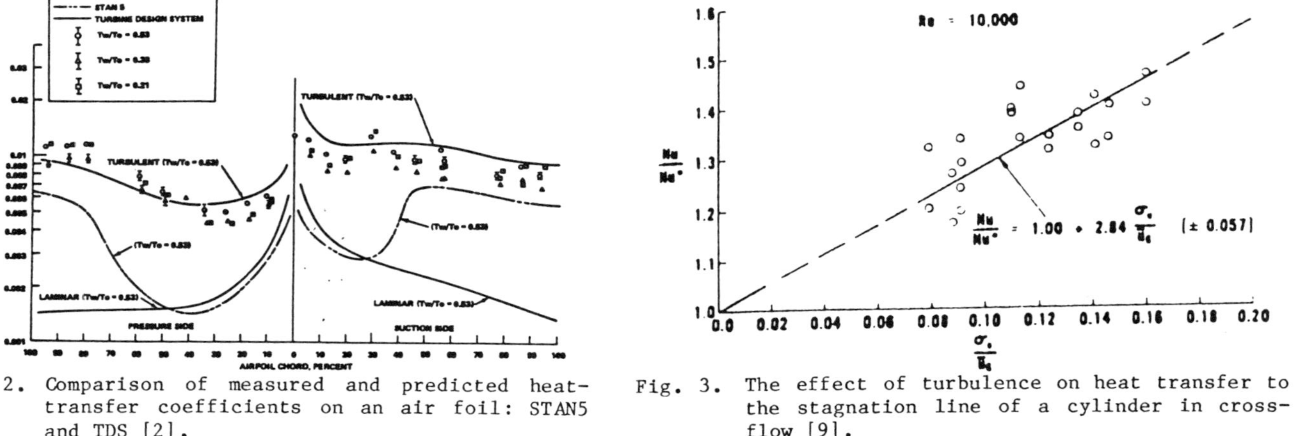

Fig. 2. Comparison of measured and predicted heat-transfer coefficients on an air foil: STAN5 and TDS [2].

Fig. 3. The effect of turbulence on heat transfer to the stagnation line of a cylinder in cross-flow [9].

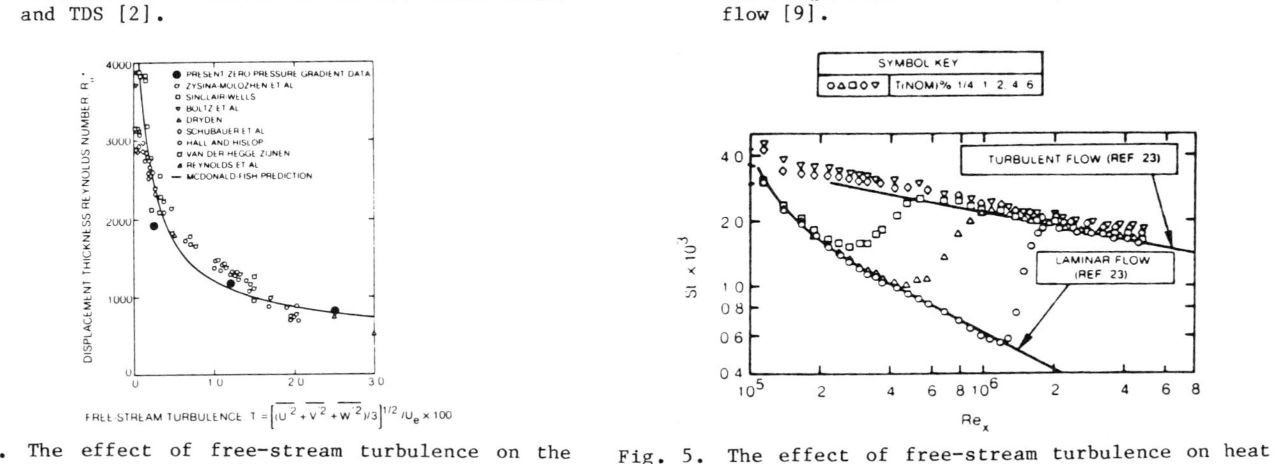

Fig. 4. The effect of free-stream turbulence on the transition Reynolds number for flow over a flat plate [11].

Fig. 5. The effect of free-stream turbulence on heat transfer to a flat plate [10].

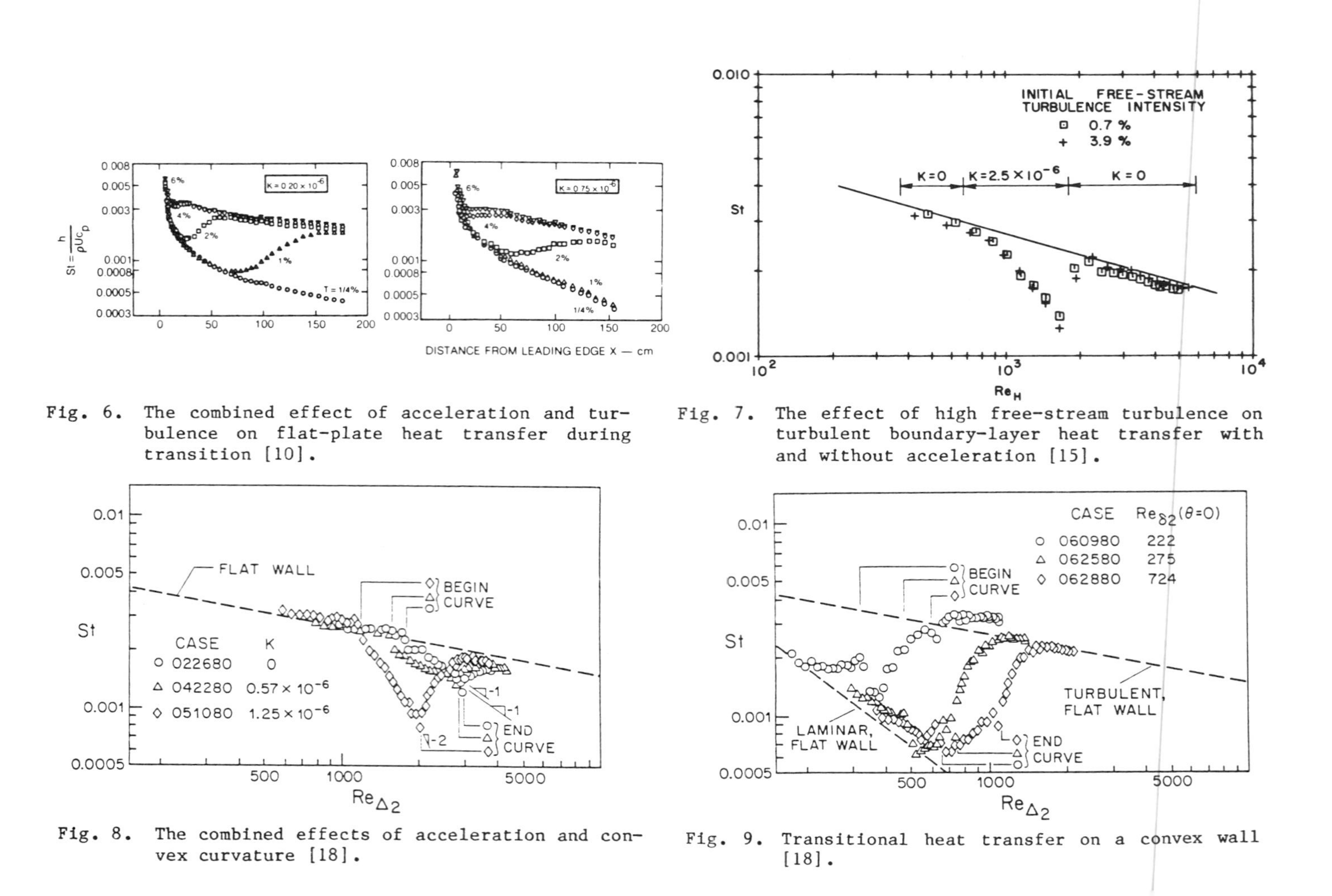

Fig. 6. The combined effect of acceleration and turbulence on flat-plate heat transfer during transition [10].

Fig. 7. The effect of high free-stream turbulence on turbulent boundary-layer heat transfer with and without acceleration [15].

Fig. 8. The combined effects of acceleration and convex curvature [18].

Fig. 9. Transitional heat transfer on a convex wall [18].

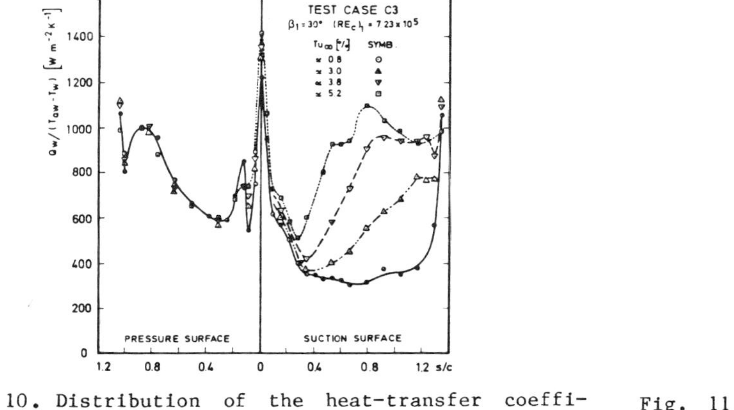

Fig. 10. Distribution of the heat-transfer coefficient around an airfoil at three levels of turbulence [20].

Fig. 11. Comparison of measured and predicted heat transfer on the pressure and suction sides of an airfoil [20].

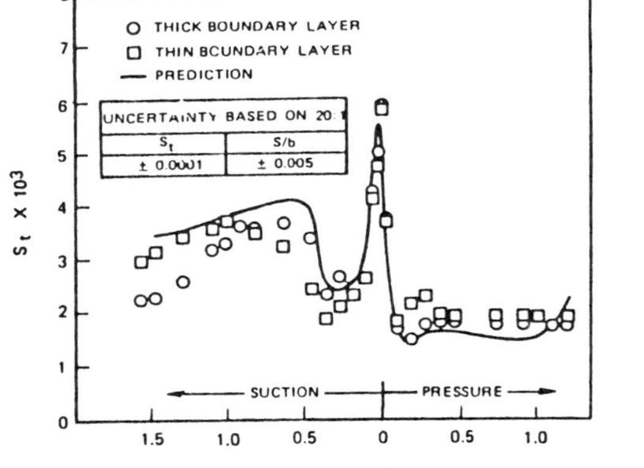

Fig. 12. Comparison of measured and predicted heat transfer on the pressure and suction sides of an airfoil [22].

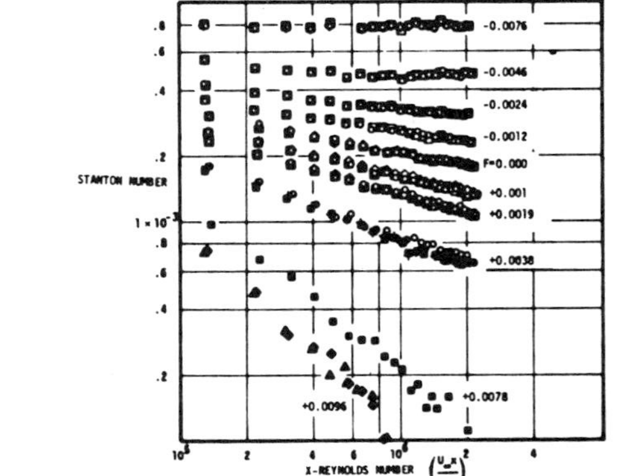

Fig. 13. Transpiration on a flat plate: blowing and suction [30].

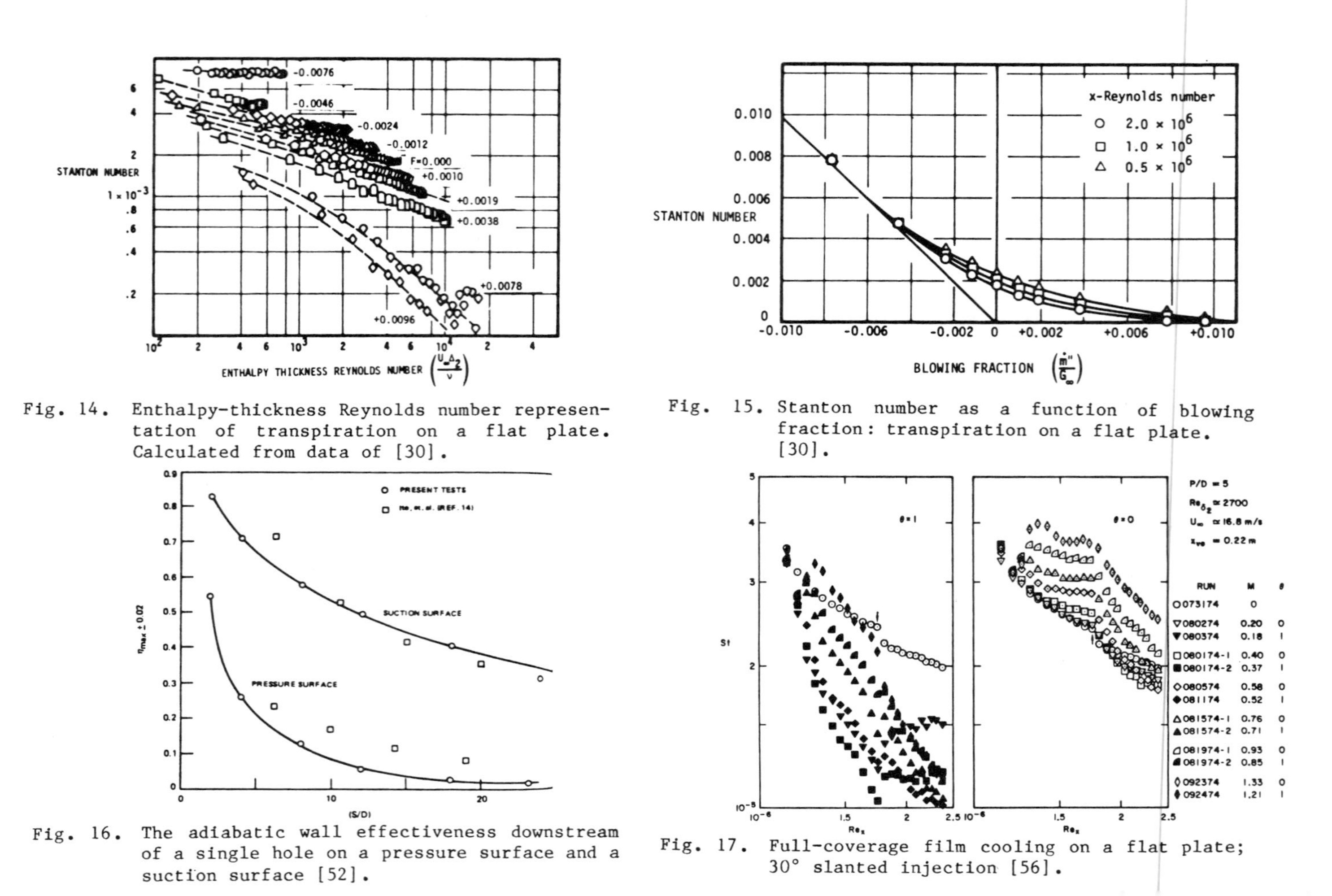

Fig. 14. Enthalpy-thickness Reynolds number representation of transpiration on a flat plate. Calculated from data of [30].

Fig. 15. Stanton number as a function of blowing fraction: transpiration on a flat plate. [30].

Fig. 16. The adiabatic wall effectiveness downstream of a single hole on a pressure surface and a suction surface [52].

Fig. 17. Full-coverage film cooling on a flat plate; 30° slanted injection [56].

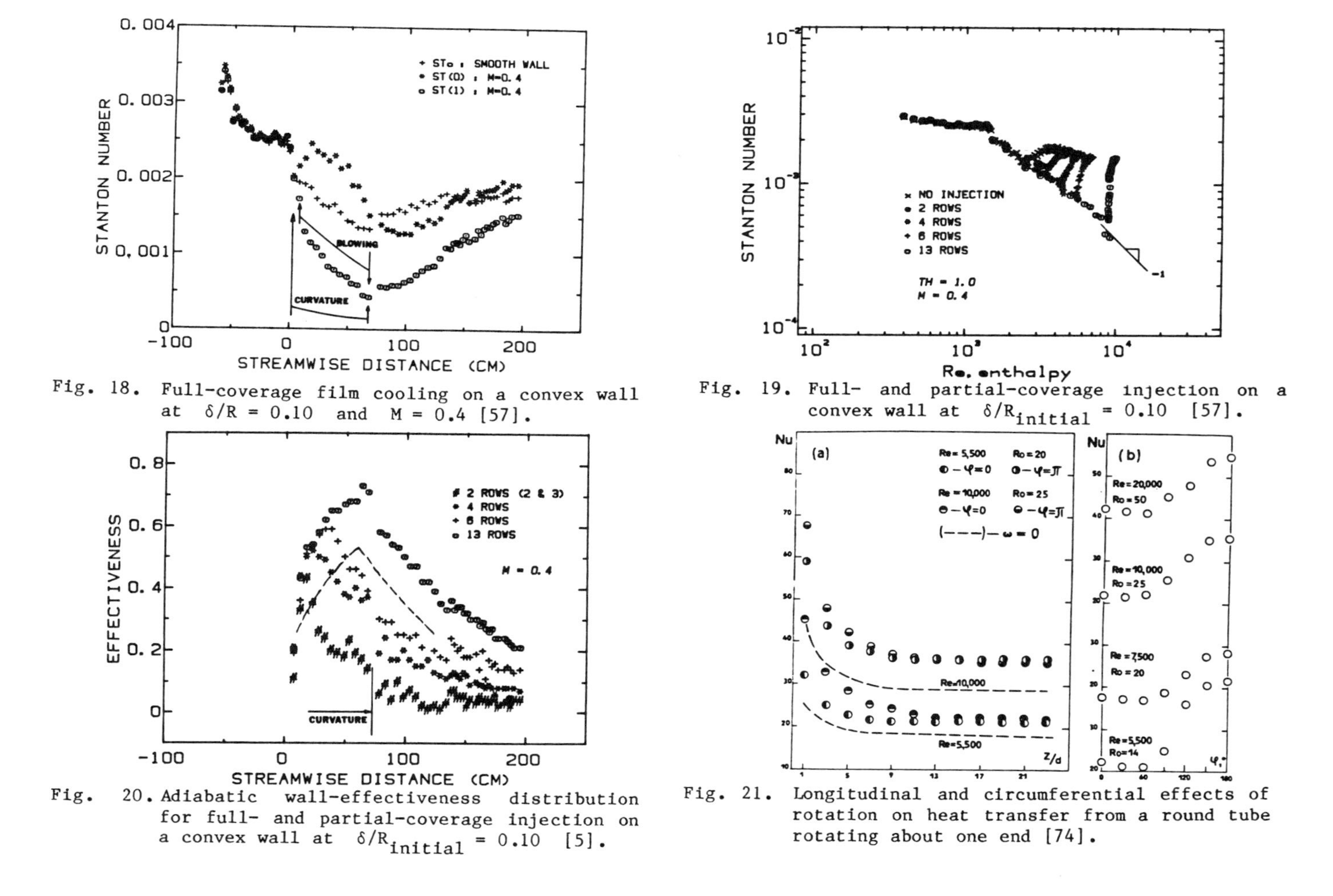

Fig. 18. Full-coverage film cooling on a convex wall at $\delta/R = 0.10$ and $M = 0.4$ [57].

Fig. 19. Full- and partial-coverage injection on a convex wall at $\delta/R_{initial} = 0.10$ [57].

Fig. 20. Adiabatic wall-effectiveness distribution for full- and partial-coverage injection on a convex wall at $\delta/R_{initial} = 0.10$ [5].

Fig. 21. Longitudinal and circumferential effects of rotation on heat transfer from a round tube rotating about one end [74].

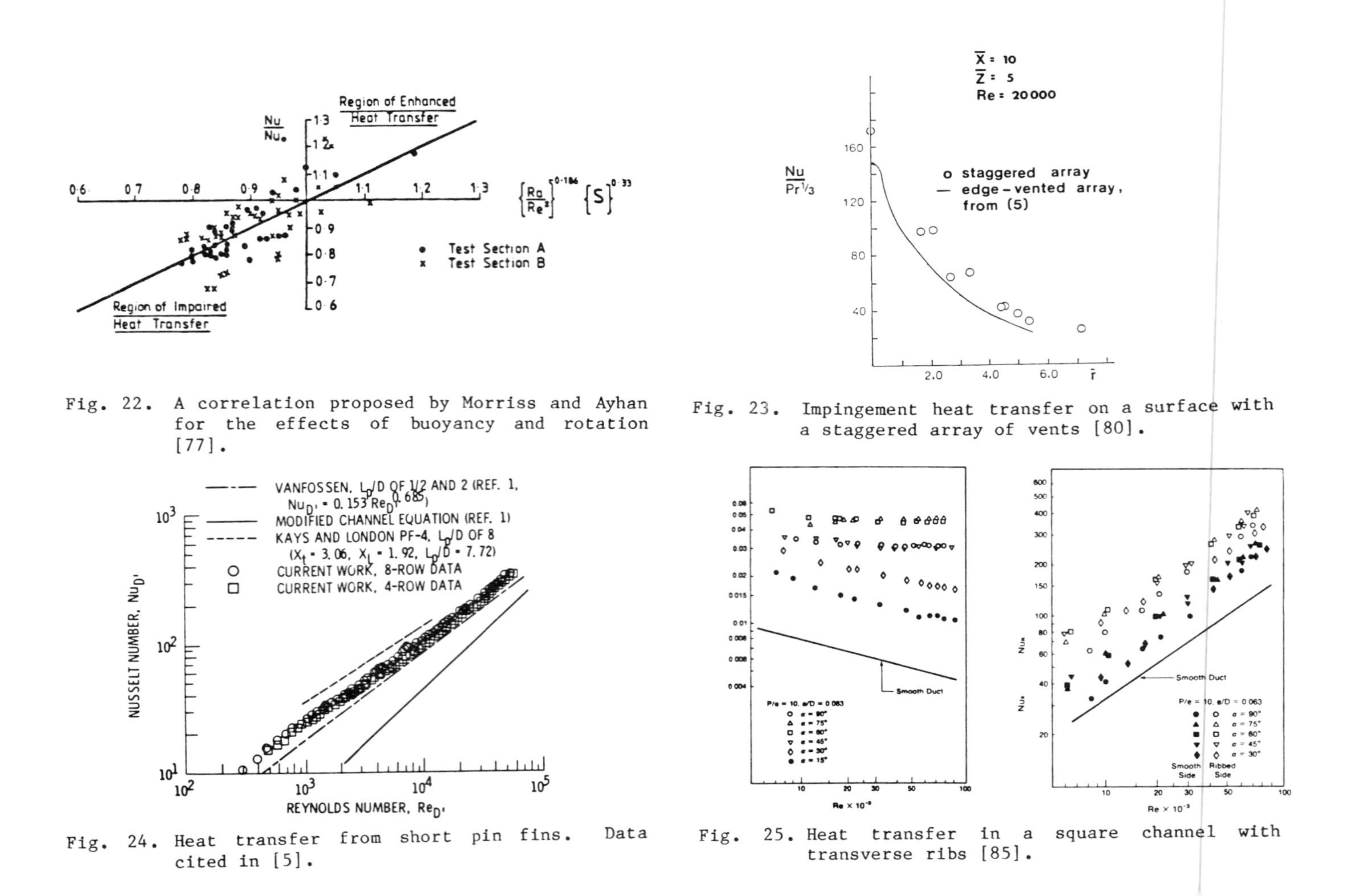

Fig. 22. A correlation proposed by Morriss and Ayhan for the effects of buoyancy and rotation [77].

Fig. 23. Impingement heat transfer on a surface with a staggered array of vents [80].

Fig. 24. Heat transfer from short pin fins. Data cited in [5].

Fig. 25. Heat transfer in a square channel with transverse ribs [85].

Experimental Investigation on Film Cooling of a Pressure Surface of a Gas Turbine Blade

SHINZO KIKKAWA and KATSUMI NAKANISHI
Department of Mechanical Engineering
Doshisha University
Kamikyo-ku, Kyoto 602 Japan

1. INTRODUCTION

One of the authors has published a paper on the experimental study of film cooling effectiveness over a gas turbine blade[1] in which coolant was two-dimensionally injected from a slot located near the leading edge. In the actual turbine cooling, however, the coolant must be three-dimensionally injected from descrete small holes because of difficulty of two-dimensional injection. The film cooling effectiveness on a concave surface is worse than on a flat or convex ones because of generation of a Görtler vortex[2]∿[4]. In order to keep the temperature throughout the pressure surface of a gas turbine blade lower than a certain value, therefore, the coolant must be injected from several rows of holes which are made in the streamwise direction. In this paper, the effects of blowing rate, injection angle, location of injection hole, geometry of injection hole, pitch-chord ratio and incidence on the film cooling effectiveness on the pressure surface of gas turbine blade are experimentally clarified and the optimum conditions for the film cooling are discussed.

2. EXPERIMENTAL APPARATUS AND PROCEDURE

The experimental apparatus used in this study was almost the same as that used in the previous work[1]. FIG.1 represents the tested cascade which is the same as that used in cited literature[1]. The secondary flow was injected with angle I_a to the surface of the blade from a rectangular hole of $2b \times s$ which is located on the midspan-line of the blade at X=7, 40 or 80mm (hereafter, these holes are named No.1, No.2 and No.3 hole as shown in FIG.1 respectively). As it is difficult to make a blade with an insulating material, the impermeable film cooling effectiveness was measured by using the analogy of heat and mass transfer, i.e. the air containing about 5% carbon dioxide was injected from a rectangular hole as the secondary flow and the gas was sampled at the rate of 3cc/min with a pump through 108 holes of 0.5mm diameter on the tested blade surface. The distribution of the film cooling effectiveness on the blade surface was obtained by analyzing the concentration of carbon dioxide in the sampled gas with an infrared analyzer. It is sure that the concentration of the sampled gas does not give the value at some distance above the blade surface, but on that surface from the experimental results made by one of the authors[5]. Prior to the experiment, the velocity distributions at the entrance of the cascade in the Y-direction and at three positions in the passage between blades in the Z-direction were measured. From the results it was made sure that the velocity distributions were uiform in the Y- and Z-directions except 20mm near the wall of the test section and the depth of the boundary layer on the end wall was nearly kept constant in the

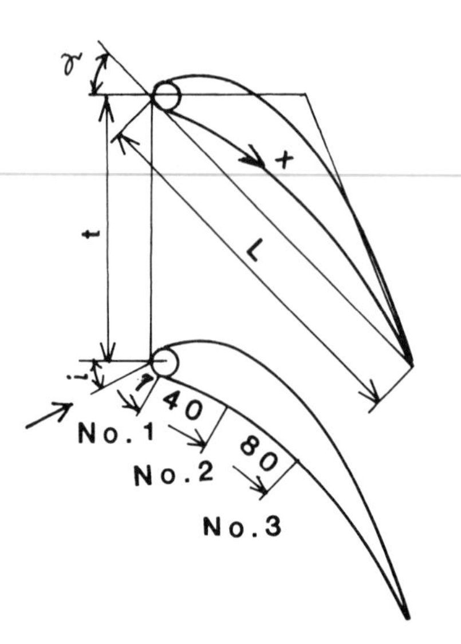

FIGURE 1. Configuration of tested cascade

streamwise direction because of the acceleration of the stream in the passage. The distance between side walls of test section was 100mm. As the standard conditions of the experiment, the following were selected; pitch-chord ratio t/L=0.7, incidence i=0, blowing rate M=0.2 and geometry of the injection hole 2bxs=6mmx2mm. To make clear the effect of a variable (e.g. blowing rate on the film cooling effectiveness, the value of the variable was varied while the other variables were kept at standard conditions. In every experiment, the staggered angle and Reynolds number based on chord length were kept constant as $\gamma=45°$ and $Re=1.2x10^5$. In the following figures, if the experimental conditions are not represented, it means the standard conditions.

3. EXPERIMENTAL RESULTS AND DISCUSSION

3.1 Effects of Injection Conditions

The profiles of the film cooling effectiveness on midspan-line η_c in the streamwise direction are shown in FIG.2, when the secondary flow is injected from injetion hole No.1 with Ia=90° and 60°. The two kinds of abscissae are used; X/L (bottom) and Xi/s (top). It is reasonable that the midspan-line effectiveness is lower especially in the downstream area than that for two-dimensional injection [1], but it must be noticeable that the difference between them becomes remarkable with increasing blowing rate M. That is, the film cooling effectiveness increases with increasing blowing rate for M<1.4 in the case of two-dimensional injection[1]; on the contrary, the midspan-line effectiveness increases with decreasing blowing rate and the optimum rate is about M=0.1∿0.2.

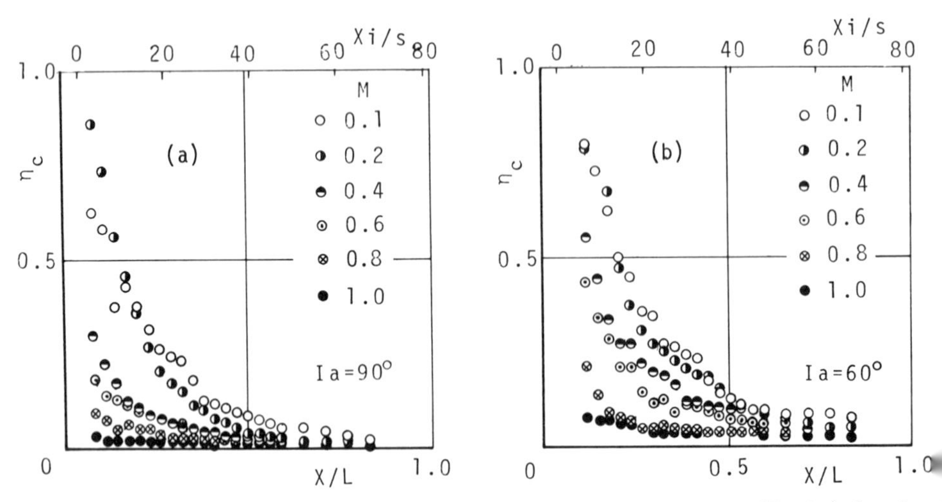

FIGURE 2. Effect of blowing rate on midspan-line effectiveness (No.1 injection hole)

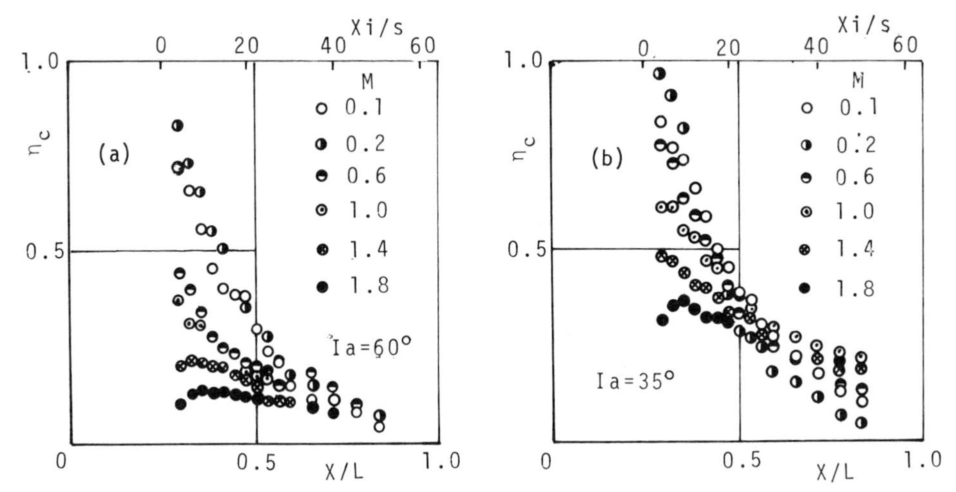

FIGURE 3. Effect of blowing rate on midspan-line effectiveness (No.2 injection hole)

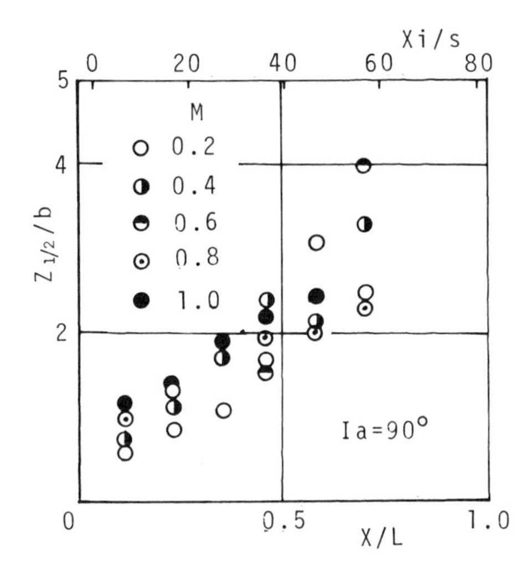

FIGURE 4. Distribution of $Z_{1/2}$ in streamwise direction (No.1 injection hole)

It is well known that the midspan-line effectiveness diminishes with blowing rate because a part of the main stream gets under the injected secondary flow and this effect may become remarkable with increasing blowing rate. This fact was verified by measuring the concentration profiles above the midspan-line downstream the injection hole. The measurement showed that the location of maximum concentration parted farther from the blade surface with increasing the blowing rate while the location of maximum concentration for M=0.2 was on the surface. This result suggests that the secondary flow injected at M≤0.2 is bent by the main stream just after the injection, and flows along the blade surface.

The distributions of the midspan-line effectiveness for injection from hole No.2 are shown in FIG.3. The optimum blowing rate for hole No.2 is about M=0.1∿0.2 as is for hole No.1.

The relation between the midspan-line effectiveness and Xi/s at M=0.1∿0.2 for hole No.2 is almost similar to that for hole No.1. However, in the downstream region, η_c at M>0.6 for hole No.2 is slightly higher than that for hole No.1. Comparing the results in FIGS.3(a) and (b), one can see that η_c increases with decreasing injection angle and the effect of injection angle becomes more remarkable as the blowing rate increases. The similar tendencies were observed in the case of hole No.3 injection. Naturally, the injected secondary flow spreads not only in the streamwise direction but spanwise as well. To make the distribution of film cooling effectiveness η in the spanwise direction clear, the distance between the midspan-line and the position where η becomes half of η_c, $Z_{1/2}$ was measured. As an example,

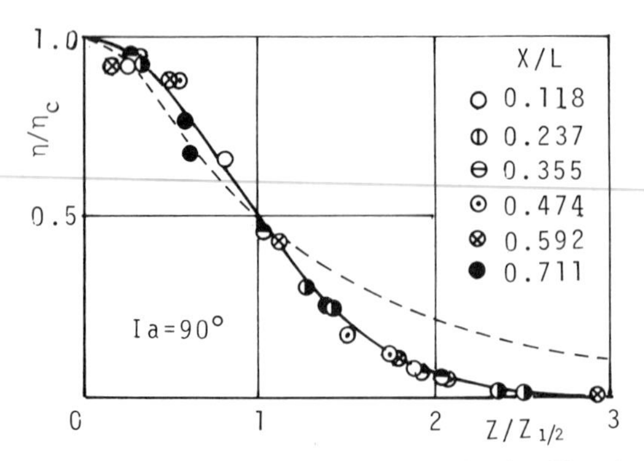

FIGURE 5. Distribution of normalized effectiveness in spanwise direction (No.1 injection hole, M=0.2)

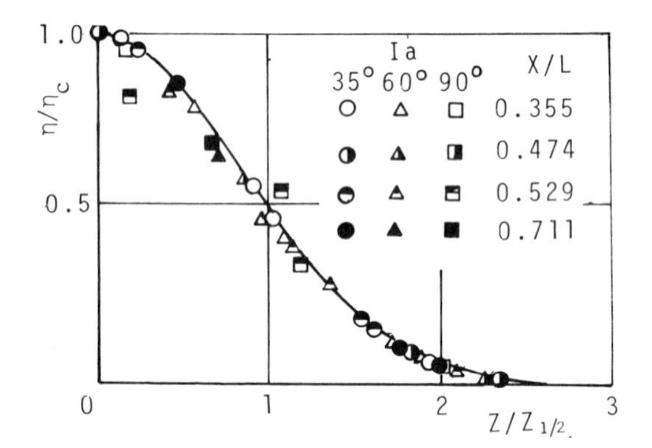

FIGURE 6. Distribution of normalized effectiveness in spanwise direction (No.2 injection hole, M=1.0)

the results for Ia=90° and hole No.1 injection are represented in FIG.4. $Z_{1/2}$ almost linearly increases with the distance from injection hole. Figures were omitted due to limited space but it was observed that the inclination of $Z_{1/2}$ toward downstream decreases with decreasing injection angle and $Z_{1/2}$ is independent of injection point if the injection angle is similar. The distribution of normalized film cooling effectiveness in non-dimensional spanwise direction is shown in FIG.5. The solid line in this figure shows a Gaussian error function given by Eq.(1).

$$\eta/\eta_c=\exp[-\ln 2(Z/Z_{1/2})^2] \quad (1)$$

It is known that the Gaussian error function agrees well not only with the non-dimensional velocity profile in the developed region of free jet but also with that of the parallel wall jet[6]. And furthermore, this Gaussian error function agrees well with the non-dimensional concentration profile of the complex jet in the present investigation. The results under the other experimental conditions showed the same behavior as shown in FIG.5. The broken line in FIG.5 shows the result obtained by Koso[7] for a three-dimensional wall jet but this curve does not agree with our results. The similar results for hole No.2 injection, M=1.0 and Ia=90°, 60° and 35° are shown in FIG.6 and the solid line represents Eq.(1). In spite of variation of injection angle, the results are in very good agreement with Eq.(1). It is imagined that the geometry of injection hole affects the film cooling effectiveness. Goldstein[8] reported that injected flow from a diverging conical pipe spreads more easily than that from a straight pipe and gives higher cooling effectiveness. Brown[9] showed that the midspan-line effectiveness increases linearly with increasing aspect ratio of the injection hole (2b/s) and coincides with the value for two-dimensional injection when 2b/s>8. The effects of the geometry of injection hole on midspan-line effectiveness are shown in FIG.7. FIG.7(a) shows the results for the case of varying streamwise width s and of constant spanwise width 2b, and FIG.7(b) for the case of varying aspect ratio and constant hole area. The injection point and injection angle are No.1 and 90° respectively. From the results in FIG.7(a), it can be seen that the midspan-line effectiveness increases with decreasing streamwise width.

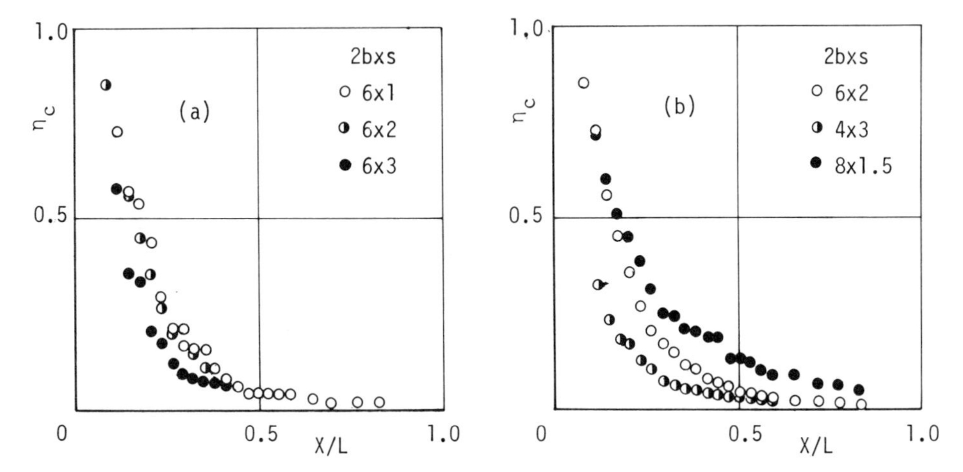

FIGURE 7. Effect of injection hole geometry on midspan-line effectiveness (No.1 injection hole, Ia=90°, M=0.2)

Goldstein[10] reported that the film cooling effectiveness increases with decreasing the momentum at the center of injected flow because the secondary flow becomes more easily bent by the main stream. It is seen that the momentum at the center of injected flow decreases more rapidly with decreasing streamwise width s. As the flow rate of injectant from a hole is proportional to the streamwise width if the spanwise width is constant, the injection from a hole having small streamwise width is very desirable from this point of view, too. The tendency showed in FIG.7(b) can be explained from the effect of streamwise width. It is worthy to note that the midspan-line effectiveness for 2bxs=8x1.5 keeps good value even in the far downstream region.

3.2 Effects of Pitch-Chord Ratio and Incidence

The effects of pitch-chord ratio and incidence of the midspan-line effectiveness in the case of hole No.1 and Ia=90° are shown in FIGS.8 and 9 respectively. In the case of three-dimensional injection, the midspan-line effectiveness increases with decreasing pitch-chord ratio and with increasing the incidence though in the case of two-dimensional injection the film cooling effectiveness is scarcely affected by these factors[1]. According to Moriya's method[11], the velocity U_1 at the outer edge of boundary layer above the position of hole No.1 increases with decreasing pitch-chord ratio and with increasing incidence. It is proper that when the velocity of main stream at injection point increases, the injectant becomes more easily bent and flow along the blade surface. On the other hand, in the case of two-dimensional injection, the injectant might be hardly affected by the velocity of the main stream, because of the large momentum of injectant (the optimum blowing rate was 1.0 for two-dimensional injection).

3.3 Prediction of Film Cooling Effectiveness by Means of Superposition

As was shown above, the three-dimensionally injected coolant on the pressure surface of gas turbine blade rapidly spreads in streamwise and spanwise directions. The behavior of the film cooling effectiveness can be calculated by using the results shown in FIGS.2,4 and 5. As an examole, a bird's eye view of the distribution of film cooling effectiveness is shown in FIG.10 under the con-

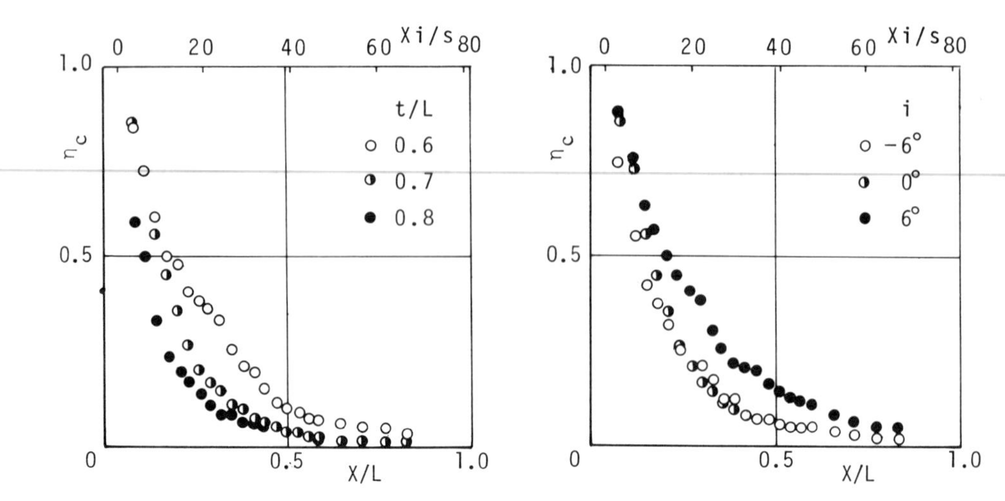

FIGURE 8. Effect of pitch-chord ratio on midspan-line effectiveness (No.1 injection hole, Ia=90°, M=0.2)

FIGURE 9. Effect of incidence on midspan-line effectiveness (No.1 injection hole, Ia=90°, M=0.2)

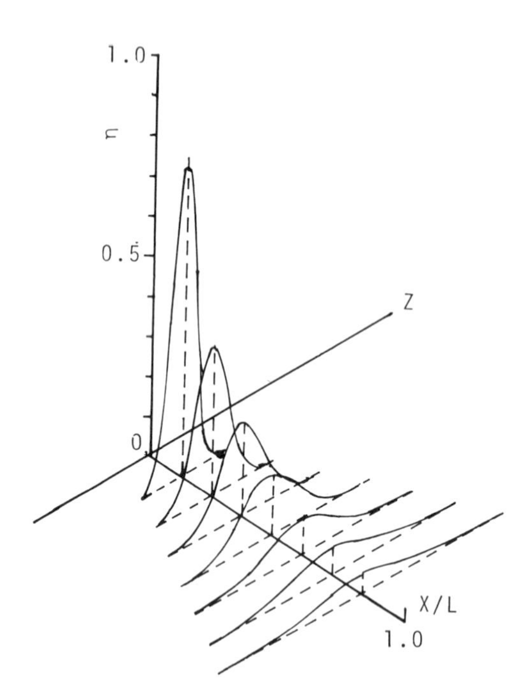

FIGURE 10. Bird's eye view profile of effectiveness
(No.1 injection hole, Ia=60°, M=0.2)

dition of hole No.1, Ia=60° and M=0.2. Using this result, the film cooling effectiveness contours can be obtained as shown in Fig.11. The arrow denotes the injection point. This figure suggests that the coolant must be injected from many holes to keep the temperature throughout the blade surface lower than a certain value.
When the multi-hole is made in the streamwise direction, the film cooling effectiveness in the downstream region can be estimated by the following expression[12].

$$\eta = \sum_{i=1}^{n} \eta_i \prod_{j=0}^{i-1} (1-\eta_j) \qquad (2)$$

$$\eta_0 = 0$$

where, η_i means the cooling effectiveness by injection from i-th hole only. This superposition is supported by experiments[13] and [14]. The calculated result by this method for the case of injection from No.1 (Ia=60°), No.2(Ia=60°) and No.3 (Ia=35°) holes is represented

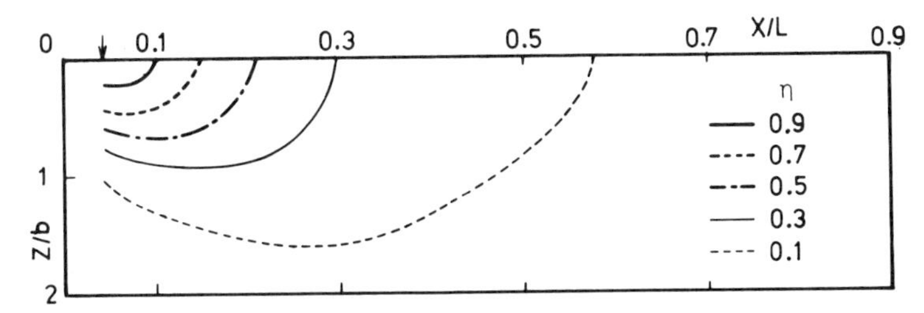

FIGURE 11. Contours of effectiveness (No.1 injection hole, Ia=60°, M=0.2)

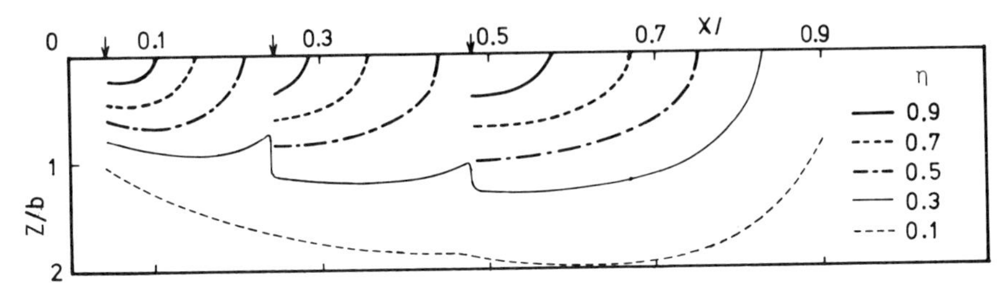

FIGURE 12. Contours of effectiveness (No.1, 2 and 3 injection holes, M=0.2)

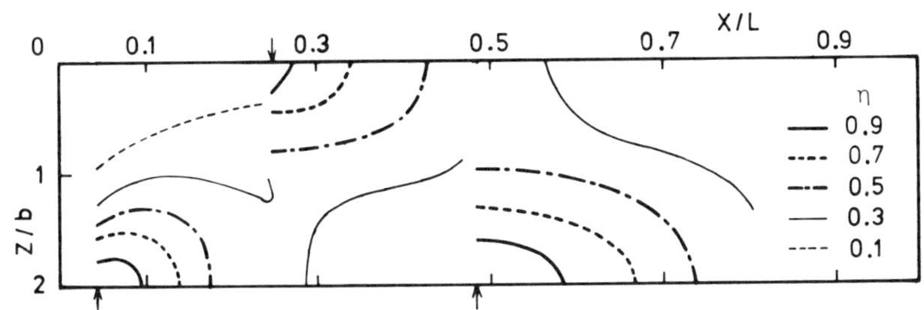

FIGURE 13. Contours of effectiveness (staggered arrangement, No.1, 2 and 3 injection holes, M=0.2)

in FIG.12. This result shows that the film cooling effectiveness in $Z>b$ is too poor and injection from the multi-hole must be arranged in the spanwise direction. As it is apparent that Eq.(2) is available for superposition of multi injection in spanwise direction, the film cooling effectiveness with injection from many staggered rows of holes at pitch of $Z/b=4$ was calculated by means of Eq.(2) and the results are represented in FIG.13. One can see that even in this figure the adequate area of $\eta<0.3$ remains. Therefore, in order to get satisfactory effectiveness throughout the blade surface, it is needed to inject coolant from more holes, especially in the streamwise direction.

As only No.1, No.2 and No.3 holes were used in the present experiment, the film cooling effectiveness with injection from other than these holes must be known to estimate the film cooling effectiveness with injections from much more holes. As the optimum blowing rate was about 0.2 for every experimental condition, the distributions of the midspan-line effectiveness for M=0.2 were re-plotted in FIG,14. In this figure $Xi_{1/2}$ on the abscissa denotes the distance from a center of injection hole along the blade surface to the position where η_c becomes 1/2. FIG.14 shows that most of all results fall on a curve, which is given by

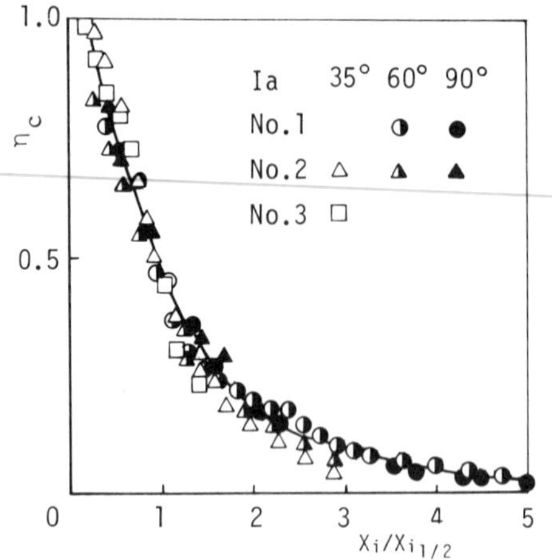

FIGURE 14. Universal profile if midspan-line effectiveness (M=0.2)

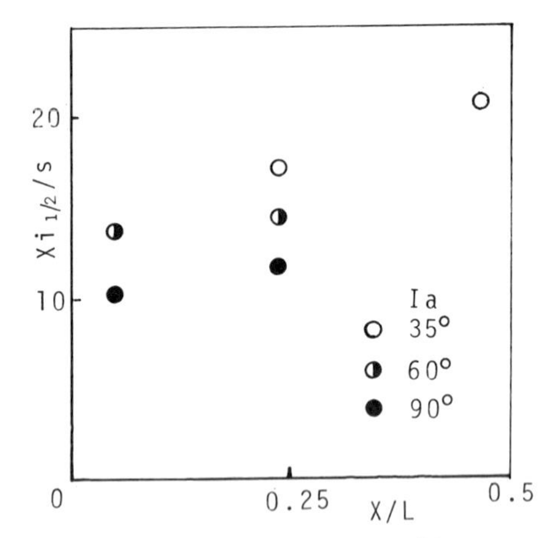

FIGURE 15. Distribution of $Xi_{1/2}$ in streamwise direction (M=0.2)

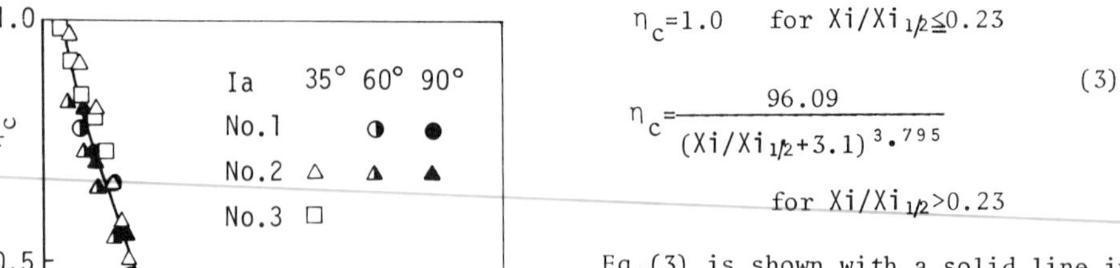

$$\eta_c = 1.0 \quad \text{for } Xi/Xi_{1/2} \leqq 0.23$$

$$\eta_c = \frac{96.09}{(Xi/Xi_{1/2}+3.1)^{3.795}} \quad \text{for } Xi/Xi_{1/2} > 0.23 \tag{3}$$

Eq.(3) is shown with a solid line in FIG.14. This equation is a universal profile of the midspan-line effectiveness in the streamwise direction, and we can calculate the value of midspan-line effectiveness for any condition if $Xi_{1/2}$ is known. The values of $Xi_{1/2}$ for six experimental conditions are presented in FIG.15. $Xi_{1/2}$ increases with decreasing injection angle and with shifting the injection point downstream. By using FIGS.4,5,14 and 15 one can estimate the distribution of film cooling effectiveness on the pressure surface of a gas turbine blade in the case which the coolant is injected from the arbitrarily arranged multi-hole with M=0.2. FIG.16 represents the film cooling effectiveness contours in the case which the coolant is injected from staggered rows of holes at X=7,20, 40, 60 and 80mm with Ia=60° (except X=80mm, Ia=35°) and M=0.2. In the region of X/L<0.8, the surface is covered by the value of η>0.5 except a very small zone.

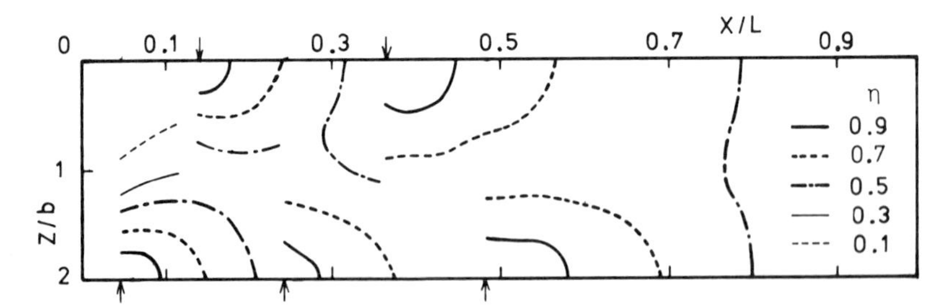

FIGURE 16. Contours of effectiveness (staggered multi-injection holes)

4. CONCLUSIONS

The film cooling effectiveness on the pressure surface of a gas turbine blade was measured. The coolant was injected from a rectangular hole. The following conclusions have been obtained.

(1) The optimum film cooling effectiveness is observed at a comparatively small blowing rate M=0.1~0.2, and the film cooling effectiveness decreases with increasing blowing rate. This tendency is opposed to that for the two-dimensional injection.

(2) The profiles of film cooling effectiveness for a low blowing rate in the streamwise direction are little affected by the location of injection and the injection angle when the dimensionless distance Xi/s is used.

(3) $Z_{1/2}$ almost linearly increases in the streamwise direction and more rapidly increases with decreasing injection angle. The behavior of $Z_{1/2}$ is nearly independent of the injection point if the injection angle is kept constant.

(4) The spanwise distribution of the film cooling effectiveness agrees well with a Gaussian error curve when the non-dimensional spanwise co-ordinate $Z/Z_{1/2}$ is used.

(5) The film cooling effectiveness increases with decreasing pitch-chord ratio and with increasing the incidence though the film cooling effectiveness is hardly affected by these factors in the case of two-dimensional injection.

(6) The injection hole having small width in the streamwise direction must be recommended from the view point of film cooling effectiveness.

(7) The streamwise distribution of the midspan-line effectiveness for M=0.2 can be expressed by an empirical formula independent of the injection point and the injection angle when the non-dimensional streamwise co-ordinate $Xi/Xi_{1/2}$ is used.

(8) By using the above results and the principle of superposition, one can predict the distribution of film cooling effectiveness on the pressure surface of a gas turbine blade on which many injection holes are laid out in arbitrary positions and injection angles.

NOMENCLATURE

b : half width of injection hole in spanwise direction
i : incidence (see FIG.1)
Ia : injection angle
L : chord length (see FIG.1)
M : blowing rate
s : width of injection hole in streamwise direction
t : pitch (see FIG.1)
U_1 : air velocity at the outer edge of the boundary layer
X : co-ordinate in streamwise direction (distance from leading edge along the blade surface (see FIG.1))
Xi : distance from center of injection hole along the blade surface
$X_{1/2}$: distance between the center of injection hole and the point at which the effectiveness is equal to 1/2
Y : co-ordinate normal to the blade surface
Z : co-ordinate in spanwise direction
$Z_{1/2}$: distance between the midspan-line and the point at which the effectiveness is equal to half the midspan-line effectiveness
γ : staggered angle (see FIG.1)
η : film cooling effectiveness

subscripts

c : value on midspan-line
w : wall

REFERENCES

1. Kikkawa,S., and Iwasaki,Y., Experimental Investigation on Film Cooling of Gas Turbine Cascade, Trans. of JSME, vol.50, no.455(B), pp1761-1768, 1984. (in Japanese, but will be published on "Heat Transfer Japanese Research" in English)

2. Nicolas,J., and LE Meur,A., Curvature Effects on a Turbine Blade Cooling Film, ASME Paper, 74-GT-156.

3. Mayle,R.E., Kopper,F.C., Blair,M.F., and Bailey,D.A., Effect of Streamline Curvature on Film Cooling, Trans. of ASME, vol.77, no.1, pp.77-82, 1977.

4. Falayan,C.O., and Whitelaw,J.H., The Effectiveness of Two-Dimensional Film Cooling over Curved Surface, ASME Paper, 76-HT-31.

5. Kikkawa,S., amd Fujii,M., Experimental and Theoretical Investigation on Two-Dimensional Film Cooling of a Flat Plate, Heat Transfer Japanese Research, vol.8, no.3, pp.52-68, 1979.

6. Kikkawa,S., and Onkura,T., Experimental and Theoretical Investigation on Three-Dimensional Film Cooling of a Flat Plate, Heat Transfer Japanese Research, vol.10, no.4, pp.40-56, 1981.

7. Koso,T., and Ohashi,H., Turbulent Diffusion of a Three-Dimensional Wall Jet, Bulletin of JSME, vol.25, no.200, pp.173-181, 1982.

8. Goldstein,R.J., and Eckert,E.R.G., Effects of Hole Geometry and Density on Three-Dimensional Film Cooling, Int. J. of H.M.T., vol.17, no.5, pp.595-607, 1974.

9. Brown,A., and Minty,A.G., The Effects of Mainstream Turbulence Intensity and Pressure Gradient on Film Cooling Effectiveness for Cold Air Injection Slit of Various Aspect Ratios, ASME Paper. 75-WA/HT-17.

10. Goldstein,R.J,, and Yoshida,T., The Influence of a Laminar Boundary Layer and Laminar Injection on Film Cooling Performance, Trans. of ASME, vol.104, no.2, pp.355-362, 1982.

11. Moriya,T., *Aerodynamics*, pp.73-104, Baifukan, Tokyo, 1962. (in Japanese).

12. Sellers,J.P., Gaseous Film Cooling with Multiple Injection Stations, AIAA J., vol.1, no.9, pp.2154-2156, 1963.

13. Muska,J.F., Fish,R.W., and Suo,M., The Additive Nature of Film Cooling from Rows of Holes, Trans. of ASME, vol.98, no.4, pp.457-464, 1976.

14. Sasaki,M., Takahara,K., Kumagai,T., and Hamano,M., Film Cooling Effectiveness for Injection from Multirow Holes, Trans. of ASME, vol.101, no.1, pp.101-108, 1979.

Film Cooling in a Plane Turbine Cascade

R. J. GOLDSTEIN and E. R. G. ECKERT
Mechanical Engineering Department
University of Minnesota
Minneapolis, Minnesota 55455

S. ITO
Mechanical Engineering Department
Ikutoku Technical University
Kanagawa-Ken, 243-02, Japan

ABSTRACT

Results of a mass transfer study show the influence of surface curvature on film cooling effectiveness. Tests using simulated turbine blades are reported for injected gas to mainstream-air density ratios near unity and two, over a range of injection rates.

INTRODUCTION

Surface curvature plays an important role in film cooling, following injection of gas from a slot [1,2] and from a single [2,3] or double [4] row of holes spaced across a surface. This influence of curvature has significant implications for the development of high-temperature gas turbines where film cooling is often employed on the suction (convex curvature) and pressure (concave curvature) sides as well as in the stagnation or leading-edge region of turbine blades and vanes. The present work provides an extension of some of the results reported earlier [3].

APPARATUS AND TEST PROCEDURE

Experiments performed on two blades held in a plane cascade in the exit section of a wind tunnel were reported in Ref. 3. A diagram of the cascade in the tunnel is shown in figure 1. Figure 2 is a sketch of the two central blades through which injection takes place--on the pressure side of one blade and on the suction side of the adjacent blade. The blades' curvature and geometry are similar to that of a high-performance gas turbine blade, although a large chord length (169 mm, or about four times the engine blade size) is used so that local measurements can be made along the blade surface.

The mainstream Reynolds number, based on the mean velocity at the cascade exit and on the chord length, is 2.32×10^5. The injection holes are spaced three diameters apart (centerline to centerline) across the span. The axis of each hole is perpendicular to a line across the span and is inclined at an angle of 35° to the blade surface. The secondary (injected) fluid density is controlled by mixing air with either a tracer gas (helium) or a heavier gas (refrigerant) to match density ratios encountered in high-

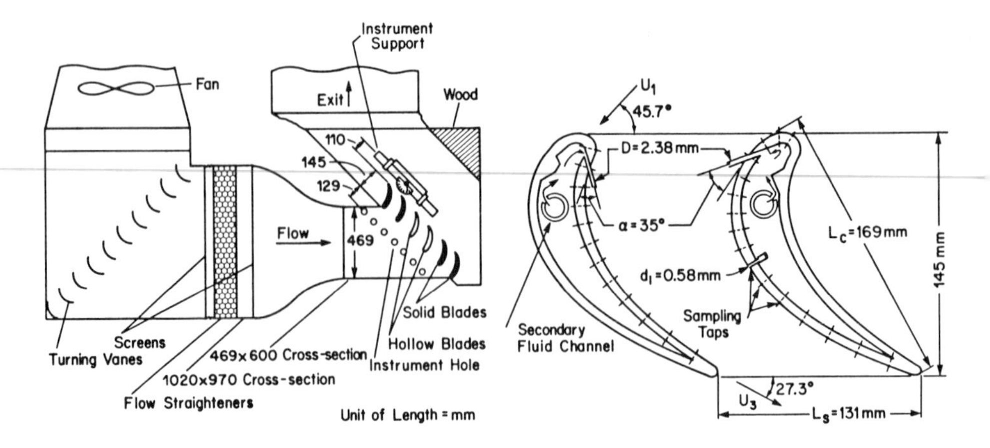

Figure 1. Schematic of wind tunnel cascade.

Figure 2. Cross section of blades with injection holes.

temperature turbine systems. A number of taps on the surface downstream of injection are used to draw off samples of the gas in the boundary layer adjacent to the wall. The concentration of the gas samples is measured in a chromatograph. The ratio of the concentration at the wall to that in the injected gas is the impermeable wall effectiveness. Applying the mass-heat transfer analogy [5], the impermeable wall effectiveness in this isothermal experiment can be related to the adiabatic wall temperature (effectiveness).

RESULTS AND DISCUSSION

Considerable differences are found between the film cooling effectiveness on the convex and concave surfaces contrasted to earlier film cooling work performed with a flat surface [5]. Figures 3,4, and 5 show comparisons of the results at three different density ratios and three different blowing rates. In these figures, the film cooling effectiveness averaged across the span is plotted over the dimensionless distance x/D for three density ratios. The parameter $I\cos^2\alpha$ in the figures expresses the ratio of the momentum flux of the secondary air jet parallel to the wall to the momentum flux in the mainstream at the location of the cooling holes.

At a relatively low blowing rate (M = 0.5, figure 3), the highest effectiveness occurs on the convex surface, with the flat surface next, and the concave surface giving the poorest, or lowest, effectiveness. At these low values of the parameter $I\cos^2\alpha$, the pressure gradient in the mainstream normal to the surface tends to push each jet on the convex side towards the blade, increasing the effectiveness. On the concave side, the jets are forced away from the surface by the pressure gradient. Note that on the flat and concave surfaces, for M = 0.5, the effectiveness increases with increasing density ratio. At the blowing rate of 0.5, the velocity ratio is near the value at which the jet lifts off the surface at the lower density ratio, yielding a lower effectiveness [3].

At higher blowing rates, figures 4 and 5, the film cooling effectiveness on the concave surface improves relative to that on the convex and plane surface. Thus at M = 2, the concave surface provides better film cooling than the other surfaces for the lowest density ratios--i.e., when $I\cos^2\alpha$ is relatively large. At the higher density ratios, the convex surface has the best film cooling performance.

The parameter $I\cos^2\alpha$ is a measure of the momentum flux of the injected fluid parallel to the wall surface compared to that of the free stream. When this ratio is very high with injection on a concave surface, the jet will tend to stay close to the wall. When the parameter is low on a concave surface, the jets will tend to move away from the wall due to the pressure gradient normal to the

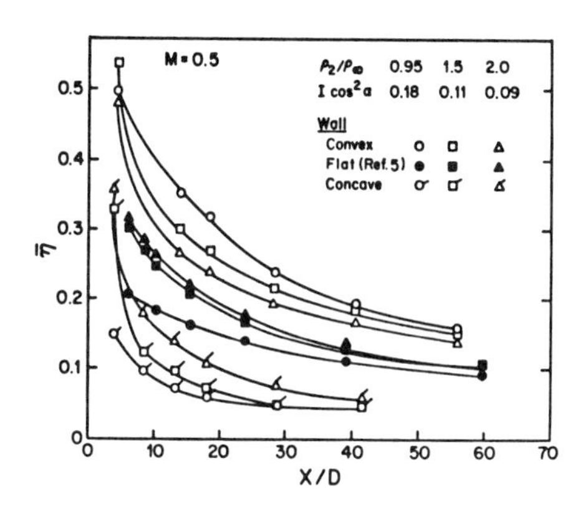

Figure 3. Lateral average effectiveness for M = 0.5

Figure 4. Lateral average effectiveness for M = 1.0

Figure 5. Lateral average effectiveness for M = 2

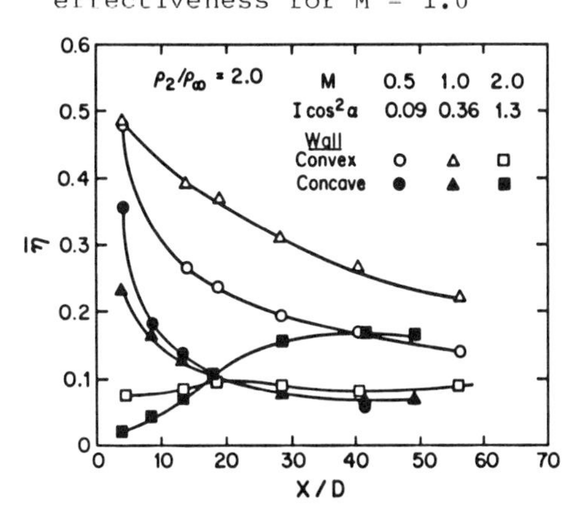

Figure 6. Lateral average effectiveness for density ratio of two--effect of M or $I\cos^2\alpha$

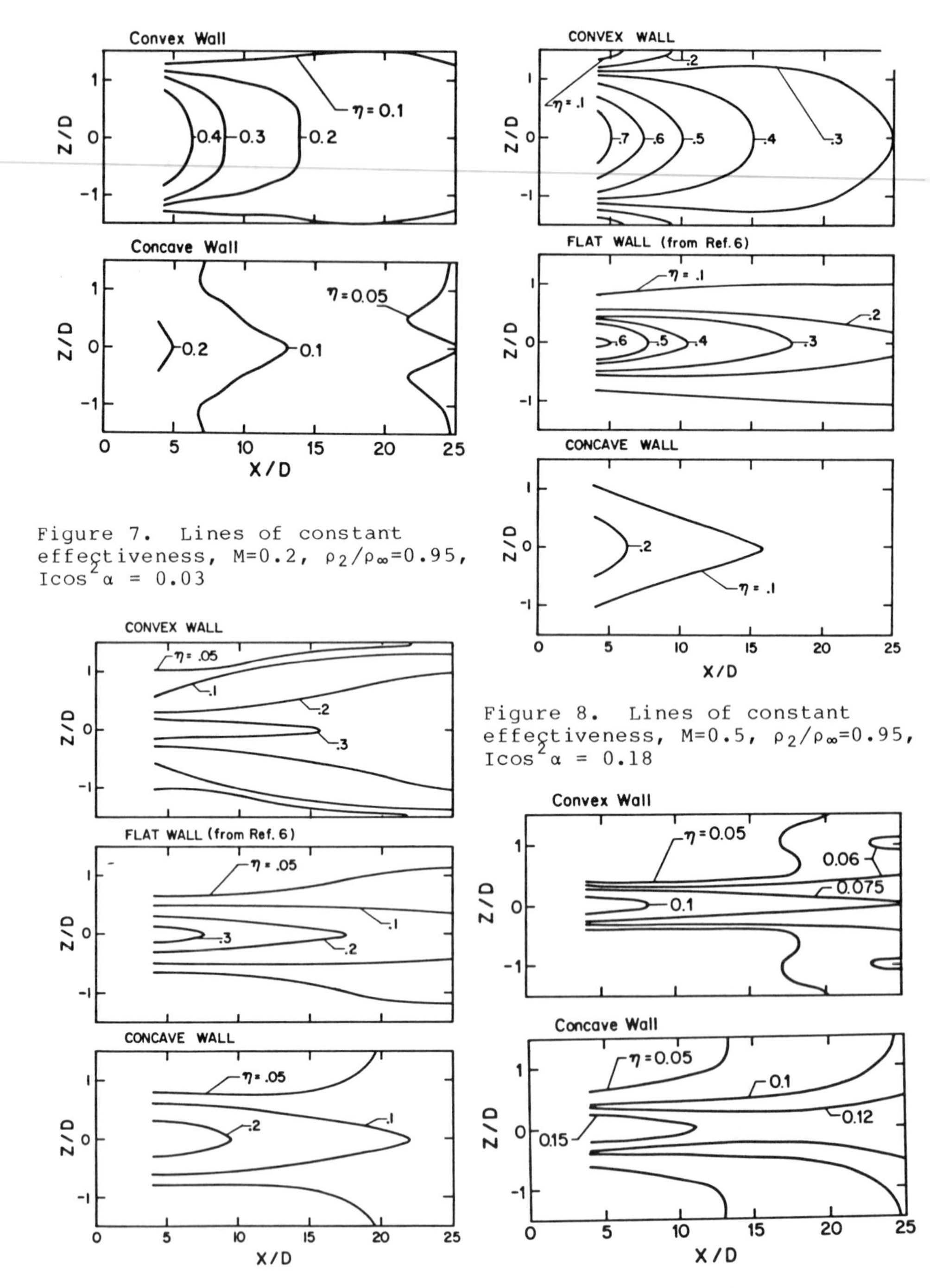

Figure 7. Lines of constant effectiveness, M=0.2, ρ_2/ρ_∞=0.95, $I\cos^2\alpha$ = 0.03

Figure 8. Lines of constant effectiveness, M=0.5, ρ_2/ρ_∞=0.95, $I\cos^2\alpha$ = 0.18

Figure 9. Lines of constant effectiveness, M=1.0, ρ_2/ρ_∞=0.95, $I\cos^2\alpha$ = 0.72

Figure 10. Lines of constant effectiveness, M=1.5, ρ_2/ρ_∞=0.95, $I\cos^2\alpha$ = 1.5

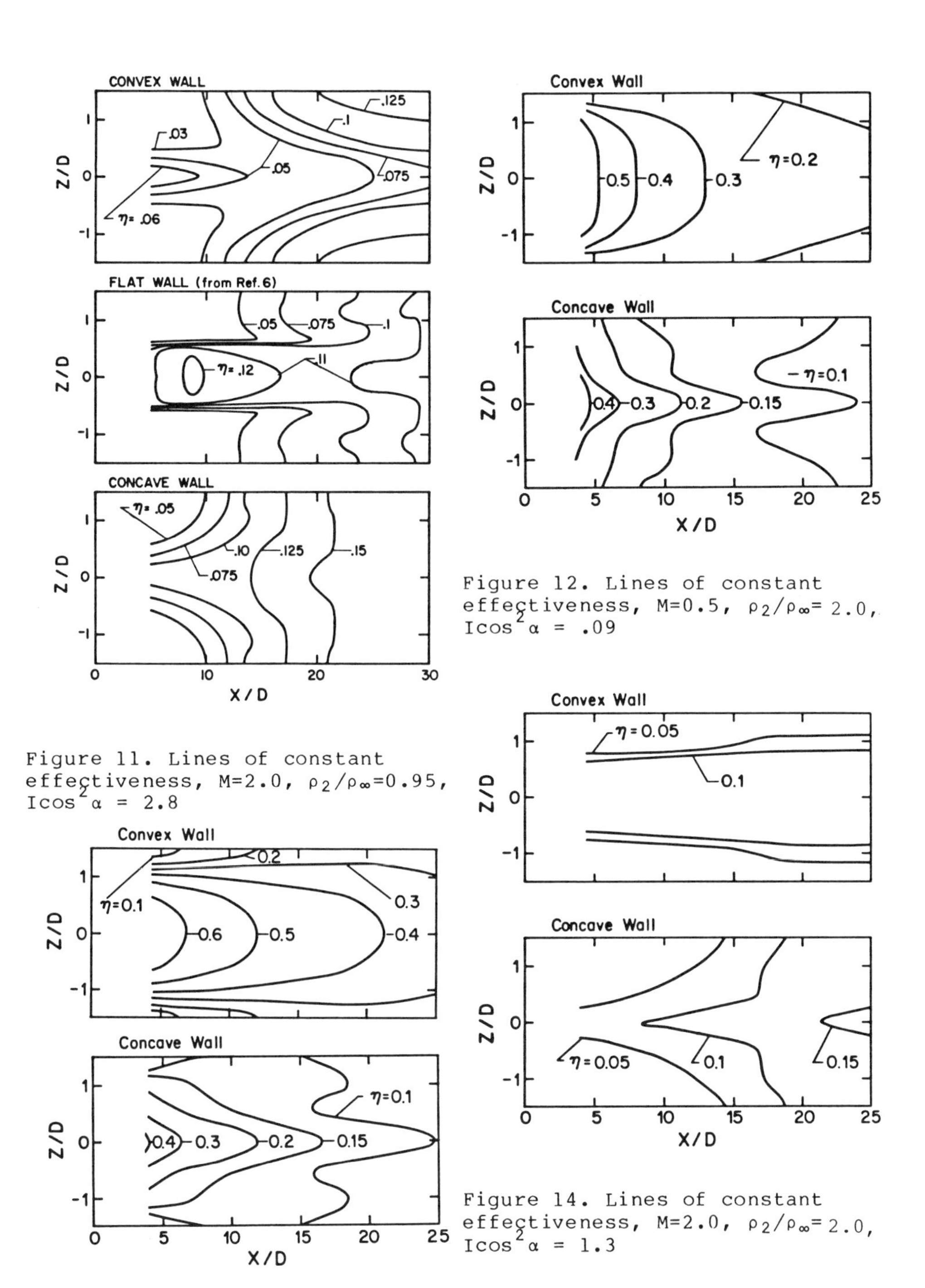

Figure 11. Lines of constant effectiveness, M=2.0, ρ_2/ρ_∞=0.95, $I\cos^2\alpha$ = 2.8

Figure 12. Lines of constant effectiveness, M=0.5, ρ_2/ρ_∞= 2.0, $I\cos^2\alpha$ = .09

Figure 13. Lines of constant effectiveness, M=1.0, ρ_2/ρ_∞= 2.0, $I\cos^2\alpha$ = .36

Figure 14. Lines of constant effectiveness, M=2.0, ρ_2/ρ_∞= 2.0, $I\cos^2\alpha$ = 1.3

surface in the curved mainstream flow. On the convex surface at low values of $I\cos^2\alpha$, the jets are pushed towards the surface. At higher values, the increased momentum of the jet essentially overcomes the pressure gradient; the jet moves away from the surface, resulting in a lower film cooling effectiveness. These effects are well-shown in Figure 6 where average effectiveness for a density ratio of two is shown for the convex and concave sides of a blade. Note in particular the crossover at higher M ($I\cos^2\alpha$) where the jets appear to impinge downstream on the concave wall (see contours below) giving a higher effectiveness.

There are also marked differences between the local values of the effectiveness and their distribution for the different surface curvatures, as can be observed in figures 7,8,9,10 and 11 for a density ratio near unity and in figures 12,13 and 14 for a density ratio of two. On these figures, contours of constant effectiveness on the convex and concave surfaces are compared to each other and in some cases to contours obtained on a flat surface [6]. Note, for example, in figure 7, that at low values of $I\cos^2\alpha$ with convex curvature, the jets are spread out on the surface in the z-direction by the pressure gradient so that the contour lines are quite broad. At a high blowing rate ($I\cos^2\alpha = 2.8$, figure 11), the broad character and high effectiveness on the concave surface give evidence of the jets impinging back on the blade. Generally similar shape contours are observed (cf. Figures 7 and 12, and Figures 8 and 13) when values of $I \cos^2\alpha$ are not too different for the different density ratios.

SUMMARY

Great care must be taken in applying the results of flat-plate film cooling experiments with a single row of holes to the design of coolant systems for gas turbine blades. Even relative trends for the improvement or loss in effectiveness depend on the momentum flux ratio as well as the sign of the radius of curvature of the surface. The influence of curvature is not as great when continuous slots or two rows of holes are used for injection[4]. Greater blockage of the mainstream decreases the relative penetration of jets into the freestream and thus decreases the importance of the pressure gradient normal to the surface.

Support for the final stages of this work by the Air Force Office of Scientific Research is gratefully acknowledged. P. H. Chen aided in the preparation of the final manuscript and figures.

NOMENCLATURE

English Symbols

di diameter of sampling tap, 0.58 mm in present study
D diameter of injection hole, 2.38 mm in present study
I momentum flux ratio, $\rho_2 U_2^2/\rho_\infty U_\infty^2$
L_c chord length, Figure 2
Ls blade pitch, Figure 2
M blowing rate or mass flux ratio, $\rho_2 U_2/\rho_\infty U_\infty$
U_1 upstream velocity
U_2 mean velocity of secondary fluid at injection hole exit
U_3 mean velocity at cascade exit
U_∞ mainstream velocity in location of injection holes
x distance along the blade wall downstream of the downstream edge of the injection hole
z distance from injection hole center in direction of blade span

Greek Symbols

α angle between injection hole centerline and surface
$\underline{\eta}$ local impermeable wall effectiveness
$\overline{\eta}$ effectiveness averaged across the blade span
ρ_2 density of secondary fluid at exit of injection hole
ρ_∞ density of mainstream fluid

REFERENCES

1. Mayle, R. E., Kopper, F. C., Blair, M. F., and Bailey, D. A., "Effect of Streamline Curvature on Film Cooling," ASME Paper No. 76-GT-90.

2. Nicholas, J., and Le Meur, A., "Curvature Effects on a Turbine Blade Cooling Film," ASME Paper No. 74-GT-156.

3. Ito, S., Goldstein, R. J., and Eckert, E. R. G., "Film Cooling of a Gas Turbine Blade", Trans. ASME, Series A, J. Engr. for Power, vol. 100, no. 3, pp. 476-481, 1978.

4. Goldstein, R. J., Kornblum, Y., and Eckert, E. R. G., "Film Cooling Effectiveness on a Turbine Blade," Israel J. Tech., 20, 193-200, 1982.

5. Pedersen, D. R., Eckert, E. R. G., and Goldstein, R. J., "Film Cooling with Large Density Differences Between the Mainstream and the Secondary Fluid Measured by the Heat-Mass Transfer Analogy", Trans. ASME, Series C, J. Heat Transfer, vol. 99, pp. 620-627, 1977. See also, Pedersen, D. R., "Effect of Density Ratio on Film Cooling Effectiveness Through a Row of Holes and for a Porous Slot", University of Minnesota, Ph.D. Thesis, March 1972.

6. Ericksen, V. L., and Goldstein, R. J., "Heat Transfer and Film Cooling Following Injection Through Inclined Circular Tubes," J. Heat Transfer, 96, 329-245, 1974.

Mist Cooling of High Temperature Gas Turbine Blades

YUKIO YAMADA
Mechanical Engineering Laboratory
Agency of Industrial Science and Technology
Tsukuba Science City, Ibaraki 305, Japan

YASUO MORI
Department of Mechanical Engineering
University of Electro-Communications
Chofugaoka, Chofu, Tokyo 182, Japan

ABSTRACT

The conventional turbine blade cooling technique using air is considered to have an upper feasible limit of about 1300°C on the turbine inlet gas temperature. Water cooling has some problems of flow instability and burnout caused by boiling in cooling holes although water cooling is supposed to give cooling performance better than steam cooling. The purpose of this study is to show a performance of mist cooling superior to conventional one-phase cooling by use of air, water or steam. Mist cooling has several advantages such as of providing high cooling performance, preventing flow instability and satisfying the condition of decreasing thermal stress of the blade by providing the appropriate cooling rate for an individual hole by adjusting the flow rate and water-air ratio for the hole. This study shows numerical results on cooling characteristics of mist cooling. As an example, a two-dimensional blade cooling with simple 12 cooling holes is studied and the temperature and stress profiles are calculated to find a better cooling condition by giving an adequate cooling rate in each hole. As a result, mist cooling is found to have cooling characteristics better than air, water or steam cooling.

1. INTRODUCTION

The combined gas turbine/steam turbine cycles have bean developed to increase the thermal efficiencies of the fossil fuel-fired power plants. The goal of the Japan project for development of a combined cycle using an advanced gas turbine has been set at about 55% thermal efficiency [1]. To attain this goal, the development of a high temperature gas turbine with turbine inlet temperatures up to 1300°C has been in progress. For further increases of thermal efficiency, higher turbine inlet temperatures are necessary and developments of new material and progresses of blade cooling technique are required to solve the strength problem of blade material at high temperatures.

Air cooling, which has been the only technique implemented by actual industrial turbine systems, is considered to have the upper limit of about 1300°C as the turbine inlet temperature because of the large amount of compressed air needed and of large power consumption. Steam cooling and liquid cooling, as well as their combinations, which use water, liquid metals or organic liquids, are considered to improve cooling performance. Especially, water cooling [2] and steam cooling [3] have been extensively studied in view of development of combined gas turbine/steam turbine cycles. Water cooling utilizes the large latent heat of evaporation of water and its high specific heat with a high heat transfer coefficient. Steam cooling has some advantages over air cooling. Steam has a specific heat twice that of air and also has a lower viscosity than air. The cooling efficiency of steam was experimentally reported to be about 50% larger than that of air cooling [4]. Furthermore, steam cooling has the capability of utilizing the latent heat of evaporation by use of steam containing moisture, although this is still just an idea at the present status, where no data are available on complicated one-component steam-water mist flow.

The present paper discusses the feasibility of blade cooling by air-water mist flow, which was proposed by one of the authors [5], from the view points of heat transfer and stress analyses. Air in the two-component mist flow absorbs the volume increase by evaporation of water droplets to keep the trend of pressure loss variation with flow rate. Therefore, the mist flow can prevent the occurence of flow instability while the large latent heat of evaporation of water improves the cooling performance. This fundamental study on mist cooling by two-component air-water mist flow can also give fundamental data of one-component steam-water mist cooling technique for combined cycles.

The experimental heat transfer results on internal mist cooling of thin tubes heated at high temperatures [5] suggest that the heat transfer coefficients at each cooling holes can be adjusted by controlling the mist flow rate and mixture ratio in each hole. The present study shows an excellent performances of mist flow using a distributed cooling rate by comparing with air and water cooling based on results from temperature and stress analyses of blades using the finite element method. Also, the effect of the size of cooling holes on the cooling performances by air, water and mist cooling is investigated.

2. FUNDAMENTAL EQUATIONS AND CONDITIONS OF CALCULATION

2.1 Temperature Profile

The main feature of the present paper is to compare the temperature and stress profiles of mist cooled blades with those of air and water cooled blades. Therefore, the following simplifications are made by leaving the detailed three-dimensional analyses to future studies.

Steady state is considered for two-dimensional turbine blades where variations in the direction of blade height are neglected. This simplification is acceptable because no serious temperature gradient in the height direction occurs generally in cooled blades, thus rendering the temperature gradient through the thickness as the major concern on temperature and stress analyses. Four kinds of rotating blades (buckets) No.1 to No.4 with cross sections as shown in Fig.1 are considered here. They all have the same airfoil and the same number of holes with different sizes of cooling holes. The cooling system is open for cooling fluid to be injected into the burned gas flow from the tips after flowing through the cooling holes.

Heat transfer coefficients at the blade surfaces α_g for both the suction and pressure sides are assumed to have the profile shown in Fig.2 [6]. When a turbine is rotating with a small load, α_g at the suction side and that at the pressure side, can be considered to have the same value and also the temperature drop of burned gas across a rotating blade can be neglected.

Heat transfer coefficients at the cooling hole surfaces are given as follows, depending on the cooling fluid.

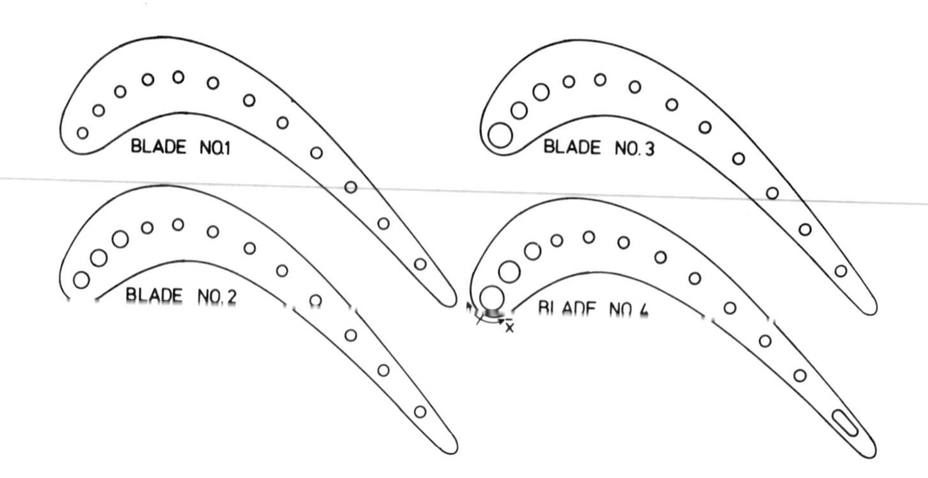

FIGURE 1. Shapes of blades and cooling holes

1. The heat transfer coefficient for air (or steam) cooling is a constant given as 1000 W/m^2K and that for water cooling as 10000 W/m^2K for any cooling holes. The variation of the heat transfer coefficient along the cooling hole periphery is neglected. Also, the temperature rise of the cooling fluid during flowing through the cooling holes is considered negligibly small.

2. Heat transfer coefficients for air-water mist flow have been obtained as follows by the experiment [5] in which Coriolis forces were simulated by centrifugal forces by use of helical thin tubes. Heat transfer coefficients for large rates of mist flow are as large as 10000 W/m^2K because of the evaporation of liquid film on the cooling surface. On the other hand, the heat transfer coefficient for flow without mist is 1000 W/m^2K, the same as that for air cooling. In between these two limiting heat transfer

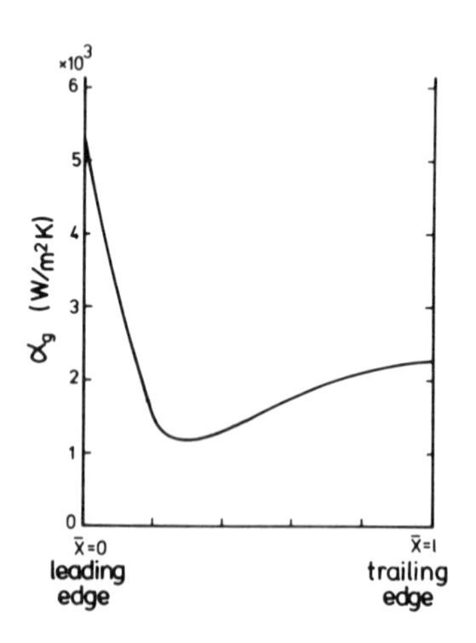

FIGURE 2. Heat transfer coefficients at the outer surfaces of the blades

coefficients, arbitrary heat transfer coefficients can be obtained by adjusting the mixture ratio in a mist flow.

Therefore, in this study, heat transfer coefficients at each cooling hole for mist cooling are assumed to be varied by controlling the mist rates in each hole. The temperatures of mist flows are considered to be constant through the cooling holes by taking mist evaporation into account. In the experiment stated above [5], circumferential variations of heat transfer coefficients in thin tubes were observed by simulated Coriolis forces. But this effect of Coriolis forces in rotating blades is left for future works and heat transfer coefficients are assumed uniform in each cooling hole, because the objective of the present study is to know the feasibility of mist cooling.

The variation of the thermal conductivity of blade material λ with temperature is neglected in the temperature calculation since the variation does not affect the results greatly and for simplicity.

On these assumptions, the equation of conduction for temperatures in a blade is solved by a finite element method with the appropriate boundary conditions at the surfaces of the blade and cooling holes. The four blades shown in Fig.1 are divided into about 700 triangle elements by about 430 nodes, and the cooling holes are approximated by hexagonals or octagonals.

2.2 Stress Profile

Thermal stresses resulting from non-uniform thermal expansion are calculated using the obtained temperature profiles. The blades are also dealt as two-dimensional, and the displacements u and v at each node for x and y direction, respectively, which are calculated from the temperature profile, give the principal stresses σ_1 and σ_2 and their direction θ_1 and θ_2 in each element. Plane strain or generalized plane strain in the two-dimensional stress field are assumed. Temperature dependences of linear expansion coefficient α, Young's modulus E and Poisson ratio ν of the blade material are incorporated in the calculation [7].

2.3 Allowable Stress

The predicted principal stresses in each element are compared with the allowable stress of the element σ_a.

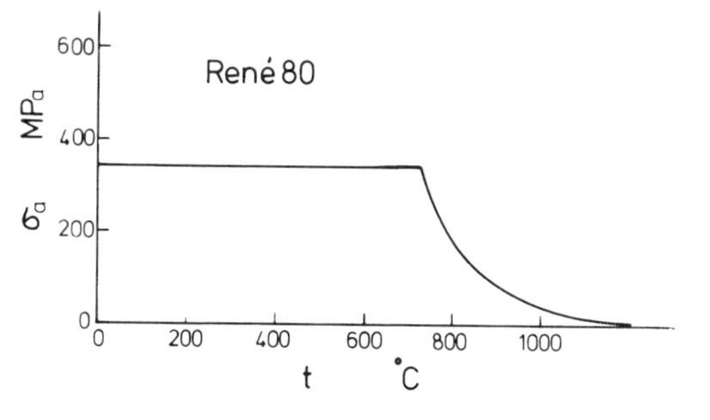

FIGURE 3. Allowable stress of René 80

René 80 is taken as the sample material for blades. Allowable stress is determined as the smaller stress of either one third of the tensile strength for short duration at the temperature or two third of the stress for 10^4 hours breaking by long duration creep tests at the same temperature. Figure 3 shows the allowable stress for René 80 [8].

3. RESULTS

Figures 4, 5 and 6 show the temperature and stress profiles of blade No.4 used for the first stage bucket of a four-stage turbine. The turbine inlet temperature is fixed as t_g= 1200°C and the temperature of cooling fluid t_c= 200°C. Figure 4 shows the result for air cooling, Fig.5 for water cooling and Fig.6 for mist cooling. In each figure, constant temperature curves of the whole blade are shown in the upper half and principal stresses which are larger than the allowable stress of each element are partially shown in the lower half by focusing our attention especially on the vicinity of the leading edge, since no over-stressed elements appears from the fifth hole and after along the chord direction for water and mist cooling. The lengths of the arrow bars in the stress profiles show the ratios of principal stresses to the allowable stresses $|r_\sigma|$, with the segments showing the ratio of 5 attached to the bottom of each figure for comparison. Outward arrows mean tensile stresses and inward arrows compression stresses. The gas temperature at the inlet of the first stage bucket is t_r= 1136°C for the turbine inlet temperature t_g= 1200°C and the heat transfer coefficients at the cooling holes are given as shown in Table 1. The stresses are obtained by a plane strain condition.

Air cooling shown in Fig.4 has poor cooling performance resulting in the number of elements with stresses larger than the allowable stresses to be N_σ=130 which is more than one fifth of the total number of the elements. The over-stressed elements are distributed through the whole blade although it is not shown. The maximum temperature is higher than 1000°C. On the other hand, water cooling in Fig.5 has good cooling performance showing no over-stressed elements downstream of the fifth hole. However, the areas in the vicinity of cooling holes are

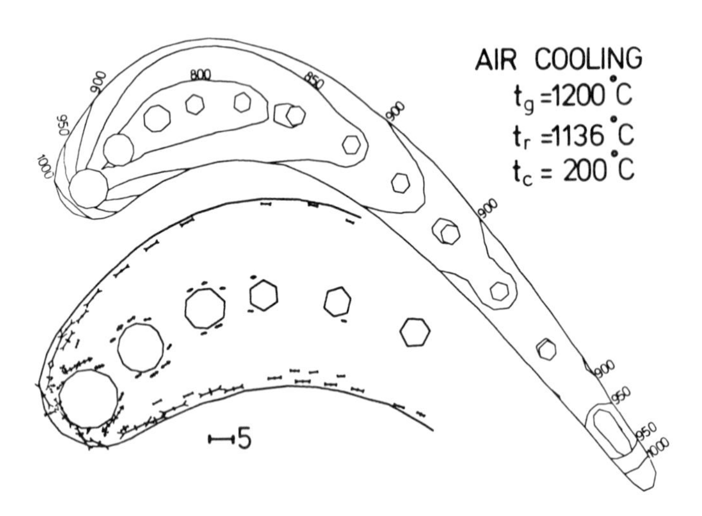

FIGURE 4. Temperature and stress profiles for air cooling

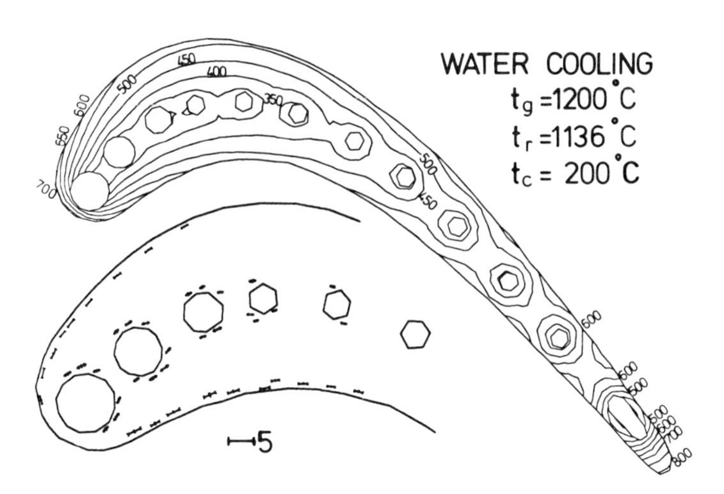

FIGURE 5. Temperature and stress profiles for water cooling

TABLE 1. Heat transfer coefficients at cooling holes

$\times 10^4$ W/m²K

No. of cooling hole	1	2	3	4	5	6	7	8	9	10	11	12
Air cooling	0.1											
Water cooling	1.0											
Mist cooling	0.8	0.2	0.2	0.1	0.1	0.2	0.3	0.4	0.5	0.6	0.8	1.0
Mist loading	.12	.03	.03	.00	.00	.03	.06	.08	.09	.10	.12	.20

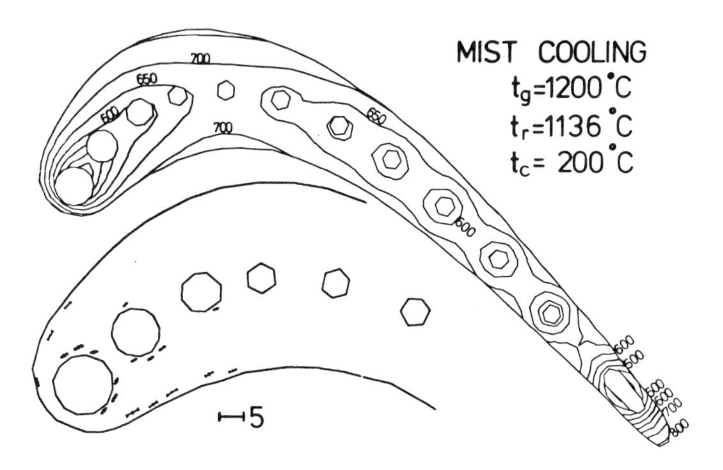

FIGURE 6. Temperature and stress profiles for mist cooling

cooled too much as seen from the temperature profile, causing sharp temperature gradients. Thus, the thermal stresses of the elements near the leading edge become larger than the allowable stresses, where the heat transfer coefficient at the blade surface is large. The number of over-stressed elements is N_σ=43 in this case.

In case of mist cooling shown in Fig.6 where heat transfer coefficients vary with cooling holes, the second to fifth cooling holes are not excessively cooled resulting in smaller temperature gradients than those for water cooling. Consequently, the number of the over-stressed elements is N_σ=19 which is less than half of N_σ=43 for water cooling. Thus, the mist cooling with controlled heat transfer coefficients is shown to be superior to air and water cooling. The distribution of the heat transfer coefficients shown in Table 1 gave the best performance among several different distributions tesyed for comparison. Table 1 also includes the values of the mist loading in each hole which shows the ratio of mist to total flow rate for that distribution estimated from the results of the reference [5]

Figure 7 shows N_σ as a function of t_g for the three cooling methods using the blade No.4. Air cooling is satisfactory up to t_g= 1000°C, but it does not suffice for higher temperatures at all. The number of over-stressed elements N_σ increases rapidly with increase in turbine inlet temperature t_g. On the other hand, the number of over-stressed elements N_σ for water cooling does not increase considerably even at the temperature of t_g= 1300°C. However, N_σ for water cooling at t_g= 1000°C and 1100°C are larger than those for air cooling at the same temperatures, since water cooling locally cools the blades down to unnecessarily low temperatures. Mist cooling to provide the appropriate heat transfer coefficient distribution has a smaller N_σ than air cooling and water cooling at any temperature, thus showing superiority to air cooling and water cooling again.

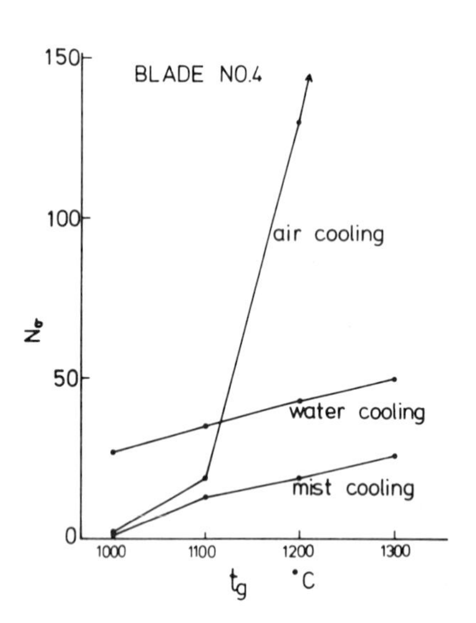

FIGURE 7. Number of over-stressed elements, N_σ

Figure 8 shows the maximum absolute values of the stress ratio $|r_{\sigma,max}|$ and the local temperatures of those elements. $|r_{\sigma,max}|$ is defined as the maximum of the absolute ratio of the local stress to the local allowable stress. For air cooling at t_g= 1300°C, $|r_{\sigma,max}|$ is more than 10 and the temperature of the element is about 1000°C meaning that the strength of the material is dependent on its creeping strength. The value for water cooling and mist cooling at t_g= 1300°C , $|r_{\sigma,max}|$, is about 2.2 and the temperature of the element is 600 to 650°C where the tensile strength for short duration should be considered as the allowable one. This maximum temperature suggests the choice of material and the direction of new materials development for mist cooling application.

The effects of hole size on N_σ are examined in Fig.9 using the four blades having different hole sizes as shown in Fig.1 at t_g=1100 and 1300°C. N_σ of mist cooling is smaller than those of air cooling and water cooling for any blades. The effect of hole size at t_g= 1100°C is not significant, but at t_g= 1300°C N_σ decreases with increasing hole size (increasing blade number) for all the three cooling methods. Among them, the decreasing rate is the most remarkable for mist cooling, indicating that mist cooling becomes more advantageous as the diameter of cooling holes in the near-leading edge region become large. Hence, mist cooling could be highly favorable to the advanced cooling blades which are supposed to have large cooling holes with thin shells.

The results shown in Fig.4 to Fig.9 were obtained in the condition of plane strain which might give too large compression stresses in the height direction. Then, stresses were calculated in the generalized plane strain condition which approximates two-dimensionality better to evaluate the stresses of the elements far enough from the blade root. The generalized plane strain approximation usually give stresses in the height direction smaller than the allowable

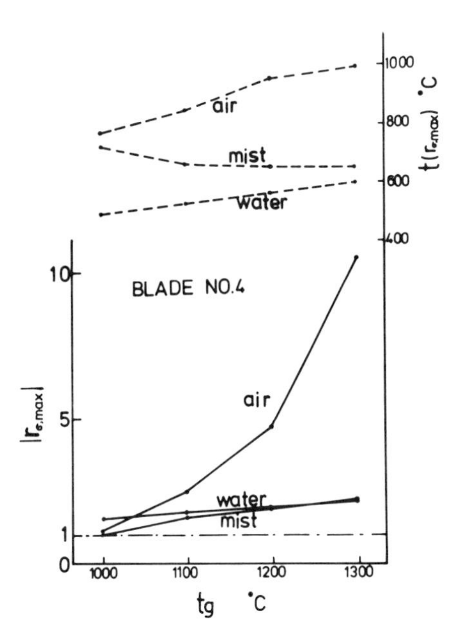

FIGURE 8. The maximum absolute values of the stress ratio $|r_{\sigma,max}|$ and the temperatures of those elements

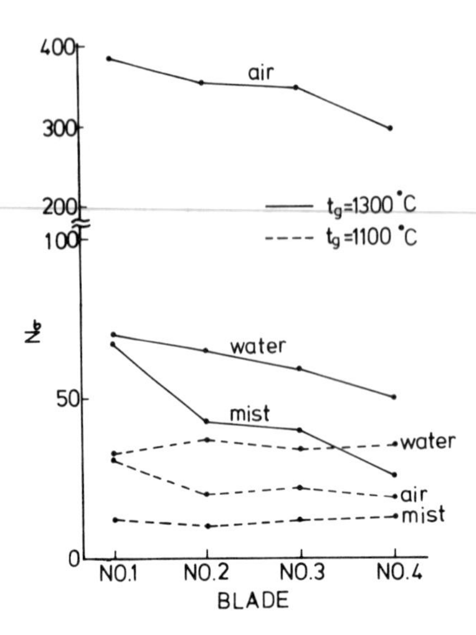

FIGURE 9. Effect of size of cooling holes

stresses and reduces N_σ to almost half of N_σ given by the plane strain condition. For the same conditions as in Fig.8, the generalized plain strain condition predicts $|r_{\sigma,max}|$ for water cooling and mist cooling as about 1.6 at t_g= 1300°C, while it was about 2.2 when the plane strain condition was used. But superiority of mist cooling to air cooling and water cooling was not altered by the change from the plane strain to the generalized plane strain condition. Consequently, even the more detailed three-dimensional analyses would show that mist cooling with controlled distributed heat transfer coefficients is preferable to air cooling and water cooling from the view point of temperature and stress analyses.

It is well expected that saturated steam-water mixtures in place of air-water mixtures can be used for the blade cooling, particularly in combined cycles. However, on the assumption that the similar cooling performance could be obtained as explained in this paper, no detailed discussion on steam-water mist cooling has been made. It is to be suggested that the mist cooling technique is also applicable to blades with many tiny holes located close to and along the blade surface, but detailed discussion is left for a future study. Mixing of water into air could be done upstream of the cooling holes of blades or on the way of cooling air flow passage, and is not considered to bring in serious construction difficulties.

4. CONCLUSION

Performances of air-water mist cooling for high temperature gas turbine blades were investigated by temperature and stress analyses. The mist cooling method featured by a controlled heat transfer coefficient at each hole was compared with air cooling or water cooling alone. As the result, mist cooling provided with appropriately distributed heat transfer coefficients has cooling performance better than air cooling and water cooling in cases studied in the present paper. Specifically, mist cooling is of great advantage to high turbine inlet temperatures and to large cooling holes. Taking the flow stability of mist flow into consideration, in addition to its high cooling performance, mist cooling using air-water mist flow or steam-water mist flow has been shown to be very promising for high temperature gas turbine blade cooling.

REFFERENCES

1. Hori, A., and Takeya, K., Outline of Plan for Advanced Research Gas Turbine, *ASME Paper*, No. 81-GT-28, 1981

2. Alff, R.K., Manning, G.B., and Sheldon, R.C., The High Temperature Water Cooled Gas Turbine in Combined Cycle with Integrated Low Btu Gasification, *ASME Paper*, No. 77-JPGC-GT-7, 1977

3. Rice, I.G., The Reheat Gas Turbine with Steam-Blade Cooling - A Means of Increasing Reheat Pressure, Output and Combined Cycle Efficiency, *ASME Paper*, No. 81-GT-30, 1981

4. Obata, M., and Taniguchi, H., Steam Cooling of Gas Turbine Blade, *J. of Japan Soc. of Mech. Eng.*, vol.87, no.788, pp.690-695, 1984 (in Japanese)

5. Mori, Y., Hijikata, K., and Yasunaga, T., Study on Inner Mist Cooling of High Temperature Small Tubes, *Trans. of Japan Soc. of Mech. Eng.*, series B, vol.47, no.419, pp.1332-1340, 1981 (in Japanese)

6. Nakayama, W., Torii, T., and Ikegawa, M., Assesment of Cooling Methods for Gas Turbine Blades in Terms of "Entropy Production", *J. of Gas Turbine Soc. of Japan*, vol.8, no.29, pp.41-50, 1980 (in Japanese)

7. Allen, J.M., Effect of Temperature Dependent Mechanical Properties on Thermal Stress in Cooled Turbine Blades, *ASME Paper*, No. 81-GT-105, 1981

8. Mechanical Properties Datacenter, DOD, *Aerospace Structural Metals Handbook*, vol.5, code 4214, Battelle Columbus Laboratories, 1980

Local Heat/Mass Transfer Distribution around Sharp 180° Turn in a Smooth Square Channel

J. C. HAN, P. R. CHANDRA, and S. C. LAU
Turbomachinery Laboratories
Mechanical Engineering Department
Texas A&M University
College Station, Texas 77843

ABSTRACT

The heat transfer characteristics of turbulent flow in a three-pass square channel were studied via the naphthalene sublimation technique. The test channel, which consisted of three square channels connected with two sharp 180° turns, resembled the internal cooling passages of gas turbine blades and vanes. The top and bottom surfaces of the test channel were two naphathalene plates. The distributions of the local heat (mass) transfer coefficient on the bottom wall of the test channel were measured from the channel entrance to just upstream of the second 180° turn for three Reynolds numbers ranging from 12,500 to 50,000. The results showed that the spanwise-averaged heat transfer decreased initially with increasing distance from the channel entrance. The spanwise-averaged heat transfer then increased sharply entering the 180° turn and reached its maximum value just before the end of the turn. The highly detailed measurements of the local heat transfer distributions at the 180° turn revealed that there existed a low heat transfer region near the inner wall and a region of high heat transfer near the outer wall. The former is believed to be due to flow separation while the latter may be caused by the increase in the turbulence intensity of the flow at the turn and/or flow reattachment. For all three Reynolds numbers investigated, the local heat transfer coefficients at the 180° turn were found to be two to three times higher than the fully developed values.

1. INTRODUCTION

In modern gas turbine blades, cooling air is circulated through multi-pass internal cooling passages to remove heat from the blade external surfaces which are directly exposed to the flow of hot gases (Figure 1). Turbulence promoters are often cast onto the two opposite active walls of the cooling channels in order to enhance the heat transfer to the cooling air [1]. Since the cooling passages are generally not very long ($X/D \sim 10$) between turns, the flow of the cooling air is not fully developed anywhere in the cooling passages. Therefore, the heat transfer and pressure drop data for conventional fully developed channel flow cannot be used in the design of such flow passages.

Earlier studies [2-4] showed that the secondary flows induced by centrifugal forces in curved tubes and channels increased both the heat transfer and the pressure drop. However, the channel geometries which were studied in [2-4] were very different from those generally encountered in turbine blade cooling passages which can better be modelled as straight rectangular channels connected by sharp 180° turns. Metzger, et. al. [5] systematically determined the effects of

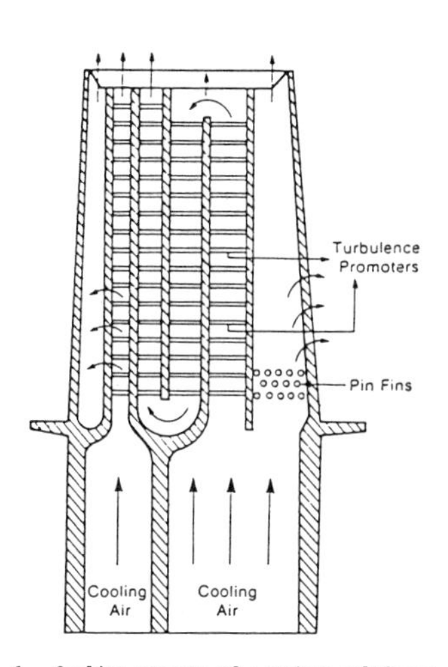

Fig. 1. Cooling concept of a modern multipass turbine blade [1]

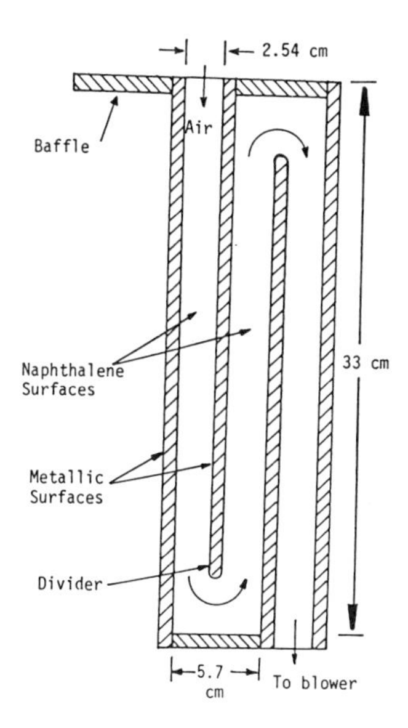

Fig. 2. Sketch of three-pass test channel

the turn geometry and the flow Reynolds number on the pressure drop in smooth rectangular channels with sharp 180° turns. The variable geometric parameters in [5] were the ratio of the upstream and downstream channel widths, the ratio of the channel depth to the channel width, and the ratio of the clearance height at the tip of the turn to the channel width. Flow visualization experiments in [5] revealed a large separated flow region downstream of the sharp 180° turn. Such a flow recirculation zone may create a low heat transfer region at the turn which is undesirable for the effective cooling of gas turbine blades.

Boyle [6] studied the heat transfer in both smooth and rib-roughened square channels with 180° turns. The top and bottom walls of the test channel were heated uniformly by allowing electric current to pass through 0.05-mm (0.002-in) thick Inconnel foils. The smooth channel results showed that the heat transfer decreased with increasing distance from the channel entrance. The heat transfer increased sharply at the turn and then decreased almost as rapidly downstream of the turn. The highest heat transfer at the turn was found to be three times the fully developed values. In the rib-roughened channel, similar trends were observed and the local heat transfer coefficients were found to be much higher than those in the smooth channel. Since the test apparatus used in [6] was only sparsely instrumented with thermocouples, the detailed distributions of the heat transfer coefficient at the sharp 180° turn could not be determined.

The determination of the detailed distributions of the heat transfer coefficient at the turns in multipass turbine blade cooling channels is important for two reasons. Firstly, the detailed heat transfer distribution data will enable researchers to understand the effect of a sharp 180° turn on the channel surface heat transfer and will provide researchers with a data base to develop numerical models to predict the flow field and heat transfer characteristics in such channels. Secondly, the detailed local heat transfer data will help engineers design effectively-cooled turbine blades which are not susceptible to structural failure due to uneven thermal stresses.

Recently, naphthalene sublimation technique was successfully used to determine the highly detailed heat/mass transfer coefficient distributions on plate-fin and tube heat exchanger surfaces [7]; on film cooling heat transfer surfaces with one row of inclined jets [8]; and on rib-roughened surfaces in high-aspect-ratio rectangular channels [9]. In the present investigation, the naphthalene sublimation technique was employed. The test section was a three-pass square channel, the top and bottom walls of which were two naphthalene plates. The naphthalene surfaces are analogous to surfaces at uniform temperature. The surfaces of the side walls of the channel were of aluminum and they are analogous to adiabatic surfaces. Highly detailed heat/mass transfer distributions at the sharp 180° turn were obtained for three Reynolds numbers between 12,500 and 50,000. The streamwise distributions of the spanwise-averaged heat/mass transfe coefficient between the channel entrance and immediately upstream of the second 180° turn were also determined.

2. EXPERIMENTAL APPARATUS AND PROCEDURE

2.1 Test Section

A schematic of the test section is shown in Figure 2. The test section was a three-pass smooth channel of square cross section. The channel width, the channel height, and the gap at the tip of the divider wall all measured 2.54 cm (1 in.). In order to simulate turbine cooling passages, the ratio of the divide thickness to the channel width (t/D) and the ratio of the upstream (and downstream) channel length to the channel width (X/D) were kept at 0.25 and 13, respectively. The top and bottom walls of the test duct were of naphthalene. The smoothness of the surface of the naphthalene plates was comparable to the highly polished stainless steel plate against which they were cast. The outer side walls and the inner divider walls were constructed of 0.63-cm (0.25-in.) thick aluminum plates. A relatively large metallic baffle was attached to the inlet o the test section to provide a sudden contraction entrance to the flow channel.

2.2 Instrumentation

The key feature of the mass transfer experiments was the instrumentation used to measure the highly detailed distribution of the mass transfer on the naphthalene surface. A Starrett electronic depth gage with an accuracy of 0.00001 in./0.0001 mm (digital readout) was used to determine the contour of the naphthalene surface before and after a test run. The depth gage consists of an electronic amplifier and a level-type gaging head. The naphthalene plate was mounted firmly on a coordinate table while the gaging head was affixed to a dial-gage-stand with a magnetic base so that the gaging head overhung the coordinate table. The coordinate table allowed the traverse of the naphthalene plate in two perpendicular directions (X-Y) tangential to the plate surface such that the surface contour of the plate could be measured.

2.3 Procedure

The naphthalene casting, tightly wrapped in sealed plastic bags to prevent sublimation, was left in the laboratory for four hours to attain thermal equilibrium. Immediately before a test run, the contour of the naphthalene surface (the bottom wall of the test section) was measured. The entire test section wa then assembled and attached to the rest of the test rig. The test rig was operated in the suction mode. Air was drawn through the test section from the naphthalene free laboratory with a blower. The air then passed through a calibrated orifice flow meter and eventually exhausted to the outside of the buildi

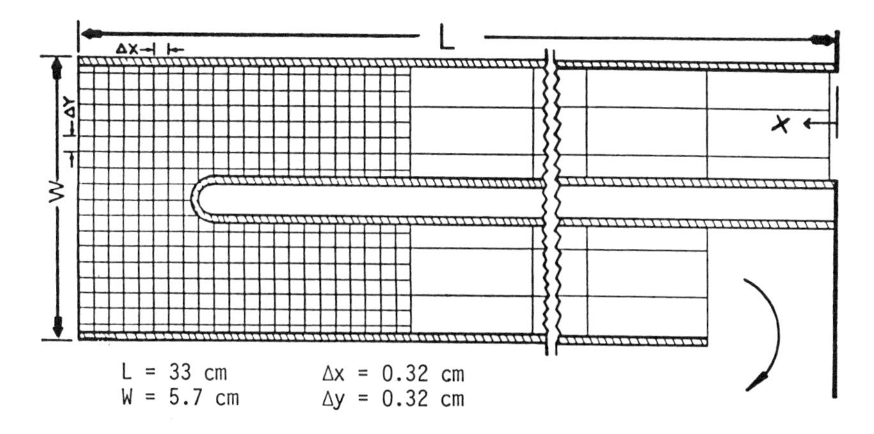

Fig. 3. Grid stations for measurements around sharp 180° turn

During the test run, the air temperature, the pressure drop across the orifice plate, the static pressure upstream of the orifice plate, and the atmospheric pressure were measured and recorded periodically. Immediately after the test run, which took about 30 minutes, the contour of the naphthalene surface (the bottom wall of the test section) was measured again. From the two contour measurements, the elevation changes at various points on the naphthalene surface were calculated. Contour measurements were made at as many as 329 discrete locations in the sharp 180° turn region and at 38 other selected locations on the rest of the naphthalene-covered bottom wall of the test channel. As shown in Figure 3, the measurement stations in the sharp 180° turn region formed a fine square grid with the distance between adjacent stations equals 0.32 cm (0.125 in.).

3. DATA REDUCTION

The local mass transfer coefficient, h_m, at any measurement point is given by the following equation

$$h_m = \dot{m}''/(\rho_w - \rho_b), \tag{1}$$

where $\dot{m}''$ is the mass flux at the measurement point; ρ_w and ρ_b are the naphthalene vapor density at the measurement point and the bulk naphthalene vapor density, respectively. The mass flux was calculated from

$$\dot{m}'' = \rho_s \cdot \Delta Z/\Delta t, \tag{2}$$

where ρ_s is the density of solid naphthalene; ΔZ is the measured change of elevation at the measurement point; and Δt is the duration of the experiment. The vapor density of naphthalene at the measurement point, ρ_w, was determined based on Sogin's vapor pressure-temperature relation in conjunction with the perfect gas law [10]. The bulk naphthalene vapor density, ρ_b, was evaluated from the ratio of the cumulative mass of naphthalene transferred from the channel walls to the air to the volumetric flow rate of air.

The local Sherwood number is defined as

$$Sh = h_m \cdot D/\mathscr{D}, \tag{3}$$

where D is the channel width and $\mathcal{D}$ is the diffusion coefficient which can be expressed in terms of the Schmidt number, Sc, as follows

$$\mathcal{D} = \nu/Sc, \quad (4)$$

where the Schmidt number is 2.5 [10] and ν is the kinematic viscosity of pure air. The local Sherwood number can be converted to Nusselt number by using the heat and mass transfer analogy as follows

$$Nu = (Pr/Sc)^r Sh, \quad (5)$$

where the Prandtl number, Pr, is about 0.71 for air at room temperature; and the exponent, r, is 0.4.

The Nusselt number for fully developed turbulent channel flow correlated by Dittus and Boelter was used as the reference for comparison

$$Nu = 0.023\ Re^{0.8}\ Pr^{0.4}. \quad (6)$$

4. DISCUSSION OF EXPERIMENTAL RESULTS

The experimental results will now be presented and discussed. The distributions of the local Sherwood number were obtained for three values of Reynolds number, namely, 12,500, 25,000, and 50,000. For each Reynolds number, two separate runs were performed under the same test conditions on different days to verify the accuracy and reproducibility of the experimental data. It was found that the local Sherwood number data for corresponding runs differed by no more than five percent. Supplementary experiments were also conducted to determine the mass transfer by natural convection from the napthalene surface during the time the contour of the naphthalene surface was measured. These experiments showed that the mass loss by natural convection was on the order of one percent of the mass transfer during test runs.

In order to examine the effect of the flow channel geometry on the streamwise distribution of the local heat/mass transfer coefficient, the local Sherwood numbers were averaged in the spanwise direction. At the sharp 180° turn, the local Sherwood numbers were averaged along lines corresponding to the various values of the turning angle (θ). The spanwise-averaged Sherwood numbers as functions of the streamwise distance from the channel entrance for all three Reynolds number investigated are shown in Figure 4. In Figure 4, the Sherwood numbers calculated from the fully developed channel flow correlation are also provided for comparison. In general, the spanwise-averaged Sherwood number distributions for all three Reynolds numbers exhibit the same trend. The values of the Sherwood numbers are always higher than the corresponding fully developed channel flow values everywhere in the test channel. For each of the three cases, the Sherwood number decreases initially with increasing distance from the channel entrance. Further downstream, the Sherwood number distribution levels off. Just before entering the turn ($\theta \cong 0°$), the Sherwood number increases sharply and reaches its maximum value near the end of the turn ($\theta \cong 180°$). Two peaks can be identified in the turn region of the Sherwood number distribution. These peaks occur immediately upstream of the streamwise stations corresponding to $\theta \cong 90°$ and $\theta \cong 180°$, respectively.

The spanwise-average Sherwood number continues to increase downstream of the turn. Further downstream of the turn, the Sherwood number decreases and then increases as the second sharp 180° turn approaches. Figure 4 also shows that the Sherwood number in the straight section of the channel downstream of

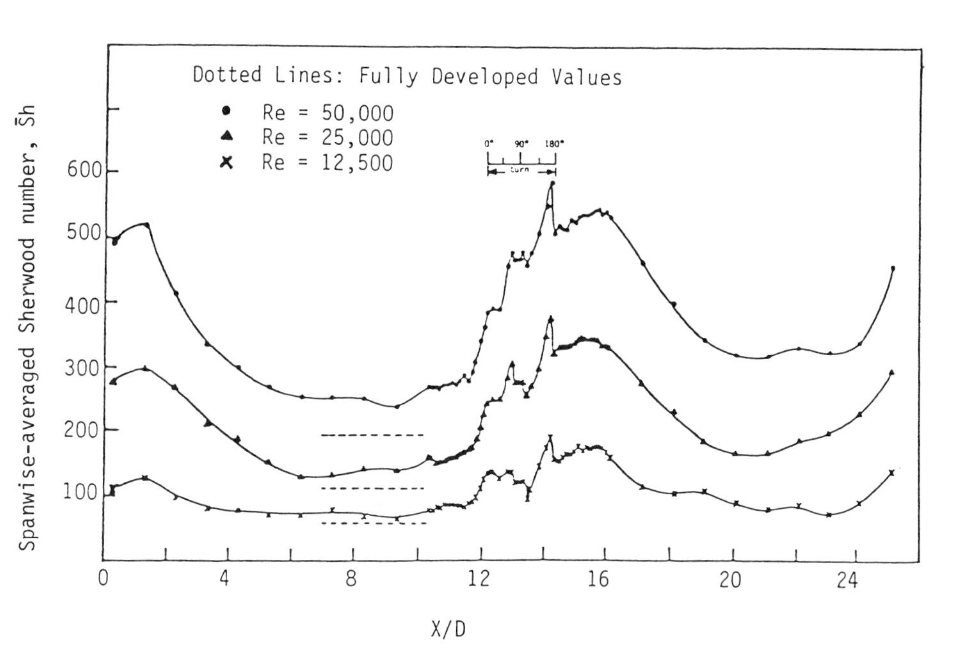

Fig. 4. Spanwise-averaged Sherwood number vs X/D ratio for varied Reynolds numbers

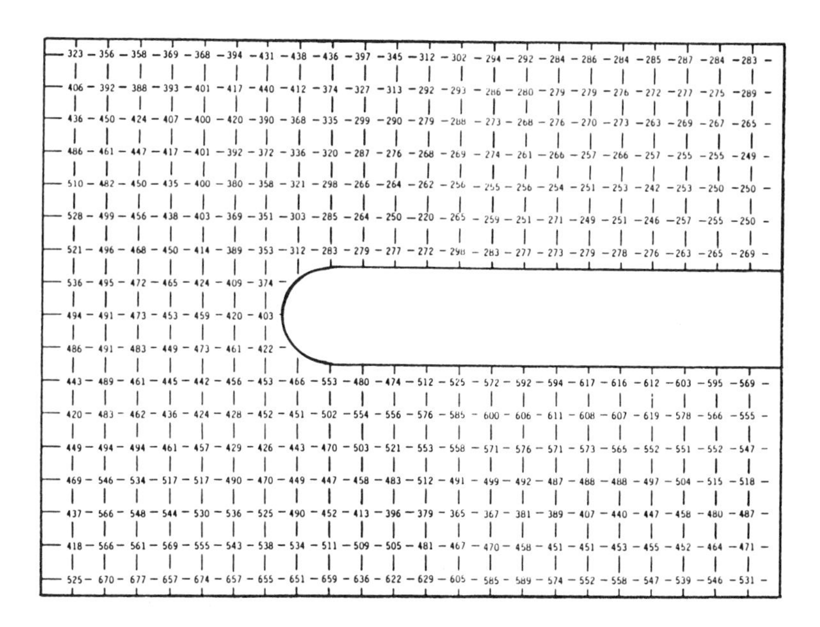

Fig. 5-(a). Local Sherwood number around sharp 180° turn for Re = 50,000

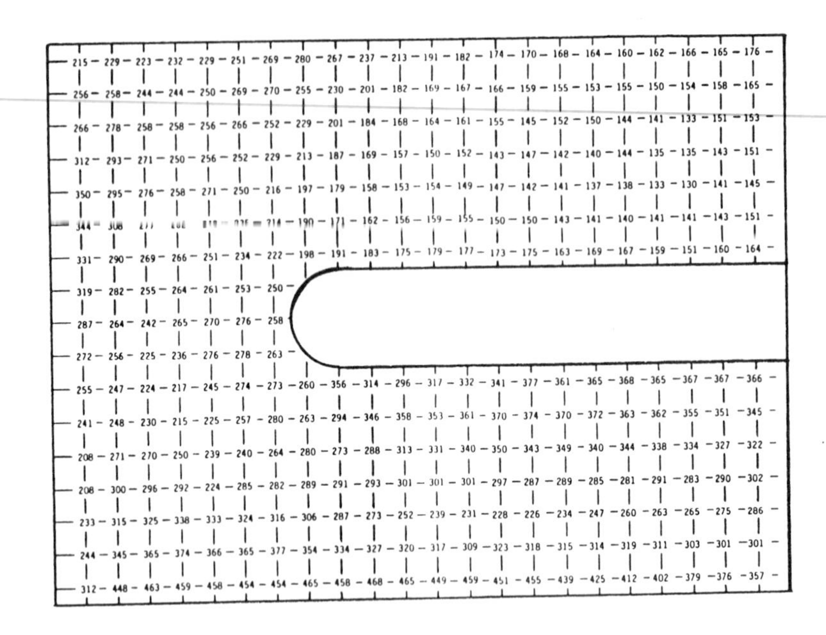

Fig. 5-(b). Local Sherwood number around sharp 180° turn for Re = 25,000

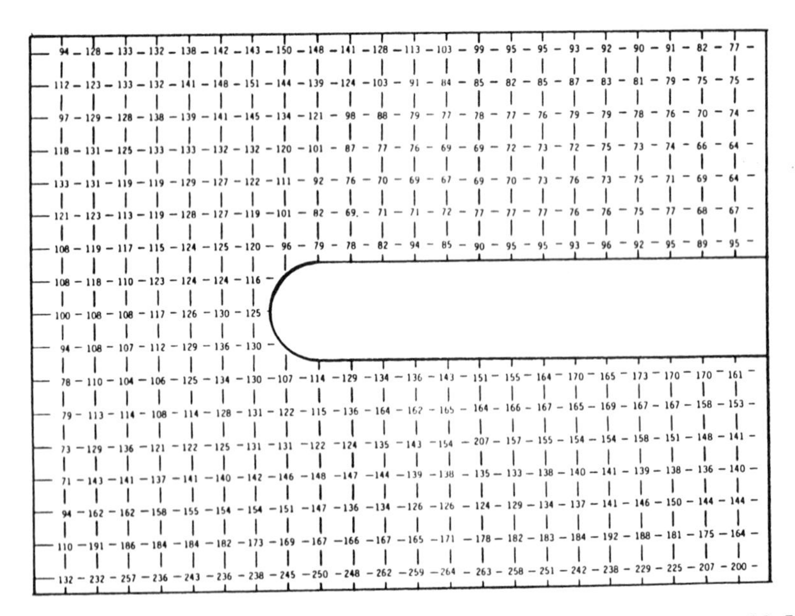

Fig. 5-(c). Local Sherwood number around sharp 180° turn for Re = 12,500

the first sharp turn is higher than that in the entrance section of the flow channel. This may be due to the fact that the flow becomes more turbulent after the sharp 180° turn. Finally, the small peak on the Sherwood number distribution which can be observed at $X/D \cong 1.5$ is probably due to flow separation at the sudden contraction entrance to the flow channel.

Highly detailed Sherwood number distributions around the sharp 180° turn (329 points) are tabulated in Figures 5-(a), (b), and (c). In order to examine closely the heat/mass transfer behavior around the turn, the ratios of the local Sherwood number to the fully developed value, Sh/Sh_0, at 45 selected measurement stations are re-tabulated in Figures 6(a) and (b) for Reynolds number of 50,000 and 12,500, respectively. At Re = 50,000, the Sherwood number ratio increases continuously from 1.3 before the turn to 3.0 after the turn along the inner wall. Along the outer wall, Sh/Sh_0 increases from 1.4 before the turn to 3.0 at $\theta \cong 135°$ and then decreases to 2.4 after the turn. The low heat/mass transfer zone (dotted lines in Figure 6(a)) near the inner wall at the turn may be due to flow separation at the tip of the divider wall. The high heat/mass transfer zone (solid lines) near the inner wall downstream of the aforementioned region is believed to be caused by flow reattachment. Around the turn, higher turbulent intensity in the flow is believed to contribute to the high heat transfer along the outer wall. Further downstream the low heat transfer along the outer wall may be due to the growth of the turbulent boundary layer.

At Re = 12,500 (Figure 6(b)), the distribution of Sh/Sh_0 exhibits trends which are slightly different from those for Re = 50,000. The low heat/mass transfer zone (flow separation) occurs at the end of the 180° turn along the inner wall, whereas the high heat/mass transfer zone (flow reattachment) is shifted further downstream along the inner wall. In addition, there is a low heat/mass transfer zone at $\theta \cong 90°$ and a high heat/mass transfer zone downstream of the turn along the outer wall. It appears that, at low Reynolds number, the flow downstream of $\theta \cong 45°$ tends to move along the inner wall, and the highly disturbed flow tends to move along the outer wall at the end of the turn. The distributions of local Sherwood number ratio in Figures 6(a) and 6(b) show that the maximum values of the local Sherwood number ratio can be as high as 3 in the turn region.

5. CONCLUSIONS

The local heat transfer characteristics in a three-pass smooth channel of square cross section were investigated via the naphthalene sublimation technique. The top and bottom of the test duct were of naphthalene. Highly localized heat/mass transfer measurements along the bottom wall of the test duct and around the sharp 180° turn region were made for three values of Reynolds numbers between 12,500 and 50,000. The following conclusions can be drawn:

1. The spanwise-averaged Sherwood number at any location of the test duct is higher than the fully developed values. The average heat/mass transfer downstream of the 180° turn is higher than that upstream of the turn. The maximum average Sherwood number is observed immediately before the end of the 180° turn.

2. The measured highly detailed heat/mass transfer distributions around the sharp 180° turn indicate that the local Sherwood number is about 2 to 3 times higher than the fully developed value for the corresponding Reynolds number.

Dotted Lines: Lower Heat Transfer Zone
Solid Lines: Higher Heat Transfer Zone

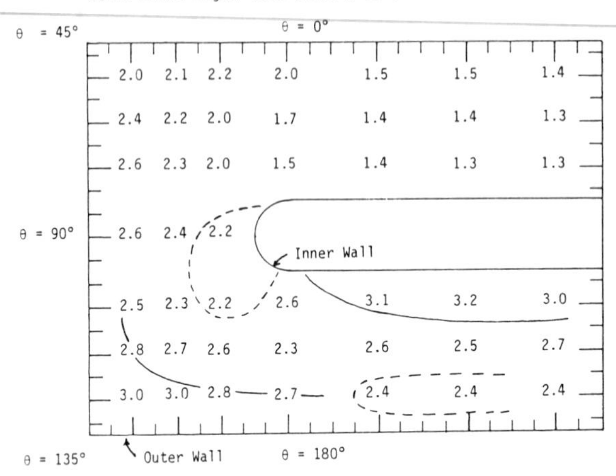

Fig. 6-(a). Local Sherwood number ratio Sh/Sh_o for Re = 50,000

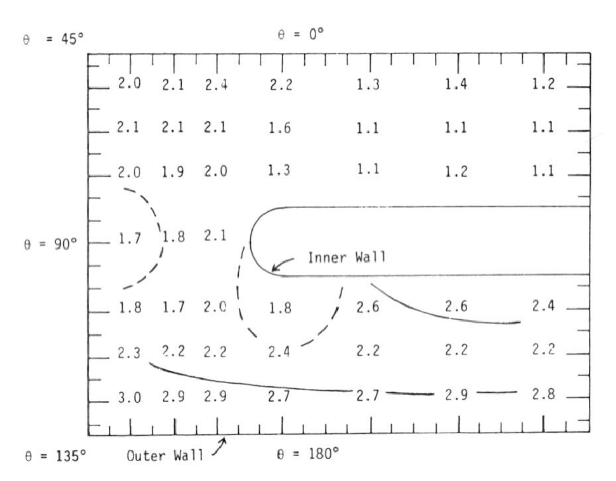

Fig. 6-(b). Local Sherwood number ratio Sh/Sh_o for Re = 12,500

3. At Re = 50,000, low heat/mass transfer zones are observed near the inne wall at $\theta = 90°$ ($Sh/Sh_o \cong 2.2$) and near the outer wall downstream of th turn ($Sh/Sh_o \cong 2.4$). At Re = 12,500, low heat/mass transfer zones are found near the inner wall downstream of the tip of the divider wall ($Sh/Sh_o \cong 1.8$) and near the outer wall at $\theta = 90°$ ($Sh/Sh_o \cong 1.7$).

ACKNOWLEDGEMENT

The first author would like to appreciate the partial financial support by the National Science Foundation under Grant MEA-8205234.

NOMENCLATURE

D	channel width
h_m	local mass transfer coefficient, eq. (1)
$\dot{m}''$	mass transfer rate per unit area, eq. (2)
Nu	Nusselt number, eq. (6)
Re	Reynolds number based on channel width, eq. (6)
Sh	Sherwood number, eq. (3)
$\bar{Sh}$	spanwise-averaged Sherwood number
Sh_o	fully developed Sherwood number
x	axial distance from duct entrance
θ	angle of turn, Figures 4 and 6

REFERENCES

1. Han, J.C., Park, J.S., and Lei, C.K., Heat Transfer Enhancement in Channels with Turbulence Promoters, ASME Paper No. 84-WA/HT-72, 1984 and to be published in J. of Engineering for Gas Turbines and Power.
2. Mori, Y., and Nakayama, W., Study on Forced Convective Transfer in Curved Pipes (2nd Report, Turbulent Region), Int. J. Heat Mass Transfer, Vol. 10, 1967, pp. 37-59.
3. Mori, Y., and Nakayama, W., Study on Forced Convective Transfer in a Curved Channel with a Square Cross Section, Int. J. Heat Mass Transfer, Vol. 14, 1971, pp. 1787-1805.
4. Seki, N., Fukusako, S., and Yoneta, M., Heat Transfer from the Heated Convex Wall of a Return Bend with Rectangular Cross Section, J. Heat Transfer, Vol. 105, 1983, pp. 64-69.
5. Metzger, D.E., Plevich, C.W., and Fan, C.S., Pressure Loss Through Sharp 180 Degree Turns in Smooth Rectangular Channels, J. of Engineering for Gas Turbines and Power, Vol. 106, 1984, pp. 677-681.
6. Boyle, R.J., Heat Transfer in Serpentine Passages with Turbulence Promoters, ASME Paper No. 84-HT-24, 1984.
7. Saboya, F.E.M., and Sparrow, E.M., Local and Average Transfer Coefficients for One-Row Plate Fin and Tube Heat Exchanger Configurations, J. Heat Transfer, Vol. 96, 1974, pp. 265-272.
8. Goldstein, R.J., and Taylor, J.R., Mass Transfer in the Neighborhood of Jets Entering a Cross Flow, ASME Paper No. 82-HT-62, 1982.
9. Sparrow, E.M., and Tao, W.Q., Enhanced Heat Transfer in a Flat Rectangular Duct with Streamwise-Periodic Disturbances at One Principal Wall, J. Heat Transfer, Vol. 105, 1983, pp. 851-861.
10. Sogin, H.H., Sublimation from Disks to Air Streams Flowing Normal to Their Surfaces, Trans. of ASME, Vol. 80, 1958, pp. 61-69.

Local Endwall Heat/Mass Transfer in a Pin Fin Channel

S. C. LAU, Y. S. KIM, and J. C. HAN
Turbomachinery Laboratories
Mechanical Engineering Department
Texas A&M University
College Station, Texas 77843

ABSTRACT

The distribution of the local endwall heat/mass transfer coefficient in a channel with a staggered array (X/D = S/D = 2.5) of pin fins was studied via the naphthalene sublimation technique. The length-to-diameter ratio was kept at 1.0 and the Reynolds number ranged from 9,000 to 33,000. Results show that, near the leading edge, the heat/mass transfer coefficient decreases as a result of the growth of the boundary layer and then increases rapidly as the flow encounters the pins on the first and the second rows. The streamwise distributions of the heat/mass transfer coefficient downstream of the first pin row are nearly periodic--the corresponding transfer coefficient distributions have similar shapes although the magnitudes of the transfer coefficients at corresponding locations may vary. Downstream of the first pin row, the heat/mass transfer coefficient is generally very high immediately upstream of a pin and is relatively high in the wake region downstream of a pin.

1. INTRODUCTION

In advanced gas turbine blades, pin fins are often cast onto the internal cooling channel near the narrow trailing edge region to increase the heat transfer to the cooling air. The cutaway view of a typical modern gas turbine blade is shown in Figure 1. The dimensions of the trailing edge of gas turbine blades and manufacturing constraints require that the length-to-diameter ratio of the pin fins be relatively small, generally between 0.5 and 4.0. Although heat transfer associated with crossflow over tube banks and flow in plate-fin and tube compact heat exchangers has been studied extensively [1, 2], data on heat transfer in pin fin channels with intermediate pin length-to-diameter ratios ($1/2 \leq L/D \leq 4$) were not available until recently. While heat is transferred primarily from the tube surfaces in tube banks ($L/D > 8$) and from the plates in compact heat exchangers ($L/D < 1/4$), heat is transferred from both the pin surfaces and the end walls in short pin fin channels such as those encountered in gas turbine blades.

VanFossen [3] studied the effect of the length-to-diameter ratio on the heat transfer from pin and endwall surfaces for a channel with four-row staggered arrays of short pin fins (L/D = 0.5 and 2). He found that the overall heat transfer coefficients were lower than data available in the literature for longer pins ($L/D \geq 8$) but about two times higher than those for a plain channel with no pins. Brigham and VanFossen [4] conducted similar experiments with staggered-arrays (four and eight rows) of slightly longer pin

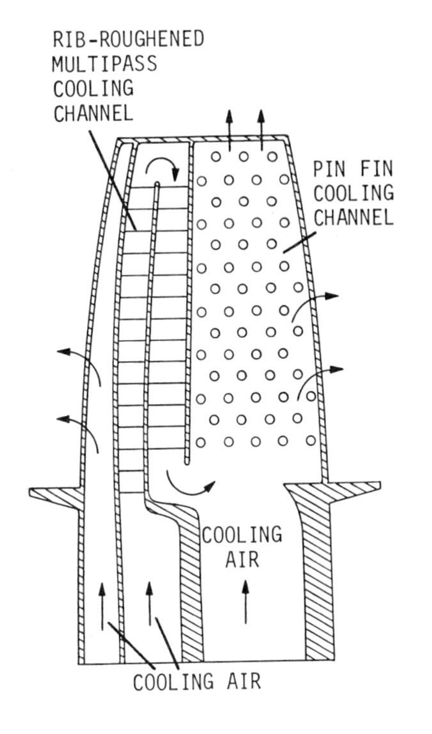

Figure 1. Conceptual Cutaway View of a Modern Internally-Cooled Gas Turbine Blade

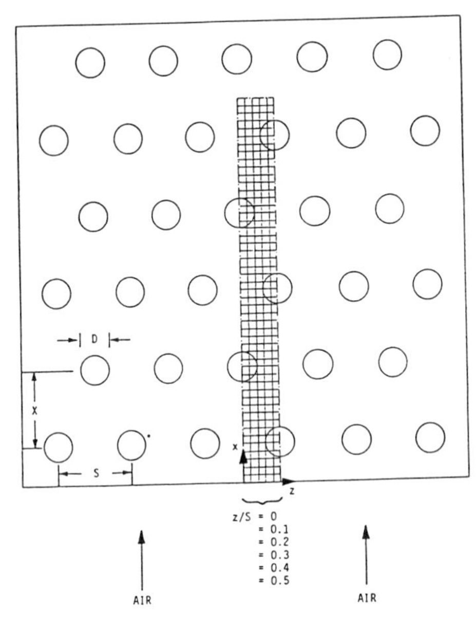

Figure 2. The Staggered Array of Pin Fins on Bottom Surface of Test Section Channel

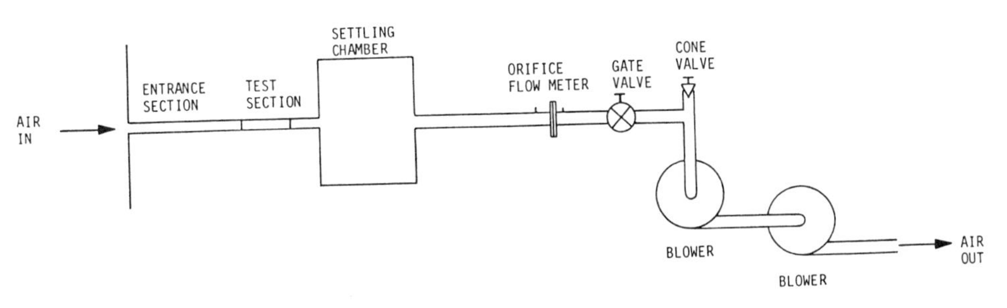

Figure 3. Schematic of Test Apparatus

fins (L/D = 4) and found that the overall heat transfer coefficients were higher than those for shorter pin fins (L/D = 0.5 and 2). Based on their results as well as those available in the literature, they concluded that, for short pin fins ($L/D \leq 2$), the heat transfer is a function of Reynolds number only, whereas the heat transfer for longer pin fins ($L/D > 2$) was a function of both the length-to-diameter ratio and the Reynolds number.

Metzger et al. [5] investigated the heat transfer for channels with ten-row staggered arrays (S/D = 2.5, X/D = 2.5 and 1.25) of short pin fins (L/D = 1). They found that, for all the cases studied, the spanwise-averaged heat transfer increased in the streamwise direction until it reached its maximum value at about the third to fifth row, then decreased gradually through the rest of the array. Simoneau and VanFossen [6] conducted experiments to measure the heat transfer from a heated pin fin (L/D = 3) placed at various locations staggered and in-line arrays of identical unheated pins in a rectangular channel. The spanwise turbulence intensity profiles immediately upstream of the row containing the heated pin were also measured. They reported that both the heat transfer from the pin and the average turbulence intensity for the staggered arrays increased then decreased with increasing number of rows of pins upstream of the heated pin. The results were consistent with those reported in [5].

In the aforementioned studies, the average of spanwise-averaged heat transfer from the pins and the endwalls in pin-fin channels were measured and reported. The only local heat transfer data available in the literature are those given by Saboya and Sparrow [7, 8]. They conducted experiments to measure the local heat/mass transfer coefficients on the endwalls of one-row and two-row pin fin channels via the naphthalene sublimation technique. However, their results may not be applied to the turbine blade trailing edge pin fin channels because they were for pin fins of a very small length-to-diameter ratio (L/D = 0.193) and for low Reynolds numbers. From a designer's point of view, local heat transfer data are needed in order to better understand the heat transfer characteristics of pin fin channels and to predict the distributions of local endwall temperatures in such channels. The need for local heat transfer data for pin fin channels motivated the present investigation.

The main objective of the present work is to study the distributions of the local endwall heat/mass transfer coefficient in a pin fin channel which resembles the cooling passage near the trailing edge of a typical gas turbine blade. The test section was a rectangular channel, of aspect ratio 15:1, in which a staggered array (S/D = X/D = 2.5) of six rows of pin fins of length-to-diameter ratio 1.0 was installed (Figure 2). The top wall of the test section was a naphthalene plate, which is analagous to a surface at uniform temperature. The pins were made of aluminum rods. The solid metal surfaces are analogous to adiabatic boundaries. Detailed local heat/mass transfer measurements were made at 300 points near the middle of the channel as shown in Figure 2 for Reynolds numbers of 9,000, 13,000, and 33,000.

2. EXPERIMENTAL APPARATUS AND PROCEDURE

Attention is now turned to the experimental apparatus which was designed and constructed for the present investigation. A schematic diagram of the apparatus is shown in Figure 3. The essential components of the open-loop flow circuit, arranged in the streamwise direction, were a rectangular channel of aspect ratio of 15:1 which included a hydrodynamic development section and the test section, a settling chamber, an orifice flow meter, a gate valve, a cone valve, and two centrifugal blowers. The apparatus was set up in a small, completely-enclosed, fully-equipped, and air-conditioned laboratory. Air, which was the working fluid, was drawn from the laboratory. Upon exiting the test apparatus, the air was ducted to the outside of the building.

The bottom wall and the two side walls of the flow channel were fabricated as one single piece from a 12.70-mm (0.50-in.) thick aluminum plate. The flow channel was a 9.53-cm (3.75-in.) wide and 6.35-mm (0.25-in.) deep rectangular slot on the top side of the plate. The top wall of the test section was a naphthalene cassette. The cassette, which measured 11.81 cm (4.65 in.) by 9.73 cm (3.83 in.), was a hollowed-out aluminum block of thickness 12.70 mm (0.50 in.). The naphthalene surface measured 9.53 cm (3.75 in.) by 9.53 cm (3.75 in.), and was framed by an aluminum border 1.02-mm (0.04-in.) and 11.43-mm (0.45-in.) wide, respectively, in the streamwise and spanwise directions. One 9.53-mm (0.375-in.) hole and three 3.18-mm (0.125-in.) holes were drilled on the back surface of the cassette to facilitate the pouring of molten naphthalene and to allow trapped air to escape during the casting of the mass transfer surface.

A staggered array of pins was installed on the bottom surface of the test section channel, as shown in Figure 2. The pins were made from 6.35-mm (0.25-in.) diameter aluminum rods. In order to ensure that all the pins were of the same height, the pins were initially cut to approximately 6.35-mm (0.25-in.) long before they were fastened with machine screws onto the bottom of the test section channel. The top of the pins were then cut off on a mill so that all the pins were of the same height. Disk-like rubber gaskets of thickness 0.64 mm (0.025 in.), which were cut with a 6.35-mm (0.25-in.) hole puncher, were cemented onto the top of the pins to prevent air leakage between the pins and the naphthalene surface during experiments.

The entrance section was 13.34-cm (5.25-in.) long, which is equivalent to 11.2 times the hydraulic diameter of the flow channel. The top wall of the entrance section was made of a 6.35 mm (0.25 in.) thick acrylic sheet. A large baffle attached to the upstream end of the entrance section provided an abrupt contraction entrance to the flow channel.

Leaving the test section, air passed through a short downstream duct and a settling chamber made of 6.35-mm (0.25-in.) thick acrylic sheets before entering an orifice flow meter.

Flow rate was controlled with a gate valve and a cone valve located between the orifice flow meter and two Cadillac Model HP33 blowers connected in series. As the air left the blower, it was ducted to the outside of the building. As a result, the laboratory was free of naphthalene vapor at any time during test runs.

The preparation of the mass transfer surface (the casting of the naphthalene plate) was done in a fume hood. The hollowed-out side of the aluminum cassette was placed against a professionally-polished stainless steel plate. Crystalline naphthalene was heated in a glass beaker until it melted. The molten naphthalene was then poured through a funnel and through the 9.53-mm (0.375-in.) hole on the back of the cassette. The naphthalene which filled the cavity in the cassette was then allowed to cool. After several minutes, the cassette was separated from the stainless steel plate with a light hammer tap on the side of the cassette in a direction tangential to the stainless steel plate surface. The resulting surface of the naphthalene was as smooth and as flat as the stainless steel plate surface against which it was cast.

A Starrett Model 812 electronic depth gage with an accuracy of 0.0001 mm (0.00001 in.) was used to measure the surface contour of the naphthalene plate. The depth gage consisted of a Model 812-13 electronic amplifier and a Model 812-1 level-type gaging head . The gaging head was mounted on a fixed standard dial gage stand with a magnetic base. In order to measure the surface contour of the naphthalene plate, the plate was mounted on a Jet Model CTSB-3 coordinate table such that the plate was directly below the gaging head of the depth gage. The contact ball on the spindle of the gaging head was then allowed to press (with a "contact pressure" of 8 to 12 grams as specified by the manufacturer) against the naphthalene surface. The coordinate table enabled the naphthalene plate to be traversed in two orthogonal directions in a plane perpendicular to the directions of movement of the gaging head spindle. For each experiment, the elevations at 300 discrete points (50 points and 6 points in the streamwise and spanwise directions, respectively, as shown in Figure 2) on the naphthalene surface were measured. Elevations were also measured on the aluminum frame surrounding the naphthalene. The latter measurements were neccessary to provide reference points to facilitate the reduction of the experimental data.

Pressure drop across the orifice plate and the static pressure upstream of the orifice plate were measured with either a mercury or an oil U-tube manometer. An Omega Model 412B temperature indicator with two copper-constantan thermocouples was used to measure the temperature of the air flowing through the test section. The atmospheric pressure was registered with a Princo Model C469 barometer. The duration of each experiment was recorded with a stopwatch.

A new naphthalene plate was prepared from fresh crystalline naphthalene for each experiment. The naphthalene plate, sealed in plastic bags, was left in the laboratory for at least twelve hours in order for it to attain the temperature of the laboratory. At the beginning of an experiment, the naphthalene plate was installed on the test section. Two large binder clips were used to hold the naphthalene plate in place. After five minutes, the naphthalene plate was removed and the contour of the naphthalene surface was measured. This procedure was necessary so that the contour measurement would include the deformation of the plate surface due to the slight pressure exerted by the pins on the plate surface.

Immediately after the contour measurement, the naphthalene plate was installed on the test section. The same two binder clips were used to hold the plate in place. Adhesive tape was used to prevent air leakage between the test section and the naphthalene plate.

Air was then allowed to flow through the test apparatus at a predetermined flow rate. During the experiment, the pressure drop across the orifice plate, the static pressure upstream of the orifice plate, the air temperature, and the atmospheric pressure were measured and recorded periodically.

A test run lasted from fifteen to thirty minutes, depending on the air flow rate. Immediately after each experiment, the contour of the naphthalene surface was measured again.

Supplementary experiments were also conducted to determine the effects of the pressure exerted by the pins on the naphthalene plate surface and sublimation by natural convection on the contour measurement data.

3. REDUCTION OF EXPERIMENTAL DATA

The reduction of the experimental data will now be described. The local mass transfer coefficient at any measurement point was evaluated from the following equation,

$$h_m = \dot{m}''/(\rho_w - \rho_b),$$

where $\dot{m}''$ is the mass transfer rate per unit area at the measurement point; ρ_w and ρ_b are the naphthalene vapor density at the measurement point and the bulk naphthalene vapor density, respectively.

The mass transfer rate per unit area at a measurement point was calculated from the measured change of elevation at the point, Δy,

$$\dot{m}'' = \rho_s \cdot \Delta y/t,$$

where ρ_s is the density of solid naphthalene; and t is the duration of the experiment.

The vapor density of naphthalene at the measurement point was evaluated from Sogin's vapor pressure-temperature relation [9] and the perfect gas law. The bulk naphthalene vapor density was determined by dividing the cumulative amount of naphthalene transferred from the naphthalene plate to the air by the volumetric air flow rate.

The local Sherwood number was defined as

$$Sh = h_m \cdot D/\mathcal{D},$$

in which $\mathcal{D}$, the diffusion coefficient, was determined by dividing the kinematic viscosity of pure air by the Schmidt number of naphthalene.

The flow Reynolds number was given by the following equation

$$Re_D = u_{max} D/\nu$$

where u_{max} and ν are the maximum air velocity in the test section, and the kinematic viscosity of air, respectively.

A computer program was written to reduce the experimental data. One of the special features of the computer program was the correction of the raw elevation data to take into account the fact that the naphthalene plate was tilted differently each time it was affixed to the coordinate table.

4. EXPERIMENTAL RESULTS

The distributions of the local endwall heat/mass transfer coefficient in a pin fin channel were obtained for a staggered array of pin fins with X/D = S/D = 2.5, L/D = 1.0, and for three Reynolds numbers: 9,000, 13,000, and 33,000. The local Sherwood numbers were calculated at 300 points on the endwall, which formed a rectangular grid between two symmetry lines near the middle of the flow channel (Figure 2). The grid extended from the leading edge (the upstream edge of the naphthalene surface) to halfway between the fifth and the sixth rows of pin fins. The distance between adjacent points was 1.59 mm (0.0625 in.), which was equivalent to 0.25 times the pin diameter. The local Sherwood numbers were plotted as functions of the distance from the leading edge (x/X) for various values of z/S. In addition, the detailed local

Sherwood number distributions near the pins on the first, the third, and the fifth rows for the highest Reynolds number run were tabulated in order to illustrate the two-dimensional row-by-row variation of the endwall heat/mass transfer.

The results will now be presented and discussed. In Figures 4, 5, and 6, the streamwise distributions of the local Sherwood number are given for z/S varying from 0 to 0.5. The figures are for Reynolds numbers of 9,000, 13,000, and 33,000, respectively. The distributions along z/S = 0 and 0.5 correspond to measurements along symmetry lines which pass through the centers of pins on alternate rows. The discussion of the figures, which will focus on the region between these symmetry lines, will begin with the first row and will proceed downstream to successive rows.

Figures 4, 5, and 6 show that, near the leading edge, the local Sherwood number decreases in the streamwise direction as a result of the growth of the concentration boundary layer. The Sherwood number then increases sharply when the flow encounters the pins on the first and the second rows. Since the pin on the first row is located very close to the leading edge, the limited number of measurement points directly upstream of the pin is not sufficient to show the U-shaped Sherwood number distributions along z/S = 0.3, 0.4, and 0.5. However, if the Sherwood numbers along the leading edge are assumed to be very large, as they should be, the sharp reversal of the slopes of the distributions upstream of the pin on the first row becomes evident. At z/S = 0 and 0.1, the Sherwood number decreases more gradually due to the larger distance between the leading edge and the pin on the second row.

The existence of the pin on the first row results in peaks on the Sherwood number distributions along z/S = 0.2. At Re = 9,000 and 13,000, a peak occurs at x/X = 0.4. The peak is believed to be caused by secondary flow created by the nearby pin on the first row. At Re = 13,000, the slope of the Sherwood number distribution changes sharply again at x/X = 1.0. This may be due to the fact that the flow is being forced away from the pin on the second row. At Re = 33,000, because of the high flow rate, the secondary flow created by the nearby pin on the first row remains attached to the pin surface longer than in the previous two cases. The effect of the secondary flow on the Sherwood number distribution along z/S = 0.2 is obvious further downstream at x/X = 1.1.

Along z/S = 0.3, a peak occurs on the distributions for the medium and high Reynolds number runs at x/X = 0.9. There is no such peak on the distribution for Re = 9,000.

The Sherwood number distributions become very nearly periodic downstream of the pin on the first row. In each of the three figures, the distribution along z/S = 0 has the same shape as that along z/S = 0.5, if the latter distribution is shifted downstream by a distance equal to the one-row width (X). In order to facilitate the comparison, the data points along the aforementioned distributions are represented by open circles in the figures. Similarly, the shapes of the Sherwood number distributions along z/S = 0.1 (squares) and z/S = 0.2 (diamonds) resemble the shapes of the distributions along z/S = 0.4 and 0.3, respectively. The peaks on the distributions along z/S = 0.2 and 0.3 clearly demonstrate the existence of secondary flow in the staggered array of pin fins investigated.

The Sherwood number is very large immediately upstream of a pin, as can be seen from the rapid rise of the Sherwood number distribution in the region

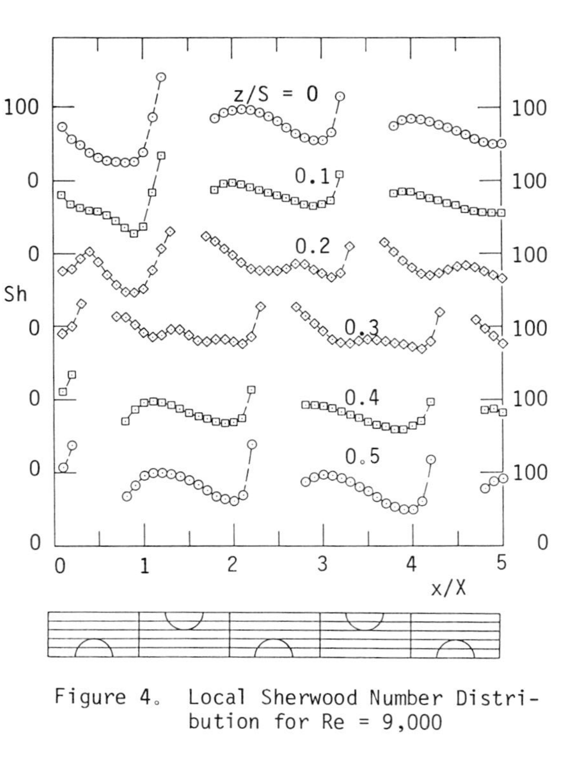

Figure 4. Local Sherwood Number Distribution for Re = 9,000

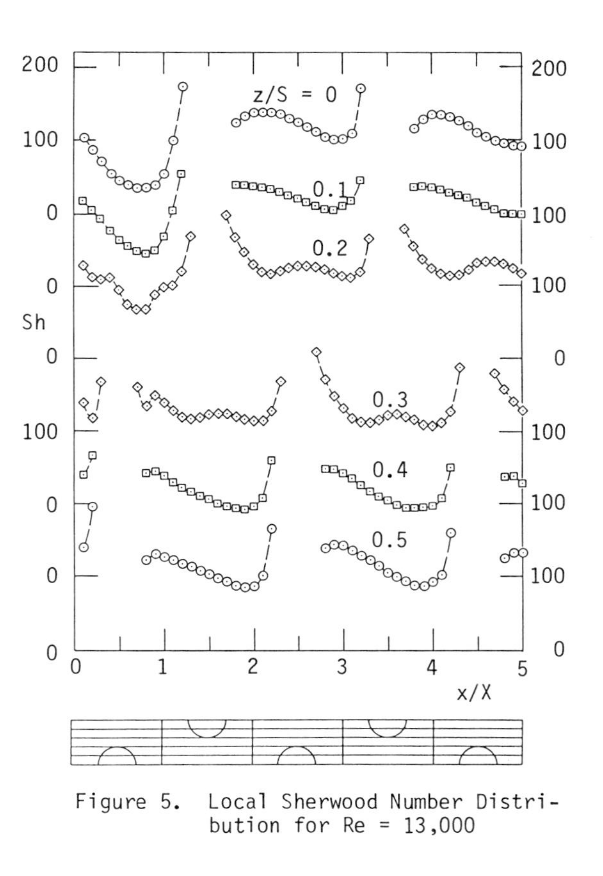

Figure 5. Local Sherwood Number Distribution for Re = 13,000

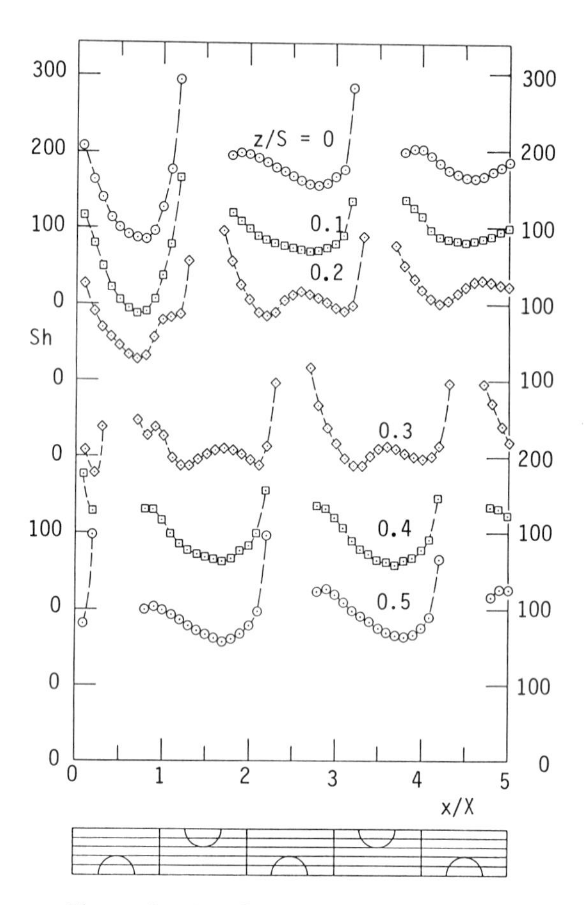

Figure 6. Local Sherwood Number Distribution for Re = 33,000

upstream of a pin. The Sherwood number is also relatively large in the region downstream of a pin, especially along $z/S = 0.2$ and 0.3. The results are consistent with Goldstein's [10] description of a horseshoe vortex system near the base of a cylinder in crossflow. The high values of the Sherwood number downstream of a pin ($z/S = 0$ and 0.5) may be caused by the high turbulence intensity in the wake region [6].

The relative magnitudes of the Sherwood numbers presented in Figures 4, 5, and 6 for the three different Reynolds numbers investigated reaffirm the results of previous investigations that the heat/mass transfer from the endwalls of a pin fin channel increases with increasing rate of mass flow through the channel.

In Figures 7, 8, and 9, the detailed two-dimensional distributions of the local Sherwood number are given for the highest Reynolds number investigated and for pin row numbers one, three, and five, respectively. Since no measurement was made along the leading edge, there are only 10 streamwise stations in the two-dimensional distribution for pin row number one (Figure 7), whereas there are 11 streamwise stations in the distributions given in Figures 8 and 9. Recognizing that the Sherwood number along the leading edge should be very large, the distribution in Figure 7 shows a decrease and then an increase in the Sherwood number in the streamwise direction. The Sherwood number is very large immediately upstream of the pin but is relatively small along $z/S = 0$, 0.1, and 0.2. The Sherwood number is also large in the wake region downstream of the pin. The maximum value in the region occurs at $z/S = 0.3$ and $x/X = 0.7$.

In Figures 8 and 9, the local Sherwood number is again very large in the region upstream and downstream of the pin. In the wake region, the maximum Sherwood numbers occur at $z/S = 0.3$ and $x/X = 2.7$ and 4.7, respectively. Along $z/S = 0$ and 0.1, the Sherwood number decreases and then increases in the streamwise direction. However, a peak is evident along $z/S = 0.2$ due to secondary flow created by the deflection of the main flow by the pin.

Finally, it should be pointed out that, since the rate of mass flow through the test channel was very high and the test section was relatively short, the calculated bulk density of naphthalene was very small compared to the vapor density on the naphthalene surface (less than one percent). The results, therefore, can be applied to a pin fin channel with two active endwalls without introducing any significant errors.

5. CONCLUDING REMARKS

Experiments were performed to study the local endwall heat/mass transfer in a pin fin channel with a staggered array ($X/D = S/D = 2.5$) of pin fins of length-to-diameter ratio 1.0. Reynolds numbers ranged from 9,000 to 33,000. The following conclusions can be drawn:

1. The overall heat/mass transfer from the endwalls of a pin fin channel increases with increasing rate of mass flow through the channel.

2. Near the leading edge, the Sherwood number decreases in the streamwise direction as a result of the growth of the concentration boundary layer, and then increases rapidly as the flow encounters the pins on the first and the second rows.

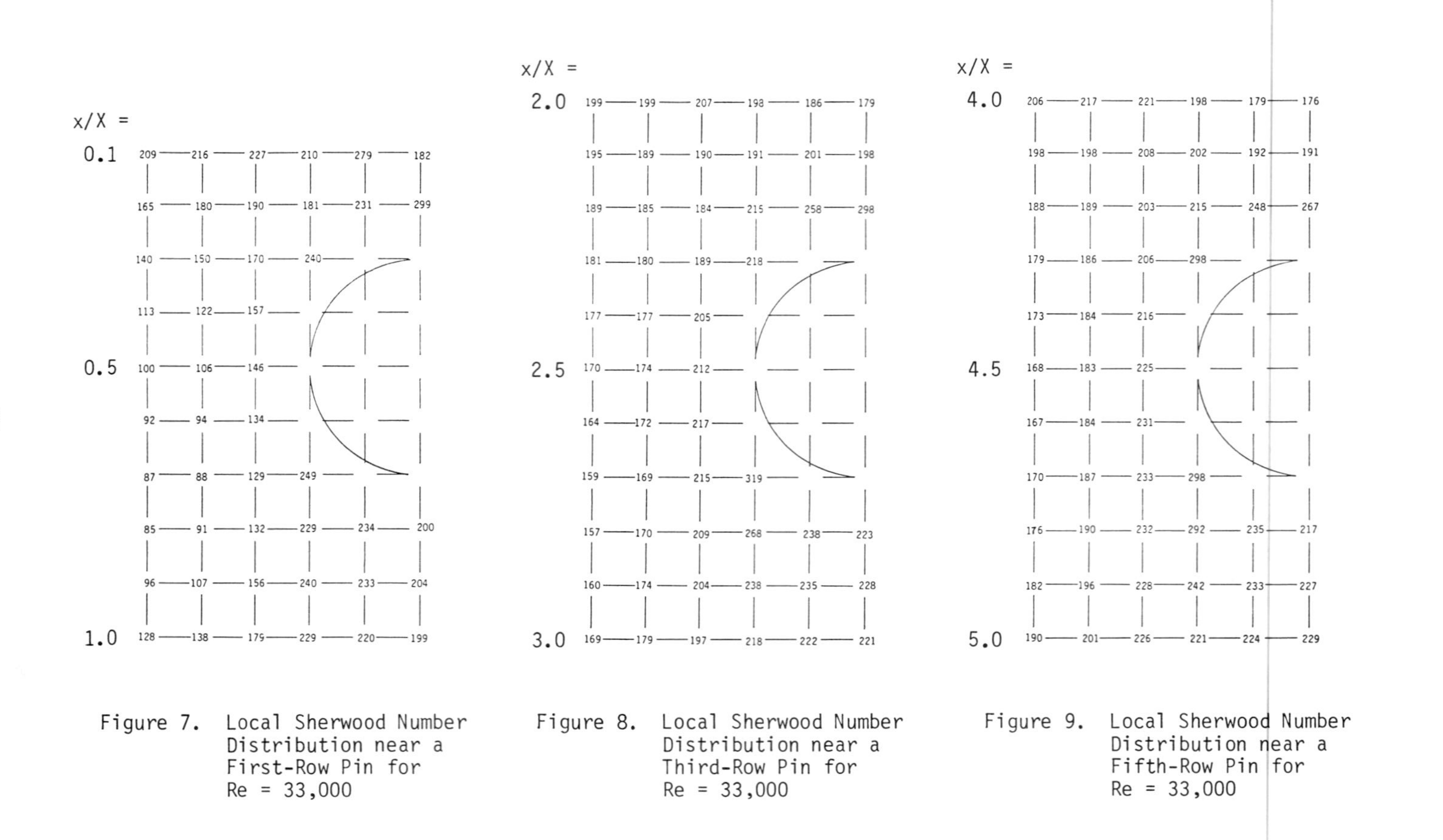

Figure 7. Local Sherwood Number Distribution near a First-Row Pin for Re = 33,000

Figure 8. Local Sherwood Number Distribution near a Third-Row Pin for Re = 33,000

Figure 9. Local Sherwood Number Distribution near a Fifth-Row Pin for Re = 33,000

3. The Sherwood number distributions become nearly periodic downstream of the first pin row--the corresponding Sherwood number distributions have similar shapes although the magnitudes of the Sherwood numbers at corresponding locations may vary.

4. The Sherwood number is generally very high immediately upstream of a pin and is relatively high in the wake region downstream of a pin.

The present investigation is part of a continuing effort to better understand the local heat/mass transfer characteristics in pin fin channels. Experiments are being conducted to study the effects of the entrance geometry, the pin fin array geometry, and the heat/mass transfer from the pin fin surfaces on the local endwall heat/mass transfer characteristics in such channels.

ACKNOWLEDGEMENT

This research was performed under the auspices of the National Science Foundation (Grant No. MEA 83-00722).

REFERENCES

1. Zukauskas, A., Heat Transfer from Tubes in Crossflow, Advances in Heat Transfer, Vol. 8, 1972, pp. 93-160.

2. Webb, R. L., Air-Side Heat Transfer in Finned Tube Heat Exchangers, Heat Transfer Engineering, Vol. 1, 1980, pp. 33-49.

3. VanFossen, G. J., Heat-Transfer Coefficients for Staggered Arrays of Short Pin Fins, J. of Engineering for Power, Vol. 104, 1982, pp. 268-274.

4. Brigham, B. A., and VanFossen, G. J., Length to Diameter Ratio and Row Number Effects in Short Pin Fin Heat Transfer, J. of Engineering for Gas Turbines and Power, Vol. 106, 1984, pp. 241-245.

5. Metzger, D. E., Berry, R. A., and Bronson, J. P., Developing Heat Transfer in Rectangular Ducts with Staggered Arrays of Short Pin Fins, J. of Heat Transfer, Vol. 104, 1982, pp. 700-706.

6. Simoneau, R. J., and VanFossen, G. J., Jr., Effect of Location in an Array on Heat Transfer to a Short Cylinder in Crossflow, J. of Heat Transfer, Vol. 106, 1984, pp. 42-48.

7. Saboya, F. E. M., and Sparrow, E. M., Local and Average Transfer Coefficients for One-Row Plate Fin and Tube Heat Exchanger Configurations, J. of Heat Transfer, Vol. 96, 1974, pp. 265-272.

8. Saboya, F. E. M., and Sparrow, E. M., Transfer Characteristics of Two-Row Plate Fin and Tube Heat Exchanger Configurations, Int. J. Heat Mass Transfer, Vol. 19, 1976, pp. 41-49.

9. Sogin, H. H., Sublimation from Disks to Air Streams Flowing Normal to their Surfaces, Trans. of ASME, Vol. 80, 1958, pp. 61-69.

10. Goldstein, R. J., and Karni, J., The Effect of a Wall Boundary Layer on Local Mass Transfer from a Cylinder in Crossflow, J. of Heat Transfer, Vol. 106, 1984, pp. 260-267.

Heat Transfer Measurement in the Full-coverage Film Cooling on a Convexly Curved Wall

MASAYA KUMADA
University of Gifu
Gifu, Japan

TERUAKI MITSUYA
Hitachi Research Laboratory
Hitachi, Japan

NOBUHIDE KASAGI and MASARU HIRATA
University of Tokyo
Tokyo, Japan

INTRODUCTION

An advanced cooling technique termed full-coverage film cooling has been studied for future high-temperature gas turbines [1-9]. The authors carried out a series of basic studies in which a numerical method was developed for prediction of the FCFC cooling performance with internal heat conduction inside the film-cooled wall [7,8]. Local heat transfer coefficients were measured on the film-cooled and the backside surfaces of a flat plate, and also on the inner surface of injection holes, with the aid of the naphthalene sublimation technique [7]. Even though these data were equivalent to those on the isothermal wall, an advantage was offered that the FCFC heat transfer coefficient with an arbitrary injection temperature was given by a simple linear function of an injection/wall temperature ratio. They were extensively used as boundary conditions in the numerical prediction of the three-dimensional temperature distribution in the non-isothermal film-cooled plate [8] and a reasonably good agreement was obtained with the result of a separate heat transfer experiment [8]. Hence, it is suggested that the heat transfer data on the isothermal wall could be useful for the general case with heat conduction inside the FCFC wall.

As a further step of FCFC heat transfer research, the present project deals with the effect of the wall curvature. Up to the present, only a few works have been reported, e.g., the heat transfer rate at each row of discrete injection holes has been measured on the convexly curved isothermal wall by Furuhama et al. [6], while the measurement of cooling effectiveness on convexly and concavely curved walls has been carried out by a separate work of the present authors [9]. Following is the discussion of the local heat transfer coefficients on the convex wall of a constant radius as well as on the flat recovery wall with three blowing mass fluxes measured by means of the mass transfer technique.

EXPERIMENTAL APPARATUS AND PROCEDURE

The experimental apparatus used and a test plate are schematically shown in Figs. 1(a) and 1(b). All the experiments have been carried out in the test duct of 300mmx400mm cross section provided at the exit of a wind tunnel. The test wall is composed of a flat plate for the developing region, a convexly curved plate with 11 rows of discrete injection holes and a flat plate for the recovery region. With turbulence promoters in the developing region, the turbulent boundary layer at the entrance of the curved section has been confirmed fully-developed. Seventy-two injection holes of 12mm ID are arranged in the staggered manner with the hole pitch of five diameters in the streamwise and lateral directions. The injection angle is slant by 30° as shown in Fig. 1(b). The profile of a flexible ceiling plate has been adjusted so that the streamwise pressure gradient should be negligibly small, i.e., the wall static pressure is constant within ±2% of the main stream dynamic pressure along the whole streamwise length. In order to avoid the influence of the secondary flows at the corner of the test duct, two auxiliary fences, 30mm in height and 2mm in thickness, are prepared on the test wall at the distance of 35mm from the sidewalls. The boundary layer flow has been confirmed two-dimensional in the central region of about 250mm in spanwise width. The main stream velocity is kept constant and is 20m/s throughout the present work, while the injection mass flux ratio is changed as M=0.3, 0.4 and 0.5.

The local mass transfer coefficient has been measured by using the naphthalene sublimation technique, which is equivalent to the heat transfer measurement on the isothermal wall. The afore-mentioned test wall is made of molded naphthalene. The amount of surface sublimation is measured after each experiment with a resolution of 0.001mm by Magne Scale (LY-101, SONY MAGNESCALE Co. Ltd.) which is attached to the 3-D automatic traversing mechanism. The output of Magne Scale is transferred to the micro-computer and is processed successively. As described in the previous paper [7], special care has also been paid in the control of system temperature. The data reduction is made by the method described in ditail in [7] using the physical properties of naphthalene by [10].

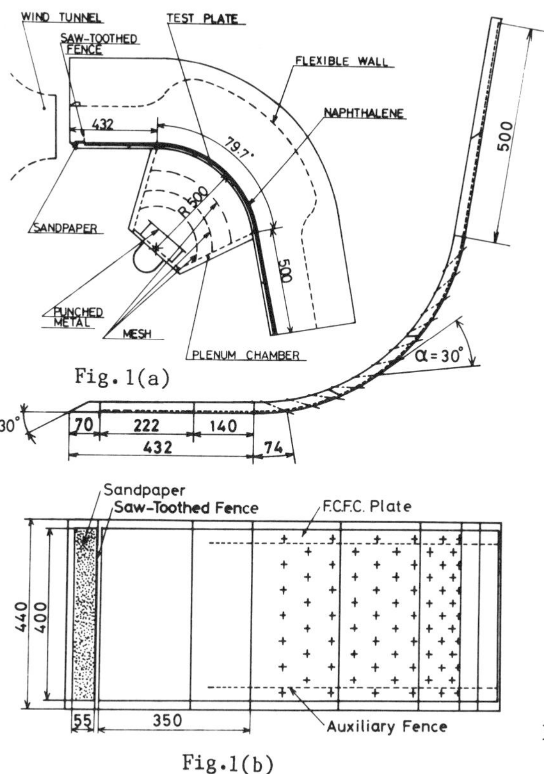

FIGURE 1. Experimental apparatus and test wall

EXPERIMENTAL RESULTS AND DISCUSSION

Stanton Number on the Convexly Curved Wall Without Injection Holes

In order to qualify the present experiment, the mass transfer measurement has been made on the convex wall without injection holes. The result is compared with the flat wall data [7] and the result of Simon et al. [11] in Fig. 2. The latter data have been modified for direct comparison with the assumption that the law of analogy to mass transfer should even be valid for the convex wall as in the case of the flat wall [7], i.e., assuming that the power index of Sc and Pr should be 0.6. As shown in this figure, the Stanton number on the convex wall drops down by a appreciable extent from the flat wall value at the inlet of the curved section and continuously decreases with the streamwise distance. The present result is in good agreement with that of Simon et al., even though a considerable difference in the experimental conditions exists (R=450mm, u_∞=14m/s and the starting length of 2000mm [11]). This suggests that, by the remarkable effect of centrifugal force, the typical convex-wall-type flow discussed by Simon et al. takes place accompanying an increase of the boundary layer thickness and a decrease of the friction factor. Thus, the mass transfer coefficient decreases mainly due to the decrease of the fluid momentum near the wall surface as well as that of the turbulent shear stress.

Local Stanton Number Under the Condition of C_2=0 (θ=0)

As described later, the local Stanton number, St(0), under the condition without naphthalene vapor in the secondary injection is used as a reference value, when the Stanton number for an arbitrary value of the dimensionless injection/wall temperature ratio, St(θ), is discussed. As a typical example, the spanwise distributions at x/d=11.33 and 12.50 are shown in Fig. 3, where the former location is right downstream of the hole in the 3rd row and the latter is the middle in between the 3rd and 4th rows of holes. This figure also includes the data of flat wall [7] for comparison. Although the distibutions are qualitatively similar, the values of St(0) on the convex wall are considerably smaller than those on the flat wall. In addition, the distance between the two highest peaks of St(0), which are caused by the secondary injection flow, is

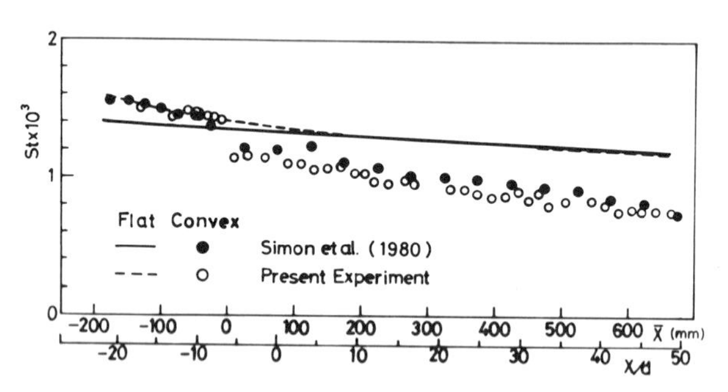

FIGURE 2. Streamwise distribution of Stanton number without injection holes

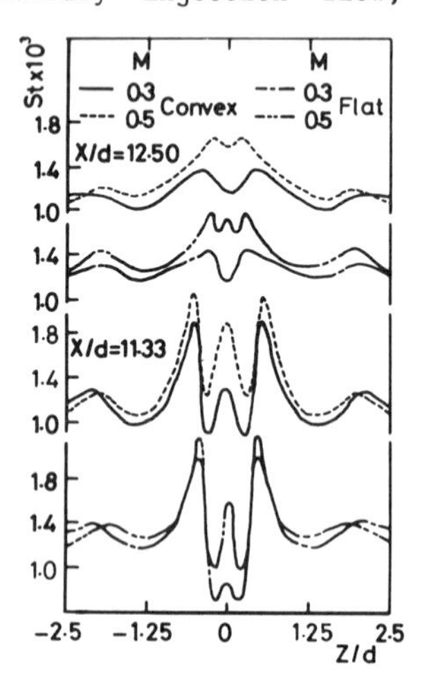

FIGURE 3. Spanwise distribtuion of local Stanton number

slightly wider, while the spanwise variation is generally smaller, if compared with the flat wall case.

The contour diagrams of constant Stanton number on the convexly curved wall in the case of M=0.3 and 0.5 with θ=0 are shown in Fig. 4(a) and 4(b). Generally speaking, the patterns of contours are not so much different from those of flat wall [7], and those for present three blowing ratios are also similar each other showing only a slight change with M. It is commonly observed in this figure that the large change of St(0) is confined to the regions around the injection holes and high Stanton numbers due to the reattachment of the secondary flow injected are also obtained in these regions. The absolute values of St(0) on the convex wall are, however, considerably small if compared with those of flat wall. The spanwise variation of St(0) is also smaller as has been pointed out in Fig. 3. These must be due to the fact that the flow trajectory of the secondary injection changes depending upon the wall curvature, yet the structure of flow field dose not change basically. According to the flow visualization by the separate work [9], the secondary injection on the convex wall tends to attach closer to the wall and expand wider over the surface than on the flat wall. These observations are qualitatively in good agreement with the present results.

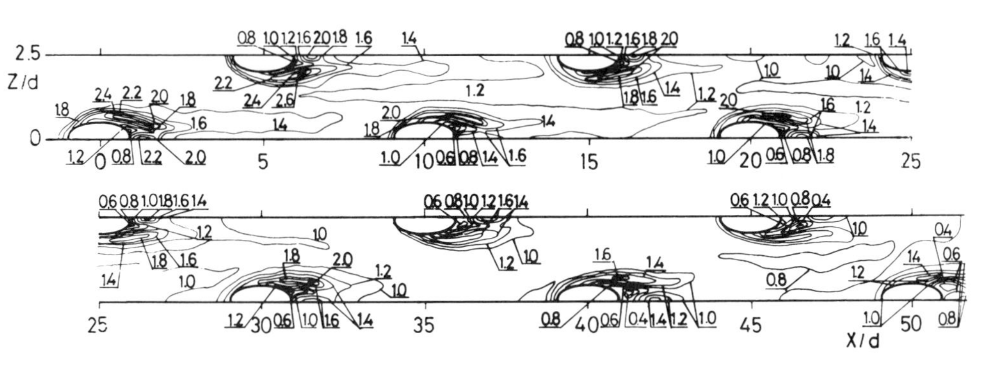

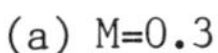
(a) M=0.3

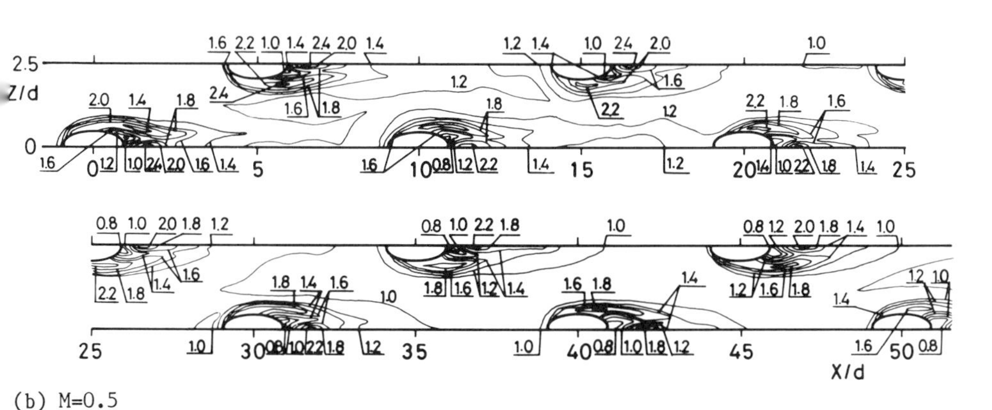

(b) M=0.5

FIGURE 4. Contour diagram of Stanton number with θ=0 (numerical values in $St \times 10^3$)

The spanwise averaged Stanton number, $\overline{St}(0)$, on the convex wall is calculated for three mass flux ratios by numerical integration of the local Stanton number as shown in Fig. 5. The results without injection holes and of the flat wall are also included for comparison. It is noted that all of the present values of $\overline{St}(0)$ decrease with the streamwise distance in contrast to the case of the flat wall and that the change of spanwise averaged Stanton number with M is smaller. In addition, the difference between the peak and valley of the jagged shape of the streamwise distributions are considerably lessened on the convex wall.

Local Stanton Number under the Condition of $C_2>0$ ($\theta>0$)

The diagram of local Stanton number with the naphthalene vapor in the secondary injection is represented in Fig. 6, where the mass flux ratio is M=0.3 and the concentration ratio $\theta=0.523$. If compared with the case of $\theta=0$ in Fig. 4, it is

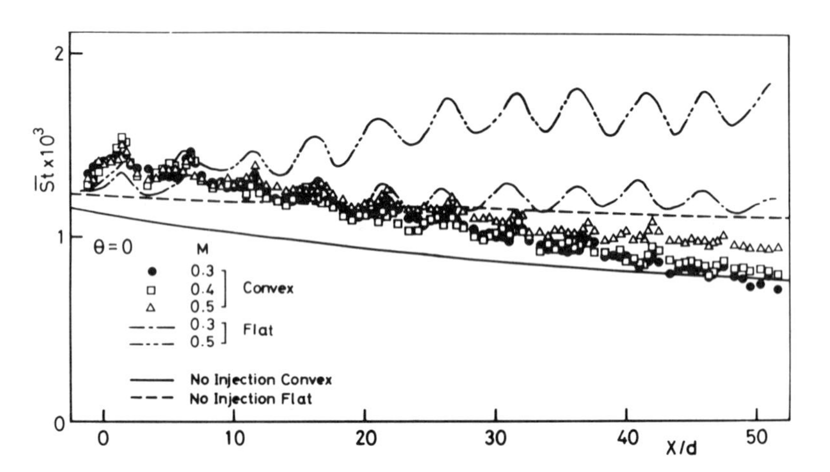

FIGURE 5. Streamwise distribution of spanwise averaged Stanton number

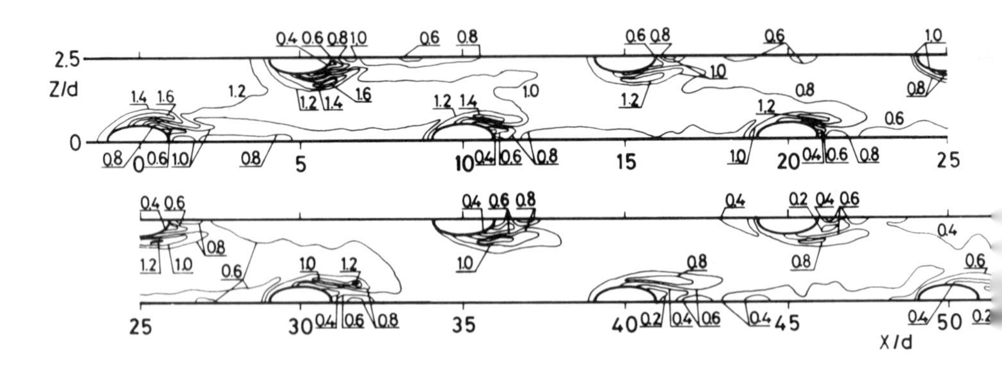

FIGURE 6. Contour diagram of Stanton number with M=0.3 and $\theta=0.523$ (numerical values in $St\times10^3$)

easily understood that the Stanton number decreases with θ particularly in the reattaching region of the secondary injection, while the profile of contour is generally similar to the case of $\theta=0$. Figure 7 shows the dependency of the spanwise averaged Stanton number, $\overline{St}(\theta)$, on θ when M=0.3. $\overline{St}(\theta)$ has a general tendency to decrease with x/d as well as θ increased, similarly to the case of the flat wall [7]. The difference between the peak and valley of the jagged shape also becomes gradually smaller with θ increased.

The linear relationship between the local Stanton number and the dimensionless injection/wall temperature ratio [3,4,7]:

$$St(\theta)/St(0) = 1 + \theta(St(1)-St(0))/St(0) = 1 + k\theta \tag{1}$$

is experimentally confirmed in Fig. 8, where St(1) and St(0) are the local Stanton numbers when $\theta=1$ and 0, respectively. As shown in the figure, the linear relation is reasonably well established. The linear constant k in Eq. (1), which is the gradient of straight line in Fig. 8, is usually negative and has a physical meaning how vigorously the film-cooling effect of secondary

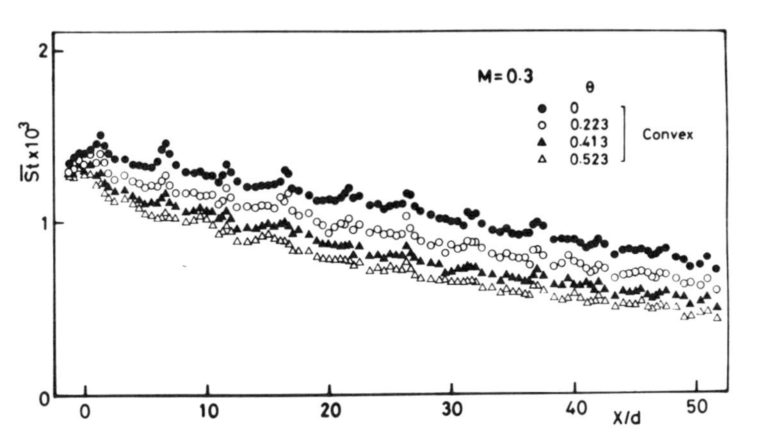

FIGURE 7. Streamwise distribution of spanwise averaged Stanton number with M=0.3

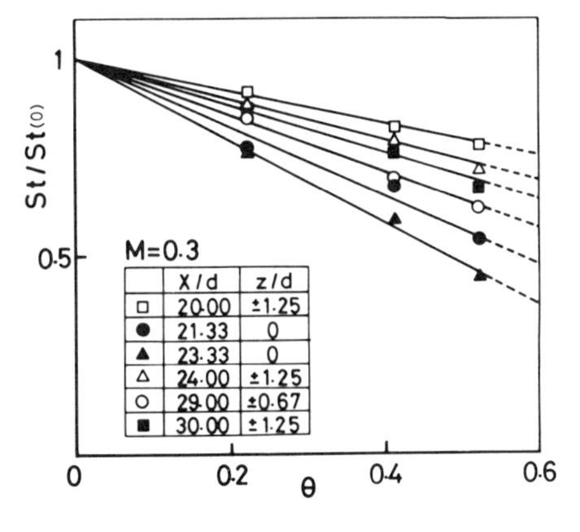

FIGURE 8. Dependency of local Stanton number on θ with M=0.3

injection takes place. As an example, figures 9(a) and 9(b) show the spanwise distribution of -k at x/d=20 and 22.5. It is commonly observed that the value of -k is large near and downstream of the holes and has the minimum at the location midway between the holes in the spanwise direction. The value of -k is the largest downstream of the holes in the case of M=0.4, while that in the case of M=0.5 is relatively larger at the both sides and right downstream of hole, but it rapidly decreases in the downstream direction. These results mean that the width of the region and the degree affected by the injection flow dose not depend only on the trajectory of the secondary injection but also on more complicated flow structure such as a separation, a reattachment and a concentration of the secondary injection flow.

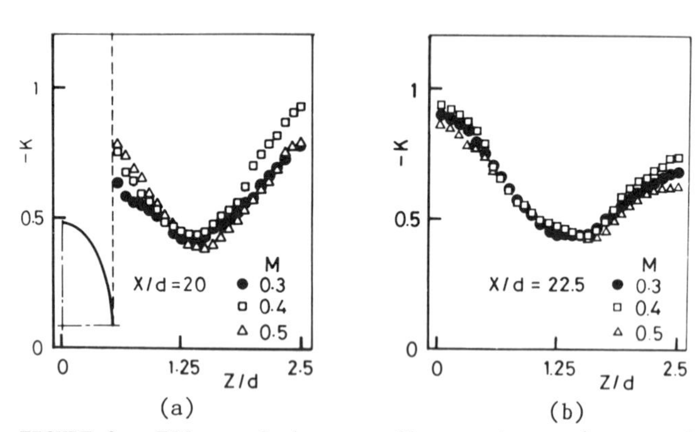

FIGURE 9. Effect of the mass flux ratio on the spanwise distribution of -k at the 5th row of holes

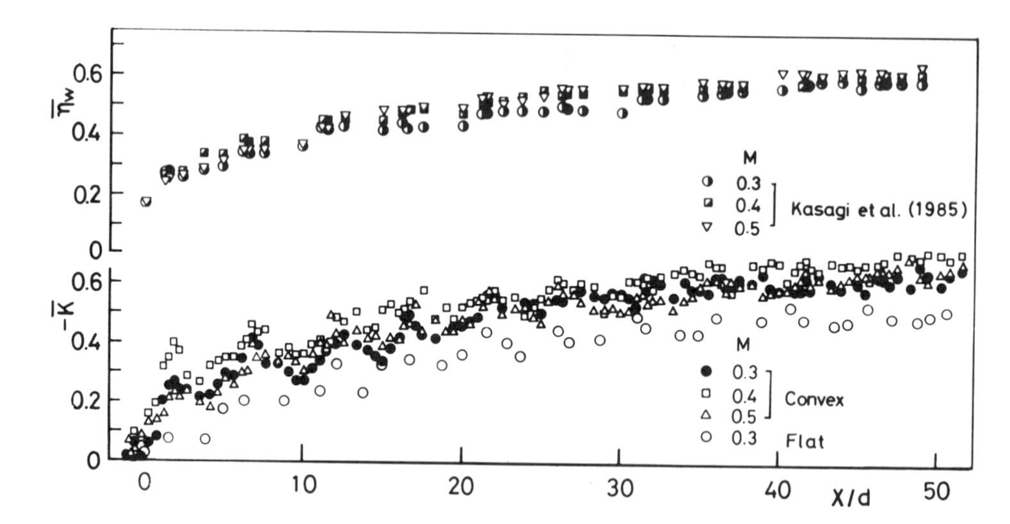

FIGURE 10. Effect of the mass flux ratio on the streamwise distribution of $\overline{-k}$

Based on the data of local k at various streamwise positions, the spanwise averaged values, $\overline{-k}$, have been calculated as shown in Fig. 10. The value of $\overline{-k}$ increases monotonously with the streamwise distance regardless of M, and a step-like change immediately downstream of each injection hole becomes smaller with x/d increased, again similarly to the case of the flat wall [7]. But the value of $\overline{-k}$ on the convex wall is considerably larger than that on the flat wall. In adition, it changes slightly with M and indicates the maximum value in the case of M=0.4, which is in agreement with the result of Furuhama et al. [6] in the case of the convexly curved wall and also with that of Crawford et al. [4] in the case of the flat wall. When an ideally adiabatic wall is considered, St(θ) must be zero in Eq. (1). Then, from the definition of θ, the following relation are obtained:

$$\theta = -1/k = 1/\eta_{w_{ad}} \tag{2}$$

where the subscript "ad" means the value on the adiabatic wall. Thus, the value of -k corresponds to the cooling effectiveness on the adiabatic wall. The spanwise averaged cooling effectiveness on the convex wall [9] is also included in Fig. 10, where the wall is made of acrylic resin. Since the thermal conductivity of acrylic resin is very small, the test wall can be approximately considered as adiabatic. As shown in the figure, the correspondence between the present result of $\overline{-k}$ and the cooling effectiveness is generally good. In the measurement of cooling effectiveness [9], the optimum value for the mass flux ratio has been evaluated to exist in the range of M=0.4-0.5, and this also agrees well with the present result.

Stanton Number in the Recovery Region

In the recovery region downstream of the curved region, the three-dimensional structure of the boundary layer flow in the spanwise direction gradually disappears and the spanwise distribution of Stanton number becomes uniform. Figure 11 shows the streamwise distribution of the spanwise averaged Stanton number in the case of θ=0. Without the wall curvature in the recovery region, the Stanton number still decreases along the considerably long streamwise distance and seems to increase beyond the position near around x/d=70, gradually approaching the value of the flat wall. This is the process that the flow field characterized by the convex wall curvature is released from the effects of the centrifugal force and recovers again to the equilibrium turbulent boundary layer flow. The streamwise behaviour of St is similar regardless of the existence of

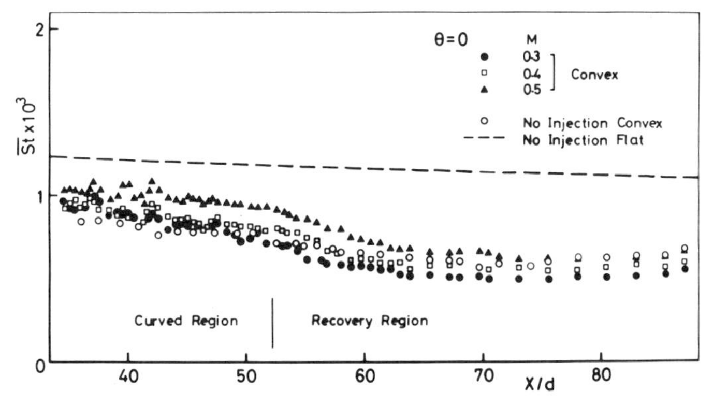

FIGURE 11. Streamwise distribution of spanwise averaged Stanton number in the recovery region with θ=0

the injection flow and the mass flux ratio.

CONCLUSIONS

By using the convexly curved FCFC wall with the same injection hole arrangement as the flat wall previously studied, the local Stanton numbers on the convex wall and on the recovery flat wall have been measured by means of the naphthalene sublimation technique which is equivalent to an isothermal wall condition. In addition, the constant included in the relationship between the heat transfer coefficient and the injection/wall temperature ratio is determined experimentally. As a result, the effects of the wall curvature on the FCFC Stanton number are clarified quantitatively under various conditions of mass flux ratio. Because of the convexly curved-wall-type flow, which takes place due to the influence of a centrifugal force, the local Stanton numbers are generally decreased on the convex wall. The Stanton number with the naphthalene vapour in the secondary injection is remarkably reduced, since the injection flow tends to attach to the wall and to expand over the downstream region due to the curvature effect. The FCFC Stanton number on the convex wall changes only slightly with the mass flux ratio, while the maximum reduction is obtained with M=0.4. In the recovery region downstream of the curved region, the Stanton number shows a considerably delay in recovering toward the value of the equilibrium turbulent boundary layer.

ACKNOWLEDGEMENTS

The authors are grateful to Messrs. N. Ito, N. Kobayashi and N. Yoneda for their contibutions in carrying out the present experiments. They also courteously thank Aero-Engine & Space Operations, Ishikawajima-Harima Heavy Industries Co. Ltd., for their manufacturing the FCFC test plates. This work was supported through the Grant-in-Aid for Special Project Research by the Japanese Ministry of Education, Science and Culture.

NOMENCLATURE

C = concentration of naphthalene
D = diffusion coefficient, m/s
d = injection-hole diameter, m
h = heat transfer coeficient, W/mK
hd = mass transfer coefficient, m/s
k = constant
M = mass flux ratio
Pr = Prandtl number, ν/κ
R = radius of wall curvature, m
Re = Reynolds number, $u_\infty x/\nu$
Sc = Schmidt number, ν/D
Sh = Sherwood number, hdx/D
St = Stanton number, hd/u_∞ or $h/\rho_\infty c p u_\infty$
x = streamwise distance from the first row of holes, m
$\overline{x}$ = streamwise distance from the inlet of the curveture section, m
z = spanwise distance, m
α = injection hole angle, °
η_w = cooling effectiveness
θ = temperature or concentration ratio
κ = thermal diffusivity, m^2/s
ν = kinematic viscosity, m^2/s

Subscripts

2 = secondary injection flow
w = wall
∞ = main stream

Superscripts

$(\bar{\ })$ = spanwise average

REFERENCES

1. Metzger, D. G., Takeuchi, D. I. and Kuenstler, P. A., Effectivenss and Heat Transafer with Full-Coverage Film Cooling, ASME Journal of Engineering for Poewr, vol. 95, pp. 180-184, 1973.

2. Mayle, R. E. and Camarata, F. J., Multihole Cooling Film Effectiveness and Heat Transfer, ASME Journal of Heat Transfer, vol. 97, pp. 534-538, 1975.

3. Crawford, M. E., Kays, W. M. and Moffat, R. J., Full-Coverage Film Cooling Heat Transfer Studies - A Summary of the Data for Normal-Hole and 30-deg Slant-Hole Injection, Report HMT-19, Thermosciences Division, Mech. Engrg. Dept., Stanford University, 1975.

4. Crawford, M. E., Kays, W. M. and Moffat, R. J., Heat Transfer to a Full-Coverage Film-Cooled Surface with 30-deg Slant-Hole Injection, Report HMT-25, Thermosciences Division, Mech. Engrg. Dept., Stanford University, 1976.

5. Bergeles, G., Gosman, A. D., and Launder, B. E., The Prediction of Three-Dimensional Discrete-Hole Cooling Processes, Part 2: Turbulent Flow, ASME Journal of Heat Transfer, vol. 103, pp. 141-145, 1981.

6. Furuhama, K., Moffat, R. J. and Frota, M. N., Heat Transfer and Turbulence Measurements of a Film-Cooled Flow over a Canvexly Curved Surface, 1983 Tokyo Int'l Gas Turbine Congress, ASME Paper No. 83-GTJ-2, 1983.

7. Kumada, M., Hirata, M. and Kasagi, N., Studies of Full-Coverage Film Cooling (Part 1; Measurement of Local Heat Transfer Coefficient by Naphthalene Sublimation Technique), submitted to ASME J. Heat Transfer, 1985.

8. Kasagi, N., Hirata, M., Kumada, M. and Ikeyama, M., Stuides of Full-Coverage Film Cooling (Part 2; Prediction of FCFC Performance and Its Comparison With the Heat Transfer Experiment), submitted to ASME J. Heat Transfer, 1985.

9. Kasagi, N., Hirata, M., Ikeyama, M., Makino, M. and Kumada, M., Effects of the Wall Curvature on the Full-Coverage Film Cooling Effectivenss, to be presented at Symposium on Transport Phenomena in Rotating Machinery, Hwawaii, 1985.

10. Schlumberger, E., Dampfdrucke des Naphthalins und dessen Analytische Bestimmung im Gereinigten Leuchtgas, J. fur Gasbeluchting u. Wasserversorgung, no. 51, pp. 1257-1260, 1912.

11. Simon, T., Moffat, R. J. and Kays, W. M., Turbulent Boundary Layer Heat Transfer Experiments: Convex Curvature Effects, Including Introduction and Recovery, Report HMT-32, Thermosciences Division, Mech. Engrg. Dept., Stanford University, 1980.

Rotational Effects on Impingement Cooling

A. H. EPSTEIN, J. L. KERREBROCK, J. J. KOO, and U. Z. PREISER
Massachusetts Institute of Technology
Cambridge, Massachusetts 02139

ABSTRACT

The effects of rotation on the heat transfer in a radially exhausted impingement cooled model of a turbine blade are discussed. Experimental results are presented for Reynolds and Rossby numbers and for blade to coolant temperature ratios typical of small gas turbine for two stagger angles, zero and -30 degrees. The experiments show an average Nusselt number some 20 to 30 percent lower than expected from previous nonrotating data. From a model which includes the buoyancy and Coriolis effects on the heat transfer it is inferred that the observed reduction is due to buoyancy. The experiments also show a radical change in the heat transfer distribution near the blade root at low Rossby number and -30 degree stagger, relative to the distribution with zero stagger or Rossby numbers larger than unity. An heuristic model is advanced which predicts that the impingement jets nearest the blade root should deflect inward due to a centripetal force resulting from their tangential velocity counter to the blade motion, while at larger radii this effect is offset by the radial outflow. It is concluded that the effects of rotation on internal heat transfer are so large as to cause potentially serious thermal stresses if not anticipated in design.

2 Associate Professor of Aeronautics and Astronautics
3 R.C. Maclaurin Professor and Head, Department of Aeronautics and Astronautics
4 Research Assistant, Department of Aeronautics and Astronautics
5 Research Staff Member, Institute for Defense Analysis
1 This research was supported by NASA Lewis Research Center under Grant NAG3-335

A. H. Epstein is an Associate Professor of Aeronautics and Astronautics.
J. L. Kerrebrock is R.C. Maclaurin Professor and Head, Department of Aeronautics and Astronautics.
J. J. Koo is a Research Assistant, Department of Aeronautics and Astronautics.
U. Z. Preiser is a Research Staff Member, Institute for Defense Analysis.
This research was supported by NASA Lewis Research Center under Grant NAG3-335.

NTRODUCTION

The design of air cooled turbine blades for gas turbines depends critically on prediction of the rate of heat transfer to the blades both on their external surfaces which are in the engine flowpath, and on their internal surfaces, which are exposed to the cooling flow. The blade metal temperature is at a level between that of the working fluid and that of the cooling flow, which may differ by as much as 1500°F, the portion of this difference on the flowpath side ordinarily being larger than on the cooling side, so that the cooling maintains a temperature difference between flowpath and blade in the order of 1000°F. Since a change in blade metal temperature as small as 20°F has a serious impact on the blade life, a high degree of precision is required in predicting the heat transfer if the deal operating condition is to be achieved.

n general, this ideal condition would be a nearly uniform blade temperature, since this would minimize the thermal stresses. Further, uniformity is desired on a length scale small compared to the blade chord or span, because the thermal conductivity of blade materials is such that at the high heat fluxes encountered in modern engines, large thermal stresses can result from heat transfer variations on length scales as small as a few percent of blade chord.

Measured against this requirement, present understanding of the heat transfer phenomena is inadequate for both the external and internal surfaces of turbine blades. On the external surfaces the location of transition is a critical factor, and it is strongly influenced by free stream turbulence, and probably by rotation, while the data base used for design does not include either in a comprehensive way. For the internal surfaces, the design base is largely empirical, derived from scaled experiments, almost all of which have been conducted in stationary (nonrotating) apparatus.

This paper is a status report on a continuing program aimed at elucidating and quantifying the effects of rotation on internal heat transfer in turbine blades. That these effects should be of first order may be inferred from examination of the dimensionless parameters which measure the importance of rotation relative to other flow parameters.

One of these is the Rossby number, which is the ratio of a characteristic flow speed to the tangential velocity of rotation at the same point in the flow, say $V/\Omega r$, where Ω is the angular velocity, r the radius, and V the flow speed. In typical internal cooling arrangements this ratio is less than unity at the design speed of the machine, a value of 0.5 being common. It is known from studies of rotating flows that at such Rossby numbers the flow exhibits phenomena not found in nonrotating flow, such as inertial waves and the tendency toward uniformity along the axis of rotation found in "Taylor columns." These phenomena stem from the centrifugal and Coriolis forces which attend flow variations in a strongly rotating base flow. It can be anticipated that these phenomena will alter the character of the flow in the cooling passages in the presence of rotation, from its character in the same passage geometry without rotation.

An additional class of phenomena due to rotation stem from the effects of buoyancy induced by temperature differences in the very strong acceleration fields due to rotation. The temperature differences may arise, for example, from the introduction of jets of cooling fluid into passages where the mean fluid temperature is higher than that of the jets, as well as in the shear layers on the passage walls. The importance of such effects relative to the centrifugal and Coriolis effects is measured by the ratio of the fractional temperature difference to Rossby number, which ratio in general will be of order unity for typical turbines.

There is not a large body of experimental data relevant to the effects of rotation on heat transfer. A review of the previous work has been given by Morris and Ayhan[1] who discuss the roles of Coriolis forces and buoyancy on the circumferentially averaged heat transfer in heated round tubes rotating about an axis perpendicular to the tube axis, with radial outflow. They point out that whereas the Coriolis effects increase the heat transfer, the effect of buoyancy is to reduce it, (for outflow and heating), so that the overall effect of rotation may be either to increase or decrease the heat transfer. Which occurs depends on the ratio of the Raleigh number to the square of the Reynolds number. If this ratio is less than unity, rotation increases the heat transfer.

Our purposes here are first to describe some results of an experiment designed to allow measurement of the spanwise and chordwise variation of heat transfer in a geometry typical of the impingement cooled leading edge of a gas turbine blade, and second to describe some qualitative models for the phenomena which occur due to rotation, and which result in large effects on both the mean (spatially averaged) heat transfer coefficient and on its spatial variation.

EXPERIMENTAL APPARATUS

The apparatus, which is described in detail in Ref (2), is shown schematically in Figs 1 and 2. The leading edge of the blade is modeled by a thin, electrically heated metal foil on the interior of which the cooling jets impinge. The blade model rotates in vacuum to eliminate heat transfer (other than radiative) on the outside of the foil. An infrared radiometer measures the temperature distribution in the metal foil through an imaging system. The ohmic heat input to the foil gives a uniform heat flux q (apart from a small correction for conduction in the foil), so that the measured foil temperature T_W gives directly a heat transfer coefficient, defined as

$$h = \frac{q}{T_W - T_c} \tag{1}$$

in terms of the cooling fluid temperature Tc.

Some additional features of the apparatus which are of general interest are; the counterflow heat exchangers mounted on the rotating system, which allow the cooling fluid to enter the blade model at a temperature scaled properly to the model temperature, after passing through the shaft seals and passages at near room temperature, and the use of Freon-12 as a cooling fluid, which allows scaling of the tip Mach number at low mechanical stress levels.

The apparatus was designed to provide realistic simulation of all the dimensionless paramet deemed important to the problem. These include the Reynolds, Rossby, Mach and Raleigh numbers, the last reflecting the temperature ratio of cooling fluid to blade skin. Because of the use of Freon, the ratio of specific heats is not quite right, but this is considered of minor importance.

This experimental concept is useful for detailed measurement of the internal heat transfer coefficient in any closed passage. In the experiments to be described here it was applied to the impingement geometry shown in Fig (2), which models the radially-exhausted imping cooling used in some small engine blades. The blade model is mounted so that it can be set at stagger angles from zero to -30 degrees, the latter being typical of a root section.

Data is acquired in the form of a set of circumferential scans of the foil radiance, at discrete radii from the model base to the tip. The radiometer is calibrated for temperatur sensitivity and for the effect of the model geometry, by scanning a calibrating body of the same geometry as the model, but at a known uniform temperature. It is mounted at the opposite end of the swing arm from the model.

Pressures in the model flow passages are measured by a series of taps carried to a speciall constructed scanning valve mounted on the shaft, via which a single silicon-diaphragm transducer measures the several pressures, referenced to vacuum. Similarly, thermocoupl are referenced to a body of known temperature, on the shaft. All electrical data are carried out via instrumentation slip rings.

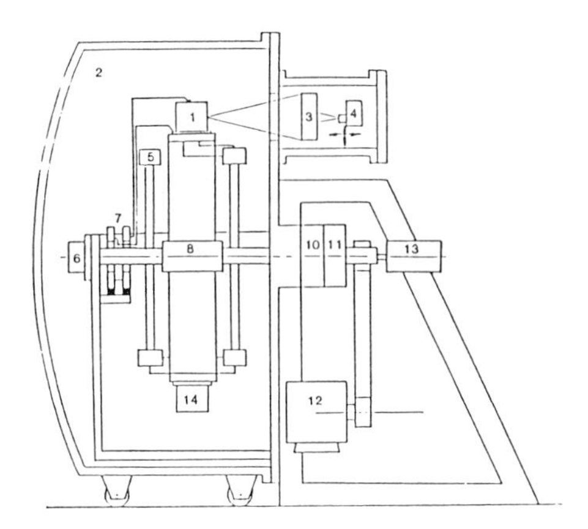

Fig. (1) Overall schematic of internal heat transfer facility, showing blade model (1) on rotating arm (8), calibrating body (14), imaging system (3) and infrared detector (4), all enclosed in vacuum chamber (2). The counter-flow heat exchangers are at (5).

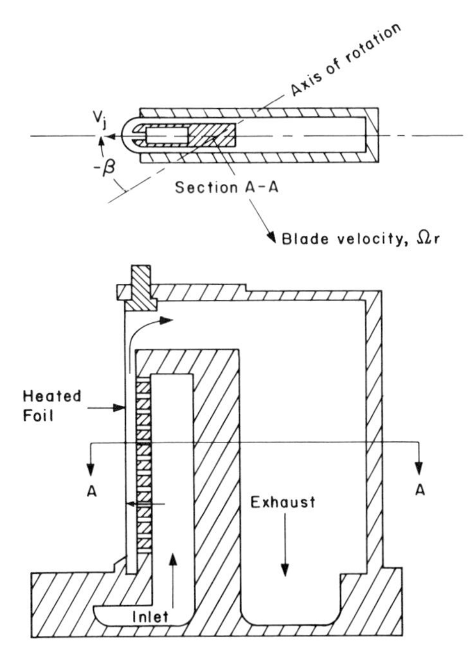

Fig. (2) Schematic of blade model, showing nominal flow pattern from inlet through impingement jets, radially to exhaust. The foil is heated by current from connection at the tip to its grounded base.

EXPERIMENTAL DATA

Data have been acquired by Kreatsoulas[2] for a range of Reynolds number (based on jet diameter and velocity) from 0.18×10^5 to $.75 \times 10^5$, for Rossby numbers (defined as $V_j/\Omega r$, where V_j is jet velocity and r is the mean radius of the blade) from .24 to 1.22 and for ratios of foil-averaged temperature to coolant temperature from 1.33 to 1.42, all at a stagger angle of -30 degrees. Preiser[3] has repeated the measurements for zero stagger angle, the conditions being otherwise essentially the same. Here we will present only enough of this data to exhibit the major findings.

Fig (3), adapted from Ref (2), shows a set of three temperature maps, all taken at a Reynolc number of about 74,000 and a temperature ratio of 1.37, but at three Rossby numbers decreasing from 1.22 to 0.53 from left to right. The cool spots formed by the impinging jets are quite noticeable, and in general reasonably close to their geometrical impingment points, indicated by the crosses. But between the second and third figures a striking change occurs near the root of the blade, with the appearance of a very cool spot very close to the root, and a hot spot just above it. We will discuss the reasons for this later.

At zero stagger, this formation of adjacent hot and cold spots at the base of the blade with decreasing Rossby number does not occur. Fig (4) from Ref (3) compares Nusselt number maps, (Nu=hd/k where d is jet diameter and k is gas conductivity) computed from the temperature maps by Eq (1) at -30 degree and 0 stagger angles, for the same Rossby number as the right map on Fig (3). There is an overall difference in Nu, but the most striking feature is the absence of the phenomenon near the blade root for zero stagger.

It can be seen that the experiment exhibits details of the heat transfer distribution which have not previously been available, in addition to the effects of rotation. Previous investig such as Chupp et. al.[4] have obtained chordwise variations of heat transfer by means of spanwise strip calorimeters, or averages over the entire passage, e.g. Morris'[1] rotating tube experiment. We believe the experiments of Kreatsoulas are the first to give resolutior both chordwise and spanwise.

The data thus offer an opportunity for comparison with experiment of a complete model for the impingement flow. Unfortunately, no such model is available as yet. In the next sections we describe some attempts at piecemeal modeling of the flow, which shed some light on the observed phenomena.

FLOW MODELING

Our initial objectives in modeling the flow are first to capture those effects of rotation on the mean Nusselt number, that is the average over the inner surface of the skin, and second to elucidate the phenomena which lead to the major redistributions of heat transfer shown in Fig (3) and Fig (4).

Channel Flow

As a first step, Koo[5] has developed a channel flow model for the flows in the coolant supply passage and in the impingement passage. This description is needed to relate the Nu measured at any point on the foil to the properties of the jet at the same radius (temperature and velocity) and to the bulk temperature, pressure and velocity in the impingement passage at the same radius. Without this connection, the Nu at any radius must be regarded as specified only in terms of the coolant supply and exhaust conditions, the blade geometry, total heat input, etc. It could happen, for example, that the flow would reverse in some of the impingement jets if the pressure drops in the supply and impingement passages were badly mismatched (an admittedly pathological happening).

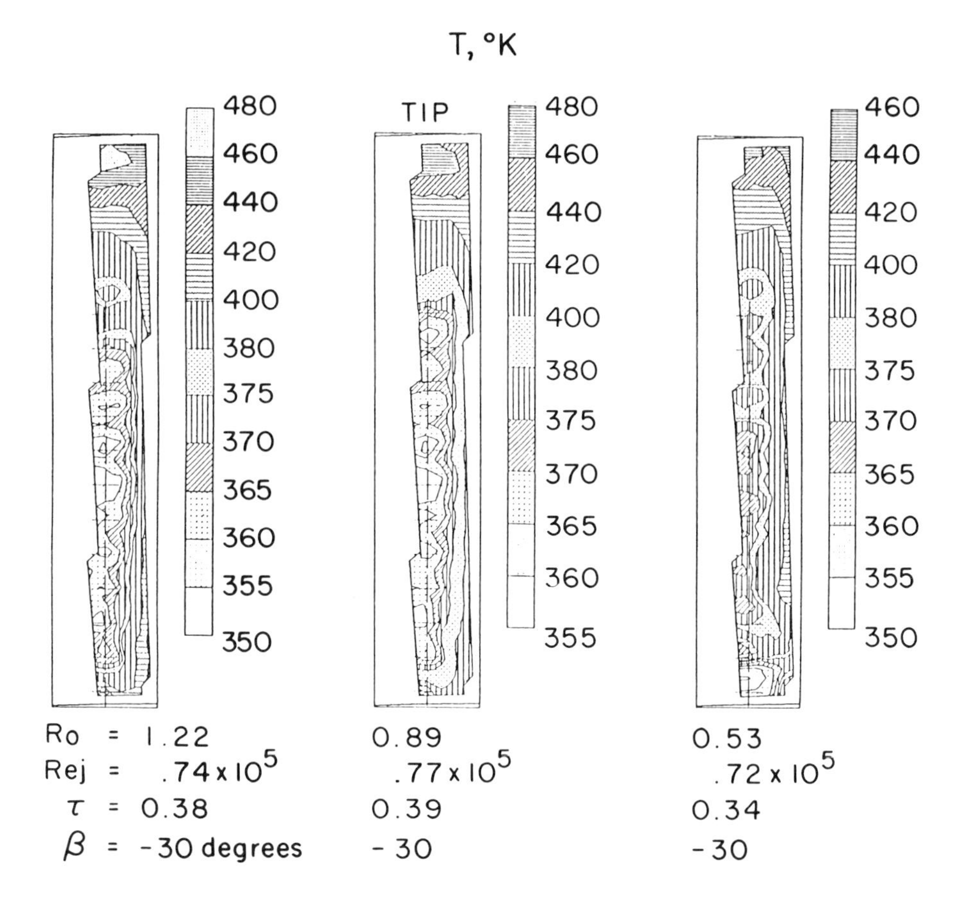

Fig. (3) Temperature maps as found by Kreatsoulas[2] for a set of Rossby numbers, decreasing from left to right, for nearly constant Reynold's number and temperature ratio. Note the change in temperature distribution near the blade root (bottom of figures) between the last two figures.

To provide this channel flow description, the equations of conservation of mass, momentum and energy have been integrated for both the supply passage and the impingement passage, in the channel flow approximation. The formalism is given in Ref (5); only its physical content will be described here.

Conservation of mass in each passage, is implemented by calculation of the flow through the jets, from the local pressures and temperatures in the two channels.

The momentum balance accounts for the zero momentum of the jets entering the impingem passage, and uses an average friction coefficient for each passage adjusted so that the computed pressure difference between root and tip in each passage matches the measured difference.

In the energy equation, the measured heat input from the foil is used, together with an estimated Nu for the other surfaces of the passage.

These three relations are integrated iteratively, by a Runga Kutta method, until a consisten solution is obtained, which yields the flow properties in each of the passages as functions of radius. With these local properties, the measured heat fluxes are then recomputed as Nusselt numbers referred to the local jet temperature, and the Reynolds Number of the jets based on their actual velocity are also computed.

Data from one experiment are displayed in this form in Figs (5) and (6), the former showing the velocities, pressures and temperatures in the flow passages,while the latter gives the Nusselt number. The data set selected is the rightmost of Fig (5), which exhibits the large Nu excursion near the blade root.

With knowledge of the local conditions, it is now possible to compare our data to that of other investigators. Two have been selected for this comparison. Morris'[1] correlation for a heated rotating tube is indicated in Fig (6) by the square points. The variation with radius is largely due to the increase of (passage) Re as the flow in the impingement passage increased radially. Of course the Nu is much lower than for our data because the effect of impingement is absent. It is satisfying to note the good agreement with our result for a radius of 50cm, which is at the end of the portion of the passage without impingement jets (see Fig (2)).

Chupps[4] correlation for an impingement geometry similar to ours, but with chordwise exhaust is shown by the triangles. The effects of rotation are absent from his experiment. Here the increase with radius is due to the increase in jet Reynolds number resulting from an increasing pressure drop across the jets, as may be seen from Fig (5). The differen in average level between Chupp's and our data will be addressed below.

The effects of rotation and stagger are clearly exhibited by comparison of results from the various conditions in the form of Fig (6). Thus Fig (7) is for conditions which are essentially identical to Fig (6) except that the Rossby number was 1.22, while for Fig (6) it was 0.53. The appearance of the large temperature gradient at the hub is clearly a result of rotation. That it is also connected with the stagger is evidenced by Fig (8) which again differs only in that it is for zero stagger, whereas Fig (6) had -30 degree stagger.

These same trends are evident at other Reynolds numbers and temperature ratios. Thus, Fig (9) shows the Nu variation at low Reynolds number ($.19\times10^5$) compared to a Reynolds number of ($.72\times10^5$) for Fig (6). The jet displacement near the blade root still appears.

A summary of the data available for -30 degree stagger angle is presented in Fig (10) on Re-Ro coordinates. The magnitude of the average foil-to-coolant temperature ratio is indicated by each point. The filled symbols denote the cases exhibiting the jet displacem near the root. From this display it is clear that low Ro (high rotational speed) is the major criterion for the jet misdirection. However, Re does seem to play a role.

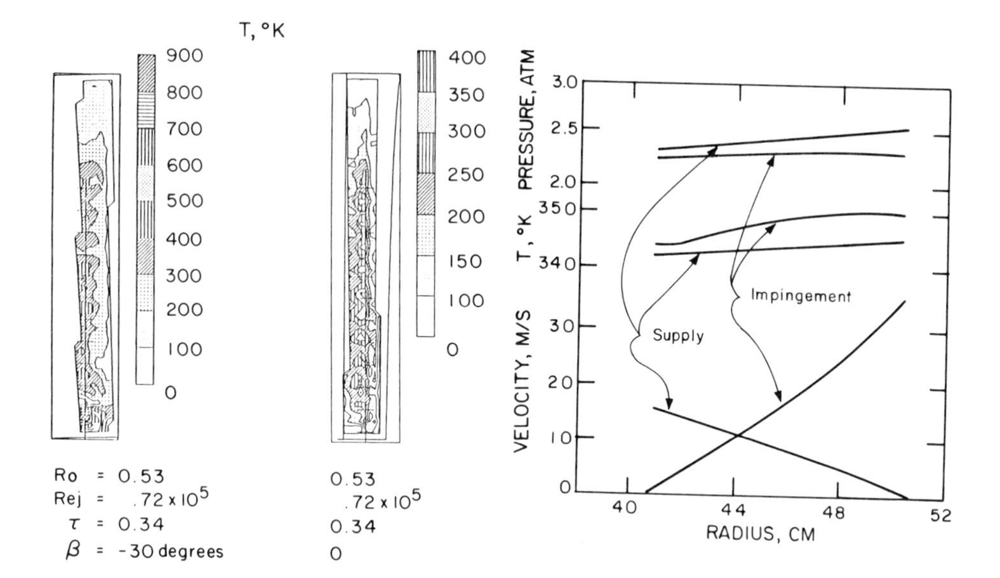

Fig. (4) Nusselt number distributions for two stagger angles, -30 and zero degrees. The figure at left corresponds to the rightmost temperature map of Fig. (3).

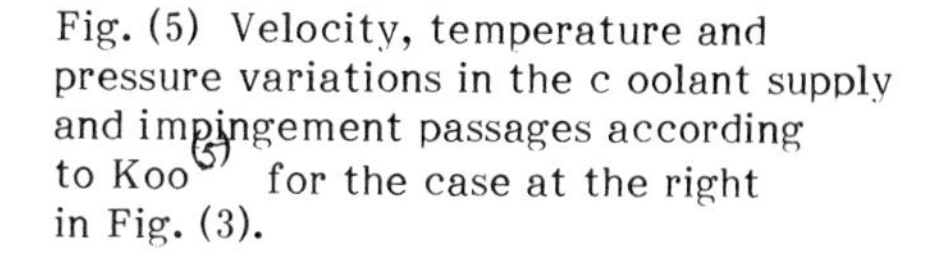

Fig. (5) Velocity, temperature and pressure variations in the c oolant supply and impingement passages according to Koo[5] for the case at the right in Fig. (3).

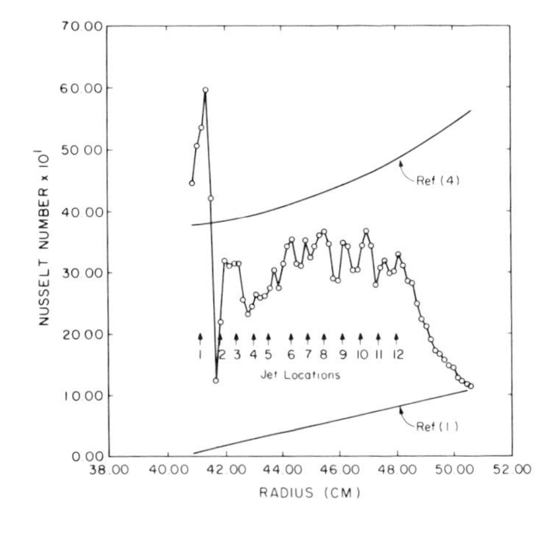

Fig.(6) The radial variation of Nusselt number based on local jet and bulk gas conditions, and averaged around the impingement passage circumference for the case at the right in Fig. (3). Note the extreme variation in Nu near the blade root, for -30 degree stagger and Roj ≃ .53. The upper curve from the correlation of Chupp[4] et. al., is for impingement cooling without rotation. The lower curve from Morris et al[1] is for a circular rotating tube.

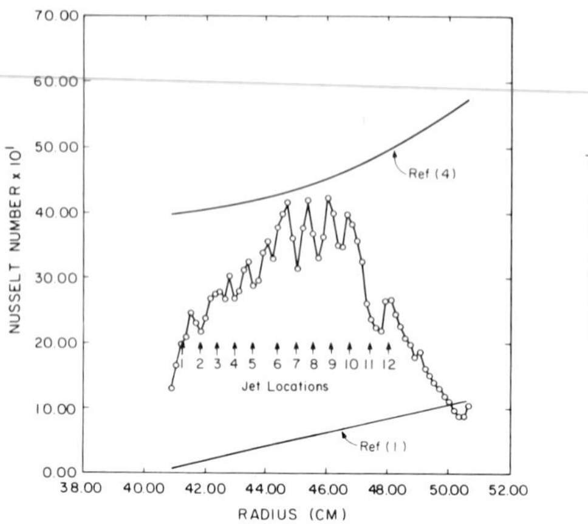

Fig. (7) The radial variation of Nusselt number as in Fig (6) for -30 degree stagger, but Roj = 1.22, showing the absence of the extreme variation at the blade root.

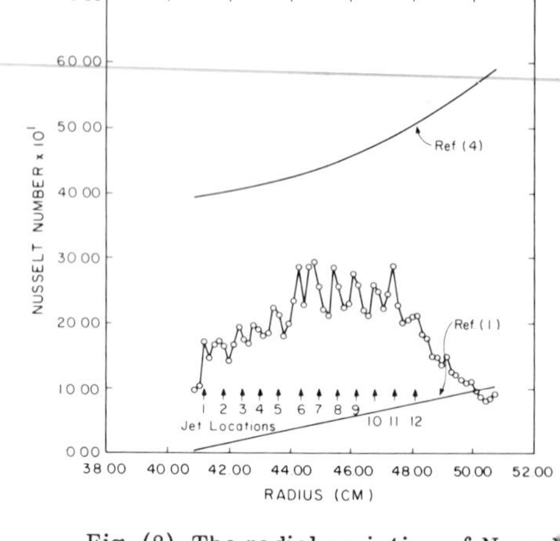

Fig. (8) The radial variation of Nussel number as in Figs (6) and (7), for Roj= but for zero stagger, again showing the absence of the extreme variation at the blade root.

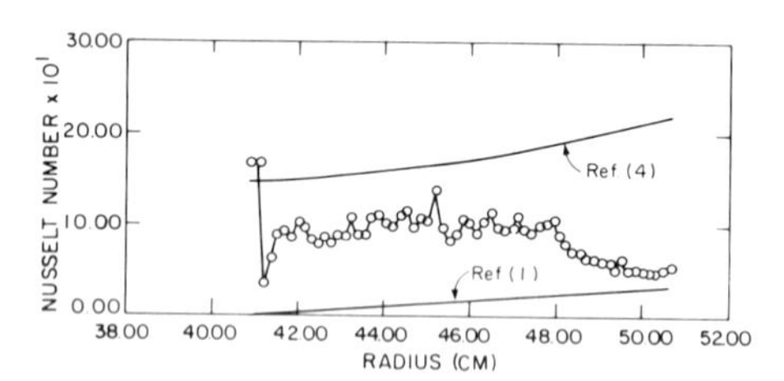

Fig. (9) The radial variation of Nusselt number as in figs. (6), (7), (8), for Roj but for Re=$.18 \times 10^5$, showing the extreme variation near the blade root.

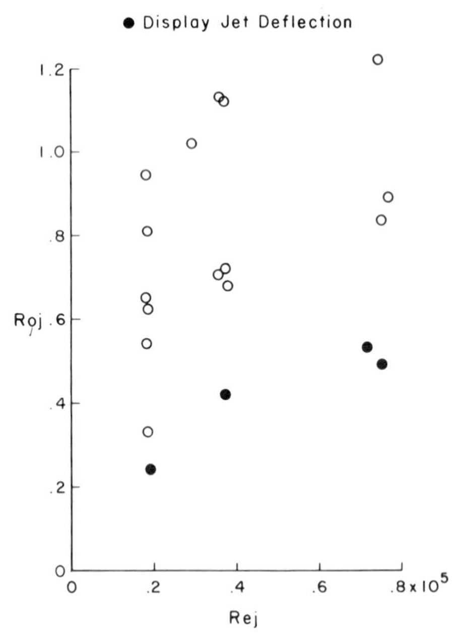

Fig. (10) Summary of the data available for -30 degree stagger, on Re-Ro coordinates. The filled symbols indicate that the extreme variation in Nu was observed near the blade root.

Buoyancy and Coriolis Effects

As a first step toward modeling of the effects of buoyancy and Coriolis accelerations on the overall (average) heat transfer rates, we shall address the problem of flow radially outward over a heated rotating surface which is normal to the axis of rotation, as shown schematically in Fig (11). We think of this surface as representative of the portion of the internal surface of the impingement passage which is tangent to a plane normal to the axis of rotation. In the core flow outside the viscous layer on the surface the flow velocity is u_b, which is radial. The surface temperature is T_W, which is different from the core temperature, T_b.

Two phenomena are of principal concern here. Due to the difference between the temperature of the surface and that of the bulk flow, the fluid in the layer near the surface will feel a buoyant body force, acting either inward for a hot surface, or outward for a cool surface. Due to the variation in the radial flow velocity in the shear layer, the fluid in the layer will experience a Coriolis acceleration in the $r\theta$ direction, which generates a tangential flow within the shear layer. This tangential flow in turn perturbs the centrifugal force on the flow in the layer, modifying the profile of u within the layer.

To obtain the simplest representation of these effects, we consider the asymptotic case where the layer is fully developed, in the sense that these effects are in balance, so that the substantial derivatives of the velocities in the shear layer are zero. The radial and tangential momentum equations are then:

radial

$$-2\rho\Omega v - \rho\Omega^2 r = -\frac{\partial p}{\partial r} + \mu\frac{\partial^2 u}{\partial z^2} \tag{2}$$

tangential

$$2\rho\Omega u = -\frac{1}{r}\frac{\partial p}{\partial \theta} + \mu\frac{\partial^2 v}{\partial z^2} \tag{3}$$

In the core flow, $u = u_b$, $v = 0$ and from these

$$-\rho_b\Omega^2 r = \frac{\partial p_b}{\partial r}$$

$$2\rho_b\Omega u_b = -\frac{1}{r}\frac{\partial p_b}{\partial \theta}$$

Taking the pressure to be uniform across the layer, (2) and (3) reduce to

$$-2\rho\Omega v + \Omega^2 r(\rho_b - \rho) = \mu\frac{d^2 u}{dz^2} \tag{4}$$

$$-2\Omega(\rho_b u_b - \rho u) = \mu\frac{d^2 v}{dz^2} \tag{5}$$

Integrating these across a layer of thickness δ, we have

$$-2\Omega\int_0^\delta \rho v\,dz + \Omega^2 r\int_0^\delta(\rho_b - \rho)\,dz = -\mu\left.\frac{du}{dz}\right|_0 \tag{6}$$

$$-2\Omega\int_0^\delta(\rho_b u_b - \rho u)\,dz = -\mu\left.\frac{dv}{dz}\right|_0 \tag{7}$$

The simplest profiles that represent the phenomena of interest are linear velocity and temperature variations and a quadratic v profile. Thus we take $u = u_b\zeta$

$$v = 4v_m\zeta(1-\zeta)\quad,\quad T = T_b[1+\tau(1-\zeta)]$$

where $\zeta = z/\delta$

and $\tau = (T_W - T_b)/T_b$

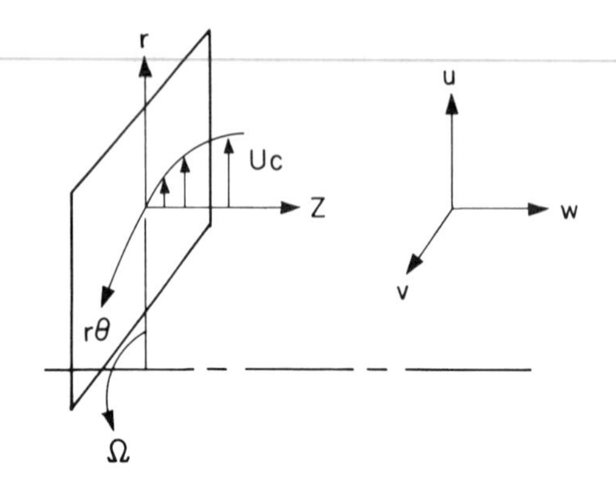

Fig. (11) Schematic of flat plate rotating about an axis perpendicular to its surface, and showing a shear layer formed in the radial flow.

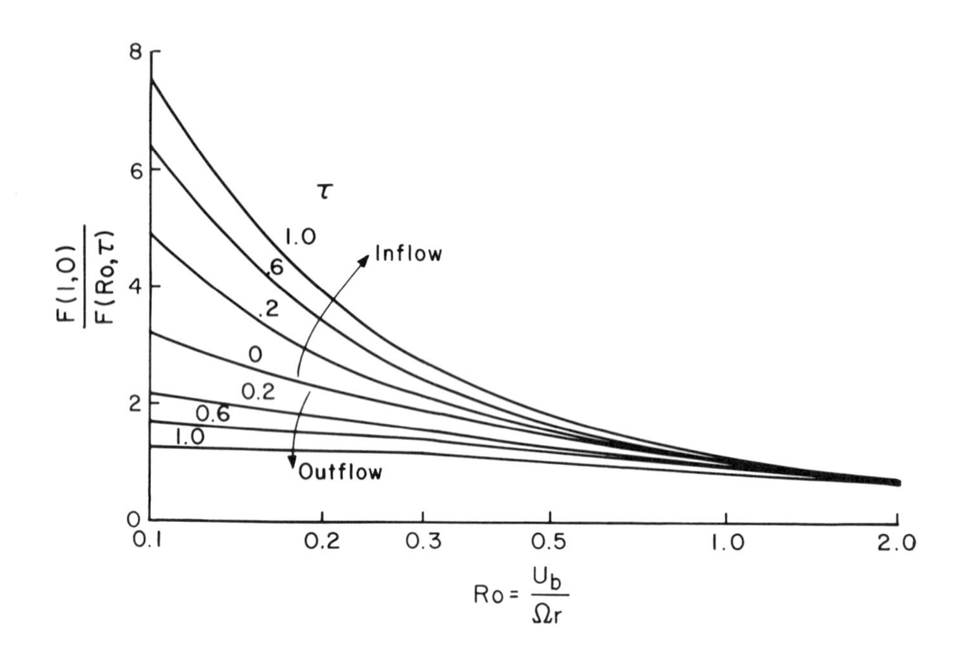

Fig. (12) The ratio of Nusselt numbers at the Rossby number Ro and temperature ratio to its value at Ro=1, τ =0, as a function of Ro and τ , showing that buoyancy increases Nu for inflow, and decreases it for outflow.

Equations (6) and (7) then reduce to

$$-8\Omega\delta\rho_b v_m I_1 + \Omega^2 r\delta\rho_b I_2 = -\mu u_b/\delta \tag{8}$$

$$-2\Omega\delta\rho_b u_b I_3 = -4\mu v_m/\delta \tag{9}$$

which constitute two relations for the shear layer thickness δ, and the peak tangential velocity v_m. The integrals I_1, I_2, I_3 are:

$$I_1(\tau) = \int_0^1 \frac{\zeta(1-\zeta)}{1+\tau(1-\zeta)}d\zeta, \quad I_2(\tau) = \int_0^1 \frac{\tau(1-\zeta)}{1+\tau(1-\zeta)}d\zeta, \quad I_3(\tau) = \int_0^1 \left[1 - \frac{\zeta}{1+\tau(1-\zeta)}\right]d\zeta$$

and these are given in Table 1.

Table 1

τ	I_1	I_2	I_3
0	.1666	0	.5000
0.1	.1588	.0469	.5159
0.2	.1518	.0884	.5304
.4	.1397	.1588	.5559
0.6	.1296	.2167	.5778
0.8	.1211	.2653	.5969
1.0	.1137	.3069	.6137

Eliminating v_m with (9) and solving for δ/r we find:

$$\frac{\delta}{r}\sqrt{\frac{\rho_b u_b r}{\mu}} = \left\{\frac{I_2}{8I_1I_3}(\mathrm{sgn}\,u_b) + \frac{1}{2}\left[\left(\frac{I_2}{4I_1I_3}\right)^2 + \left(\frac{u_b}{\Omega r}\right)^2\frac{1}{I_1 I_3}\right]^{\frac{1}{2}}\right\}^{\frac{1}{2}} \tag{10}$$

which is of the form

$$\frac{\delta}{r} = \frac{F(Ro,\tau)}{\sqrt{\rho_b u_b r/\mu}}$$

or if we introduce the hydraulic diameter of the impingement passage, D, then

$$\frac{\delta}{D} = \left(\frac{r}{D}\right)^{\frac{1}{2}}\frac{F(Ro,\tau)}{\sqrt{\rho_b u_b D/\mu}} \tag{11}$$

If we assume $Nu \propto 1/\delta$, this relation gives us immediately the functional variation of Nu. Referring it to $Ro = 1$ and $\tau = 0$, we have

$$Nu(Ro,\tau)/Nu(1,0) = F(1,0)/F(Ro,\tau) \tag{12}$$

This relationship is shown in Fig (12). It displays several interesting results. First as expected, the effect of buoyancy, for a given Ro, is always to decrease Nu for inflow, and increase it for outflow, the magnitude of the effect increasing as decreases. Second, for the case of outflow, which applies to our experiment, the net effect of variation of Ro decreases as τ increases, the rotational effect of buoyancy tending to cancel that of the Coriolis acceleration.

By comparison to our data, summarized on Fig (13), we see that the model either overestima the Coriolis effects or underestimates the buoyancy, since the experiment shows almost no effect of Rossby number on Nu, for the values of $\tilde{\epsilon}$ found in the experiment, which all lie between 0.3 and 0.38. We incline to the view that the Coriolis thinning is overestimat since in the experimental channel only about half the surface is normal to the axis of rotation.

Jet Deflections

The jets impinging on the concave internal surface of the leading edge can give rise to rather complex flow phenomena. In nonrotating experiments with jets entering a two dimensional "blind pass" it has been demonstrated that the jet can experience a transverse oscillation (in the plane depicted in Fig (2)), this being evidenced by peaks of heat transfer to either side of the geometrical line of impingement. There is however, no evidence that this sort of oscillation in the θ, z plane occurs in our experiment, either at high or low Rossby numbers.

There is also no evidence that oscillations of the jet direction occur in the r-z plane; indeed the fairly distinct pattern of cool spots at approximately the geometrical impingemen points of the jets argues against fluctuations, which would tend to smear out this pattern.

On the other hand, we believe the shift in the heat transfer pattern near the blade root with decreasing Rossby number is explained by a deflection of some of the jets toward the blade root, as a result of the defect of centrifugal force produced by the tangential component of the jet velocity. We have not yet developed a quantitative model for this phenomenon, but the following plausibility argument seems to connect the experimental observations in a useful way.

We consider a jet issuing into the impingement space with velocity Vj; the tangential component of this is $v_j = V_j \sin\beta$ where β is the stagger angle, defined such that v_j is in the direction of blade motion for positive β.

The fluid in the jet relative to the background fluid experiments a centrifugal force perturbation $2\rho_j \Omega v_j$ and a buoyancy force perturbation $(\rho_j - \rho)\Omega^2 r$. To account for the effects of crossflow (in the radial direction) we postulate a drag force acting on the jet due to the mean radial flow. Then the substantial derivative of the radial velocity of the jet fluid, along the jet axis, s, is

$$V_j \frac{du_j}{ds} = 2\Omega v_j + \left(1 - \frac{\rho}{\rho_j}\right)\Omega^2 r + \frac{2C_D}{\pi}\frac{u^2}{d_j}$$

We relate u to V_j by noting that the fluid which enters the impingement passage through jets located from the root radius r_0 to the radius r, must flow through the impingement passage, of area A, at r. Approximating the array of jets, which have flow area $\pi d_j^2/4$ and radial spacing s, by a uniform distribution of flow,

$$\frac{d(\rho u A)}{dr} = \frac{\rho_j V_j \pi d_j^2}{4s}$$

and we find

$$\frac{u}{V_j} = \frac{\rho_j}{\rho}\frac{\pi d_j^2}{4A}\left(\frac{r - r_0}{s}\right)$$

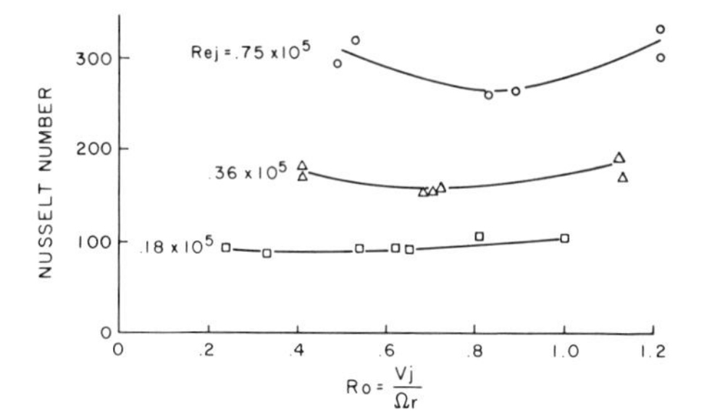

Fig. (13) Observed variation of surface - averaged Nu with jet Rossby number Roj and jet Reynold's number Rej. Compare variation with Roj to Fig. (12).

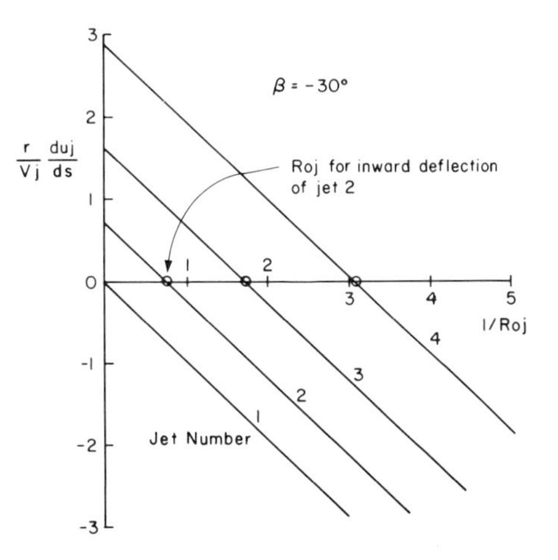

Fig. (14) Deflection of impingement jets as a function of 1/Roj. The jets are numbered 1,2,3,4, beginning with the innermost/ Where the line for each jet crosses the axis is the lowest speed at which inward deflection should occur.

Thus (13) becomes, putting $\rho/\rho_j = 1$,

$$\frac{r}{V_j}\frac{du_j}{ds} = 2\,\mathrm{Sen}\,\beta\left(\frac{\Omega r}{V_j}\right) + \left(1-\frac{\rho}{\rho_j}\right)\left(\frac{\Omega r}{V_j}\right)^2 + \frac{2C_D}{\pi}\left(\frac{\pi d_j^2}{4A}\right)^2 \frac{r}{d_j}\left(\frac{r-r_0}{s}\right)^2$$

To a first approximation, since the blade span is small compared to the radius, r, we can regard $\Omega r/V_j$ as proportional to Ω, and independent of r, hence the Rossby number, just characterizes the rotational speed. Similarly in the last term r/d_j is nearly constant over the blade span, the principal variation coming from $(r-r_0)/s$, which varies from zero at the blade root to 11 in the experiments discribed above.

Substituting values for our experiment, and putting $Ro_j = \frac{V_j}{\Omega r}$, we have

$$\frac{r}{V_j}\frac{du_j}{ds} = 2\,\mathrm{Sen}\,\beta/Ro_j + (1-\rho/\rho_j)/Ro_j^2 + 0.18\left(\frac{r-r_0}{s}\right)^2$$

According to the channel flow model, (see Fig (6)), the ratio of jet temperature to bulk temperature is .985 for that case, so $(1-\rho/\rho_j) = .015$ and the buoyancy term is relatively small.

Two points are quite clear from Eq (15)

a) The inward deflection should only occur for $\beta < 0$, since the other two terms are positive for all conditions in our experiment. This agrees with the available data, i.e., the inward deflection did not appear in Preiser's experiment, for which $\beta = 0$ and it did in Kreatsoulas', for which $\beta < 0$.

b) For $\beta < 0$, there should be inward deflection of the jets when Ro_j becomes small enough to offset the drag term. This is shown graphically in Fig (14), from which it can be seen that the first jet should always be deflected inward, that the second should deflect inward for $1/Ro_j > .75$, the third for $\frac{1}{Ro_j} > 1.75$ etc. For the outermost jets (there are 12 total) the effect of crossflow should be dominant for all Rossby numbers encountered in our experiments, so they should deflect outward.

Referring to Fig (10), we see that the largest value of Ro_j attained in Kreatsoulas' experiments (Preiser's data covers the same range) is about 1.2, so we would expect the first jet to be deflected inward in all experimental cases with -30 degree stagger and it would not be surprising if the second were as well. As Ro_j is reduced by increasing speed, we should expect more of the jets near the hub to deflect inward.

On the other hand, for the zero stagger data of Preiser we should expect all jets to deflect outward if at all.

The magnitude of the jet deflection predicted by (15) is not large for the inner jets since the slope of the jet at impingement would be approximately $\Delta u_j/V_j$ which is

$$\frac{\Delta u_j}{V_j} \approx \frac{\Delta s}{r}\left(\frac{r}{V_j}\frac{du_j}{ds}\right)$$

where Δs is the distance from impingement insert to leading edge, so the ratio $\Delta s/r \approx .01$.
For the outer most jets, where $[(r-r_0)/s]^2 \doteq 121$ we find $\Delta u_j/V_j \approx 1$ so the outward displacemen could be a significant fraction of Δs which is 0.4cm.

These trends do conform reasonably well to the data in hand. To establish a baseline, consider first the zero stagger case shown in Fig (8) for which $Ro_j = .5$, $Re_j = .72 \times 10^5$

and $\tau = .34$. We see that there are 12 identifiable peaks in the Nu curve, and each lies close to the nominal impingement point of one of the jets. Thus there is no evidence for any strong deflections of the jets, with the possible exception that the outermost peak is much weaker than its near neighbor, suggesting that the outermost jet may deflect outward, due to crossflow, and the absence of another jet on its outer side.

Turning now to Fig (6), for which the flow conditions are substantially the same as for Fig (8) but for which the stagger is -30 degrees, we note several changes. First, if we identify the outermost peak with the outermost jet, it is displaced nearly one jet spacing outward, as are the next four jets, through 8. Jets 7, 6 and 5 are about one half space outward of their nominal impingement points. There is then a gap of low Nu, which we suggest is due to inward deflection of jets 3 and 4, which produce the small peak near a 42 cm radius. Apparently jets 1 and 2 combine, with perhaps some contribution from 3 and 4 as well, to produce the Nu peak just at the blade root.

Of course this evidence may yield to other interpretations when a more complete understanding of the complexities of the flow is available. A more complete model is needed which incorporates viscous effects and the effects of rotation in a consistent way. Flow visualization would also be very helpful, but is technically difficult if all the parameters are to be correctly modeled, as in these experiments.

CONCLUSIONS

The experiments and heuristic models presented here support the following conclusions:

1) That the effects of rotation on the internal heat transfer coefficients of air cooled turbine blades are so large as to cause shifts in the heat transfer distributions with rotative speed large enough to introduce damaging thermal stresses.

2) For the radially exhausted, closed impingement geometry investigated here, the mean Nusselt number is some 20 to 30 percent lower than would be expected from correlations of data from nonrotating chordwise exhausted but otherwise geometrically similar models.

3) From an asymptotic shear layer model, we infer that the reduction is probably due to buoyancy effects on the shear layer in the radial outflow. If this is true, the effect would diminish with chordwise exhaust, and in radial inflow would enhance the heat transfer.

4) At low Rossby numbers $V_j/\Omega r$ based on jet velocity and blade tangential velocity, the innermost jets deflect inward from their geometrical points of impingement when their direction (due to blade stagger) gives them a tangential velocity component against the blade rotation. The result is closely adjacent regions of very high and very low heat transfer near the blade root, which would result in very severe thermal stresses.

REFERENCES

1. Morris, W.D. and Ayhan, T., "Observations on the Influence of Rotation on Heat Transf in the Coolant Channels of Gas Turbine Rotor Blades," Proceedings of the Institu of Mechanical Engineers, Vol. 193, 1979, pp. 303-311

2. Kreatsoulas, J.C., "Experimental Study of Impingement Cooling in Rotating Turbi Blades," Ph.D. Thesis, MIT Department of Aeronautics and Astronautics, Septemb 1983.

3. Preiser, Uriel Z., "Stagger Angle Effects on Impingement Cooling of a Rotating Turbi Blade," M.S. Thesis, MIT Department of Aeronautics and Astronautics, May 1984.

4. Chupp, R.E., Helms, H.E., McFadden, P.W. and Brown, T.R., "Evaluation of Intern Heat Transfer Coefficients for Impingement Cooled Turbine Airfoils," Journal of Aircraf Vol. 6, 1969, pp. 203-208.

5. Koo, J.J., "Channel Flow Modeling of Impingement Cooling of A Rotating Turbine Blade MIT GT&PDL Report No. 181, December, 1984.

Effects of the Wall Curvature on the Full-coverage Film Cooling Effectiveness

NOBUHIDE KASAGI and MASARU HIRATA
University of Tokyo
Tokyo, Japan

MASAKI MAKINO
University of Tokyo
Tokyo, Japan

MASATAKA IKEYAMA
Ishikawajima-Harima Heavy Industries Co. Ltd.
Tokyo, Japan

MASAYA KUMADA
University of Gifu
Gifu, Japan

INTRODUCTION

As one of the most feasible methodologies utilizing thermal energy effectively, the combined-cycle plants have been paid much attention in recent years. Among them, the use of high-temperature gas turbine as a topping cycle of the combined-cycle plant has to be of practical importance, while an effort to raise the turbine inlet temperature is desirable for higher total utilization efficiency. Related with the design of the turbine components, further basic research of gas turbine heat transfer with a particular emphasis on the turbine blade cooling must be still required.

As an advanced cooling technique protecting hot part components of high-temperature gas turbines, full-coverage film cooling has been intensively studied both experimentally and analytically [1-9]. This technique utilizes a number of small holes for coolant injection over the component surface exposed to hot gas stream. Hence, serious difficulties arise in dealing with FCFC heat transfer from the complex three-dimensional boundary layer flow strongly disturbed by discrete hole injections. In addition, the heat conduction effect inside the film-cooled wall is appreciable and it must also be properly taken into account. The present authors carrid out a series of basic studies to develop a numerical technique for prediction of the FCFC cooling performance [8,9]. Basic data sets of local heat transfer coefficients were obtained under the isothermal wall condition by using the naphthalene sublimation technique, which was based on the heat and mass transfer analogy, by Kumada et al. [8]. These data of heat transfer coefficients were used in the computational prediction of FCFC cooling performance by Kasagi et al. [9]. The predicted cooling effectiveness on the thermally conductive walls agreed well with the experimental results.

Related with the FCFC heat transfer in the real gas turbines, there are several hydrodynamically influential factors, such as compressibility and variable properties of fluid, main stream turbulence and oscillation, flow acceleration and deceleration, and streamline curvature. In the present project, the effect of wall curvature has been studied by using the concave, flat and convex test plates. It is generally known that only a small curvature of streamline may cause a drastic change in the turbulence activity in turbulent wall shear flows [10]. In the heat transfer aspect [11,12], convex wall curvature suppresses heat transfer rate, while concave curvature augments it with the three-dimensional Gortler-vortex-type turbulence structure. In the case of the full-coverage film cooling, the effect of convex wall curvature was only studied by Furuhama et al. [7]. From the measurement of heat transfer coefficient on the isothermal convex wall, they reported that the stabilization effect took place and contributed appreciably to the increase of the FCFC cooling effectiveness and that the optimum blowing mass flux ratio existed in the range around 0.4.

In the present study, the flow field behaviour influenced by the centrifugal force is analyzed by the smoke-wire flow visualization with an emphasis on the change of the trajectory of secondary injection with the wall curvature. The cooling effectiveness is obtained on the curved walls, which are made of acrylic resin, in the heat transfer experiment, and are compared with that on the flat plate previously studied [9]. In addition, the effect of mass flux ratio is discussed.

PRELIMINARY CONSIDERATION

The effect of streamline curvature on the trajectory of the secondary coolant injected is considered preliminarily by using a simple flow model as shown in Fig. 1. If the streamwise pressure gradient is negligibly small, the centrifugal force acting on the fluid element flowing along a curved wall must be balanced with the pressure gradient in the radial direction (the direction normal to the wall) [13]. If the velocity profile in the boundary layer is assumed to obey the 1/n power law, the mean pressure gradient in the boundary layer is evaluated as follows:

$$\overline{dp/dr} = (n/(n+2))(\rho_\infty u_\infty^2/R) \tag{1}$$

The momentum flux of the secondary injection is:

$$J = (4/\pi d^2)\int_0^{2\pi}\int_0^{d/2}\rho_2 u_2^2 r'dr'd\theta \tag{2}$$

$$= k\rho_2\overline{u}_2^2$$

where k and $\overline{u}_2$ are a constant and the bulk mean velocity of injection, respectively. The value of k is about 2.3 according to the measurement of the velocity profile of injection. The centrifugal force acting on the injection is given as:

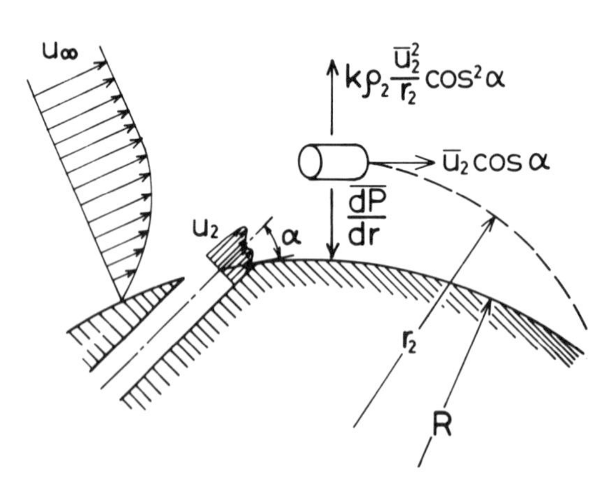

FIGURE 1. Discrete hole injection on a curved wall

$$F = k\rho_2\bar{u}_2^2\cos^2\alpha/r_2 \quad (3)$$

where the injection hole angle, α, is 30° presently. Considering the balance between the forces of Eqs. (1) and (3) and using Eq. (2), the following relationship between the radius of trajectory, r_2, and the wall curvature, R, is obtained.

$$r_2/R = k((n+2)/n)I\cos^2\alpha \quad (4)$$

$$I = \rho_2\bar{u}_2^2/\rho_\infty u_\infty^2 = (\rho_\infty/\rho_2)M^2 \quad (5)$$

$$M = \rho_2\bar{u}_2/\rho_\infty u_\infty \quad (6)$$

If the power index of n is assumed as n=7, then the value of the left hand side of Eq. (4) is evaluated as about 0.22-0.98 and is always less than 1 under the present experimental conditions of mass flux ratio, M=0.3-0.65. This suggests that the secondary injection tends to attach to the convex wall and depart away from the concave wall and this tendency is more enhanced for smaller mass flux ratio.

EXPERIMENTAL CONDITIONS AND PROCEDURE

Basic experimental conditions and flow parameters are summarized in Table 1. In addition to the flat plate, convexly and concavely curved plates with constant radii of curvature, R=±500mm, have been manufactured. As a test plate material, acrylic resin (λ=0.15W/mK) is used as an approximation to an ideally adiabatic wall. Strictly speaking, the effect of heat conduction inside the FCFC wall must still exist, although it is very small, and computational treatment is necessary to quantitatively evaluate this effect [9]. One of three different test sections for each test plate has been attached to the exit of a wind tunnel [8,9]. As an example, the setup for the concave test plate is shown in Fig. 2. The secondary air is heated up to a prescribed temperature and injected through the injection holes into the main stream at a room temperature. The mass flux ratio, M, is changed as M=0.3, 0.4, 0.5, and 0.65 for each curved wall.

TABLE 1. Basic experimental conditions

Radius of Wall Curvature	±R=500mm
Injection Hole Diameter	d=12mm
Test Plate Thickness	t=25mm
Injection Hole Arrangement	Staggered Array Streamwise 11 rows
Hole Pitch	p/d=s/d=5
Injection Hole Angle	α=30°
Main Stream Velocity	u_∞=20m/s
Boundary Layer Parameter ($)	
Displacement Thickness	δ =2.11mm
Momentum Thickness	θ =1.64mm
Shape Factor	H =1.28

($; at x/d=-7.5 on the flat plate)

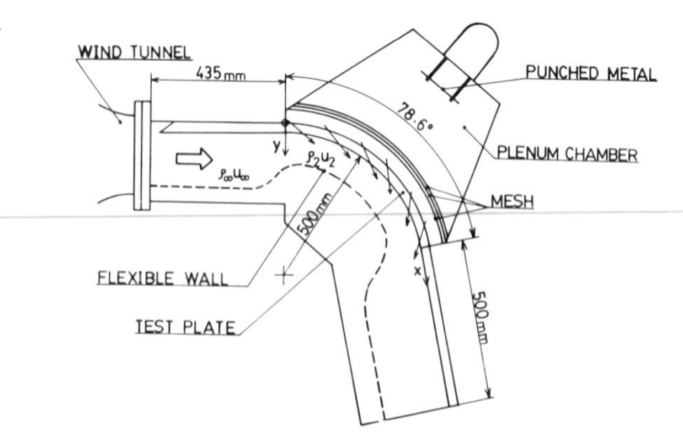

FIGURE 2. Experimental apparatus of the concave test plate

Accordingly, the momentum flux ratio is I=0.1, 0.17, 0.26, and 0.44, respectively. The streamwise pressure gradient has been kept very small by adjusting the profile of a ceiling plate in the curved section, and the resultant change in the wall static pressure is within ±2% of the main stream dynamic pressure along whole length of the test section.

For the measurement of local surface temperature on the FCFC plate, temperature sensitive liquid crystals [14] are utilized by the technique described by Kasagi [9,15,16]. The liquid crystals are painted uniformly in a small thickness of about 0.1mm on the test plate. The use of monochromatic light source of 5890A in wavelength makes it possible to take an instantaneous photograph of an isotherm on the plate under a certain condition of the secondary injection temperature. Thus, the local cooling effectiveness defined as:

$$\eta_w = (Tw(x,z)-T_\infty)/(T_2-T_\infty) \tag{7}$$

can be obtained. Note that a series of experiments with various secondary injection temperature produce a contour diagram of cooling effectiveness as shown later. The experimental uncertainty associated with the cooling effectiveness measured is evaluated less than ±10% at 20:1 odds, referring to Kline and McClintock [17].

VISUALIZATION OF 3-D TURBULENT FLOW FIELD WITH DISCRETE HOLE INJECTION

In order to investigate the effect of wall curvature on the secondary flow behavior, the smoke-wire flow visualization [18] has been applied using a 0.1mm heating wire at the location 5mm away from the wall and downstream of the 7th row of holes. Typical examples of photographs are represented of the three test plates in Fig. 3, where plane and end pictures have been taken simultaneously for each case. On the convex wall, the smoke trace of secondary flow injected remains almost at the same elevation as the smoke-wire and is seen to reattach over the surface at some streamwise distance. In the plane view, in addition, the smoke is seen swept away from the region just downstream of the injection hole. On the concave wall, the smoke tends to concentrate into the streamwise path downstream of the holes.

These observations by the smoke-wire technique can be interpreted as the change of the trajectory of secondary injection, if recalling the existence of a pair

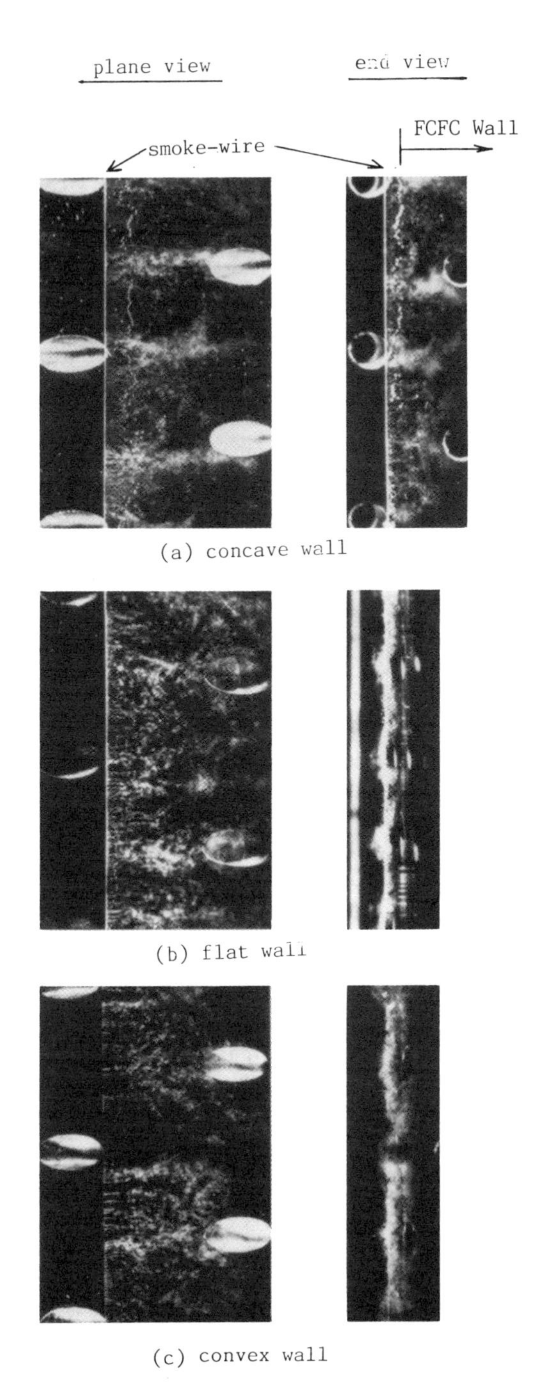

FIGURE 3. Flow visualization by the smoke-wire technique

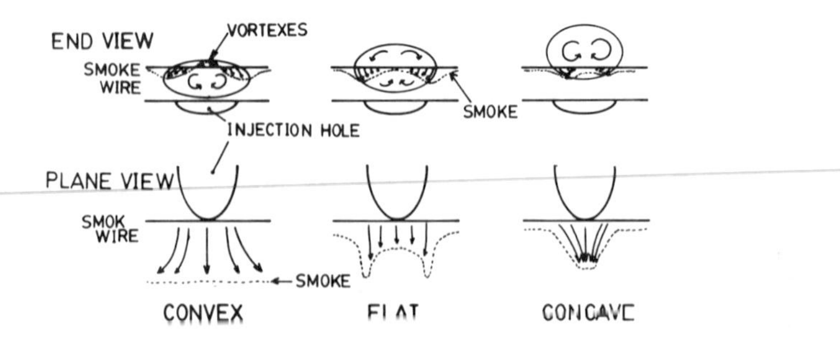

FIGURE 4. Smoke-pattern generated by the secondary injection on the curved walls

of counter-rotating streamwise vortices in the secondary injected flow [19]. That is to say, although the location of the smoke-wire is kept at a fixed distance from the wall, its location relative to the secondary flow has been changed and the smoke-wire must have sliced the upper and lower parts of the secondary injection on the convex and concave walls, respectively, as schematically shown in Fig. 4. These understandings are qualitatively in good agreement with the picture given by the preliminary consideration of the trajectory of the secondary coolant injected.

COOLING EFFECTIVENESS ON THE CURVED WALLS

The contour diagrams of cooling effectiveness on the concave, flat and convex plates, for typical two values of blowing mass flux ratio, are represented in Fig. 5. On the convex wall, the streamwise extent of constant cooling-effectiveness line is generally smaller than that on the concave wall. In addition, a tendency is observed for these contour lines to expand rapidly in the spanwise direction over the surface immediately downstream of injection and, hence, the spanwise distribution of cooling effectiveness is comparatively flatter than that on the concave wall. It is also seen that the smaller mass flux ratio of M=0.3 gives higher cooling effectiveness than M=0.65. When M=0.3, the contour line right downstream of each injection hole shows a small dent and this suggests the reattachment of secondary injection. On the concave wall, there is formed a streamwise long path in between streamwise rows of holes, where the cooling effectiveness remains very low. The patterns of contour lines for two mass flux ratios in Fig. 5 are similar each other with a little change in the region downstream of the injection hole. Since the effect of heat conduction inside the wall should be very small, these results of cooling effectiveness distributions can be recognized as wall trace of heated secondary injection and they are qualitatively consistent with the results of smoke-wire flow visualization.

The streamwise distributions of the spanwise averaged cooling effectiveness are shown in Fig. 6. The cooling effectiveness has a general tendency to increase with the streamwise distance due to the accumulated effects of the repeated injections. But, on the concave wall, it seems almost constant in the downstream region beyond the third row of holes and remains at a comparatively low value of about 0.35. The spanwise averaged cooling effectiveness on the convex wall is considerably higher than that on the concave wall, reaching about 0.6 at the 11th row of injection holes. These results are consistent with the cooling effectiveness diagrams shown in Fig. 5. The effect of mass flux ratio, M, is also clearly seen in Fig. 6. There is no appreciable change in cooling effectiveness with M on the concave wall. On the convex wall, the highest

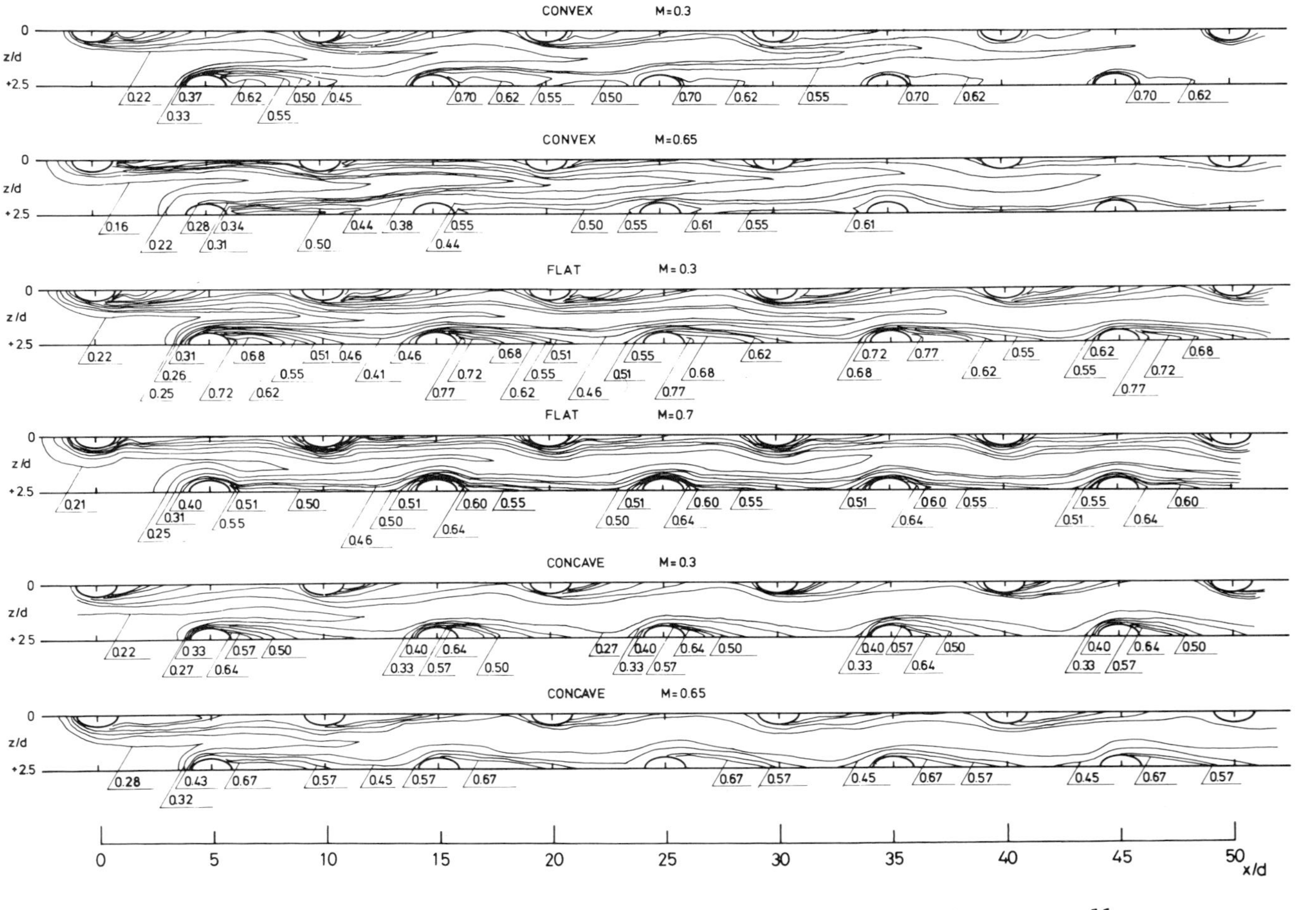

FIGURE 5. Cooling effectiveness diagram on the convex, flat and concave walls

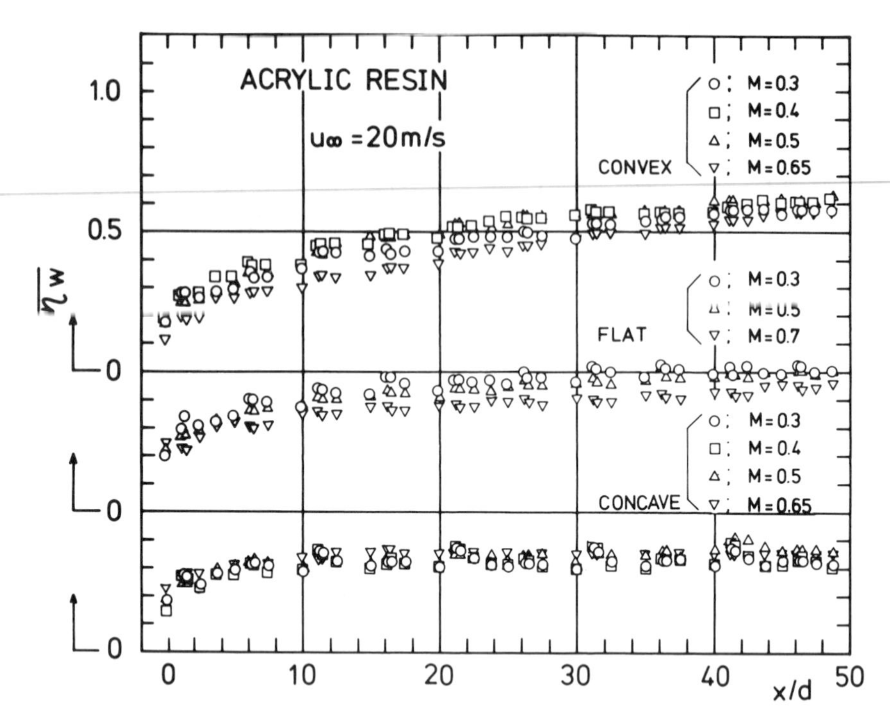

FIGURE 6. Spanwise averaged cooling effectiveness

cooling effectiveness can be obtained when the mass flux ratio, M, is in the range of 0.4-0.5 and this fact is in good agreement with the result of Furuhama et al. [7]. Note that the case of M=0.65 also gives high values in the downstream region. This is due to the thermal barrier effect caused by the thick boundary layer flow, which entrains a larger quantity of secondary fluid injected.

CONCLUSIONS

As an advanced cooling technique protecting hot-part components of high-temperature gas turbines, full-coverage film cooling has been experimentally studied focusing the effect of wall curvature on the cooling performance by using concave, flat and convex walls made of acrylic resin. Two curved plates have radii of curvature of ±500mm. The effect of coolant mass flux has also been investigated. A preliminary consideration, based on the balance of the centrifugal force associated with an injected fluid element and the pressure gradient normal to the wall, leads to the understanding that the trajectory of the injected fluid tends to depart away from the concave wall and attach to the convex wall. This has been experimentally confirmed by the smoke-wire flow-visualization, which can produce an instantaneous stereoscopic picture of complex turbulent flow field. In accordance with the flow structure observed, the cooling effectiveness measured on the convex and concave walls are higher and lower than that on the flat wall. In addition, the mass flux ratio has little effect on the cooling effectiveness on the concave wall, while it is optimum in the range of 0.4-0.5 on the convex wall.

ACKNOWLEDGEMENTS

The authors courteously thank Aero-Engine & Space Operations, Ishikawajima-Harima Heavy Industries Co. Ltd., for their manufacturing the FCFC test plates. This work was supported through the Grant-in-Aid for Special Project Research by the Japanese Ministry of Education, Science and Culture.

NOMENCLATURE

d = injection hole diameter, mm
H = shape factor
I = momentum flux ratio
J = momentum flux of secondary injection, kg/ms^2
K = constant
M = mass flux ratio
p = streamwise hole pitch, mm
R = radius of wall curvature, mm
s = spanwise hole pitch, mm
t = thickness of test plate, mm
T = temperature, K
u = velocity, m/s
x = streamwise distance from the first row of holes, mm
y = distance from the wall, mm
z = spanwise distance, mm
α = injection hole angle, °
δ = displacement thickness, mm
η_w = cooling effctiveness
θ = momentum thickness, mm
λ = thermal conductivity, W/mK

Subscripts

2 = secondary injection flow
w = wall
∞ = main stream

Superscripts

$\overline{(\;)}$ = spanwise average

EFERENCES

. Metzger, D. E., Takeuchi, D. I., and Kuenstler, P. A., Effectiveness and eat Transfer with Full-Coverage Film Cooling, ASME J. Engrg. Power, vol. 95, p. 180-184, 1973.

. Mayle, R. E., and Camarata, F. J., Multihole Cooling Film Effectiveness and eat Transfer, ASME J. Heat Transfer, vol. 97, pp. 534-538, 1975.

. Bergeles, G., Gosman, A. D., and Launder, B. E., The Prediction of Three-imensional Discrete-Hole Cooling Processes, Part 2: Turbulent Flow, ASME J. eat Transfer, vol. 103, pp. 141-145, 1981.

. Sasaki, M., Takahara, K., Kumagai, T., and Hamano, M., Film Cooling ffectiveness for Injection from Multirow Holes, ASME J. Engrg. Power, vol. 101, p. 101-108, 1979.

5. Crawford, M. E., Kays, W. M., and Moffat, R. J., Full-Coverage Film Cooling Heat Transfer Studies - A Summary of the Data for Normal-Hole and 30-deg Slant-Hole Injection, Report HMT-19, Thermosciences Division, Mech. Engrg. Dept., Stanford University, 1975.

6. Eckert, E. R. G., Analysis of Film Cooling and Full-Coverage Film Cooling of Gas Turbine Blades, 1983 Tokyo Int'l Gas Turbine Congress, ASME Paper no. 83-GTJ-2, 1983.

7. Furuhama, K., Moffat, R. J., and Frota, M. N., Heat Transfer and Turbulence Measurements of a Film-Cooled Flow Over a Convexly Curved Surface, 1983 Tokyo Int'l Gas Turbine Congress, Paper no. 83-Tokyo-IGTC-16, 1983.

8. Kumada, M., Kasagi, N., and Hirata, M., Studies of Full-Coverage Film Cooling (Part 1; Measurement of Local Heat Transfer Coefficient by Naphthalene Sublimation Technique), submitted to ASME J. Heat Transfer, 1985.

9. Kasagi, N., Hirata, M., Kumada, M. and Ikeyama, M., Studies of Full-Coverage Film Cooling (Part 2; Prediction of FCFC performance and Its Comparison With the Heat Transfer Experiment), submitted to ASME J. Heat Transfer, 1985.

10. Bradshaw, P., Effects of Streamline Curvature on Turbulent Flow, AGARDograph no. 169, 1973.

11. Mayle, R. E., Blair, M. F. and Kopper, F. C., Turbulent Boundary Layer Heat Transfer on Curved Surface, ASME J. Heat Transfer, vol. 101, pp. 521-525, 1979.

12. Simonich, J. C. and Moffat, R. J., Liquid Crystal Visualization of Surface Heat Transfer on a Concavely Curved Turbulent Boundary Layer, 1983 Tokyo Int'l Gas Turbine Congress, Paper no. 83-Tokyo-IGTC-1, 1983.

13. Ito, S., Goldstein, R. J. and Eckert, E. R. G., Film Cooling of a Gas Turbine Blades, ASME J. Engrg. Power, vol. 100, pp. 476-481, 1978.

14. Fergason, J. L., Cholesteric Structure - I Optical Properties, Molecular Crystals, vol. 1, pp. 293-307, 1966.

15. Kasagi, N., Liquid Crystal Applications in Heat Transfer Experiments, Report IL-27, Thermosciences Division, Mech. Engrg. Dept., Stanford University, 1980.

16. Kasagi, N., Visualization of Temperature Field - Application of Temperature-Sensitive Liquid Crystals, J. JSME, vol. 87, no. 783, pp. 145-151, 1984.

17. Kline, S. J. and McClintock, F. A., Describing Uncertainties in Single-Sample Experiments, Mechanical Engineering, vol. 75, pp. 3-8, 1953.

18. Kasagi, N., Hirata, M. and Yokobori, S., Visual Studies of Large Eddy Structures in Turbulent Shear Flows by Means of Smoke-Wire Method, Proc. Int'l Symp. Flow Visualization, Tokyo, pp. 169-174, 1977.

19. Liess, C., Experimental Investigation of Film Cooling With Ejection From a Row of Holes for the Application of Gas Turbine Blades, ASME J. Engrg. Power 97, pp. 21-27, 1975.

FLOW VISUALIZATION AND FLOW MEASUREMENTS

Flow Visualization of Secondary Flow in Curved, Heated and Rotating Channels, and Internal Flow in Rotating Machinery

K. C. CHENG
Department of Mechanical Engineering
University of Alberta
Edmonton, Alberta, Canada

YASUKI NAKAYAMA
Department of Mechanical Engineering
Tokai University, Kitakaname
Hiratsuka, Kanagawa, Japan

ABSTRACT

Centrifugal instability problems for Taylor, Görtler and Dean vortices and convective instability problems for longitudinal vortices revealed by flow visualization are compared considering analogy. Recent flow visualization results for secondary flows caused by such body forces as centrifugal, buoyancy and Coriolis forces in channel or ducts are presented. Representative recent photographic results obtained by flow visualization techniques applicable to internal flow in rotating machinery are briefly reviewed.

1. INTRODUCTION

Since the famous flow visualization experiment [1,2] of Reynolds in 1883 on transition from laminar to turbulent flow by dye injection into a horizontal water flow in tube, flow visualization studies have been carried out by many investigators and the related literature is now very extensive. Photographs showing many basic flow phenomena for laminar or turbulent flows and internal flow in turbomachinery are now available. It is of interest to observe that both Prandtl [3,4] and Taylor [5] utilized flow visualization experiments extensively in their pioneer works. Prandtl's books are illustrated with clear and physically revealing photographs and Taylor vortices [6] between rotating cylinders are still under investigations by many investigators. Flow visualization also played a decisive role in the development of hydrodynamic stability theory. Probably the most well-known photographs in hydrodynamic stability are Taylor vortices between rotating cylinders and Bénard cells due to thermal instability presented in [7], for example. In the field of natural convection, the Schlieren pictures of temperature fields near heated bodies obtained by Schmidt [8] in 1932 are considered to be classic examples of flow visualization.

The published photographs of fluid motion are scattered throughout the literature. The collected photographic results for various flow phenomena can be found in [11-14]. Recently, Eckert [15] points out a need for a collection of photographs illustrating various heat transfer processes (conduction, convection, radiation) obtained by visualization as well as velocity and temperature fields in convection revealed by computer simulation. Currently two books [16,17] and one handbook [18] on flow visualization are available explaining various flow visualization techniques showing photographic results. Many review articles on flow visualization techniques are also available [17,19-30]. Several recent books on fluid mechanics [31-35] are also illustrated with flow visualization photographs.

The purpose of this paper is to present a series of photographic results on secondary flow patterns caused by such body forces as centrifugal, Coriolis and buoyancy forces in laminar flow channels and some typical flow visualization results for internal flow in rotating machinery revealed by various flow visualization techniques. Recent hydrodynamic instability results for centrifugal and convective instabilities revealed by flow visualization with emphasis on analogy between physical phenomena and photographic results obtained by recent flow visualization techniques for flow in turbomachinery will also be presented. It is noted that the internal flow inside the rotating machinery is characterized by three-dimensional flow, turbulence and unsteady flow and a theoretical analysis from the inlet to the outlet of a turbomachinery is not possible due to extremely complicated flow passages involved. The centrifugal and Coriolis forces inside the turbomachinery cause secondary flow but the secondary flow is rather difficult to observe by flow visualization. Curved passages are also used extensively in the design of rotating machinery. Thus, the simpler problems of secondary flow in channels caused by centrifugal and Coriolis forces may be of some interest.

The annual flow visualization symposium in Japan was started in 1973 and a new Journal of the Flow Visualization Society of Japan was first published in 1981. In the meantime, a large number of photographic results revealed by flow visualization were published in Japan and a brief literature review on related problems may be of interest. The first symposium on flow visualization was organized by ASME in 1960 [36] and the international symposium was held in 1977 [37], 1980 [83] and 1983 [39]. The various flow visualization techniques and photographic results in many areas can be found in the published proceedings.

2. SECONDARY FLOW DUE TO BODY FORCES IN CHANNELS AND HYDRODYNAMIC INSTABILITY PROBLEMS

The secondary flows caused by such body forces as centrifugal forces ($\rho v^2/R$) in curved tubes, Coriolis forces ($2\,\rho\bar{\omega} \times \bar{u}$) in radial rotating tubes and buoyancy forces $\rho g \beta (T - T_w)$ in heated horizontal tubes are known to be analogous [40]. The effects of secondary flow on friction factor and laminar forced convection are also known to be similar. The methods of solution for this class of fluid flow and convection heat transfer problems are also known to be similar. The analogy is clear when one considers the following dimensionless parameters [41]:

$$\text{Taylor number, } Ta = (Ud/\nu)(d/R)^{1/2}, \text{ Görtler number, } G_\delta = (U\delta/\nu)(\delta/R)^{1/2} \tag{1}$$

$$\text{Dean number, } K = Re(a/R)^{1/2} \quad \text{and} \quad \text{Rayleigh number, } Ra = g\beta\Delta T\, h^3/\nu x$$

The analogy between thermal and hydrodynamic (viscous) instabilities are also well-known. Thus, the centrifugal instability for Taylor-Görtler vortices between two rotating cylinders or in the laminar boundary layer along concave walls and thermal instability for Bénard cells between two horizontal rigid plates at different temperatures heated from below are analogous [41]. It can be shown that [41]

$$Ta^2_{crit.} = (41.3)^2, \; Ra_{crit.} = 1707 \tag{2}$$

and one has a complete agreement of the smallest eigenvalues. An examination of the dynamical parameters, Taylor, Görtler and Dean numbers reveals that these longitudinal vortices belong to the same family and the centrifugal instability problems are qualitatively similar. It is of interest to note that Dean vortices also appear in curved pipe laminar flow [42-44]. The centrifugal instability problems become much more complicated when buoyancy forces appear.

When body forces act in a direction normal to the main flow in ducts or channels, secondary flow arises. After the classical work of Dean [45,46] on fully developed laminar flow in curved pipes, many theoretical and experimental studies on flow and forced convective heat transfer in curved pipes have been carried out. The related literature is well surveyed in [47-51]. A general review of secondary flows and enhanced heat transfer in curved ducts, and rotating pipes and ducts is also available [52-54]. Further literature on heat transfer and fluid flow in rotating coolant channels can be found in a recent monograph [55].

Probably one of the most striking examples of the utilization of flow visualization is the confirmation of Taylor vortices in two concentric cylinders with inner cylinder rotating and outer cylinder at rest [6,7]. It is of interest to note that theoretical prediction precedes the flow visualization experiment in this case. In contrast, the flow visualization experiment [7] of Bénard cells [1900] precedes a theoretical prediction [7] of Rayleigh [1916]. Many reviews on Bénard convection and related problems are available [7,56]. From these classical studies, one can see clearly the importance of flow visualization experiment in understanding the complicated flow phenomena relating to the onset of instability and the post-critical flow regime and in confirming the theoretical prediction. Some recent photographic results will be represented to illustrate the qualitative analogy relating to hydrodynamic instability problems caused by such body forces as centrifugal, buoyancy and Coriolis forces. Similarily, selected recent photographic results on secondary flow patterns in ducts or channels caused by such body forces as centrifugal, buoyancy and Coriolis forces will be presented. The emphasis here is the physical phenomena revealed by flow visualization and the flow visualization techniques will not be discussed. Because of extensive literature on the related topics and time constraint, only those photographs which are readily available to the authors will be used for illustrations and are not representative of the photographic results available in the literature at present.

3. INTERNAL FLOW IN ROTATING MACHINERY

Because of rotation, curvature effect, high velocity, rather complicated flow passages and many other factors, internal flow in fluid machinery represents one of the most complicated fluid mechanics problems and one cannot attempt a theoretical analysis from the inlet to the outlet of the rotating machinery in the same way as the entrance region problem for internal flow in ducts or channels. The secondary flow caused by centrifugal and Coriolis forces is also a complicating factor. Thus, measurements, flow visualization and computer simulation are useful tools in understanding the flow phenomena inside the rotating machinery.

Various flow visualization methods [17,18] have been applied to internal flow in rotating impellers in recent years and review articles [28,57-67] are available explaining flow visualization techniques and presenting recent photographic results. Many flow visualization studies relating to internal flow in turbomachinery are reported in the proceedings of the past three international symposiums on flow visualization. The Japanese works on this subject are reported frequently in the Journal of the Flow Visualization Society of Japan and Turbomachinery. A brief literature survey will be presented with emphasis on recent Japanese works. Representative photographs obtained by various flow visualization techniques will also be presented.

4. PHOTOGRAPHIC RESULTS ON HYDRODYNAMIC INSTABILITY AND SECONDARY FLOW PATTERNS IN CHANNELS

4.1 Centrifugal and Thermal Instabilities

The photographs of Bénard cells shown in Fig. 1 (a), (b), (c) [68] and Taylor vortices shown in Fig. 2 [69] are the two classical examples of hydrodynamic instability phenomena revealed clearly by flow visualization. The growth and decay of Görtler vortices are shown in Fig. 3 (a) [70], (b), (c), (d) [71]. The Görtler vortices also appear on turbine blade pressure surface [72]. Dean vortices due to centrifugal instability appear in fully developed laminar flow in a curved rectangular channel with large aspect ratio (width/height) at a certain Dean number and the development and decay are shown in Fig. 4 [73]. The photographic results for other aspect ratios are given in [74-78].

The computer-generated secondary flow streamlines in an 8 x 1 channel caused by Coriolis forces introduced by steady rotation about an axis perpendicular to the plane of mean flow are shown in Fig. 5 [79] as a function of time. Similarly, the development of secondary-flow streamline patterns in the thermal entrance region of a horizontal rectangular channel with aspect ratio 10 heated from below is presented in Fig. 6 [80] to show the similarity between the hydrodynamic instability due to Coriolis forces and convective instability due to buoyancy forces. Some recent flow visualization results obtained by smoke injection method for the roll-cell instability in a radial rotating laminar flow in rectangular channel with aspect ratio 20 are shown in Fig. 7 for reference. The definitions of Reynolds number Re and rotation number Ro are similar to those used in [79]. The right-hand edge of the channel represents the center of the channel.

The flow visualization results obtained by smoke injection method on the onset of longitudinal vortices in laminar forced convection between horizontal plates heated from above or below are shown in Fig. 8 (a) for developing secondary flow patterns and in (b) for fully developed secondary flow [81]. The fully developed secondary-flow pattern is also shown in [82]. A flow visualization study of convective instability in the thermal entrance region of a horizontal parallel-plate channel heated from below is presented in [83]. The onset, development and decay of longitudinal vortices (cross-sectional view) in natural convection flow along an isothermally heated horizontal rectangular plate are shown in Fig. 9. The experiment was conducted in a suction low speed wind tunnel designed for flow visualization studies. The smoke was injected at the leading edge $x = 0$ and Grashof number Gr_x is defined using the temperature difference between the plate and air, and a distance x from the leading edge. The photographic results presented serve to show the qualitative similarity between hydrodynamic instability problems due to centrifugal, Coriolis and buoyancy forces. It is expected that the instability problem becomes more difficult if two or more kinds of body forces appear simultaneously.

4.2 Secondary Flow in Curved, Heated and Rotating Channels

Fully developed laminar forced convection with secondary flow in curved pipes, heated horizontal tubes and rotating ducts is well reviewed in [53]. The general features of secondary flow pattern and distributions of axial velocity and fluid temperature are shown in Fig. 10 [53]. The secondary flow pattern depends on the intensity of secondary flow which in turn is governed by such physical parameters as Dean number or Rayleigh number. With a very strong secondary flow, numerical solution is not possible and one must employ the boundary layer approximation for analytical solution. The flow visualization is useful in confirming the secondary flow pattern obtained by numerical solution.

The photographs of secondary flow patterns revealed by smoke injection method are presented in [84] for fully developed flow in curved circular and semicircular pipes. The secondary flow patterns shown in Fig. 11 are arranged in the order of increasing Dean number. The highest Dean number is $K = 642$ and 370 for curved circular and semicircular tubes, respectively. It is seen that the secondary flow pattern for $K = 642$ is quite different from the theoretical model shown in Fig. 10. For curved semicircular tubes, Dean vortices caused by centrifugal instability appear in the form of one or two pairs of vortices near the curved flat outer wall and the flow phenomenon agrees with numerical prediction [85].

The secondary flow patterns in a straight tube section downstream of 180° degree bend at $Re = 403$ and Dean number $K = 127$ are shown in Fig. 12 where a downstream distance x is expressed in terms of tube diameter d [44]. It is noted that the pictures on the right-hand side were obtained by disturbing the flow using a hypodermic needle at 90° position from the start of the 180° bend and those on the left-hand side represent natural disturbances. A pair of Dean vortices is clearly seen near the outer curved wall for the artificially disturbed flow.

The secondary flow patterns observed at the exit of an isothermally heated 180° bend with parabolic entrance flow and entrance air temperature 25°C are shown in Fig. 13 (a) for upward flow with wall temperature at 57°C, (b) for horizontal flow with wall temperature at 56°C (outer curved wall to the left) and (c) for downward flow with wall temperature at 91°C [86]. The Dean number and the parameter Re Ra are also shown for reference. Because of the superimposed centrifugal and buoyancy effects, the secondary flow patterns are seen to be very complicated. In the downward flow configuration, the Dean vortices appear naturally near the lower outer curved wall at Dean number $K = 215$, 182 and 149. The mixed convection problems in the entrance region of curved circular tubes are studied both numerically and experimentally in [87].

he developing secondary flow patterns in the entrance region of a curved pipe with fully developed parabolic entrance velocity profile are shown in ig. 14 for Dean number $K = 364$ (left-hand side photographs) and $K = 520$ right-hand side photographs) with entrance angle from the start of bend = 45°, 90°, 135° and 180° [86]. The developing secondary flow patterns n the laminar entry region of a curved pipe obtained by flow visualization, measurements and numerical solution are shown in Fig. 15 for Dean umber $K = 333$ and $Re = 10^3$ [88].

he secondary flow patterns in the thermal entrance region of an isothermally heated horizontal tube and inclined tube with inclination angle ϕ = 0° from the horizontal direction and upward flow are shown in Figs. 16

(wall temperature 74°C) and 17 (wall temperature 78°C), respectively [84]. The dimensionless axial distance z = (l/d)/RePr, Rayleigh number based on temperature difference between wall temperature and entrance air temperature, Ra_1, Rayleigh number based on temperature difference between wall temperature and average of centerline air temperatures at thermal entrance and exit, Ra_2, and $\Gamma = Gr_1/Re^2$ are shown for reference. As expected, the secondary flow will disappear eventually as the bulk temperature approaches the wall temperature. The transient secondary flow patterns observed at the end of a horizontal tube after the start of heating by flowing hot water at high velocity in annular duct are shown in Fig. 18 [86]. The entrance air is at room temperature. Secondary flow patterns for forced convection heat transfer in uniformly heated horizontal tubes are also shown in [89].

The secondary flow can be generated in a heated or cooled tube rotating about its own axis with main flow. A Pyrex tube (inside dia. = 4.29 cm) heated by a single band heater (525 W) was cooled by air. After confirming the steady state, the power input to the heater was then turned off and the rotation of the tube started. The secondary flow patterns were observed at some distance from the heater by smoke injection method. The results are shown in Fig. 19 for Re = 178. The left-hand side photograph represents secondary flow patterns at 5 cm downstream from heater and the right-hand side photographs are obtained at 10 cm from the heater.

The secondary flow patterns of a rotating radial tube (inside dia. = 3.81 cm) observed at a distance 41.1 cm from the axis of rotation are shown in Figs. 20 and 21 for N = 36 rpm, Re = 68 and N = 96 rpm, Re = 826, respectively. The series of photographs in each case represent the changing secondary flow patterns after the start of air flow. The last photograph can be considered to be at steady state. The photographs were obtained in a dark room with camera lens open and by synchronizing the stroboflash with tube outlet position.

5. FLOW VISUALIZATION EXAMPLES FOR INTERNAL FLOW IN TURBOMACHINERY

The Visualization of flow in fluid machinery is realized with one technique or a combination of techniques. For pumps and hydraulic turbines, oil film (wall trace) method is used for flow near surface and tuft method, tracer method and smoke-wire method are used for internal flow. The oil film method provides flow field near the surface but differs considerably from internal flow when secondary flow is significant. It is not suitable for unsteady phenomena since it takes several minutes before a pattern is formed. Tuft method and dye injection method are not suitable for high velocity and turbulence. Because of dye diffusion, dye injection method must be used at low velocity.

For the blade cascade experiments of compressor, blower and turbine, optical method and spark tracing method are used for air, and tracer method and water table visualization method are used for water. For rotating experiments, tracer method, tuft method, spark tracing method and smoke-wire method are used for air, and tracer method and hydrogen bubble method are used for water. Flow visualization experiments are usually performed at lower velocity than the actual operating condition. Optical method is used at high velocity but is usually concerned with air flow around stationary blades. It is seen that examples of flow visualization experiments of actual fluid machinery at operating conditions are rather limited.

5.1 Hydraulic Machinery

Fig. 22 shows path lines revealed by surface coating on the impeller surface of an axial flow pump [59]. The application of tuft method for flow inside the impeller of a centrifugal pump is shown in Fig. 23 [12] where the upper picture is for large flow rate and the lower one is for small flow rate. The surface coating method is also applied to volute casing wall [12]. Prescale film method [59] is used for both visualization and measurement of pressure distribution due to cavitation.

5.2 Fluid (Air) Machinery

Schlieren picture of the flow (Freon gas) in a rotating supersonic annular cascade and shock wave pattern deduced from this visualization is shown in Fig. 24 [64,90]. The flow around a single blade visualized by spark tracing method is shown in Fig. 25 [59]. A photographic study of the three-dimensional flow in a radial compressor is reported in [91]. The rotating stall of a centrifugal impeller is visualized in Fig. 26 [59] by using plastic pellets as tracer particles for water table visualization method. The time lines in blade passage visualized by spark tracing method are shown in Fig. 27 [92]. The unsteady flow phenomena in rotating centrifugal impeller passages are studied in [93] using hydrogen bubble technique and a flow phenomenon revealed at the exit of the shrouded impeller with zero flow is shown in Fig. 28.

The flow around the impeller of a crossflow fan revealed by suspending polystyrene beads in water is shown in Fig. 29 [59]. Similar flow revealed by Moiré method showing an eccentric vortex is presented in Fig. 30 [59]. For flow visualization of both the main flow and the secondary flow near the walls, the hydraulic model using coloured liquid or air injection can be used. Visualization of secondary flows near the bottom walls in a turbine blade cascade with coloured filmaments is shown in Fig. 31 [64]. Other examples using hydraulic models are given in [64].

5.3 Heat Engine (Turbine)

It appears that flow visualization study of combustion process in actual engine has not been carried out so far. Flow visualization has been realized by installing a partial model to the engine using external drive. For jet engine and gas turbine, flow visualization is limited to components. For steam turbine, flow visualization is also limited to blade cascade experiment and blades are usually kept stationary. Flow visualization techniques in diesel and gasoline engines are reviewed in [59]. The flows inside the vortex chamber of a four-cycle engine and around suction valve visualized by spark-tracing method are shown in Figs. 32 and 33, respectively [59].

For steam and gas turbines, optical methods are used for high speed flow in nozzles and rotor blades by using mainly two-dimensional blade cascade in wind tunnel. Flow visualization under rotating conditions has not been reported. Figs. 34 and 35 show the equal density interference fringe pattern obtained by Mach-Zehnder interferometer and Schlieren method, respectively, for flow in the nozzle of a supersonic turbine under overloading condition [59]. The exit Mach number is 2.5 and the shock waves leaving the trailing edge on the pressure surface of the blade suggest that the flow is deficient. From the interference fringe pattern, one can understand the expansion process and the growth of boundary layer. By counting the number of fringes, one can compute the local density, pressure and velocity. From Schlieren picture, one can determine the flow exit direction and the flow direction change caused by the interference of the shock waves. The pressure distribution along the blade surface obtained from the interference fringe pattern agrees with theoretical calculations [59]. These flow visualization methods have also been applied to the wind tunnel test of blade cascade for steam turbine [94].

Fig. 36 shows the equal density interference fringe pattern for flow around the long blades in the low pressure stage of a steam turbine. A color Schlieren photograph is shown in Fig. 37 [95] for transonic flow around the turbine blades. By using hydraulic analogy, the flow around steam turbine rotor blade can be simulated in the free-surface water table as shown in Fig. 38 [59]. The film cooling by air for high temperature turbine blade of a jet engine can be visualized by flowing paraffin mist as shown in Fig. 39 [59]. A color Schlieren photograph for air intake flow of an engine of a supersonic plane is shown in Fig. 40 [59]. The swirling flow in the continuous combustor of a gas turbine can be visualized by spark-tracer method using air [59]. Fig. 41 shows the swirling flow pattern inside a cylindrical combustor (inside dia. = 60 mm) photographed simultaneously by two cameras from front and side. It is noted that flow direction can be detected by mixing air with AlN particles (size, 1.5 ~ 6 μm) giving out luminous tails without disturbing the spark lines [67,96].

For actual turbomachines with compressibility effects, shock waves in absolute or relative flow can be visualized by some modifications of the conventional shadowgraph or Schlieren techniques. The examples of application of shadow and Schlieren techniques to various turbomachines are reviewed in [64]. The use of holography techniques for turbomachinery flow diagnostic is reviewed briefly in [64].

Since the literature on flow visualization in rotating machinery is very extensive, it is difficult to present a complete catalogue of applications and references. Methods such as hydrogen bubble method, spark tracing method, smoke-wire method and Mach-Zehnder interferometer yield quantitative information. Other subjects such as laser holography, holographic interferometry [17], image recording system, image processing and computer aided flow visualization are not discussed. A good summary of flow visualization for secondary flows in turbomachines is given in [97]. Flow visualization provides an overall, qualitative understanding of the flow field. It provides understanding on flow phenomena such as flow separation, shock waves, rotating stall, and surging phenomenon. It can be used to check numerical solution and theoretical flow modelling. It is expected that flow visualization techniques will be used extensively in conjunction with measurements and theoretical works in future research and development relating to advanced turbomachines.

6. CONCLUDING REMARKS

Recent flow visualization results for secondary flow patterns in ducts or channels caused by such body forces as centrifgual, buoyancy and Coriolis forces are presented. Flow visualization examples are presented to illustrate the applications of various visualization techniques to internal flow in turbomachinery. Flow visualization techniques are not discussed. The details can be found in publications in the form of technical reports or papers. The references given are incomplete and further references can be found in books and review articles on flow visualization in each field.

6.1 Secondary Flows in Channels or Ducts

For hydrodynamic instabiity problems involving longitudinal vortices, flow visualization also provides quantitative results on the lowest eigenvalue and wavelength. For complicated secondary flow problems involving two or more body forces, flow visualization provides information on flow structure and can be used for comparison with numerical predictions and checking of theoretical models such as boundary layer approximation. The smoke injection method is restricted to low speed flow and is similar to the present applicability of numerical solution. Optical methods should be used for flow regime with higher dynamical parameter. Such fundamental flow processes as relaminarization due to secondary flow or the delay or occurrence of turbulent flow in curved pipes, for example, should be studied by flow visualization experiment. Many secondary flow problems remain to be clarified by flow visualization.

It is of considerable interest to observe that temperature boundary development in a tube (thermal entrance region problem) with fully developed laminar flow and constant wall temperature was solved by Graetz in 1885 and the boundary layer concept for flow on a flat plate was proposed by Prandtl in 1904. The developments of temperature and velocity boundary layers are similar in nature (parabolic problem). Thus it is useful to visualize temperature boundary layer development directly in ducts or channels somewhat similar to time lines for velocity developments.

6.2 Internal Flow In Turbomachinery

Smoke filaments, hydraulic models and optical methods (Shadow, Schlieren, Mach-Zehnder interferometer and holography techniques) are used for flow visualization in turbomachinery. The flow visualization for turbomachinery flow phenomena is at least one order of magnitude more difficult than the visualization of flow phenomena in channels or ducts due to rotation and complicated flow passages. Such flow phenomena as boundary layer separation, shock waves, wake flow, rotating stall, surging phenomenon, recirculation and secondary flow and other unsteady phenomena are visualized by various flow visualization techniques. This review is restricted to photographic results of example problems. Flow visualization techniques, devices, evaluation procedures and actual applications are not considered.

ACKNOWLEDGEMENT

This review was prepared using operating grant from the Natural Sciences and Engineering Research Council of Canada. The authors wish to thank graduate students, F.P. Yuen and Y.W. Kim for providing the photographs used and T. Villett, Research Laboratory at the University of Alberta for technical assistance.

REFERENCES

1. Reynolds, O., Phil. Trans., Vol. 174, pp. 935-982, 1883.
2. Eckert, E.R.G., Proc. 7th Int. Heat Transfer Conf., Munchen, Vol. 1, pp. 1-8, 1982.

3. Prandtl, L., Essentials of Fluid Dynamics, Hafner Pub. Co., 1952.
4. Prandtl and Tietjens, Applied Hydro- and Aeromechanics, Dover Pub., 1934.
5. Batchelor, G.K. (Ed.), Scientific Papers of G.I. Taylor, Vol. 4, Cambridge Univ. Press, 1971.
6. Taylor, G.I., Phil. Trans. 223A, pp. 289-343, 1923.
7. Stuart, J.T., Hydrodynamic Stability, Rosenhead, L. (Ed.), Laminar Boundary Layers, Oxford Univ. Press, Chapter 9, 1963.
8. Schmidt, E., Forsch. Ing. Wesen, Vol. 3, pp. 181-189, 1932.
9. Shapiro, A.H., Shape and Flow, Anchor Books, New York, 1961.
10. The NCFMF Book of Film Notes, Illustrated Experiments in Fluid Mechanics, MIT Press, 1972.
11. Van Dyke, M., An Album of Fluid Motion, The Parabolic Press, 1982.
12. JSME - An Album of Fluid Flow, Maruzen Co., 1984.
13. The Flow Visualization Soc. of Japan, An Album of Flow Visualization, No. 1, 1984.
14. Lugt, H.J., Vortex Flow in Nature and Technology, John Wiley & Sons, 1983.
15. Eckert, E.R.G., ASME Paper 84-WA/HT-31.
16. Reznicek, R., Visualisace Proudeni (Flow Visualization), Academia, Praha, 1972.
17. Merzkirch, W., Flow Visualization, Academic Press, 1974.
18. Asanuma, T. (Ed.), Handbook of Flow Visualization (in Japanese) Asakura Shoten, Tokyo, 1977.
19. Clayton, B.R. and Massey, B.S., J. Sci. Instrum. Vol. 44, pp. 2-11, 1967.
20. Hauf, W. and Grigull, U., Advances in Heat Transfer, Vol. 6, pp. 133-366, 1970.
21. Hanawa, J. and Okamoto, Y., J. of JSME, Vol. 70, pp. 1793-1801, 1967.
22. Taneda, S., J. of the Physical Soc. of Japan, Vol. 23, pp. 430-447, 1968.
23. Asanuma, T., J. of JSME, Vol. 72, pp. 1370-1377, 1969.
24. Werlé, H., Annu. Rev. Fluid Mech., Vol. 5, pp. 361-382, 1973.
25. Asanuma, T., J. of JSME, Vol. 77, pp. 567-574, 1974.
26. Sakagami, H. and Taneda, S., Meteorological Research Note, No. 124 (Japan), 1975.
27. Nakayama, Y., J. of JSME, Vol. 81, pp. 636-642, 1978.
28. Akashi, K., J. of JSME, Vol. 81, pp. 663-669, 1978.
29. Mueller, T.J., Fluid Mechanics Measurements (Ed. R.J. Goldstein), Hemisphere Pub. Corp., Chapter 7, 1983.
30. Goldstein, R.J., Fluid Mechanics Measurements, Hemisphere Pub. Corp., Chapter 8, 1983.
31. Batchelor, G.K., An Introduction to Fluid Dynamics, Cambridge Univ. Press, 1967.
32. Greenspan, H.P., The Theory of Rotating Fluids, Cambridge Univ. Press, 1969.
33. Turner, J.S., Buoyancy Effects in Fluids, Cambridge Univ. Press, 1973.
34. Tritton, D.J., Physical Fluid Dynamics, Van Nostrand Reinhold, 1977.
35. Nakayama, Y., Fluid Mechanics (in Japanese), Yokendo, Tokyo, 1979.
36. Kilne, S.J. (Ed.), ASME Symposium on Flow Visualization, 1960.

37. Asanuma, T. (Ed.), Flow Visualization, Hemisphere Pub. Corp., 1979.
38. Merzkirch, W. (Ed.), Flow Visualization II, Hemisphere Pub. Corp., 1982.
39. Yang, W.J. (Ed.), Flow Visualization III, Hemisphere Pub. Corp., 1985.
40. Trefethen, L., Proc. 9th Int. Congress of Applied Mechanics, Vol. 2, pp. 341-350, 1957.
41. Zierep, J., Convective Transport and Instability Phenomena, Zierep, J. and Oertel, H. (Eds.), G. Braun, Karlsruhe, pp. 25-37, 1982.
42. Dennis, S.C.R. and Ng, M., Q. J. Mech. Appl. Math., Vol. 35, pp. 305-324, 1982.
43. Nandakumar, K. and Masliyah, J.H., J. Fluid Mech., Vol. 119, pp. 475-490, 1982.
44. Cheng, K.C. and Yuen, F.P., ASME Paper, 84-HT-62.
45. Dean, W.R., Phil. Mag., Vol. 4, pp. 208-223, 1927.
46. Dean, W.R., Phil. Mag., Vol. 5, pp. 673-695, 1928.
47. Ito, H., J. of JSME, Vol. 66, pp. 1368-1375, 1963.
48. Akiyama, M., Thermal Entrance Region Heat Transfer and Hydrodynamic Stability in Curved Channels, Ph.D. Thesis, Univ. of Alberta, 1973.
49. Berger, S.A., Talbot, L. and Yao, L.S., Ann. Rev. Fluid Mech., Vol. 15, pp. 461-512, 1983.
50. Masliyah, J.H. and Nandakumar, K., Adv. in Transport Processes, Vol. 4, 1985.
51. Ito, H., Trans. JSME, Vol. 50, pp. 2267-2274, 1984.
52. Mori, Y., Nakayama, W. and Uchida, Y., J. of JSME, Vol. 70, pp. 1188-1196, 1967.
53. Mori, Y. and Nakayama, W., Heat and Mass Transfer in Rotating Machinery, Metzger, D.E. and Afgan, N.H. (Eds.), Hemisphere Pub. Corp., pp. 3-24, 1984.
54. Kreith, F., Adv. in Heat Transfer, Academic Press, Vol. 5, pp. 129-251, 1968.
55. Morris, W.D., Heat Transfer and Fluid Flow in Rotating Channels, John Wiley & Sons, 1981.
56. Koschmieder, E.L., Adv. in Chemical Physics, Vol. 26, pp. 177-212, 1974.
57. Ito, H., Science of Machine, Yokendo, Japan, Vol. 17, pp. 1230-1236, 1965.
58. Murai, H. and Watanabe, H., Turbomachinery, Japan, Vol. 5, pp. 661-668, 1977.
59. Nakayama, Y., Science of Machine, Japan, Vol. 33, pp. 981-985, pp. 1094-1098, 1981.
60. Akashi, J., J. Flow Visual. Soc. of Japan, Vol. 1, pp. 29-35, 1981.
61. Nakayama, Y., Turbomachinery, Vol. 11, pp. 173-184, 1983.
62. Ohki, H., Yoshinaga, Y. and Tsutsumi, Y., J. Flow Visual. Soc. Japan, Vol. 3, pp. 330-337, 1983.
63. Fabri, J., Flow Visualization Techniques for Radial Compressors, VKI Lecture Series 66, 1974.
64. Paulon, J., Optical Measurements in Turbomachinery, AGARDograph No. 207, pp. 123-139, 1975.
65. Sieverding, C., Starken, H., Lichtfuss, H. and Schimming, P., AGARDograph No. 207, pp. 1-76, 1975.
66. Kraft, H., ASME Symposium on Unsteady Flow, 68-FE-41, 1968.

67. Nakayama, Y., Yamamoto, T., Aoki, K. and Ohta, H., Trans. JSME, Vol. 51B, pp. 325-332, 1985.
68. Chandra, K., Proc. Roy. Soc. London, 164A, pp. 231-242, 1938.
69. Zierep, J., Z. Flugwiss. Weltraumforsch, Vol. 2, pp. 143-150, 1978.
70. Wortmann, F.X., Proc. 11th Int. Congress of Applied Mechanics, pp. 815-825, 1964.
71. Ito, A., J. Japan Soc. Aero. S[ace Sciences, Vol. 28, pp. 327-333, 1980.
72. Han, L.S. and Cox, W.R., J. Eng. for Power, ASME Trans., Vol. 105, pp. 47-52, 1983.
73. Urai, I. and Akiyama, M., J. Flow Visual. Soc. Japan, Vol. 2, 1982.
74. Cheng, K.C., Nakayama, J. and Akiyama, M., Flow Visualization I, pp. 181-186, 1979.
75. Sugiyama, S., Hayashi, T. and Yamazaki, K., Bulletin JSME, Vol. 26, pp. 964-969, 1983.
76. Akiyama, M., Cheng, K.C., Urai, I. Suzuki, M. and Nishiwaki, I., J. Flow Visual. Soc. Japan, Vol. 2, pp. 553-558, 1982.
77. Akiyama, M., Kikuchi, K., Nakayama, J., Suzuki, M., Nishiwaki, I., Cheng, K.C., Trans. JSME, Vol. 47B, pp. 1705-1714, 1981.
78. Akiyama, M., Kikuchi, K. Suzuki, M., Nishiwaki, I., Cheng, K.C., Nakayama, J., Trans. JSME, Vol. 47B, pp. 1960-1970, 1981.
79. Speziale, C.G. and Thangai, S., J. Fluid Mech., Vol. 130, pp. 377-395, 1983.
80. Cheng, K.C. and Ou, J.W., Proc. 7th Int. Heat Transfer Conf., Vol. 2, pp. 189-194, 1982.
81. Akiyama, M., Hwang, G.J. and Cheng, K.C., J. Heat Transfer, Trans. ASME, Vol. 93, pp. 335-341, 1971.
82. Mori, Y. and Uchida, Y., Int. J. Heat Mass Transfer, Vol. 9, pp. 803-817, 1966.
83. Hwang, G.J. and Liu, C.L., Canadian J. Chem. Eng., Vol. 54, pp. 521-525, 1976.
84. Cheng, K.C., Inaba, T. and Akiyama, M., Flow Visualization III, pp. 531-536, 1985.
85. Masliyah, J.H., J. Fluid Mech., Vol. 99, pp. 469-479, 1980.
86. Yuen, F.P., Flow Visualization Experiments on Secondary Flow Patterns in Curved and Heated Pipes, M.Sc. Thesis, Univ. of Alberta, 1985.
87. Akiyama, M., Suzuki, M., Cheng, K.C., Suzuki, M. and Nishiwaki, I., Trans. JSME, Vol. 50B, 1197-1204, 1984.
88. Akiyama, M., Hanaoka, Y., Cheng, K.C., Urai, I. and Suzuki, M., Flow Visualization III, pp. 526-530, 1985.
89. Mori, Y. and Futagami, K., Int. J. Heat Mass Transfer, Vol. 10, pp. 1801-1813, 1967.
90. Veret, C., Philbert, M., Surget, J. and Fertin, G., Flow Visualization I, pp. 335-340, 1979.
91. Senoo, Y., Yamaguchi, M. and Nishi, M., J. Eng. for Power, Vol. 90A, pp. 237-244, 1968.
92. Fister, W., BWK, Vol. 18, pp. 425-429, 1966.
93. Lennemann, E. and Howard, J.H.G., J. Eng. for Power, Trans. ASME, Vol. 92A, pp. 65-72, 1970.
94. Ukeguchi, M. and Kuramoto, Y., Proc. 2nd Int. JSME Symp., Tokyo, Vol. 2, p. 131, 1972.

95. Ikeda, T. and Suzuki, A., Proc. Heat and Fluid Flow in Steam and Gas Turbine Plant, Inst. Mech. Engr., p. 46, 1974.

96. Nakayama, Y., Okitsu, S., Aoki, K. and Ohta, H., Flow Visualization I, pp. 239-244, 1976.

97. AGARD Conf. Proc., No. 214, Paris, 1977.

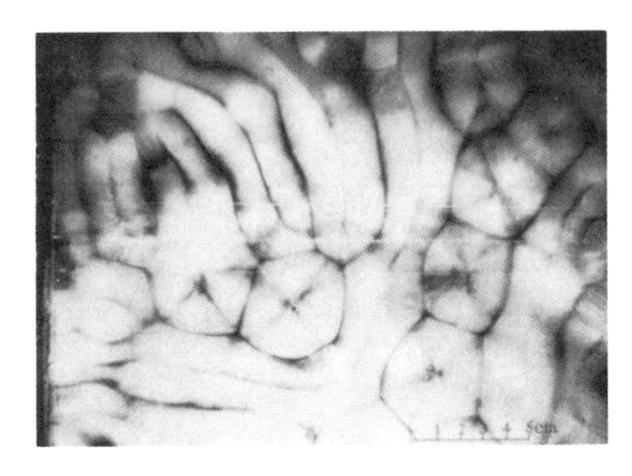

(a) 10 mm depth, with temperature difference 117°C.

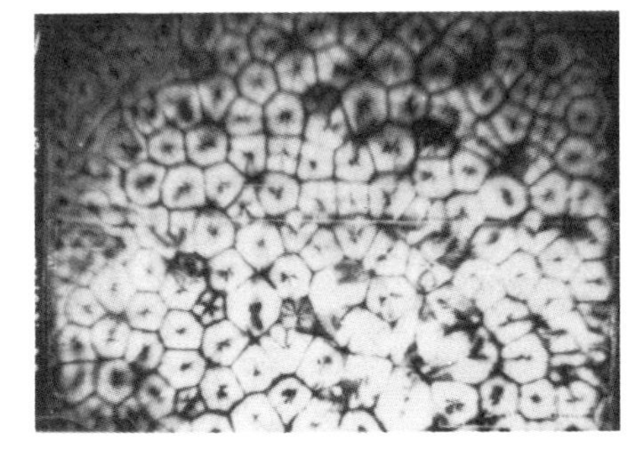

(b) 7 mm depth, with temperature difference 130°C.

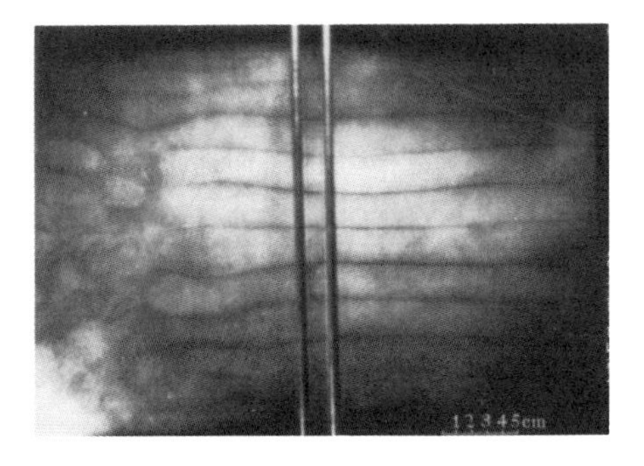

(c) 16 mm depth, with temperature difference 29°C, and shear 10 cm/sec.

Fig. 1 Benard convection cells [68].

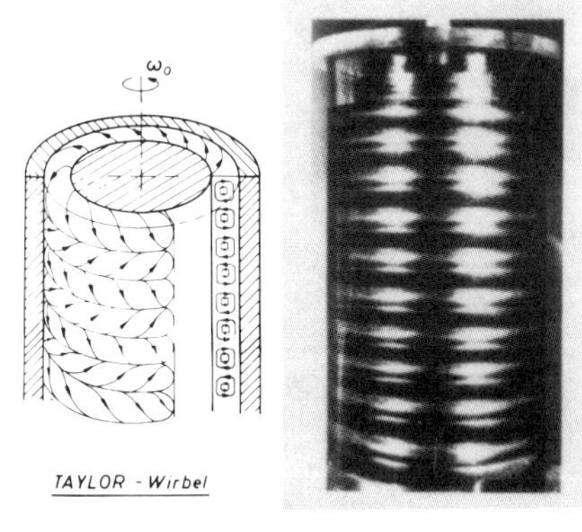

Fig. 2 Taylor vortices in post-critical state. Outer cylinder is at rest, aluminum powders in silicon oil [69].

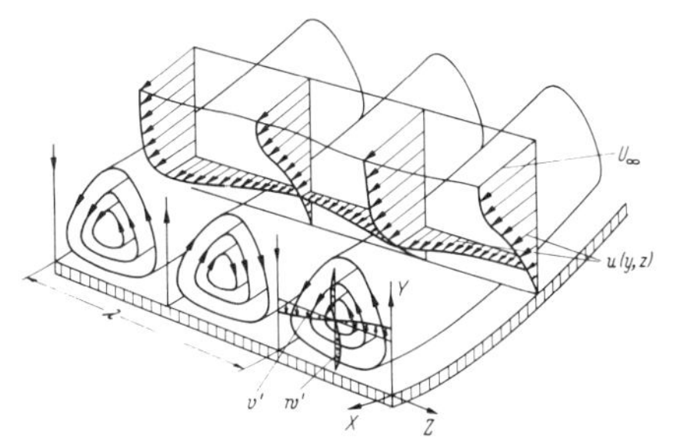

(a) Görtler vortices on a concave wall [70].

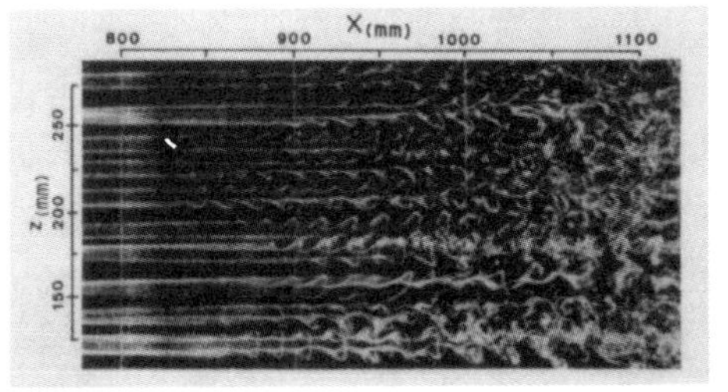

(b) Formation of Görtler vortices [71].

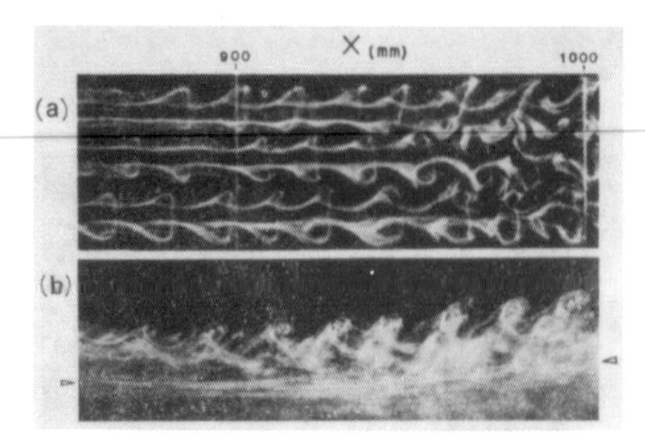

(c) Top and side views of vortices [71].

(d) Decay process of longitudinal vortices [71].

Fig. 3 Development of Görtler vortices.

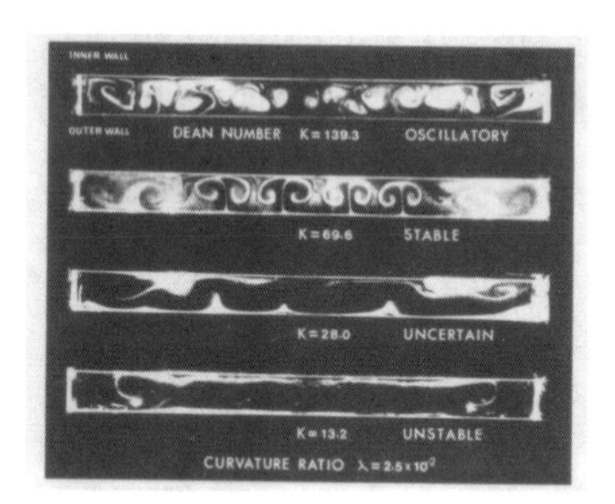

Fig. 4 Development of Dean vortices in a curved rectangular channel [73].

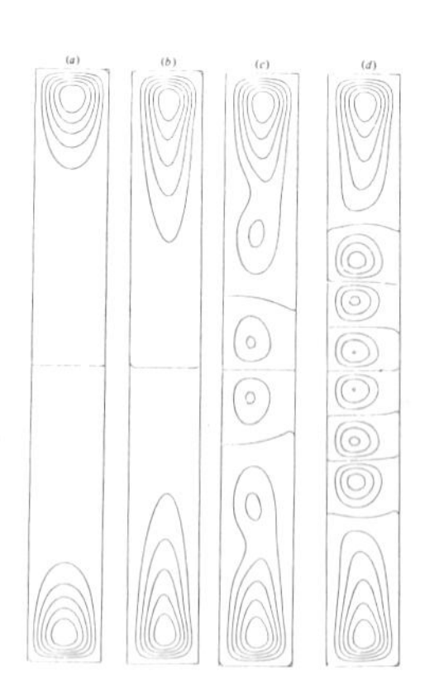

Fig. 5 Secondary-flow streamlines in a rotating radial rectangular channel, (a) 10 s, (b) 500 s, (c) 1,600 s, (d) fully developed.

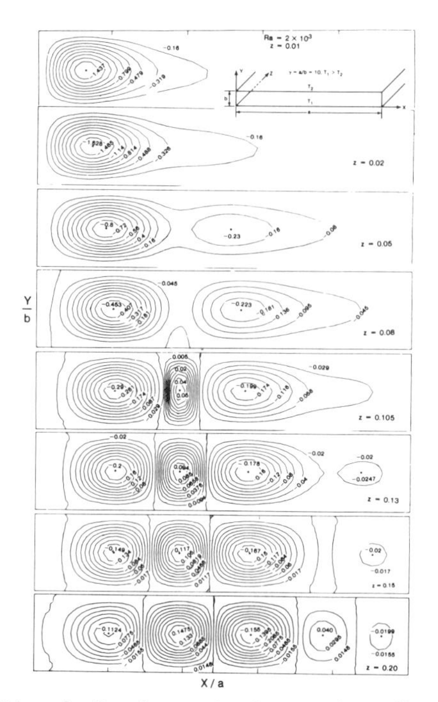

Fig. 6 Development of secondary-flow streamlines in the thermal entrance region of a rectangular channel heated from below.

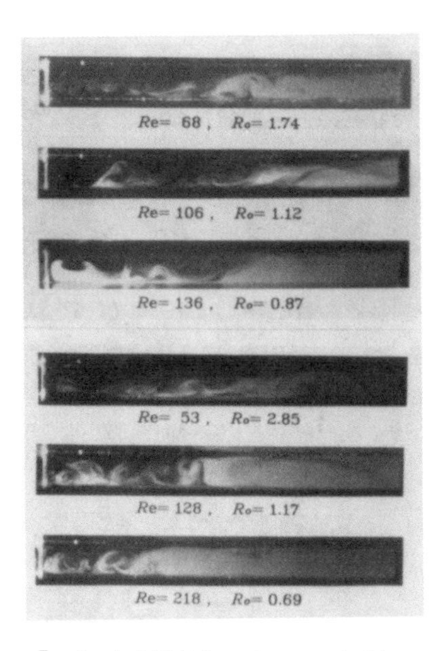

Fig. 7 Instabilities in a rotating radial rectangular channel.

(a) Convective instability in a horizontal rectangular channel with negative Rayleigh number.

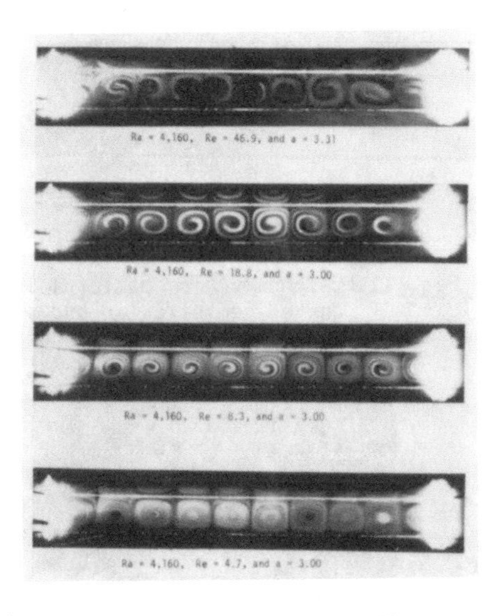

(b) Fully developed secondary flow streamlines in a horizontal rectangular channel heated from below.

Fig. 8 Convective instability in a horizontal rectangular channel with fully developed laminar flow [81].

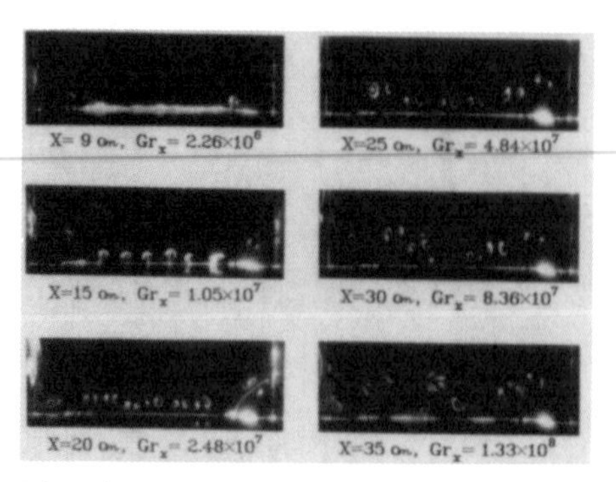

Fig. 9 Vortex instability of free convection flow over an isothermally heated horizontal plate.

body force

secondary flow pattern

U

distribution of axial velocity

T

distribution of fluid temperature

(a) case of weak secondary flow

(b) case of strong secondary flow

Fig. 10 Secondary flow pattern and distributions of axial velocity and temperature [53].

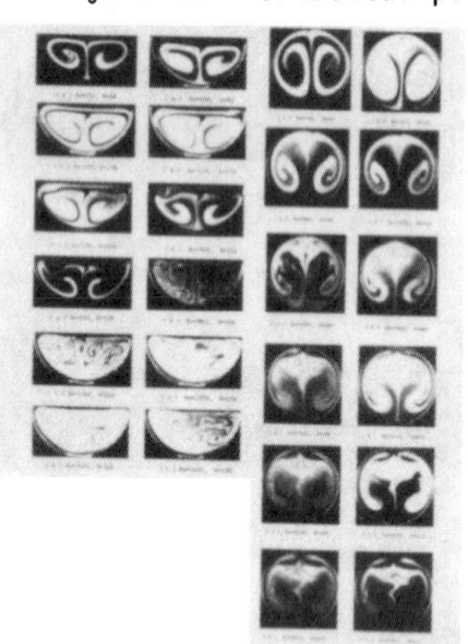

Fig. 11 Secondary flow patterns in curved semicircular and circular tubes [84].

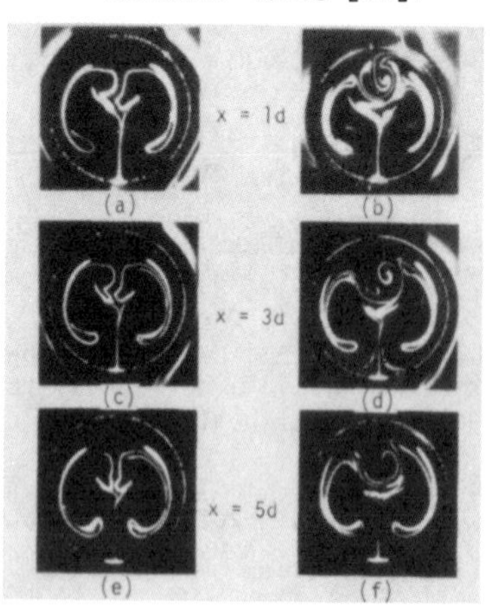

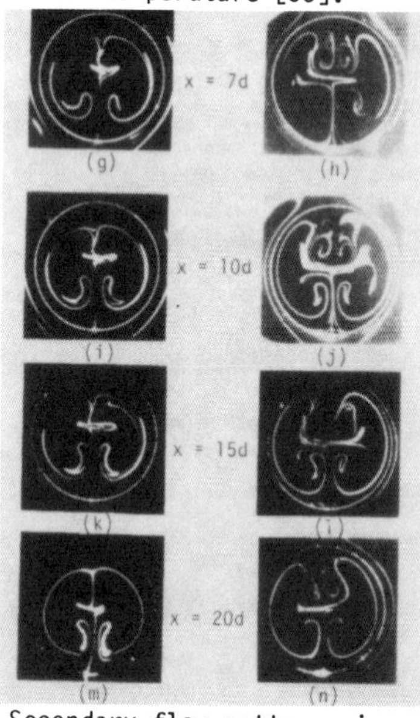

Fig. 12 Secondary flow patterns in a straight tube downstream of 180° bend [44].

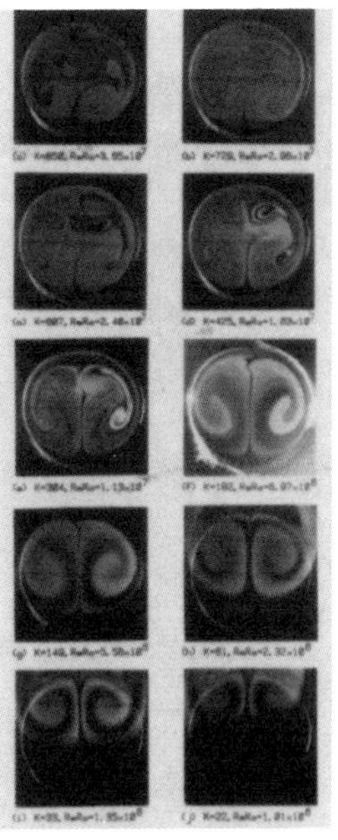

(a) Upward flow.

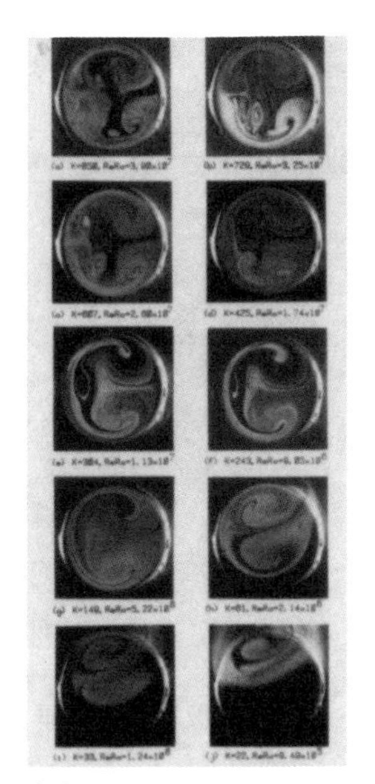

(b) Horizontal flow.

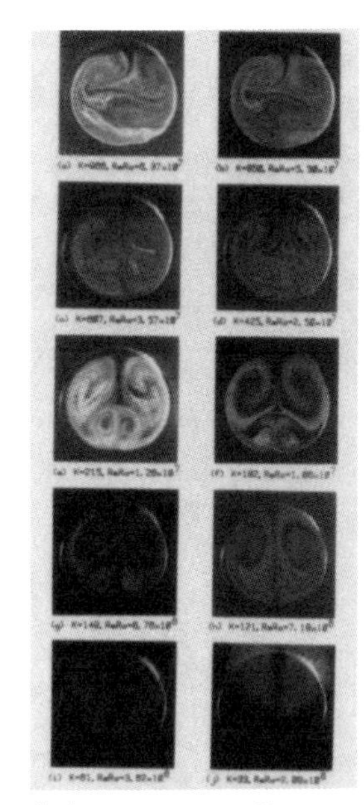

(c) Downward flow.

Fig. 13 Secondary flow patterns at the exit of an isothermally heated 180° bend [44].

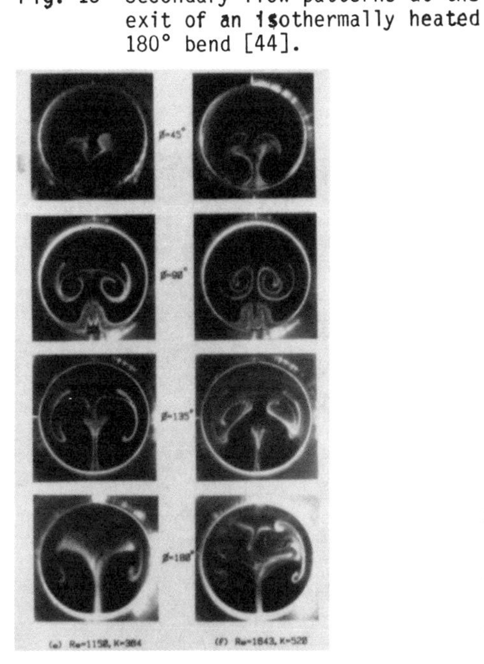

Fig. 14 Developing secondary flow patterns in the entrance region of a curved pipe [86].

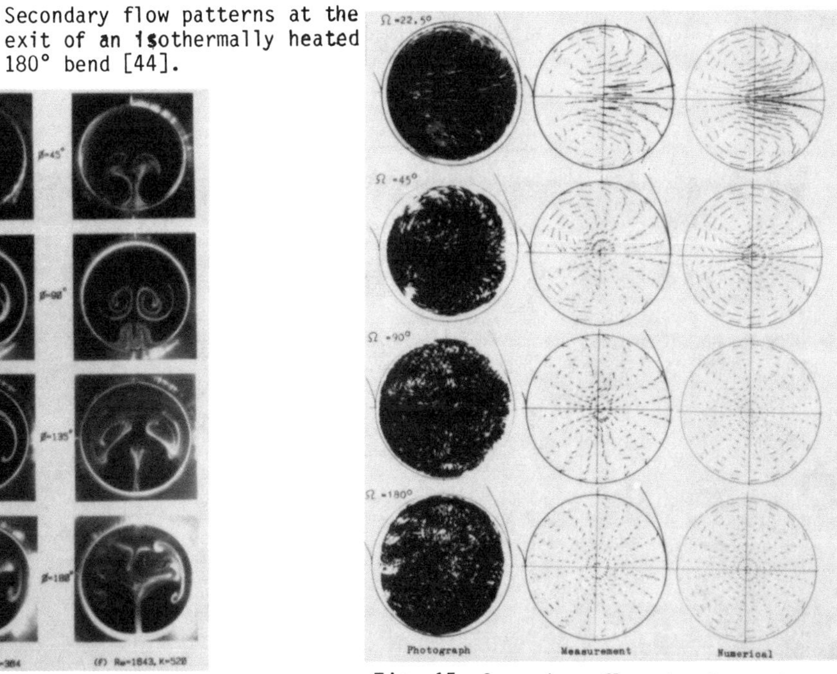

Fig. 15 Secondary flow development revealed by flow visualization, measurement and numerical solution [88].

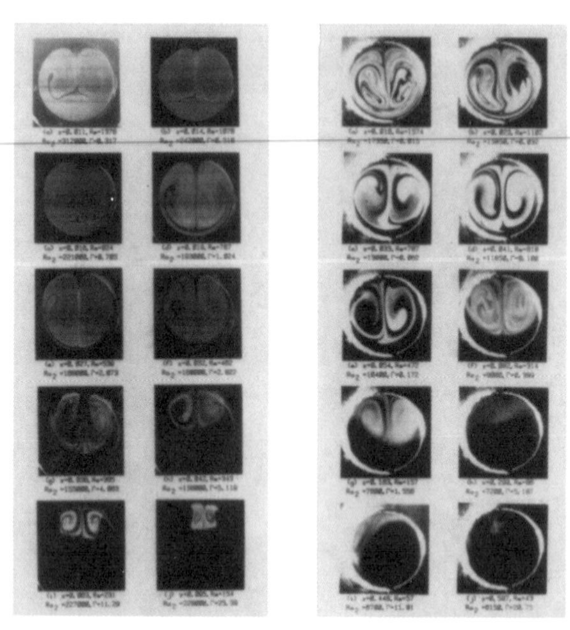

Fig. 16 Secondary flow patterns in the thermal entrance region of an isothermally heated horizontal tube [86].

Fig. 17 Secondary flow patterns in the thermal entrance region of an isothermally heated inclined tube with inclination angle 60 from horizontal direction [86]

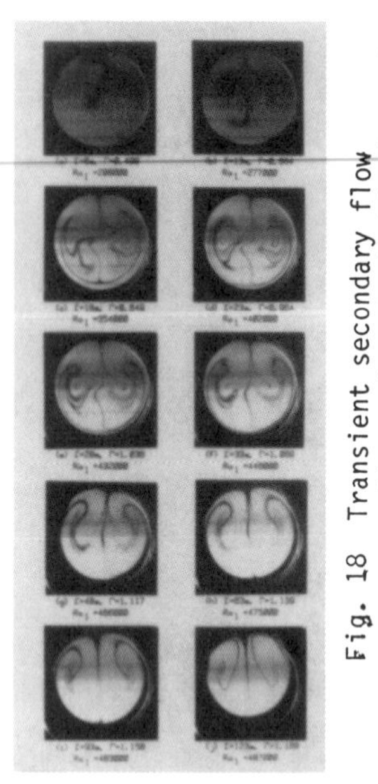

Fig. 18 Transient secondary flow patterns in a heated horizontal tube [86].

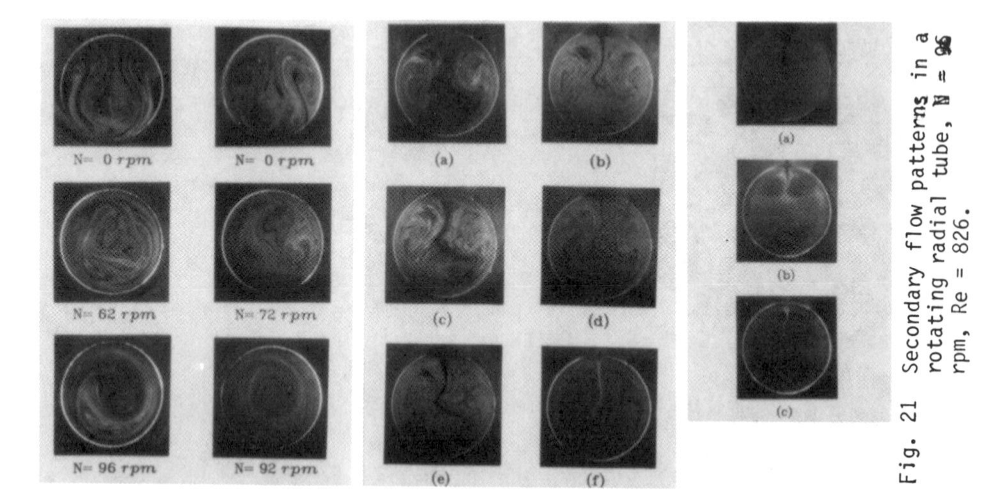

Fig. 19 Secondary flow patterns in a cooled Pyrex tube.

Fig. 20 Secondary flow patterns in a rotating radial tube, N = 36 rpm and Re = 68.

Fig. 21 Secondary flow patterns in a rotating radial tube, N = 96 rpm, Re = 826.

Fig. 22 Path lines on the impeller surface of an axial flow pump, oil film method, left, pressure surface, right, suction surface rpm 400, lapse time 30 s ~ 3 min [59].

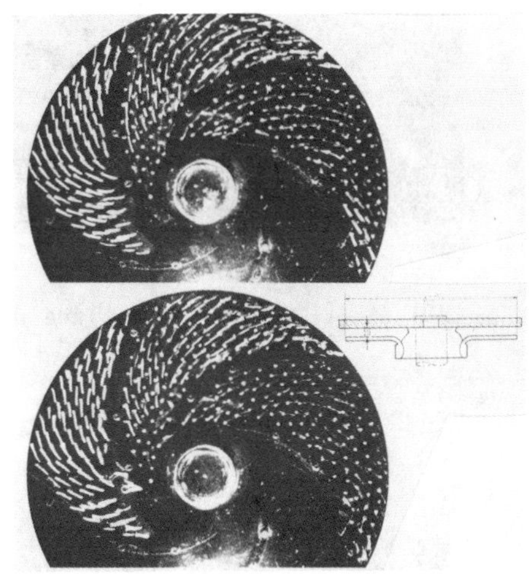

Fig. 23 Flow inside the impeller of a centrifugal pump, tuft method [12].

Fig. 24 Schlieren picture and deduced shock wave pattern in a rotating supersonic annular cascade (Freon) 12,000 rpm [64-90].

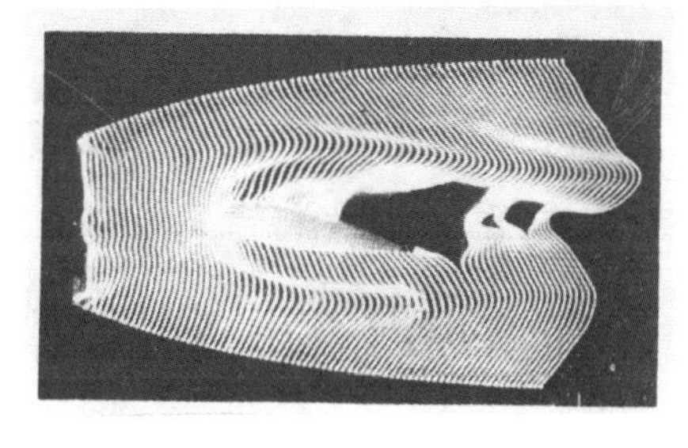

Fig. 25 Flow around a blade, spark tracing method, air, velocity 28 m/s, Re = 7.4 x 10^4, angle of attack 10° [59].

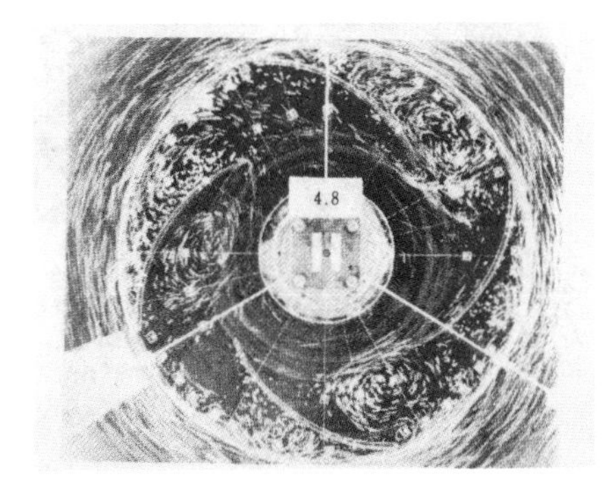

Fig. 26 Stall vortices, water table visualization method, 20 rpm [59].

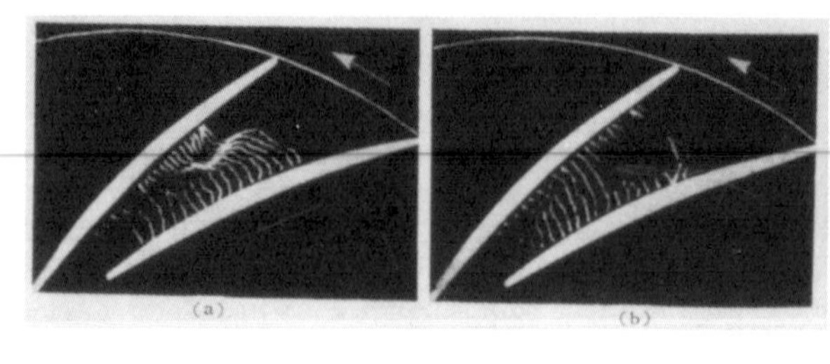

Fig. 27 Time lines in a blade passage, spark tracing method [92].

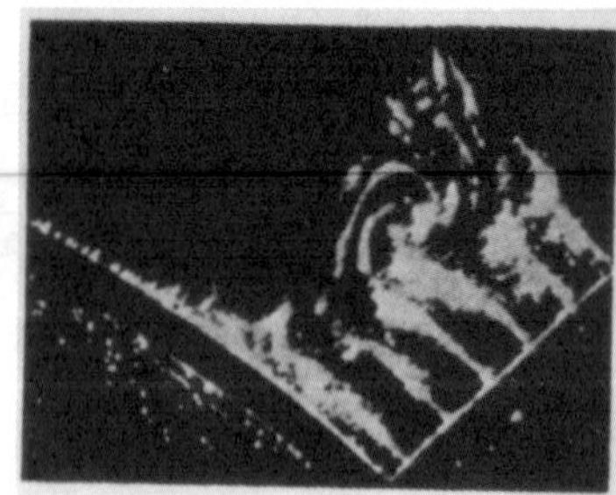

Fig. 28 Flow at the exit of the shrouded impeller of a centrifgual compressor, hydrogen bubble method [93

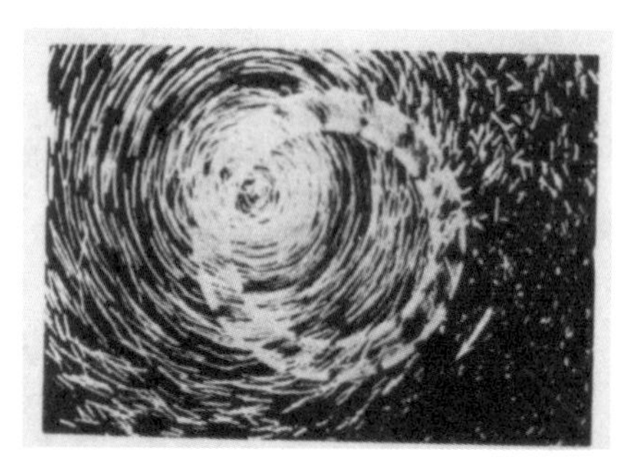

Fig. 29 Flow pattern around radial impeller of a crossflow fan, water table visualization method, impeller peripheral velocity 2.51 cm/s, rpm 30 [59].

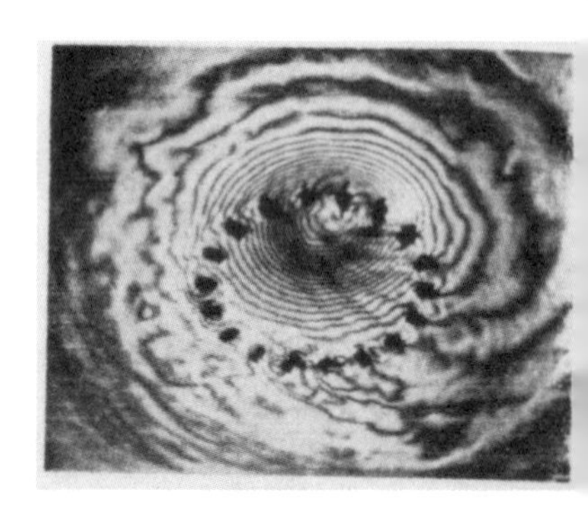

Fig. 30 An eccentric vortex formed by impeller of a crossflow fan, Moiré method, water, impeller peripheral velocity 25 cm/s, = 2,400 [59].

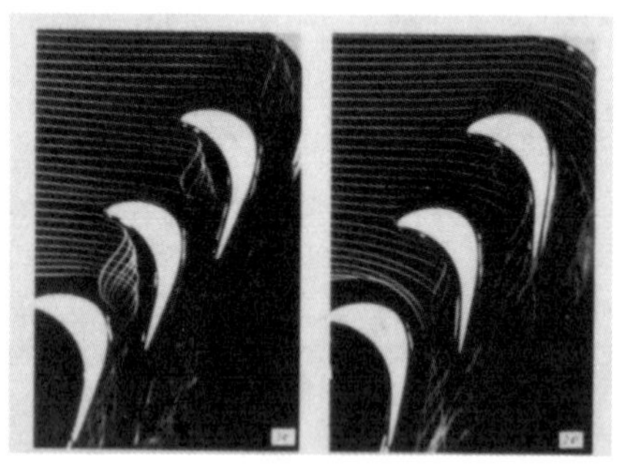

Fig. 31 Secondary flows near the bottom wall in a turbine blade cascade with coloured filaments, hydraulic analogy method [64].

Fig. 32 Flow in a vortex chamber, spar tracing method, air, voltage 100 kV, frequency 30 kHz [59].

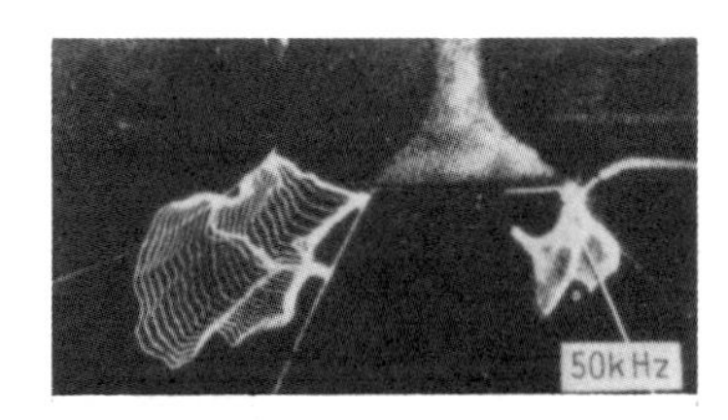

Fig. 33 Flow around a suction valve, spark tracing method, air, cylinder inside dia. 60 mm [59].

Fig. 34 Interferometer flow picture of supersonic flow through nozzle of a turbine, Mach-Zehnder interferometer method, air, inlet angle 90°, exit Mach number 2.56 [59].

Fig. 35 Schlieren picture of supersonic flow through nozzle of a turbine, air, inlet angle 90°, exit Mach number 2.45 [59].

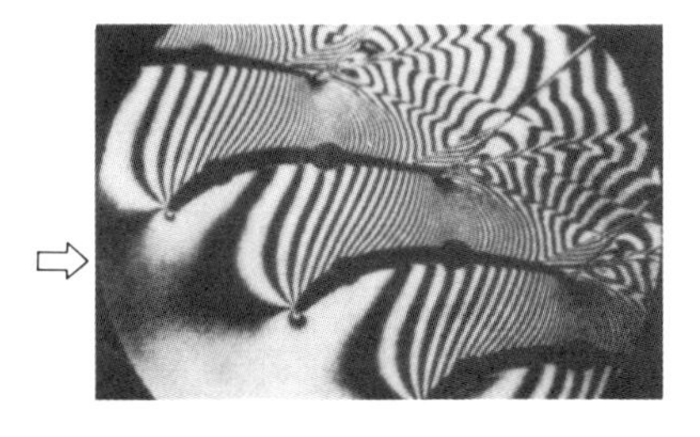

Fig. 36 Interferometer flow picture of long blade cascade of a low pressure turbine, Mach-Zehnder interferometer method, exit Mach number 1.43 [59].

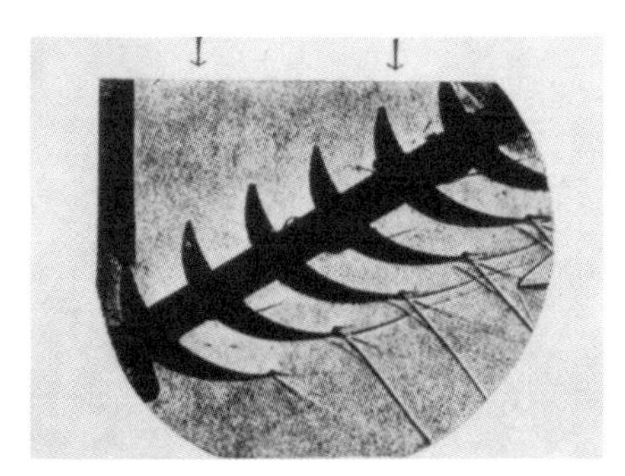

Fig. 37 Colour Schlieren picture of transonic flow in turbine blade cascade under over expansion condition, superheated steam, inlet temperature 175°C, Mach number 1.6, $Re = 3.5 \times 10^5$, chord length 50 mm, exit angle 60° [95].

Fig. 38 Hydraulic jump near the trailing edge of turbine blades, hydraulic analogy method, exit Froude number 1.90 [59].

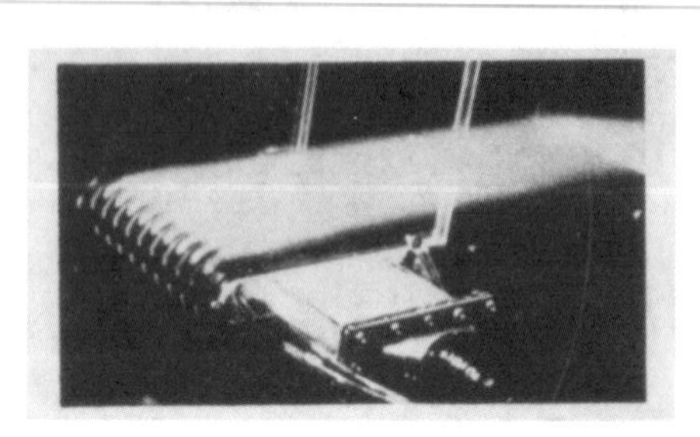

Fig. 39 Air film cooling of high temperature turbine blade, mist tracer method, uniform velocity 14 m/s [59].

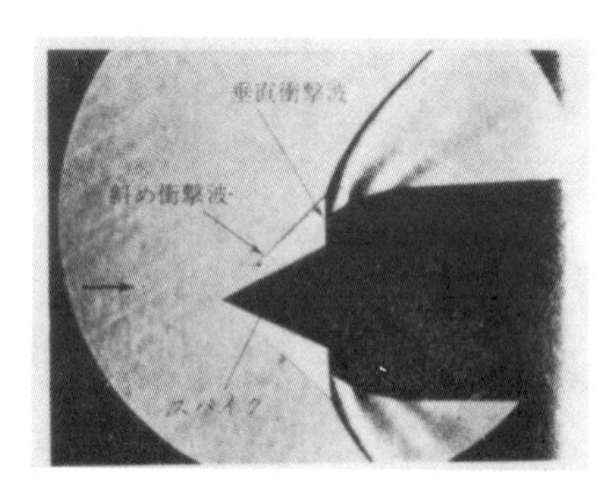

Fig. 40 Colour Schlieren picture of engine air intake flow of a supersonic airplane, Mach number 2.0, Re = 1.0 x 10^7 [59].

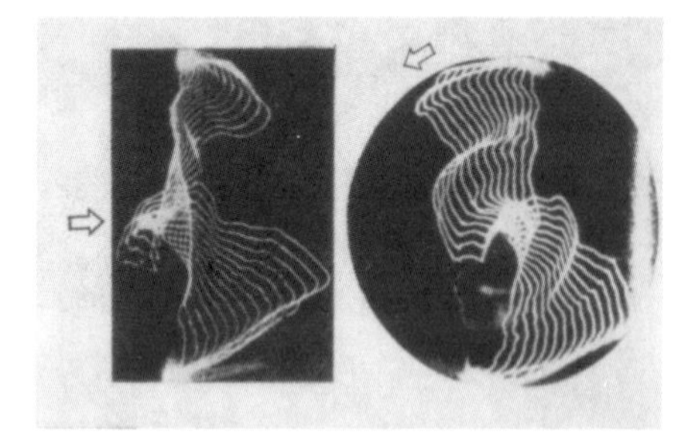

Fig. 41 Flow in a continuous combustor, spark tracing method, air, cylinder dia. 60 mm, exit mean velocity 10 m/s [59].

Flow Visualization Studies on Secondary Flow Patterns at the Outlet and in the Downstream Region of a Centrifugal Blower

K. C. CHENG and F. P. YUEN
Department of Mechanical Engineering
University of Alberta
Edmonton, Alberta, Canada T6G 2G8

ABSTRACT

Photographs are presented for secondary flow patterns and axial flow patterns in a downstream duct section after the circular outlet (d=3.18 cm) of a small centrifugal fan with fan speed range n=0~640 rpm. The outlet ducts tested are a straight tube (d=3.18 cm), a straight tube with abrupt change in diameter (d=5.11 cm), and two conical diffusers with included angles β=6.0° and 14.1°. The flow visualization was realized by a smoke injection method. The secondary flow patterns are also presented at the free outlets of two small centrifugal fans with circular and rectangular outlets. The varying complicated secondary flow patterns and flow separation phenomena in the conical diffusers are of interest. The photographic results provide some insight into three-dimensional, unsteady, turbulent flow inside a centrifugal fan.

1. INTRODUCTION

Blowers (centrifugal fans) and fans are used extensively in many applications and represent the heart of air distribution systems. The fluid motion inside the blower is characterized by three-dimensional flow, turbulence and unsteady flow and is extremely complicated. Because of scroll-type or volute-type housing and the effects of rotating blades inside centrifugal fan, one expects a strong secondary flow at the outlet and in the downstream region of the blower due to centrifugal and Coriolis forces effects. Since theoretical analysis of the internal fluid motion from the inlet to the exit of a centrifugal fan is extremely difficult, it is believed that the flow visualization studies of the secondary flow patterns after the discharge are of considerable practical interest.

The purpose of this paper is to present a series of photographic results on secondary flow patterns at the free outlet and in the downstream region (straight, sudden enlargement and conical diffuser) of a small centrifugal fan with circular outlet in the fan speed range n=0~640 rpm. The secondary flow patterns at the free outlet of a small centrifugal fan with rectangular outlet are also presented.

Flow visualization of internal flow in rotating machinery has been studied by many investigators and review articles are available [1-4]. Recently, flow visualizations of Sirocco fan [5] and cross flow fan [6-8] were also reported. Although the present flow visualization technique using smoke injection method is limited to a relatively low exit velocity, the secondary flow patterns to be presented provide some insight into a three-dimensional unsteady fluid motion after the discharge of a centrifugal fan. The decay of secondary flow in a straight downstream pipe and conical diffusers with included angles β=6.0° and 14.1° will also be studied.

2. EXPERIMENTAL APPARATUS AND PROCEDURE

The experimental setup for flow visualization is very simple and a schematic diagram is shown in Fig. 1. Two small shaded pole centrifugal blowers shown in Fig. 2(a) were chosen for flow visualization studies and some specifications are given in Table 1.

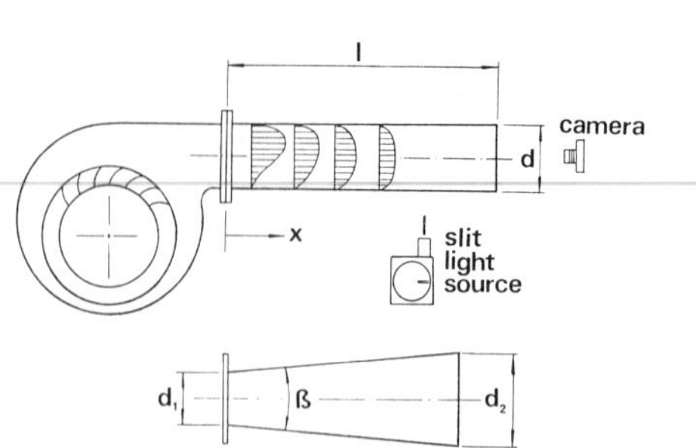

Fig. 1 Schematic diagram of experimental apparatus.

TABLE 1. Specifications of two centrifugal blowers for experiment

Make	Dayton 2C782	Dayton 4C723
Power Rating	1/250 hp	1/45 hp
Free Air RPM	3160	2300
Outlet Size	3.18 cm dia.	7 x 4.5 cm
Inlet Size	4.76 cm dia.	11.9 cm dia.
Casing	Scroll-Type	Volute-Type
CFM Free Air	15 (.42 m^3/min.)	82 (2.32 m^3/min.)
Cut-Off SP	0.22" (0.56 cm)	1.35" (3.43 cm)
No. of Blades	24	26

The downstream ductings were made of plexiglas with different lengths to make visualization possible. Two different sizes of conical diffuser with length ℓ=24.21 cm, included angle β=6.0° and ℓ=7.58 cm, β=14.1° were also machined from plexiglas. The fan motor speed can be adjusted by speed controller. The rpm of centrifugal fan was measured by variable flash rate stroboscope and local centerline velocity (V_c) was measured by TSI air velocity meter (Model 1650).

The flow visualization was facilitated by burning a bundle of Chinese incense sticks near the free inlet of a centrifugal fan with forward-curved blading. Air enters the inlet axially and is discharged tangentially. A slit light source was provided by a 300 w slide projector. The secondary flow patterns were observed at the outlet (x=0) of a centrifugal fan and at various downstream sections. The plan (top) and side views of the axial flow in the downstream section were also photographed. Over 1,200 photographs were taken using a Nikon FM2 SLR camera with 55 mm microlens and Kodak Tri-X Pan film using f 3.5 aperture and 1/2 ~ 1/15 sec shutter speed settings.

3. EXPERIMENTAL PARAMETERS

In this investigation, the flow rate of the centrifugal fan was not measured. The centerline velocity at the outlet of the fan or downstream tube was measured for reference only. Because of smoke diffusion at high exit velocity, the smoke injection method for flow visualization was restricted to relatively low rpm range. For the centrifugal fan with circular outlet, the measured maximum rpm was 3,000 and the centerline exit velocity was V_c=4.8 m/s. The experimental ranges for flow visualization are n=0~640 rpm and V_c= 0~0.77 m/s. For the centrifugal fan with rectangular outlet, the measured maximum rpm was 2,400 with V_c = 15 m/s. The experimental ranges are restricted to n=0~305 rpm and V_c= 0~1.7 m/s. Flow visualization in downstream region was performed only for the centrifugal fan with circular outlet. Two conical diffusers with d_1=3.17 cm, d_2=5.04 cm, ℓ=7.58 cm, β=14.1°, ℓ/d_1=2.52 and d_1=3.20 cm, d_2=5.75 cm, ℓ=24.21 cm, β=6.0°, ℓ/d_1=7.57 were used. For the downstream tube with sudden enlargement at the fan outlet d_1=3.18 cm, d_2=5.11 cm and d_2/d_1=1.61. For the centrifugal fan with circular outlet, the Reynolds number (Re) based on V_c=0.77 m/s and outlet diameter d_1 was 1,480 and good photographs were obtained up to Re=826 (V_c=0.43 m/s). For the centrifugal fan with rectangular outlet, the Reynolds number based on V_c=1.7 m/s and hydraulic diameter d_e=5.48 cm was 5,628 and good photographs were obtained up to Re=1,655(V_c=0.5 m/s).

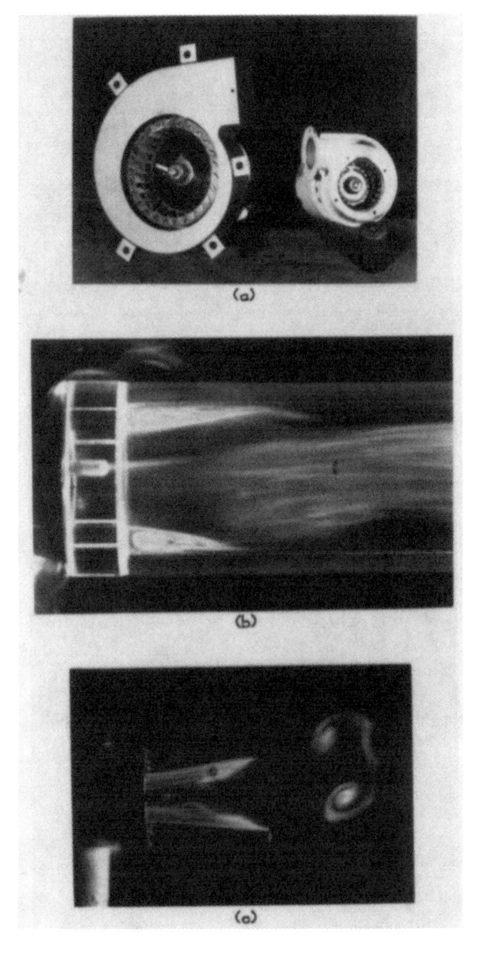

Fig. 2 Close-up views of two centrifugal fans and axial flow patterns.

4. RESULTS AND DISCUSSION

4.1 Photographs at the exit with discharge into atmosphere.

The velocity profile at the outlet of a centrifugal fan is not uniform. Typical secondary flow patterns as seen at the exit of a centrifugal fan are shown in Figs. 3 and 4 for circular and rectangular outlets, respectively. At fan rpm n=0, one has a natural flow of smoke through the fan. A pair of Dean-type vortices is clearly seen near the top in Fig. 3(a) in addition to a pair of larger vortices. The secondary flow pattern is different at different rpm and several vortices can be observed. At n=640 rpm, the present smoke injection method does not yield a clear secondary flow pattern due to smoke diffusion at higher exit velocity. For rectangular outlet, a pair of vortices appear near

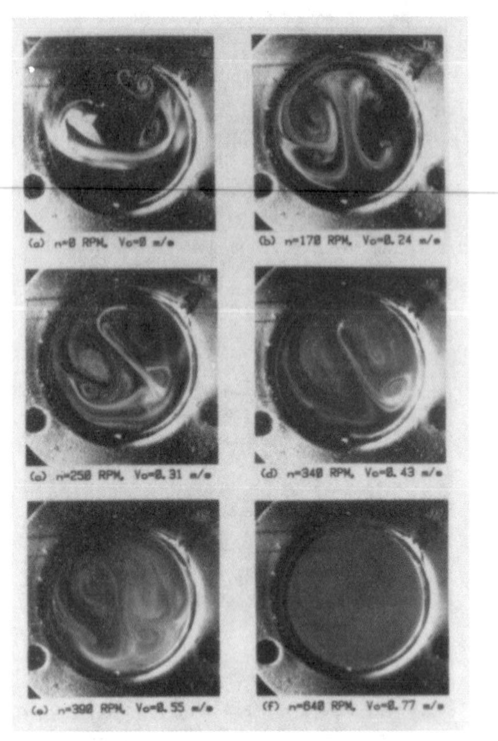

Fig. 3 Secondary flow patterns at the circular outlet.

the top at n=0 due to natural flow of smoke through the fan. At n=85 rpm, two corner vortices are seen near the lower wall and persist up to n=155 rpm. At n=220 rpm, the corner vortices cannot be detected.

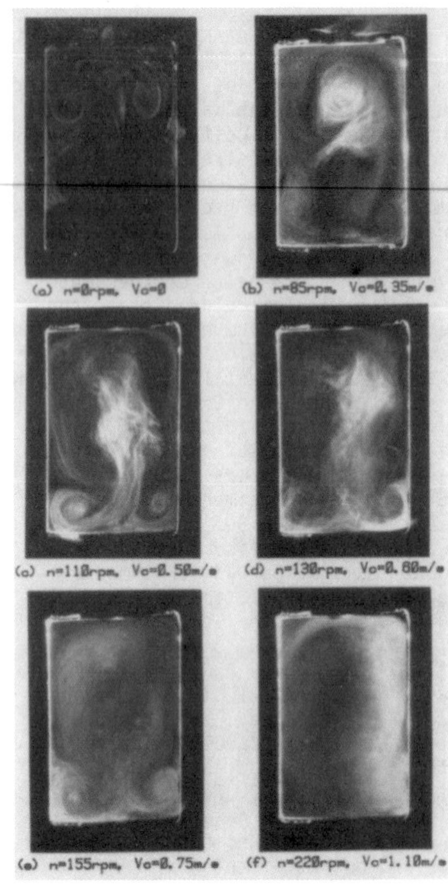

Fig. 4 Secondary flow patterns at the rectangular outlet.

4.2 Photographs in the ducted outlet with constant inside diameter (d=3.18 cm).

The secondary flow patterns in the outlet tube with constant diameter (d=3.18 cm) are shown in Fig. 5 for n=0~390 rpm and downstream distance x from the fan outlet x=0~10 d. At x=0, one observes four or more vortices depending on fan rpm. At n=0, the secondary flow pattern is fairly symmetric. For n>0, the symmetry cannot be maintained. It is noted that a mounting screw (about 1 cm long) is protruding at the inside casing wall near the outlet. The impeller offset in the casing may also be a contributing factor.

The secondary flow is unsteady and the photographs shown represent the instantaneous results. Although not shown here, the secondary flow was confirmed up to n=640 rpm. Due to the upward moving secondary flow in the core region, the maximum axial velocity is located near the upper wall. Since the Reynolds num-

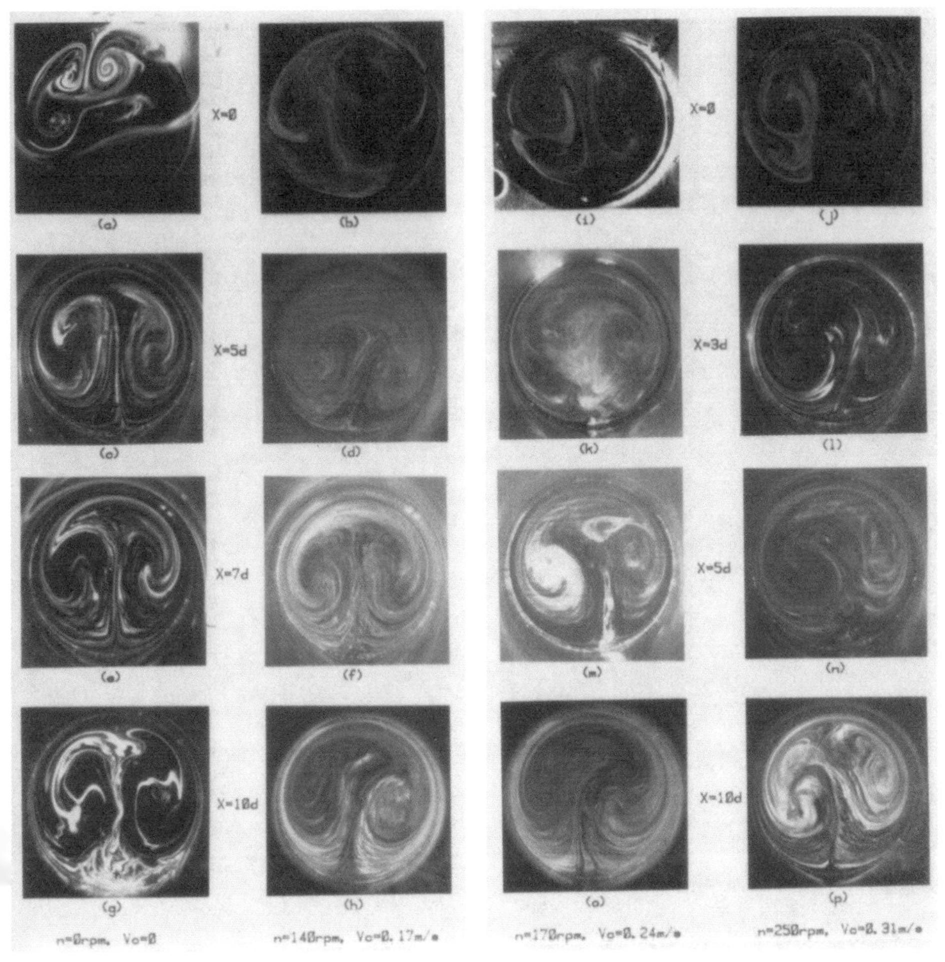

Fig. 5 Secondary flow patterns in the downstream straight section d=3.18 cm.

ber is less than 2,300, the secondary flow and turbulence generated by impeller are expected to decay after a sufficiently long outlet tube. The present result seems to suggest that the secondary flow will persist for some distance before reaching a fully developed parabolic profile for low rpm range. At x=10 d, the secondary flow appears to be more symmetric in each case.

4.3 Photographs in downstream region after sudden enlargement.

The secondary flow patterns in the downstream region after a sudden enlargement of diameter from d=3.18 cm to d=5.11 cm (diameter ratio = 1.61) at the fan outlet are shown in Fig. 6 for the case n=250 rpm. For $x \leq 3d$ (d=3.18 cm), the effect of a sudden enlargement of diameter appears to be appreciable and many vortices are seen. At x=10 d, only a pair of vortices occupies the whole region. It is interesting to note a gap near the lower wall at x=5d and 7 d. This may be an indication of flow separation there.

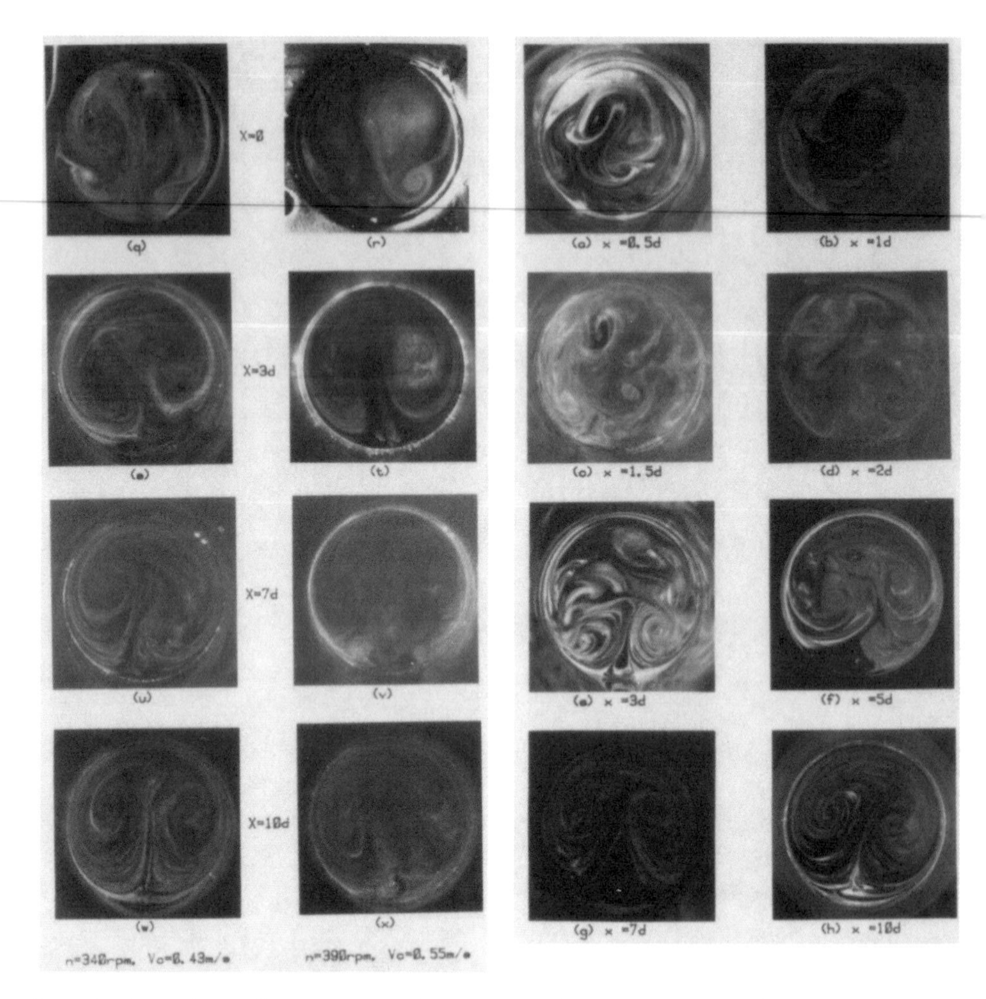

Fig. 5 Secondary flow patterns in the downstream straight section d=3.18 cm.

Fig. 6 Secondary flow patterns in the downstream section with abrupt change in diameter d=5.11 cm.

The behavior of axial flow in the downstream region is also of interest. By illuminating a central vertical or horizontal plane with a sheet of light, a side or plan view of the main flow in the downstream section can be obtained. The results are shown in Figs. 7 and 8 for n=0 to 250 rpm. Fig. 7 reveals that at n=140 and 250 rpm, a wave-like flow exists in the central region. Furthermore, the flow pattern at n=250 rpm seems to confirm the flow separation at x=5 d and 7 d in Fig. 6. Figs. 7 and 8 also confirm the appreciable change of flow pattern after the sudden enlargement of cross-sectional area. Fig. 2(b) shows the top view of recirculating zone near the shoulder.

4.4 Photographs in conical diffusers (ℓ=7.58 cm, β=14.1° and ℓ=24.21 cm, β=6.0°).

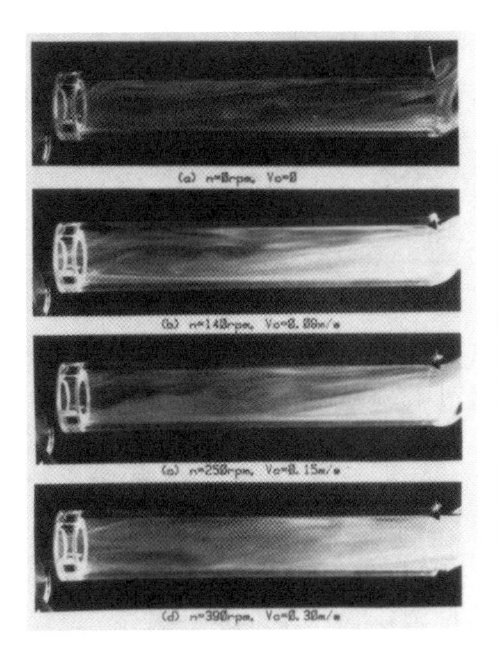

Fig. 7 Side view of main flow in tube with d=5.11 cm.

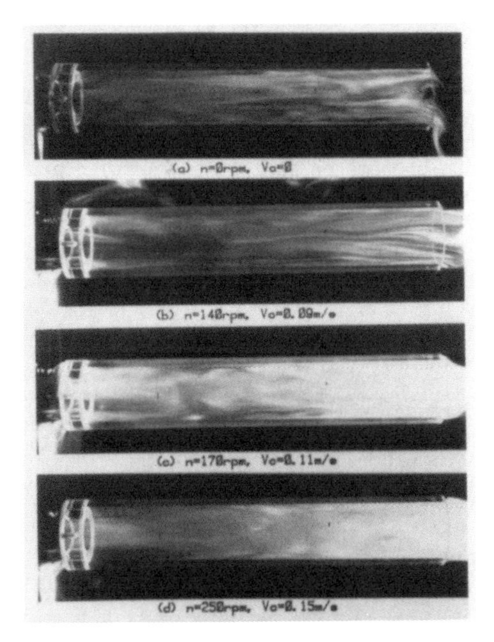

Fig. 8 Plan view of main flow in tube with d=5.11 cm.

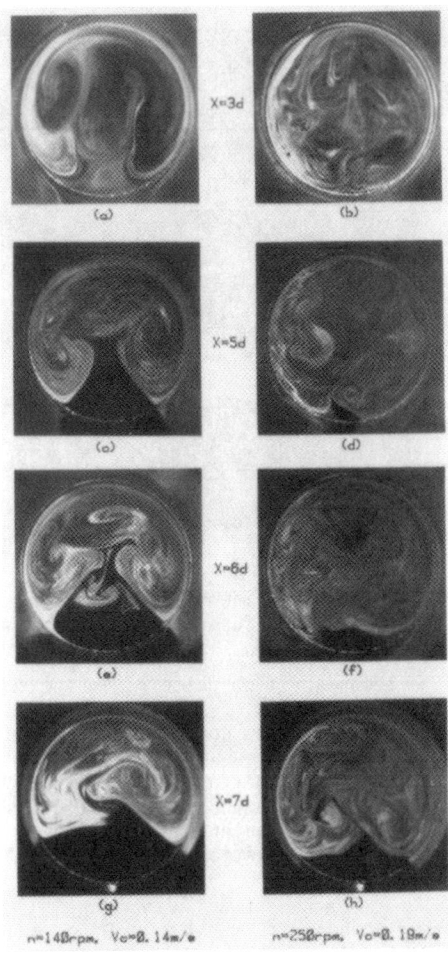

Fig. 9 Secondary flow patterns in diffuser with β=6.0°.

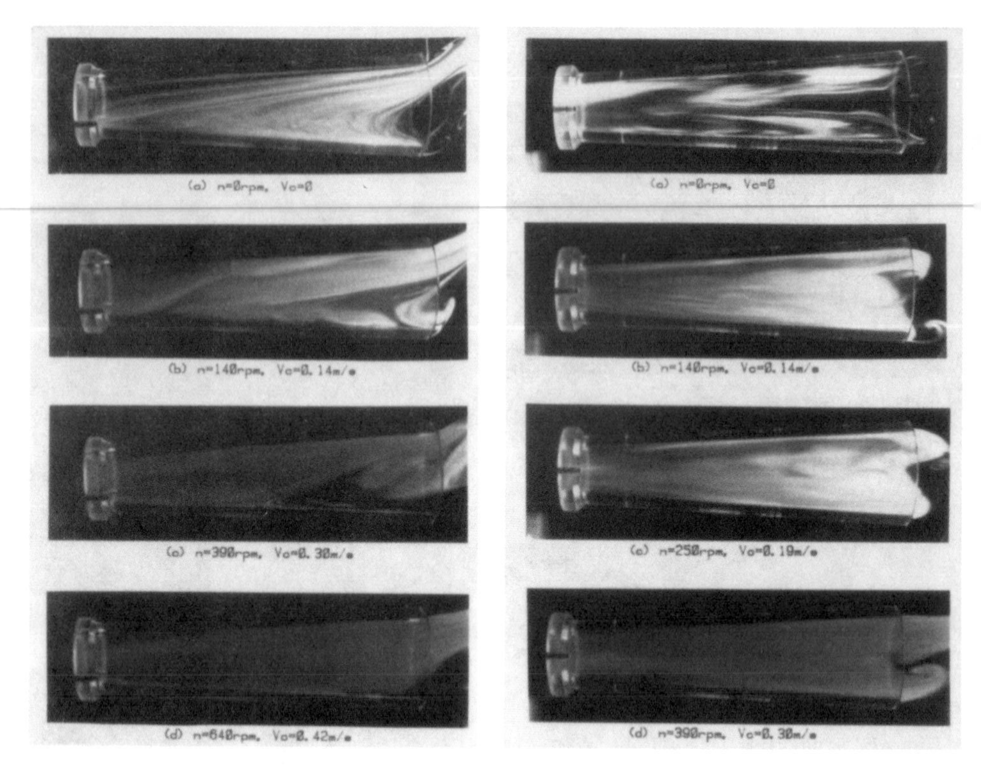

Fig. 10 Side view of main flow in diffuser with β=6.0°.

Fig. 11 Plan view of main flow in diffuser with β=6.0°.

The secondary flow patterns in the conical diffuser with ℓ=24.21 cm and β=6.0° are shown in Fig. 9 for n=140 and 250 rpm. The side and plan views of the axial flow in the conical section are shown in Figs. 10 and 11 respectively. Fig. 9 clearly reveals the flow separation phenomenon in the lower region and can be confirmed from axial flow pattern in Fig. 10. At n=140 rpm, one sees laminar-like flow but at n=250, the secondary flow is more complicated and appreciable turbulence can be seen. One notes that the plan view in Fig. 11 reveals no separation phenomenon in the upper half region of the conical diffuser.

The secondary flow patterns observed at the exit of the two conical diffusers are shown in Fig. 12 with the left-hand side for the case ℓ=7.58 cm, β=14.1° and the right-hand side for the case ℓ=24.21 cm, β=6.0°. The side and plan views of the axial flow in the conical diffuser with ℓ=7.58 cm, β=14.1° are shown in Fig. 13. It is instructive to compare the photographs on the left-hand side of Fig. 12 with the side views of Fig. 13 to understand the flow separation phenomenon. The secondary flow patterns shown in Fig. 12 are characterized by rather complicated multiple vortices of different sizes and flow separation near the lower wall. Fig. 2(c) shows the top view of axial flow in conical diffuser (β=14.1°) with a pair of decaying vortices leaving the diffuser.

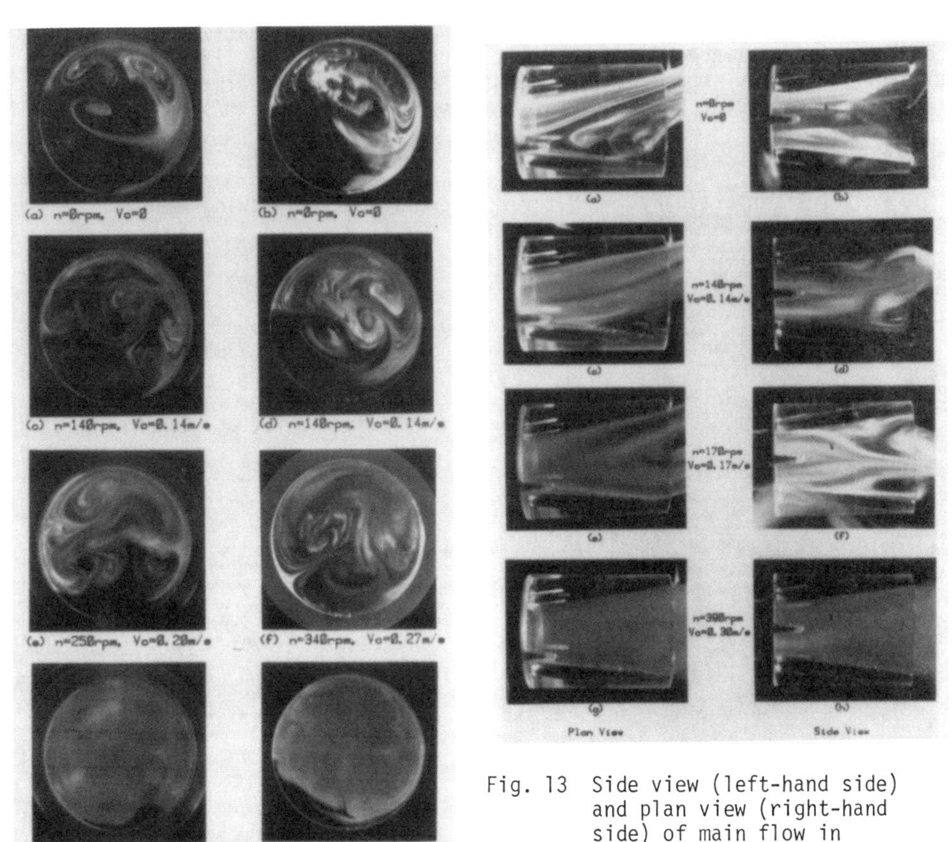

Fig. 13 Side view (left-hand side) and plan view (right-hand side) of main flow in diffuser with β=14.1°.

Fig. 12 Secondary flow patterns at the exits of diffusers, left-hand side β=14.1°, right-hand side β=6.0°.

5. CONCLUDING REMARKS

The secondary flow at the outlet of a centrifugal fan represents an integrated result of the complicated flow from the inlet to the outlet of the fan and a theoretical analysis would be extremely difficult. A simple flow visualization experiment using smoke injection method yields readily the secondary flow patterns downstream of the outlet for low rpm range n=0~640 rpm of a small centrifugal fan.

The secondary flow patterns in the downstream region are presented for a straight tube, a straight tube with an abrupt change in diameter and two conical diffusers with included angle β=6.0° and 14.1°. The flow patterns for the axial flow along the vertical and horizontal central planes are also obtained for the outlet sections. The flow separation phenomena in the conical diffusers are clearly identified.

The present flow visualization study for very low fan speeds shows that the entry flow problem from the fan outlet to a fully developed laminar flow with parabolic profile is extremely complicated and cannot be readily approached theoretically.

The present flow visualization study does not yield quantitative result but confirms the existence of the unsteady secondary flow. The limit of the applicability of the smoke injection method is confirmed and other flow visualization methods should be used for future work with $v_c > 0.6$ m/s.

ACKNOWLEDGEMENT

This work was supported by the Natural Sciences and Engineering Research Council of Canada through operating grant.

REFERENCES

1. Ito, H., Secondary flow Problems in Fluid Mechanics, J. of the Japan Society of Mechanical Engineers, Vol. 66, No. 537, pp. 1368-1375, 1963.

2. Murai, H. and Watabe, H., Flow Visualization Methods for Internal Flows in Turbomachinery, Turbomachinery (Japan), Vol. 5, No. 11, pp. 21-28, 1977.

3. Akashi, K., Application of Flow Visualization to Machinery, J. of the Japan Society of Mechanical Engineers, Vol. 81, No. 716, pp. 663-669, 1978.

4. Nakayama, Y., Flow Visualization in Rotating Impèllers, Turbomachinery (Japan), Vol. 11, No. 3, pp. 173-184, 1983.

5. Suzuki, M., Flow Visualization in a Sirocco Fan Blower for Automotive Air Conditioning System, J. of the Flow Visualization Cos. of Japan, Vol. 2, No. 6, pp. 469-472, 1982.

6. Bush, E.H., Crossflow Fans - History and Recent Developments, Conference on Fan Technology and Practice, Institution of Mechanical Engineers, pp. 50-65, 1972.

7. Tuckey, P.R., Holgate, M.J. and Clayton, B.R., Performance and Aerodynamics of a Cross Flow Fan, Int. Conf. on Fan Design and Applications, BHRA Fluid Engineering, pp. 407-424, 1982.

8. Takahashi, K. and Daikoku, T., Flow Visualization of Cross-Flow Fan, J. of the Flow Visualization Soc. of Japan, Vol. 2, No. 6, pp. 473-478, 1982.

MS9001E Transition Piece Cold-Flow Visualization Test

LI C. SZEMA
Combustion Mechanical Unit
Gas Turbine Division
General Electric Company
Schenectady, New York 12345

ABSTRACT

The MS9001E combustion transition piece cold-flow visualization test was conducted at General Electric's Research and Development Center in December 1984. The test focused on the local flow phenomena at the air gap between the combustion transition piece body and the first-stage nozzle side wall.

Conclusions drawn from the results of this cold-flow visualization test are:

1. The gap width of 0.315 inch appears to be an acceptable design
2. Hot-gas recirculation may exist when the gap width is increased to 0.555 inch by removing the transition piece lip

INTRODUCTION

General Electric is committed to improving the efficiency and quality of its gas turbines. As part of this commitment, a cold-flow visualization test was performed on a model MS9001E simple-cycle, single-shaft, heavy-duty gas turbine (Figure 1) to determine the effect of removing the downstream lip of the transition piece on the potential recirculation of hot gases into the resulting slot. The experiments were conducted at General Electric's Research and Development Center in December 1984. The tests were performed on the water table, a tool used to visualize complex flows within turbomachinery. Figure 2 shows a close-up view of the area investigated.

In this series of experiments, the local air gap in the original design of the MS9001E combustion transition piece (Figure 3) was compared to a modified version (Figure 4) at the joint between the transition piece body and the first-stage nozzle inner side wall.

In operation, the lip of the transition piece tends to wear, apparently during transients, against the first-stage nozzle. This wearing results in a shortening of the lip in the transition piece frame and an increase in gap width during steady-state operation.

In order to prevent metal-to-metal contact, the downstream lip on the transition piece opposite the inner side wall of the first-stage nozzle was removed. Based on the results of previous combustion laboratory tests on a similar configuration, there was concern that hot gases might recirculate in the slot and overheat the transition piece inner seal.

TEST PROCEDURES

The gas flow at the transition piece exit of the MS90001 has a Mach number of 0.26. At this flow rate, the gas stream can be treated as an incompressible flow, and modeled by using the flow of water rather than gases. Since the intent of the tests was to show general flow patterns,

FIGURE 1. Model series 9001E simple-cycle, single-shaft heavy-duty gas turbine

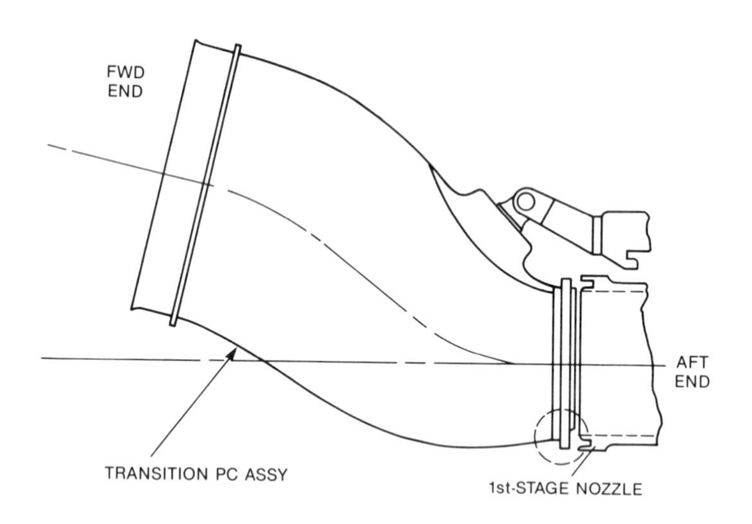

FIGURE 2. MS9001E T/P — closeup view

rather than detailed mixing effects caused by the viscous interaction between jets and mainstream flow, this cold-flow model using water is adequate.

For the visualization experiments, a cold-flow visualization model was built so that all dimensions, including two axial gaps between the transition piece and the first-stage nozzle, were four times the actual size. Figure 5a is the top view of this model; the side view can be seen in Figure 5b. Secondary water flows from two cooling jets located on the floating seal were constantly on, and the dye was separately delivered by the dye injection tubes to each cooling hole.

The original design was tested first on the water table. To make the fluid's circulation readily visible, dye was injected into the main water flow at the wall. The dye showed that there was no recirculation of hot gas into the slot (Figure 6). Using the same design, dye was injected into the secondary flow to illustrate the effect of the cooling jets (Figure 7).

For the second test, the original transition piece model was removed from the water table, and the modified design was put in place. This design was tested next by using the same dye injection procedures of the first test. It was evident that with the larger gap, the water representing the hot gas recirculated into the slot (Figure 8). When dye was injected into the secondary flow (Figure 9), the dye partially filled the slot, and some hot gas flow was still present. This could have a negative effect on the first-stage nozzle side wall and the floating seal if the actual machine conditions cause this phenomenon.

TEST CONDITIONS AND RESULTS

Two approaches were used to simulate machine operating conditions:

1. The momentum ratio of cooling jets versus mainstream hot-gas flow
2. The velocity ratio of cooling jets versus mainstream flow

The results of the calculations (see "Nomenclature and Test Parameters") indicate that the manometer height ratio for Condition 1 is 4.3 and the ratio for Condition 2 is 2.0. During these tests, there was no obvious difference after the color dye injection between these two conditions. However, the momentum ratio due to the flow mixing at the vicinity of the air gap seems to be a proper representation (Figure 10).

CONCLUSIONS

Several conclusions can be drawn from the results of this cold-flow visualization test:

1. The gap width of 0.315 inch appears to be an acceptable design (Figures 6 and 7)
2. A chamfer or smoooth transition at the first-stage nozzle side wall apparently is a necessary design condition (Figure 6). During actual machine operating conditions, there might be a mismatching of the inner surface between the transition piece and the first-stage nozzle. The current design at the first-stage nozzle side wall has a smooth corner, which could be beneficial for hot flow re-attachment, and minimizes the potential for recirculation.
3. Hot-gas recirculation may exist when the gap width is increased to 0.555 inch by removing the transition piece lip. This points out the need to monitor the hardware in service to determine if actual machine conditions are producing the hot gas recirculation predicted by the flow model.

NOMENCLATURE AND TEST PARAMETERS

Main Flow (Hot-Gas Side) – General Electric Combustion Transition Piece

$V_{\text{hot flow}}$ = transition piece exit velocity = 91.75 m/s

$\rho_{\text{hot flow}}$ = flow density at transition piece exit = 3.044 kg/m^3

ΔP (System pressure drop) = 4.545×10^4 N/m^2

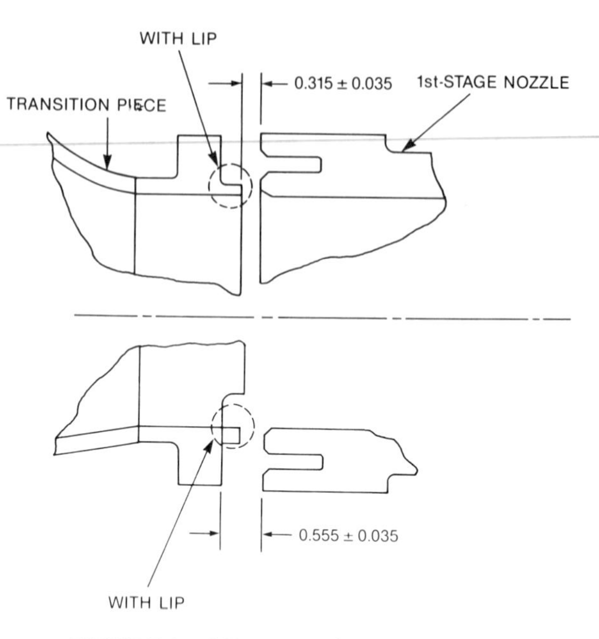

FIGURE 3. MS9001E T/P original design

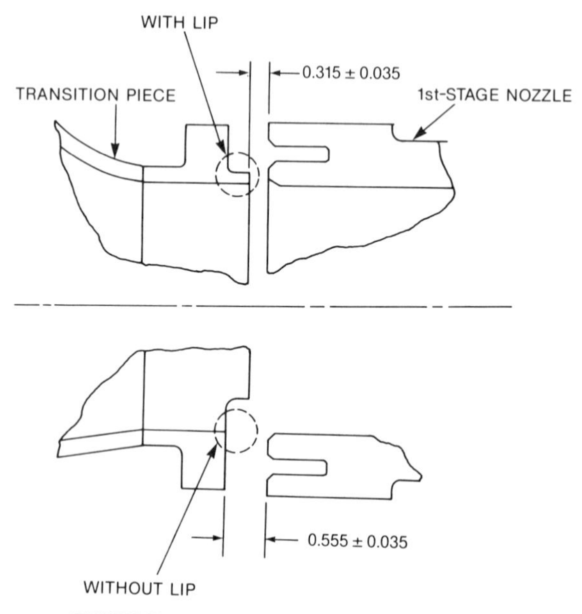

FIGURE 4. MS9001E T/P modified design

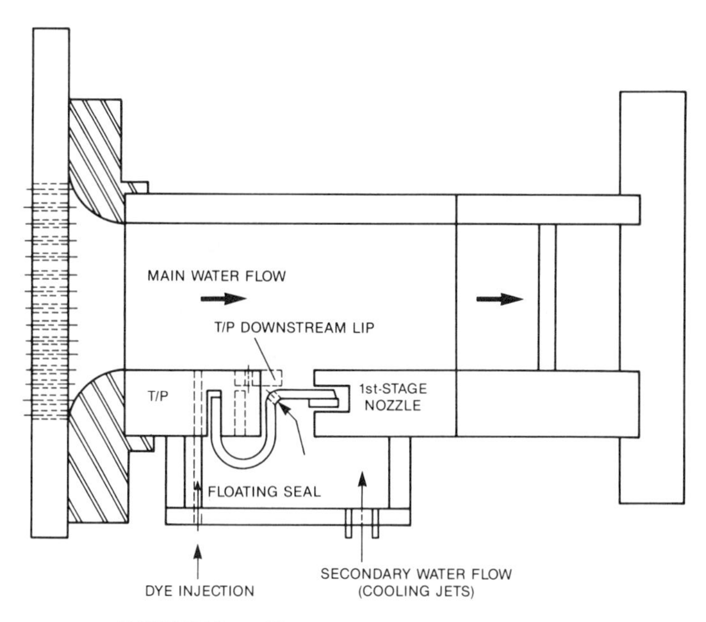

FIGURE 5a. Flow visualization model – top view

FIGURE 5b. Flow visualization model – side view

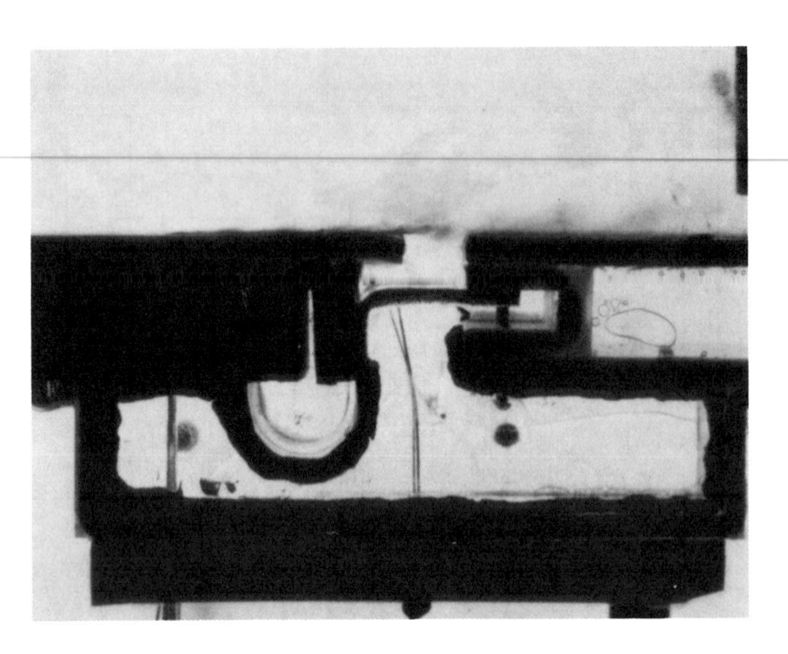

FIGURE 6. Transition piece with lip (air gap = 0.315 inch). Dye injected into main flow; no recirculation of hot air is present.
Note: no camfer or smooth surface at nozzle side wall

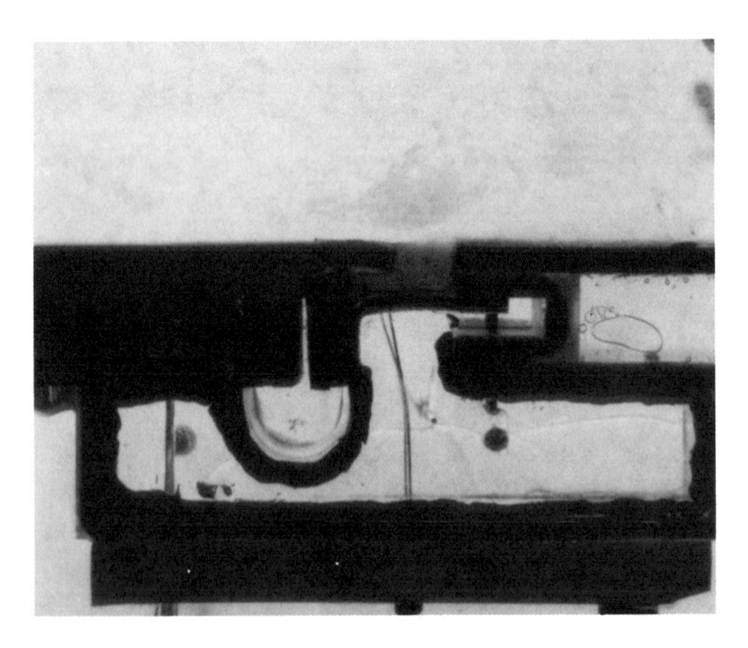

FIGURE 7. Transition piece with lip (air gap = 0.315 inch). Dye injected into secondary water flow; slot is filled with the dye.

FIGURE 8. Transition piece without lip (air gap = 0.555 inch). Dye injected into main water flow; hot gas recirculation occurred at the slot.

FIGURE 9. Transition piece without lip (air gap = 0.555 inch). Dye injected into secondary water flow.

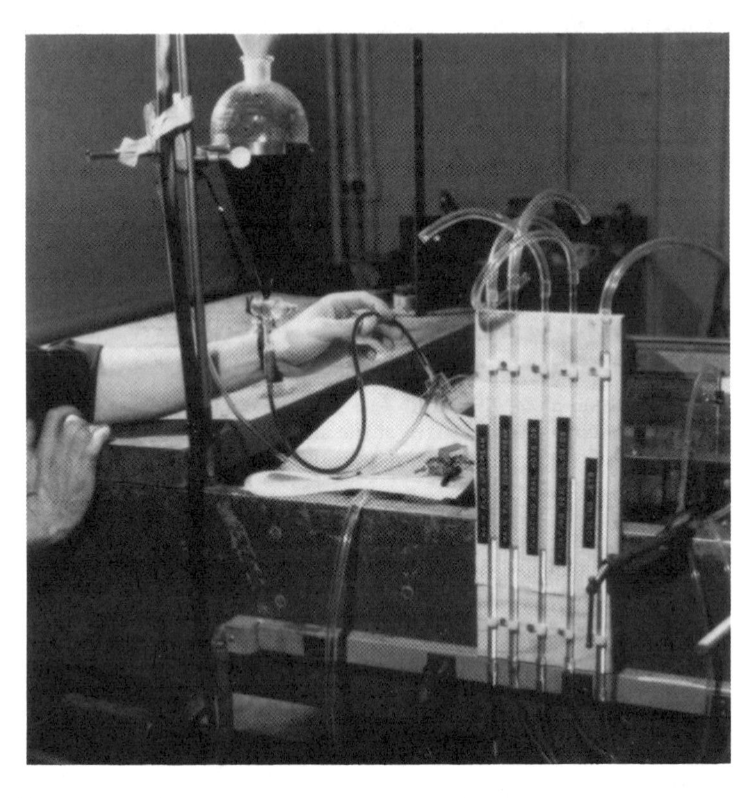

FIGURE 10. Test condition setting (momentum ratio). Manometer height ratio = (h_{jet} / $h_{main\ flow}$ = 4.3)

Secondary Flow (Compressor Discharge)

$$\frac{P_s}{P_t} = \frac{\text{static pressure of main flow near slot}}{\text{total pressure of secondary flow}} = 0.954$$

$$\gamma = \frac{C_p}{C_v} = \frac{\text{specific heat at constant pressure}}{\text{specific heat at constant volume}} = 1.4$$

M_N = Mach number = 0.26

P_t = total pressure

= 1.193×10^6 N/m^2

T_t = total temperature = 342.5°C

G_t = total impingement jet flow = 0.6286 kg/s

V_{jet} = jet flow velocity = 128.0 m/s

$\rho_{air\ jet}$ = air jet density = 6.727 kg/m^3

Machine Condition 1

$$M = \text{blowing rate parameter} = \frac{(\rho V)_{jet}}{(\rho V)_{hot\ flow}} = \frac{(6.727)\ (128.0)}{(3.044)(91.75)} = 4.3$$

In Cold-Flow Visualization Test Condition 1

$$\rho_{jet} = \rho_{hot\ flow} = \rho_{water}$$

Because $V^2 = 2gh$ (water height), the ratio of water jet with respect to main flow is

$$h_{jet}/h_{main\ flow} = 4.3$$

In Cold-Flow Visualization Test Condition 2

$$V_{jet}/V_{main\ flow} = \frac{128}{91.75} = 1.4$$

By the same approach, the ratio of water jet with respect to main flow is

$$h_{jet}/h_{main\ flow} = (1.4)^2 = 2.0$$

ACKNOWLEDGEMENTS

The author is greatly indebted to Mr. Ivan Edelfelt, Research Engineer, General Electric Corporate Research and Development, and Mr. Dick Schiefer, Senior Engineer of the Combustion Sub-section, General Electric Gas Turbine Engineering, Gas Turbine Division for their assistance.

Surface Indicator and Smoke Flow Visualization Techniques in Rotating Machinery

H. DAVID JOSLYN and ROBERT P. DRING
United Technologies Research Center
East Hartford, Connecticut 06018

INTRODUCTION

Historically, the observation of fluid motion has unveiled complex flow phenomena, verified concepts and aided in the development of analytical models. Flow visualization was the first and, in many instances, the only experimental technique available to study complex fluid dynamics problems. Engineering applications of surface indicator and smoke or vapor flow visualization techniques have been in use since the advent of manned flight and are still widely used today. Recent reviews[1] and [2] indicate the broad useage of these flow visualization techniques in a wide variety of fluid dynamic research problems. In turbomachinery research, these techniques have been applied mainly to stationary(non-rotating) wind tunnel or cascade models, e.g., [3],[4] and [5]. An exception was the use of smoke visualization by Phillips and Head[6] to study the tip region of a rotating blade row in a low speed axial flow research compressor. Currently, there is little information in the open literature to demonstrate the value of flow visualization in rotating turbomachinery research.

Since 1977 the United Technologies Research Center Large Scale Rotating Rig(LSRR-1) has been committed to studying in detail the complex, three dimensional flow fields of rotating axial flow compressors and turbines. As will be shown in this paper, the capability to conduct flow visualization studies on the rotating airfoil and endwall surfaces and in the rotating airfoil passages is a valuable aide in not only identifying the aerodynamic mechanisms present, but also in understanding their impact on the various parameters used by engineers to characterize turbomachinery flow fields.

Specific applications of the surface indicator technique include studies of turbine film cooling, three dimensional flows on turbine airfoils, analytical modeling of turbine pressure surface flows and compressor suction surface flows, and compressor hub corner stall/blockage effects. Smoke flow visualization results from a preliminary study of wake-airfoil interaction are also presented.

EXPERIMENTAL FACILITY AND FLOW VISUALIZATION TECHNIQUES

The United Technologies Research Center Large Scale Rotating Rig (LSRR-1) test section is 5 ft(1.52m) in diameter and test rotors can run at speeds up to 900 rpm. Inlet flow is drawn from ambient (out-of-doors) air and the flow through the facility is essentially incompressible. Test conditions are set by the model inlet flow coefficient(ϕ) based on the area average inlet axial velocity(Cx) and the test rotor wheel speed at midspan (Um). To date, 0.8 hub to tip radius ratio models of a 1 and 1/2 stage turbine, an isolated compressor rotor, and a two stage compressor have been studied in detail in this facility. Detailed descriptions of the test facility, models, aerodynamic test conditions and the dedicated online data acquistion and reduction system can be found in [7] through [11].

Surface Indictor Technique

Surface indicator techniques used in wind tunnels or cascades typically consist of applying an array of tufts or pigmented oil or ink dots, e.g.,[3], to the model surface. In general, these approaches can not be used on the rotating components of compressor or turbine models because the ratio of aerodynamic to centrifugal force acting on a tuft or oil/ink dot is typically low. As a result, the surface indicators tend to align themselves somewhere between the actual surface flow direction and the radial centrifugal force field. An adaptation of Ruden's [12] ammonia/Ozalid paper technique solves this problem. Also, with this technique, a permanent record of the flow pattern remains on the Ozalid paper which can be removed from the model for later study.

The ammonia/Ozalid paper surface indicator technique has been used extensively in the LSRR-1 to obtain airfoil and endwall flow visualization for both rotors and stators. Typically, it consists of seeping a trace amount of anhydrous ammonia out of selected airfoil and endwall surface static pressure tap locations. For surface film coolant studies, a trace amount of ammonia is mixed with the coolant prior to injection at the blowing sites. Ozalid paper is attached smoothly to the surfaces immediately downstream of the pressure taps or coolant blowing sites. The ammonia is swept along with the surface flow and leaves a dark blue streak in the flow direction on the Ozalid paper. A high intensity strobe light(Strobotac Model 1538A) synchronized to the rotor permits one to observe the trace developing on the Ozalid paper while the model is running. This capability permits direct observation of flow behavior that might otherwise be lost if the ammonia flow is left on too long and the Ozalid paper is over exposed.

Smoke Flow Visualization Technique

Recently, a smoke flow visualization technique has been demonstrated at UTRC that permits visualization of the flow in turbine and compressor passages. An Elvin Precision Ltd. System B smoke generator and NPL-type injector probe were used to inject smoke into the flow at various locations upstream of a single stage turbine model. With this system a mineral oil, Shell "Ondina 17", is pumped to the NPL probe tip where an electrical coil heats it to produce a dense white vapor(smoke) at the injection site. This is an advantage over other types of smoke generator/injectors that produce smoke in a chamber and transport the smoke through a feed line to the injection site. Usually, with the chamber type generator, the resulting smoke stream is less dense and tends to become invisible in high speed flows. To achieve high resolution flow visualization in the present study, where the maximum flow speed was 200 feet per second(60.9 m/s), it was necessary to inject smoke from two System B units with the NRL probes installed in tandem upstream of the stator row. This technique produced a smoke streamer over 6 feet long which was visible to the eye and easily photographed.

A high intensity strobe light(Strobotac Model 1538A) synchronized to the rotor through a variable time delay controller illuminated smoke stream in a selected rotor blade passage at different circumferential positions of the rotor blade passage relative to the the stationary smoke injector or the upstream stator row. The path of the smoke trace through the rotor passage was observed with a Panasonic Model WV1500 black and white video(TV) camera and recorded on video tape with a SONY Model VO5850 recorder. Mulitiple exposure(ensemble average) photographs were also obtained on Polaroid 600 color film in a Polaroid Model 680SF camera. This was done by setting the strobe light to trigger at a selected rotor/stator position and then opening the camera shutter for fifteen seconds. This procedure resulted in approximately 100 exposures of the smoke stream in the same rotor passage per picture.

APPLICATIONS OF THE SURFACE INDICATOR TECHNIQUE

Turbine Film Cooling

Film cooling has been applied to many turbine rotor blade configurations and many parameters affecting film cooling have been extensively studied,e.g.,[13] and [14]. However, until recently[8] there has been no information at all on the impact of rotation and radial (spanwise) flow on the film coolant trajectory or film effectiveness. Typically, design systems have been based on the assumption that the radial velocity on the blade pressure surface and hence in the film is negligible. Any shortcoming of this approach is typically compensated for by calibrating the design system on engine experience.

To provide some insight in this area, a film cooling study was conducted on the rotor blade of a large scale model of a high pressure turbine first stage [8]. Film coolant was discharged from a single blowing site on the pressure surface of the rotor. The coolant to freestream mass flux ratio (M) and density ratio (R) were varied from 0.5 to 1.5 and 1.0 to 4.0 respectively. Both surface flow visualization and local film cooling adiabatic effectiveness data were obtained. Flow visualization results on the pressure surface for a M of 1.0 and a R of 4.0 are shown in Fig. 1.

The pressure surface trajectory displays noticable curavature and has a significant radial component(roughly 30 degrees radially outward). Reducing the density ratio to 1.0 or varying M from 0.5 to 1.5 resulted in no significant change to the flow visualization pattern(Fig. 1). Based on the flow visualization results, thermocouple arrays were designed and installed downstream of the blowing site on the turbine airfoil model also shown in Fig. 1. These arrays were used to obtain the surface film effectiveness patterns (Fig. 1). In this study [8], twelve combinations of M and R were examined. The surface patterns,e.g., Fig. 1, obtained from the separate flow visualization and film effectiveness tests were consistent for all conditions. Although the maximum effetiveness and footprint width were observed to vary markedly over the range of M and R tested, the location of the centerline of the effectiveness and the flow visualization footprint appeared to be insensitive to these variables.

Three Dimensional Flow on Turbine Rotors

Radial flows on a turbine rotor airfoil and in the blade-to-blade passage

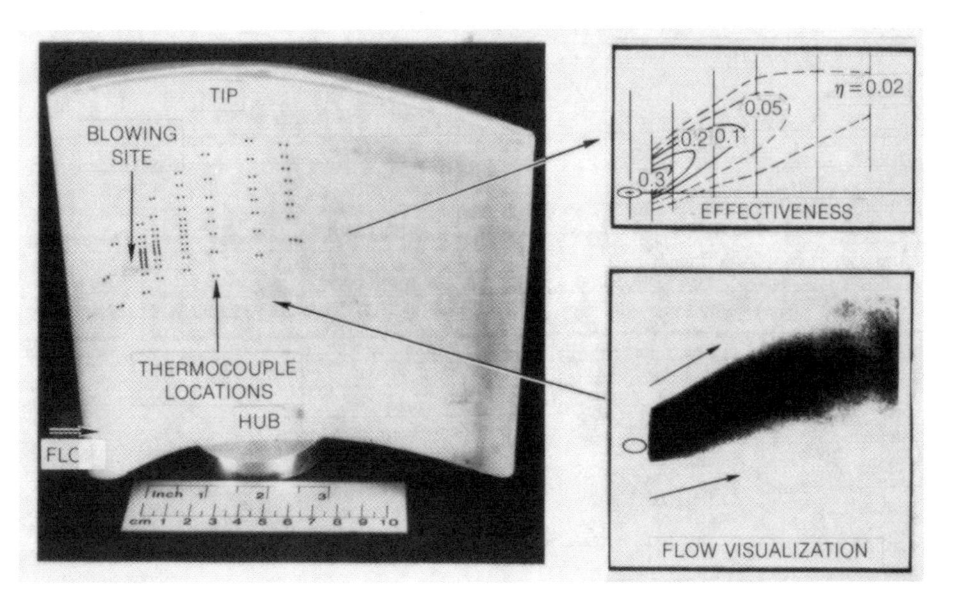

FIGURE 1 Film coolant trajectory flow visualization, thermocouple instrumentation locations and pressure surface effectiveness contours (M = 1.0, R = 4.0)

not only affect film coolant trajectories, but also affect boundary layers, loss distributions, and heat transfer coefficients. Radial flows also affect the local free stream temperature by distorting carefully tailored combustor exit temperature profiles as the flow passes through the turbine airfoil rows. The flow visualization results obtained downstream of a single coolant hole (Fig. 1) indicated the need to better understand the surface flow on the entire turbine rotor airfoil. The flow visualization results subsequently obtained over the remainder of the airfoil surface are presented in Fig. 2. The surface flows on both the pressure and suction surfaces of this airfoil are highly three dimensional.

On the pressure surface there is a strong radial outflow near the leading edge that tends to turn toward axial as the flow proceeds aft. There is relatively little radial outflow near the hub. Near the leading edge the radial component increases rapidly with span. This is in marked contrast to what is typically observed in turbine cascades (without rotation) where the spanwise component of flow on the pressure surface is very small. It was demonstrated[8] that the radial flow on the turbine rotor pressure surface was not due to centrifugal pumping in the boundary layer. By employing the ammonia/Ozalid technique it was shown[15] that leakage flow(Fig. 2) over the unshrouded tip contributes to radial flows only in the immediate vicinity of the rotor tip. Subsequently, it was analytically shown[16] that an inviscid mechanism, the relative eddy, was the source of most of the radial flow on the turbine rotor pressure surface. This mechanism does not occur in stationary cascades but occurs only in rotating turbomachinery passages.

On the suction surface(Fig. 2) there is no significant radial component to

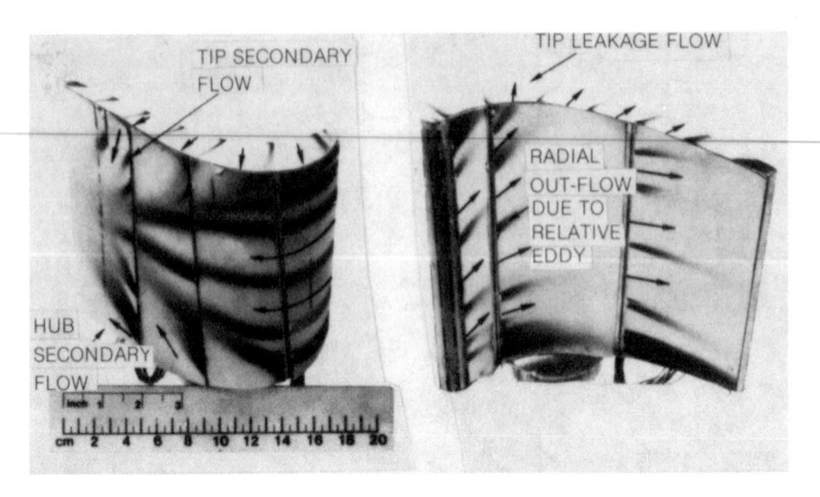

FIGURE 2 Flow visualization on a turbine rotor blade to illustrate the three-dimensionality of the surface flow

the flow in the leading edge region. The high flow speed(approximately five times that on the pressure surface) dominates the influence of the relative eddy. In the trailing edge region the suction surface flows do have a large radial flow component directed toward midspan. These radial flows are a result of the hub and tip endwall secondary flow vorticies that are symmetric about the 45 percent span location. The behavior of the suction surface flows are similair to what one might expect from stationary cascade results such as those of Langston[3].

Analytical Model of Turbine Pressure Surface Flow

The flow visualization results of Fig. 2 illustrate the complex three dimensional nature of the flow on a turbine rotor. The prediction of such a flow field is extremely difficult using analyses which are based on inviscid and/or viscous two dimensional theories. There is a need to locally enhance current predictive techniques by including three dimensional viscous effects and rotation. A prediction method[17] based on three dimensional boundary layer theory was evaluated with the pressure surface flow visualization results shown in Fig. 3. The flow visualization results were obtained by injecting the ammonia from a stationary frame probe located upstream of the turbine rotor blade row. The surface of the airfoil was covered with Ozalid paper as before(Fig. 2).

A fully turbulent boundary layer was assumed throughout the computation domain. Free stream crossflow velocities were estimated using the relative eddy approximation[16]. A comparison of the calculated limiting streamline flow with the flow visualization is shown in Fig. 3. The calculation clearly shows the radially outward flow near the leading edge increasing toward the tip as was seen in the surface flow visualization(Figs. 2 and

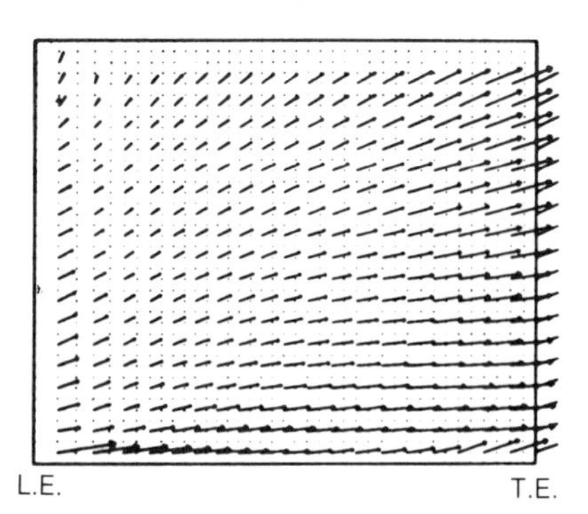

FIGURE 3 Flow visualization and calculated limiting streamlines on the pressure surface of a rotating turbine blade ($C_x/U_m = 0.78$)

3). Both the calculation and the flow visualization show the surface flow to turn toward axial as it proceeds aft. A detailed comparison of the calculated flow angle for the streamline starting at midspan with the flow visualization results is shown in Fig. 4. The measured and calculated flow angles are in reasonably good agreement. This favorable comparison with surface flow visualization results indicates that a zonal application of three dimensional boundary layer theory can be formulated to predict the flow on the pressure surface of turbine rotor blades. This computational technique will be a valuable tool permitting the turbine designer to make better choices in orienting airfoil film coolant holes to acheive optimum effectiveness and adequate coverage.

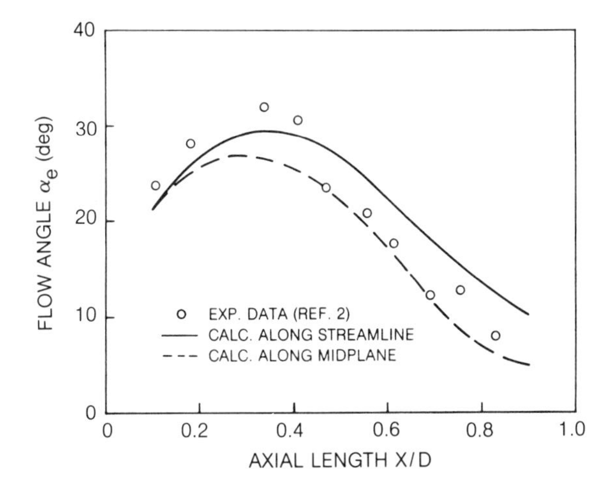

FIGURE 4 Comparison of calculated and measured flow angle on rotating turbine blade

Analysis of Compressor Suction Surface Flow

Analytical tools for the prediction of blade-to-blade potential and boundary layer flow over the airfoils of axial compressors and turbines are in common use today. The substantiation of these tools has historically been based on cascade tests. Until recently, there has been little in the way of actual rotating rig data to compare with. A recent study[9] of axial compressor aerodynamics used the ammonia /Ozalid flow visualization technique to assess the ability of a boundary layer analysis[18] to predict the locations of boundary layer transition and separation at the midspan of the test rotor. Flow visualization results for the design flow coefficient(Cx/Um) of 0.85 and for a lower value(0.65) are shown in Fig. 5. At 0.85 there is flaring in the streakline at midspan which appeared to be due to transition. Following the flare region, the flow remained attached all the way to the trailing edge. With the airfoil at more positive incidence(Cx/Um = 0.65), the location of the flaring (transition) moved forward and there is a fullspan boudary layer separation in the trailing edge region beginning at the 70 percent chord location. It was observed that the location of the flaring always occurred slightly down stream of the minimum pressure on the suction surface where transition would be expected. The flaring location is indicated by the "t" on the measured and predicted[19] airfoil pressure (Cp) distributions (Fig. 6). The predicted pressure distributions were based on measured rotor inlet and exit conditions. At 0.85 the agreement of the prediction with the measured pressure was excellent and at 0.65 it was good up to separation.

Shown below the pressure distributions are the computed suction surface velocity(V/V1) distributions used as input to the boundary layer calculation. The resulting skin friction coefficiant(C_f) distributions are also shown. The location of separation indicated by the "s" is based on the flow visualization(Fig. 5). Two freestream turbulence levels were used in the calculations. The first(Tu = 1/2 percent) is typical for the test model environment and the second(Tu = 4 percent) is typical of a multistage test rig or engine environment. At these turbulence levels, the results

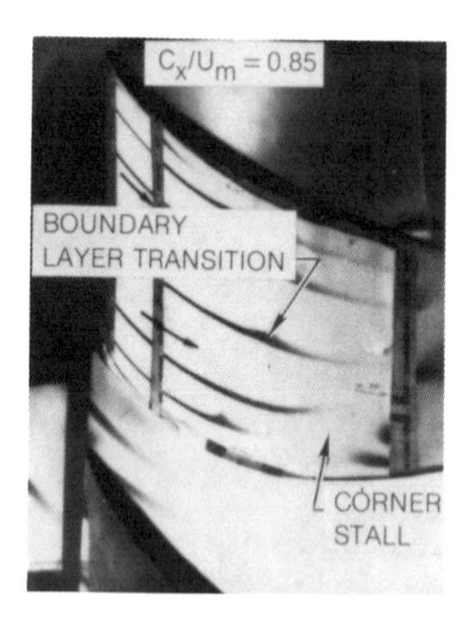

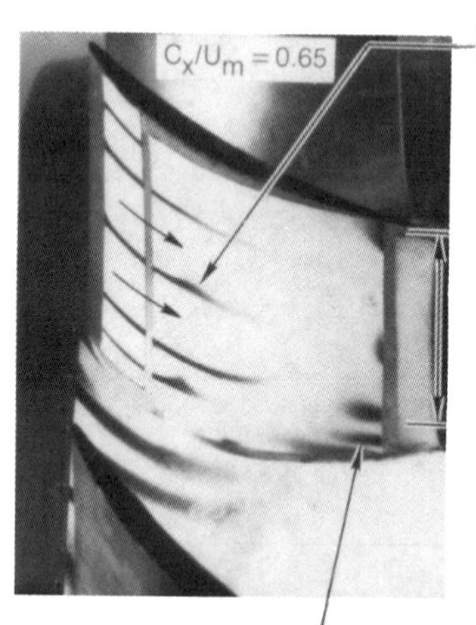

FIGURE 5 Flow visualization on a compressor rotor blade to illustrate boundary layer transition, hub corner stall and full span stall (separation)

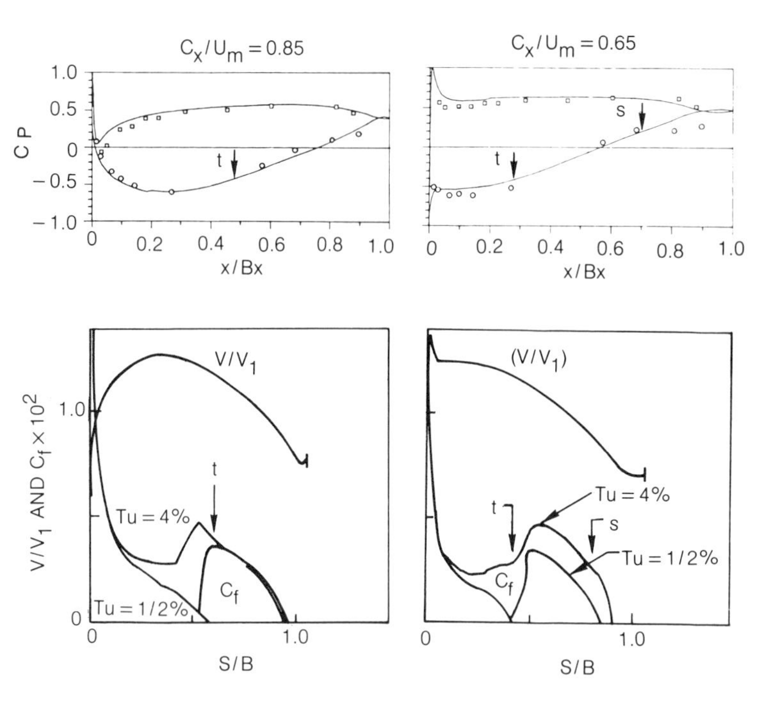

FIGURE 6 Measured compressor rotor airfoil pressure distributions and calculated suction surface velocity and skin friction coefficient distributions to determine transition and separation

for both 0.85 and 0.65 are similar in that the predicted and measured transition locations were in good agreement. Also, for the 0.65 case the predicted and measured trailing edge separation locations closely matched. Therefor, at least for the flow conditions examined, the approach of using potential flow and boundary layer analyses for the design of compressor airfoils has been partially substantiated by the ammonia/Ozalid surface flow visualization technique.

Surface flow visualization(Fig. 5) was also an indispensible aide in clarifying the phenomena observed in the rotor exit traverse data[9]. At 0.85 there is a hub corner stall on the suction surface and considerable back flow. As the loading was increased ($Cx/Um = 0.65$) the suction surface boundary layer near the trailing edge and aft of the 70 percent chord separated over the entire span. There was radial outflow in this region. Until the suction surface flow reached the separated region there was no other significant radial flow on the suction surface.

It can be anticipated that at 0.85 the hub corner stall should show up as a region of high loss near the hub and that at 0.65 the radial flow in the fullspan suction surface boundary layer separation would result in a significant radial redistribution of high loss fluid toward the tip. These results were observed in the data[9, Figs. 15 and 16]. At 0.85 the spanwise distribution of total pressure loss was roughly symmetric about midspan. However, at 0.65 there was a shift in high loss fluid from the hub to the tip. The radial redistribution of loss at 0.65 is peculiar to a rotating turbomachinery environment where centrifuging can occur in the separated region. If the only evidence one had was the loss distribution at 0.65 one might conclude that the blade tip was generating high loss; however, as seen in the flow visualization, this was not the case. The origin of the high loss fluid was at the hub and it was being centrifuged out to the tip.

Hub Corner Stall in a Multistage Compressor

Tests conducted in a two stage compressor model[10] and [11] showed that corner stall occurred on both the second stage rotor and stator suction surfaces and was the dominant endwall mechanism influencing total pressure loss, deviation and blockage. Surface flow visualization on the airfoil surfaces and the endwalls aided in understanding the impact of this mechanism on the quantitative measurements. Flow visualization results obtained on the stator suction and pressure surfaces and on the hub and tip endwalls at the design flow coefficient(0.51) are shown in Fig. 7. Also shown are the total pressure (C_{pt}) contours obtained from radial-circumferential traverse data [11] acquired downstream of the stator.

On the pressure surface the flow is nearly two dimensional. The hub and tip endwall flow visualization show no significant evidence of inlet boundary layer skewing or cross flow toward the suction surface due to secondary flow. However, as seen in Fig. 7, the hub corner stall does force the flow away from the suction surface at the hub endwall and radially outward on the suction surface. In the total pressure contours (Fig. 7) the low total pressure region near the hub is due to the corner stall and not to secondary flow in the endwall boundary layer.

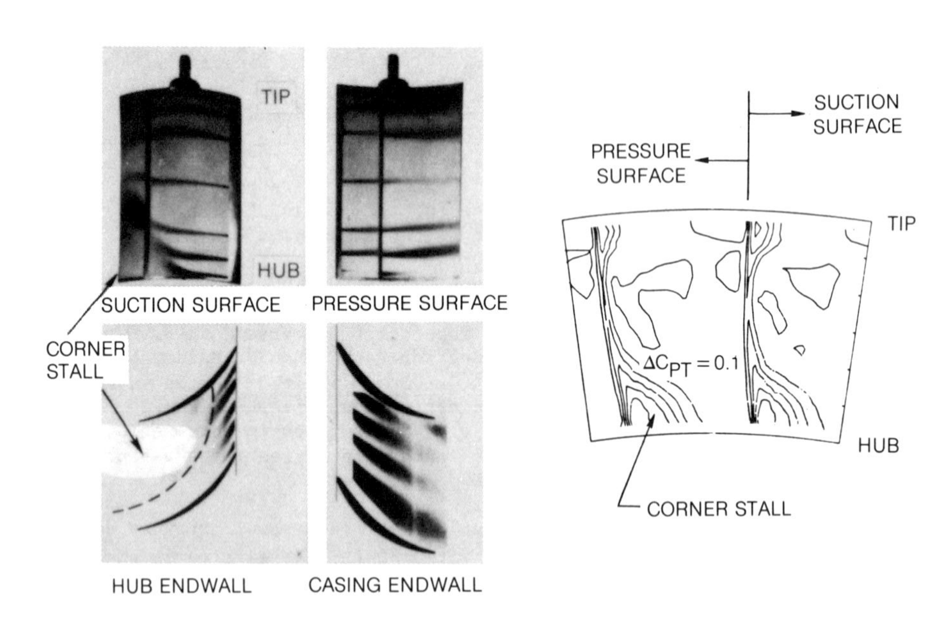

FIGURE 7 Surface flow visualization to interpret second stage compressor stator exit plane total pressure contours at design flow coefficient ($C_x/U_m = 0.51$)

The concept of blockage has been introduced by compressor analysts to account for both endwall boundary layers(endwall blockage) and non-axisymmetric effects(tangential blockage). Based on the flow visualization one would expect the corner stall to generate flow blockage in the stator passage. This effect is seen in the spanwise distributions of tangential blockage(Fig. 8) calculated from measured traverse data[11] acquired at 0.51 and near stall(0.45). The highest levels of blockage are in the hub region and are due to low total pressure in the hub corner stall region. Suction surface flow visualization results[11] showed that as the flow coefficient was reduced from 0.51 to 0.45, the hub corner stall grew dramatically in spanwise and chordwise extent. The effect of this growth is reflected in the increase in blockage near the hub region as shown in Fig. 8. In this study, the surface flow visualization provided a direct means to observe and identify the physical mechanism (corner stall) that resulted in the high loss and blockage seen in the quantitative (measured) results.

APPLICATION OF SMOKE FLOW VISUALIZATION

Wake-Airfoil Interaction

The problem of unsteady wake-airfoil interaction is of importance in nearly all fluid mechanical devices that involve rotating machinery. Unsteady wake-airfoil interactions significantly affect turbomachinery component aerodynamic efficiency, airfoil heat transfer and unsteady loading. The degree of impact is related to the manner in which upstream wake fluid is transported through the downstream rotor airfoil passage.

To provide the basis for the development of analytical wake-airfoil interaction and wake transport models, a smoke stream flow visualization technique was demonstrated in a single stage turbine model. In this case the smoke stream corresponds to a streamline in the stationary frame of reference (for example, a stator wake with a very small velocity defect).

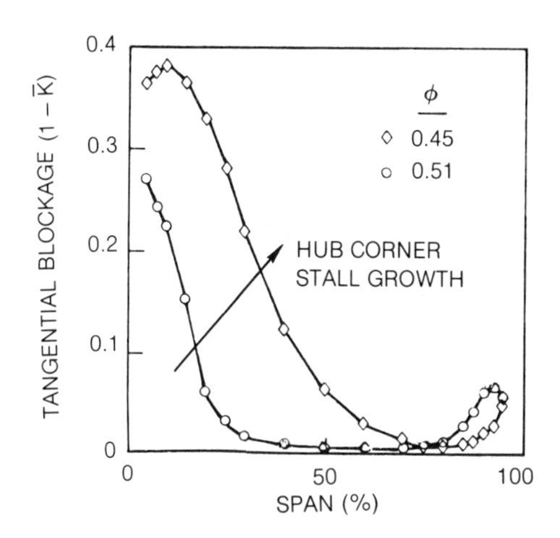

FIGURE 8 Spanwise distributions of axial compressor second stage stator tangential blockage

The results at midspan from this exploratory study are shown in Fig. 9. Flow visualization of the smoke flow through the rotor passage with the rotor passage at different circumferential positions relative to the stationary smoke injector and upstream stator row illustrates the different stages of wake chopping and subsequent distortion through the rotor passage.

The corresponding results obtained from an inviscid wake trajectory calculation[20] are also shown. The inclined line labeled "t = 0" in Fig. 9 corresponds to the location of the centerline of wake from the stator row at a specific reference time or rotor/stator position. The other numbered lines represent the position and shape of the initial wake centerline at subsequent time intervals(rotor/stator positions) as it is convected through the rotor passage. It is clear from the flow visualization and the calculated results that there is considerable distortion(bowing) to the wake. This distortion is due to retardation of the incoming flow near the leading edge stagnation region of adjacent rotors and the cross passage variation in the relative frame velocity. Calculated results[20] for both a sharp leading edge turbine rotor and a compressor rotor did not exhibit any flow retardation in the leading egde region and as a result there was significantly less wake distortion. For the compressor, the calculated wake distortion was also confirmed with the smoke flow visualization[20].

CONCLUSIONS

This paper has shown the value of flow visualzation in rotating turbomachinery research. In particular, the surface indicator(ammonia/Ozalid paper) technique has been shown to be a valuable aide in specifying instrumentation arrays in film cooling studies, in interpreting quantitative aerodynamic measurements, and in detecting the presence of three dimensional phenomena (the relative eddy, endwall secondary flow, tip leakage flow, corner stall, boundary layer transition and separation, and radial fluid transport) in rotating airfoil rows. It has also been used to assess

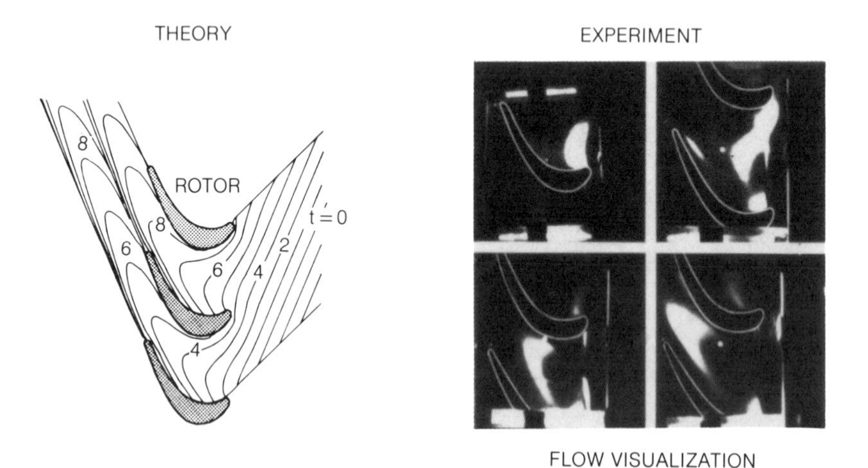

FIGURE 9 Flow visualization of turbine rotor blade passage compared to theory

analytical models. The smoke flow visualization technique has the potential to lead to a better understanding of wake-airfoil interaction and also to provide a basis for the assessment of analytical models that are developed.

ACKNOWLEDGMENTS

The development and demonstration of these techniques was done under United Technologies Research Center Corporate funding.

NOMENCLATURE

B = airfoil chord

C_f = skin friction coefficient

C_p = static pressure coefficient

$(P-P_1)/(\frac{1}{2}\rho V_1^2)$

C_{pt} = total pressure

$(P_t-P_{T1})/(\frac{1}{2}\rho U_m^2)$

C_x = axial component of flow velocity

K = blockage factor

M = mass flux ratio

$(\rho V)_c/(\rho V)_f$

P = pressure

Q = dynamic heat $\frac{1}{2}\rho V$

R = density ratio (ρ_c/ρ_f)

S = surface arc length

T = temperature

T_u = freestream turbulence

U_m = rotor wheel speed at midspan

V = flow velocity

η = film cooling effectiveness

$\frac{(Ts-Tf)}{(Tc-Tf)}$

ϕ = flow coefficient (C_x/U_m)

Subscripts

1 = inlet conditions

c = coolant

f = freestream

s = surface

REFERENCES

1. Yang, W. J.," Survey of Recent Flow Visualization Studies in U.S.A. and Canada," General Lecture 4, Proceedings of the International Symposium on Flow Visualization, October 1977.

2. Sedney, R., Kitchens, C. W. Jr., and Bush, C. C., "Combined Techniques for Flow Visualization," AIAA Paper 76-55, presented at AIAA 14th Aerospace Sciences Meeting, Washington, D.C., January 26-28, 1976.

3. Langston, L.S., and Boyle, M.T., "A New Surface Flow Visualization Technique," Journal of Fluid Mechanics, Vol. 125, Dec. 1982, pp 53-57.

4. Severding, C. H., and Voon Den Bosche, P., "The Use of Colored Smoke to Visualize Secondary Flows in a Turbine-Blade Cascade" Journal of Fluid Mehanics, Vol. 134, 1983, pp. 85-89.

5. Gaugler, R. E., and Russell, L. M., "Comparison of Visualized Turbine Endwall Secondary Flows and Measured Heat Transfer Patterns", ASME Journal of Engineering for Gas Turbines and Power, Vol. 106, Jan. 1984, pp. 168-172.

6. Phillips, W. R. C., and Head, M. R., "Flow Visualization in the Tip Region of A Rotating Blade Row", Int. J. Mech. Sci., Vol. 22, 1980, pp. 495-521.

7. Joslyn, H. D., Dring, R. P., and Sharma, D. P., "Unsteady Three-Dimensional Turbine Aerodynamics," ASME Journal of Engineering for Power, Vol. 105, No. 2, April 1983, pp. 322-331.

8. Dring, R. P., Blair, M. F., and Joslyn, H. D., "An Experimental Investigation of Film Cooling on a Turbine Rotor Blade," ASME Journal of Engineering for Power, Vol. 102, No. 1, Jan. 1980, pp. 81-87.

9. Dring, R. P., Joslyn, H. D., and Hardin, L. W., "Experimental Investigation of Compressor Rotor Wakes," AFAPL-TR-79-2107, Air Force Aero Propulsion Laboratory, Technology Branch, Turbine Engine Division (TBX), Wright-Patterson AFB, Ohio.

10. Dring, R. P., Joslyn, H. D., and Wagner, J. H., "Compressor Rotor Aerodynamics." AGARD/PEP Specialists' Meeting 61-A, Viscous Effects in Turbomachines, Copenhagen, Denmark, June 1-3, 1983.

11. Joslyn, H. D., and Dring, R. P., "Axial Compressor Stator Aerodynamics", To be published in the ASME Journal of Engineering for Power, 1985

12. Ruden, P., "Investigation of Single Stage Axial Fons," NACA RM No. 1062, April 1944

13. Goldstein., R. J., "Film Cooling". Advances in Heat Transfer. Academic Press, New York and London, Vol. 7, 1971, p. 321.

14. Ifo, S., Goldstein, R. J., and Eckert, E. R. G., "Film Cooling of a Gas Turbine Blade." ASME Journal Engineering for Power, Vol. 100, July 1978, pp. 476-481.

15. Dring, R. P. and Joslyn, H. D., "Measurements of Turbine Rotor Blade Flows", ASME Journal of Engineering for Power, Vol 103, No. 2, April 1981, pp. 400-405.

16. Dring, R. P., and Joslyn, H. D., "The Relative Eddy in Axial Turbine Rotor Passages", ASME Paper 83-GT-22, 1983.

17. Anderson, O. L., "Assessment of a 3-D Boundary Layer Code to Predict Heat Transfer and Flow Field in a Turbine Passage," submitted to be presented at the 1985 ASME Winter Annual Meeting.

18. McDonald, H., and Fish, R. W., "Practical Calculation of Transitional Boundary Layer," International Journal of Heat and Mass Transfer, Vol. 16, No. 9, 1973.

19. Caspar, J. R., Hobbs, D. E. and Davis, R. L., "Calculation of Two-Dimensional Potential Cascade Flow Using Finite Area Methods," AIAA Journal, Vol. 18, No. 1, Jan. 1980, pp. 103-109.

20. Joslyn, H. D., Caspar, J. R., and Dring, R. P., "Unsteady Wake-Airfoil Interaction and Flow Distortion", to be presented at the 21st AIAA Joint Propulsion Conference, Monterey, CA., June 1985.

Flow Visualization Study of the Effect of Injection Hole Geometry on an Inclined Jet in Crossflow

FREDERICK F. SIMON and MICHAEL L. CIANCONE
National Aeronautics and Space Administration
Lewis Research Center
Cleveland, Ohio 44135

ABSTRACT

A flow visualization study using neutrally buoyant, helium-filled soap bubbles was conducted to determine the effect of injection hole geometry on the trajectory of an air jet in a crossflow and to investigate the mechanisms involved in jet deflection. Experimental variables were the blowing rate (M = 0.53, 1.1, 1.6, 4.1, and 6.2) and the injection hole geometry (cusp facing upstream (CUS), cusp facing downstream (CDS), round, swirl passage, and oblong). Results indicate that jet deflection is governed by both the pressure drag forces and the entrainment of free-stream fluid into the jet flow. The effect of the pressure drag force is that a jet presenting a larger projected area to the crossflow will be deflected initially to a greater extent. Thus for injection hole geometries with similar cross-sectional areas and similar mass flow rates, the jet configuration with the larger aspect ratio (major axis perpendicular to the crossflow) experienced a greater deflection. Entrainment arises as a result of lateral shearing forces on the sides of the jet, which set up a dual vortex motion within the jet and thereby cause some of the main-stream fluid momentum to be swept into the jet flow. This additional momentum forces the jet nearer the surface. Of the jet configurations examined in this study, the oblong, CDS, and CUS configurations exhibited the largest deflections. These results correlate well with film cooling effectiveness data, suggesting the need to determine the jet exit configuration of optimum aspect ratio to provide maximum film cooling effectiveness.

SYMBOLS

A	cross-sectional area
C_d	flow discharge coefficient
D	effective diameter of jet at exit
g_c	Newton's constant
M	blowing rate, $(\rho U)_j/(\rho U)_\infty$
m	mass flow rate
P	static pressure
R	ratio of jet to main-stream velocity

Re	Reynolds number
T	temperature
U	velocity
X	axial distance from downstream edge of jet exit
X/D	dimensionless distance based on effective jet diameter at exit
Y	vertical distance from wall
Y/D	dimensionless vertical distance based on effective jet diameter at exit
η	adiabatic film cooling effectiveness, $(T_\infty - T_{aw})/T_\infty - T_j)$
ρ	density

Subscripts

aw	adiabatic wall
c	centerline
j	jet
L	lower jet boundary
p	plenum
∞	tunnel air, crossflow or free stream

INTRODUCTION

A jet in crossflow is of great practical significance for many engineering applications. The ratio of the jet mass flux to the main-stream mass flux (the blowing rate M) determines to a great extent the application to be considered. Applications range from the film cooling of turbine blades and the injection of jets into combusters to the transition flight of V/STOL aircraft or the disposal of wastes into the atmosphere. In dealing with these phenomena, it is important to know the flow field or jet trajectory that results from a given value of M. Although the results reported herein have general application, the motivation for the present study was the need to maintain a coolant film (film cooling) as close as possible to the surface of turbine blades exposed to high-temperature gases.

Papell (Ref. 1) compares the film cooling efficiencies of a jet emanating from either a cusp-shaped hole or a standard discrete round hole into a crossflow for a range of blowing rates ($0.2 \leq M \leq 2.05$) and an injection angle of 30°. His visual evidence indicates that the cusp-shaped hole has a higher film cooling efficiency because its lower coolant jet trajectory deflects closer to the surface than the trajectory from a round hole at comparable conditions. Papell further postulates that the cusp-shaped hole produces a secondary flow consisting of a pair of counterrotating vortices that enhances the deflection of the jet trajectory. He supports this concept by using neutrally buoyant helium-filled bubbles to delineate the jet flow region. Papell thus establishes the advantage of some noncircular holes in film cooling and hence the need for a greater understanding of the mechanisms involved in determining the trajectory of a jet in crossflow.

There is an abundance of information on jet trajectories for round injection holes (Refs. 2 and 3) but little information for noncircular holes. Reference 2 alone contains 24 references of experimental investigations of round holes in crossflow. Reference 4 reports on the penetration of air jets from circular, square, and elliptical orifices at a known distance from the orifice. The long axis of the elliptical orifice was placed parallel to the crossflow. Reference 5 studies the temperature profile in the dilution zone of a combustion chamber created by jets flowing from "bluff"-shaped[1] slots and slots or orifices of other shapes. References 6 and 7 investigated the effect of a normal jet on the pressure distribution on a flat surface with round- and oblong-shaped injection holes (crossflow parallel and perpendicular to major axis). References 5 and 6 provide some general information on the effect of orifice configuration on jet flow but no detailed information on jet trajectories. Rather than focusing on film cooling effectiveness, the present study goes beyond the preliminary flow visualization results of Ref. 1 and attempts to develop a sound relationship by providing experimental jet trajectory data for circular and noncircular holes for several values of M.

Previously it had been demonstrated that greater film cooling efficiency could be obtained by using a curved-tube inlet channel (Ref. 8), "shaped" holes (Ref. 9) (which decrease the jet momentum and employ the Coanda effect[2] to decrease the penetration of the coolant jet into the main stream), and compound-angle injection (Ref. 10) (to keep the jet attached to the surface). An analysis of the existing literature on jet trajectories indicates three general categories of jet/crossflow interactive mechanisms:

1. Only entrainment of the free stream by the jet governs the interaction (Refs. 11 to 13).

2. Only pressure forces acting on the jet govern the interaction (Ref. 14).

3. Both entrainment and pressure forces are considered in the interaction (Refs. 15 to 20).

From these studies it became apparent that an understanding of the interaction between the jet and the crossflow is crucial to determining the trajectory of the deflected jet.

A jet, in terms of its history as it penetrates into the main stream, can be described in terms of three regions: (1) the potential core region, region I; (2) the developed turbulent flow region, region II; and (3) the far downstream region, region III.

In region I, the fluid jet penetrating the crossflow forms a potential core of essentially constant velocity. This retards the main stream along the upstream side of the jet and increases the pressure. On the downstream side of the jet a rarefaction, or wake region, occurs. Coupled with the upstream pressure, it produces a pressure differential that deflects the jet toward the surface. Kamotani's (Ref. 21) experimental results indicate that this deflection begins

---

[1] Oblong-shaped orifice with long axis perpendicular to crossflow.
[2] Henri Coanda (1932) observed that a free jet emerging from a nozzle will follow a nearby curved or inclined surface or will come in contact with the surface. This effect is caused by jet stream entrainment, which creates a partial vacuum.

very close to the jet exit. Platten (Ref. 22) found that for low values of the ratio of jet to main-stream velocity R the deflection of the potential core by the pressure gradient normal to the jet begins to become appreciable.

Viscous entrainment of the main-stream fluid denotes the beginning of the developed turbulent flow region (region II, approximately three diameters downstream). Lateral shearing action sweeps main-stream fluid around the sides of the jet and into the central jet region via entrainment through the underside of the jet. The overall effect is the creation of a counterrotating pair of vortices within the jet that tend to deform the jet cross section into a kidney shape (Fig. 1). This secondary motion enhances the entrainment of the main-stream fluid, along with its corresponding momentum, into the jet. This further deflects the jet toward the surface. In addition, as the jet proceeds downstream, the kidney-shaped cross section presents a greater drag surface to the main stream, thereby enhancing the deflection due to pressure forces. Reference 23 suggests that this secondary vortex motion is influenced by the velocity profile within the jet flow passage and that it may be possible to enhance the secondary motion caused by the interaction of the jet and the main stream by changing the shape of the passage. In the developed turbulent flow region (region II) jet deflection is due to both the entrainment of main-stream fluid and the pressure forces induced by the main stream interacting with the jet.

Figure 61 of Ref. 13 shows that for a ratio of jet length to diameter greater than approximately 18 the effects of main-stream entrainment dominate the jet trajectory. This is the far downstream region (region III), where pressure forces no longer play a significant role in determining the jet trajectory.

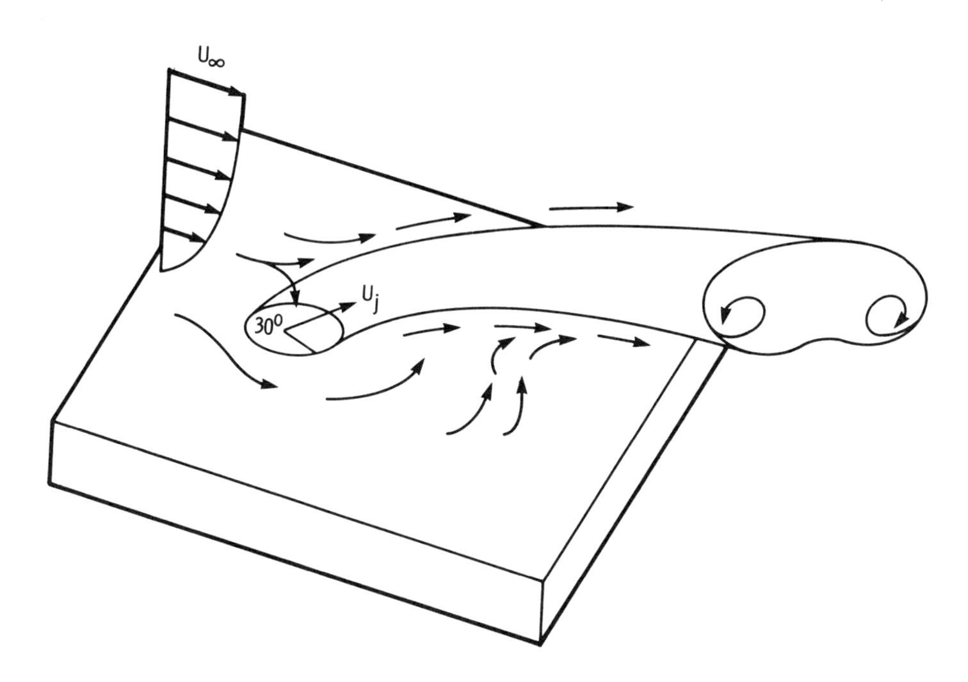

FIGURE 1. Representations of a jet in crossflow.

In the present work jet trajectories were photographed for five injection hole geometries at blowing rates M of 0.53, 1.1, 1.6, 4.1, and 6.2 for a jet injection angle of 30° with respect to the upstream horizontal. Relevant data were extracted from these photographs and used in a comparative analysis of the effect of injection hole geometry on jet trajectory to assess the mechanisms contributing to jet deflection.

APPARATUS

The flow visualization test rig (Fig. 2) consisted of a transparent plastic tunnel through which air was drawn into a vacuum exhaust line. This simple construction provided flexibility for testing a large number of injection hole geometries appropriate to turbine and combustor cooling applications. The test configuration for this report consisted of a zero-pressure-gradient, free-stream flow over a flat surface containing an injection hole. The tunnel section containing the test plate was positioned so that there was about 1.3 m of tunnel length (not including contoured inlet) upstream of the jet exhaust. Thus at the point of jet exit the injection surface boundary layer was fully turbulent as determined in a previous investigation with this tunnel (Ref. 8). The three separate ambient airflow sources were the primary free-stream airflow, the bubble generator airflow, and the secondary jet air (Fig. 3).

The jet air was supplied to the plenum by means of a Hilsch tube connected to a 827-kPa (120-psi) dry air source. This source, which incorporated a vortex generator element, separated the inlet air into hot and cold streams. This separation resulted from the forced vortex, or wheel, type of angular velocity imparted to the air entering the device. Conservation of the total energy of the inner region of the contained vortex caused heat to be transferred to the outer region of the vortex. Consequently a relatively cold inner core of air and a warm outer ring of air were available. In the Hilsch tube design the warm and cold air discharge ports are on opposite ends of the tube. Cold-end temperatures of 0 °C were available with this device, so some variation in the jet density was possible.

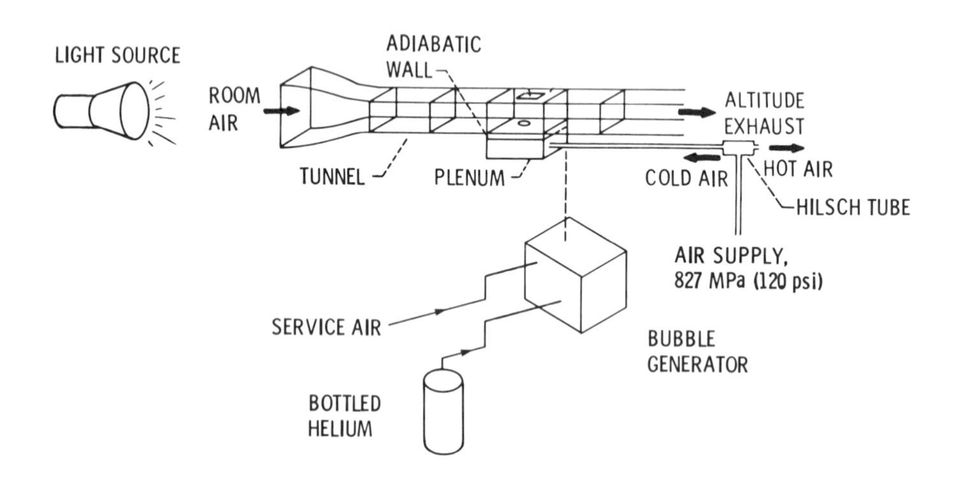

FIGURE 2. Schematic of test plate and plenum air supply.

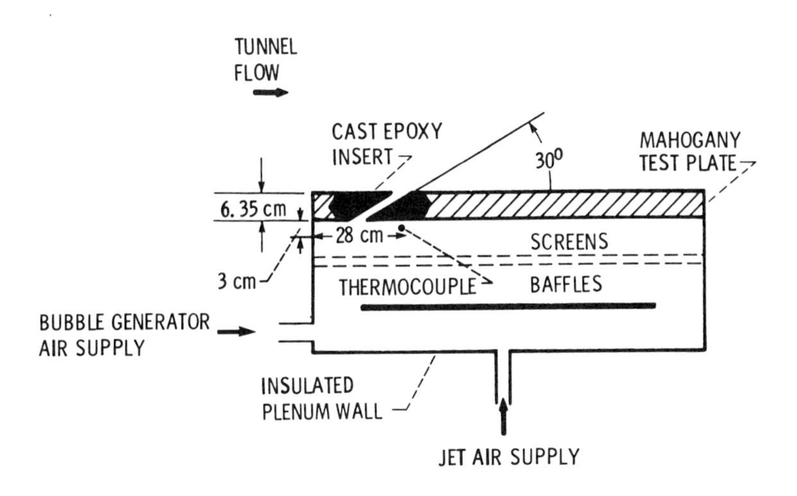

FIGURE 3. Flow visualization rig.

The tunnel air temperature was measured with a thermocouple mounted in the contoured inlet. The coolant air temperature was measured with a thermocouple mounted in the plenum between the screens and the mahogany test plate. The coolant airflow rate was measured with a turbine type of flowmeter installed between the Hilsch tube and the coolant plenum. The tunnel velocity was determined from pitot-static pressure readings taken upstream of the test section.

The bubble generator system employed in the present study was used in the visual study of Ref. 10 and consisted of a head (which formed the bubbles) and a console (which controlled the flow of helium, bubble solution, and air to the head). A drawing illustrating the basic features of the head is shown in Fig. 4. Neutrally buoyant, helium-filled bubbles, about 1 mm in diameter, formed on the tip of the concentric tubes and were blown off the tip by air flowing through the shroud passage. Bubble solution flowed through the annular passage and was formed into bubbles at the tip. These bubbles were inflated with helium passing through the inner concentric tube. The desired bubble size and neutral buoyancy were achieved by proper adjustment of air, bubble solution, and helium flow rates. For this study a setting was established to produce the largest number of bubbles possible that were small enough to survive passage through the plenum and the jet exit channel. As many as 300 bubbles per second can be formed by this device.

The neutrally buoyant, helium-filled bubbles were injected into a plenum, which served as a collection chamber for the bubbles and the jet air. The air, seeded with the bubbles, then passed through the jet passage and into the test region. The small quantity of air used by the bubble generator to blow the bubbles off the tip of the annulus as they formed ended up as part of the jet air in the plenum. Consequently the mass contribution of the bubble generator was measured by a rotameter. This small, but not negligible, correction to jet mass flow was subsequently accounted for in calculating the blowing rate M.

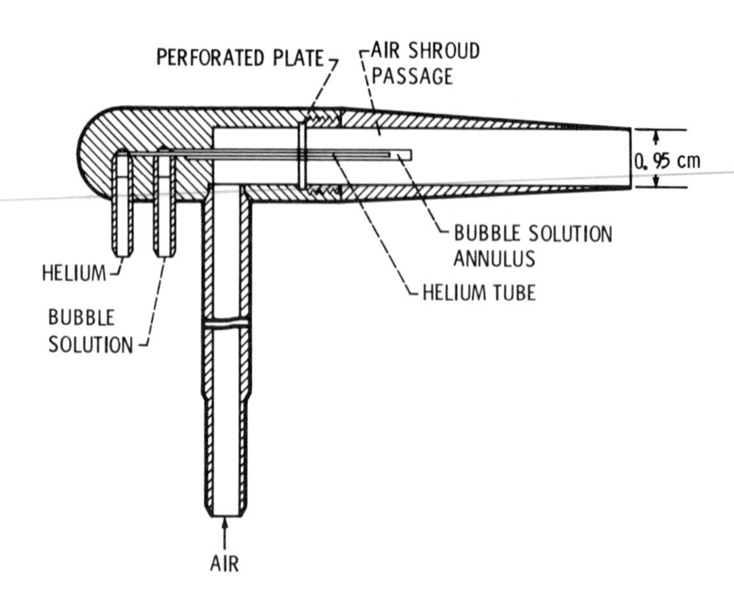

FIGURE 4. Bubble generator head.

The plenum box was clamped onto the bottom of the test section for easy removal when another test plate with a different injection hole geometry was to be tested. The configurations were cast in epoxy as inserts to be installed in the flat plate. Schematic drawings of the cross-sectional areas of the various configurations investigated are displayed in Fig. 5.

The 0.38- by 0.61-m floor of the test section, which contained the jet injection hole, was easily removable to allow bottom plates with different hole configurations to be installed without affecting the rest of the test section or the plenum chamber. The jet flow passage length of 6.35 cm provided a ratio of jet flow passage length to diameter of 5.0 (typical of aircraft turbine applications). All configurations had an equivalent diameter of 1.27 cm based on a constant cross-sectional area of 5.07 cm^2 to allow the mass flow rate to be independent of jet configuration. The floor and back side of the test section were made of wood and had a glossy black finish to give maximum contrast with the bubble streaklines.

A high-efficiency 300-W xenon quartz arc lamp provided sufficient light intensity for photographing the bubbles (Fig. 6). A metal plate with a rectangular slot cutout was placed between the light source and the lens to shape the light beam, and an infrared reflecting filter was used to prevent heating of bubbles passing through the beam. The beam was then focused through a 300-mm lens to form a sharply defined rectangular pattern of collimated light (7.7 by 15.2 cm) through which the bubbles passed as they exited the jet flow passage.

EXPERIMENTAL PROCEDURE

Typical test procedure consisted of filming the test section from the lower surface level for a variety of injection hole geometries and blowing rates as

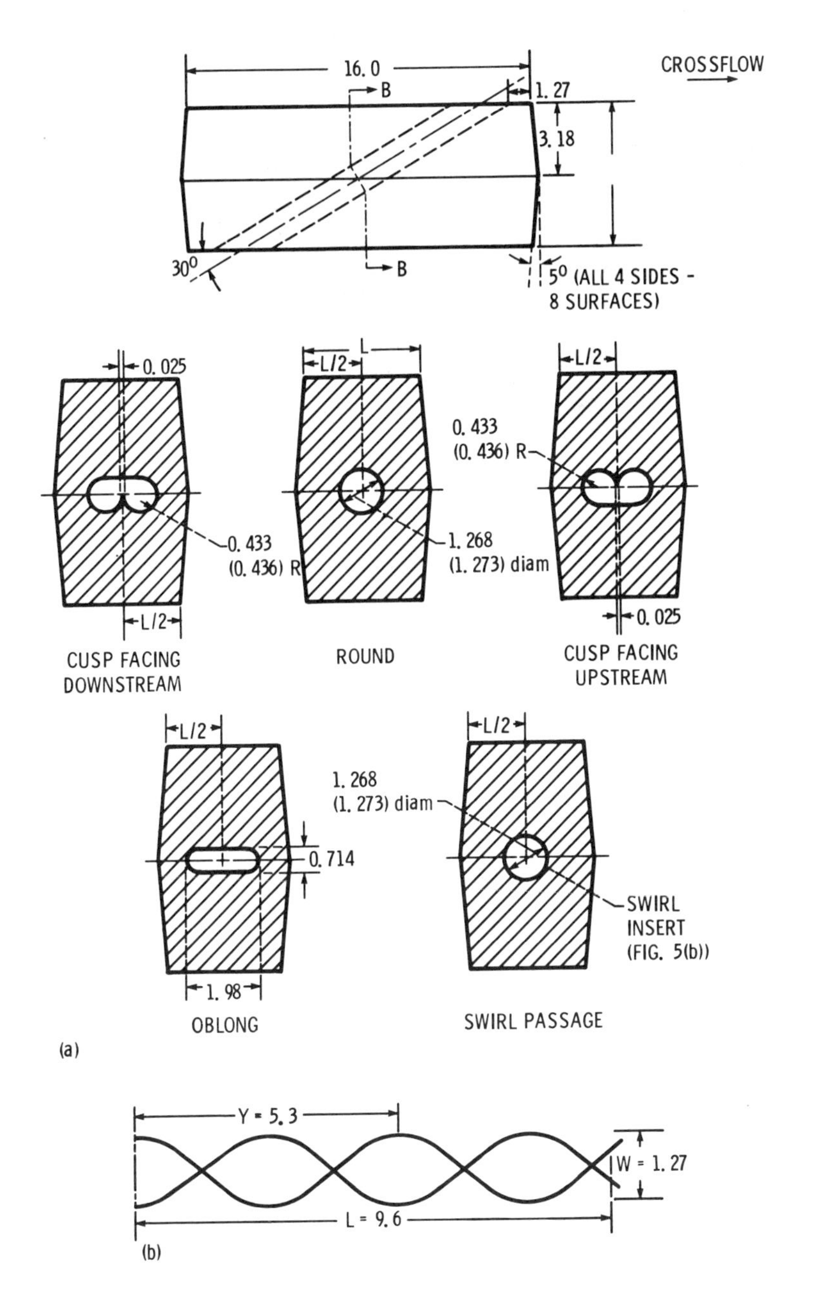

(a) Jet flow passage.
(b) Swirl insert (twisted tape) for film cooling hole. Twist is in clockwise direction when viewed from end of tape; Y is period of revolution (i.e., distance over which a complete 360° twist is completed); t = 0.16 cm.

FIGURE 5. Geometry. (Dimensions are in centimeters.)

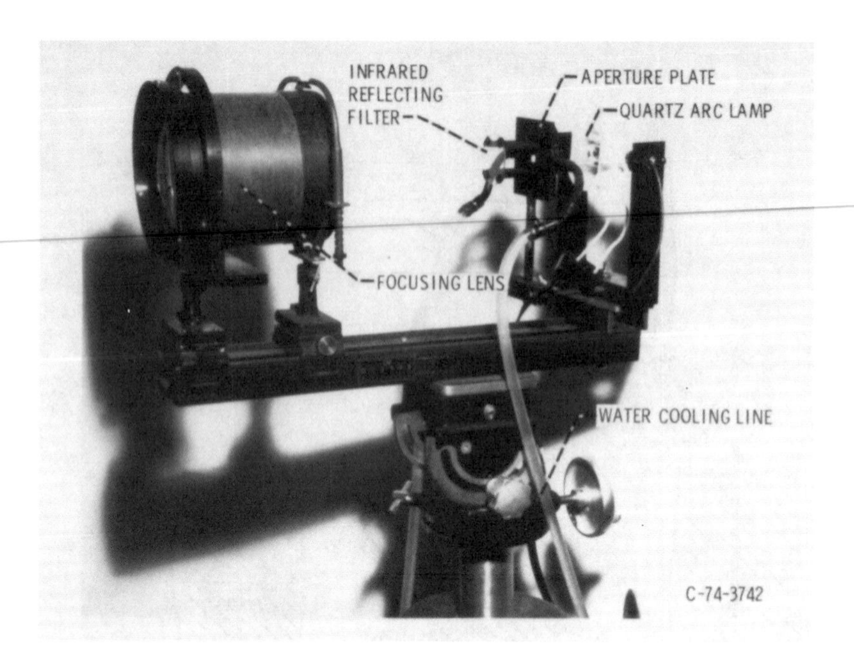

FIGURE 6. Light source assembly.

the bubble-delineated jet flow interacted with the free stream. Upon their passage through the jet exit channel and entrance into the test section, the bubbles contained within the jet flow were illuminated by the light beam emitted by a high-intensity xenon arc lamp situated upstream of the flow and directed parallel to the free stream.

As the soap bubbles passed through the illuminated region of the test section, their movement was recorded on film as a series of streaklines caused by light reflecting from the surface of the bubbles as they passed through the illuminated portion of the photographic field. A sequence of photographs exposed for different times was produced to achieve the optimum setting for each set of test conditions. Too few streaklines would result in a photograph lacking definition; too many streaklines would tend to wash out the entire frame. Generally speaking exposure times ranged from 20 to 80 sec at an aperture setting of f5.6 for a film speed of ASA 400.

This attention to exposure time was a direct result of the need for a statistically significant number of streaklines in each photograph, in addition to the desire for quality flow visualization. Assuming the bubbles would faithfully and accurately follow the jet flow as it entered the test region and mixed with the free stream, this nonetheless dictated the statistical nature of the bubble movement as a function of each bubble's departure point from the jet orifice - hence the randomness associated with each streakline location. It was desirable to establish a large number of streaklines in each photograph to ensure that the jet region was delineated realistically as it began to mix, and eventually merge, with the free stream. This procedure provided sufficient data to identify the effect of hole geometry on jet trajectory in a crossflow.

The high and low extents of the jet flow region were identified in each photograph at axial downstream distances X/D of 0.5, 1.0, 1.5, 2.0, 3.0, 4.0, 5.0, 8.0, 10.0, and 12.0. From these, the jet centerline height was calculated by averaging the values of the high and low vertical positions of the jet boundary. Measured data were stored in a data set to be graphically output via the ZETA12 graphics capabilities of the IBM 370 computer system.

RESULTS AND DISCUSSION

Figures 7 and 8 represent the jet trajectory centerline data and the lower jet boundary data for the five injection hole geometries investigated in this study. Representative values of blowing rate M and density ratio ρ_j/ρ_∞ are indicated in each figure. Abramovich's empirical prediction (Ref. 24) based on round holes is included as a basis for comparison with the jet centerline data. There is general agreement with the round-hole data. In general, the round hole with a swirl passage insert produced the highest jet trajectory. The cusp facing upstream (CUS), cusp facing downstream (CDS), and oblong holes produced the lowest trajectories. The lowering of the jet trajectory by an oblong hole is also suggested by the temperature profile results of Ref. 5 and by the limited jet trajectory information of Ref. 6. Photographs of jets are shown in Fig. 9.

According to Ref. 25, the use of a swirl tape insert (Fig. 5(b)) should markedly increase the entrainment of main-stream fluid into the jet. In the present experiments the spreading rate of the jet produced by the swirl configuration was greater than those of the other configurations, an indication that entrainment would also be greater for the swirl configuration. It had been expected that the additional entrainment of main-stream fluid would effectively increase jet deflection toward the surface. However, inasmuch as this was not the case in the present study, it was conjectured that the swirl component of jet velocity diminished the effect of drag and entrainment on the jet centerline trajectory. The swirl-produced jet (Fig. 9) appeared relatively unaffected by the main-stream flow to an axial distance of several diameters downstream of the jet exit. Apparently the main-stream fluid entrained by the jet lost some of its axial momentum to the swirl component of the jet. In addition, the small effect of pressure drag forces on jet trajectory in the potential core region was probably due to the swirl component of the jet preventing the crossflow shearing action across the jet, which would otherwise set up the pressure differential necessary to deflect the jet. This is analogous to the method given in Ref. 26 for reducing pressure drag through the use of a moving surface that effectively reduces the relative velocity at the shear interface. Therefore it appears that in the case of the swirl configuration, decreased drag and decreased axial momentum combined to produce a jet that was deflected least among the configurations within the scope of this study.

Figure 10 points out the effect of the initial jet cross section on pressure drag, and hence on the deflection trajectory, by comparing the deflection trajectories for round and oblong holes. The oblong hole produces an initial jet shape that has a drag coefficient at least twice that of the round hole. The effect that increased pressure drag can have on jet trajectory, namely a greater deflection of the jet, is illustrated in Fig. 10 for blow rates M of 1.6 and 0.5 for the analysis of Ref. 14 and the data of the present experiments. The analysis of Ref. 14, which predicts jet centerline trajectory based solely on drag, is insufficient to describe the jet centerline trajectory. Because of the lack of difference in entrainment for M of 1.6 and 0.5 (based on inspection of jet expansion) for both the round and oblong holes, the difference in jet trajectories observed in Fig. 10 was assumed to be a function

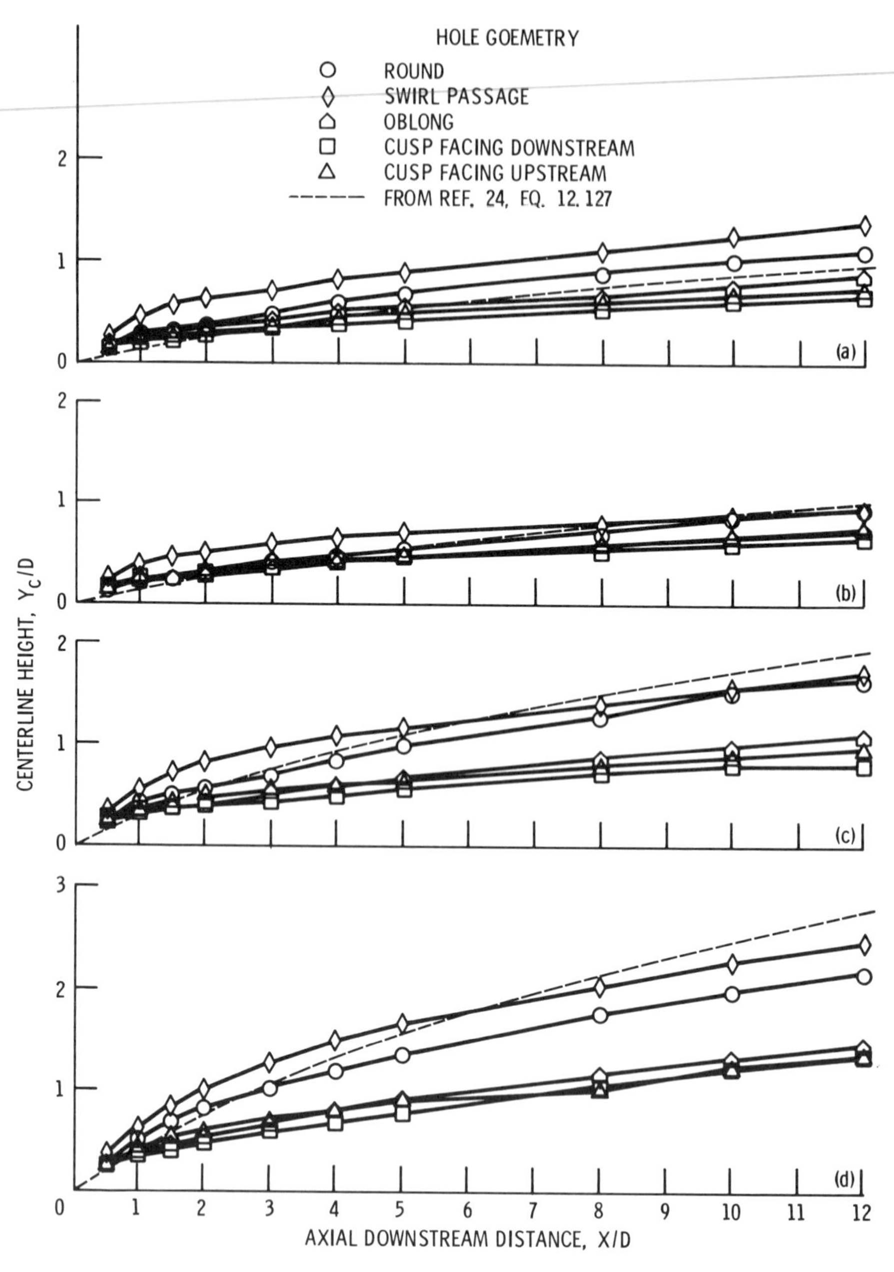

(a) M = 0.53; ρ_j/ρ_∞ = 1.02; U_∞ = 16 m/sec.
(b) M = 0.53; ρ_j/ρ_∞ = 1.02; U_∞ = 31 m/sec.
(c) M = 1.1; ρ_j/ρ_∞ = 1.03; U_∞ = 16 m/sec.

FIGURE 7. Jet centerline height as a function of axial distance.

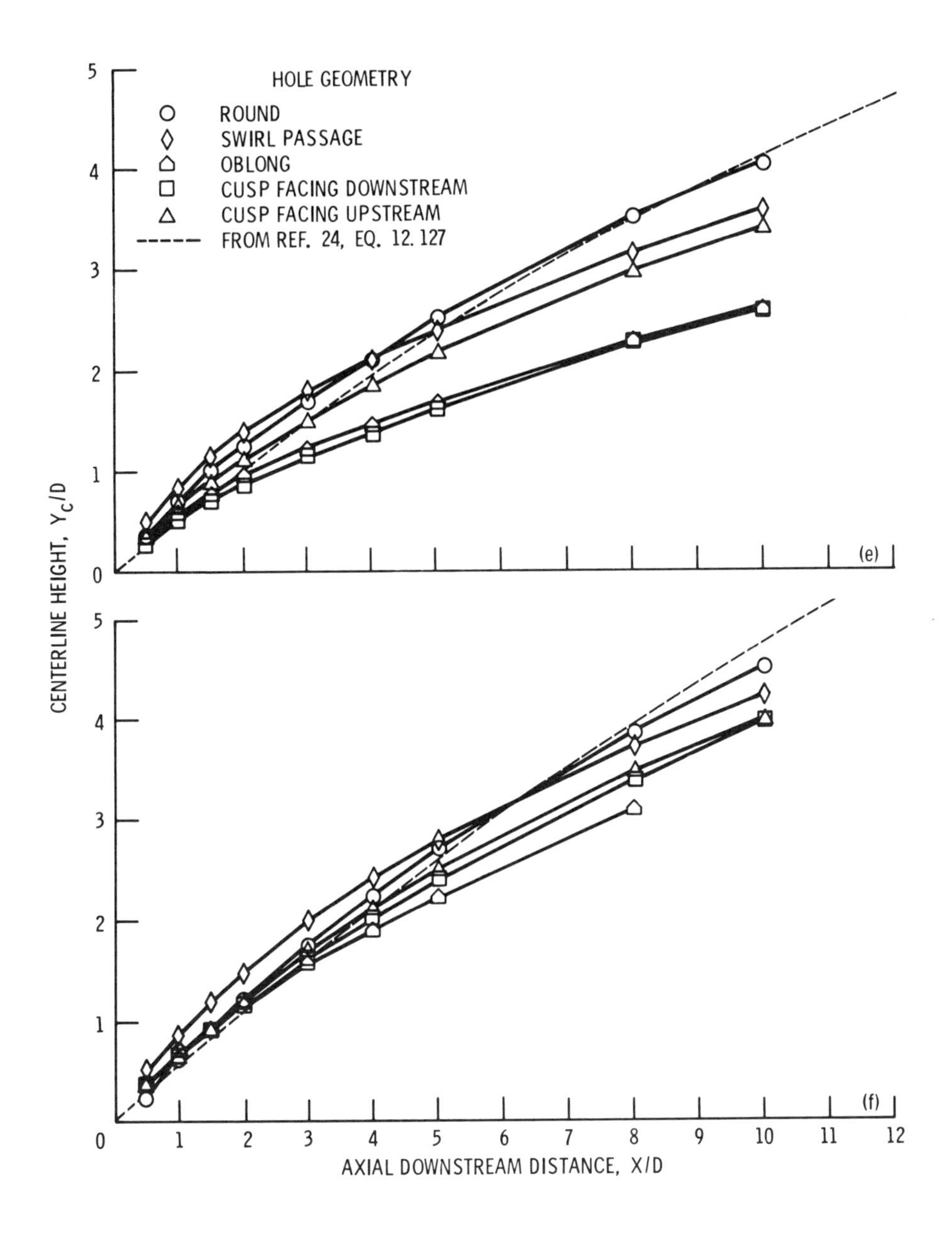

(d) M = 1.6; ρ_j/ρ_∞ = 1.04; U_∞ = 16 m/sec.
(e) M = 4.1; ρ_j/ρ_∞ = 1.03; U_∞ = 5.6 m/sec.
(f) M = 6.2; ρ_j/ρ_∞ = 1.04; U_∞ = 5.6 m/sec.

FIGURE 7. Concluded.

of pressure drag only. The difference in the experimental curves would be slightly greater if the projected area diameter (D = 1.98 cm) were used for the oblong hole. The use of a projected area diameter is consistent with the drag analysis of Ref. 14. This difference is also seen in the theoretical curves of Fig. 10. An example of how drag plays a key role in determining jet trajectory is provided by injecting jets at angles lateral to the direction of the main stream as a means of increasing film cooling efficiency by forcing the jet nearer to the surface (Ref. 10). A jet that attempts to laterally penetrate the main stream presents a much greater projected area to the main-stream flow than an aligned jet with an aspect ratio that increases with lateral angle. The greater projected area results in increased pressure drag. This pressure drag and the accompanying entrainment of main-stream fluid into the jet keeps the jet flow near the surface.

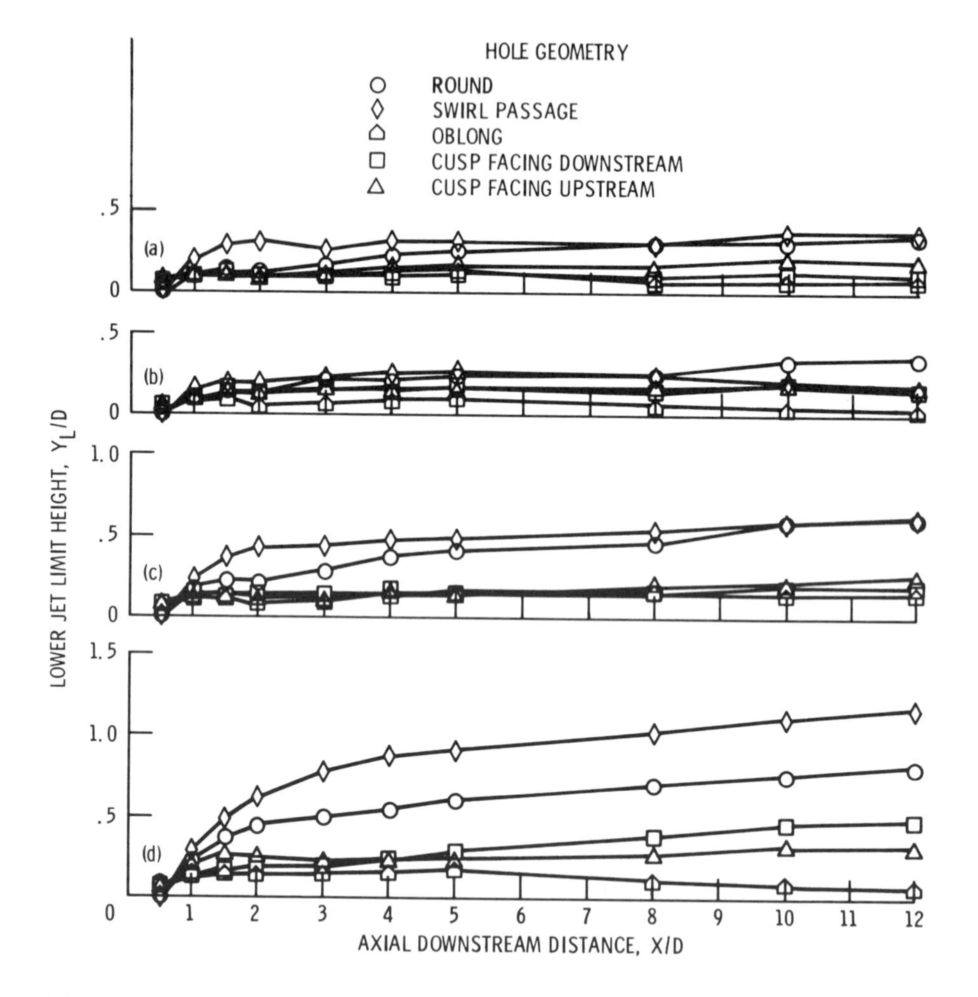

(a) M = 0.53; ρ_j/ρ_∞ = 1.02; U_∞ = 16 m/sec.
(b) M = 0.53; ρ_j/ρ_∞ = 1.02; U_∞ = 31 m/sec.
(c) M = 1.1; ρ_j/ρ_∞ = 1.03; U_∞ = 16 m/sec.

FIGURE 8. Lower jet boundary height as a function of axial distance.

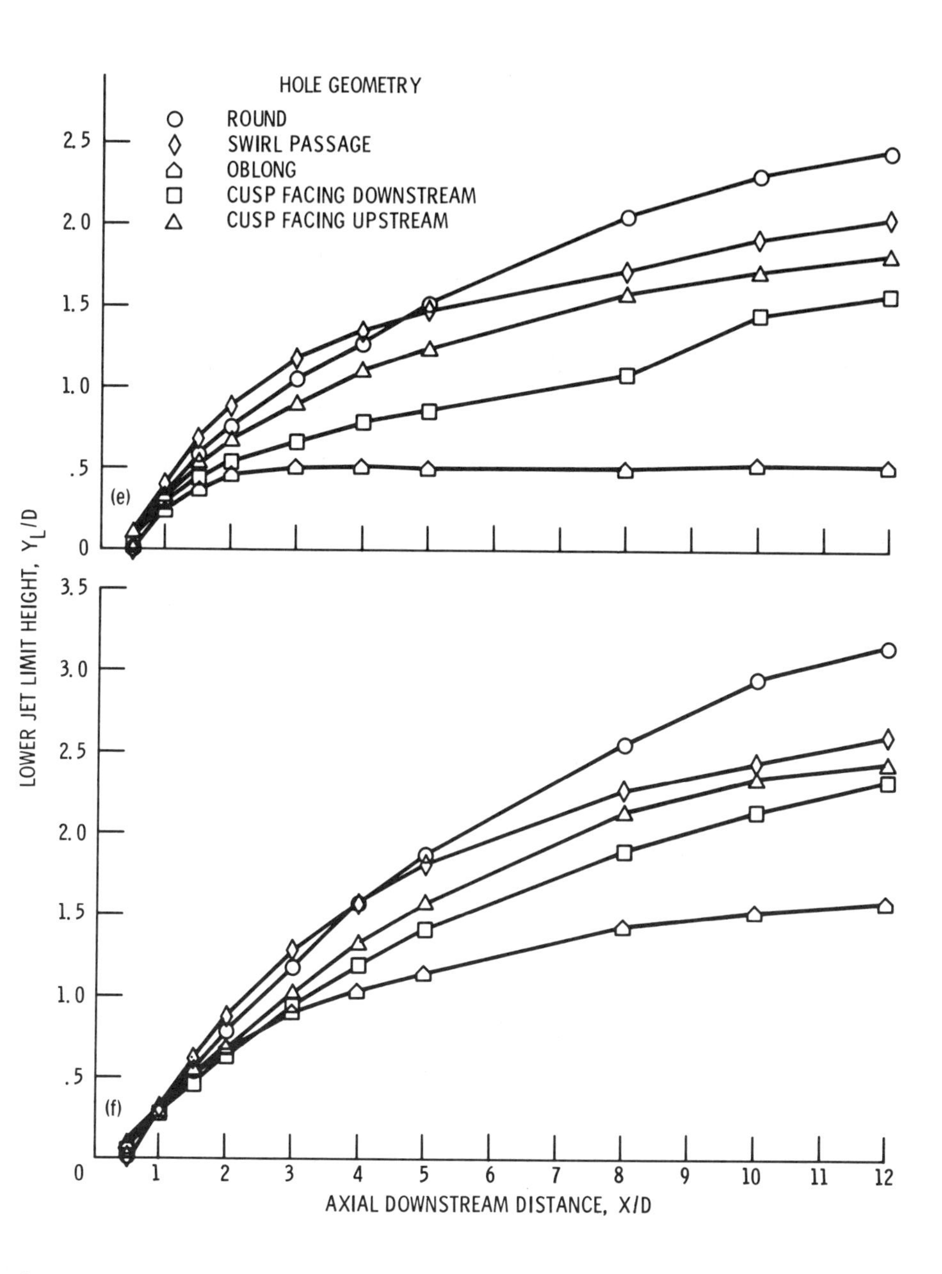

d) M = 1.6; ρ_j/ρ_∞ = 1.04; U_∞ = 16 m/sec.
e) M = 4.1; ρ_j/ρ_∞ = 1.03; U_∞ = 5.6 m/sec.
f) M = 6.2; ρ_j/ρ_∞ = 1.04; U_∞ = 5.6 m/sec.

IGURE 8. Concluded.

From a comparison of the experimental and theoretical curves of Fig. 10, it appears that entrainment begins to play a major role in determining the jet trajectory at relatively small axial downstream distances. This is consistent with the length of the potential core region (region I) being only of the order of one diameter at low values of M. It is expected that greater drag in region I will increase both the drag and the entrainment of main-stream fluid into the jet in region II. This additional entrainment of fluid should deflect the jet closer to the wall.

For all the blowing rates considered, the jet produced by the CDS configuration generally exhibited a larger deflection than that produced by the CUS configuration. This suggested that the jet surface facing the cross stream had higher pressure drag properties for the CDS configuration than for the CUS configuration. Some confirmation of this is suggested by the measurements of jet flow discharge coefficients.

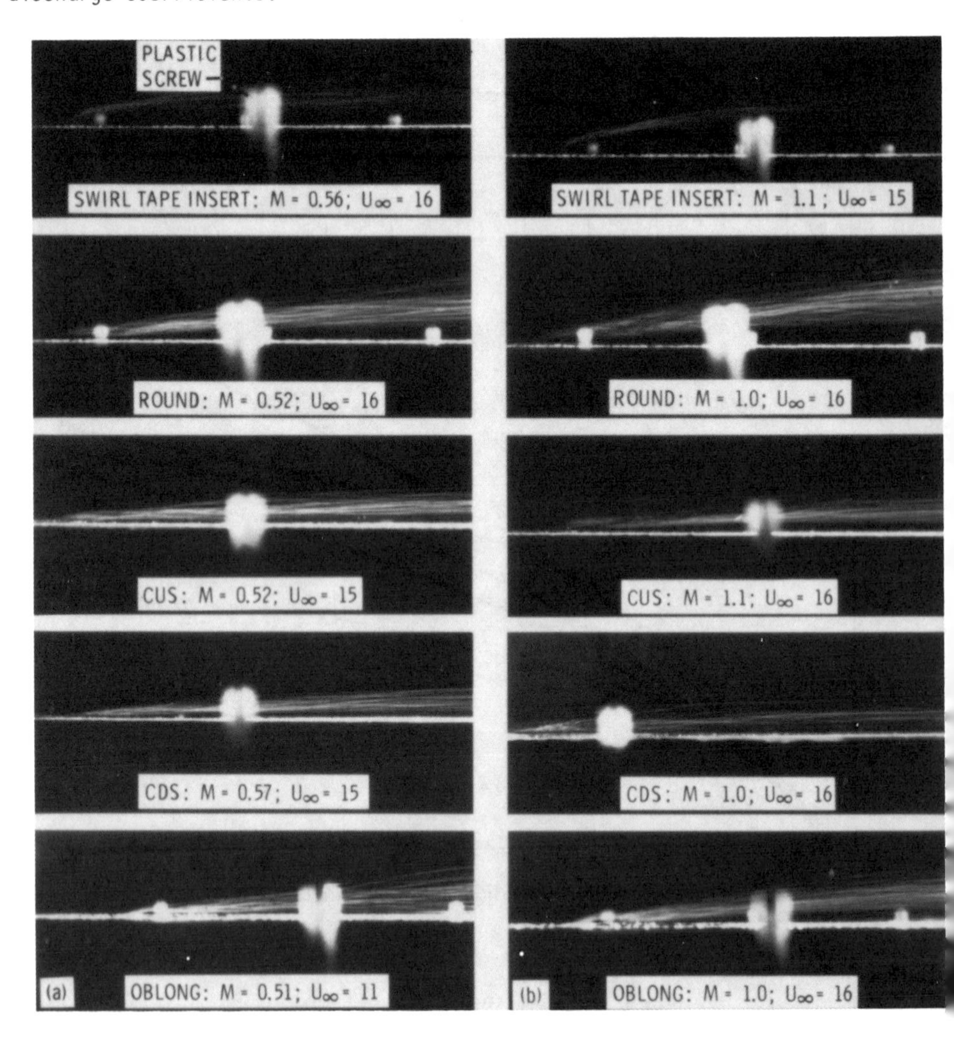

FIGURE 9. Side views of jets in crossflow for various hole configurations.

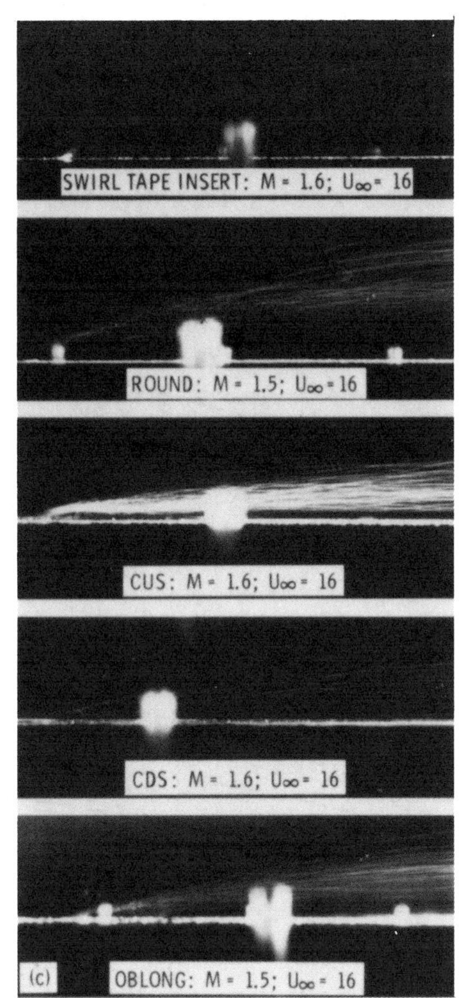

FIGURE 9. Concluded.

The jet flow discharge coefficient, which is a measure of the jet frictional losses, is defined as follows:

$$C_d = U_j \Bigg/ \sqrt{\frac{2g_c(P_p - P_\infty)}{\rho_j}} \tag{1}$$

where

$U_j = m_j/\rho_j A_j$

The discharge coefficient is plotted as a function of the jet Reynolds number based on hydraulic diameter in Fig. 11 for the round, CUS, and CDS injection hole geometries. Discharge coefficients were measured by exhausting a jet into

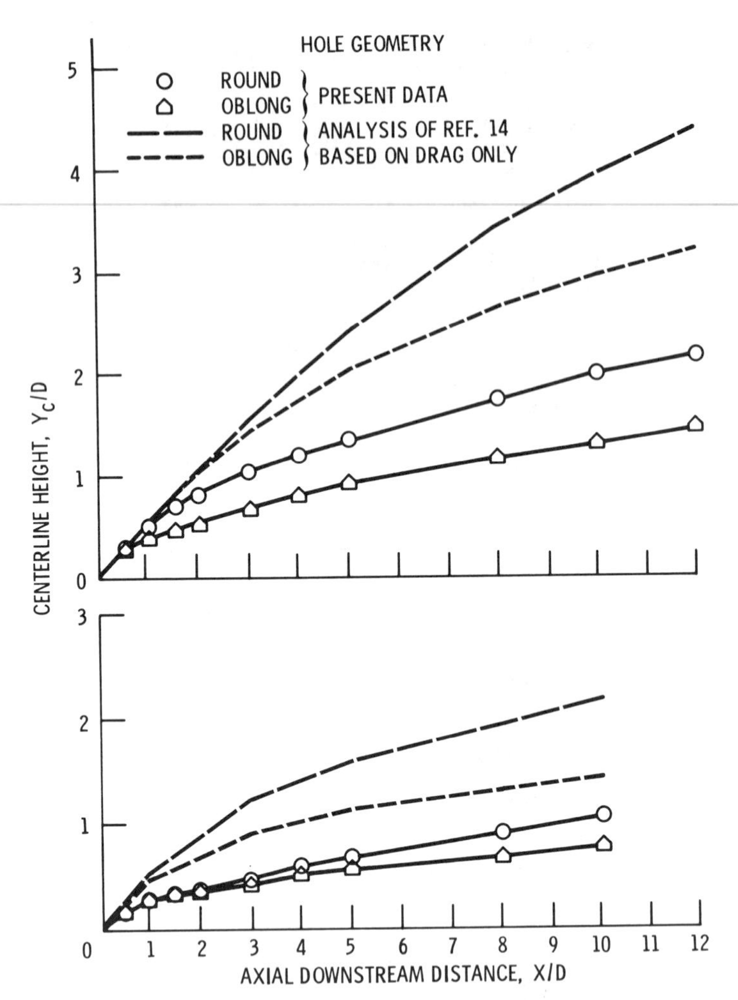

(a) M = 1.6; U_∞ = 16 m/sec. (b) M = 0.5; U_∞ = 16 m/sec.

FIGURE 10. Theoretical and experimental effect of drag coefficient on jet trajectory.

a relatively slow-moving cross stream so as to minimize the entrainment effect of free-stream axial momentum and thus leave a relative measure of frictional losses. Since the same test section was used for both CUS and CDS hole configurations, a difference in discharge coefficient could not be expected on the basis of frictional losses incurred within the jet channel. However, Eq. (1) states that the discharge pressure is the free-stream pressure; therefore the flow coefficient includes frictional losses created by the jet-crossflow interaction, which could differ between the injection hole geometries in question. It was expected from jet trajectory curves that the CDS hole would produce the highest frictional loss and the lowest discharge coefficient since it produces a higher pressure drag. The results of Fig. 11 support this assumption.

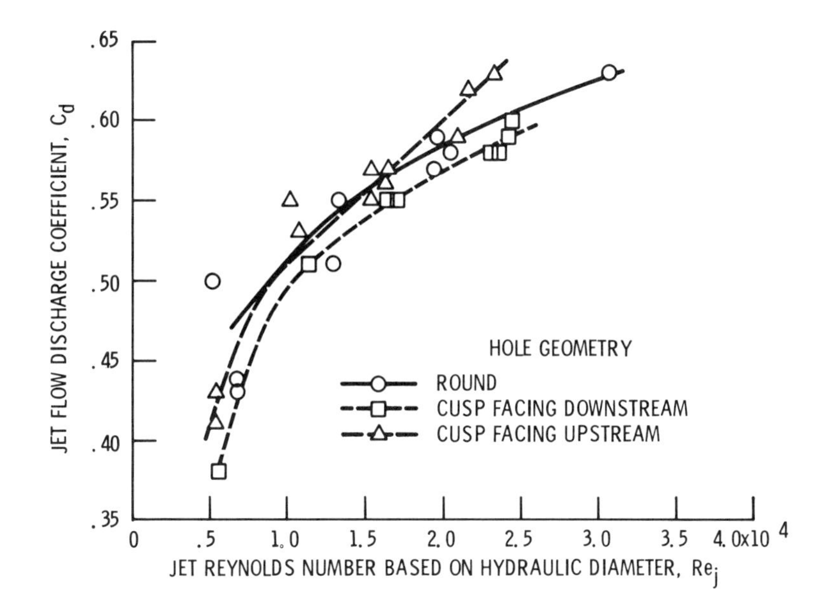

FIGURE 11. Jet flow discharge coefficient as a function of jet Reynolds number.

The aspect ratios were 2.78 for the oblong hole and 2.0 for the CUS and CDS holes. It was inferred from this difference that the aspect ratio of a jet injection hole is important in determining the jet trajectory, particularly when drag forces dominate. However, as indicated in Ref. 23, the jet velocity profile is also of importance. The differences in velocity profile produced by the CUS and oblong holes may account for the respective differences in jet trajectories. Papell (Refs. 1 and 8) postulates that the production of secondary flows in the jet before its injection into the free stream will result in greater deflection of the jet toward the wall. Although the effects of secondary motions or velocity profiles were not considered in this study, there is a need for these effects to be studied in the future.

FILM COOLING APPLICATION

Since the same facility and some of the same jet injection hole geometries (CUS, CDS, and round) were used in this study and in Papell's investigation (Ref. 1), it was of interest to compare the jet deflection trajectory data of this study with the film cooling effectiveness data of Ref. 1. Some of these data are reproduced in Fig. 12. Although the jet centerline data of Fig. 7 give some correlation with film cooling effectiveness, a better correlation is obtained by using the height of the lower jet boundary as a function of axial distance downstream X/D (Fig. 8).

Figure 13 shows that a general relationship exists between the film cooling effectiveness (from Fig. 12) and the height of the lower jet boundary Y_L (Figs. 8(a), (c), and (d)). This set of curves illustrates, as expected, that as Y_L decreased, film cooling effectiveness increased. In addition, the film cooling effectiveness near the jet exit (X/D = 1, Fig. 13(a)) is quite sensitive to the location of the jet with respect to the wall. Depending on

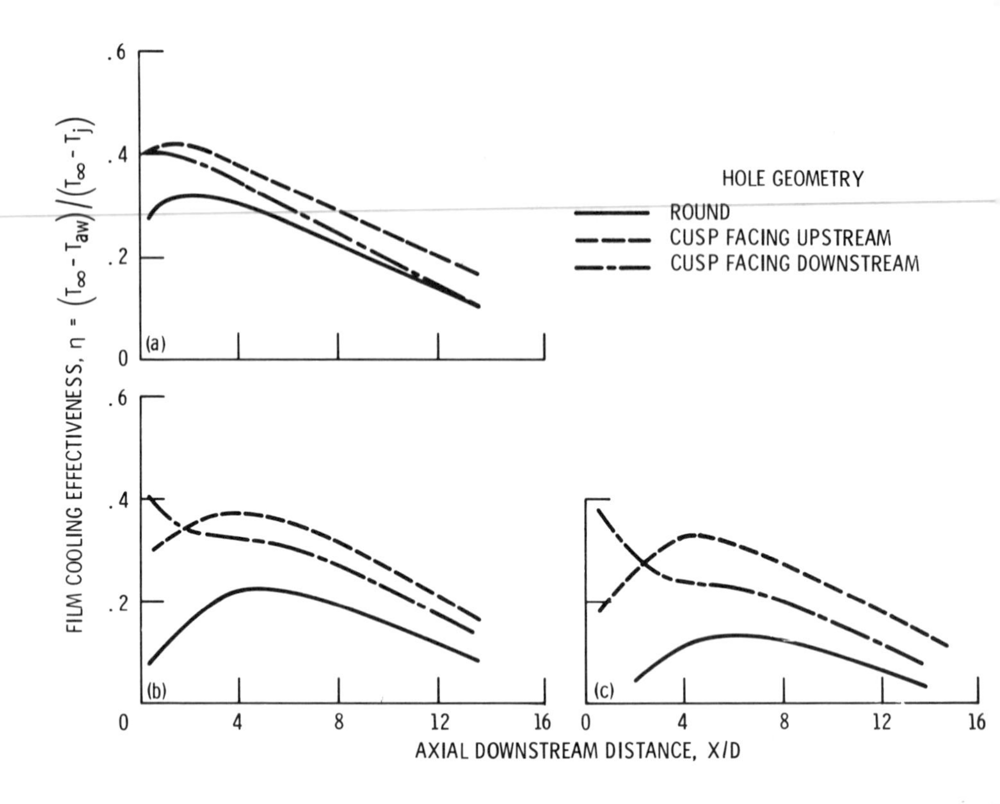

(a) M = 0.5. (b) M = 1.0. (c) M = 1.5.

FIGURE 12. Centerline film cooling effectiveness as a function of axial distance for blowing rates of 0.5, 1.0, and 1.5. Injection angle, 30°, free-stream velocity, U_∞, 15.5 m/sec. (From Ref. 1.)

the value of Y_L, the cooling of the downstream exit area is either efficient or relatively inefficient (Fig. 13(c)). Based on the results of Ref. 27, this indicates little or no recirculation of main-stream fluid about the exit location as the jet touched or was very close to the wall. In the case of separation, inferred from larger values of Y_L, the jet turned toward the surface and reattached. Reattachment represents the maximum in film cooling effectiveness (Fig. 12).

Comparing the film cooling effectiveness for M = 1.5 (Fig. 12(c)) with Y_L at a blowing rate of M = 1.6 (Fig. 8(d)) illustrates the correlation of heat transfer and jet location. At X/D = 1, the relative positions from Fig. 8(d) in decreasing order of vertical height are round, CUS, and CDS. This corresponds to the greater film cooling effectiveness shown in Fig. 12(c) for M = 1.5 and X/D = 1. Film cooling effectiveness increased in the order of round, CUS, and CDS. The order of film cooling effectiveness changed beyond X/D = 2

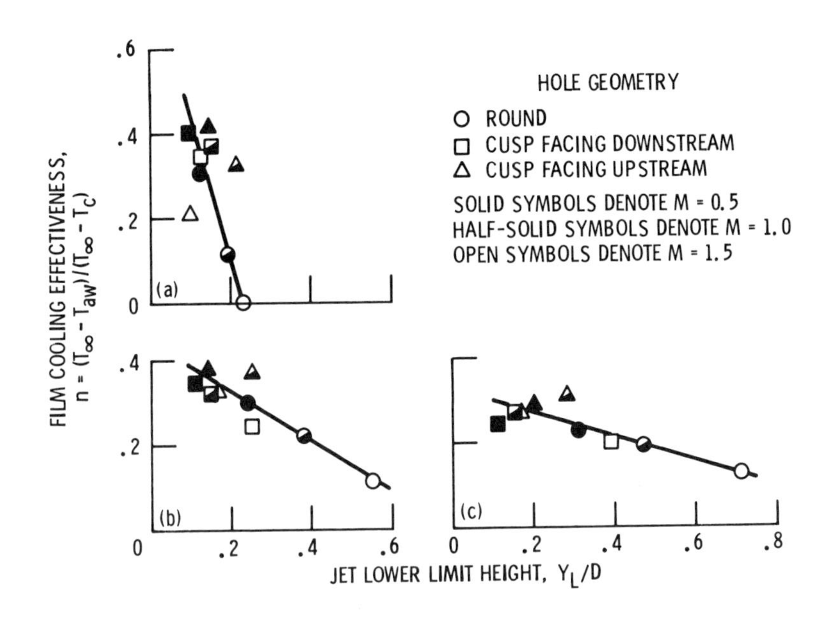

(a) X/D = 1. (b) X/D = 4. (c) X/D = 8.

FIGURE 13. Film cooling effectiveness as a function of lower jet limit height for axial distances of 1, 4, and 8.

(Fig. 12(c)), with the new order being round, CDS, and CUS. This change is reflected in the change in order of decreasing lower jet boundary Y_L beyond X/D = 4 of round, CDS, and CUS. Therefore the crossover in film cooling effectiveness (Fig. 12(c), M = 1.5) appears to be supported by the visual evidence. At lower values of M (1.0 or 0.5), the jet trajectory data grouped quite closely together (and well within the level of experimental error). Hence a correspondence to the order of film cooling effectiveness is more difficult to determine.

The visual (Fig. 9(b)) and heat transfer (Fig. 12) evidence suggests that there is initially a greater turning of the jet toward the surface for the CDS configuration than for the CUS configuration. This turning is probably due to the greater pressure drag created by the CDS hole, as mentioned in our earlier discussion on jet centerline trajectory, and is best seen for the oblong hole (Fig. 9). Drag near the jet exit has the additional effect of increasing the dual vortex motion and entrainment of the free-stream fluid into the jet (a further aid in deflecting the jet). This increased entrainment should decrease the coolant potential of the jet as noted in Fig. 12 at M of 1.0 and 1.5 for $X/D > 2$. In addition, the visual evidence of Fig. 9(b) for M = 1.0 indicates that after the jet reattaches to the wall it proceeds further and further from the surface. This effect was originally observed in the visual experiments of Ref. 26.

The lower jet boundary information (Figs. 8(a), (c), and (d)) and the visual information (Fig. 9) for the oblong hole suggests that it has a high potential for providing effective film cooling. If we use Y_L as a measure of film cooling effectiveness, the oblong hole should be as effective as both cusp holes. An additional factor to be considered is the amount of free-stream fluid entrained into the jet created by the oblong hole. Indeed, there is probably an optimum aspect ratio of such a hole since increasing the aspect ratio increases both drag and entrainment and the entrainment effectively diminishes the jet film cooling effectiveness. However, this increased jet dilution could be advantageous in combustor applications. For information on dilution jet experiments, refer to Refs. 5 and 29.

The following are recommendations for future work:

1. Experimentally determine the optimum aspect ratio for an oblong or elliptical jet injection hole that results in maximum film cooling effectiveness.

2. Perform a general analysis of a jet in crossflow that takes into consideration the effects of jet cross section, jet velocity profile, and secondary flows on entrainment and on such jet characteristics as jet deflection and spreading, both horizontally and vertically.

CONCLUSIONS

Trajectory information was obtained for jet in a crossflow for five jet injection hole geometries at ratios of jet mass flux to main-stream mass flux (the blowing rate M) of 0.53, 1.1, 1.6, 4.1, and 6.2. From this information, the following conclusions were drawn:

1. The nature and extent of pressure drag forces and the entrainment of free-stream fluid into the jet play an important role in determining the extent of jet deflection toward the injection surface.

2. Increasing the aspect ratio of the jet injection hole, with the long axis measured perpendicular to the main flow direction, increases jet deflection toward the injection surface.

3. Visual evidence confirms that film cooling effectiveness increases with increasing deflection of the jet toward the injection surface.

REFERENCES

1. Papell, S.S., "Vortex Generating Flow Passage Design for Increased Film-Cooling Effectiveness and Surface Coverage," NASA TM-83617, 1984.

2. Crabb, D., Durao, D.F.G., and Whitelaw, J.H., "A Round Jet Normal to a Crossflow," ASME Journal of Fluids Engineering, Vol. 103, No. 1, Mar. 1981, pp. 142-153.

3. Margason, R.J., "The Path of a Jet Directed at Large Angles to a Subsonic Free Stream," NASA TN D-4919, 1968.

4. Ruggeri, R.S., Callaghan, E.E., and Bowden, D.T., "Penetration of Airjets Issuing from Circular, Square and Elliptical Orifices Directed Perpendicularly to an Air Stream," NACA TN-2019, 1950.

5. Holdeman, J.D., and Srinivasan, R., "Experiments in Dilution Jet Mixing - Effects of Multiple Rows and Non-Circular Orifices," AIAA/ASME/SAE 21st Joint Propulsion Conference, 8-11 July 1985, Monterey, CA.

6. McMahon, H.M. and Mosher, D.K., "Experimental Investigation of Pressures Induced on a Flat Plate by a Jet Issuing into a Subsonic Crosswind," Analysis of a Jet in a Subsonic Crosswind, NASA SP-218, 1969, pp. 49-62.

7. Wu, J.C. and Wright, M.A., "A Blockage-Sink Representation of Jet Interference Effects for Noncircular Jet Orifices," Analysis of a Jet in a Subsonic Crosswind, NASA SP-218, 1969, pp. 85-100.

8. Papell, S.S., Wang, C.R., and Graham, R.W., "Film-Cooling Effectiveness with Developing Coolant Flow Through Straight and Curved Tubular Passages," NASA TP-2062, 1982.

9. Goldstein, R.J., and Eckert, E.R.G., "Effects of Hole Geometry and Density on Three-Dimensional Film Cooling," International Journal of Heat and Mass Transfer, Vol. 17, No. 5, May 1974, pp. 595-607.

10. Colladay, R.S., and Russell, L.M., "Streakline Flow Visualization of Discrete-Hole Film Cooling with Normal, Slanted, and Compound Angle Injection," NASA TN D-8248, 1976.

11. Braun, G.W., and McAllister, J.D., "Cross Wind Effects on Trajectory and Cross Sections of Turbulent Jets," Analysis of a Jet in a Subsonic Crosswind, NASA SP-218, 1969, pp. 141-164.

12. Fearn, R,L., "Mass Entrainment of a Circular Jet in a Crossflow." Analysis of a Jet in a Subsonic Crosswind, NASA SP-218, 1969, pp. 239-248.

13. McAllister, J.D., "A Momentum Theory for the Effects of Cross Flow on Incompressible Turbulent Jets," Ph.D. Thesis, Univ. of Tennessee, Aug. 1968.

14. Hanus, G.J., and L'Ecuyer, M.R., "Turbine Vane Gas Film Cooling with Injection in the Leading Edge Region from a Single Row of Spanwise Angled Holes," TSPC TR-76-1, Purdue Univ., West Lafayette, IN, Apr. 1976, (NASA CR-147160).

15. Kamotani, Y., and Greber, I., "Experiment on a Turbulent Jet in Cross Flow," FTAS/TR-71-62, Case Western Reserve Univ., Cleveland, OH, June 1971, (NASA CR-72893).

16. Wooler, P.T., Burghart, G.H., and Gallagher, J.T., "Pressure Distribution on a Rectangular Wing with a Jet Exhausting Normally into an Airstream," Journal of Aircraft, Vol. 4, No. 6, Nov.-Dec. 1967, pp. 537-543.

17. Sucec, J., and Bowley, W.W., "Prediction of the Trajectory of a Turbulent Jet Injected into a Crossflowing Stream," ASME Journal of Fluids Engineering, Vol. 98, No. 4, Dec. 1976, pp. 667-673.

18. Adler, D., and Baron, A., "Prediction of a Three-Dimensional Circular Turbulent Jet in Crossflow," AIAA Journal, Vol. 17, No. 2, Feb. 1979, pp. 168-174.

19. Stoy, R.L., and Ben-Haim, Y., "Turbulent Jets in a Confined Crossflow," ASME Journal of Fluids Engineering, Vol. 95, No. 4, Dec. 1973, pp. 551-556.

20. Wang, C.R., Papell, S.S., and Graham, R.W., "Analysis for Predicting Adiabatic Wall Temperatures with Single Hole Coolant Injection into a Low Speed Crossflow," NASA TM-81620, 1981.

21. Kamotani, Y., and Greber, I., "Experiments on a Turbulent Jet in a Cross flow," AIAA Journal, Vol. 10, No. 11, Nov. 1972, pp. 1425-1429.

22. Platten, J.L., and Keffer, J.F., "Deflected Turbulent Jet Flows," ASME Journal of Applied Mechanics, Vol. 38, No. 4, Dec. 1971, pp. 756-758.

23. Gibeling, H.J., et al., "Computation of Discrete Slanted Hole Film Cooling Flow Using the Navier-Stokes Equations," R83-910002-F, Scientific Research Associates, Inc., Glastonbury, CT, Sept. 1983, (AFOSR-83-1288TR, AD-A137022).

24. Abramovich, G.N., The Theory of Turbulent Jets. MIT Press, Cambridge, MA, 1963.

25. Pratte, B.D., and Keffer, J.F., "Swirling Turbulent Jet Flows - Part 1: The Single Swirling Jet," UTME-TP-6901, Univ. of Toronto, Toronto, Canada, 1969.

26. Hoerner, S.F., Fluid-Dynamic Drag, Midland Park, NJ, 1965.

27. Bergeles, G., Gosman, A.D., and Launder, B.E., "Near-Field Character of a Jet Discharged Through a Wall at 30° to a Mainstream," AIAA Journal, Vol. 15, No. 4, Apr.1977, pp. 499-504.

28. Colladay, R.S., Russell, L.M., and Lane, J.M., "Streakline Flow Visualization of Discrete Hole Film Cooling with Holes Inclined 30° to Surface," NASA TN D-8175, 1976.

29. Holdeman, J.D., Srinivasan, R., and Berenfeld, A., "Experiments in Dilution Jet Mixing," AIAA Journal, Vol. 22, No. 10, Oct. 1984, pp. 1436-1443.

Numerical Visualization of the Elliptic Nature of Laminar Flow in Square Bend Ducts

K. C. CHENG
Department of Mechanical Engineering, Faculty of Engineering
The University of Alberta
Edmonton, Alberta, Canada T6G 2G8

MITSUNOBU AKIYAMA, YUICHI ENDO, MICHIYOSHI SUZUKI, and ICHIRO NISHIWAKI
Department of Mechanical Engineering, Faculty of Engineering
Utsunomiya University
Utsunomiya 321, Japan

ABSTRACT

This paper dealts with a numerical study on developing laminar flow in curved square channels. It was solved by using a finite-difference approximation adopted for the system of full N-S equations written in the form of toroidal and Cartician coordinates. The flow intensity is of 200 and 500 in Reynolds number and the channel consists of 90° and 180° bends with curvature ratio of four connected with an inlet and an outlet straight ducts. Typical flow field solutions in steady conditions are shown in terms of main velocity contour, secondary flow vector, pressure value, shearing stresses and some other physical values related to centrifugal force effects. Stress is paid to the graphical expressions of three-dimensional elliptic nature of the phenomena. In the case of Reynolds number 500,for example,the outer side of the curved channel wall located in the downstream will make a inward secondary flow on the upcoming stream in the straight section up to 2.0 hydraulic diameter, de. In the range of 0.4de to 2.0de from the curved channel inlet, it is found to have flow separation in the direction to the main flow at the two corners of the outer side wall. The secondary flow becomes maximum in intensity at 2.8de having 40% of the bulk flow velocity, and begins to form a Dean's type vorticity pair in addition to the original vortex pair at 4.9de(70°). The computed friction factor for a fully developed region was found to agree very well with the values of a previous solution of a parabolic approximation in the main flow direction.

INTRODUCTION

Numerical solutions for laminar flow in a square ducts with a bend section are obtained by using a finite difference approximation applied for the time-dependnet three dimensional N-S equation. In the previous study (1,2), our problem was limited to the smoothly curved duct flow with no flow separation, the calculations were performed on a parabolic type partial differential equations in the main flow direction and only elliptic in the secondary flow direction. In reality, depending on the magnitudes of curvature and

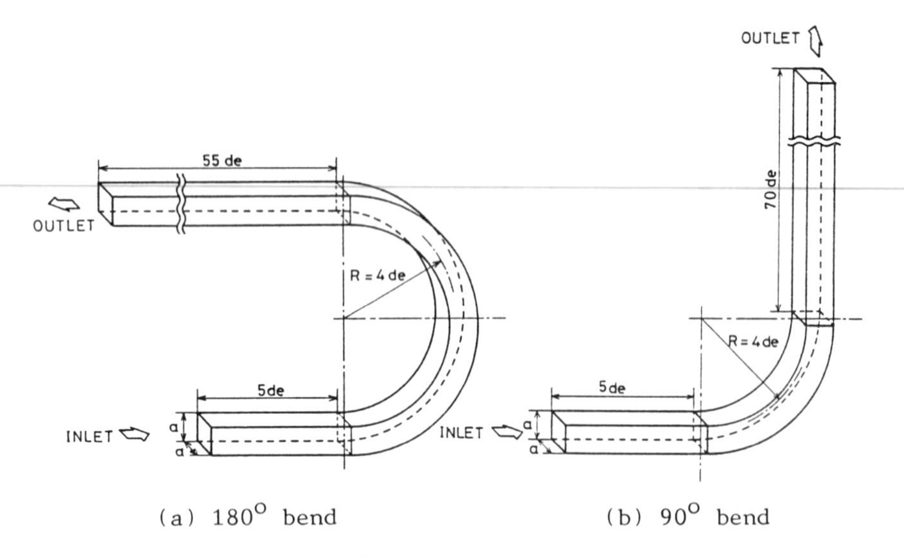

Fig.1 Test section

flow, there must be a chance of forming a flow separation, reverse flow and time dependent swaying in addition to two types of secondary flows observed in (1,2). Therefore, it is natural to think about the flow situation where the downstream condition affects on to the upcoming flow . It is where parabolic assumption in the main flow direction may not be suitable and the effect of elliptic nature should be examined (3,4). Based on the above viewpoint, a detailed numerical calculation was carried out focusing on a downstream effect of the curved section on an upcoming flow of straight duct section and a possible flow separation in curved section. The N-S equation retains diffusion terms in three directions and naturally parabolic in time wise direction. The channels used are 90° and 180° bends with connected smoothly with straight inlet and outlet ducts.

CALCULATION MODEL

The geometry of the ducts used in this analysis are shown in Fig.1. The symbol de is the hydraulic diameter defined as $de=a\times a/(4a)/4$. Reynolds number, Re is given by $w\times de/\nu$ where w is mean value of the main flow velocity and ν is kinematic viscosity. The values of Reynolds number used are 200 and 500 which correspond to the values of Dean number to be 100 and 250, respectively. Dean number, De is made up with $Re\times(de/R)^{1/2}$. The coordinate system used is of the toroidal type with a geometrical length R^* to be $1+x/R$ where x is the distance measured from the center plane of the curvature R along with R-direction. When R approaches infinity, the term $1+x/R$ becomes 1, and the system will be Cartesian. A fully developed velocity profile of a Dirichlet condition at the duct inlet and the outlet was used. A Neumann condition is applied for a pressure field. Only a half domain divided by a central symmetric plane having the radius of curvature is treated. Nodal points are 20X10 in the duct section and 200 in the main flow direction. Hybrid modification with donor cell approximation is used for special directions and solved by the A.D.I. method. In the time-wise direction, an implicit scheme is applied.

RESULTS AND DISCUSSION

The results for Re=500 in terms of main velocity contour, secondary flow vector and isobaric contours are shown in Fig.2 for six typical stages of the development. Fig.2(a) indicates that the concave wall of the curved duct in downstream makes inward secondary flow in the up-stream straight section. The straight duct itself dose not have inner and outer walls, thus the word "inward" is named related to the straight wall section which is smoothly connected to the inner wall of the curved duct. The intensity of this secondary flow becomes about 2.8% of the main flow velocity. Therefore, the main velocity contour shifts towards the " inner " side wall and the pressure exhibits negative gradient towards this wall. At $\theta=30^{\circ}$ in Fig.2(c), the maximum intensity of the secondary flow becomes about 45% of the mean main flow velocity. A Dean's type secondary flow or an additional vortex pair appears in Fig.2(d) and $\theta=135^{\circ}$ of Fig.2(e) shows a fully developed stage in a curved duct.The decaying process of the secondary flow in straight section can be seen in Fig.2(f).

The main flow separation is observed at the two corners of the outer side wall. Fig.3 indicates such a separation at the A-A' section near the side wall. The distance from the side wall to the A-A' line is de/40. The separation region ranges from $\theta=8^{\circ}$ to 18° in the main flow direction. The maximum intensity of separation observed is about 2%.

It is noted that results for the 90° bend duct does not allow enough for full development in the bend.

For the flow with Re=200 or De=100, we could not obtaine Dean's type vortex pair. This fact is reasonable, since the critical Dean's number Dec for the existance Dean's type additional vortex is found to be 110 (1).

Development of pressure drop is shown in Fig.4. $\bar{P}$ is representing averaged, Po for outer and Pi is for inner.

The shearing stress can be calculated for main flow τ_{ww}, for secondary flow τ_{ws} and for total flow τ_w. Fig.5 shows the results. It can be noted that there is some difference between an usual friction factor τ_{ww} and a total friction τ_w. Also noted is a slight difference in characteristics between development of pressure loss and the friction loss indicating energy accumulation process into the third term of secondary flow energy. For

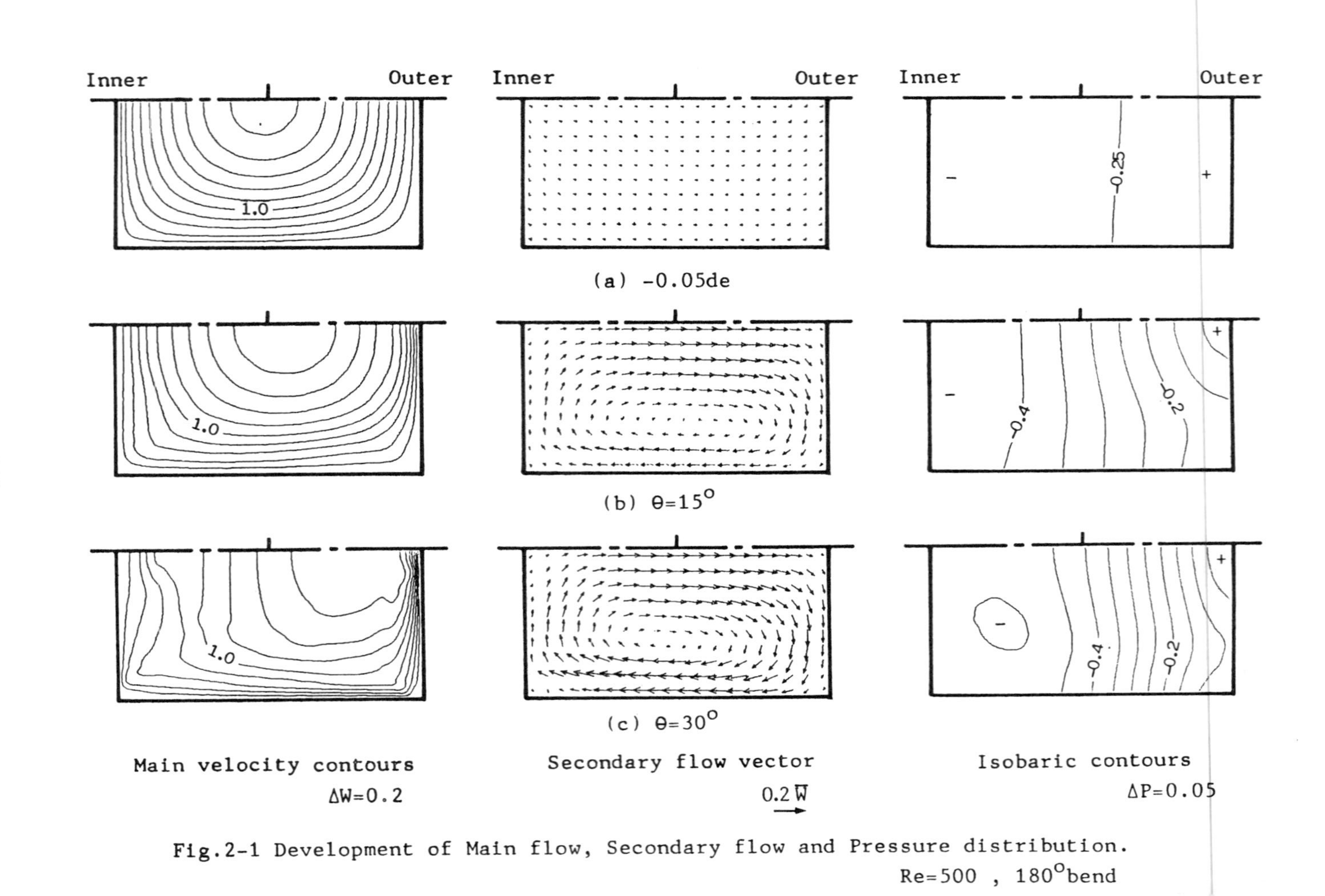

Fig.2-1 Development of Main flow, Secondary flow and Pressure distribution.
Re=500 , 180°bend

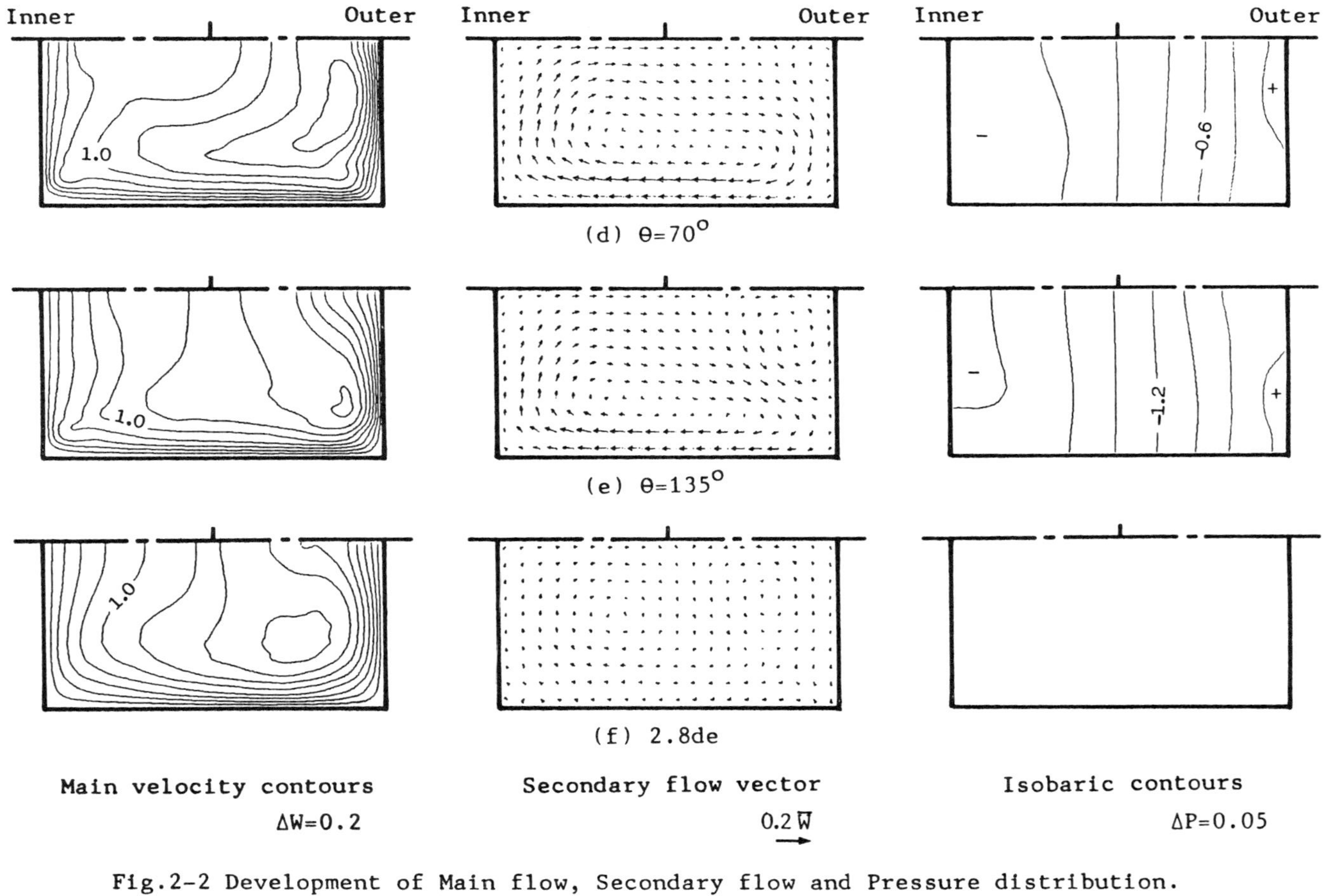

Fig.2-2 Development of Main flow, Secondary flow and Pressure distribution. Re=500 , 180°bend

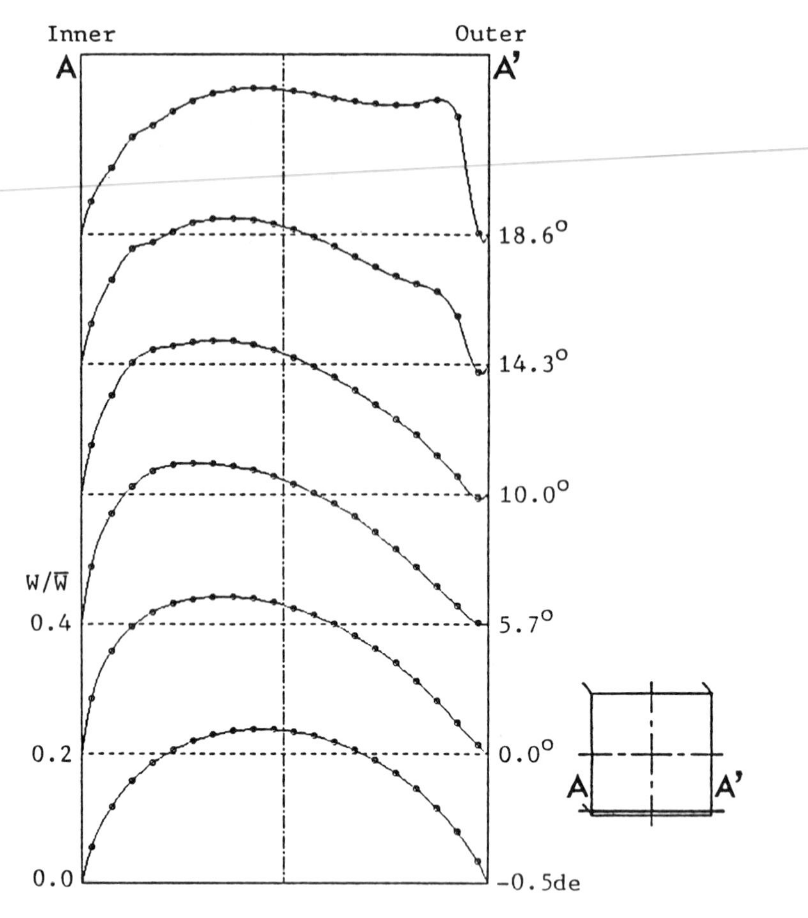

Fig.3 Development of main velocity distribution in the recirculating region. Re=500, Rc/de=4, 180°bend

reference, an averaged secondary-velocity intensity is shown in Fig.6. These results will serve to understand the force balance among the individual terms in N-S equation.

Fig.7 indicates superiority of the hybrid difference over the upwind difference. This scheme contains central difference aproximation for the small velocity region. The results by the hybrid difference agree well with the experimental results (1).

References (5) and (6) are interesting to note for comparison.

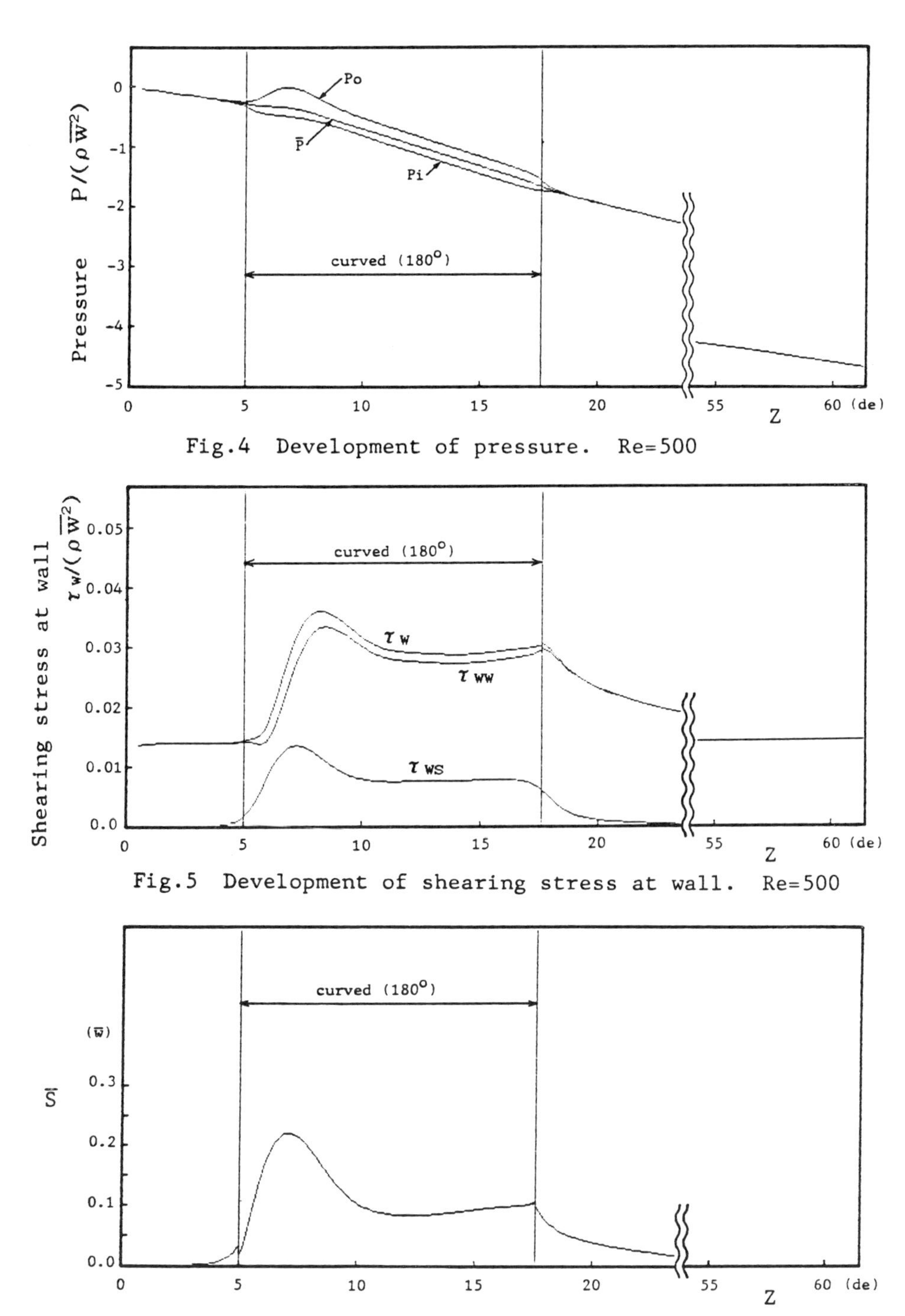

Fig.4 Development of pressure. Re=500

Fig.5 Development of shearing stress at wall. Re=500

Fig.6 Development of averaged secondary-velocity component. Re=500

CONCLUSION

The elliptic nature of the laminar flow field in the main flow direction for square bend ducts is quantitatively explained for moderate Reynolds number.

We acknowledge the support of Mr.I.Urai and Mr.T.Serizawa in preparing the present manuscript.

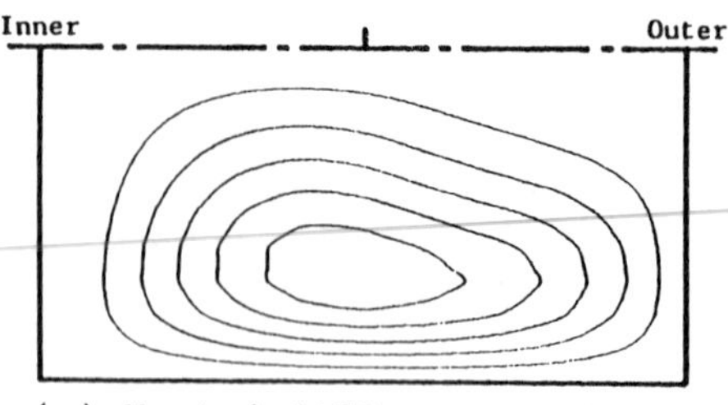

(a) Upwind difference scheme

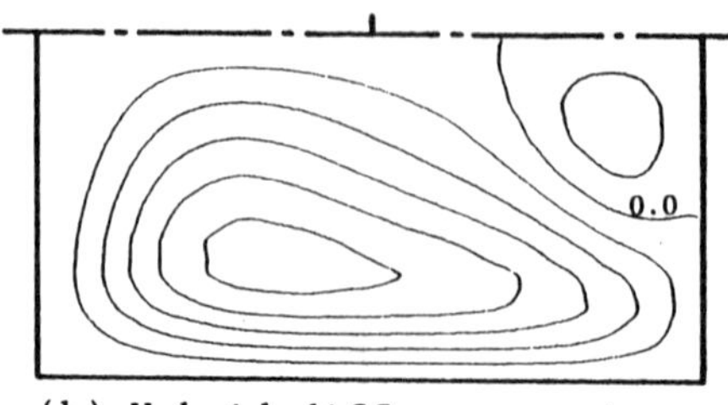

(b) Hybrid difference scheme

Fig.7 Comparison of difference schemes (Stream function) Re=500, 180°bend, $\theta=135^{\circ}$

REFERENCE

1. M.Akiyama, K.Kikuchi, M.Suzuki, I.Nishiwaki, K.C.Cheng, and J.Nakayama, "Numerical Analysis and Flow Visualization on the Hydrodynamis Entrance Region of Laminar Flow in curved Square Channels," Transactions of the Japan Society of Mechanical Engineers, Vol.47,No.422, Oct.1981,pp.1960-1970.

2. M.Akiyama, K.Kikuchi, J.Nakayama, M.Suzuki, I.Nishiwaki, and K.C.Cheng, "Two stage Developments of Entry Flow with an Ineraction of Boundary-wall and Dean's-Instability Type Secondary Flows," Transactions of the Japan Society of Mechanical Engineers, Vol.47, No.421, Sept.1981,pp.1705-1714.

3. M.Akiyama, Y.Endo, M.Suzuki and I.Nishiwaki, "Numerical Experiment on the Elliptic Nature of Laminar Flow in Curved Square Channels," Flow Visualization, Vol.4,Supp.,1984,pp.85-88

4. M.Akiyama, Y.Endo, M.Suzuki and I.Nishiwaki, "Numerical Experiment on the Elliptic Nature of Developing Flow in 180 Bend Square Duct," Flow Visualization, Vol.4,No.14, 1984,pp.327-330

5. J.A.C.Humphrey, A.M.K.Taylor and J.H.Whitelaw, "Laminar flow in a square duct of strong curvature," J.Fluid Mech., Vol.83, part3, 1977 ,pp.509-527

6. A.M.K.P.Taylor, J.H.Whitelaw and M.Yianneskis, "Curved Ducts With Strong Secondary Motion: Velocity Measuewments of Developng Laminar and Turbulent Flow," ASME Journal of Fluids Engineering, Vol.104, Sept.1982,pp.350-359

Compressible Flow in a Diffusing S-Duct with Flow Separation

A. D. VAKILI, J. M. WU, M. K. BHAT, and P. LIVER
The University of Tennessee Space Institute
Tullahoma, Tennessee 37388

SUMMARY

Accurate measurements have been made of secondary flows in a diffusing 30°-30° S-Duct with circular cross section. Turbulent flow was entering the duct at Mach number of 0.6, the boundary layer thickness at the duct entrance was ten percent of the duct inlet diameter. Flow parameters were measured at six streamwise stations along the length of duct. These measurements were made using a five port cone probe. Local flow velocity vector as well as static-and total pressures along ten radial traverses at six stations were obtained. Strong secondary flow was measured in the first bend which continued into the second bend with new vorticity produced in there in the opposite direction. Surface oil flow visualization indicated a region of separated flow starting at $\theta \simeq 22^o$ on the inside of the first bend and ending at $\theta \simeq 44^o$ on the outside of the second bend. Wall static pressure measurements were made along three azimuth angles of $10^o, 90^o$ and 170° along the duct. Contour plots presenting the transverse velocity field as well as total and static pressure contours have been obtained. As a result of the secondary flow and the separation, significant total pressure distortion was observed at the exit of the duct.

I. INTRODUCTION

Performance of engines and compressors depends significantly on the entering flow characteristics. Flow unsteadiness and non-uniformities at the inlet of such systems not only reduce the overall efficiency, but also could result in flutter and stall. Due to many design criteria at the upstream of these systems, diffusers are made short in length. Also, in many cases where misalignment of centerlines is required, short S-shaped diffusers are employed. In such cases the presence of secondary flow could result in flow separation and significant flow distortion. Design of S-shaped diffusers is complex and needs careful attention. Computer codes are available to predict the flow fields in such ducts. However, predictions are not accurate where separation may be present.

Secondary flows formed in these ducts are due to the pressure gradients generated by the streamline curvature. The fluid near the centerline having a higher velocity is acted upon by larger centrifugal forces than the slower fluid near the walls. Therefore, the faster moving fluid moves outwards, pushing the fluid in the boundary at the outer wall around the side toward the inner wall. Fresh fluid is continually being brought into the neighborhood of the outer wall and then forced round toward the inner wall.

In straight duct flow, there is no static pressure gradient in the transverse plane. In transition from a straight duct to a curved duct flow, at the outer wall the static pressure increases and at the inner wall the static pressure decreases. The increase of pressure at the outer wall results in the thickening of the boundary layer. The direction of the pressure gradient along the duct is reversed and in combination with the centrifugal forces present the overall flow becomes three dimensional and complex.

In bending diffusers, the longitudinal adverse pressure gradient due to the diffusion works against the low kinetic energy secondary flow and could enhance the possibility for flow separation. Flow separation adds to the level of total pressure distortion and flow non-uniformity at the duct exit. When such a duct is used as the inlet to a compressor, the exit flow must ideally be fairly uniform and steady.

Early studies of secondary flow were mainly concerned with the losses associated in bending pipes. Other studies have been made to provide detail description of the flow in curved pipes. More recent investigations have concentrated on in-depth studies of flows in bending pipes for better understanding of the physics of the secondary flows and collecting data necessary for computational comparison and verification (Ref. 1–5). Most of these studies, however, have been performed at low speeds, low Reynolds numbers and in incompressible flow.

The present study is part of a systematic benchmark experimental effort to provide an accurate set of measurements of secondary flows in S-shaped ducts. In the first phase of this work (Ref. 5) detail measurements were made in a 30°-30° S-duct with constant area circular cross section. These measurements were performed for inlet Mach number of 0.6, therefore the compressibility effects are present in the data. This study is the second phase of stated efforts, which includes measurements of secondary flows in a $30^o - 30^o$ S-duct with variable area with similar flow parameters.

The measurements were made using a five port cone probe to obtain mean value flow parameters relevant to the development of secondary flow. The advantages associated with cone probe measurement in this case are discussed in Reference 5. This probe provides accurate measurement of flow total and static pressure which is used to obtain the three dimensional velocity field. The cone probe was traversed in radial directions at six stations along the duct. Total and static pressure contours have been plotted presenting the pressure data at each station. The secondary flow velocity vector plots are discussed. In addition all of the data are tabulated in Reference 5.

Surface oil flow visualization revealed a relatively large region of separated flow in the duct. This was absent in the constant area $30^o - 30^o$ S-duct studied previously. The oil flow technique was used to provide details of the flow on the duct boundaries.

II. EXPERIMENTAL APPROACH

Duct Set Up

The circular cross section variable area S-duct shown in Figure 1 was made from two symmetric sections. Each was manufactured from fiberglass, using a mold made from wood. Extreme care was taken to ensure symmetry and good surface finish quality. The fabrication

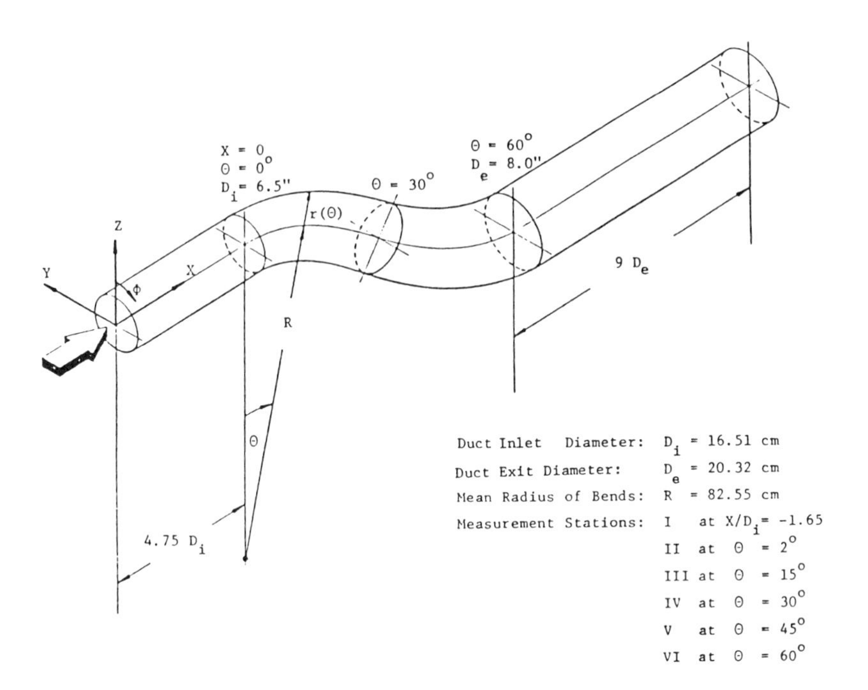

Figure 1. Geometry and Coordinates of the $30^o - 30^o$ Diffusing S-Duct.

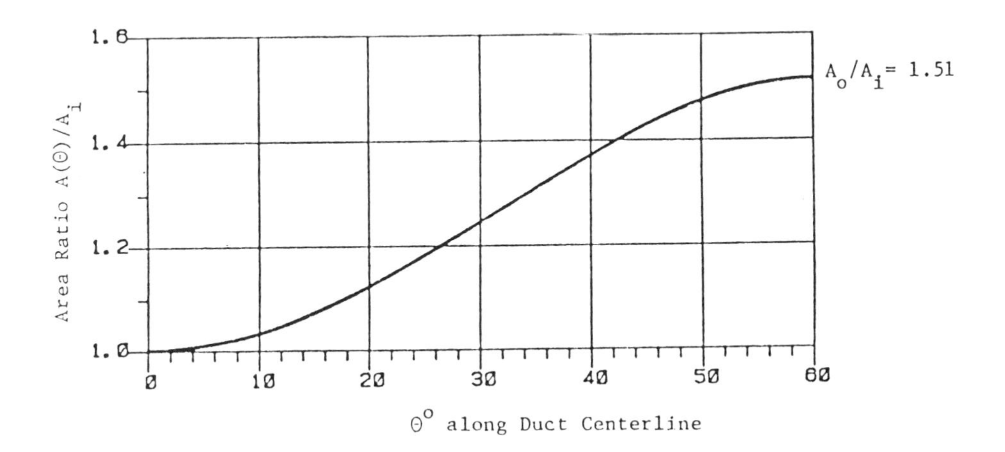

Figure 2. Diffusing Duct Area Variation with Bending Angle.

tolerances were $\pm.16$ cm ($\pm1/16$") maximum. The two symmetric sections were bolted together and the junctions were filled and smoothed to a circle each time the duct was opened for visual observation. Figure 1 shows the duct geometry with the notations and coordinates. The duct centerline was defined by $30^o - 30^o$ circular arcs, the local radius as a function of θ is described by:

$$r = -1.3061871\,\theta^3 + 2.051754\,\theta^2 + 3.25$$

where $\quad 0 \leq \theta\ [\text{radian}] \leq \pi/3$

The duct area ratio was $(A_{exit}/A_{inlet}) = 1.51$, as shown in Figure 2. At the upstream a section of constant area duct, 78. cm long (4.75 D_{inlet}), was installed to allow for the development of the turbulent boundary layer to the desired thickness. At the exit, the duct was attached to a long pipe, 183 cm (nearly 9 D_{exit}), to minimize any end effect.

Instrumentation

All the pressure transducers were calibrated before the measurements and the pressure measurements are accurate to $\pm1\%$ of reading. The angles were measured with respect to the horizontal and the vertical to a precision of one second, however, deflections of up to $\pm0.2^o$ under load could be present. The cone probe manufacturing misalignments were identified in the straight section and a uniform correction was applied to all of the cone probe data. The total pressure in the tunnel varied nearly ±0.5 psi during the runs. Since this does not significantly influence the flow angularity no corrections have been applied for this.

All the measurements were made for one nominal free stream Mach number of 0.60 at the reference measurement station in the straight section, $x = -27.3$ cm. The flow parameters at this point were used as reference conditions for non-dimentionalizing downstream flow data. The unit Reynolds number at this station was 3.25×10^6 per ft. The boundary layer thickness was 0.64 cm, and the Dean number, $De = (\frac{D_i/2}{R})^{1/2} Re$, was in the order of one million.

Measurement Technique

In order to measure local flow angularity a five-port cone probe was used. The five orifices on the cone probe include one total pressure and four static pressures at 90° azimuthally apart on the cone surface. The flow static pressure was obtained from the calibration data while the flow angularity was calculated using theoretical relationships (Ref. 6). This approach was chosen, since the determination of static pressure is less sensitive to errors in angular orientation during the calibration and also that the differential pressures between symmetrically placed orifices on the cone are used to determine the flow angularity.

The five port cone probe allowed the determination of the local flow conditions, i.e., Mach number, static- and total-pressures and the flow angularity both in pitch and yaw. A suitable model for the velocity-temperature relation using boundary layer concepts (Ref. 5) with the aid of cone theory (Ref. 6) then allowed the determination of the complete local velocity vector. The limitations in the use of cone probes in flows with gradients in the direction normal to the flow have been discussed in Ref. 7. Details of the cone probe data reduction is discussed in Reference 5.

The cone probe was traversed along radial directions. Ten azimuth angles, approximately 20° apart, were traversed at each station. The traversing mechanism uses a computer controlled stepper motor, preprogrammed to obtain more data near the boundaries than in the center region. At least seventy data points were measured along each traverse. Wall static pressures were also measured along the duct on three azimuth angles of $10^o, 90^o$ and 170°.

III. FLOW VISUALIZATION

The duct was symmetric with respect to the vertical plane passing through its centerline. This symmetry was verified through surface oil flow visualization and cone probe measurements. Oil was injected through small holes upstream of regions of interest which was carried along by the local flow during the run. A trace of the oil remained on the surface which was visible after the flow was stopped. The pattern shown in the photographs of Figure 3 shows typical surface flow visualization obtained. Figure 4 shows the separation details in one half of the duct. A similar pattern exists in the other half.

This flow visualization was used to identify and locate regions of possible flow separation within the duct. As shown on these photographs, a region of strong flow separation was present in the duct. The surface flow pattern clearly defines the separation boundaries. Separation starts at $\theta \simeq 22^o$ and the surface flow pattern shows the dividing streamline at $\theta \simeq 44^o$.

IV. DISCUSSIONS ON THE MEASUREMENTS

Flow entering the duct is exposed to adverse pressure gradients at the outer wall in the first bend. This pressure is due to the centrifugal forces exerted on the flow. At the outer wall the adverse pressure thickens the boundary layer, and due to the transverse pressure the boundary layer is pushed around toward the inner wall. Under combined influence of the pressure recovery due to the area increase and the secondary flow, the boundary layer grows at a rapid pace in the inner wall region. Under present duct boundary conditions (geometry), the boundary layer could not negotiate the pressure gradient and therefore was separated.

The resulting three dimensional separation was a large region as shown in the surface flow visualization photographs. Separated flows are, in general, unsteady. In the present study no attempt was made to identify the dynamics of the separation region and its influence on the flow unsteadiness at the duct exit. References 8 and 9 discuss typical dynamics of separation in a diffusing S-duct. In a separate study various techniques were successfully used to eliminate this flow separation. The resulting influence on the secondary flow is discussed in Reference 10.

Wall static pressure measurements made along three azimuth angles are plotted in Figure 4. Azimuth angles of 0° and 180° could not be measured since flanges were interfering with the static pressure tubings. The wall pressure along 10° and 170° azimuth lines are expected to be nearly similar to the 0° and 180° angles. The separation region is indicated on this plot, with the actual locations identified from the surface oil flow visualization patterns.

The five port cone probe was traversed at six stations, these include one station in

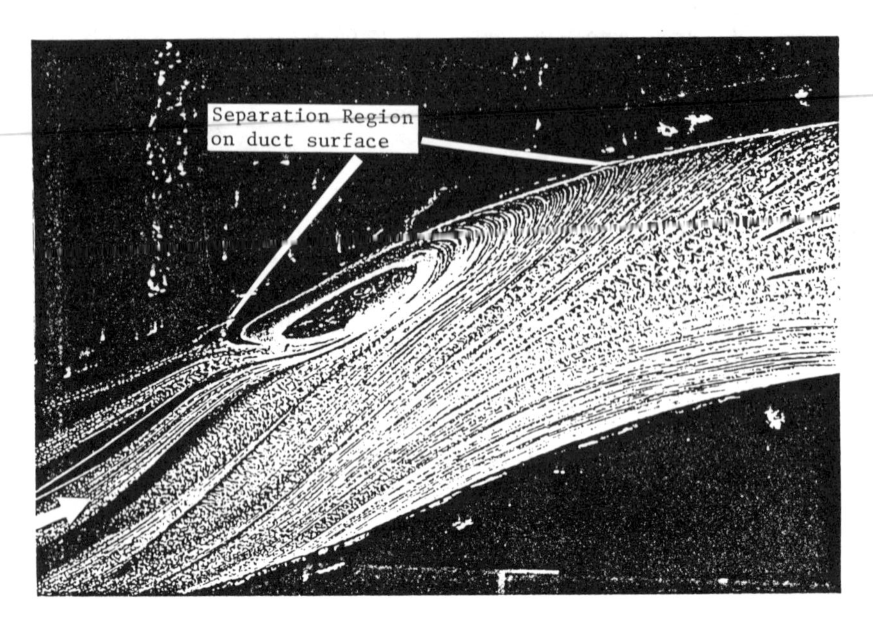

Figure 3. Surface Oil Flow Patterns Showing the Separation in the Diffusing S-Duct.

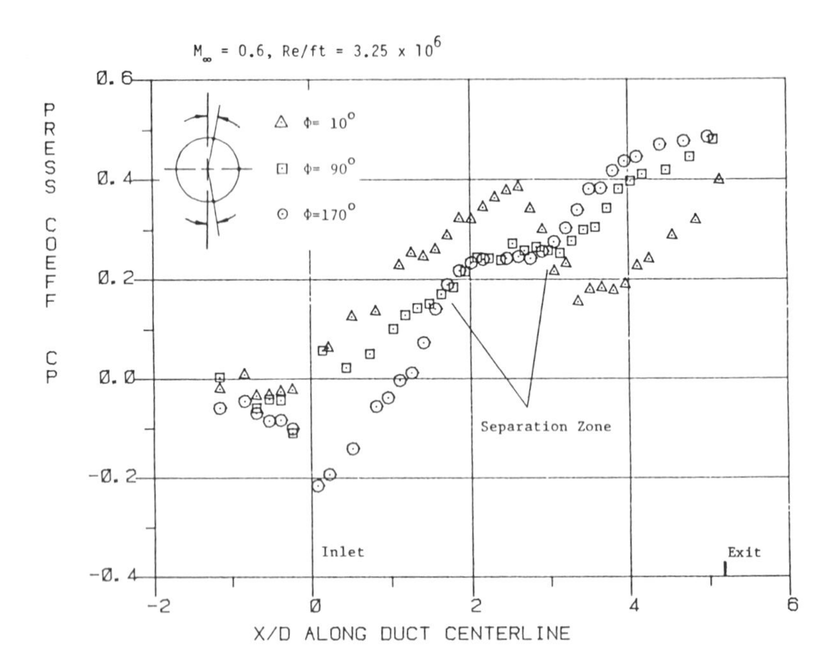

Figure 4. Wall Static Pressures Along the Length of the Diffusing S-Duct.

the straight section upstream of the duct entrance at X/D_{inlet} = -1.65 and stations $\theta = 2^o, 15^o, 30^o, 45^o$ and 60°. At each station, a total of ten radial trasverses were made. Analysis of the cone probe data resulted in the determination of total- and static-pressure contours as well as the local flow velocity vector.

Figure 5 shows the transverse velocity vectors at the six stations along the duct. Since the flow is symmetric with respect to a vertical plane passing through the centerline, only half of the plots are shown in the figures to save space. The development of secondary flow along the duct is made clear in these plots. In the first bend, $\theta \leq 30^o$, strong vortex flow is developed which moves downstream into the second bend. The presence of separated flow, beginning on the inside of the first bend, effectively increases the duct radius of curvature. Therefore, the secondary flow vortices develop further into the second bend up to $\theta \simeq 45^o$. The formation of additional vorticity (secondary flow) in the second bend is small and at the duct exit there is only a small indication of vorticity in the opposite direction. Even though there are two pairs of counterrotating vortices at the duct exit, the vorticity formed in the second bend is small compared to the constant area duct (Ref. 5). This is believed to be caused by the presence of flow separation. The observations of surface flow visualization agree with the trends on the transverse velocity vector plots.

Contours are obtained using linear interpolation both in radial and circumferential directions. Total pressure contours at the six stations are shown in Figure 6. Flow remains nearly axisymmetric up to the middle of the first bend. Inside the first bend the potential core was moved toward the outer bend, especially corresponding to the location where the flow separation begins on the inner wall. The flow entered the second bend with the potential core near the inner wall. In the second bend, the development of flow separation pushed the core flow further towards the inner boundary. The region near the outer wall in the second bend identifies the large extent of low total pressure flow. The distortion at the exit, as shown in Figure 7, $\theta = 60^o$, covers more than half of the exit plane.

Static pressure contour plots are presented in Figure 7. These plots also indicate the increase in static pressure toward the outer wall, in the region of transition from the straight section into the first bend. This is true in the first bend up to somewhere between $15^o \simeq \theta \simeq 30^o$, where the gradient direction is reversed. That is, the static pressure gradient increases toward the inner wall in the first bend and the outer wall in the second bend. At the station $\theta = 30^o$, the low pressure cores of secondary flow vortices are identified. The secondary flow combined with the three dimensional flow separation at $\theta = 45^o$ and 60° resulted in a large region of low energy flow in the outer half of the second bend. All the data measured in this work are listed in tabulated form in Reference 11.

V. CONCLUDING REMARKS

Benchmark quality data has been obtained for compressible flow in a circular cross section diffusing S-duct. The duct area ratio was 1.51, with the inlet diameter of 16.51 cm. Turbulent flow was entering the duct at $M_\infty = 0.6$, the boundary layer thickness at the duct entrance was nearly ten percent of the duct inlet diameter. Flow parameters were measured at six streamwise stations. Using a five port cone probe, total and static pressure contours in addition to the secondary velocity vector were obtained. A complete listing of the data is given in Reference 11. Strong three dimensional flow separation was observed which extended from $\theta \sim 22^o$ to $\theta \sim 44^o$. Due to this region of flow separation, the duct

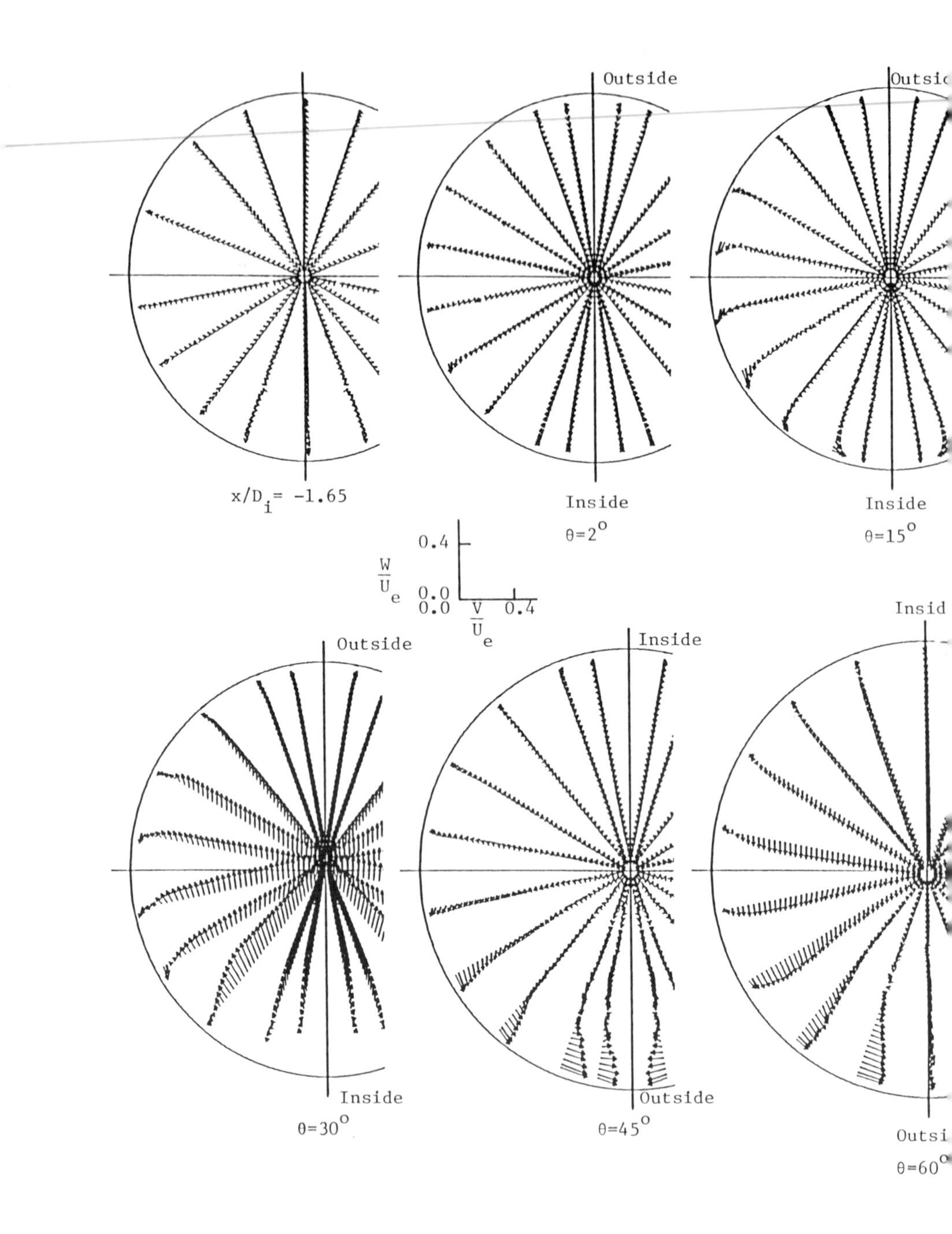

Figure 5. Secondary Velocity Components V, W Measured in Transverse Planes at Six Streamwise Stations.

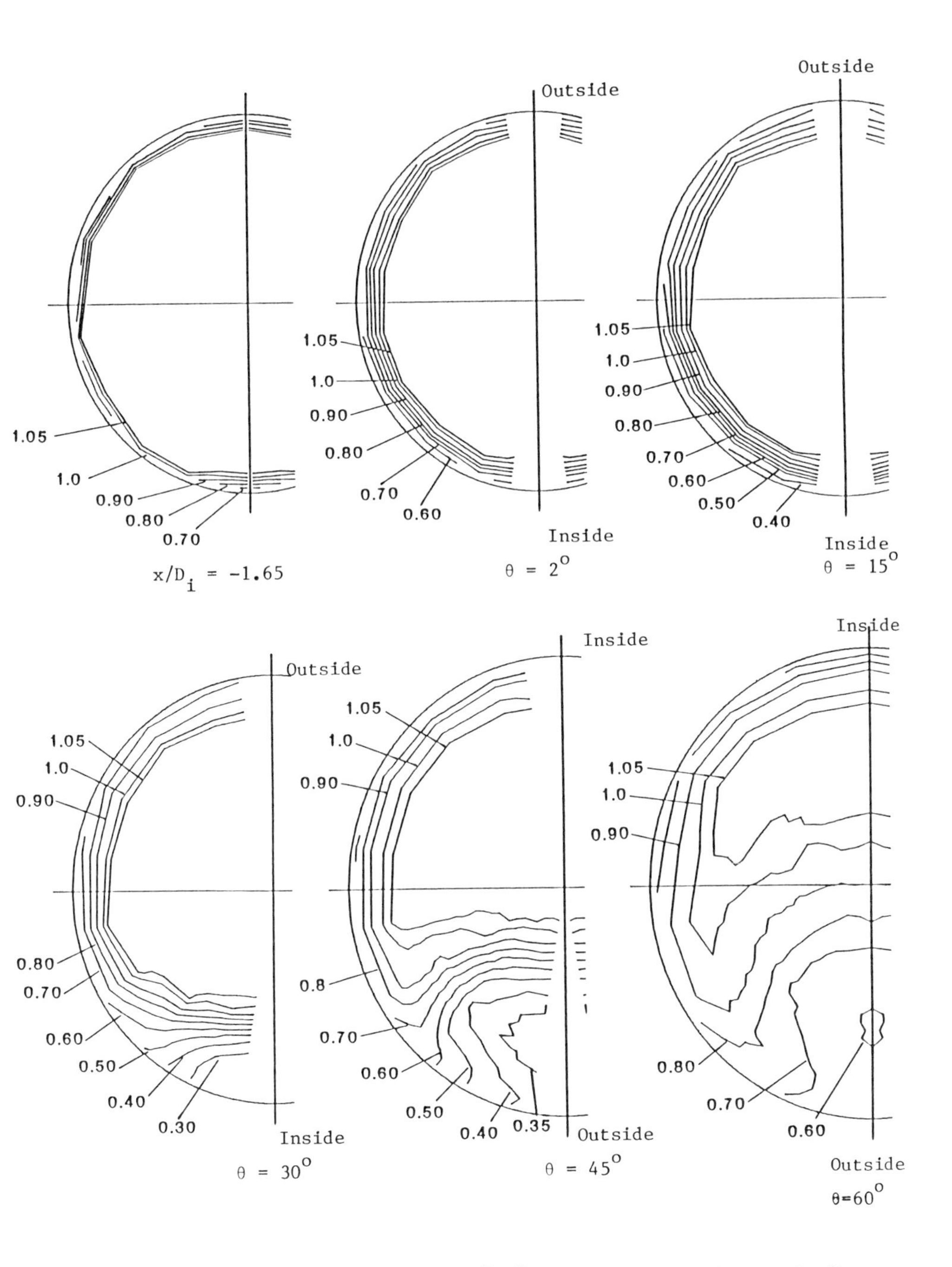

Figure 6. Total Pressure ($C_{Pt} = \frac{P_t - P_\infty}{q_\infty}$) Contours at Six Streamwise Stations.

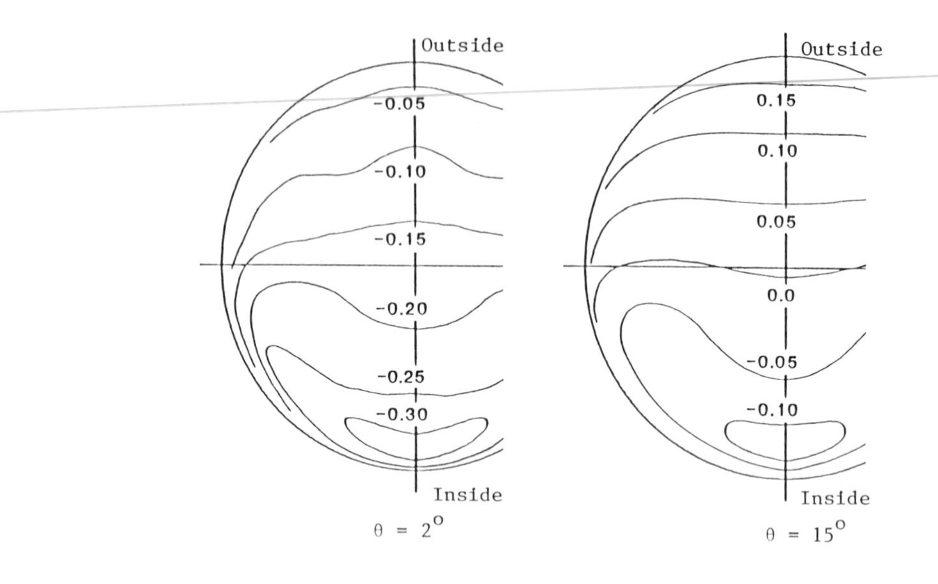

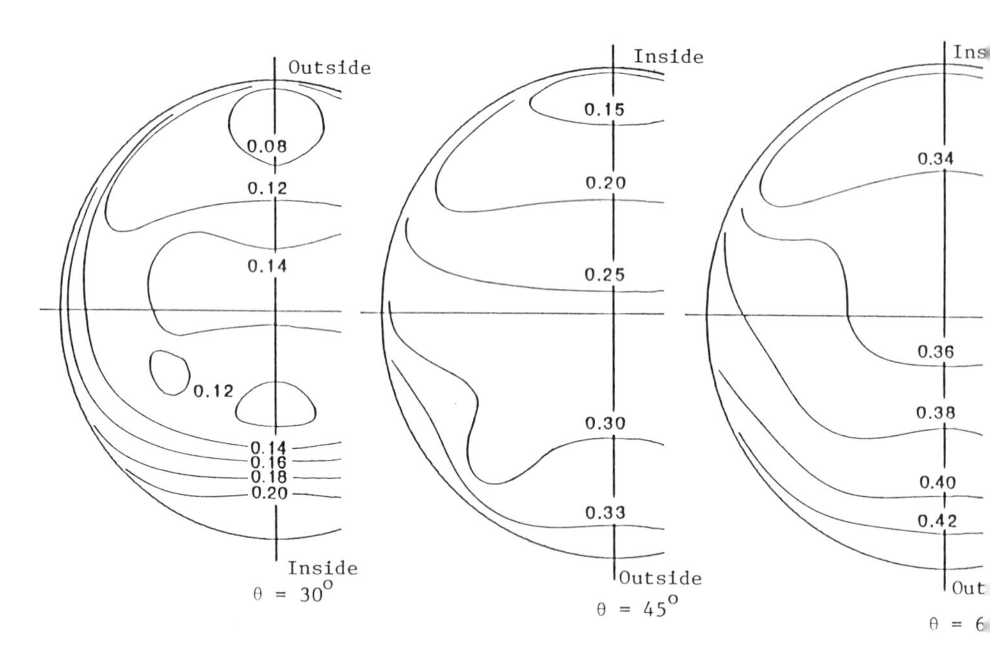

Figure 7. Approximate Static Pressure ($C_{Ps} = \frac{P-P_\infty}{q_\infty}$) Contours at Six Streamwise Stations.

curvature (r/R) was effectively changed. Also due to this flow separation, the pressure recovery along the duct was reduced significantly. The exiting flow contained a region of flow distortion larger than half of the exit area.

Acknowledgement. This work was supported by NASA Lewis Research Center with Dr. Warren R. Hingst as the technical monitor. The continuous support and advice of Dr. Hingst is greatly appreciated.

REFERENCES

1. Squire, H.B. and Winter, K.G. 1951, "The Secondary Flow in a Cascade of Airfoils in a Non-Uniform Stream," J. Aero. Science 18, pp. 271–277.

2. Rowe, M. "Measurements and Computations of Flow in Pipe Bends," J.F.M. (1970), Vol. 43, Part 4, pp. 771–783.

3. Taylor, A.M.K.P., Whitelaw, J. H. and Yianneskis, M. 1982. "Developing Flow in S-Shaped Ducts," Part I: Square Cross-Section Duct. Final Report Prepared for NASA Lewis Research Center.

4. Taylor, A.M.K.P., Whitelaw, J.H. and Yianneskis, M. "Developing Flow in S-Shaped Ducts," Part II: Circular Cross-Section Duct. Final Report Prepared for NASA Lewis Research Center.

5. Vakili, A.D., Wu, J. M., Liver, P., Bhat, M.K., "An Experimental Investigation of Secondary Flows in a S-Shaped Circular Duct" NASA Lewis Final Report 1983.

6. Wu, J.M. and R. C. Lock, "A Theory for Subsonic and Transonic Flow Over a Cone-With and Without Small Yaw Angle," U.S. Army Missile Command, Technical Report RD-74-2, December 1973.

7. Reed, T. D., T. C. Pope and J. M. Cooksey, "Wind Tunnel Calibration Procedure Manual," NASA CR 2920, 1977.

8. Neumann, H.E., L.A., Povinelli and R.E. Coltrin, "An Analytical and Experimental Study of a Short S-Shaped Subsonic Diffuser of a Subsonic Inlet", AIAA 80-0386.

9. Stumpf, R., H.E. Neumann and C.C. Giamati," Dynamic Distortion in a Short S-Shaped Subsonic Diffuser with Flow Separation". NASA TM-83412.

10. Vakili, A. D., J. M. Wu, P. Liver, M. K. Bhat. "Flow Control in a Diffusing S-Duct." AIAA Paper 85-0524, AIAA Shear Flow Control Conference, Boulder, Colorado, March 12 - 14, 1985.

11. Vakili, A. D., J. M. Wu, P. Liver and M. K. Bhat, "Experimental Measurements in a Variable Area 30° - 30° S-Duct." Final Report prepared for NASA Lewis Research Center (to be published), 1985.

LDA Measurement within a Vortex Boundary Layer Generated in a Rotating Drum

B. C. KHOO and S. H. WINOTO
Department of Mechanical & Production Engineering
National University of Singapore
Kent Ridge, Singapore 0511

INTRODUCTION

Atmospheric vortices like tornadoes, whirlwinds and watersprouts have attracted the interest of many people because of their destructive nature. Nevertheless, very little is known about the flow dynamics, and data are often deduced from indirect measurement using radar, instrumented aircraft and photogrammetric technique. The essential data regarding distribution of velocity, pressure and energy are lacking as tornadoes destroy all standard measuring instruments. A detailed study is usually confined to a laboratory simulated natural vortex. Even then, there are disagreements about the dynamics used in producing the vortex.

Perhaps the simplest conceptual model of the vortex interaction problem is that of a potential vortex with its axis vertical and coaxial with a stationary horizontal disk. Such a model has possible applications to geophysical phenomena and produces one of the simplest form of a three-dimensional boundary layer. The azimuthal velocity, due to the potential vortex, gives rise to a radial pressure gradient as the flow is in cyclostropic balance (the centripetal force is in equilibrium with the force due to the radial pressure gradient). This pressure gradient induces a radial flow in the boundary layer towards the axis. Near the axis, the radial flow fluid changes direction and move away from the base into the core of the vortex, in order to preserve the continuity of flux of fluid in the boundary layer and the core. The flow of fluid into the vortex core is called effusion.

The objective of the present work is to study the flow field of such a three-dimensional vortex boundary layer and effusing core at various Reynolds numbers Re (based on disk radius and free stream velocity of the edge of disk) which range from 5,000 to 30,000. This is done by measuring the three velocity components (tangential, radial and axial) in the flow field using a non-disturbing laser Doppler anemometer (LDA). From these measurements, a quantitative assessment of the flow characteristics and flow development is possible.

THE TEST RIG

The test rig is schematically shown in Fig. 1. The potential vortex is generated in a rotating perspex drum of 630 mm diameter and 250 mm height. The drum is supported by stainless steel ball bearings placed in an aluminium ring bracket with inverted V-shaped grooves machined underneath. A perforated

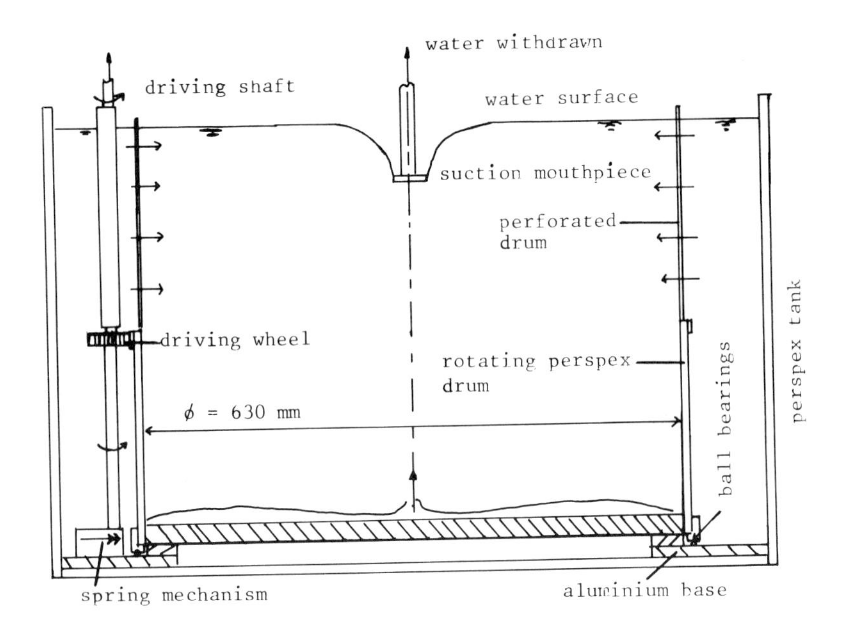

FIGURE 1. Schematic view of the test rig.

stainless steel drum of 250 mm height is mounted on top of the perspex drum to allow the working fluid (water) to flow in the inner surfaces of the perspex and perforated drums are flushed.

The base is a 12 mm thick aluminium plate with a 570 mm diameter hole in the centre. A 25 mm thick circular perspex base is placed over the hole. Concentric with the hole is a V-shaped groove having a ring diameter of 640 mm; this lower groove with the upper groove of the aluminium ring bracket acts as a channel for the ball bearings to run. This bearing system allows the rotation of the drum from 0.2 rpm to 40 rpm (equivalent to the range of Re from 2,000 to 500,000). The vibration associated with the ball bearing system is damped out by the water medium.

The drum is rotated by a drive wheel acting firmly against a section of rubber interface which is attached to the perspex drum. To prevent any slippage during rotation, the surface of the drive wheel is knurled while the rubber interface is made of soft rubber with roughened surface. At the base near the drive shaft, is a casing containing a compressed spring which always exert a centripetal force on the shaft to ensure contact between drive wheel (which is attached to the shaft) and rubber interface.

A d.c. motor with a 40:1 speed reducer is used to turn the driving wheel. The speed reducer in turn is coupled directly to a 530 mm long driving shaft. Both motor and reducer are mounted on an L-shaped aluminium plate. These are supported on a separate supporting system away from the main rotating drum. The vibration caused by the motor and its bearing is not transmitted to upset the vortex flow inside the rotating drum. Because the driving shaft is rotating just outside the main drum, a stationary aluminium sleeve is used to enclose the shaft, shielding any vortices generated from interfering with the main flow.

The main perspex drum and its aluminium base are housed inside a perspex tank whose two side walls are 25 mm thick and the other two are 10 mm thick. The thicker side wall faces the laser source of the LDA; the maximum deflection of the wall is less than 1.0 mm for a 500 mm height of water. The side walls are secured to the base by screws and sealing is ensured by applying silicone sealant to the inner edges.

A Torishima made low head, high capacity pump (Model ETA 32-20) is used to remove the vortical fluid at the centre. The water is replaced by the flow in action through the perforated cylindrical drum. The suction side of the pump is connected to a suction tube placed vertically in line with the axis of the vortex. At high Re flow, the water surface dipped significantly towards the axis and hence a specially shaped mouthpiece (designed to remove the separation bubbles formed between the edge of a disk-shaped mouthpiece and the pipe inlet) is attached to the suction tube and positioned just on the water surface.

Since the base pressure gradient at the centre of vortex is very steep, any slight movement of the vortex would precipitate a large variation in pressure readings. The location of the vortex is determined by the position of the suction mouthpiece; hence the suction tube is mounted on an x, y, z traversing mechanism capable of fine adjustments. On the pressure side of the pump, an orifice plate is installed to measure flow rate. The circulating water is passed through a water filter before being returned to the perspex tank which housed the rotating drum. The water filter has dual purposes: to remove dirt and impurities in the water, and to dissipitate the returned water into tiny streams instead of a strong jet.

THE LASER DOPPLER ANEMOMETER

The velocity components (tangential, axial and radial) of the vorted boundary layer and its effusing core were measured by using a laser Doppler anemometer (LDA) since it introduces no disturbances to the flow.

The LDA consists of a 5 mW He-Ne coaxial laser source (NEC, GLG 5350), a laser adaptor (DISA, 55X19), a single colour beam splitter (DISA, 55X26) with 60 mm beam separation, an optical support (DISA, 55X23), a beam translator (DISA, 55X32), a beam expander of ratio 1.9:1 (DISA, 55X12), a lens mounting ring (DISA, 55X33) to mount a front objective planoconvex lens of 600 mm focal length (DISA, 55X52), a photomultiplier (DISA, 55X08) and its optics (DISA, 55X34) which is supported by a tripod and connected to a frequency tracker (DISA, 55N20), a monitoring oscilloscope and an r.m.s. meter (DISA, 55D35). The optical equipments and photo multiplier were mounted on a 1200 mm long optical bench (DISA, 55X42) which was lengthened to 1700 mm. The optical bench was attached to a traversing mechanism capable of providing fine movements in three-perpendicular directions. The LDA set-up is shown in Fig. 2.

The half beam intersection angle was $\tan^{-1}$ (7/120) which produced a measuring volume of 5.5 mm in length and 0.3 mm in diameter (at mid-length cross section).

VELOCITY MEASUREMENT

In this flow condition, calculations of velocities from LDA readings (Doppler frequency) require correction because refraction of the laser beams at cylindrical interfaces changes the paths of the beams thus moving both their point of intersection and the angle between the beams. In this case, the

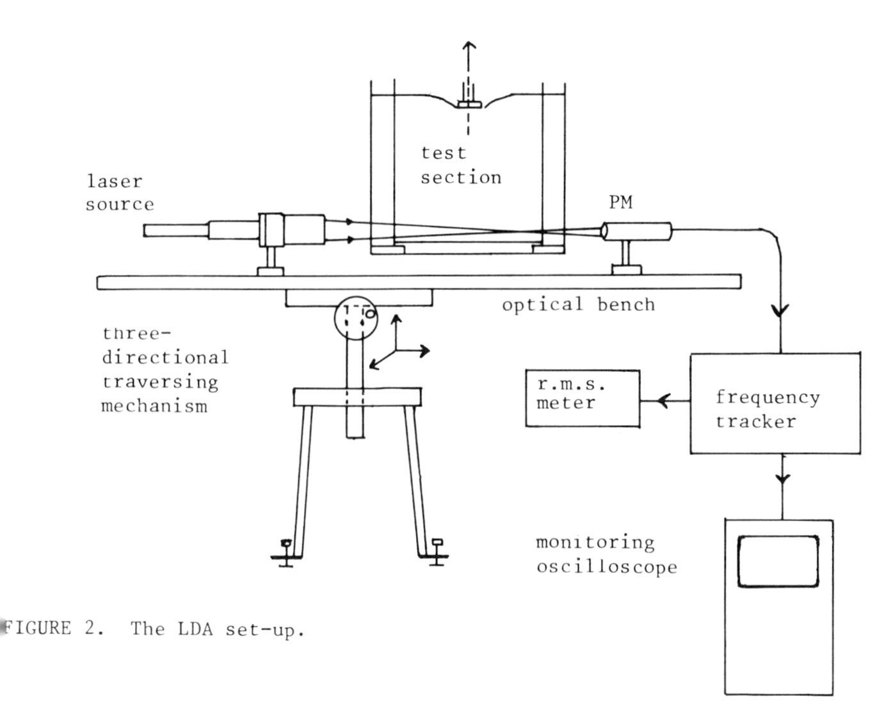

FIGURE 2. The LDA set-up.

orrection methods used for measurements of tangential, axial and radial elocity components are those suggested by Boadway and Karahan (1981).

angential Velocity Measurements

o measure the tangential velocity component v_θ, the optical system was riented so that the beams bisector was perpendicular to, and passes through, he axis of the cylinder with the plane containing both beams perpendicular to he axis (Fig. 3(a)).

ince a region of high velocity gradient is present near the vortex axis, as ell as in the boundary layer near the base, the measurement of tangential elocity in these regions has a large margin of uncertainty, as a result of a ide velocity gradient broadening. (This is one of the major disadvantages of n LDA).

his configuration was also used to "calibrate" the LDA; the measuring volume as positioned very near to the inner wall of the perspex drum. A stop watch as used to time the rotation rate which was varied from about 0.15 rpm to 50 pm. For each rotation, the timing of five cycles were taken and the average inear velocity was calculated. The LDA readings were taken over the duration nd its mean velocity was computed. A variation of at most 4% was found in he comparison between the measured velocities and LDA readings.

xial Velocity Measurements

o measure the axial velocity component V_z, the optical system was oriented so hat both beams are coplanar with the cylinder axis with the bisector between he beams at right angle to this axis.

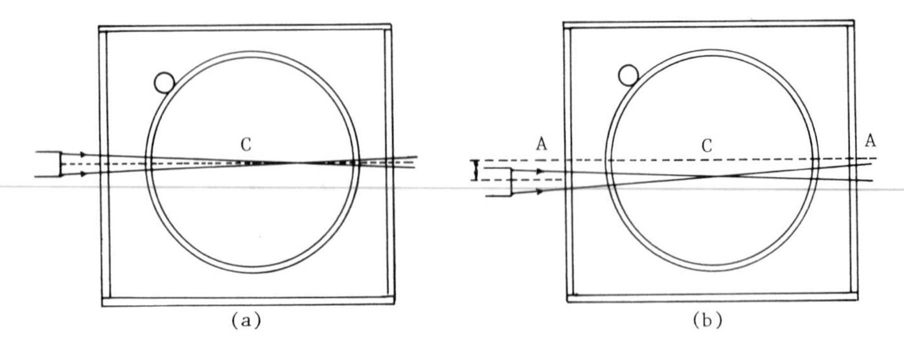

FIGURE 3. Plan views of LDA optical system:
(a) for measuring tangential velocity v_θ (the LDA optical axis passes through drum centre C),
(b) for measuring radial velocity v_r (the LDA optical axis is parallel) to line AA passing through drum centre C and perpendicular to perspex tank walls).

Radial Velocity Measurements

To measure the radial velocity component v_r, the optical system, was oriented as for the tangential velocity measurement except that the beams bisector did not pass through the axis of the drum (Fig. 3(b)). The beams and bisector however still lie in a plane perpendicular to the axis. Unlike in the tangential and axial velocity measurements, the velocity measured consists of two components; in the radial and tangential directions. By resolving it in the respective directions, radial velocity was computed.

The measurement of radial velocity has a larger error bound than the tangential or axial components since it depends on several factors such as the optical and geometrical calculations of the angle of intersection of beams, the angle subtended at the centre by the perpendicular made with the bisector of the two beams, and the tangential velocity at that particular position. It is more sensitive at positions further away from the base as the tangential velocity component is larger.

RESULTS AND DISCUSSION

The results of velocity measurements at various Re values are presented in Figs 4-7. For the boundary layer region, the variations of radial and tangential velocities across the layer are shown in Fig. 4 while for the effusing core region the variations of tangential and axial velocites are shown in Figs. 5-6.

All the heights z and radial positions r are non-dimensionalised by z_n (r*) which is defined as the height at the disk radius position $r^*(= r/r_o)$ where its radial velocity is half of the maximum radial velocity, (r_o is the disk radius). This non-dimensional scheme was adopted by some researchers like Newman (1967) and Irwin (1973) in investigating turbulent wall jets. In the boundary layer region, the corresponding velocities are non-dimensionalised by the local free stream velocity, whereas the free stream velocity at z_n (r*), denoted by $U(z_n (r^*))$, is used in the effusing core region.

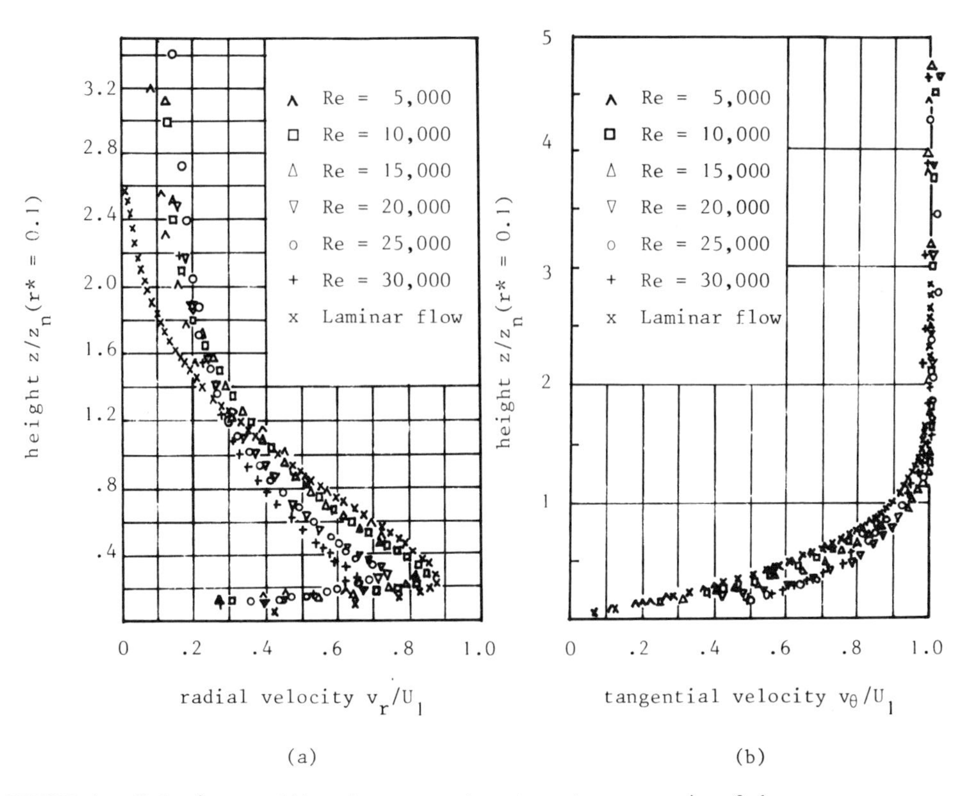

FIGURE 4. Velocity profiles in vortex boundary layer at r* = 0.1.

Measurements of v_r across the boundary layer were made at r* = 0.1 (Fig. 4(a)) and 0.2 (not presented here since the profiles are similar in shapes and characteristics as those presented in Fig. 4(a)). The laminar profile is plotted by reproducing the numerical results of Prahlad and Head (1976) independently. For both values of r* the lower the Re value, the closer the experimental results approach the laminar profile. Agreement is good except at greater height above the base. For increasing Re, the maximum value of v_{rn}, the non-dimensionalised v_r (= v_r/U_1 where U_1 is the local free stream velocity decreases gradually from 0.9 to 0.65 for the case of r* = 0.1, and from 0.8 to 0.72 for r* = 0.2.

The non-dimensionalised tangential velocity, $v_{\theta n}$ (= v_θ/U_1) is plotted against z/z_n (r*) for various Re values at r* = 0.1 (Fig. 4(b)) and 0.2 (not presented here since the profiles are similar in shapes and characteristics as those presented in Fig. 4(b)). The laminar velocity profile is again obtained from Prahlad and Head (1976). It can be seen that the profile at Re = 5,000 is close to the laminar profile. The higher the Re, the higher the velocity gradient of the profiles at z = 0.

The non-dimensionalised tangential velocity, $v_\theta/U(z_n$ (r*)) is plotted against non-dimensionalised lateral distance r/z_n (r*) for various Re values at a

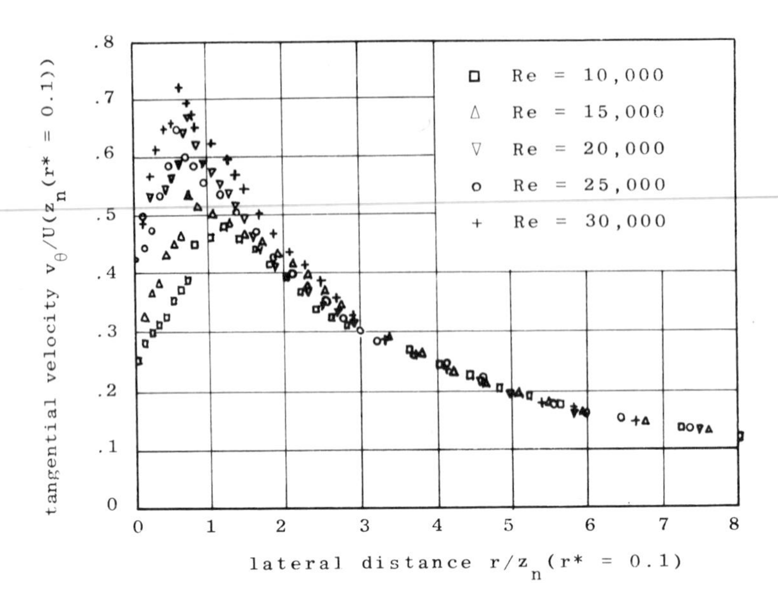

FIGURE 5. Tangential velocity profiles in vortex effusing core at height of 20 mm above the base.

constant height. The plots for $r^* = 0.1$ at height = 20 mm are presented in Fig. 5. It can be seen that the plots follow the potential velocity variation for r/z_n (0.1) greater than 3.3. As r/z_n (0.1) decreases, $v_\theta/U(z_n(0.1))$ is higher for respectively larger Re. The values of r/z_n (0.1) which correspond to the peak of $v_\theta/U(z_n(0.1))$ also become smaller for increasing Re. It is interesting to note that the measurements do not go to zero as r/z_n (0.1) approaches zero. Closer examination of the nature of the vortex shows that it is always wandering about the centre with a very high frequency. The plots for $r^* = 0.1$ at height = 40 mm, for $r^* = 0.2$ at height = 20 mm and 40 mm were also obtained but not presented here since they are very similar in shapes and have the same characteristics as those shown in Fig. 5.

The non-dimensionalised axial velocity, $v_z/v_{z\ max}$ (where $v_{z\ max}$ is the maximum axial velocity) is plotted against non-dimensionalised lateral distance r/z_n (r^*) for various Re values at a constant height. The plots for $r^* = 0.1$ at height = 20 mm are shown in Fig. 6. It can be seen that all the plots collapse onto a single profile. The plots for $r^* = 0.1$ at height = 40 mm, for $r^* = 0.2$ at height = 20 mm and 40 mm were also obtained but not presented here since they are very similar in shapes and characteristics as those shown in Fig. 6. Depending on the use of z_n ($r^* = 0.1$) or Z_n ($r^* = 0.2$) to normalise the radial displacement r, the axial flow profiles could be represented by the equation:

$$v_z/v_{z\ max} = \exp. - [0.52\ \{r/z_n\ (0.1)\}^2] \quad (1)$$

or,

$$v_z/v_{z\ max} = \exp. - [0.63\ \{r/Z_n\ (0.2)\}^2] \quad (2)$$

The above equations are applicable irrespective of height (20 mm and 40 mm above the base) and Re.

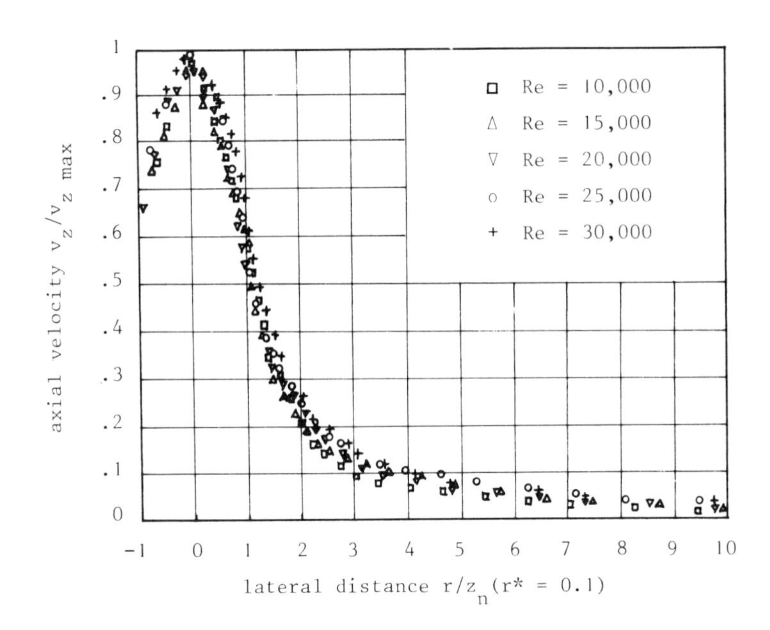

FIGURE 6. Axial velocity profiles in vortex effusing core at height of 20 mm above the base.

The maximum axial velocity $v_{z\ max}$ is non-dimensionalised by the free stream velocity at z_n (r^*). $v_{z\ max}/U(z_n(r^*))$ is then plotted against Re for r^* = 0.1 and 0.2 at height = 20 mm and 40 mm (Fig. 7). It can be seen that all the plots decrease with increasing Re in an almost linear fashion.

SUMMARY AND CONCLUSION

A vortex boundary layer has been successfully generated in a rotating perspex drum. The tangential, axial, and radial velocity profiles in the vortex have been revealed by using non-disturbing laser Doppler anemometry at Reynolds number values range from 5,000 to 30,000.

Radial and tangential velocity profiles in the vortex boundary layer (Fig. 4) follow the trend of the laminar flow profiles of Prahlad and Head (1976).

In the vortex effusing core, the tangential velocity profiles (Fig. 5) follow the potential velocity variation, starting at certain distance from the core. The non-zero values at the core centre are caused by high frequency vortex wandering about the centre. The axial velocity profiles (Fig. 6) collapse onto a single profile which can be represented by an exponential function as in equation (1) or (2). The maximum axial velocity, however, decreases almost linearly, with increasing Reynolds number (Fig. 7).

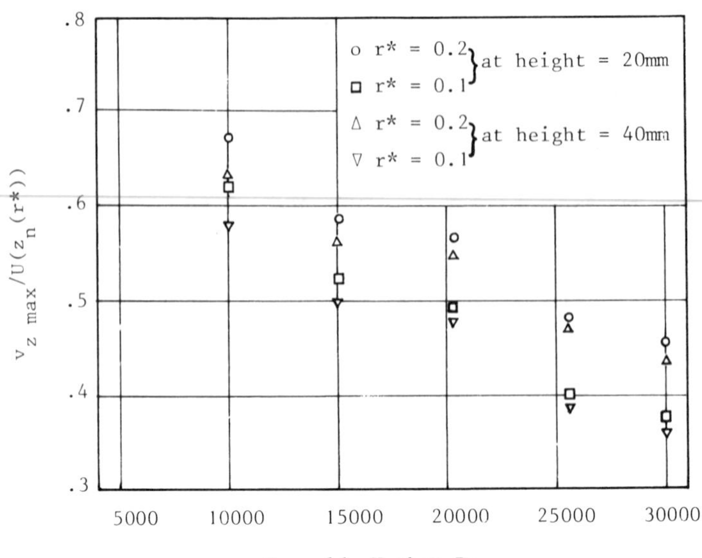

FIGURE 7. Variations of maximum axial velocity with Reynolds number at height of 20 mm and 40 mm above the base.

ACKNOWLEDGEMENTS

Dr W.R.C. Phillips is thanked for his suggestion of the problem and valuable advice. The assistance of Messrs. C.S. Yap, S.W. Looi, Mokhtar, C.P. Ng and Miss Jaleha with constructional and experimental, drafting and typing aspects is gratefully acknowledged. The work was funded by a research grant from the National University of Singapore.

REFERENCES

Boadway, J.D., and Karahan, E., "Correction of Laser Doppler Anemometer Readings for Refraction at Cylindrical Interfaces", DISA Information, No. 26, pp. 4-6, 1981.

Irwin, H.P.A.H., "Measurements in Self-Preserving Plane Wall Jet in a Positive Pressure Gradient", Journal of Fluid Mechanics, Vol. 61, pp. 33-63, 1973.

Newman, B.G., "Turbulent Jets and Wakes in a Pressure Gradient", Fluid Mechanics Internal Flow, Gen. Motors Conf. Elsevier, 1967.

Prahlad, T.S., and Head, M.R., "Numerical Solutions for Boundary Layer Beneath a Potential Vortex", Computers and Fluids, Vol. 4, pp. 157-169, 1970.

Measurements of Isothermal Film-cooling Effectiveness Using a Short Duration Flow Facility

S. J. CHEN, D. Y. LIU, and FU-KANG TSOU
Department of Mechanical Engineering and Mechanics
Drexel University
Philadelphia, Pennsylvania 19104

ABSTRACT

Experiments were performed in an 11 m long Ludwieg tube wind tunnel to investigate the effects of injection ratios and injection stream turbulence on the film cooling process downstream of a two-dimensional film cooling slot. The ranges of these parameters are: the injection ratios (M) from 0.41 to 1.88 and injection turbulence intensities (ε_i) from 3% to 14.5% at locations of x/s from 0 to 82.8. The dimensionless heat transfer coefficients (h/h_o) indicate good linear relations with the dimensionless temperatures (θ). With the superposition method, the adiabatic film cooling effectiveness (η_{aw}) and the adiabatic heat transfer coefficients (h_f) were obtained by interpolation and extrapolation respectively.

Results demonstrate that increases in the injection-stream turbulence intensities reduce the film cooling effectiveness. In the near-slot region, the heat transfer coefficients (h_f) are increased with the increase in the injection ratios or in the injection turbulence intensities.

1. INTRODUCTION

Film cooling has been used to protect solid surfaces from high temperature environments, such as those in gas turbine components including blades, combustors, and afterburners. In the present paper, an experimental program is conducted to study the effects of the injection ratios (M) and the injection flow turbulence intensity (ε_i) on the film cooling process downstream of a two-dimensional slot.

In most laboratory experiments, the injection turbulence is kept minimum by the well designed injection system. However, in an actual design [1], the cold air first enters through holes for mixing and expansion to take place before the cold air exits from the slot. Such a geometry gives rise to a high value of turbulence and non-uniformity, resulting in a lower adiabatic film cooling effectiveness (defined in Eq. 2). Sturgess [1] reported a significant decrease in the adiabatic film-cooling effectiveness with the high injection turbulence induced by the practical slots. Best [2] found the same effects for M greater than unity. However, the effect of ε_i on η_{aw} was reported very small by Kacker and Whitelaw [3]. The present study is therefore initiated to reevaluate such an effect.

Supported by National Science Foundation

In film cooling studies, it is customarily adopted an adiabatic wall temperature (T_{aw}) as the reference temperature in the definitive expression of the heat transfer coefficient, i.e.,

$$q_f = h_f(T_w - T_{aw}) \tag{1}$$

where T_w is the wall temperature. For constant property flows in film cooling, the heat transfer coefficient (h_f) is independent of the temperature difference ($T_w - T_{aw}$) based on a superposition method [4]. For the sake of convenience, the dimensionless representation of T_{aw} called adiabatic film-cooling effectiveness is introduced in most experimental work, or

$$\eta_{aw} = \frac{T_{aw} - T_m}{T_i - T_m} \tag{2}$$

where the injection fluid temperature (T_i) and the mainstream temperature (T_m) are generally constant.

An observation of Eq. (1) reveals that T_{aw} and h_f have to be known in order to predict the temperature of the film-cooled wall (T_w). In the region away from the slot where the flow is of boundary-layer type, h_f is equal to the heat transfer coefficient without film cooling. Thus, η_{aw} is the only quantity that has to be determined. Emphasis has been placed to determine η_{aw} in the earlier film cooling literature [5]. Since the most important region in film cooling is the one near the slot, determination of both T_{aw} and h_f becomes the center of interest in later studies.

This study utilizes a transient technique associated with a short duration flow facility (Ludwieg Tunnel) for acquiring film cooling data [6]. The flow prevails in the tunnel in a period of about 20 ms. The advantages are that the isothermal film-cooling effectiveness (defined in Eq. 5) can be measured with ease. With this information available, the adiabatic film-cooling effectiveness and the heat transfer coefficient (h_f) are then determined conveniently from the superposition method. No heating or cooling the test plate is needed in conducting experiments.

2. EXPERIMENTAL APPARATUS

The overall experimental apparatus is sketched in Fig. 1, which shows the various components including the Ludwieg Tube, the injection system, the instrumentation and the data acquisition system. Further description of the components is given below:

2.1 Ludwieg Tube Wind Tunnel

The Ludwieg tube consists of three sections: a 127 mm I.D. x 3 m high pressure supply tube, a 89 mm I.D. x 7 m low pressure dump tube, and a 127 mm x 50.4 mm rectangular test section (700 mm long) placed between the tubes. A mylar diaphragm, located downstream of the test section, can be broken by a pin to start the flow. The operations and the wave diagrams of the tunnel were discussed in detail elsewhere [7].

As seen in Fig. 2, the test section is 0.7 m long, 50.4 mm high, and 127 mm wide. The two-dimensional film cooling slot has a slot height (s) of 2.3 mm and a lip thickness (t_1) of 1.0 mm. The details of the test section were reported earlier [6]. At the end of the test section, a nozzle portion is installed to choke the flow at the nozzle throat in order to produce subsonic flows in the test section.

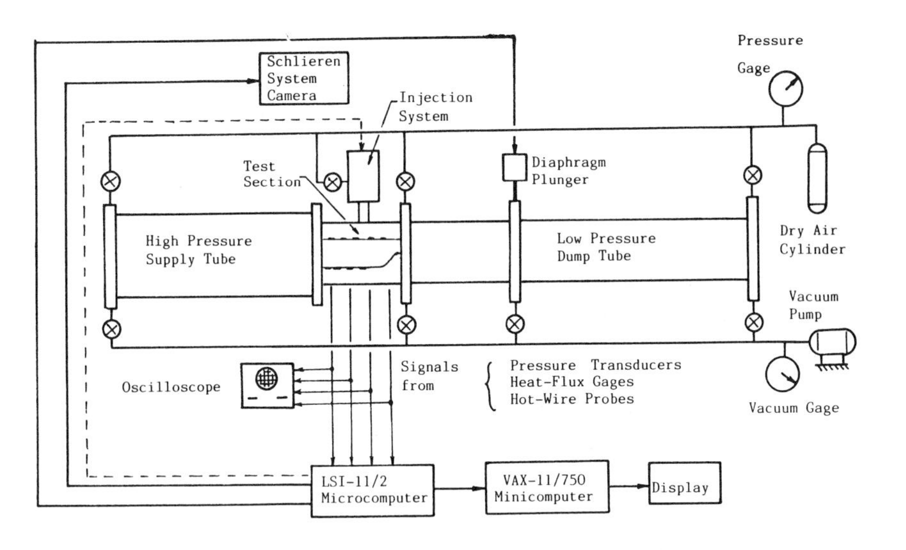

Fig. 1 Ludwieg Tube Wind Tunnel and Instrumentation

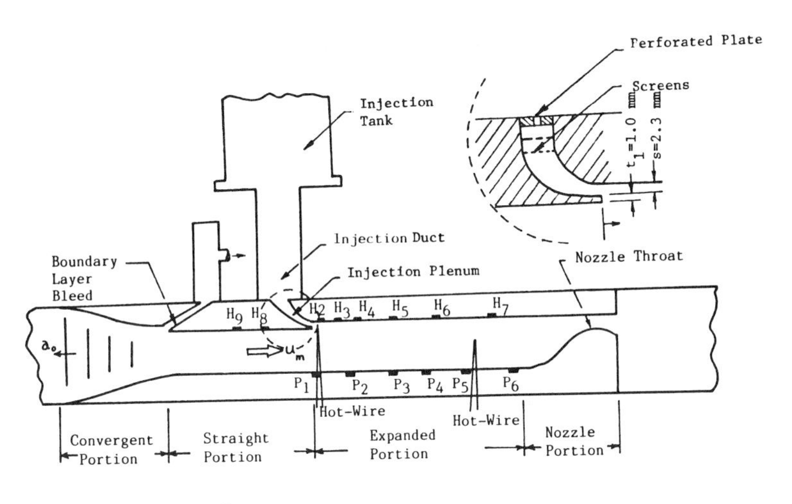

Fig. 2 Test Section and Injection System

2.2 Injection System

The injection system was fed from a large tank so that the pressure drop is negligible during each run (dropping less than 2% in 20 milliseconds); hence the injection temperature and the injection flow rate can be kept constant. The injection system (Fig. 2), consists of a tank with a volume of 0.0444 cubic meters, placed on a rectangular duct with a cross section of 114.3 mm x 63.5 mm and a height of 317.5 mm, and finally an injection plenum with a span of 114.3 mm, which converges and bends 90 degrees from 12.7 mm to 2.3 mm and keeps a constant slot height for a distance of 6.35 mm before the injection flow exits from the injection plenum. The injection plenum is designed to produce uniform velocity and turbulence at the slot exit both in spanwise and vertical directions. Air injection is produced by leaving the injection system open to the test section prior to the start of the main flow. The initial pressures are thus in equilibrium. As the expansion waves pass over the injection slot, the decreasing main flow pressure causes injection air to flow from the injection system to the test section through the tangential slot. If the test section is placed in a shock tunnel, the injection system may be started before the main flow by a computer-controlled solenoids (dashed line in Fig. 1).

The injection ratios and the injection turbulence are controlled by inserting different perforated plates and screens [8] between the injection plenum and the rectangular duct. The screens with various mesh sizes are used to reduce the turbulence intensities, which are generated behind the perforated plates. The injection temperature is varied by heating the injection duct. The injection tank is not heated, because only a small fraction of the injection air (less tha 2%) is used during the tunnel run. For each injection ratio and injection turbulence, five injection temperatures are used to obtain the heat fluxes. The first run is taken with the initial temperature in the injection duct (T_{id}) bein equal to the initial temperature of the high pressure tube (T_o). Next, the injection duct is heated to about 20 to 25 degrees Celsius above T_o for the second run. The remaining three runs are then taken about 15 minutes apart to let the temperature in the injection duct cool down. Consequently, five sets of θ value (defined in Eq. 6) ranging from 0.7 to 2.5 are obtained. The injection ratio fo each case varies slightly due to the temperature variations in the injection system, but the effect is less than 2% and an average value of M may be taken fo correction.

2.3 Instrumentation

Measurements of pressure as functions of time are made using Kistler Model 603B1 piezoelectric pressure transducers with 503D charge amplifiers. These transducers, P_1 through P_6 located at x = 0, 25.35, 50.80, 76.22, 114.3, and 165.1 mm respectively.

Heat transfer rates are measured using the thin platinum film heat flux gauges. The response time of the gauge is less than one microsecond. With such a rapid response, the gauge is able to detect the nature of the boundary layer (laminar, transitional, or turbulent). Theories, calibrations, installations of the heat-flux gauges, and the electric circuits were discussed in detail elsewhere [6]. In the present work, eight gauges at x/s=1.96, 5.22, 6.96, 10.9, 19.6, 32.6, 58.7, and 82.8 were used to obtain the wall heat fluxes. Two heat flux gauges in front of the slot were used to check the turbulent boundary layer heat transfer.

A 5 μm diameter hot-wire probe, with a TSI Model 1050 constant temperature hot-wire anemometer are used to determine the average velocity and turbulence intensity. The calibration was done before and after the experiments. The hot-wire probe is also used to measure the flow temperature by using the constant current (set to 1.0 to 1.5 mA) mode. As compared to 500 khz as a velocity sensc the hot wire has a frequency response of 3 khz as a temperature sensor at flow velocity of 70 m/s based on the report [8]. This is adequate for flow temperature measurements since the steady state main flow lasts about 15 milliseconds. The hot-wire probe can be translated along a vertical direction using a model 430-2M (made by Newport Corp.) translation stage with a resolution of 0.01 millimeter.

In order to improve the data collection and processing procedures, an 8-channel A/D, 4-channel D/A data-acquisition and control system was developed to collect the data and control several events using a LSI-11/2 microcomputer. Eventually, the data was processed and the results were displayed using a VAX-11/750 mini-computer. The system description and software were given in [6, 8]. For measurements of pressures, temperatures, and mean velocities, sampling rates of either 5 khz or 10 khz are used. The maximum time for data collection does not exceed 30 milliseconds in the present flow facility. The reproducibility of the pressure, temperature, and velocity are all within 3%. For the measurements of turbulence intensities, it is necessary to use the highest sampling rate and the maximum number of data points because the turbulence intensities are evaluated from the root mean squares of the fluctuation components of the velocities. At present, 40 khz sampling rates which result in 400 to 600 data points during the steady state flow are used to measure the turbulence intensities. Under these conditions, the turbulence intensities can be reproduced within 5%.

3. EXPERIMENTAL RESULTS

3.1 Flow Field

The main stream properties were reported earlier in the absence of injection [6]. With and without injection, the main stream properties vary slightly. In the present study, the main flow properties obtained are: (1) the temperature is 11.5 C below the ambient for Ma = 0.21, (2) the temperature is 17.5 C below the ambient for Ma = 0.30. The turbulence intensity of the main stream is 0.4%.

The temperature, velocity, and turbulence intensity of the injection flow are measured at the center line of the injection slot exit, and they are assumed constant across the slot height. The profiles are checked in several cases and good flat profiles are obtained [8]. The two-dimensionality of the flow field is checked within 3% at two spanwise locations which are 6.35 mm apart. A signal of u_i (the subscript "i" refers to the injection flow) from the hot-wire anemometer, as shown in Fig. 3, was recorded on the computer at a 40 khz sampling rate. It was found that the injection ratio is 0.41 and the turbulence intensity is 14.5%

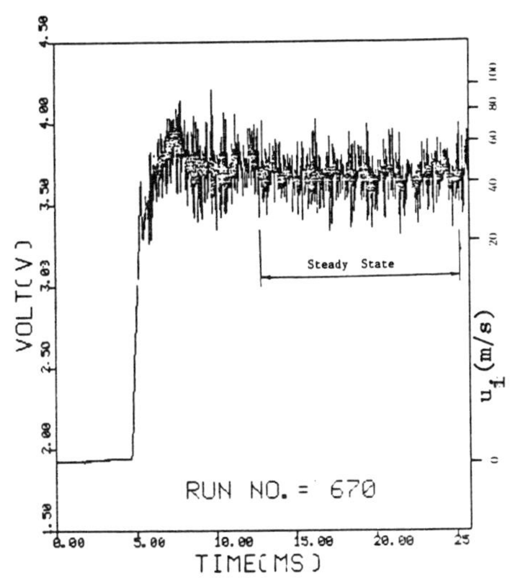

Fig. 3 Output of u_i from the Hot-Wire Anemometer for Ma=0.30, M=0.41, ξ_i=14.5%

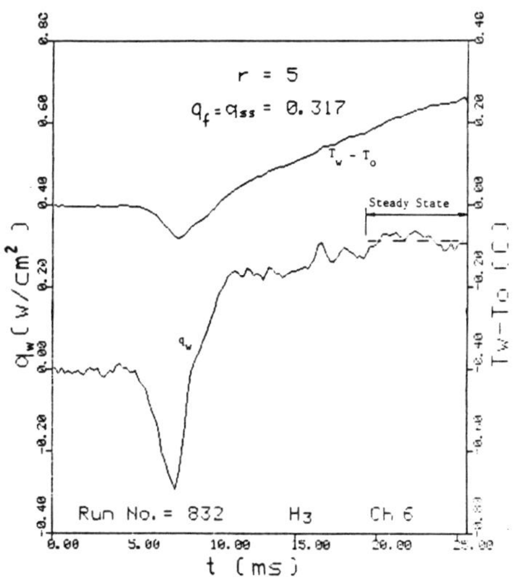

Fig. 4 Output of (T_w-T_o) and q_w obtained from the Least Square Method (128 data points, 0.2 ms time intervals)

Various combinations of plates and screens are summarized in Table 1, in which 11 cases with different injection ratios (M) and injection turbulences (ε_i) are obtained. These cases can be divided into two groups: (A) five cases from A1 to A5 with injection-stream turbulence larger than 7%, in which one plate is used without screens, (B) six cases from B1 to B6 with injection-stream turbulence less than 5%, in which either the combination of a plate with screens or no plate is used.

Table 1 Eleven Cases for Different Injection ratios (M) and Injection Turbulence Intensities (ε_i)

Case No.	M	ε_i(%)	$Re_i (= \frac{\rho_i u_i s}{\mu_i})$	Ma (Mach Number)
A1	1.35	7.2	13120	0.21
A2	1.05	8.2	10210	0.21
A3	1.10	8.0	10690	0.21
A4	0.80	9.7	7780	0.21
A5	0.41	14.5	5320	0.30
B1	1.88	3.6	18270	0.21
B2	1.83	3.0	17790	0.21
B3	1.80	3.8	17500	0.21
B4	1.63	4.5	15850	0.21
B5	1.27	3.3	12340	0.21
B6	1.03	3.0	10050	0.21

$T_i - T_m$ = -4 to 22 C ρ_i/ρ_m(density ratio) = 0.94 ∿ 0.98

θ = 0.7 to 2.5 (The subscripts "i" and "m" refer to the condition of injection flow and main flow, respectively)

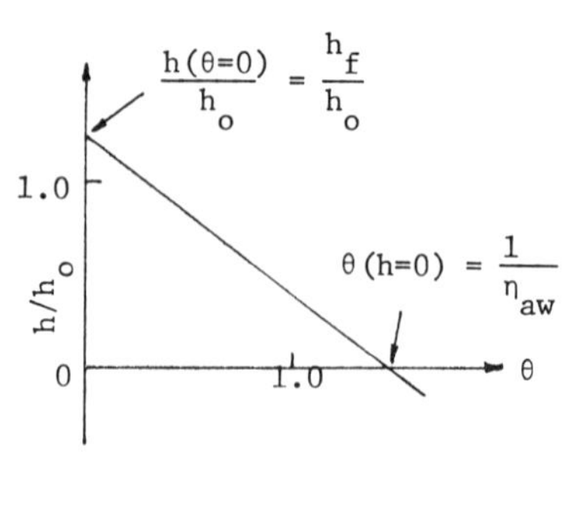

Fig. 5 h/h_o Versus θ at a Given Location for a Given Flow Field

3.2 Measurements of Heat Flux With and Without Injection

An estimation of the time required to reach steady state in the turbulent boundary layer gives less than one millisecond in the present tunnel [6]. The stead state heat flux can therefore be obtained. The heat flux in the absence of injection was reported earlier [6]. The heat flux with injection (q_f) is measured for several values of θ by changing the injection temperatures. The least squar method of data smoothing [6, 8] is applied to obtain the steady state heat flux. A sample output from heat flux gauge H_3 of run 832 is shown in Fig. 4. The stea state heat flux is taken between t=19.5 ms and t=25.5 ms by considering the time lag for the turbulent mixing between the main flow and the injection flow to tak place. As seen in Fig. 4, the top curve ($T_w - T_o$) is less than 0.25 C and hence can be considered as a constant wall temperature boundary condition. In all cases for the present film cooling study, ($T_w - T_o$) is in the range of -0.3 C to +0.5 C when the initial supply pressure (p_o) of 101 Kpa is used.

The injection temperature reaches a steady state at slightly different time stages for various injection ratios. Thus, the steady state values of heat fluxes for various values of M are taken at different time intervals. The shortest steady state time for heat flux is about 5 milliseconds for the case of M=0.41.

3.3 Isothermal Film Cooling Effectiveness

Since the main stream and the wall temperatures are constant during the experiments, it is convenient to define:

$$q_f = h(T_w - T_m) \tag{3}$$

$$q_o = h_o(T_w - T_m) \tag{4}$$

where the subscripts "f" and "o" refer to the cases with and without injection respectively. The dimensionless form of the heat flux,

$$\eta_{iso} = \frac{q_o - q_f}{q_o} \tag{5}$$

called isothermal wall effectiveness, is often used for data presentation in short duration flow facilities [8, 9, 10]. The name "isothermal" is used here because the test wall is essentially at a constant wall temperature. In the turbine blade or combustor cooling, the surface is close to the isothermal wall condition due to its internal and longitudinal cooling.

It should be noted that while the η_{aw} and h_f/h_o are functions of the flow parameters and the film cooling geometries only, η_{iso} has an additional dependence on the wall temperature. More precisely, η_{iso} depends on the dimensionless parameter,

$$\theta = \frac{T_i - T_m}{T_w - T_m} \tag{6}$$

Using Eq. (1) to (7), this dependence can be shown as [8]:

$$1 - \eta_{iso} = \frac{h}{h_o} = \frac{h_f}{h_o}(1 - \eta_{aw}\,\theta) \tag{7}$$

An observation of the equation shows either η_{iso} or h/h_o is linear with θ. The latter linear relation is plotted in Fig. 5, where the intercept along the vertical axis ($\theta = 0$) gives $h = h_f$ and that along the horizontal axis ($h = 0$) indicates $\eta_{aw} = 1/\theta$. Some experimental verification of this linear relation is available [8, 9, 10].

Due to the slight change in the temperature difference $(T_w - T_m)$ for cases with and without injection, it is more accurate to employ the ratio of h to h_o than the ratio of q_f/q_o. The values of h_o are chosen as if there were no backward-facing step and no injection, i.e., a turbulent boundary layer on a flat plate. Therefore, the curves which fit the data in the absence of injection at $x/s < 0$ and $x/s > 15$ [8] are used to obtain the values of h_o.

In Fig. 6(a), 6(b), and 6(c), three selected cases of h/h_o versus θ are presented for different values of x/s. For the sake of clarity, only five selected locations of x/s are presented in each figure, but all eight locations are shown at least once. As seen in Fig. 6(a), h/h_o versus θ for $M = 0.80$ and $\varepsilon_i = 9.7\%$ indicates a good linear relationship between h/h_o and θ for each x/s. The least square fit of a straight line was then performed for each location (x/s) for a given flow field (M and ε_i). The adiabatic wall heat transfer coefficients (h_f) are obtained by extrapolation to the values of $\theta = 0$, and the adiabatic film cooling effectiveness (η_{aw}) are obtained by interpolation at the values of $h/h_o = 0$.

Fig. 6(b) and 6(c) also show good linear relationship between h/h_o and θ for $M = 1.03$ and $M = 1.80$ respectively. Here, some discussions are given using Fig. 6(b). At $\theta = 0.9$, $0 < h/h_o < 1$ indicates that the heat fluxes from the wall to the fluid are reduced (remembering that film heating is used for the present work), and the isothermal effectiveness ($\eta_{iso} = 1 - h/h_o$) is between zero and unity. At $\theta = 1.37$, the values of h/h_o are smaller than zero at x/s smaller than 58.7, h/h_o is equal to zero at x/s equal to 58.7, and h/h_o is greater than zero at

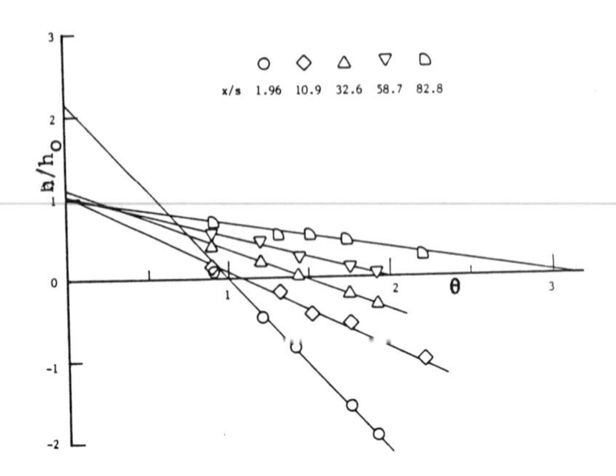

Fig. 6(a) Isothermal Heat Transfer Coefficient (h) Versus the Injection Temperature Parameter (θ) for M=0,80, ε_i=9.7%.

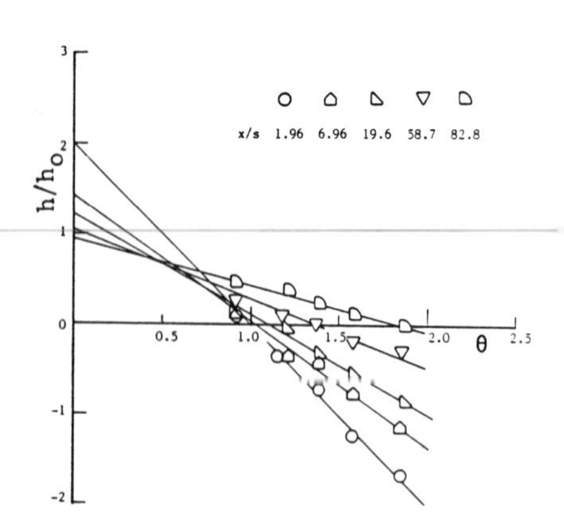

Fig. 6(b) Isothermal Heat Transfer Coefficient (h) Versus the Injection Temperature Parameter (θ) for M=1.03, ε_i =3.0%.

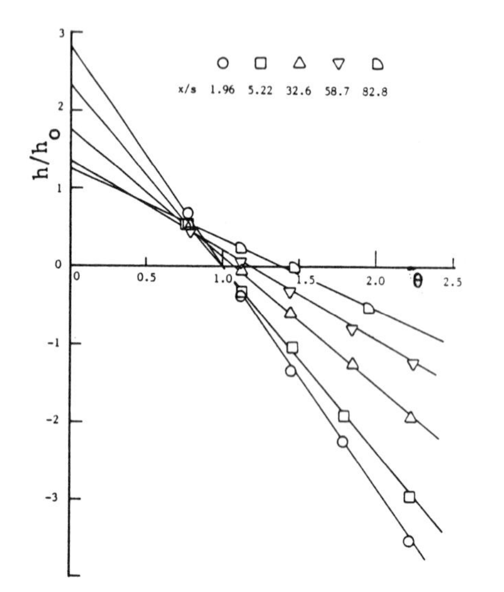

Fig. 6(c) Isothermal Heat Transfer Coefficient (h) Versus the Injection Temperature Parameter (θ) for M=1.80, ε_i=3.8%.

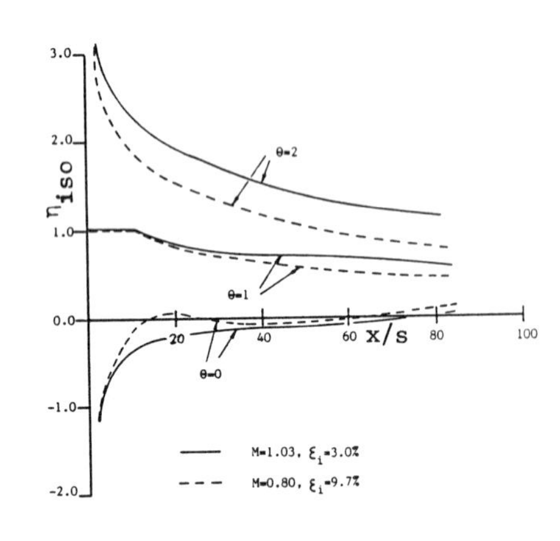

Fig. 7 Isothermal Film Cooling Effectiveness (η_{iso}) Versus the Dimensionless Distance (x/s).

x/s larger than 58.7. The negative values of h/h_o represent the heat fluxes from the fluid to the wall rather than from the wall to the fluid. The value of h/h_o as zero represents that the wall is adiabatic and the corresponding value of θ is the reciprocal of η_{aw}. At values of θ larger than 1.9, all values of h/h_o become negative and result in values of η_{iso} being greater than unity.

With the values of h/h_o available, the isothermal film cooling effectiveness can then be obtained ($\eta_{iso} = 1-h/h_o$). The results of two cases for M=1.03 and M=0.80, as shown in Fig. 7, are obtained from the curves in Fig. 6(a) and (b). It can be observed that the values of η_{iso} range from -1 to +3, which depends strongly on the values of θ.

The values of θ are probably in the range of 1 to 5 in practical applications. For values of θ less than 1, it may be convenient to perform experiments in laboratories. For instance, h_f/h_o can be obtained by measuring h/h_o at θ = 0; without any heating or cooling device, values of θ from 0.7 to 1 can be obtained in the present Ludwieg tube wind tunnel.

3.4 Adiabatic Film Cooling Effectiveness (η_{aw})

Film cooling data are customarily presented using η_{aw}. For the purpose of comparison, η_{aw} is converted from η_{iso} (or h/h_o) and its plot for different injection ratios (M) is shown in Fig. 8 using the location x/s as a parameter (x/s=19.6, 32.6, 58.7, and 82.8). The values of η_{aw} at x/s < 11 are all above 0.90; they are hardly distinguishable and thus are not presented in this figure.

The data show that increases in the injection ratios (M) increase the effectiveness (η_{aw}), but η_{aw} tends to level off at M > 2. Thus, further increase in M above 2 will not increase η_{aw} further. The data in Fig. 8 can be divided into two groups as shown in Table 1. It can be seen that the values of η_{aw} in group B (lower ε_i with open symbols) are higher than those in group A (higher ε_i with solid symbols) for the same x/s. The differences in η_{aw} range from 10% to 25% at x/s=58.7 and 82.8, and 5% to 10% at x/s=32.6 and 19.6.

3.5 Adiabatic Heat Transfer Coefficients (h_f)

The adiabatic heat transfer coefficients with film cooling are extrapolated using the h/h_o versus θ curves. For the purpose of clarity, h_f/h_o for eleven cases are shown in three separate figures. As shown in Fig. 9(a), h_f/h_o is plotted against x/s for M=1.80, 1.27, 1.03, and 0.41. The largest value of h_f/h_o is obtained at the highest injection ratio (M=1.80) in which h_f/h_o drops from 2.8 at x/s=1.96 to 1.25 at x/s=82.8. A good agreement is indicated by comparing the data of M=1.27 with that measured by Hartnett et al. [11] for M=1.23 (dash line). Fig. 9(b)

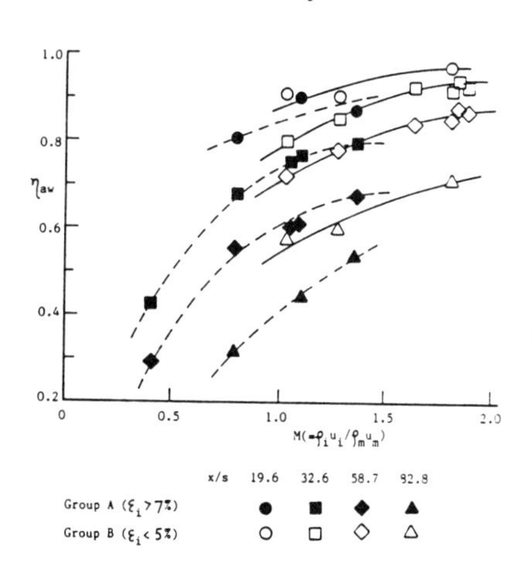

Fig. 8 Adiabatic Film Cooling Effectiveness (η_{aw}) Versus the Injection Ratio (M) for x/s < 15.

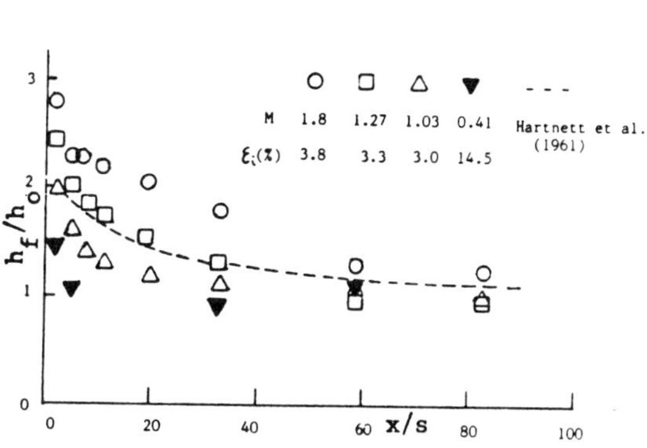

Fig. 9(a) h_f/h_o Versus x/s for Selected Cases.

shows h_f/h_o for five values of M in which solid symbols represent cases with higher turbulence. From the comparisons of data with higher turbulence (M=1.35 and M=1.10) to data with lower turbulence (M=1.27 and M=1.03), it is observed that higher injection turbulence results in higher heat transfer rates. This effect is very strong for small x/s but decays as x/s increases. The remaining cases of h_f/h_o are presented in Fig. 9(c). By comparisons of data for M=1.10 (ε_i=8.0%), M=1.05 (ε_i=8.2%), and M=1.03 (ε_i=3.0%), it can be concluded that the heat transfer coefficients with and without injection are equal far downstream n matter what the initial values of injection stream turbulence intensities are.

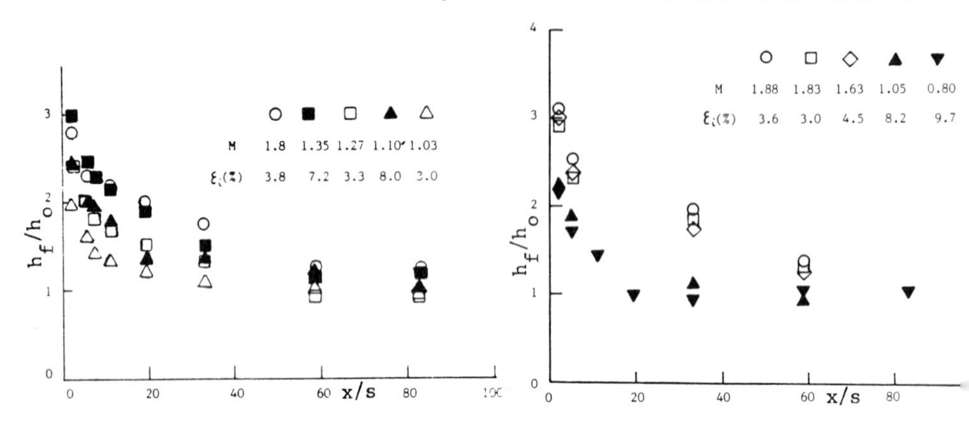

Fig. 9(b) h_f/h_o Versus x/s for Selected Cases.

Fig. 9(c) h_f/h_o Versus x/s for the Remaining Cases.

4. CONCLUSIONS

a. The steady state values of the flow field can be reached and maintained in the order of 10 milliseconds. The time required for the mixing of two streams t take place downstream of the injection slot is very short, such that steady stat values of wall heat transfer rates can be obtained.

b. The linear relation between the isothermal film-cooling effectiveness (η_{iso}) and the adiabatic film-cooling effectiveness (η_{aw}) as obtained from the superposition method are demonstrated using the present experiments.

c. As the injection ratio (M) increases, the effectiveness (η_{aw}) increases but levels off at M > 2. An increase in M also enhances the heat transfer coefficient in the near-slot region.

d. The effect of the injection stream turbulence intensity on film cooling effectiveness is to reduce the effectiveness resulting from the enhanced mixing and higher entrainment of the main flow into the boundary layer. The enhancement effect of the injection-stream turbulence on the heat transfer coefficient is significant in the near-slot region. The effect decays downstream and finall vanishes sufficiently downstream.

e. In real applications, the values of θ are greater than unity. As discussed in Section 3.3, there is a small region close to the injection slot where the heat transfer with injection are in the direction opposite to the heat transfer without film cooling, i.e., the isothermal effectiveness (η_{iso}) greater than one T_w is close to the constant wall temperature condition in real applications. Therefore, the present study of heat transfer distributions is close to the real situations.

5. REFERENCES

[1] Sturgess, G. J., "Account of Film Turbulence for Predicting Film Cooling Effectiveness in Gas Turbine Combustors," *ASME J. Engr. for Power,* Vol. 102, pp. 524-534, July 1980.

[2] Best, R., "The Influence of Coolant Turbulence Intensity on Film Cooling Effectiveness," AGARD-CP-229, *High Temperature Problems in Gas Turbine Engines,* pp. 17/1-17/14, 1978.

[3] Kacker, S. C. and Whitelaw, J. H., "The Effect of Slot Height and Slot-Turbulence Intensity on the Effectiveness of the Uniform Density, Two-Dimensional Wall Jet," *ASME J. Heat Transfer,* Vo. 90, pp. 469-475, Nov. 1968.

[4] Eckert, E. R. G., "Analysis of Film Cooling and Full-Coverage Film Cooling of Gas Turbine Blades," *ASME J. Engr. for Gas Turbines and Power,* Vol. 106, pp. 206-213, Jan. 1984.

[5] Goldstein, R. J., "Film Cooling," *Advances in Heat Transfer,* Vol. 7, pp. 321-379, ed. Irvine, T. F. Jr. and Hartnett, J. P., Academic Press, 1971.

[6] Chen, S. J. and Tsou, F. K., "Computer-Based Data Acquisition of Heat Transfer in Short-Duration Flow over a Backward-Facing Step," ASME/AIChE National Heat Transfer Conference, Niagara Falls, New York, Aug. 1984.

[7] Tsou, F. K., Smith, L. T., and Chen, S. J., "Experimental Determination of Transition Reynolds Number for Unsteady Film Cooling," ASME paper 81-GT-94, Mar. 1981.

[8] Chen, S. J., "The Effect of Injection-Stream Turbulence on Film Cooling Effectiveness and Heat Transfer Coefficients," Ph.D. Thesis, Drexel University, Feb. 1985.

[9] Metzger, D. E., Carper, H. J., and Swank, L. R., "Heat Transfer with Film Cooling Near Nontangential Injection Slots," *ASME J. Engr. for Power,* Vol. 90, pp. 157-163, Apr. 1968.

[10] Richards, B. E., "Heat Transfer Measurements Related to Hot Turbine Components in the Von Karman Institute Hot Cascade Tunnel," AGARD-CP-281, *Testing and Measurement Techniques in Heat Transfer and Combustion,* pp. 6/1-6/13, 1980.

[11] Hartnett, J. P., Birkebak, R. C., and Eckert, E.R.G., "Velocity Distributions, Temperature Distributions, Effectiveness and Heat Transfer in Cooling of a Surface with a Pressure Gardient," *International Development in Heat Transfer,* Part 4, ASME, pp. 682, 1961.

ROTATING TUBES, CHANNELS, AND HEAT PIPES

Influence of Internal Axial Fins on Condensation Heat Transfer in Co-Axial Rotating Heat Pipes

P. J. MARTO and A. S. WANNIARACHCHI
Department of Mechanical Engineering
Naval Postgraduate School
Monterey, California 93943

INTRODUCTION

Co-axial rotating heat pipes are being utilized to effectively cool rotating machinery components [1-6]. Marto [7] described the concept of these devices and reviewed the existing literature pertaining to their operating principles, thermal limitations and applications.

Figure 1(a) shows the co-axial rotating heat pipe in its simplest form, which is nothing more than an evacuated hollow shaft rotating about its axis. In this configuration, the thermal performance of the device is very sensitive to the volume charge of liquid and the speed of rotation. The operation of this device at various liquid fills and operational speeds has been thoroughly studied by Nakayama et al. [8] and more recently by Katsuta et al. [9]. When the heat pipe is operated at very high rotational speeds, the liquid will form an annulus in solid-body rotation. Figure 1(b) shows a stepped-wall heat pipe operating at high

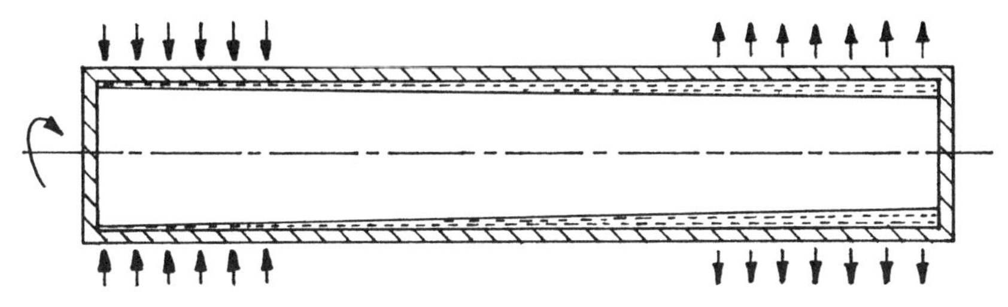

(a) Simple cylindrical heat pipe

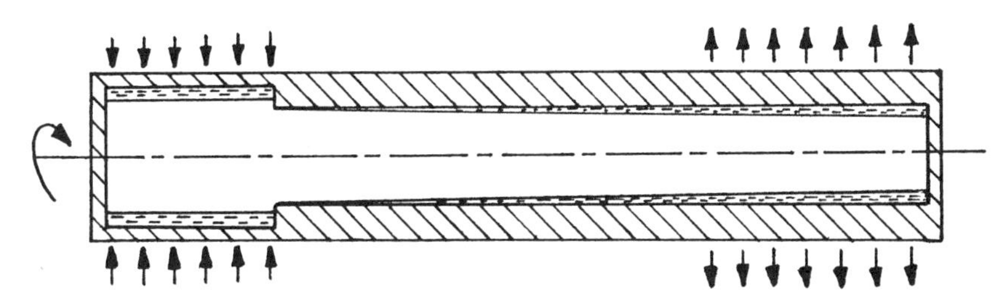

(b) Stepped cylindrical heat pipe

FIGURE 1. Schematics of rotating co-axial heat pipes.

rotational speed. The presence of the stepped wall and the high rotational speed can improve upon thermal performance although care must be exercised to ensure that the condenser wall is not too thick [10]. It is easy to see that, in this situation, the thermal behavior is almost completely independent of the volume charge of liquid, provided enough liquid is present to form a complete annulus in the evaporator and a thin liquid film in the condenser.

Various investigations have occurred to study the heat-transfer process in the condenser section of these stepped, co-axial rotating heat pipes. Marto and Wagenseil [11] showed that the thermal performance of a smooth, cylindrical condenser section could be dramatically improved upon by tapering the condenser walls or by using internal, helical fins along the cylindrical condenser surface. Marto and Weigel [12] studied several economical ways to improve upon the condenser performance by using off-the-shelf, internally finned tubing for the condenser sections. Vasiliev and Khrolenok [13] demonstrated that straight, longitudinal grooves in the condenser wall can also significantly improve upon thermal performance.

The purpose of this paper is to report on some additional experimental work where the influence of straight, axial fins in the condenser section is examined in relation to a smooth cylindrical surface. Using wall-mounted thermocouples to estimate the inside wall temperature, a detailed comparison between the condensation heat-transfer coefficients was possible, yielding more insight into the important mechanisms occurring in these devices.

EXPERIMENTAL EQUIPMENT AND PROCEDURES

The experimental apparatus is shown schematically in Figure 2. The evaporator was an electrically-heated copper cylinder 100 mm in diameter and 70 mm in length. A glass end-window was provided to view the boiling and condensing processes. Two copper condensers were tested during this investigation. Each condenser was 295 mm long with an effective length of 250 mm. They were flanged on each end for easy installation and removal. One condenser was a smooth-walled cylinder with a

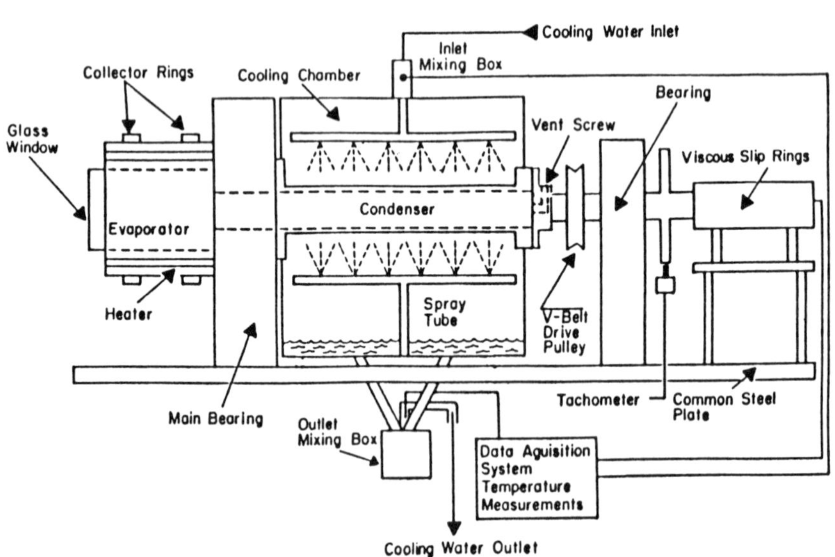

FIGURE 2. Schematic of heat pipe apparatus.

FIGURE 3. Photograph of axially-finned condenser section.

inside diameter of 23 mm and a wall thickness of 2.0 mm. The other was a sample of "Forge-Fin" tubing which was manufactured by Noranda Metal Industries. It had an inside diameter of 26 mm and a wall thickness of 1.5 mm, and contained twenty-two straight axial fins as shown in Figure 3. Each fin was 1.35 mm high and 1.2 mm thick. The internal surface area of this finned condenser was estimated to be 1.44 times larger than the internal surface area of the smooth cylinder. Cooling of the condenser sections took place by spraying tap water along the length of the condensers using numerous jets issuing from four distribution tubes placed 90 degrees apart around the condensers. This spray mechanism was located within an insulated chamber so that an accurate heat measurement could be obtained.

All temperatures were measured using 30-gage, type-E (i.e., chromel-constantan) thermocouple wire encased in Teflon and plastic insulation. Two vapor-space thermocouples were mounted inside 1.6 mm diameter stainless steel wells, which projected approximately 50 mm into the vapor space as shown in Figure 4. Eight thermocouples were used to measure the condenser wall temperature distribution. The locations of these thermocouples are also shown in Figure 4. They were placed at various axial and circumferential positions and were flush-mounted in the wall to measure the average wall temperature during condensation. Care was taken to ensure that accurate temperature measurements could be obtained. Each welded thermocouple junction was epoxied into a very small groove (approximately 1 mm wide, 1 mm deep and 10 mm long) which had been milled into the copper wall. The thermocouple was placed into the groove and a small amount of epoxy was added

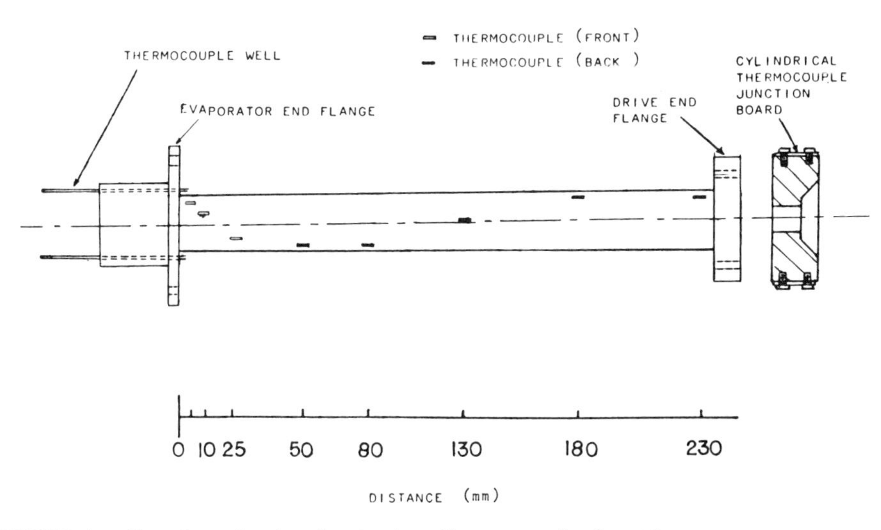

FIGURE 4. Drawing showing heat pipe thermocouple locations.

to fill the groove. A lubricated brass foil was then placed tightly over the epoxy for approximately four hours, after which the foil was removed. This left smooth cylindrical surface with little excess epoxy.

The cooling water inlet temperature was measured using a single thermocouple, whereas the outlet temperature was measured using five thermocouples wired in parallel to ensure a good representation of an average value. The outlet thermocouples were mounted in the discharge line from the cooling chamber after the water had flowed through an insulated mixing chamber. All the thermocouples were read by a Hewlett Packard 3054A data acquisition system controlled by an HP-9826 computer. The revolving thermocouples were wired to the data acquisitio system through mercury slip rings. All the thermocouples were calibrated using Rosemount constant-temperature bath and an accurate mercury-in-glass thermometer (0.05-K resolution) as a standard. The condenser wall thermocouples were calibrated both before and after installation and no significant difference was observed. A single calibration curve was generated for both sets of condenser wall and vapor space thermocouples, with an estimated uncertainty of ± 0.2 K. T cooling water inlet and outlet temperatures were calibrated to better than ± 0.1 K. Cooling water flow rate through the cooling chamber was measured with rotameter which had been calibrated using a weigh tank and stop watch. Addition experimental details may be found in [14].

Since it is well known that noncondensable gases can significantly reduce condensation heat-transfer rates in rotating heat pipes [15], care was exercised to eliminate as many system leaks as possible. Repeated pressure and vacuum checks were made to ensure a tight system. The heat pipe was accepted for operation if the leak rate, at an absolute pressure of about 10 mm Hg, was less than 1.0 mm Hg in 10 hours. Each of the condenser surfaces was cleaned chemical to ensure wettability of water [14]. A similar filling and venting procedure as used successfully by Marto and Weigel [12] was followed during these experiments Operation took place with the heat pipe in a horizontal orientation. Data were taken at different rotational speeds and with different input power settings. I took approximately 10 minutes to reach steady-state conditions for each data point. During steady-state conditions, all thermocouple readings were recorded along with the cooling water flow rate. The condenser heat-transfer rate was determined using the measured values of the cooling water flow rate and temperature increase:

$$Q = \dot{m}\, c_p\, (T_{co} - T_{ci}) \quad . \qquad ($$

A correction was made to this result to take into account frictional heating effects of the bearings and seals, and viscous dissipation effects within the cooling water during rotation. The average inside heat-transfer coefficient of the condenser section $\overline{h_i}$ was calculated using the calculated value of the heat-transfer rate Q and the measured values of the heat pipe vapor temperature T_s an the average condenser wall temperature $\overline{T}_{wo}$. For one-dimensional radial outflow heat, the heat-transfer rate may be written as:

$$Q = (T_s - \overline{T}_{wo})/(R_i + R_w) \quad , \qquad ($$

where R_i represents the thermal resistance on the inside of the condenser and R_w represents the wall resistance. Re-arranging eq'n (2), the average inside heat-transfer coefficient may be expressed as:

$$\overline{h}_i = \left[A_i\left(\frac{T_s - \overline{T}_{wo}}{Q} - \frac{\ln(r_o/r_i)}{2\pi L k_w}\right)\right]^{-1} \quad , \qquad ($$

where A_i is the inside surface area, r_i and r_o represent the inner and outer rad respectively, k_w is the wall thermal conductivity and L is the effective length of the condenser.

RESULTS AND DISCUSSION

Each of the test condenser surfaces was operated at rotational speeds of 700, 1400 and 2800 RPM using water as the test fluid. Figure 5 shows the overall heat-transfer results for the smooth-walled cylindrical condenser. The solid lines represent the data of Marto and Weigel [12] for a condenser of nearly the same size. Agreement between the two sets of data is excellent considering the uncertainty in the experimental measurements.

The film condensation process inside a co-axial rotating heat pipe whose condenser section is a smooth-walled cylinder should be similar to film condensation on the inside of a rotating drum, or to film condensation on a finite horizontal plate. In each of these cases, the condensate that builds up on the condenser surface will flow over the edge of the condenser under the action of hydrostatic forces on the condensate film created by either centrifugal force or the earth's gravitational force. Nimmo and Leppert were the first to analyze this problem [16-18]. Later, related studies were conducted by Marto [19] and Roetzel and Newman [20]. Nimmo and Leppert [18] analyzed film condensation on a finite horizontal surface, assuming laminar flow conditions in the condensate film and negligible liquid-vapor shear. They arrived at an expression for the average Nusselt number given by:

$$\overline{Nu} = \frac{\bar{h}L}{k} = 0.82 \left(\frac{g \rho^2 h_{fg} L^3}{k \mu (T_s - T_w)} \right)^{1/5} , \tag{4}$$

where the dimensionless grouping in the parentheses is known as the Sherwood number, Sh. Their experimental data fell approximately 20 percent lower than their theory, so that the recommended coefficient in eq'n (4) was reduced from 0.82 to 0.64. Notice that when their analysis is applied to condensation on the inside of a rotating cylinder, the gravitational acceleration g in eq'n (4) should be replaced by $\omega^2 r_i$.

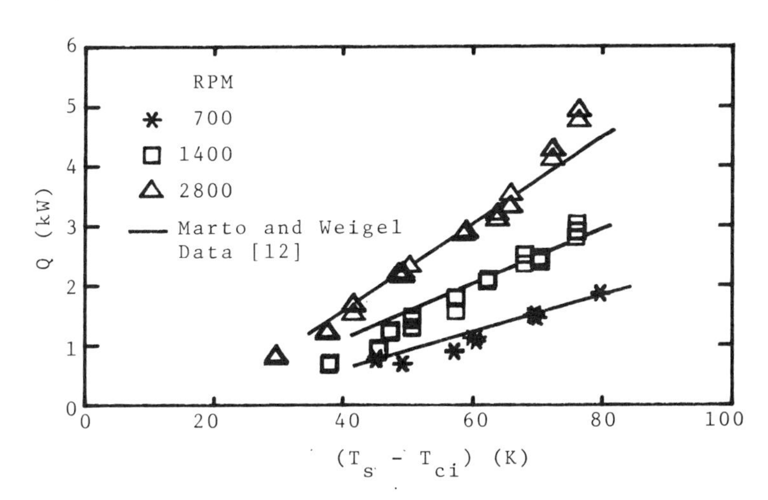

FIGURE 5. Thermal performance of smooth-walled cylindrical condenser.

When the present data for the smooth-walled cylinder are utilized to find the inside heat-transfer coefficient using eq'n (3), and these data are plotted as t average Nusselt number ($\bar{h}_i L/k$) versus Sherwood number, Figure 6 results. Equati (4) with the lower coefficient of 0.64 is plotted in the figure for comparison. Notice that the three sets of data at different rotational speeds appear to foll the Sherwood number trend given by eq'n (4), but at a given rotational speed, th data have considerable spread, indicating that the correlation as proposed is no suitable. The lower data points, taken at low heat fluxes, are in better agreement with eq'n (4) than the data points taken at high heat fluxes which hav correspondingly thicker condensate films. A possible explanation of this discrepancy is that the condensate film was not laminar, but perhaps was experiencing turbulence due to inherent vibrations in the rotating heat pipe.

It is well known that, on a vertical surface, as the condensate film gets thicke and the film Reynolds number Re_f increases, it is possible for turbulent eddy motions to occur within the film. Usually, film Reynolds numbers of 1600-1800 a required for this characteristic change to occur. Further, once turbulence has begun, the average heat-transfer coefficient increases, rather than decreases, with film Reynolds number [21]. When the data of Figure 6 are re-plotted versus film Reynolds number, a successful turbulent-type correlation was found as shown in Figure 7 and expressed below:

$$\frac{\bar{h}_i}{k}\left(\frac{\nu}{\omega}\right)^{1/2} = 0.027\, Re_f^{0.328} \quad . \qquad ($$

The above expression correlates the data at all three rotational speeds to ± 12 percent. Notice that the data span a film Reynolds number range of 25 to 300 which is far below the critical Reynolds number of 1600-1800 mentioned above. However, in the presence of interfacial shear, Rohsenow et al.[22] have shown th for condensation on a flat plate, the critical film Reynolds number can be reduc to 50. Carpenter and Colburn [23] also reasoned that the condensate layer could become turbulent at much lower film Reynolds numbers when interfacial shear was present. Thus, with this information and realizing that the rotating heat pipe experiences high-frequency vibrations during operation, it seems plausible to expect turbulence to have occurred.

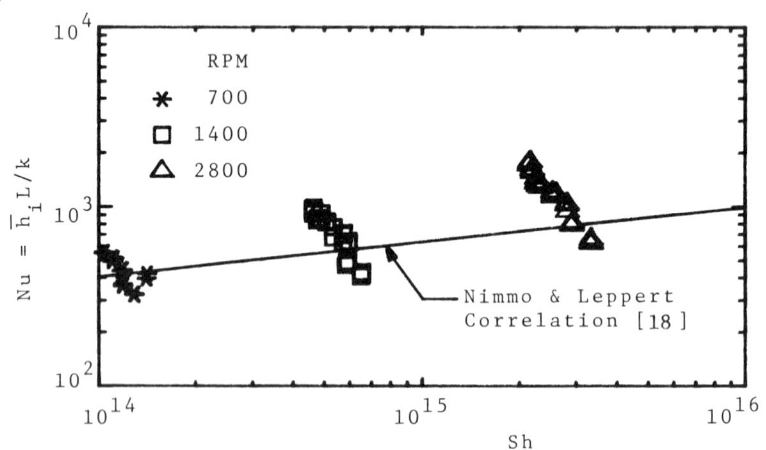

FIGURE 6. Comparison of smooth-walled condenser data with laminar film condensation theory.

Equation (4) (with a coefficient of 0.64) for laminar flow is also plotted in Figure 7 and it clearly shows an opposite trend to the data. An interesting observation can be made regarding the Nimmo and Leppert analysis. In deriving eq'n (4), they assumed that the condensate overfall is well rounded rather than sharp edged. If the condensing surface is terminated with a sharp edge, then surface tension effects, which were neglected, may become very important. Nimmo and Leppert, using an earlier analysis of Bankoff [24] which included surface-tension effects, arrived at an approximate correction factor for the average Nusselt number when a sharp edged overfall exists:

$$\frac{Nu_c}{Nu} = \frac{0.61\ Bo^{1/2}}{Sh^{1/5}\ \sin(\phi/2)} \tag{6}$$

In the above expression, Bo is the Bond number given by $\rho g L^2/\sigma$ and ϕ is the liquid contact angle. Using the conditions of this experiment and substituting this information into eq'n (6) yields a corrected Nusselt number which is plotted for 700 RPM in Figure 7. The corrected Nusselt number for the other rotational speeds would be larger since the Bond number increases with rotational acceleration.

Figure 8 shows the overall heat-transfer results for the finned condenser. The solid lines represent the least-squares fits to the data, while the dashed lines represent the smooth-walled data referred to earlier in Figure 5. Clearly, the presence of the axial fins enhances the heat transfer at all rotational speeds. For example, at a temperature difference (T_s-T_{ci}) of 60 K and a rotational speed of 700 RPM, the finned condenser carries away approximately 3 kW of heat whereas at the same conditions the smooth-walled condenser carries away only approximately 1 kW. Since these two condensers have the same approximate outside diameter and wall thickness, the three-fold gain in heat transfer must be due primarily to a significant enhancement process on the inside surface. Figure 9 shows a plot of the condensation heat-transfer coefficient ratio for the finned tube to the smooth tube as a function of the vapor to wall temperature difference (T_s-$\overline{T}_{wi}$). The curves show that at low temperature differences, corresponding to low heat fluxes and thin condensate films, enhancements of 4-7 are possible. Increased enhancement occurs with an increase in rotational speed. Clearly, as the

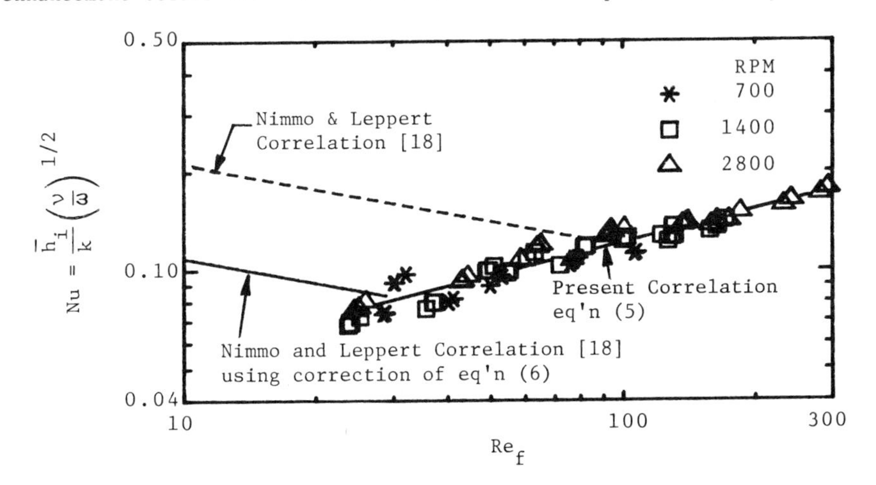

FIGURE 7. Correlation of smooth-walled condenser data.

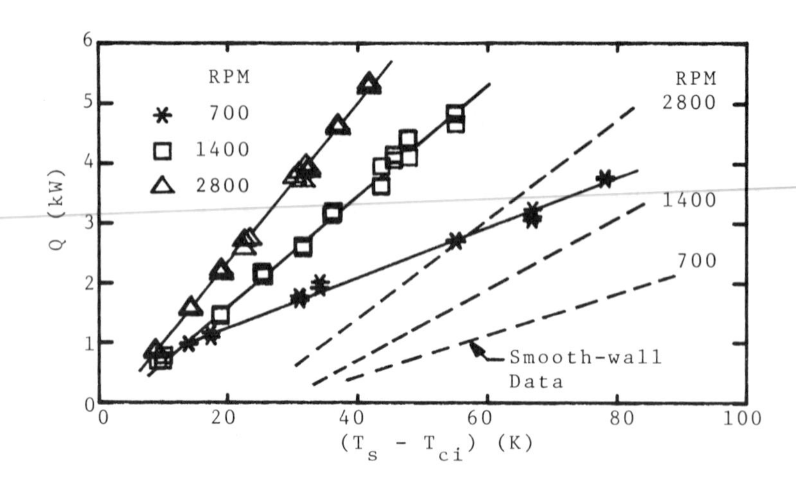

FIGURE 8. Thermal performance of internally-finned condenser.

centrifugal force increases, the condensate formed on the fins is quickly removed keeping the condensate film on the fin very thin and leading to very high heat-transfer coefficients. At large temperature differences, as the condensate begins to build up in the trough regions between fins, the enhancement is not as dramatic. Based upon this experimental information, it will be important to mode this process to investigate what fin geometries will give the best performance. An analysis to describe this problem is now underway, and results will be reporte in the future.

CONCLUSIONS

From the above-described experimental results, several important conclusions can be reached:

1. In a co-axial rotating heat pipe, the use of a stepped condenser section can provide efficient thermal transport of heat.

2. When using a smooth-walled cylindrical condenser, turbulence may occur in the condensate film. This may be unavoidable for long slender heat pipes where sizeable interfacial shear exists between the liquid and the vapor. Vibrati may also influence the onset of turbulence. The geometry of the step betwee the evaporator and condenser could influence the heat transfer as well.

3. The use of straight axial fins in the condenser can lead to significant heat transfer enhancement since the presence of the fins in a strong centrifugal force field will generate very thin condensate films on the fins.

4. It would be very fruitful to model the condensation process in the presence axial fins. Such a model should include the effects of interfacial shear. Optimum fin dimensions could thus be generated for given heat pipe operating conditions.

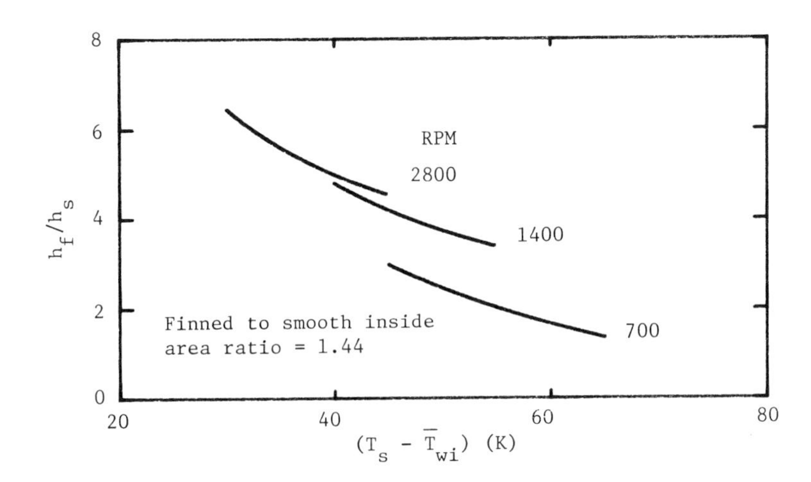

FIGURE 9. Heat-transfer enhancement with internally-finned condenser.

REFERENCES

1. Fries, P. Experimental Results with a Wickless Centrifugal Heat Pipe, Int. J. Heat Mass Transfer, vol. 13, pp. 1503-1504, 1970 (in German).

2. Pokorny, B., Polasek, F., Schneller, J., and Stulc, P., Heat Transfer in Co-axial and Parallel Rotating Heat Pipes, Proc. 5th Int. Heat Pipe Conf., Tsukuba, Japan, vol. 4, pp. 90-98, 1984.

3. Vasiliev, L. L., Kukharsky, M. P., and Khrolenok, V. V., Centrifugal Heat Pipes and Heat Exchangers, Energy and Mass Transfer Processes in Porous Media with Phase Change, ed. A. B. Lykov, pp. 1-31, BSSR Academy of Science, Minsk, 1982 (in Russian).

4. Sattler, P. K. and Thoren, F. , Totally Enclosed Heat Pipe Cooled Induction Motor, Proc. Int. Conf. Electric Machines, Laussane, Switzerland, 1984.

5. Brost, O., Unk, J. and Canders, W. R.,Heat Pipes for Electric Motors, Proc. 5th Int. Heat Pipe Conf., Tsukuba, Japan, vol. 4, 151-153, 1984.

6. Furuya, H., Heat Pipe Applications to Electronic Devices, Electric Equipment and Machine Tools, The Institute of Space and Astronautical Science Research Note, Tokyo, Japan, pp. 18-39, April, 1983.

7. Marto, P. J., Rotating Heat Pipes, Heat and Mass Transfer in Rotating Machinery, eds. D. E. Metzger and N. H. Afgan, pp. 609-632, Hemisphere, Washington, D. C., 1984.

8. Nakayama, W., Ohtsuka, Y., Itoh, H., and Yoshikawa, T., Optimum Charge of Working Fluids in Horizontal Rotating Heat Pipes, Heat and Mass Transfer in Rotating Machinery, eds. D. E. Metzger and N. H. Afgan, pp. 633-644, Hemisphere, Washington, D.C., 1984.

9. Katsuta, M., Kigami, H., Nagata, K., Sotani, J. and Koizumi, T., A Study on Performance and Characteristics of the Rotating Heat Pipe, Proc. 5th Int. Heat Pipe Conf., Tsukuba, Japan, vol. 4, pp. 106-112, 1984.

10. Nakayama, W., Ohtsuka, Y., and Yoshikawa, T., The Effects of Fine Surface Structures on the Performance of Horizontal Rotating Heat Pipes, Proc. 5th Int. Heat Pipe Conf., Tsukuba, Japan, vol. 4, pp. 99-103, 1984.

11. Marto, P. J. and Wagenseil, L. L., Augmenting the Condenser Heat-Transfer Performance of Rotating Heat Pipes, AIAA Journal, vol. 17, no. 6, pp. 647-652, 1979.

12. Marto, P. J. and Weigel, H., The Development of Economical Rotating Heat Pipes, Advances in Heat Pipe Technology, ed. D. A. Reay, pp. 709-724, Pergamon Press, New York, 1982.

13. Vasiliev, L. L., and Khrolenok, V. V., Study of a Heat Transfer Process in the Condensation Zone of Rotating Heat Pipes, Heat Recovery Systems, vol. 3 no. 4, pp. 281-290, 1983.

14. Nefesoglu, A., Heat Transfer Measurements of Internally Finned Rotating Hea Pipes, M.S. Thesis, Naval Postgraduate School, Monterey, California, 1983.

15. Daniels, T. C. and Williams, R. J., Experimental Temperature Distribution a Heat Load Characteristics of Rotating Heat Pipes, Int. J. Heat Mass Transfe vol. 21, pp. 193-201, 1978.

16. Leppert, G. and Nimmo, B., Laminar Film Condensation on Surfaces Normal to Body or Inertial Forces, J. Heat Transfer, vol. 90, pp. 178-179, 1968.

17. Nimmo, B., Laminar Film Condensation on a Finite Horizontal Surface, Ph.D. Thesis, Stanford University, California, 1968.

18. Nimmo, B. and Leppert, G., Laminar Film Condensation on a Finite Horizontal Surface, Heat Trnasfer 1970, eds. V. Grigull and E. Hahne, pp. Cs 2.2, Elsevier Publishing Co., Amsterdam.

19. Marto, P. J., Laminar Film Condensation on the Inside of Slender, Rotating Truncated Cones, J. Heat Transfer, vol. 95, pp. 270-272, 1973.

20. Roetzel, W. and Newman, M., Uniform Heat Flux in a Paper Drying Drum with a Non-Cylindrical Condensation Surface Operating Under Rimming Conditions, In J. Heat Mass Transfer, vol. 18, pp. 553-557, 1975.

21. Butterworth, D., Film Condensation of a Pure Vapor, Heat Exchanger Design Handbook, vol. 2, pp. 2.6.2-3 to 2.6.2-7, Hemisphere, 1982.

22. Rohsenow, W. M., Webber, J. H. and Ling, A. T., The Effect of Vapor Velocit on Laminar and Turbulent Film Condensation, Trans. ASME, vol. 78, pp. 1637-1643, 1956.

23. Carpenter, E. F. and Colburn, A. P., The Effect of Vapor Velocity on Condensation Inside Tubes, Proc. General Discussion on Heat Transfer, Trans ASME, pp. 20-26, 1951.

24. Bankoff, S. G., The Contortional Energy Requirement in the Spreading of La Drops, Phys. Chem., vol. 60, pp. 952, 1956.

Generation and Evolution of Turbulence in an Annulus between Two Concentric Rotating Cylinders

KUNIO KATAOKA and TETSUYA DEGUCHI
Department of Chemical Engineering
Kobe University
Rokkodai, Nada, Kobe 657, Japan

1. INTRODUCTION

The objective of the present work is to observe the generation and spectral evolution of time-dependent wavy disturbances in the Taylor-Couette flow. It is well known that as the Reynolds number Re = $R_i \Omega d/\nu$, based on the rotation speed (Ω: angular velocity) of the inner cylinder, is gradually increased, the following five dynamical transitions occur stepwise in sequence: laminar Couette flow → laminar Taylor vortex flow → wavy vortex flow → quasi-periodic wavy vortex flow → weakly turbulent wavy vortex flow → turbulent vortex flow. Time-dependent wavy disturbances appear when the transition to wavy vortex flow occurs as a result of instability of the laminar Taylor vortex flow. The disturbances are regularly periodic because it results from the azimuthally traveling waves. The next transition to the quasi-periodic wavy vortex flow is accompanied by the amplitude modulation of the wave motion. The first fundamental frequency f_1 comes from the passing frequency of the azimuthally traveling waves and the second fundamental frequency f_2 from the modulation frequency. When the transition to the weakly turbulent wavy vortex flow occurs, chaotic turbulence first appears. A spectral analysis is made to analyze the temporal variation in the local velocity gradient measured on both the inner and outer cylinder walls by using an electrochemical technique [1]. FIGURE 1 shows a schematic diagram of wavy Taylor vortex flow.

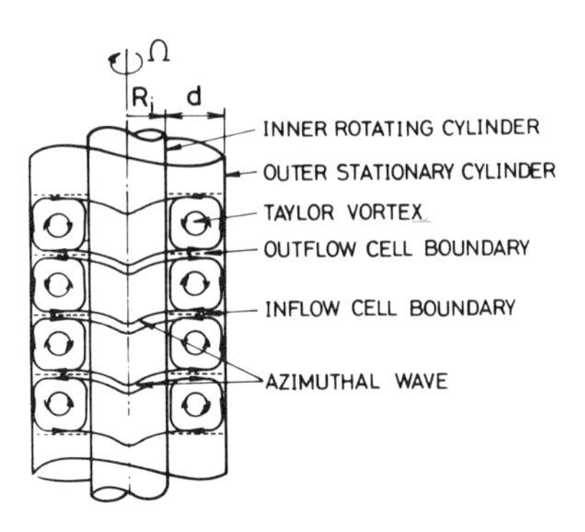

FIGURE 1. Wavy Taylor vortex flow.

2. EXPERIMENT

In a way similar to the previous work [2-6], the experiment was designed to meas ure local, time-dependent values of velocity gradient on both the inner and oute cylinder walls by making use of the following electrolytic redox reaction:

$$Fe(CN)_6^{3-} + e^- \rightarrow Fe(CN)_6^{4-} \qquad \text{at cathode}$$

$$Fe(CN)_6^{4-} \rightarrow Fe(CN)_6^{3-} + e^- \qquad \text{at anode}$$

FIGURE 2 shows the experimental apparatus consisting of two vertically mounted concentric cylinders of stainless steel. The outer cylinder was 122 mm inner diameter and 422 mm long. The inner cylinder was 75 mm outer diameter and 406 mm long. The upper and lower fluid surfaces were fixed by rigid flat flanges. The radius ratio $\eta = R_i/R_o = 0.615$ and the aspect ratio $\Gamma = 18.0$. A motor controlled by a thyristor cycloconverter was used to directly rotate the inner cylinder.

The liquid used was an electrolytic aqueous solution of 0.01 $kmol/m^3$ $K_3Fe(CN)_6$, $K_4Fe(CN)_6$ and 1 $kmol/m^3$ NaOH. Water glass (sodium silicate) was added to increas the fluid viscosity. The experiment was performed using a diffusion-controlled cathode reaction of ferri-cyanide ions.

Sixteen circular nickel anodes of 4 mm diameter were arrayed every 10 mm in the axial direction and made flush with the outer cylinder wall. These anodes were fabricated by inserting 4 mm diameter nickel rods concentrically into 5 mm diameter holes drilled in the wall of the outer cylinder and polishing them flush with the surrounding wall of the outer cylinder. As shown in FIGURE 2, five isolated rectangular nickel cathodes (numbered 01, 02, ---, 05 from above) were arrayed at intervals of 15 mm in the axial direction on the outer cylinder wall.

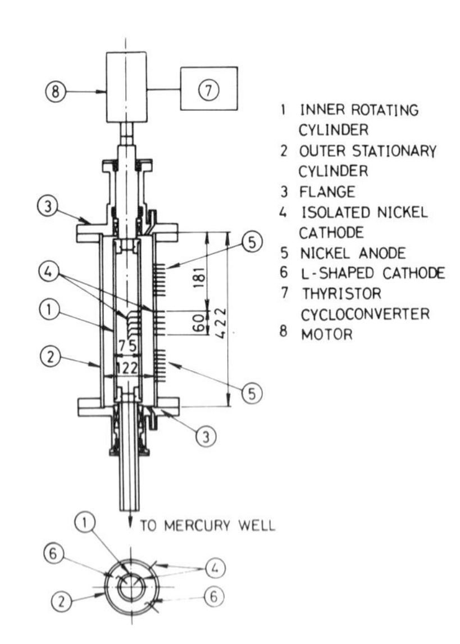

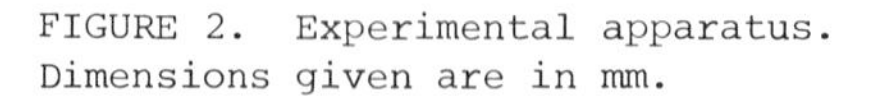

FIGURE 2. Experimental apparatus. Dimensions given are in mm.

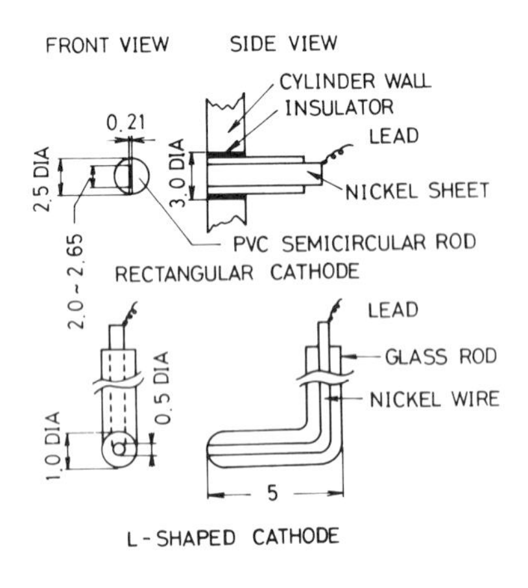

FIGURE 3. Isolated rectangular cathode and L-shaped cathode.

Similarly five isolated rectangular nickel cathodes (numbered I1, I2, ---, I5 from above) were arrayed at intervals of 15 mm in the axial direction on the inner cylinder wall. Those cathodes (I1, I2, ---, I5) of the inner cylinder confront the corresponding ones (O1, O2, ---, O5) of the outer cylinder. As shown in FIGURE 3, those electrodes were fabricated by inserting about 2 mm wide and 0.21 mm thick nickel sheet, sandwiched between two PVC semicircular cylinders, into 3 mm diameter drilled holes. The rectangular surface of each cathode exposed to the flowing fluid has an active area of 0.21 mm x 2.0 to 2.65 mm. The longer side of the rectangular surface was directed perpendicularly to the main rotating flow.

The limiting current I from each rectangular cathode (surface area A) gives local velocity gradient s at the inner or outer cylinder wall:

$$s = 1.90 \left(\frac{I}{F\ Cb\ A} \right)^3 \frac{L}{D^2} \quad (1)$$

where F is the Faraday constant, Cb is the bulk concentration of Fe^{3+} ions, L is the cathode length in the flow direction, and D is the diffusivity of Fe^{3+} ions.

Two L-shaped electrodes, shown in FIGURE 3, were used for ascertaining the azimuthal wave motion at two radial positions in the annulus. One electrode (named L1) was inserted into the annulus from the inner rotating cylinder. The circular active surface (about 0.5 mm diameter) normal to the main rotating flow was positioned at a radial position 1.0 mm apart from the inner cylinder wall and 173 mm from the top flange. The other electrode (named L2) was inserted into the annulus from the outer stationary cylinder. The circular active surface (about 0.5 mm diameter) normal to the main rotating flow was positioned at a radial position 3.5 mm apart from the outer cylinder wall and 201 mm from the top flange. Similarly to the hot-wire anemometry, these L-shaped electrodes give local values of azimuthal velocity component [1].

At a given Reynolds number, there were many distinct stable spatial states depending upon the flow history. The spatial state is characterized by the number N of cellular vortices and the number m of azimuthal waves. However, other flow properties (e.g. torque, velocity, and velocity gradient) show no substantial difference in their time-averaged values among those accessible spatial states. Therefore the main experiment was performed by rapidly accelerating the inner cylinder from rest to a specified Reynolds number. In addition, in order to observe the spectral evolution of wavy disturbances, the Reynolds number was gradually increased and then decreased over the region of the wavy vortex and the quasi-periodic wavy vortex flows.

3. EXPERIMENTAL RESULTS AND DISCUSSION

According to the linear stability theory [7], the critical Reynolds number for the present radius ratio can be considered to be $Re_c = 72.6$.

FIGURE 4 shows the axial variation of local velocity gradient at the wall of the outer stationary cylinder. FIGURE 5 shows the axial variation of local velocity gradient at the wall of the inner rotating cylinder. It is clear that the periodicity results from the axial array of Taylor cells. According to the previous work [5], the velocity gradient on the outer cylinder wall becomes maximal at the outflow cell boundaries and minimal at the inflow cell boundaries.

FIGURE 6 shows the power spectrum of the velocity-gradient fluctuation measured by the O3 electrode. The frequency was made dimensionless with respect to the frequency f_r of the inner cylinder rotation. As can be seen from FIGURE 4, this electrode was located near an outflow cell boundary. The flow showed the wavy

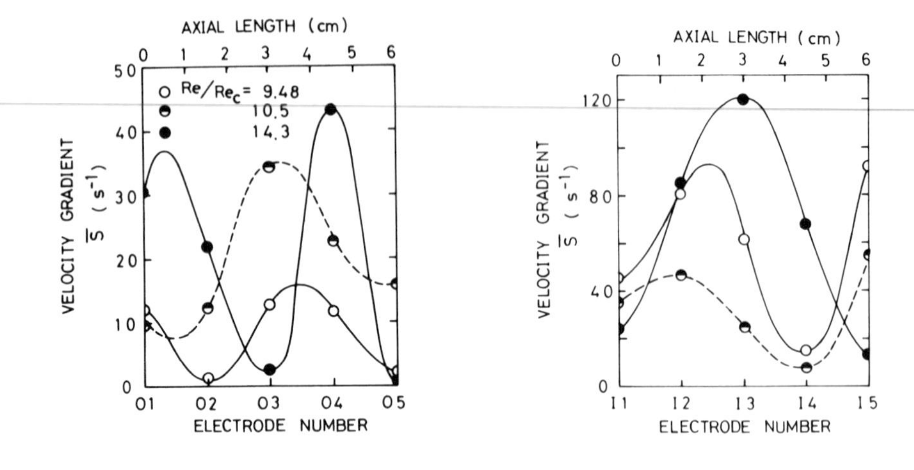

FIGURE 4. Axial variation of time-averaged velocity gradient on the wall of outer stationary cylinder.

FIGURE 5. Axial variation of time-averaged velocity gradient on the wall of inner rotating cylinder.

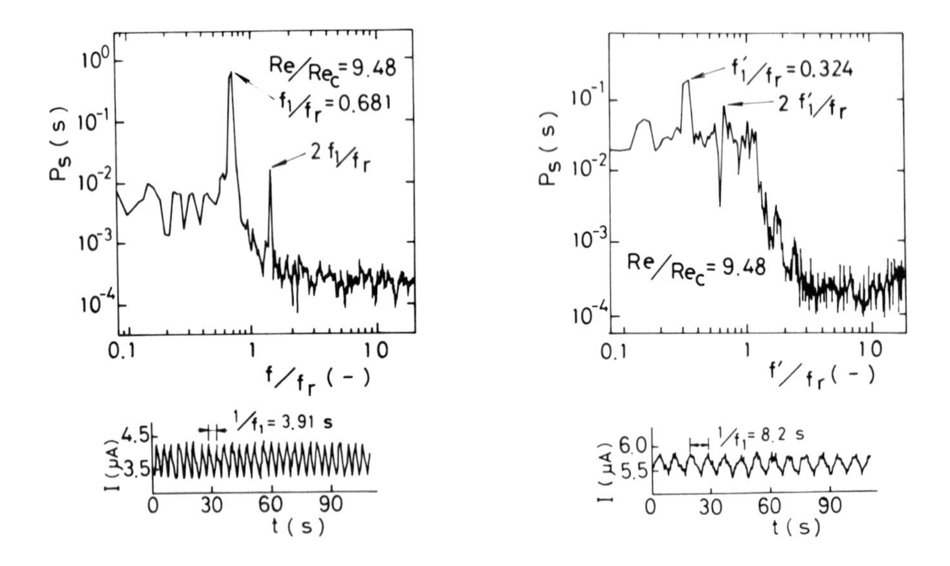

FIGURE 6. Time trace and power spectrum of velocity-gradient fluctuation measured on the wall of outer stationary cylinder. (Wavy vortex flow).

FIGURE 7. Time trace and power spectrum of velocity-gradient fluctuation measured on the wall of inner rotating cylinder.

vortex flow pattern. The time trace of the limiting current from the electrode, also shown in the same figure, has regular periodical oscillation with a frequency of f_1 = 0.256 Hz. The power spectrum indicates only one sharp peak at the same frequency (i.e. f_1/f_r = 0.681). This frequency can be considered as that of the velocity-gradient fluctuation due to the passage of azimuthal waves observed at a fixed point of the outer stationary cylinder wall.

FIGURE 7 shows the power spectrum of the velocity-gradient fluctuation measured by the I3 electrode. The electrode was also located at the same height as the O3 electrode. The power spectrum indicates a high sharp peak at a frequency f'_1/f_r = 0.324, which is coincident with the main frequency (f'_1 = 0.12 Hz) of the time trace. This frequency can be considered as that of the velocity-gradient fluctuation observed from a point rotating at the same angular velocity as the inner cylinder. It has been ascertained that neither f_1 nor f'_1 appears for the laminar Taylor vortex flow regime. The wavy vortex flow can be regarded as a singly periodic flow with a single fundamental frequency f_1.

The experimental apparatus had a peripheral array of isolated rectangular cathodes but the number of azimuthal waves could not be obtained clearly from the peripheral correlation measurements owing to the short spacing between the electrodes. The recent King et.al. observations [8] for large aspect ratios reported that the wave speed $f_1/m\ f_r$ should be about 0.13 for the present radius ratio (η = 0.615). This implies that m should be five because of f_1/f_r = 0.681. On the other hand, m is usually one or two for the wavy vortex flow regime established in wide gap annuli [9]. In addition, as pointed out by King et.al. [8], the wave speed tends

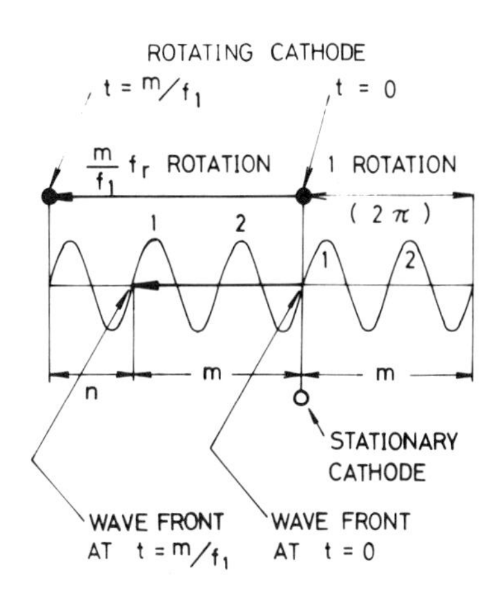

FIGURE 8. Schematic diagram of the relation of azimuthally traveling waves with azimuthally rotating cathode: m = number of waves passing the stationary cathode for time m/f_1; n = number of waves passed by the rotating cathode for time m/f_1; and n = m (m f_r/f_1 - 1).

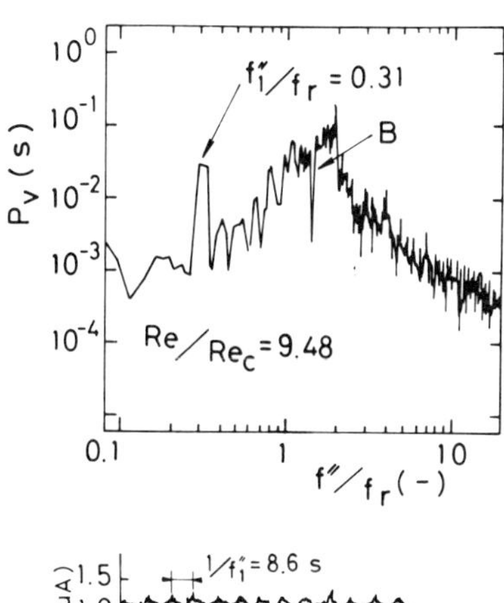

FIGURE 9. Time trace and power spectrum of velocity fluctuation measured by L1 electrode.

to increase considerably as Γ decreases lower than 20. It has been found from the flow-visualization observations [9] that the wave speed can become very large when the axial wavelength is considerably smaller than the annular gap width d. It has been concluded after careful examination that m should be two for this flow state.

As shown schematically in FIGURE 8, the rotating electrodes (I1, I2, ---, I5) should have had a large component at a frequency $m - f_1/f_r$ on the power spectra if the azimuthal wave motion could be detected by those rotating electrodes. However this is not the case. It has been found that there is another relation between f_1 and f'_1:

$$f_1 = m\, f'_1 \tag{2}$$

FIGURE 9 shows the power spectrum of the velocity fluctuation obtained by the L1 electrode. The electrode located 1 mm apart from the inner cylinder wall was rotated at the same angular velocity as the inner cylinder. The power spectrum indicates a high peak at $f''_1/f_r = 0.31$. This frequency is nearly equal to $f'_1/f_r = 0.324$. The broad component B appears in the higher frequency range due to the effect of the wake flow issuing from the inserted electrode itself. It can be

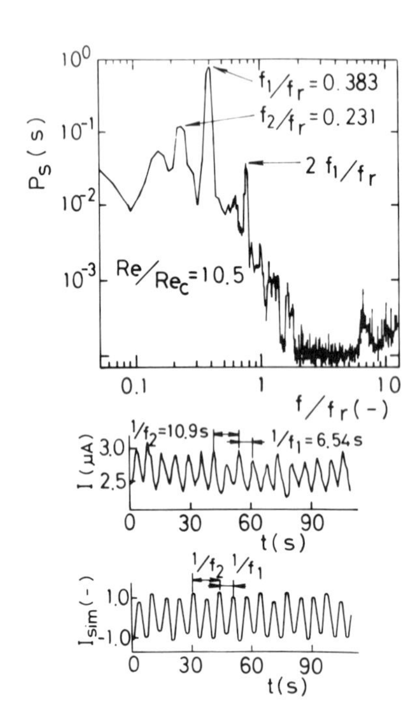

FIGURE 10. Time trace and power spectrum of velocity-gradient fluctuation measured on the wall of outer stationary cylinder. (Quasi-periodic wavy vortex flow).

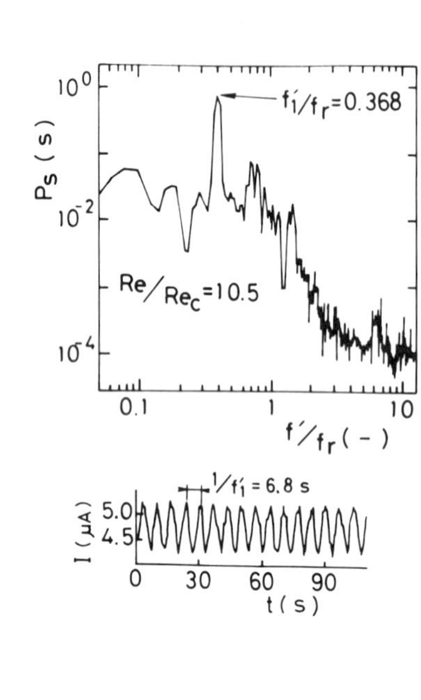

FIGURE 11. Time trace and power spectrum of velocity-gradient fluctuation measured on the wall of inner rotating cylinder.

conjectured from these results that wavy disturbance energy is supplied into the azimuthal wave motion due to the periodical bursting from the wall of the inner rotating cylinder. This periodical bursting seems to occur due to the centrifugal shedding of fluid elements near the inner rotating cylinder. This phenomenon is remarkable at the outflow cell boundaries. The power spectrum obtained by the L2 electrode also indicates the fundamental frequency equal to f_1.

FIGURE 10 shows the power spectrum of the velocity-gradient fluctuation obtained by the O3 electrode. The flow showed the quasi-periodic wavy vortex flow pattern. The electrode was located at an outflow cell boundary. The time trace of the limiting current from the electrode (shown in the same figure) has regular periodicity with small amplitude modulation. The first fundamental frequency f_1/f_r = 0.383 is coincident with the main frequency (f_1 = 0.153 Hz) of the time trace. The frequency f_2/f_r = 0.231 obtained from the second highest peak of the power spectrum is also coincident with that of the amplitude modulation. The same figure shows an oscillation curve simulated using these two frequencies f_1, f_2 and their power levels in the amplitude modulation equation:

$$I_{sim} = (1 + 0.3 \sin 2\pi f_2 t) \sin 2\pi f_1 t \tag{3}$$

The simulated curve is quite similar to the actual time trace.

FIGURE 11 shows the power spectrum of the velocity-gradient fluctuation obtained by the I3 electrode. A single sharp peak appears at f_1'/f_r = 0.368. It has been assumed that m should be one for this flow state. The relation $f_1 = m f_1'$ holds in this flow state. The power spectrum does not show a peak at the wave frequency $m - f_1/f_r$. The quasi-periodic wavy vortex flow can be regarded as a doubly-periodic flow with two fundamental frequencies f_1 and f_2.

TABLE 1 suggests that the main frequency f_1' of the wavy time-dependent disturbance produced owing to the periodical bursting is about 1/3 of the rotating frequency f_r of the inner cylinder. In addition, the wave speed relation $f_1/m f_r$ = 1/3 proposed by Coles [10] still holds in spite of small radius ratio. This is attributable to the small aspect ratio of the present cylinder system.

TABLE 1. Experimental results for the case of abrupt acceleration.

Re/Re_c	f_r(Hz)	f_1(Hz)	$f_1/m f_r$	f_1'/f_r	m	f_2(Hz)	f_2/f_r	Flow Regime
9.48	0.376	0.256	0.341	0.324	2	—	—	W
10.5	0.399	0.153	0.383	0.368	1	0.092	0.231	Q
10.5	0.399	0.269	0.337	vague	2	0.086	0.216	Q
10.7	0.352	0.244	0.347	0.296	2	—	—	W
11.0	0.365	0.256	0.351	0.301	2	—	—	W
14.3	0.222	0.348	0.314	vague	5	0.079	0.358	Q
15.6	0.518	0.354	0.342	0.307	2	0.147	0.284	Q
15.6	0.618	0.989	0.320	vague	5	0.098	0.159	Q
17.7	0.294	0.311	0.353	0.354	3	0.110	0.374	Q
17.7	0.294	0.238	0.405	0.207	2	0.043	0.145	Q
17.7	0.294	0.208	0.354	0.333	2	0.092	0.313	Q

W; wavy vortex flow   Q; quasi-periodic wavy vortex flow

FIGURE 12 shows a model of the generation and propagation of time-dependent wavy disturbance in the azimuthal wave motion. The time-dependent wavy disturbance is generated at about $f'_1 = f_r/3$ owing to the periodical bursting on the wall of the inner rotating cylinder and transferred into the wave motion. It seems that the periodical bursting occurs in resonance with the azimuthal wave motion.
The first fundamental frequency f_1 of the wave motion is in proportion to the number m of the azimuthal waves. This is similar to the fact that the frequency of string oscillation is inversely proportional to the string length.

The time-dependent periodical disturbance is propagated according to the following equation of the wave height y:

$$y = \sin (2\pi f_1 t - m \theta) \tag{4}$$

where $f_1 = m f'_1$.

The same figure shows a schematic diagram of the azimuthal wave motion calculate by this wave height equation. It can be seen from the figure that this model holds for the wave motion with m = 1, 2, and 3. It should be kept in mind that this model is based on the assumption of m.

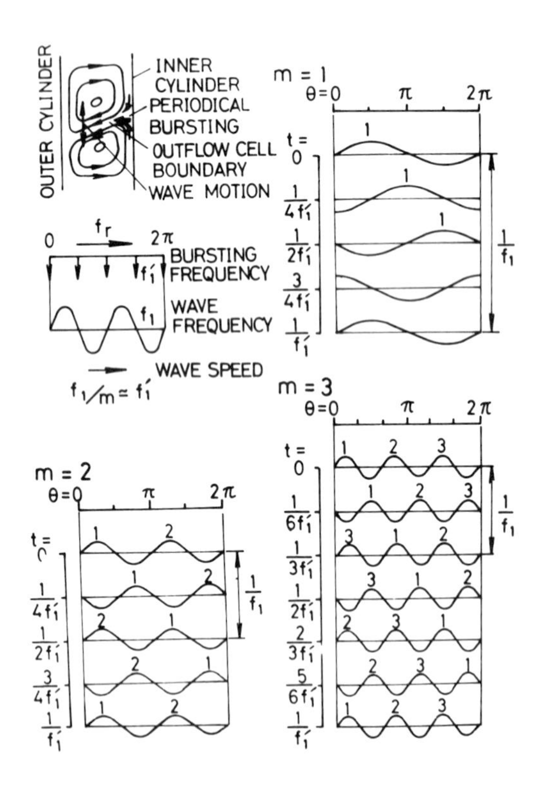

FIGURE 12. Model of generation and propagation of wavy disturbances.

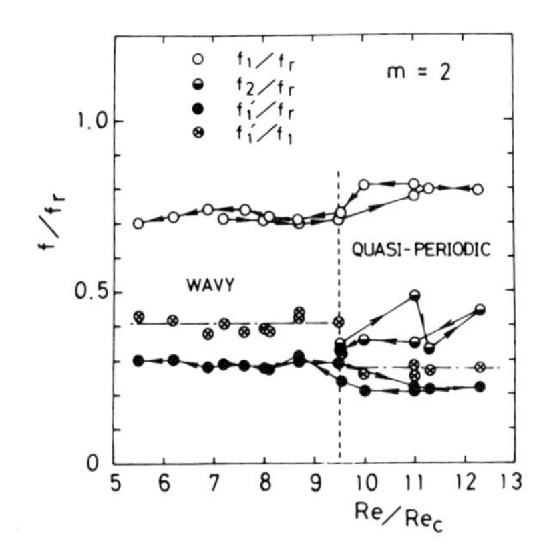

FIGURE 13. Variation of fundamental frequencies with Reynolds number for gradual acceleration and deceleration of inner cylinder.

It is found from the spectral analysis that the spectral evolution of the time-dependent wavy disturbance generated from near the inner cylinder wall is suppressed by the stable azimuthal wave motion with one or two fundamental frequency components over the regions of wavy vortex and quasi-periodic wavy vortex flow.

FIGURE 13 shows the variation of three frequencies with the Reynolds number. This result was obtained by gradually increasing or decreasing Re. The arrows designate the direction of the Re-variation. The first fundamental frequency f_1/f_r remains constant. This implies that f_1 increases in proportion to Re as long as the fluid properties are kept constant. The bursting frequency f_1'/f_r also remains constant. The experimental values are slightly lower than 1/3. The transition from wavy vortex to quasi-periodic wavy vortex flow is accompanied by a stepwise increase in f_1/f_r and decrease in f_1'/f_r. The frequency ratio f_1'/f_1 tends to be kept constant at 0.4 over the wavy vortex flow region and at about 0.28 over the quasi-periodic wavy vortex flow region. It can be considered from the figure that the periodical bursting occurs in resonance with the azimuthal wave motion.

4. CONCLUDING REMARKS

The spectral evolution of the wavy disturbances generated from near the inner rotating cylinder wall is suppressed due to the stable azimuthal wave motion with one or two fundamental frequency components over the whole regions of wavy vortex and quasi-periodic wavy vortex flow. The periodical bursting occurs very close to the wall of the inner rotating cylinder in a definite relation with the azimuthal wave motion. A simple model has been proposed for the generation and propagation of the wavy disturbance.

REFERENCES

1. Mizushina, T., The Electrochemical Method in Transport Phenomena, in Advances in Heat Transfer, vol.7, pp.87 - 161, Academic Press, NY., 1971.

2. Bouabdallah, A. and Cognet, G., Laminar-Turbulent Transition in Taylor Couette Flow, in I.U.T.A.M.Symp. on Laminar-Turbulent Transition, Stuttgart, F.R.G., Sept. 16 - 22, 1979.

3. Cognet, G., Bouabdallah, A., and Aider, A.A., Laminar-Turbulent Transition in Taylor Couette Flow, Influence of Geometrical Parameters, in Stability in the Mechanics of Continua, ed. F.H. Schroeder, pp.330 - 340, Springer-Verlag, Berlin Heidelberg, 1982.

4. Kataoka, K., Doi, H., and Komai, T., Heat/Mass Transfer in Taylor Vortex Flow with Constant Axial Flow Rates, *Int. J. Heat Mass Transfer*, vol.20, pp.57 - 63, 1977.

5. Kataoka, K., Bitou, Y., Hashioka, K., Komai, T., and Doi, H., Mass Transfer in the Annulus between Two Coaxial Rotating Cylinders, in Heat and Mass Transfer in Rotating Machinery, ed. D.E. Metzger and N.H. Afgan, pp.143 - 153, Hemisphere, Washington, D.C., 1984.

6. Kataoka, K., Bitou, Y., Deguchi, T., and Minakuchi, T., Spectral Analysis of Time-dependent Taylor Couette Flow, in Synopsis of 3rd Taylor Vortex Flow Working Party Meeting, Nancy, France, April 5 - 7, pp.70 - 73, 1983.

7. DiPrima, R.C., and Swinney, H.L., Instabilities and Transition in Flow between Concentric Rotating Cylinders, in Hydrodynamic Instabilities and the Transition to Turbulence, ed. H.L. Swinney and J.P. Gollub, Chap.6, pp.139 - 180, Springer-Verlag, Berlin Heidelberg, 1981.

8. King, G.P., Li, Y., Lee, W., Swinney, H.L., and Marcus, P.S., Wave Speeds in Wavy Taylor-Vortex Flow, *J. Fluid Mech.*, vol.141, pp.365 - 390, 1984.

9. Kataoka, K., Mizusugi, T., Ueno, H., and Ohmura, N., Hysteresis of Dynamical Transitions in Taylor-Couette Flow, in Symp. on Transport Phenomena in Rotating Machinery, Honolulu, Hawaii, April 28 - May 3, 1985.

10. Coles, D., Transition in Circular Couette Flow, *J. Fluid Mech.*, vol.21, pp.385 - 425, 1965.

Experimental Investigation of Boiling Water Films in Radial Rotating Channels

I. A. MUDAWWAR
School of Mechanical Engineering
Purdue University
West Lafayette, Indiana 47907

M. A. EL-MASRI
Department of Mechanical Engineering
Massachusetts Institute of Technology
Cambridge, Massachusetts 02139

ABSTRACT

One proposed method for water-cooling the blades of large utility gas turbines is the open-loop system. Cooling water is injected into a groove on the turbine rotor disc from which it flows into radial passages within the blades. The passages are vented to local gas pressure both at the rotor and the blade tips. The water forms a thin film on the trailing walls of the passages due to Coriolis forces and is driven radially by centrifugal forces. The blade-cooling load is absorbed by boiling a portion of the water flow and the remaining water, together with the steam produced, exit from the blade tips into the gas path.

An experimental facility has been constructed to obtain boiling heat transfer and critical heat flux data for water films under such conditions. A description of the experimental facility and the measurement techniques follows. Boiling curves and critical heat flux data are presented for a wide range of flow rates, rotational speeds up to 1775 rpm, and pressures ranging from one to 5.41 atmospheres. A correlation for the critical heat flux is also presented.

1. INTRODUCTION

Recent developments in utility gas turbine power plants have led to significant improvements in their overall cycle efficiency. For a given plant configuration and pressure ratio, the thermal efficiency of the cycle is a strong function of the turbine firing temperature. For this reason, blade cooling plays a very important role in any such improvements. Present industrial blade cooling techniques such as internal air cooling or film cooling permit maintaining the surface temperature to about 870°C with firing temperatures up to 1200°C.

Water cooling has been recently projected as a possible alternative for blade cooling. An advanced program was developed by the U.S. Department of Energy for studying the feasibility of water cooling turbine blades. Several studies by General Electric Company (G.E.) [1-3] have shown that combined-cycle coal pile to busbar efficiencies close to 42 percent may be obtained with water-cooled turbines (compared to 37 percent in present power plants) if the firing temperature is increased to about 1600°C. Two important goals could be achieved with such a cycle. The first is the obvious improvement in the thermal efficiency. The second is the utilization of coal-derived and residual petroleum fuels. Hot corrosion of the blade surface prevents conventional cooling techniques from protecting the blades against such highly corrosive environments. This is primarily due to the fact that hot corrosion becomes critical if the blade surface temperature exceeds 540°C, a temperature that can be obtained only with water cooling.

Water cooling is superior to conventional air or steam cooling for several reasons. To start with, water has higher thermal conductivity and specific heat which lead to a significant enhancement of heat transfer. Furthermore, evaporation heat transfer coefficients are heavily dependent on centrifugal forces. These forces are more important in the case of water since its density is three orders of magnitude higher than steam. Still higher heat transfer coefficients could also be obtained if boiling were to occur inside the water-cooled passages.

Two major water cooling schemes can be employed. Closed-loop cooling is achieved by circulating water between the blades and an external heat exchanger. Thermosyphons using water or liquid metals have been considered as possible configurations of closed circuit blade cooling. In the open-loop scheme, which is the subject of this paper, water enters the rotor near the turbine axis and flows freely through near radial blade-cooling passages under the action of the centrifugal forces (see Figure 1). Inside these passages a large portion of water is vaporized into steam. Steam mixes with the combustion products at the blade tips while most of the non-vaporized water is captured by a number of circumferential ports in the static casing.

Open-loop cooling has several advantages over other cooling techniques. Perhaps the most important of these is the simplicity of the open-loop design. As shown in Figure 2, water flows as a very thin film covering only one side of the cooling passage due to the Coriolis force. Steam, on the other hand, fills the rest of the cross section. For this reason, the centrifugal force drives the liquid film at fairly constant pressure and leak problems become quite insignificant. Nevertheless, when the steam mixes with the combustion products a penalty is assessed on the overall thermal efficiency. Another advantage of open-loop cooling, as compared to closed-loop cooling, is that less water is needed on board the blade at any instant of time. Therefore, mechanical unbalancing of the rotor is less critical.

The analysis of momentum and heat transfer inside the water cooling passages is complicated by the combined effect of the centrifugal and Coriolis forces. Furthermore, the wetted fraction of the coolant passage is unknown for the case of circular tubes. These effects were studied analytically [4-6] and simulated experimentally [6,7] at G.E. in an attempt to understand evaporation and boiling in rotating liquid films. Since measurement of the coolant coverage of the cross section was not possible, heat transfer coefficients were based on a pre-estimated area. Dakin [4] introduced an analytical model to predict this wetted

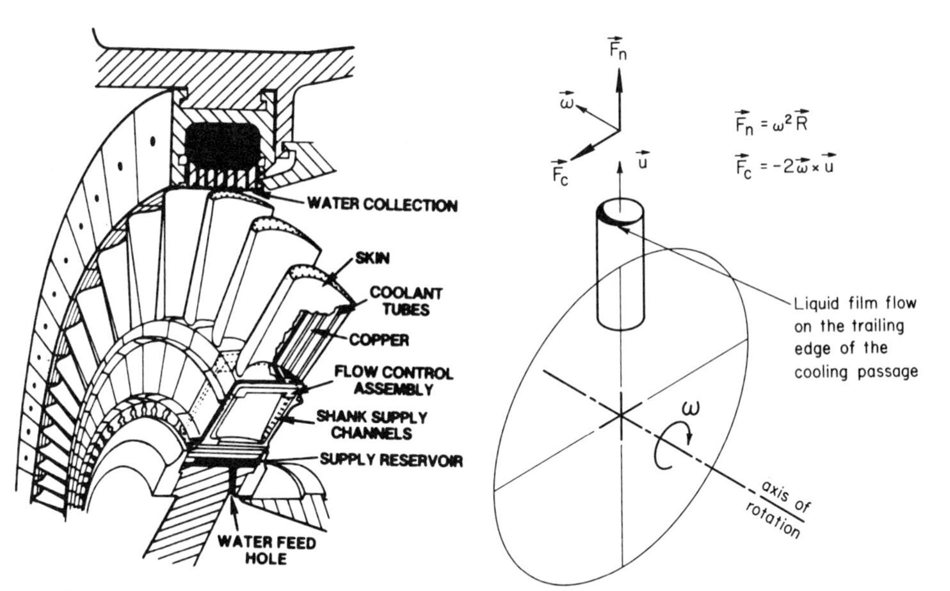

FIGURE 1. Schematic representation of open-loop water cooling of turbine blades [7].

FIGURE 2. Rotating film motion on the trailing edge of the cooling passage.

fraction for laminar film flow. For turbulent flow, Dakin [6,7] deduced this fraction using Dukler's [8] theoretical evaporation heat transfer analysis of freely-falling films in stationary coordinates. Recent efforts at G.E. [9,10] focussed on the critical heat flux (CHF) during severe boiling inside the blade cooling passage. This limit is encountered when the heat flux to the water film is increased to a level where intense vapor generation at the boiling surface prevents the liquid from reaching that surface to provide the necessary cooling. The CHF is accompanied normally by severe temperature overshooting which results in melting of the surface. Sundell et al. [9] presented several boiling heat transfer data for rotating films. Their experiments, however, failed to produce reliable estimates of the CHF since the coolant passage area covered with the boiling film was unknown.

The CHF problem in rotating liquid films was addressed theoretically by El-Masri and Louis [11]. They based their analysis on Kutateladze's [12] boundary-layer separation model of the CHF in stationary boiling systems. This model assumes that CHF is reached when the vapor stream normal to the wall acquires sufficient kinetic energy to eject the liquid remaining between the tightly packed bubbles at the heated surface. The numerical results of El-Masri and Louis suggest that, at low pressures, the CHF will occur spontaneously as soon as boiling starts at the surface. High pressure films ($p > 5$ bar), on the other hand, were found to sustain considerable boiling before reaching the separation point.

Heat transfer in rotating liquid films was simulated experimentally by Mudawwar et al. [13,14] using a rectangular, rather than circular, cooling passage. The wetted area in their experiments was known since the film covered the entire surface normal to the Coriolis force vector. They showed that turbulent boundary-layer separation is not a sufficient physical reason for the CHF in high-speed liquid films. Re-wetting by liquid droplets from the shattered boiling film could still provide sufficient cooling for the passage wall even after separation. The critical flux level was modelled as occurring when the severe vapor drag normal to the boiling surface overcame the Coriolis forces on these droplets.

Based on their experimental findings, the authors predicted [15] that open-circuit water cooling of turbine blades would impose substantial heat transfer constraints. These are the necessity of avoiding the CHF inside the cooling channel, and of providing an adequate conduction path around the cooling passage in order to maintain the blade suction surface temperature below an acceptable limit. These constraints make the open-loop cooling technique less competitive with air cooling for higher turbine pressures (in excess of 20 bars) and temperatures. Improved performance of open-loop water cooled blades was expected with composite blades of a high-conductivity matrix, which might be made of copper, covered by a skin of a protective turbine alloy material.

This paper focuses on the experimental hardware and measurement techniques used by the authors in investigating the boiling heat flux and the CHF in rotating liquid films.

2. EXPERIMENTAL METHODS

In utility gas turbines, convective heat at the blade surface is transferred by conduction to the liquid film through the wetted side of the circular cooling passages (see Figure 3). Proper simulation of this complicated heat transfer process can only be achieved by an identical cooling geometry. The disadvantage of such an effort is the absence of any prior knowledge about the wetted fraction of the coolant passage wall. Curvature of the wall becomes an additional variable whose impact on the heat transfer data might obscure the individual effects of other more significant parameters such as static pressure, coolant flow rate or rotational speed. Furthermore, recommendations have been made by the authors [15] to replace the circular passages with oval cross sections (see Figure 3) in order to spread the film and increase the wetted area. For this reason, all the simulation experiments were carried out with two-dimensional film flow characterized by a known wetted heat transfer area. In addition, the CHF data were obtained with saturated water films. Subcooling both increases the CHF and alters the effects of other transport parameters. The authors believe, therefore, that the data and correlations of this investigation provide safe CHF estimates for gas turbine applications for a given pressure, coolant flow rate, rotational speed, and blade radius.

Simulation of blade surface heating of the film was achieved by

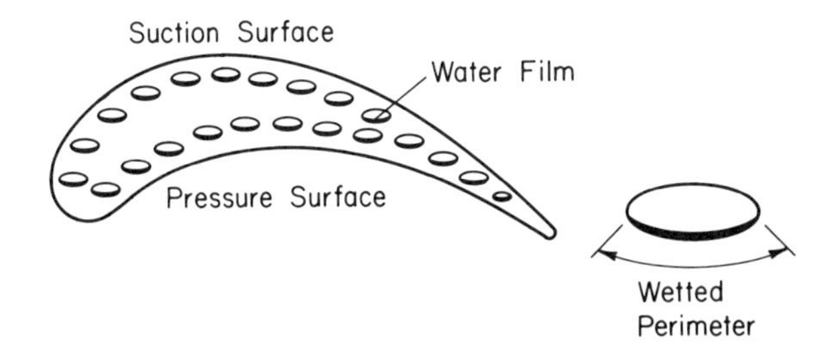

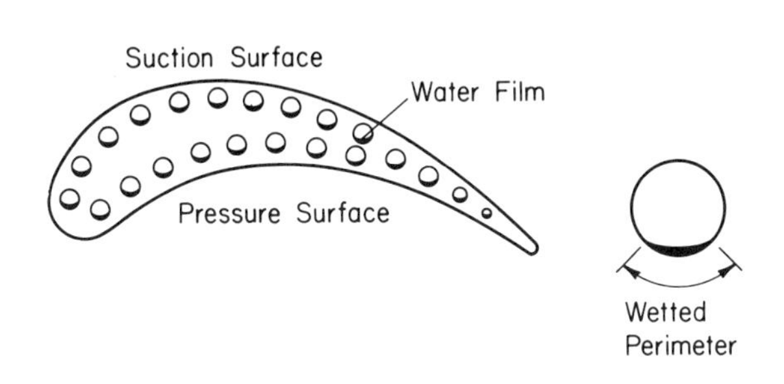

FIGURE 3. Enhancement of the wetted perimeter using oval passages in comparison with standard circular passages.

an electric heater inserted inside the rotating channel. Initial tests involved direct contact of water with a resistive metallic ribbon. As the electric current was passed through the ribbon, thermal expansion led to buckling of its surface. This resulted in nonuniform boiling of the film and local transient overheating which caused pre-mature burnout. A thermally optimized composite-wall heat transfer channel which hid the metallic ribbon from the boiling surface was developed in an attempt to overcome this problem. As shown in Figure 4, the film was forced through a rectangular nozzle over a copper plate which provided an indirect conduction path for the electric power. The copper surface was electrically insulated from the ribbon with a high thermal conductivity boron nitride plate. This ceramic material combines high thermal conductivity with excellent electrical insulation. A specially prepared paste of fine boron nitride powder and thermally conducting silicon grease was applied between the various plates to minimize thermal contact resistances. Electric power was supplied to the ribbon through two stainless steel terminals. Nichrome and stainless steel ribbons (0.075 mm thick) were fastened to the terminals by high-temperature oven brazing. The entire heater probe was then securely packaged inside a high temperature silicon-base fiberglass module which was characterized by very low thermal conductivity ($k = 0.2$ W/m K) and considerable mechanical strength. This module was then mounted on a rotating disc which was flanged to the end of a drive shaft.

Precise control of the water temperature was extremely critical in the measurement of the boiling heat flux since any subcooling would result in a higher CHF. This was actually a major source

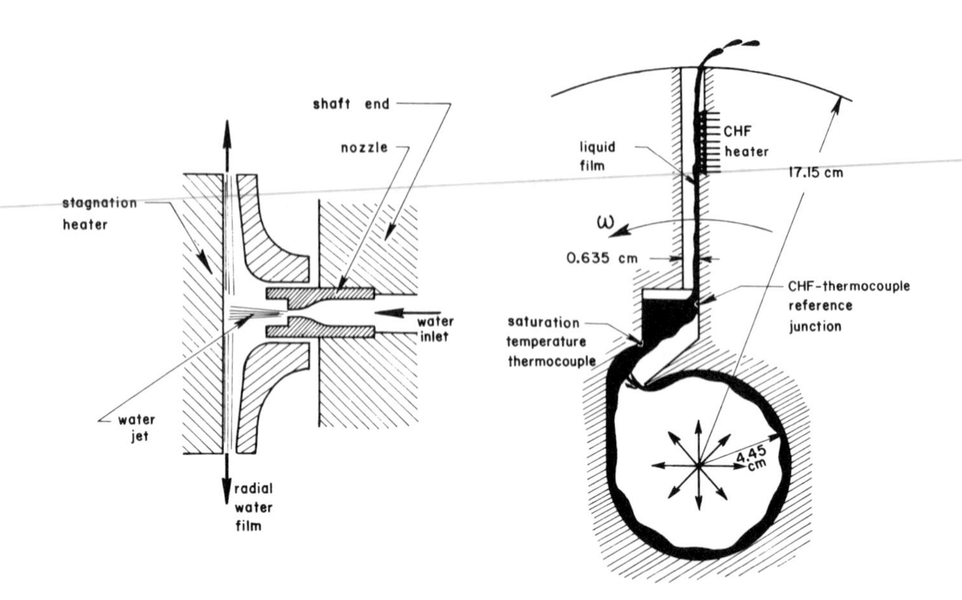

FIGURE 4. Impingement jet flow against the surface of the stagnation heater and the spiral film flow inside the rotating disc [14].

of error in earlier studies [7,8], where high degrees of subcooling (up to 75°C) were reported with some experimental data. For this reason, efforts were made by the authors to insure saturated film flow upstream of the rotating channel. Most of the sensible heat addition occurred inside a high-capacity step-up heater capable of supplying up to 10 KW of electric power. As shown in Figure 5, additional heating was provided by three 1-KW temperature-controlled heaters. Careful control was also necessary to achieve maximum preheating while avoiding in-line vapor generation. Hot water was introduced axially to the shaft through a small mechanical seal. The flow was forced through a nozzle onto a flat heated surface which was mounted at the center of the rotating disc. The impinging jet was then diverted radially outward (see Figure 4) to the test channel. The stagnation heater was essential for overcoming the heat losses from water to the stainless shaft. Saturated water flow was thus achieved simultaneously in two different ways. For a jet speed of 8-35 m/s, the pressure drop across the nozzle was in the range 5-30 psi. The saturation temperature of the water just upstream of the nozzle was 3 to 34°C above the saturation temperature corresponding to the chamber pressure. Therefore external preheating increased the water temperature without the danger of vapor generation. Furthermore, the axial jet was diverted over the surface of the rotating stagnation heater in the form of a thin high-speed radial film. This flow configuration enhanced the convective heat transfer coefficient and resulted in a highly compact stagnation heater design. The stagnation heater was a composite substrate very similar in construction to the main channel heater. Due to the enhanced heat transfer at the surface, it was capable of dissipating up to 2 KW of electric power over an area of 38.6×10^{-4} m^2.

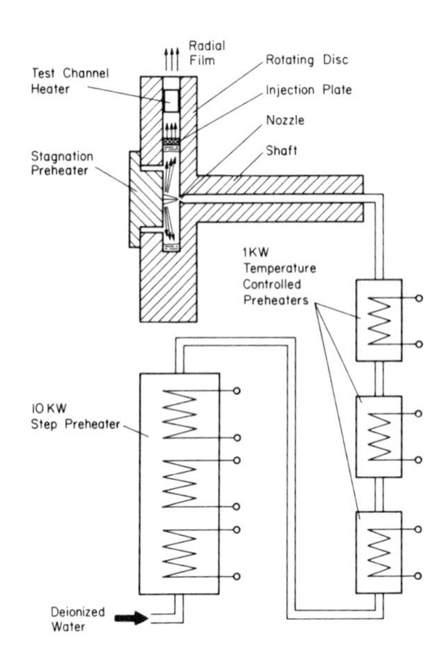

FIGURE 5. Successive stages of water preheating.

Figure 4 shows the liquid trajectory from the nozzle to the probe. During operation, a small amount of water was flashed at the nozzle exit or evaporated over the stagnation heater surface to insure saturated water flow. The radial film was accumulated in a circumferential pocket, where it maintained a continuous water layer suspended by centrifugal forces. The net pressure due to these forces, combined with a tangential swirl created by Coriolis forces at the heater surface, directed the flow into a small reservoir located upstream the channel injection plate. The pressure inside the annular space was equilibrated with the gaseous mixture of the pressure chamber, both around the stagnation heater and inside the channel by a number of small holes drilled on the surface of the rotating disc.

Sizing the injection plate was a critical part of the probe design. The initial film thickness affects the development of the film over the heated wall. Depending on this thickness the film might thin out (see Ref.[11]) in the radial direction, as it is the case with gas turbine blades, or diverge in the form of a hydraulic jump. For this reason the injection plate spacing had to be kept below the critical thickness required to maintain stable convergent film flow. On the other hand, if the injection nozzle was very restricted, the probe reservoir would flood and water would escape through the breathing holes of the stagnation heater. Proper operation was maintained by using a different injection plate spacing (0.5-2.5 mm) depending on the water flow rate or the rotational speed.

Electric power was provided to the test channel by a current-controlled 0-115 Amps, 0-20 volts DC power supply. The voltage drop was measured directly across the heater by signal wires connected through the shaft to an electronic slip-ring assembly. Two instrument amplifiers received voltage signals proportional to the heater current (across an external shunt), and voltage (directly across the heater), respectively. The heater power was deduced directly by a voltage multiplier circuit and plotted on a chart recorder.

Two thermocouple junctions were embedded inside the boiling copper surface of the channel heater. Each of these junctions was referenced with respect to another junction located in the reservoir upstream the injection plate of the test channel. This technique provided precise temperature measurement of the wall superheat ΔT_{sat} especially for small temperature differences. The thermocouple signals were both recorded and used to protect the test heater against burnout. It was estimated that the temperature of the copper plate could increase several hundred degrees within the first second after the CHF. A fast-acting electronic circuit was built for protecting the heater. The circuit received thermocouple signals from the test channel, stagnation heater, and several other external components of the experimental facility. As soon as overheating was detected in any part of the system, a pre-calibrated voltage comparator in the circuit triggered several relay switches which cut off the electric power to the heater as well as other critical components within 30 m sec.

Figure 6 shows an axial view of the heat transfer test section mounted on the disc which was made of aluminum 6061. As shown in the same figure, an identical mass of fiberglass was mounted diametrically opposite to the probe module for mechanical balancing. The instrumented disc was rotated inside the 49.5 cm O.D. aluminum-6061 pressure chamber shown in Figure 7. Access to the probe was provided by a removable cover. A Lexan window was mounted on the cover for optical studies. Air was injected continuously through the chamber during rotation to maintain uniform pressure. The chamber was equipped with a 6.67 cm I.D. double-face self-aligning seal.

The primary drive shaft was made of 7.62 cm O.D. stainless steel shaft 1.52 m in length. As shown in Figure 8, a 7.5 hp constant-torque eddy-current clutch motor was used to spin the disc. The rotational speed was monitored through an optical tachometer. Motor torque was transmitted to the shaft via three V-belts using a pair of reduction pullies. The electric current was transmitted from water-cooled carbon brushes through two copper rings which were electrically insulated from the shaft. Electric power to the disc was supplied through teflon-insulated copper rods inside axial grooves which were drilled through the shaft. The carbon brushes carried up to 125 Amps DC at 1775 rpm.

3. EXPERIMENTAL RESULTS

Boiling curves for the rotating film experiments were generated on logarithmic plots of heat flux q versus wall superheat $(T_w - T_{sat})$. The main flow parameters were the static pressure, flow rate and centrifugal acceleration. For each pressure, attempts were made to cover a matrix of five flow rates (0.5,

FIGURE 6. Axial view of the rotating disc showing the test channel (top white module), and the back side of the stagnation heater (at the center of the disc). The balancing mass is also shown diametrically opposite to the test channel.

FIGURE 7. Side view of the pressure chamber.

1.09, 1.58, 2.08 and 2.67 kg/m s) at four different rotational speeds (500, 1000, 1300 and 1775 rpm). Three different pressure conditions were investigated. At 1.0 and 3.24 atm., the complete matrix of conditions was covered. This was not possible, however, at the upper pressure limit of 5.41 atm. because of the persistent burnout of the stagnation heater and the test channel. Some operating conditions were tested more than once to confirm measurement accuracy. The heater geometry was characterized by a heat transfer area of $12.0 \times 6.35\ mm^2$ and a 13 cm radius of rotation (to the center of the heater). Experimental results spanned a centrifugal acceleration range of 36.5-460 g.

As shown in Figure 9, each boiling plot starts with a fairly linear variation of the heat flux q with ΔT_{sat}. This part of the

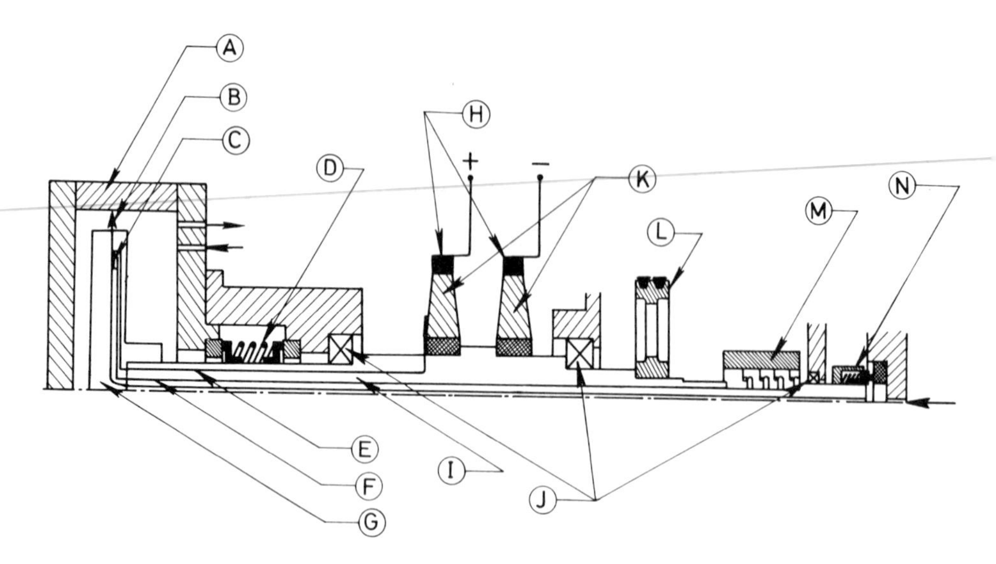

FIGURE 8. Schematic view of the rotating system. A: pressure chamber; B: liquid film; C: CHF heater; D: seal; E: CHF powerline; F: stagnation heater electric cables and sensor wires; G: rotating disc; H: carbon brushes; I: shaft; J: bearings; K: copper rings; L: pulley; M: slip-ring assembly; N: inlet seal [14].

curve corresponds to a nearly constant convective heat transfer coefficient. At higher values of q, the plots reveal a gradual departure to higher slope as a result of surface boiling. Fully developed nucleate boiling is then established, and the slope increases well beyond unity. Ultimately, the boiling curve completes an S-shaped variation leading to the CHF (represented by an arrow parallel to the temperature axis). No attempts were made to monitor the heat variation beyond that point since CHF represents the upper design limit for gas turbine blade cooling. As shown in Figure 9, the effect of centrifugal acceleration was to enhance the convective heat transfer coefficient. This resulted in a higher incipient boiling heat flux for higher rotational speeds. The CHF limit also followed a monotonically increasing path with acceleration.

Figure 10 shows similar boiling curves obtained with various mass flow rates. Note that prior to boiling, the convective heat transfer coefficient is quite sensitive to flow rate variation. The marked increase in the heat transfer coefficient with flow rate, however, does not seem to have a strong influence on fully developed boiling or the CHF.

The effect of pressure was most evident with the CHF limit (see Figure 11). Following our model of Ref.[14] based on the physical arguments given in the introduction, the CHF limit was correlated within a ± 25 percent accuracy by the following equation:

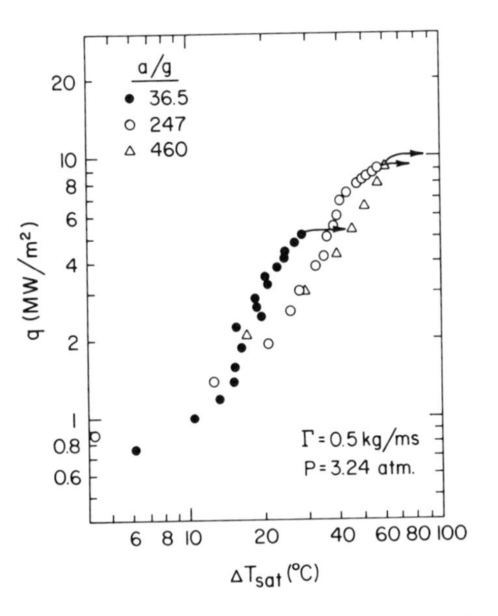

FIGURE 9. Effect of centrifugal acceleration on the boiling characteristics of the rotating film.

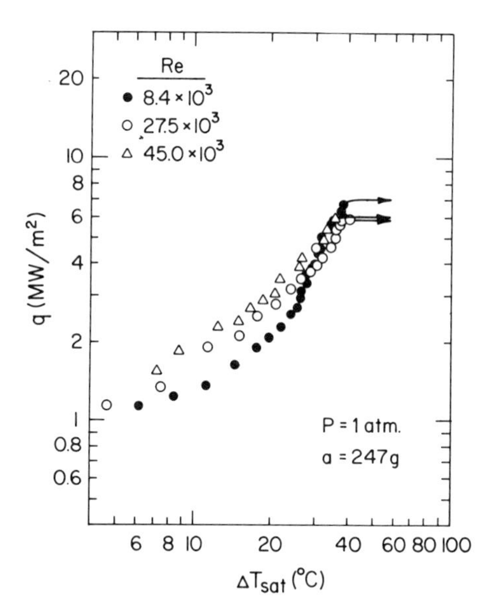

FIGURE 10. Effect of flow rate on the boiling characteristics of the rotating film.

$$q_M = 0.69 \; \rho_g h_{fg} \left[\frac{(\rho_f - \rho_g)\sigma\omega(\nu a)^{1/3}}{\rho_g^2} \right]^{1/4} \qquad (1)$$

This correlation represents one of two important thermal design constraints [15] on the performance of open-loop water-cooled turbine blades.

Note that equation (1) does not include any dependence on the flow rate. This equation can be re-written as:

$$q_M = 0.69 \quad F \, \omega^{5/12} \, R^{1/12} \qquad (2)$$

where F is a fluid property parameter which is uniquely determined by the static pressure. The additional terms of equation (2), are independently determined by the mechanical design of the gas turbine. Figure 12 shows that the CHF is a strong function of pressure below 20 atm. At higher pressures, the CHF dependence decays asymptotically as the critical pressure is approached. Since heat transfer to a turbine blade is roughly proportional to pressure, the CHF becomes a more serious design limit for reheat cycles, where the gas turbine would typically operate above 40 atm. Figure 13 illustrates the effect of centrifugal acceleration on the CHF for a turbine blade radius of 126 cm (which was recommended by G.E. [7]). Note that higher

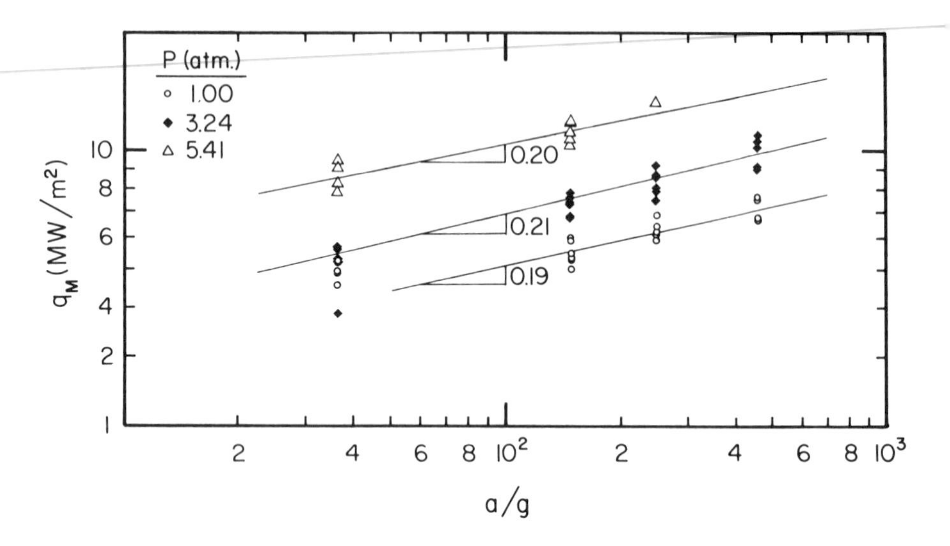

FIGURE 11. Critical heat flux variation with centrifugal acceleration for various pressures and flow rates.

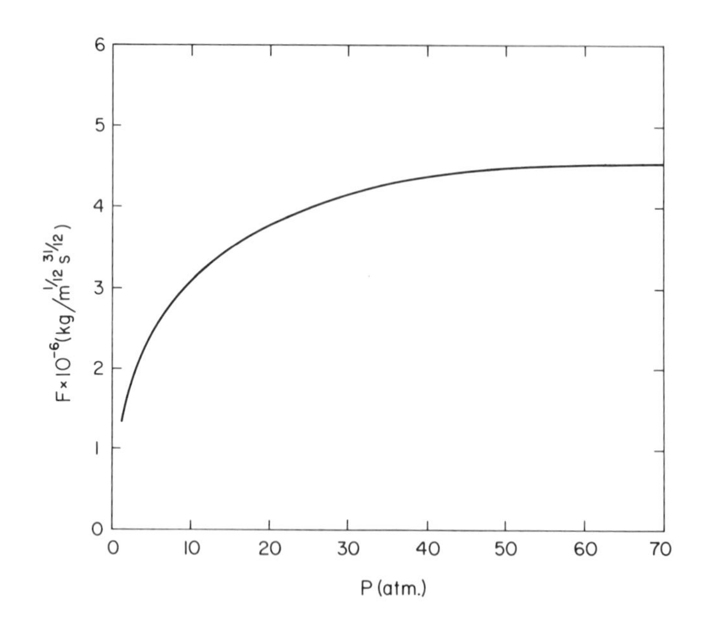

FIGURE 12. Variation of the CHF property parameter with pressure

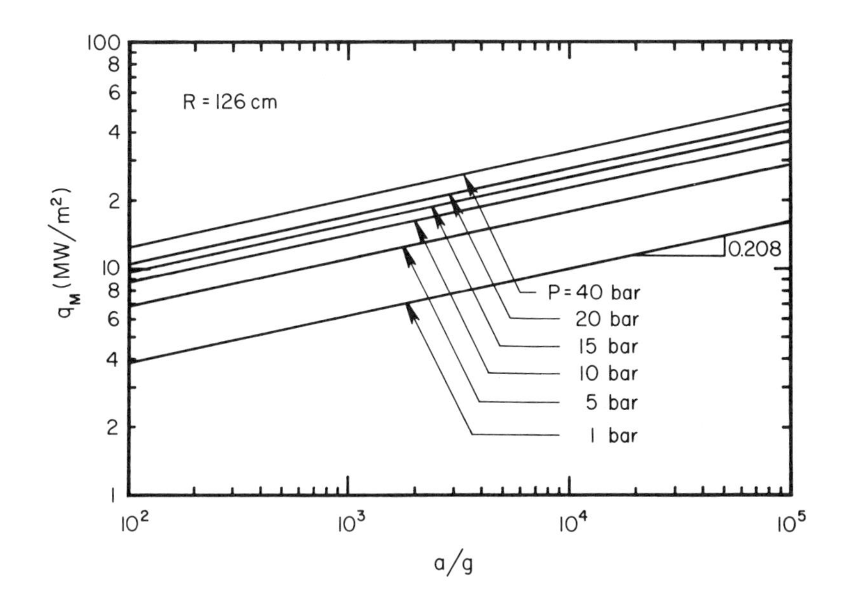

FIGURE 13. Critical heat flux variation with centrifugal acceleration for a pressure range of 1-40 bar.

centrifugal acceleration is required to avoid the CHF for reheat cycles. From equation (2), the CHF is more strongly dependent on the rotational speed ω than the radius R. The risk of the CHF at higher pressures can be reduced for smaller turbines of higher rotational speeds.

CONCLUSIONS

A description of the experimental facility and measurement techniques was presented for an investigation of heat transfer in rotating liquid films. Data on boiling curves and the critical heat flux have been obtained for saturated water films at various flow rates, pressures and rotational speeds. A correlation for the critical heat flux has been presented and its implication for water-cooled turbine design discussed.

ACKNOWLEDGEMENTS

The authors acknowledge the partial support provided by each of the following organizations: U.S. Department of Energy, contract E(49-18)-2295; NSF, grant MEA-8017416; and the MIT Energy Laboratory.

NOMENCLATURE

Latin

a centrifugal acceleration, $\omega^2 R$

F	liquid property parameter, $\rho_g^{1/2} (\rho_f - \rho_g)^{1/4} h_{fg} \sigma^{1/4} \nu_f^{1/12}$
g	acceleration due to gravity
h_{fg}	latent heat of vaporization
p	static pressure
q	wall heat flux
q_M	critical heat flux (CHF)
R	radius of rotation
Re	film Reynolds number, $4\Gamma/\mu_f$
T_{sat}	saturation temperature corresponding to the operating static pressure
T_w	wall temperature

Greek

Γ	mass flow rate per unit film width
ΔT_{sat}	wall superheat, $(T_w - T_{sat})$
μ_f	viscosity of the saturated liquid
ν_f	kinematic viscosity of the saturated liquid
ρ_f	density of the saturated liquid
ρ_g	density of the saturated vapor
σ	surface tension
ω	rotational speed.

REFERENCES

1. Kydd, P.H., and Day, W.H., An Ultra High Temperature Turbine for Maximum Performance and Fuels Flexibility, ASME paper no. 75-GT-81, 1975.

2. Horner, M.W., Day, W.H., Smith, D.P., and Cohn, A., Development of a Water-Cooled Gas Turbine, ASME paper no. 78-GT-72, 1978.

3. Caruvana, A., Manning, G.B., Day, W.H., and Sheldon, R.C., Evaluation of a Water-Cooled Gas Turbine Combined Cycle Plant, ASME paper no.78-GT-77, 1978.

4. Dakin, J.T., Viscous Liquid Films in Non-Radial Rotating Tubes, report no. 77CRD133, General Electric Company, Schenectady, New York, 1977.

5. Dakin, J.T., and So, R.M.C., The Dynamics of Thin Liquid Films in Rotating Tubes: Approximate Analysis, report no. 77CRD043, General Electric Company, Schenectady, New York, 1977.

6. Dakin, J.T., Vaporization of Water Films in Rotating Radial Pipes, International Journal of Heat and Mass Transfer, vol.21, pp.1325-1332, 1978.

7. Dakin, J.T., Horner, M.W., Piekarski, A.J., and Triandafyllis, J., Heat Transfer in the Rotating Blades of a Water-Cooled Gas Turbine, ASME book no. H00125, Gas Turbine Heat Transfer, 1978.

8. Dukler, A.E., Fluid Mechanics and Heat Transfer in Vertical Falling Film systems, Chemical Engineering Progress, vol.56, pp.1-10, 1960.

9. Sundell, R.E., Goodwin, W.W., Kercher, D.M., Dudley, J.C., and Triandafyllis, J., Boiling Heat Transfer in Turbine Bucket Cooling Passages, ASME paper no. 80-HT-14, 1980.

10. Dudley, J.C., Two-Phase Heat Transfer in Gas Turbine Bucket Cooling Passages: Part I, in Heat and Mass Transfer in Rotating Machinery, ed. D.E. Metzger, and N. Afghan, Hemisphere, Washington, D.C., 1984.

11. El-Masri, M.A., and Louis, J.F., On the Design of High-Temperature Gas Turbine Blade Water-Cooling Channels, Journal of Engineering for Power, vol.100, no.4, pp.586-591, 1978.

12. Kutateladze, S.S., and Leont´ev, A.I., Some Applications of the Asymptotic Theory of The Turbulent Boundary Layer, Proc. 3d Int. Heat Transfer Conf., Chicago, vol.3, pp.1-6, 1966.

13. Mudawwar, I.A., Boiling Heat Transfer in Rotating Channels with Reference to Gas Turbine Blade Cooling, Ph.D. Thesis, Department of Mechanical Engineering, Massachusetts Institute of Technology, 1984.

14. Mudawwar, I.A., El-Masri, M.A., Wu, C.S., and Ausman-Mudawwar, J.R., Boiling Heat Transfer and Critical Heat Flux in High-Speed Rotating Liquid Films, International Journal of Heat and Mass Transfer, 1985, in press.

15. Mudawwar, I.A., and El-Masri, M.A., Thermal Design Constraints in Open-Loop Water-Cooled Turbine Blades, ASME paper no. 84-WA/HT-68, ASME Winter Annual Meeting, New Orleans, 1984.

Hysteresis of Dynamical Transitions in Taylor-Couette Flow

KUNIO KATAOKA, TETSUYA MIZUSUGI, HIROTSUGU UENO, and NAOTO OHMURA
Department of Chemical Engineering
Kobe University
Rokkodai, Nada, Kobe 657, Japan

1. INTRODUCTION

This paper deals experimentally with the transition phenomena occurring in the fluid flow between two concentric cylinders with the inner one rotating. When the rotation speed of the inner cylinder exceeds a certain critical value, laminar Couette flow becomes unstable and leads to the first transition to Taylor vortex flow. This flow consists of cellular toroidal vortex rings stacked in the annulus, as shown in FIGURE 1. The cellular vortex structure is preserved even after time-dependent disturbances have been generated due to higher instabilities.

The principal dynamical parameter for the flow system is the Reynolds number defined by $Re = R_i \Omega d/\nu$, where R_i is the inner cylinder radius, Ω is the angular velocity of the inner cylinder, d is the annular gap width, and ν is the kinematic viscosity. It is well known that as Re is gradually increased from a small value, the following five dynamical transitions occur stepwise in sequence: laminar Couette flow → laminar Taylor vortex flow → wavy vortex flow → quasi-periodic wavy vortex flow → weakly turbulent wavy vortex flow → turbulent vortex flow.

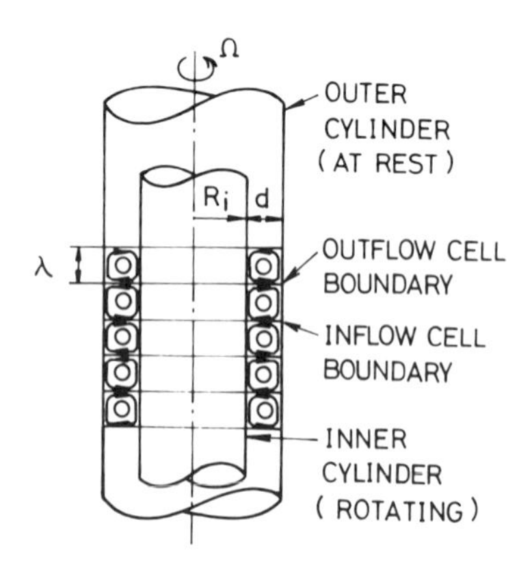

FIGURE 1. Taylor vortex flow.

It has been revealed by our previous work [1, 2] that local coefficient of mass transfer to the outer stationary cylinder varies periodically in the axial direction depending on an array of Taylor vortices. The mass transfer coefficient on the outer cylinder wall becomes maximal at the outflow cell boundaries and minimal at the inflow cell boundaries. It is very important to establish a model which permits to predict local variation of heat/mass transfer. The following fluid-dynamical questions should be considered for that purpose:

1. What flow regime will be obtained at a given Reynolds number ?
2. How many vortices will be accommodated in the annulus ?
3. What spatial state will be selected (in a sense of bifurcation) by the history of the Reynolds number ?

The objective of the present work is to observe the dynamical transition with hysteresis effect by a flow-visualization technique.

2. EXPERIMENT

FIGURE 2 shows the experimental apparatus consisting of two vertically mounted concentric cylinders. The inner cylinder made of stainless steel was 87.2 mm outer diameter and 990 mm long. The outer cylinder made of transparent acrylic resin was 120 mm inner diameter and 1010 mm long. The upper and lower fluid surfaces were fixed by flat flanges. The radius ratio $\eta = R_i/R_o = 0.727$ and the aspect ratio $\Gamma = L/d = 61.59$. A motor controlled by a thyristor cycloconverter was used to directly rotate the inner cylinder at a very small acceleration.

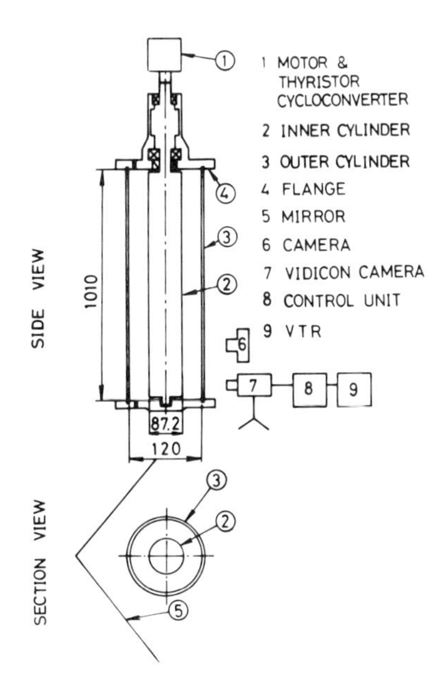

FIGURE 2. Experimental apparatus.
Dimensions given are in mm.

The liquid used was several aqueous solutions of glycerol. The concentration of glycerol was changed intentionally, so that the viscosity ranged from 1.66×10^{-3} to 5.80×10^{-3} kg/m s with the fluid density from 1.05 to 1.12 g/cm^3. A small amount (less than 0.05 vol%) of fine metallic platelet powder (aluminum paint pigment; nominal length 40 μm) was suspended to the liquid for visualization. Since the platelets are aligned along streamlines, the fluid appears light (silvery) where the fluid is moving parallel to an observer.

The visual data taken by a vidicon camera were stored into a video tape recorder The frame frequency was 30 Hz; the real time signals were displayed on the scree of the picture tube by a video timer. In each run of the experiment, the Reynold number was gradually increased from zero to a value high enough to establish the weakly turbulent wavy vortex flow regime and then decreased to a value low enoug to extinguish the Taylor vortices. Each flow pattern was observed with an adequate settling time (longer than 20 min) after Re was fixed.

3. EXPERIMENTAL RESULTS AND DISCUSSION

In the present work, the following three flow properties were measured from vide observations: N = number of Taylor vortices accommodated in the annulus; m = number of azimuthal waves; and f = passing frequency of azimuthally traveling waves.

Five photographs, shown in FIGURE 3 a to e, are representative of the flow regimes accessible in this flow system. FIGURE 3a indicates that all cell boundaries are flat and horizontal in the laminar Taylor vortex flow regime. The firs transition (from laminar Couette flow to laminar Taylor vortex flow) occurs rather gradually. The first critical Reynolds number Re_c was decided to be 82. from the solution of the linear stability theory [3].

There are many stable states with different numbers of vortices i.e. different axial wavelengths. FIGURE 4 indicates the probability of appearance of N-vortic mode. The number of vortices N is always even for the case of fixed end surface It can be seen that the 60-vortice mode appears most frequently and that those vortices are slightly larger (i.e. $\lambda_m/d = 1.03$) than the annular gap width d. It can be considered from the figure that the axial wavelength λ/d are distributed statistically around the preferred wavelength λ_m/d with separated unstable regions. As distinct from the observations of Coles [4], N remains always constant irrespective of Re during each run of the present experiment. This may be attributable to the gradual Re-variation of the present experiment.

When Re goes beyond the second critical value Re_{c2}, the transition to wavy vort flow (FIGURE 3b) occurs as a result of instability of the laminar Taylor vortex flow. Both the outflow and inflow cell boundaries frequently have azimuthal wav motion. As will be described later, Re_{c2} is considerably high owing to the wid annular gap of the present cylinder system. This tendency is consistent with Cole's observation [5]. For fixed geometrical and dynamical conditions (η, Γ, and Re), it is possible to have many spatial states different in m and N. Only in the wavy vortex flow regime, the inflow cell boundaries are usually wav in phase with the outflow cell boundaries unless N becomes very large (i.e. $N \geq$ 66).

When Re goes further beyond the third critical value Re_{c3}, the wavy vortex moti is accompanied by amplitude modulation [6]. The third transition is usually acc panied by an increase in m. The quasi-periodic wavy vortex flow (FIGURE 3c) wa confirmed from the amplitude modulation of the azimuthal wave motion. However the modulation frequency was not measured because this frequency was nc directly necessary for the present purpose. Strictly speaking, this flow is not fully turbulent. Another interesting feature is that Görtler-type vortices appear at the inflow cell boundaries when Re approaches the next critical Reynolds number Re_{c4}.

When Re exceeds the fourth critical value Re_{c4}, chaotic turbulence appears but the wave motion is still alive. As can be seen from FIGURE 3d, the weakly turbulent wavy vortex flow has small-scale turbulent motion. It is too difficult to measure precisely m in this flow regime. When Re becomes extremely large, the transition to fully-developed turbulent vortex flow occurs with the disappearance of azimuthal wave motion (FIGURE 3e). Still the cellular vortex structure remains persistently.

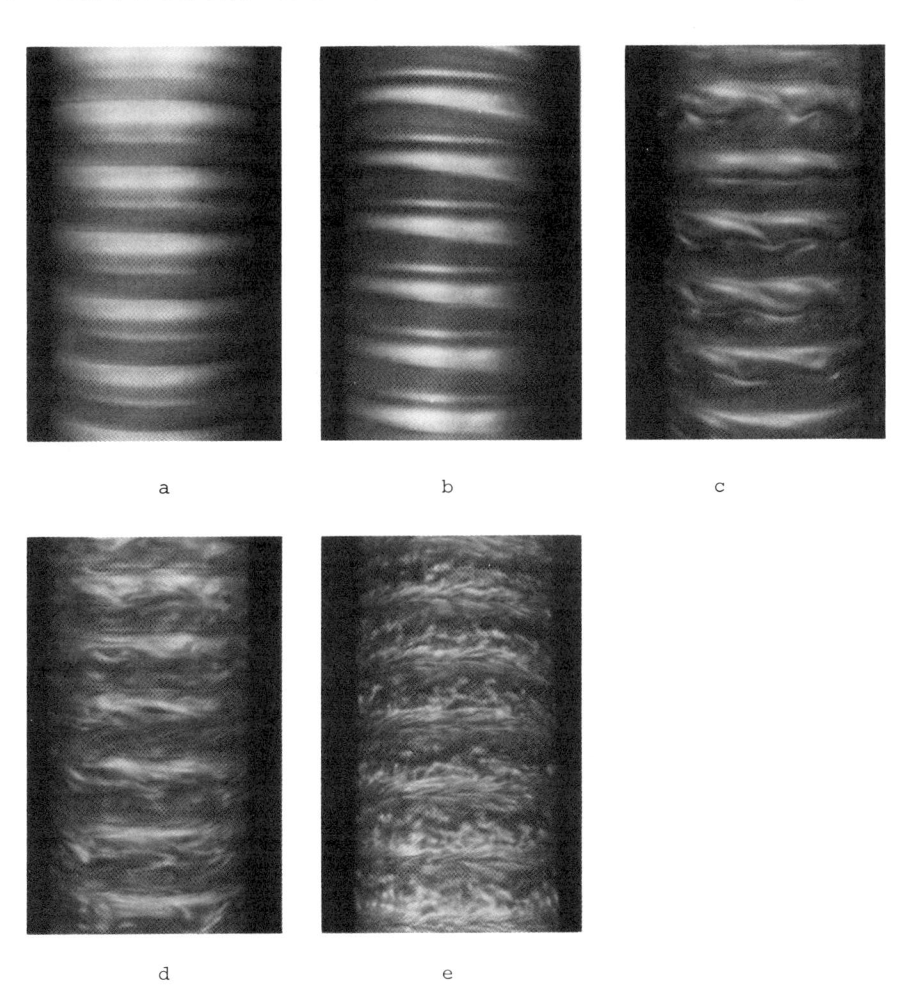

FIGURE 3.a-e. Photographs of the Taylor-Couette flow. (a) $Re/Re_c = 2.4$; laminar Taylor vortex flow. (b) $Re/Re_c = 10$; wavy vortex flow. (c) $Re/Re_c = 22$; quasi-periodic wavy vortex flow. (d) $Re/Re_c = 38$; weakly turbulent wavy vortex flow. (e) $Re/Re_c = 86$; turbulent vortex flow.

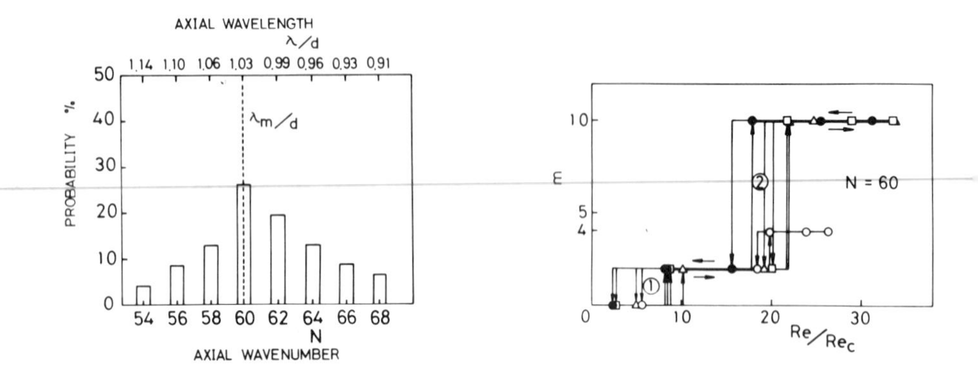

FIGURE 4. Probability of appearance of N-vortice mode.

FIGURE 5. Hysteresis loops for the most accessible axial wavenumber: (1) between laminar Taylor vortex flow and wavy vortex flow; (2) between wavy vortex flow and quasi-periodic wavy vortex flow. The keys given designate the difference in experimental run number.

As distinct from the observations of Coles [4], N remains constant during each run of the present experiment. No mode jumping is induced within each flow regime region. Depending upon the cellular mode established at the outset, however, many different routes to turbulence may be followed by gradually increasing Re. Hysteresis effects are represented by a trajectory of the sequential transitions in the $(m, Re/Re_c)$ plane with constant N. FIGURE 5 indicates the hysteresis loop for the most accessible axial wavenumber (N = 60). It is clear that each transition boundary has a band of width in Re by the hysteresis effect. In the region of the weakly turbulent wavy vortex flow, m could not be measured. The 60-vortice mode has the same jumps with respect to m on both the Re-increasing and Re-decreasing ways: m becomes always 2 at the transition from the laminar Taylor vortex flow to wavy vortex flow, but m has two choices (m = 4 or 10) at the transition to quasi-periodic wavy vortex flow. If the turning back point is moved into the region of turbulent vortex flow, the Re-decreasing trajectory will hardly follow backward the Re-increasing trajectory.

FIGURE 6 shows the demarcation lines of flow regimes, where the hatched areas imply the transition zones. Each right-side boundary with open circles indicates the highest accessible critical Reynolds number when it is approached from lower Re. Each left-side boundary with black circles indicates the lowest accessible critical Reynolds number when it is approached from higher Re. It can be seen from the figure that chaotic turbulence appears at considerably lower Reynolds numbers when $N \geq 66$. Regarding the Re-increasing way, the 56- and 64-vortice modes are stable to the transition to chaotic turbulence. The axial wavelength for these modes are larger or smaller by 0.04 d, respectively, than the annular gap width. The Görtler-type vortices appear at the inflow cell boundaries and are analogous to ones which form in the boundary layer on a concave wall. There is hysteresis in the generation of Görtler-type vortices. In the same figure, open triangles indicate the Reynolds number at which Görtler-type vortices appear on the Re-increasing way and black triangles indicate the Reynolds number at which they disappear on the Re-decreasing way. While the Görtler-type vortices exist, the inflow cell boundaries are not wavy.

FIGURE 7 shows the transition curves above Re_c by three-dimensional representation. It should be kept in mind that N always remains to be an even number and that m is integer. In the present experiment, any transition trajectory does not cross other trajectories because N does not change during each run. The horizontal arrows indicate an abrupt and discrete change in m. The thick lines of the figure indicate the most accessible trajectory. Each rectangular dotted area implies a transition zone with hysteresis loop. It can be considered that any trajectory obtained by an arbitrary acceleration must go through one of those transition zones.

FIGURE 8 shows the variation of wave speed of the outflow cell boundaries with the Reynolds number. In the wavy vortex flow regime, the inflow cell boundaries are usually wavy in phase with the outflow cell boundaries. The wave amplitude is considerably large. This is the case when N is less than 66. But the inflow cell boundaries are stationary independent of the wavy outflow cell boundaries when $N \geq 66$. This suggests that the azimuthal wave motion is related with the axial wavelength. In this work, the dimensionless wave speed is defined by the passing frequency of azimuthally traveling waves as $f_1/m\,f_r$, where f_r is the rotation frequency of the inner cylinder. In the quasi-periodic wavy vortex flow regime, the inflow cell boundaries are always stationary irrespective of N.

In the wavy vortex flow regime, m becomes 2 very frequently and the wave speed is settled down to 0.20. This value is equal to that reported for small radius ratios and large aspect ratios by King et.al.[7]. When the 68-vortice mode was

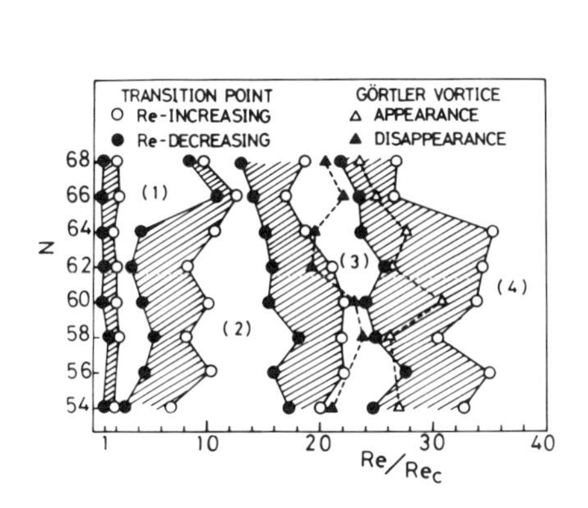

FIGURE 6. Demarcation lines of flow regimes: (1) laminar Taylor vortex flow; (2) wavy vortex flow; (3) quasi-periodic wavy vortex flow; (4) weakly turbulent wavy vortex flow. Hatched areas designate the transition zones.

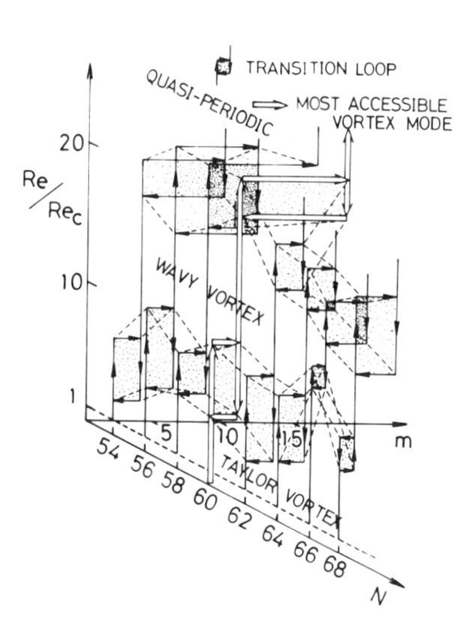

FIGURE 7. Three-dimensional diagram of dynamical transitions in Taylor-Couette flow.

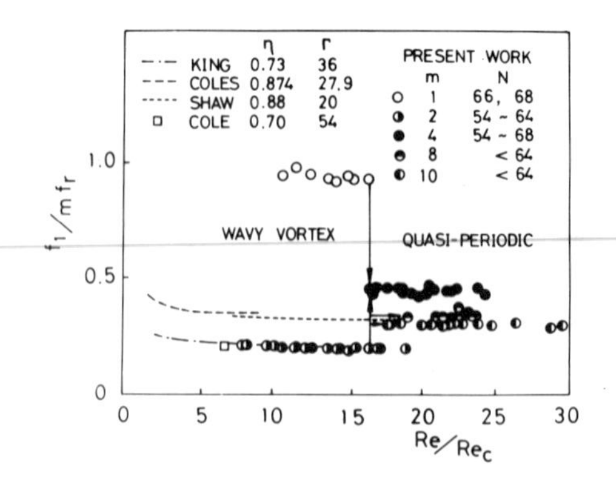

FIGURE 8. Wave speed of outflow cell boundary.

produced intentionally, the wave speed of the 1-wave mode became an extremely large value ($f_1/m\ f_r = 0.95$) in the region of wavy vortex flow. This implies that the waves move azimuthally much faster than the 2-wave mode.
In the quasi-periodic wavy vortex flow regime, m becomes 4, 8, and 10 in the present experiment and the wave speed decreases stepwise with increasing m. It has been found that the wave speed of the quasi-periodic wavy vortex flow regime is dispersionless in each group classified by m [8]. For the case of wavy vortex flow, the wave speed is dispersionless in each group classified according to whether the inflow cell boundaries are wavy or not.
This result is distinct from Shaw et.al. model [9] in that the wave speed $f_1/m\ f$ is still a function of m.

4. CONCLUDING REMARKS

The dynamical transitions have been confirmed to have those respective wide tran-sition zones. Once the axial wavenumber N has been selected at the transition to laminar Taylor vortex flow, it remains constant as long as the inner cylinder is gradually accelerated or decelerated. The hysteresis loops are formed in the (m, Re/Re_c) plane. Once the azimuthal wavenumber m has been selected at the transition to wavy vortex flow, it remains constant over the same flow region. The transition to quasi-periodic wavy vortex flow is accompanied by an abrupt increase in m. The azimuthal wavenumber remains constant over the quasi-periodic wavy vortex flow region. The dimensionless wave speed tends to be constant with respect to Re but still a function of m.

REFERENCES

1. Kataoka, K., Doi, H., and Komai, T., Heat/Mass Transfer in Taylor Vortex Flow with Constant Axial Flow Rates, *Int. J. Heat Mass Transfer*, vol.20, pp.57 - 63, 1977.

2. Kataoka, K., Bitou, Y., Hashioka, K., Komai, T., and Doi, H., Mass Transfer in the Annulus between Two Coaxial Rotating Cylinders, in Heat and Mass

Transfer in Rotating Machinery, ed. D.E. Metzger and N.H. Afgan, pp.143 - 153, Hemisphere, Washington, D.C., 1984.

3. DiPrima, R.C., and Swinney, H.L., Instabilities and Transition in Flow between Concentric Rotating Cylinders, in Hydrodynamic Instabilities and Transition to Turbulence, ed. H.L. Swinney and J.P. Gollub, Chap.6, pp.139 - 180, Springer-Verlag, Berlin Heidelberg, 1981.

4. Coles, D., Transition in Circular Couette Flow, *J. Fluid Mech.*, vol.21, pp.385 - 425, 1965.

5. Cole, J.A., The Effect of Cylinder Radius Ratio on Wavy Vortex Onset, in Synopsis of 3rd Taylor Vortex Flow Working Party Meeting, Nancy, France, April 5 - 7, pp.1 - 4, 1983.

6. Gorman, M., and Swinney, H.L., Spectral and Temporal Characteristics of Modulated Waves in the Circular Couette Systems, *J. Fluid Mech.*, vol.117, pp.123 - 142, 1982.

7. King, G.P., Li, Y., Lee, W., Swinney, H.L., and Marcus, P.S., Wave Speeds in Wavy Taylor Vortex Flow, *J. Fluid Mech.*, vol.141, pp.365 - 390, 1984.

8. Kataoka, K., and Deguchi, T., Generation and Evolution of Turbulence in an Annulus between Two Concentric Rotating Cylinders, in Symp. on Transport Phenomena in Rotating Machinery, Honolulu, Hawaii, April 28 - May 3, 1985.

9. Shaw, R.S., Andereck, C.D., Reith, L.A., and Swinney, H.L., Superposition of Traveling Waves in the Circular Couette System, *Phys. Rev. Lett.*, vol.48, pp.1172 - 1175, 1982.

ROTATING SURFACES AND ENCLOSURES

Performance Comparison between Parallel Disk Assemblies and Plate-Fin Compact Surfaces

S. MOCHIZUKI
Department of Mechanical Engineering
Tokyo University of Agriculture & Technology
Koganei, Tokyo, Japan

WEN-JEI YANG
Department of Mechanical Engineering & Applied Mechanics
University of Michigan
Ann Arbor, Michigan 48109

ABSTRACT

Multiple parallel-disk assemblies and high-performance compact surfaces have similar modes of performance enhancement. Heat transfer and friction loss performance of stationary and rotating disk systems are compared with those of high-performance plate-fin surfaces of plain, wavy, and strip types. Based on the same hydraulic diameter, the rotating disk system is superior to the plain type and comparable to the wavy and strip types. The study concludes that the rotating parallel disk system is a high-performance heat transfer device.

INTRODUCTION

Three distinct modes of performance enhancement are discovered in all types of high-performance, compact heat transfer surfaces (1-3). These three modes are also exhibited in the stationary parallel-disk assemblies (4). The similarity in the mechanisms of performance enhancement between the two systems has motivated the study to compare the performance of rotating parallel-disk assemblies and plate-fin type compact surfaces.

Compact surfaces are characterized by high surface area for heat transfer per unit volume. They have important applications where space is limited and weight must be reduced, such as in vehicular power plants. There are numerous kinds of compact surfaces. Among them, plate-fin surfaces of plain, wavy, and strip types are renowned for their high performance (5). The rotating parallel-disk system may be considered an idealized model of a turbomachinery unit (6). Heat transfer and pressure drop performance in stationary and rotating parallel-disk assemblies were experimentally studied in references 7-9.

The high performance of plate-fin type surfaces is attributed to two important factors: an increase of heat transfer area through an extended surface and a reduction of convective heat transfer resistance through repeated disruption of thermal boundary layers. The high performance of the disk system is attributed to the occurrence of self-sustained, vortex-induced oscillating flow in the radial passage and the effect of centrifugal and Coriolis forces. However, the effect diminishes with an increase in Re. When the system is in rotation at a sufficiently high speed, the f factor falls sharply with a reduction in Re and eventually becomes negative. The sign change (from positive to negative) in f signifies the occurrence of stall propagation, namely a self-excited, large amplitude reverse flow through the disk periphery into the rotor core.

Various methods have been utilized for performance comparison of heat transfer surfaces. One may employ the heat transfer factor j and the friction factor f versus the Reynolds number Re. Another may use a plot of the area goodness factor j/f versus Re. In the present study, a volume goodness factor comparison is employed. It is in the form of h_{STD} versus E_{STD}, where the heat transfer power h_{STD} and the flow friction power E_{STD} are defined as

$$h_{STD} = \left(\frac{Ca\,\mu}{Pr^{2/3}}\right)_{STD} \frac{jRe}{D_H} \; ; \; E_{STD} = \frac{f}{2}\left(\frac{\mu^3}{\rho^2}\right)_{STD} \left(\frac{Re}{D_H}\right)^3 \qquad (1)$$

The subscript STD denotes standard conditions of dry air at 293 K and 1 atmospheric pressure, Table 1.

RESULTS AND DISCUSSION

Figure 1 shows multiple parallel disks with openings in the center, which are assembled as a unit. Each disk was made of aluminum 0.5 mm in thickness; d_1 and d_2 denote the inner and outer diameter, respectively, while s is the disk spacing. d_1 was fixed at 160 mm, while d_2 and s were varied. Heat transfer and friction loss performance of the disk system were determined by a modified single-blow method with the aid of a computer-assisted data-reduction system. Details of the experimental setup and procedure are available in reference 9, which is included in the present proceedings. In the interest of brevity, only representative results are cited here for performance comparison with high-performance, plate-fin compact surfaces of plain, wavy, and strip types. They have approximately the same hydraulic diameter (3 mm).

(i) Stationary disk assembly

Figure 2 shows that the heat transfer power h_{STD} in stationary disk assemblies enhances consistently with the flow friction power E_{STD}. At the same E_{STD}, heat transfer performance improves with a reduction in the disk spacing s, which implies an increase in L/D_H for a fixed value of d_1 and d_2. With s and d_1 fixed, an enlargement in the disk size (d_2) leads to heat transfer augmentation only in the high E_{STD} range. The effect of d_2 on h_{STD} diminishes as the value of E_{STD} decreases. The stationary disk system is superior to the plain plate-fin type surface and comparable to the strip and wavy types when operated at low values of E_{STD}. At a higher E_{STD} range, however, its performance becomes lower than both the strip and wavy fins but still comparable to the plain plate fin.

(ii) Rotating disk assemblies

When the disk assembly was set in rotation, the effect of centrifug

TABLE 1. Geometric properties of parallel disk assembly

CORE NO.	1	2	3	4	5	6	7	8	9	10	11	12
Disk thickness, mm	0.5											
Inside diameter d_1, mm	160											
Outside diameter d_2, mm	240			280			310			345		
Disk spacing s, mm	1.5	2.5	3.5	1.5	2.5	3.5	1.5	2.5	3.5	1.5	2.5	3.5
Flow length L, mm	40			60			75			92.5		
Mean diameter d_m, mm	200			220			235			253		
Hydraulic diameter D_H, mm	3	5	7	3	5	7	3	5	7	3	5	7
L/D_H	13.3	8	5.71	20	12	8.57	25	15	10.7	0.8	18.5	13.2
Number of disks N	9	6	4	9	6	4	9	6	4	9	6	4

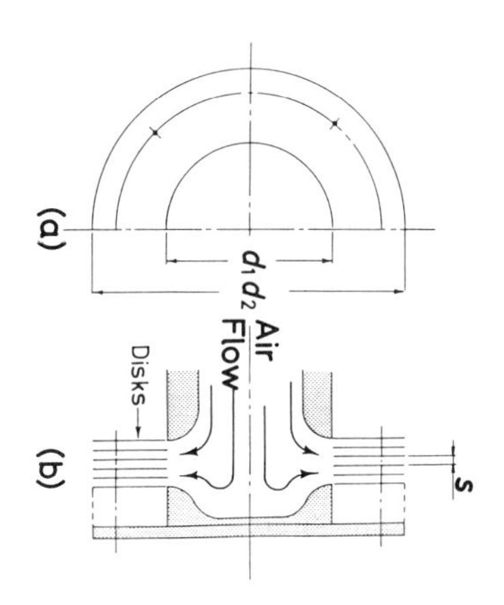

Fig. 1 A schematic of test core

and Coriolis forces augments heat transfer performance. Figures 3, 4, and 5 plot h_{STD} versus E_{STD} for the cores of the same disk size, d_1 = 160 mm and d_2 = 345 mm, but s = 3.5, 2.5, and 1.5 mm, respectively. The value of h_{STD} increases with the rotational speed. The enhancement diminishes at a higher E_{STD} range. A comparison of Fig 3, 4, and 5 indicates that the effect of s, namely D_H, is very important. It is seen that the performance of the rotating disk syst with large disk spacing (s = 3.5 mm and D_H = 7 mm) in Fig. 3 is abo the same as that of the stationary disk assemblies in Fig. 2. The rotating core with small disk spacing (s = 1.5 mm and D_H = 3 mm) is superior to the plain plate-fin surface and comparable to the wavy and strip-fin surfaces over the entire range of E_{STD}. This observation bears special importance in that the rotating disk system an the compact surfaces have about the same hydraulic diameter of 3 mm Even when the disk spacing is enlarged to 2.5 mm, Fig. 4 shows the superiority of the rotating disk system over the plain plate-fin surfaces.

CONCLUSIONS

The heat transfer and friction loss performance of multiple paralle disk assemblies have been compared with those of high-performance, plate-fin compact surfaces of plain, wavy, and strip types in vehicular power applications. It is concluded that based on the same hydraulic diameter, the rotating disk system in the range of 600 to 1800 RPM is superior to the plain plate-fin surface and comparable to both the wavy and strip-fin surfaces. The study has concluded that the rotating parallel disk system is a high performance heat transfer device.

ACKNOWLEDGEMENT

The study was partially supported by the Japanese Ministry of Education under Grant C 1981 and by the U.S. National Science Foundation under Grant Number ME 80-18031.

NOMENCLATURE

C_a	specific heat of air at constant pressure, J/(kg·K)
D_H	hydraulic diameter = 2s, m
d	disk diameter, m; d_1, inner; d_2, outer; d_m, mean
E_{STD}	flow friction power as defined by equation (1), kW/m^2
f	Fanning friction power
h_{STD}	heat transfer power as defined by equation (1), W/(m^2·K)
j	heat transfer factor
L	flow (disk) length = $(d_2 - d_1)/2$, m
N	number of disks in a test core
Pr	Prandtl number
Re	Reynolds number = $u_m D_H/\nu$
s	disk spacing, m
u_m	mean radial velocity at midpoint, m/s

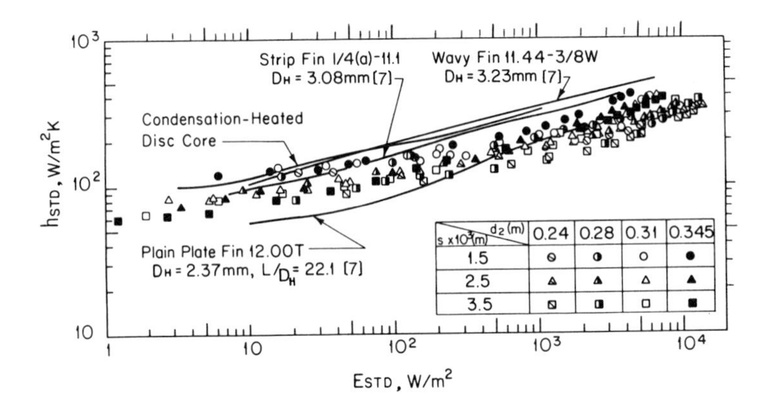

Fig. 2 Comparison of heat transfer power versus flow friction power between stationary disk assemblies and plate-fin compact surfaces

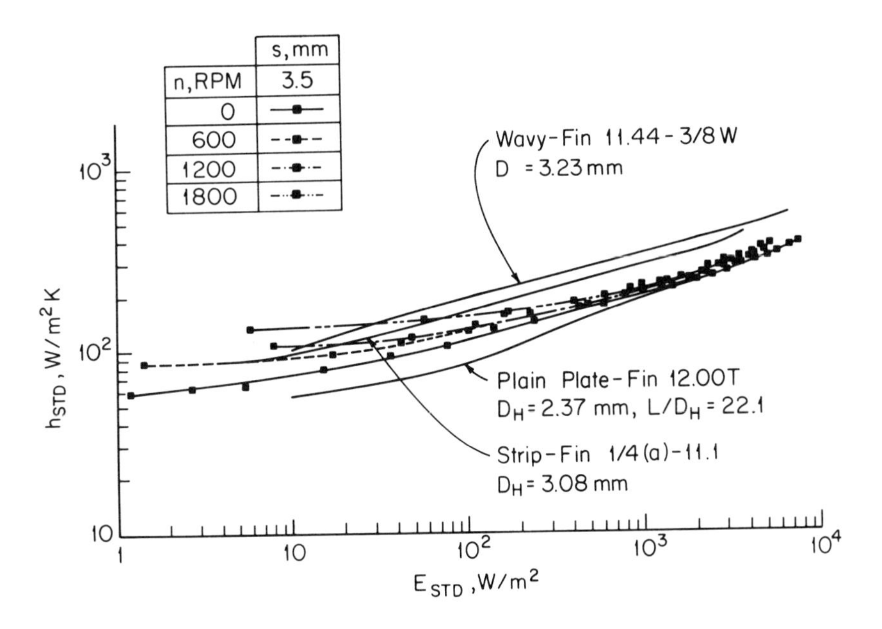

Fig. 3 Comparison of heat transfer power versus flow friction power between rotating disk assemblies with s= 3.5 mm and plate-fin compact surfaces

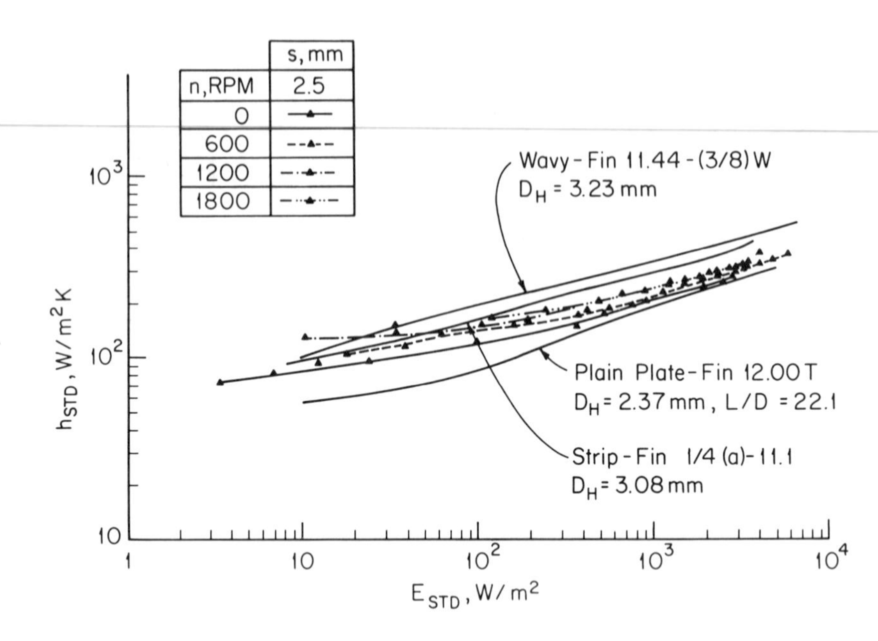

Fig. 4 Comparison of heat transfer power versus flow friction power between rotating disk assemblies with s= 2.5 mm and plate-fin compact surfaces

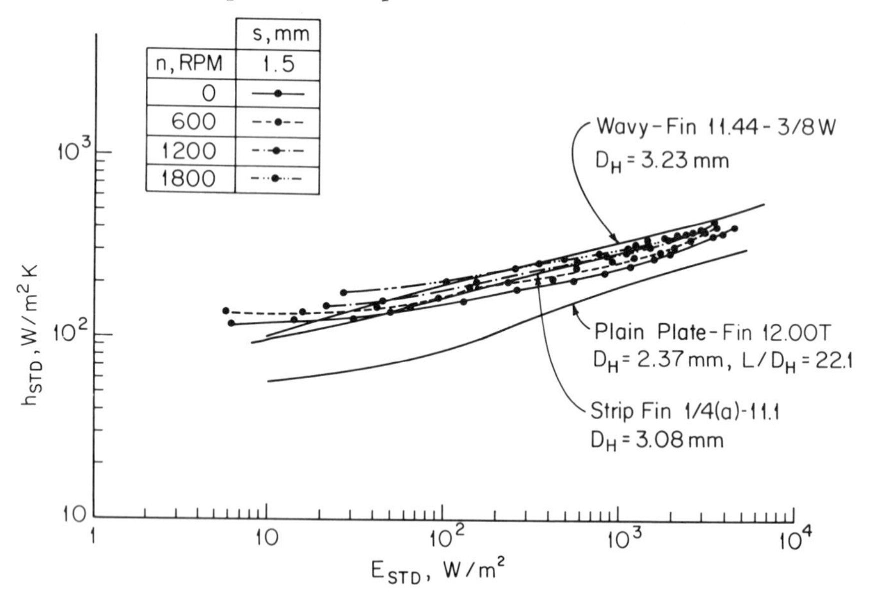

Fig. 5 Compariaon of heat transfer power versus flow friction power between rotating disk assemblies with s= 1.5 mm and plate-fin compact surfaces

ν dynamic viscosity, Pa·s
ρ fluid density, kg/m^3

Subscripts

1 inlet to test core
2 exit from test core
m mean value
STD standard condition at 293 K and 1 atm

REFERENCES

1. Lee, C.P. and Yang, Wen-Jei, "Augmentation of Convective Heat Transfer from High-Porosity Perforated Surfaces," Heat Transfer 1978, Vol. 2, pp. 589-594, Toronto, 1978.

2. Yang, Wen-Jei, "Three Kinds of Heat Transfer Augmentation in Perforated Surface," Letters Journal in Heat Transfer, Vol. 5, pp. 1-10, 1978.

3. Yang, Wen-Jei, "Forced Convective Heat Transfer in Interrupted Compact Surfaces," Proceedings of 1983 ASME-JSME Thermal Engineering Joint Conference, Vol. 3, pp. 105-111, Honolulu, March 20-24, 1983.

4. Mochizuki, S. and Yang, Wen-Jei, "Three Mechanisms of Convective Enhancement in Stationary Disk Systems," Manuscript in Preparation, 1985.

5. Kays, W.M. and London, A.L., Compact Heat Exchangers, McGraw-Hill, New York, 1964.

6. Owens, J.M., "Fluid Flow and Heat Transfer in Rotating Disc Systems," in Heat and Mass Transfer in Rotating Machinery (edited by D.E. Metzger and N.H. Afgan), Hemisphere, Washington, D.C., pp. 81-103, 1984.

7. Mochizuki, S. and Yang, Wen-Jei, "Heat Transfer and Friction Loss in Laminar Radial Flows through Rotating Annular Disks," Journal of Heat Transfer, Vol. 103, 1981, pp. 212-217.

8. Mochizuki, S., Yang, Wen-Jei, Yagi, Y., and Ueno, M., "Heat Transfer Mechanisms and Performance in Multiple Parallel Disk Assemblies," Journal of Heat Transfer, Vol. 105, 1983, pp. 598-604.

9. Yang, Wen-Jei, Mochizuki, S. and Sim, Y. S.,"Theoretical and Experimantal Studies on Transport Phenomena in Rotating Disks", Transport Phenomena in Rotating Machinery (edited by Wen-Jei Yang), Hemisphere, Washington, D. C. (1986).

Analysis of Experimental Shaft Seal Data for High-Performance Turbomachines—As for Space Shuttle Main Engines

R. C. HENDRICKS
National Aeronautics and Space Administration
Lewis Research Center
Cleveland, Ohio 44135

M. J. BRAUN
University of Akron
Akron, Ohio 44325

R. L. MULLEN
Case Western Reserve University
Cleveland, Ohio 44106

R. E. BURCHAM and W. A. DIAMOND
Rockwell International
Rocketdyne Division
Canoga Park, California 91304

ABSTRACT

High-pressure, high-temperature seal flow (leakage) data for nonrotating and rotating Rayleigh-step and convergent-tapered-bore seals have been characterized in terms of a normalized flow coefficient. The results for normalized Rayleigh-step and nonrotating tapered-bore seals were in reasonable agreement with theory, but the data for rotating tapered-bore seals were not. The tapered-bore-seal operational clearances estimated from the flow data were significantly larger than calculated. Although clearances are influenced by wear from conical to cylindrical geometry and by errors in clearance corrections, the problem was isolated to the shaft temperature - rotational speed clearance correction. The geometric changes support the use of some conical convergence in any seal. Under these conditions rotation reduced the normalized flow coefficient by nearly 10 percent.

INTRODUCTION

Few data are available in the literature to guide seal designing for high-performance turbomachines. Although Ref. 1 discusses several examples of instabilities, along with some design recommendations, it gives few data for turbomachines characteristic of the space shuttle main engine (SSME). A seals study program was undertaken (Ref. 2) to determine the leakage and wear performance of a Rayleigh-step and a tapered-bore seal. The Rayleigh-step seal's close tolerances and calculated low restoring forces (as designed) made it undesirable for this application. The tapered-bore seal, designed according to Ref. 3, appeared to have sufficient stiffness, acceptable flow leakages, and tolerable wear characteristics. Tests were conducted at various pressures, mostly above 24 MPa (3500 psi). Nominal initial temperatures were 500 K (440 °F), but some room-temperature checks were made for calibration and comparison. The nominal rotational speed was 483 Hz (29 000 rpm), but some checks were made without rotation. The working fluid was gaseous nitrogen.

The tabulated results of Ref. 2 could be analyzed by using the normalized flow coefficient method presented in Ref. 4. The working fluids of Ref. 4 were ambient-temperature gases. The seal configurations tested were nonrotating, of similar diameter to those of Ref. 2 but with nominal design clearances nearly three times larger. Using the method of Ref. 4 could therefore present a scaling problem. However, the Reynolds numbers for the data of both studies were estimated to be of the same order.[1]

The tapered-bore seal data of Ref. 2 were reanalyzed by using the real-gas theory of Ref. 4 to determine the applicability of such techniques and to assess the wear performance. First, the real-gas theory was shown to give results in reasonable agreement with those of other theories for flow rates in a Rayleigh-step seal. The theory was then compared with the tapered-bore-seal data of Ref. 2. For no rotation, with and without clearance corrections, there was reasonable agreement. However, for rotation with clearance corrections the experimental flow ratio differed significantly from the theoretical (to 1.6). These differences had to arise either from wear or from rotation or temperature clearance corrections. The wear rates are shown to be in reasonably good agreement with theory, on the basis of flow data. Therefore the problem must rest in the rotation or temperature clearance corrections. A 10 to 20 percent change in these values is shown to be sufficient to account for much of the deviation.

APPARATUS AND INSTRUMENTATION

The materials in this section are adapted from Ref. 2 and are presented here to clarify the nature of the experiment and the types of measurement available for analysis. For detailed information see Ref. 2. The turbine seal test assembly (Fig. 1) could be driven to 3142 rad/s (30 000 rpm) by the apparatus motor. High-pressure nitrogen gas (to 25 MPa; 3600 psia) was heated in a counterflow heat exchanger to 533 K (500 °F) before it passed through the seal assembly. The seal inlet temperature decreased over the course of the run because of the high mass flow rates through the heat exchanger. The basic measurements were temperature, pressure, and flow rate. Typical instrumentation locations are illustrated in Fig. 2, with the flowmetering system not shown. A sketch of the tapered-bore seal is shown in Fig. 3 and the actual seal, in Fig. 4.

RESULTS AND ANALYSIS

The flow rate (leakage) analysis of Ref. 4 (without rotation) was applied to the Rayleigh-step and tapered-bore seals described in Ref. 2 (with and without rotation). The contribution of the friction terms was small relative to that of the inertia terms and for gas flows may be related by using the normalized flow coefficient

$$\frac{G_r \sqrt{T_{r,0}}}{P_{r,0}} = C_f = \frac{C_d}{5} \tag{1}$$

For the tests conducted in Ref. 4

---

[1] $Re = \dot{w}D_h/A$; for a shaft seal, $D_h = 2h$ and $D_h/A = 2/\pi\ (D + h) \sim 1/\pi R$. Reynolds similarity criteria requires $Re_1 = Re_2$ or $(\dot{w}_1/\dot{w}_2)(\eta_2/\eta_1)(D_2 + h_2)/(D_1 + h_1) = Y$. Here $D_1 = D_2 >> h_1 > h_2$ and $\eta_2 = \eta_1 \sqrt{T_2/T_1}$. With typical values of $\dot{w}_1/\dot{w}_2 = 1/2$ and $T_2/T_1 = 533/300$, $Y \rightarrow 2/3 = [O]1$.

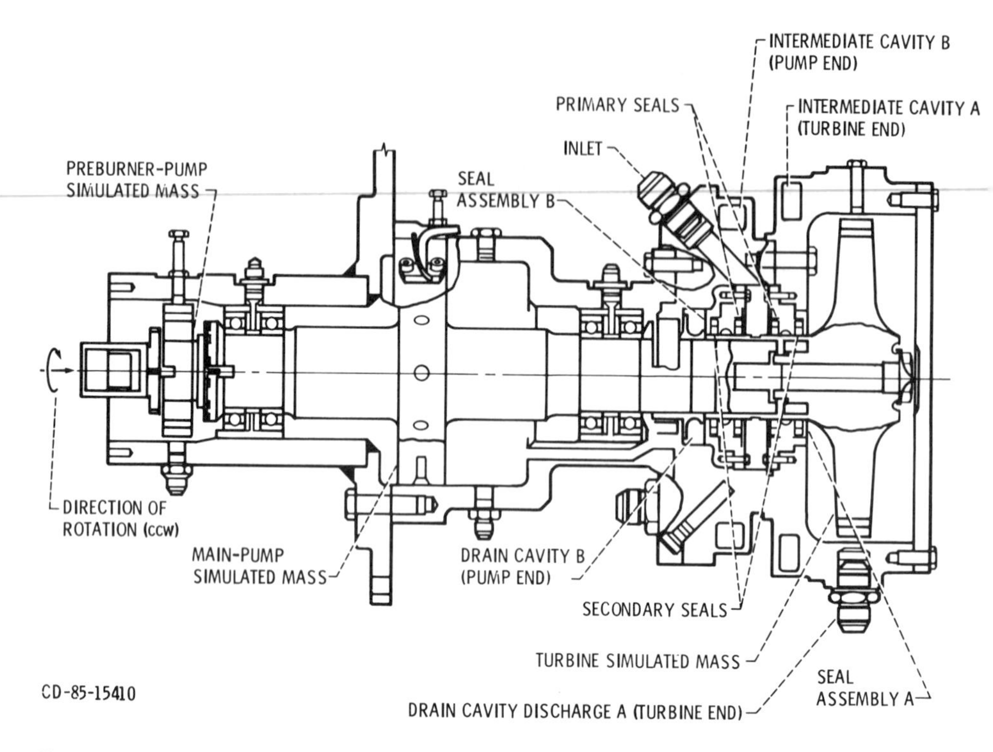

FIGURE 1. Schematic of turbine seal test apparatus. (From Ref. 2.)

$$C_f = 0.153 \quad \text{for concentric position (no rotation)}$$

$$= 0.158 \quad \text{for fully eccentric position (no rotation)} \tag{2}$$

which implies that $C_d = 0.8$ for those similar (geometric and kinematic) configurations.

In the following sections we demonstrate the application of Eqs. (1) and (2). Theoretical comparisons were made for the Rayleigh-step seal, and several cases were investigated for the tapered-bore seal, with and without rotation and correlation of flow data, to isolate a clearance correction problem.

Rayleigh-Step Seal - Theoretical Comparisons

The flow rate (leakage) analysis of Ref. 4 was compared with the theoretical results of Ref. 2 (p. 24) for the design of the Rayleigh-step seal.[2] For the primary seal with a design pressure of 27.6 MPa (4000 psia) and a design temperature of 533 K (500 °F), the parameters used in Eq. (1) became[3] $P_{r,0} = 8.07$, $T_{r,0} = 4.22$, $A_e = 0.103\ \text{cm}^2$, $G^* = 6010\ \text{g/cm}^2\ \text{s}$, $C_d = 0.6$, $G = 2833\ \text{g/cm}^2\ \text{s}$ and the flow rate became $\dot{w} = GA_e = 292$ g/s (0.64 lb/s, or

[2]The seal passage was slightly divergent.
[3]For fluid nitrogen, $P_c = 3.417$ MPa (495.5 psia); $T_c = 126.3$ K (-232.3 °F); $G^* = 6010\ \text{g/cm}^2\ \text{s}$ (85.5 lb/in^2 s).

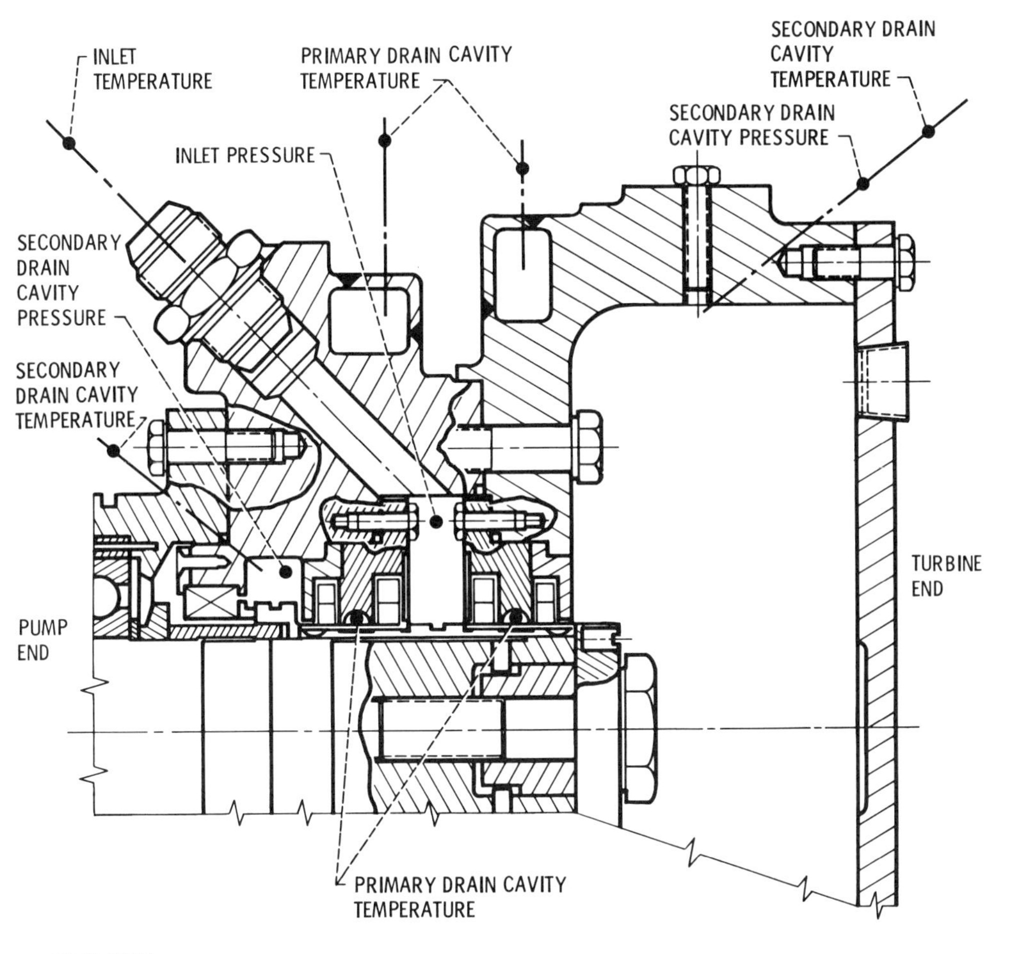

FIGURE 2. Schematic of test instrumentation locations. (From Ref. 2.)

495 ft^3/min). For the secondary seal with a design pressure of 0.059 MPa (8.5 psia) and a design temperature of 533 K (500 °F), the flow rate became $\dot{w}$ = 6.21 g/s (0.014 lb/s, or 10.5 ft^3/min). Flow rates predicted by the methods of Refs. 4 to 6 (Table 1) are in reasonable agreement.

Tapered-Bore Seal - Flow Rate Comparisons

Since the theoretical agreement between the results of Refs. 4 to 6 is reasonably good, we now propose to apply the normalized flow coefficient method to the tapered-bore-seal data of Ref. 2. The seal was designed for operation at 483 Hz (29 000 rpm) with an inlet-to-exit clearance ratio of 1.8 for a nominal exit diametral clearance of 0.051 mm (0.002 in) at 26.2 MPa (3800 psia) and 533 K (500 °F). The average flow area for a conical tapered-bore seal is (Ref. 4)

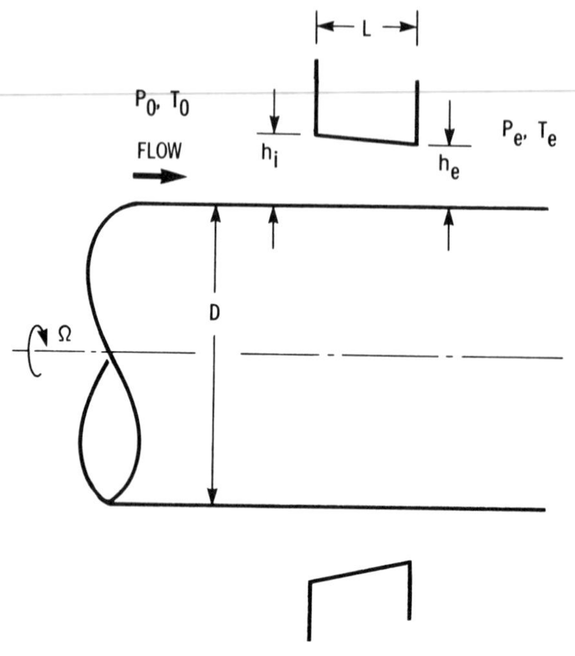

FIGURE 3. Sketch of tapered-bore seal configuration.

FIGURE 4. Tapered-bore seal. (From Ref. 2.)

TABLE 1. Comparison of Theoretical Flow Rates for a Rayleigh-Step Seal (Design clearance, h = 0.051 mm (0.002 in))

Theory	Leakage				Percent deviation	
	Primary		Secondary		Primary	Secondary
	g/s	ft^3/min	g/s	ft^3/min		
Isentropic (Ref. 5)	15.06	532	0.32	11.3	-7	-8
QUASC (Ref. 6)	14.16	500	.303	10.7	-1	-2
Real gas (Ref. 4)	14.00	495	.297	10.5	--	--

$$\bar{A} = \frac{A_i + 2A_e}{3} \tag{3a}$$

or in terms of the inlet and exit clearances (Fig. 3)

$$\frac{\bar{A}}{A_e} = \frac{h_i + 2h_e}{3h_e} = 1.27 \tag{3b}$$

In turbomachinery, radial clearance h and tolerance corrections e due to temperature, pressure, manufacturing, alignment, and operations are significant and must enter into the analysis. The equation for h can be written as

$$h = h_{inst} + (e_{\Delta T} + e_{\Delta P})_s - (e_{\Delta T} + e_{\Delta rpm})_{sh} \tag{4}$$

Flow rate comparison without rotation. For the data given in Table 12 of Ref. 2 (build 10 - pretest 108,6) without rotation (0 rpm), we find for 20.1 MPa (2915 psia), 299 K (78 °F), and the geometry of Fig. 3 that $T_{r,0} = 2.36$ and $P_{r,0} = 5.88$. The nominal clearance can be estimated by using the diametral clearances presented in Table 2, assuming that $e_{\Delta T}$ and $e_{\Delta P}$ vary linearly with temperature and pressure, respectively, and that half of the shaft growth is due to temperature and is linearly dependent on both temperature and rotational speed. Since the test (build 10 - pretest 108,6) was run at ambient temperature (300 K; 78 °F) and 20 MPa (2900 psia) without rotation, the diametral clearance corrections were estimated (Table 3). Substituting these estimated clearance corrections into Eq. (4), the diametral clearance $2h$ becomes 0.1956 mm (0.0077 in). From Eq. (3) the effective flow area becomes 0.227 cm^2, and from Eq.(1) the estimated flow rate, $\dot{w} = 1044 \times C_d$ g/s ($2.30 \times C_d$ lb/s), is about 20 percent higher than the measured value of 700 g/s (1.54 lb/s) for $C_d = 0.8$.[4] For this run, $C_d = 0.7$ would bring satisfactory agreement.[5]

---

[4]Without clearance correction and without rotation, from equations.
[5]At high pressure the seal face is strained, rotates, and may change C_d.

TABLE 2. Diametral Clearances for Tapered-Bore Primary Seal with Minimum Clearance (Ref. 2, p. 29)
(Nominal seal outlet diameter, 6.4668 cm (2.5460 in); nominal inlet-to-exit clearance ratio, 1.8; nominal seal inlet diameter, 6.4720 cm (2.5480 in); design pressure, 26.2 MPa (3800 psia); design temperature, 533 K (500 °F))

	Diametral clearance	
	mm	in
Design	0.508	0.002
Installed	.254	.0100
Seal:		
Temperature effect	.112	.0044
Pressure effect	-.0787	-.0031
Shaft - temperature and speed effect	.2286	.009

TABLE 3. Estimated Diametral Clearance Corrections for Selected Data of Ref. 2, With and Without Rotation

	Build 10 - pretest 108,6		Build 12 - test 120	
	mm	in	mm	in
Seal:				
Pressure correction, $2e_{\Delta P}$	-0.061	-0.0024	-0.074	-0.0029
Temperature correction, $2e_{\Delta T}$	.018	.0007	.10	.004
Shaft:				
Temperature correction, $2e_{\Delta T}$	.015	.0006	.10	.004
Rotational speed correction $2e_{\Delta rpm}$	0	0	.117	.0046

Flow rate comparison with rotation. We now apply the same method to the rotational data (build 12 - test 120), where the rotational speed is 469 Hz (28 150 rpm); pressure, 24.8 MPa (3590 psia); and temperature, 510 K (458 °F). With the clearance corrections given in Table 3 the average flow area is determined from Eqs. (3) and (4) with $2h_i = 0.114$ mm (0.0045 in) and $2h_e = 0.064$ mm (0.0025 in) as $A = 0.082\ cm^2$. From Eq. (1), the mass flow rate becomes $\dot{w} = 356 \times C_d$ g/s ($0.785 \times C_d$ lb/s), which is a factor of 1.6 lower than the experimental value of 584 g/s (1.29 lb/s) for $C_d \rightarrow 1$ and a factor of 2.3 lower for $C_d \rightarrow 0.7$. So how can one account for being nearly correct at no rotation and a factor of 2.3 low with rotation. It would appear that the clearance tolerances of Tables 2 and 3 and Eqs. (3) and (4) require further investigation to determine whether the disparity is solely the effect of rotation.

Tapered-Bore Seal Operational Flow Area and Rotation Effects

As an initial effort to determine the effects of operational flow area and rotation on flow rate, some of the data of Ref. 2 (p. 69, Table 12) were recalculated in terms of the normalized flow coefficient. Rewriting Eq. (1) gives

$$C_f = \frac{\dot{w}_{exp}}{\dot{w}_{calc}} = \frac{\dot{w}_{exp}}{G^* \bar{A} P_{r,0} / \sqrt{T_{r,0}}} \tag{5}$$

and, with $\bar{A} = 1.27\ A_e$, the values are given as a function of $P_{r,0}$ in Fig. 5.

To make this comparison, the clearance tolerances of Eq. (4) were held fixed (i.e., constant clearance for expediency). As is clear from Fig. 5, there is considerable scatter for the pump-end primary seal. Similar results were found for the turbine-end primary seal. However, the trend with pressure was predictable since the clearance was held constant. This further indicates the necessity of properly incorporating Eq. (4) in the design analysis. It is possible that the experimental flow area for these tests differs from the design area by a factor of 3.[6]

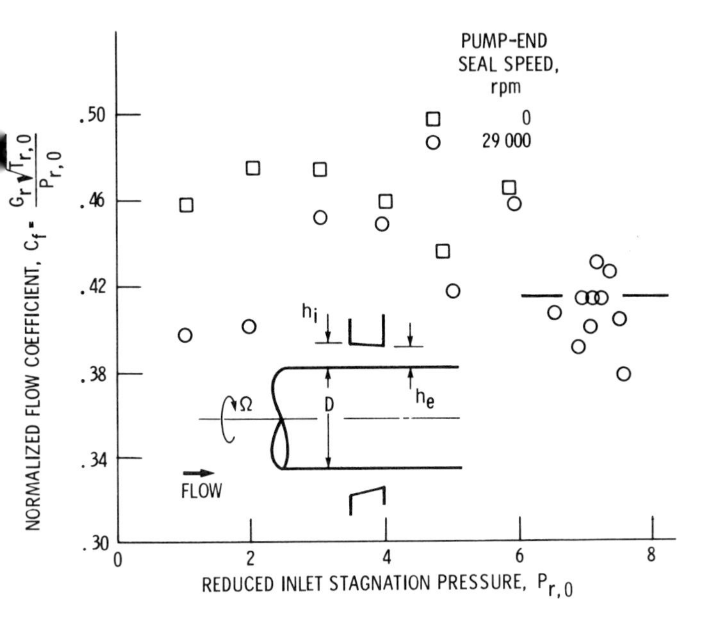

FIGURE 5. Comparison of normalized flow coefficients for pump-end primary seals with and without rotation.

Effect of rotation. The effect of rotation (Fig. 5) can be estimated as follows:

$$\frac{C_{f,\text{rotation}}}{C_{f,\text{no rotation}}} = \frac{0.137}{0.15} = 0.91 \tag{6}$$

This establishes, experimentally, about a 10-percent decrease in flow rate due to rotation independent of flow area.

Effect of wear on flow rate. If the tapered-bore seal wore cylindrical, the flow area ratio would be

---

[6]To place it in better perspective, a 10-percent error in design diametral clearance $2h$ is 5 μm. (Recall diametral clearance is 0.0508 mm (0.002 in).)

$$\frac{\bar{A}_{cyl}}{\bar{A}_{des}} = \frac{1.84\ A_e}{1.27\ A_e} = 1.45 \tag{7}$$

Because the seals did not exhibit such extensive wear and the area increase cannot account for the disparity in flow rates, the problem must reside in the shaft rotational speed clearance correction.

Operational flow area. To resolve the flow area problem, it is first necessary to establish confidence in the flow data of Ref. 2. A statistical analysis of the 14 high-pressure, high-temperature data points with rotation for the pump-end and turbine-end primary seals is given in Fig. 6 and summarized in Table 4. In general, although the scatter is apparent, it is much less than would be anticipated from the Monte Carlo sampling of a normally distributed sample space of 14 members. This gives confidence in the reproducibility of the flow data.

TABLE 4. Statistical Values[a] of Normalized Flow Coefficient for Selected Data of Ref. 2

Seal	Averaged normalized flow coefficient, C_f	Standard deviation	Correlation coefficient
Turbine end	0.414	0.0069	0.985
Pump end	.414	.0066	.977

[a]For a more definitive measure of the normalized flow coefficient, a much larger, well-controlled sample space will be required.

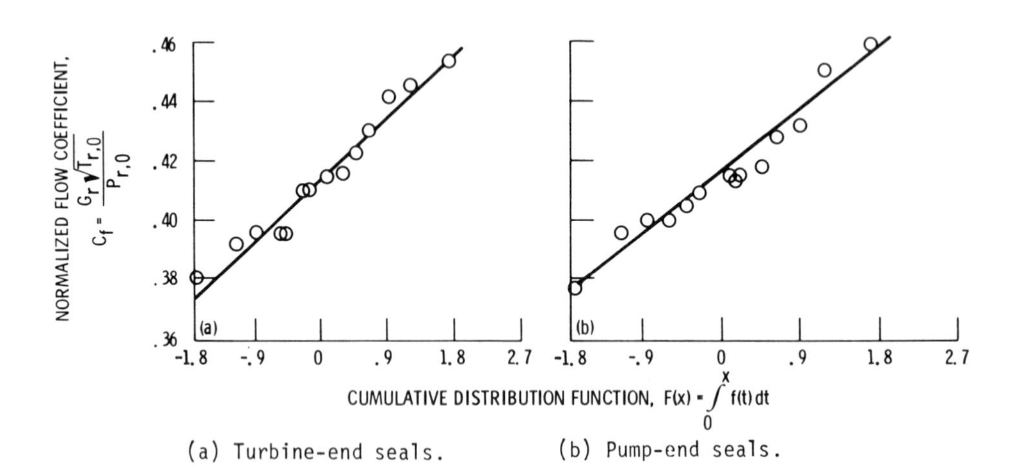

(a) Turbine-end seals. (b) Pump-end seals.

FIGURE 6. Statistical variation of normalized flow coefficient for pump-end and turbine-end seals. Pressure range, 21 - 28 MPa (3000 - 4000 psi); temperature range, 460 - 533 K (370 - 500 °F); speed, ~29 000 rpm.

If the flow data with rotation were in agreement with the results of Ref. 4, as they are for no rotation with a $C_d = 0.7$ instead of 0.8, the ratio of the operational flow clearances would be defined in terms of the flow coefficients of Table 4 and Eq. (2).

$$\frac{\bar{A}_{op}}{\bar{A}_{e,des}} = \frac{C_{f,op}}{C_{f,des}} = \left(\frac{0.8}{0.7}\right)\left(\frac{0.41}{0.153}\right) = 3.1 \tag{8}$$

From Eqs. (3) and (8) the clearance becomes

$$3.1 = \left\{D_i^2 - (D_s + 2h_0)^2 + 2\left[D_e^2 - (D_s + 2h_0)^2\right]\right\}\frac{\pi}{12} \tag{9}$$

Solving gives $2h_0 = 0.117$ mm (0.0046 in) and

$$h_e = 0.083 \text{ mm } (0.00325 \text{ in}) \tag{10}$$

or about three times the design value.[7] This suggests an error in rotation or temperature design clearance (Table 2).

For a cylinder the radial displacement u_r can be approximated as

$$u_r = \frac{\left(1 - \nu^2\right)\rho_{sh}D_{sh}^3\Omega^2}{64E} \tag{11}$$

and solving gives $u_r = 0.018$ mm (0.0007 in), which is significantly less than cited in Table 3. For a 50-percent decrease in both the rotation and temperature shaft clearance corrections, the predicted and experimental flow rates would agree, but such an error is difficult to verify and remains only bracketed. We now turn our attention to the problem of wear.

WEAR ANALYSIS

Wear - from Geometry

Combining the definition of the effective flow area $\bar{A}_1$ for a tapered-bore seal (Eq. (3)), the optimum taper recommended (Ref. 3), and the geometry (Fig. 3) gives a pretest, or initial, clearance of

$$h_{i,1} = 1.8\ h_{e,1} \tag{3c}$$

and a flow area of

$$\bar{A}_1 = 1.27\ A_{e,1} \approx 1.27\ \pi D h_{e,1} \tag{3d}$$

---

[7]It must also be recognized that during the blowdown test the full effect of the temperature may not be felt. Furthermore, some immediate rub-in is imminent.

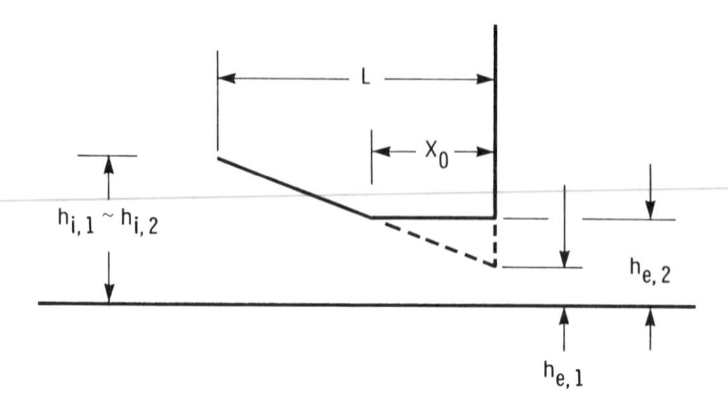

FIGURE 7. Sketch of tapered-bore seal with wear.

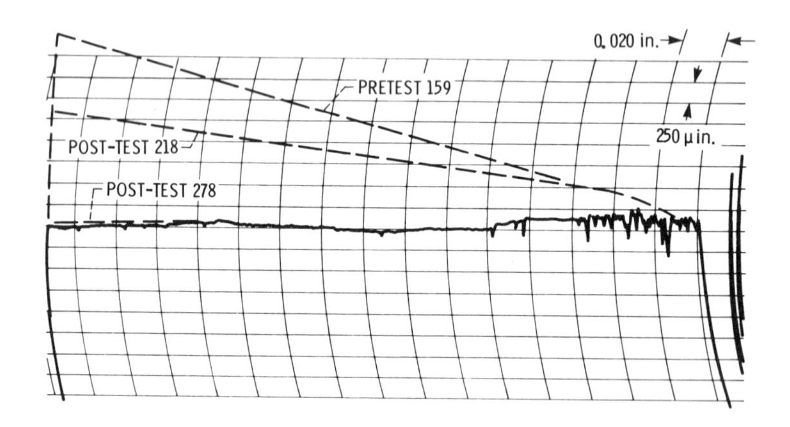

FIGURE 8. Profilometer traces for turbine-end primary seal ring. (From Ref. 2.)

The post-test flow area $\bar{A}_2$ is not so easy to establish as it varies from conical to cylindrical; with the aid of Fig. 7, it can be expressed as

$$\bar{A}_2 = \left[\left(\frac{h_{i,2} + 2h_{e,2}}{3h_{e,2}}\right)(1 - X_0) + X_0\right] A_{e,2} \tag{12}$$

where $h_{i,1} = h_{i,2}$ and X_0 represents the fraction of the seal length that is cylindrical. The values of h_i, h_e, and X_0 are determined from profilometer data (e.g., Fig. 8).

Wear - from Flow Data

Changes in the pretest and post-test flow rates provide a relative estimate of wear. From Eq. (1) we have

$$\frac{C_{f,1}}{C_{f,2}} = \frac{\left(\frac{G_r\sqrt{T_{r,0}}}{P_{r,0}}\right)_1}{\left(\frac{G_r\sqrt{T_{r,0}}}{P_{r,0}}\right)_2} = \left(\frac{\dot{w}_1}{\dot{w}_2}\right)\left(\frac{\bar{A}_2}{\bar{A}_1}\right)\left(\frac{P_{0,2}}{P_{0,1}}\right)\sqrt{\frac{T_{0,1}}{T_{0,2}}} = \chi\left(\frac{\bar{A}_2}{\bar{A}_1}\right) \tag{13}$$

Although χ is readily determined from the flow measurements, changes in C_f are difficult to assess.

Wear - from Profilometer

We can now make wear estimates from the profilometer data (Fig. 8) and the flow data for pretest run 159 and post-test run 218 (Fig. 108 and Tables 10 and 12 of Ref. 2).

Geometric wear. The average diametral wear is 0.038 mm (0.0015 in) at the seal outlet and 0.013 mm (0.0005 in) at the seal inlet (Ref. 2, p. 179) in agreement with Fig. 8. Assume the operating clearances to be $2h_{e,1}$ = 0.165 mm (0.0065 in) (see Eq. (10)) and $2h^*_{e,2}$ = 0.203 mm (0.008 in). As both profiles are tapered ($X_0 = 0$), $h^*_{i,1} \approx h^*_{i,2} \approx$ 0.216 mm (0.0085 in).

$$\frac{\bar{A}_2}{\bar{A}_1} = 1.10 \tag{14}$$

Wear by flow parameters. With $(L/2h) \rightarrow 40$, assume that the passage transition from conical to cylindrical lowers C_f by 5 percent. The operating conditions at 483 Hz (29 000 rpm) are run 159: P_1 = 22.8 MPa (3305 psia), T_1 = 464 K (375 °F), $\dot{w}_1$ = 608 g/s (1.340 lb/s) and $C_{d,1} = 0.8$; and run 218: P_2 = 23.4 MPa (3390 psia), T_2 = 533 K (500 °F), $\dot{w}_2$ = 610 g/s (1.346 lb/s) and $C_{d,2} = 0.76$.

The ratio $\bar{A}_2/\bar{A}_1$ becomes

$$\frac{\bar{A}_2}{\bar{A}_1} = \frac{C_{f,1}}{\chi C_{f,2}} \tag{15}$$

and from the flow parameters

$$\frac{\bar{A}_2}{\bar{A}_1} = \sqrt{\frac{533}{464}}\left(\frac{3305}{3390}\right)\left(\frac{1.346}{1.340}\right)\left(\frac{0.8}{0.76}\right) = 1.1 \tag{16}$$

The agreement between measured and calculated wear is reasonable. However, rub-in is common and, if the initial rub were 10 percent of the total wear,

$h^*_{e,1} = 0.1689$ mm (0.00665 in), calculated and measured wear would not agree.[8] Even so, for these results it is difficult to justify agreement to better than 10 to 15 percent.

STABILITY CONSIDERATIONS

As established in Ref. 4 and in experimental tests with similar low-L/D tapered- and straight-bore and high-L/D straight-bore seals (three seal configurations, Ref. 7), the convergent-tapered-bore seal is at least as stable as the straight-bore seal and more tolerant to clearance problems. Thus a tapered-bore seal worn cylindrical gains dynamic stiffness from the increase in flow leakage.

The wear characteristics for a cylindrical passage could produce a divergent (or convergent) flow passage, leading to decreased (or enhanced) stability. If both divergence and convergence were produced, the leakage rate would increase because $C_f \rightarrow 1$, similar to a venturi with a conjectured loss of stability. Such geometric changes are the subject of another study.

SUMMARY OF RESULTS AND RECOMMENDATIONS

Primary and secondary seals for the pump and turbine ends of an advanced turbomachine have been experimentally studied in Ref. 2. The high-pressure, high-temperature nitrogen gas flow data of Ref. 2 have been characterized in terms of a normalized flow coefficient

$$\frac{G_t \sqrt{T_{r,0}}}{P_{r,0}} = C_f$$

The data for normalized Rayleigh-step and nonrotating tapered-bore seals were in reasonable agreement with theory, but the data for rotating tapered-bore seals were not. The operational flow area had to be enlarged by a factor of 3 in order to match the data. The problem was isolated to the shaft temperature - rotational speed clearance correction. Rotation was found to reduce the normalized flow coefficient by nearly 10 percent independently of the flow area problems.

The flow area increase measured from the wear data and calculated from the flow data was 10 percent. However, rub-in is commonplace and agreement cannot be justified to better than 10 to 15 percent.

Even when normalized in terms of flow rate, the conical convergent seal appears to be more stable than the straight-bore seal. As it wears, it usually becomes more cylindrical; at the same time the flow rate (leakage) increases and the system stiffness can increase.

---

[8]Table 10 of Ref. 2 (p. 201) gives total wear values with an average of 0.028 mm (0.0011 in) for system post-test 278, but clearly, from Fig. 8, post-test 278 values give a wear of 0.057 mm (0.00225 in), a factor of 2 higher than the tabulated data. We feel that Fig. 8 is correct.

SYMBOLS

A	area
D	diameter
D_h	hydraulic diameter, $4A/p$
C_d	flow loss coefficient (e.g., entrance)
C_f	flow coefficient
E	Young's modulus
e	clearance tolerance correction
G	mass flux
G^*	mass flux normalizing parameter, $\sqrt{\rho_c P_c / Z_c}$ ($G^* = 6010$ g/cm^2 s (85.5 lb/in^2 s) for nitrogen
G_r	normalized mass flux, G/G^*
h	clearance
h^*	clearance based on profilometer data
L	seal length
R	radius
$\mathscr{R}$	gas constant
Re	Reynolds number
P	pressure
P_r	reduced pressure, P/P_c
p	"wetted" perimeter
T	temperature
T_r	reduced temperature, T/T_c
u_r	radial displacement
$\dot{w}$	mass flow rate
X	area ratio parameter
Y	similarity measure parameter
Z	compressibility, $P/\rho \mathscr{R} T$

η viscosity

ν Poisson's ratio

ρ density

χ flow parameter, Eq. (17)

Ω angular velocity

Subscripts

c thermodynamic critical point

calc calculated

cyl cylinder

des design

e exit

exp experimental

inst manufacturing, alignment, and other clearance tolerances

i inlet

op operational

ΔP pressure clearance tolerance

Δrpm rotational speed clearance tolerance

s seal; static component

sh shaft

ΔT temperature clearance tolerance

0 stagnation or reference

Superscript

average

REFERENCES

1. "Rotordynamic Instability Problems in High-Performance Turbomachinery," NASA CP-2133, 1980; see also NASA CP-2250, 1982; and NASA CP-2338, 1984.

2. Burcham, R.E., and Diamond, W.A., "High-Pressure Hot Gas Self-Acting Floating Ring Shaft Seal for Liquid Rocket Turbopumps," RI/RD80-186, Rocketdyne, Canoga Park, CA, 1981. (NASA CR-165392.)

3. Fleming, D.P., "Stiffness of Straight and Tapered Annular Gas Path Seals," ASME Paper 78-Lub-18, 1978.

4. Hendricks, R.C., "Some Flow Characteristics of Conventional and Tapered High-Pressure-Drop Simulated Seals," ASLE Trans., Vol. 24, No. 1, pp. 23-28, 1981.

5. Zuk, J., Ludwig, L.P., and Johnson, R.L., "Design Study of Shaft Face Seal with Self-Acting Lift Augmentation, I - Self-Acting Pad Geometry," NASA TN D-5744, 1970.

6. Zuk, J., Ludwig, L.P., and Johnson, R.L., "Quasi-One-Dimensional Compressible Flow Across Face Seals and Narrow Slots. I - Analysis," NASA TN D-6668, 1972.

7. Hendricks, R.C., "A Comparison of Flow Rates and Pressure Profiles for N-Sequential Inlets and Three Related Seal Configurations," Advances in Cryogenic Engineering, Vol. 29, Plenum, New York, 1983.

Fluid Flow Resistance on Surface of Rotating Cylindrical Drums in Enclosure

KINICHI TORIKAI and KOHOICHI SUZUKI
Department of Mechanical Engineering
Science University of Tokyo
Noda-shi, Chiba-ken, Japan

1. INTRODUCTION

We need to know fluid flow resistance of rotating cylindrical drums in a fixed enclosure for multi stages of chemical industrial apparatuses or centrifuges in design or maintenance in order to be reduced power. A concept on system of multi cylindrical drums in one fixed enclosure is presented by the authors for reducing the power of those apparatuses.

If we have two rotating cylindrical drums with the same velocity and direction on counter surfaces of the rotating drums, the fluid flow resistance in the narrow gap between one surface and another surface of the rotating drums will become smaller, because the fluid flow in the gap has the same velocity and direction as the moving of the drum surfaces, while the outside surfaces of the rotating drums located near the inside wall of the fixed enclosure can be considered to be similar conditions to the outside surface of single rotating cylindrical drum in one fixed enclosure for the fluid flow resistance as shown in FIGURE 1.

As the results of above mentioned articles, it is considered that the total fluid flow resistance between single rotating cylindrical drum and one fixed

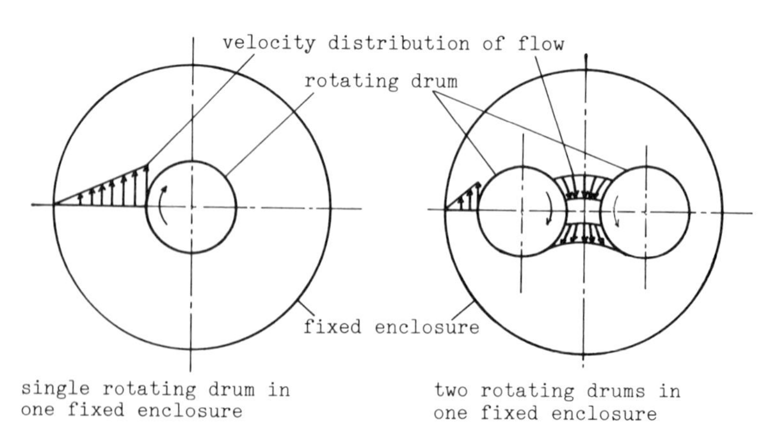

FIGURE 1. Outline of two rotating drums and one rotating drum in a fixed enclosure.

enclosure would be more than the divided fluid flow resistance between the surfaces of rotating drums and the one fixed enclosure by the number of drums. And the fluid flow resistance of the outside surfaces on rotating drums in the system of multi rotating drums in one fixed enclosure was obtained experimentally by a simulation method using rotating drums in water basin.

2. RESISTANCE OF FLUID FLOW

A theoretical treatment for fluid flow resistance on the outside surface on a single rotating cylindrical drum in one fixed enclosure is shown as follows.

The shearing stress τ on the surface of a rotating drum is generated by fluid flow friction in the gap δ between a rotaing drum and a fixed enclosure. And the friction factor f has been given by Eq.(1).

$$f \equiv \frac{\tau}{\frac{1}{2}v^2\rho} \tag{1}$$

And much research has been conducted for the friction factor f.

Stuart[1], Davey[2] and Batchelor[3] have performed theoretical analysis and Taylor[4], Wendt[5], Donnelly[6] and Kanagawa[7] have experimented for the friction factor on turbulent flow in the system of single rotating drum with a fixed enclosure, and especially Kanagawa[7] has showed the friction factor f for special Reynolds number $Re^* = [(1+\delta^*/2)/(1+\delta^*)^2]Re$, $(Re = v\delta/\nu,\ \delta^* = \delta/r)$ as result of summation of various research.

A new equation on power for fluid flow resistance without the relation for δ in this system was considered by the authors as follows; if it is considered that the above mentioned gap δ between a rotating drum and a fixed enclosure is a fluid flow channel, the hypothetical pressure drop Δp in unit length of this channel is shown in Eq.(2), and the hydrodynamic diameter in the channel is $4\delta/2 = 2\delta$ at circular flow.

$$\Delta p = f_1 \frac{1}{4\delta} v^2 \rho \tag{2}$$

Frictional shearing stress τ on the surface of the rotating cylindrical drum by fluid flow resistance in the channel, i.e. gap δ, is given by Eq.(3) and Eq.(4) from Eq.(2).

$$2\tau = \Delta p \cdot \delta \tag{3}$$

$$\tau = \frac{1}{4} f_1 \frac{v^2}{2} \rho \tag{4}$$

because $\Delta p \cdot \delta$ is total resistance in the channel. And τ is not explicitly related to the gap δ as shown in Eq.(4). A relation between f and f_1 is shown in Eq.(5) from Eq.(1) and Eq.(4).

$$f_1 = 4f \tag{5}$$

And the requested power P_1 for the fluid flow resistance on the outside surface of a rotating drum in unit height of a drum is shown in Eq.(6) and Eq.(7) from Eq.(4).

$$P_1 = \tau \cdot 2\pi r \cdot v \tag{6}$$

$$P_1 = f_1 \frac{v^3}{4} \rho \cdot \pi r \tag{7}$$

Now if multi stages of same rotating cylindrical drums are necessary for practical use, the power for the fluid flow resistance of rotating cylindrical drums will be proportional to the number of stages. But if multi rotating cylindrical drums with the same direction and velocity of the surfaces on rotating drums in part of the most narrow gap between one surface and another surface of rotating drums each other are assembled in one fixed enclosure, the fluid flow resistance in the narrow gap might be smaller.

Then four rotating cylindrical drums in a fixed enclosure are presented by the authors in order to reduce fluid flow resistance of rotation of drums as shown in FIGURE 2. These four cylindrical drums have same direction and velocity of rotating surface in part of the most narrow gap between one surface and another surface of the rotating drums each other, and they are placed with same pitch each other and symmetrically to the center of the fixed enclosure.

The fluid flow resistance on the surface of each rotating drum in this case might be reduced in two parts of the most narrow gap between one surface and another surface of the drums. And the power P_4 for fluid flow resistance of the outside surface of one rotating drum in this system are given by Eq.(8) from Eq.(7) and FIGURE 2.

$$P_4 = F_4 f_1 \frac{v^3}{4} \rho \pi r \qquad (8)$$

where F_4 is a empirical factor for reducing fluid flow resistance on the surface, determined by the experiment and is less than unity.

3. THE EXPERIMENTAL APPARATUS

The experimental apparatus is classified into two types, one concerning the system of a single rotating drum, and the other concerning the system of multi rotating drums. We have many experiments on single rotating drum in a fixed enclosure to obtain basic data for the system of multi rotating cylindrical drums in one fixed enclosure.

The experimental apparatus concerning of a single rotating drum is shown in

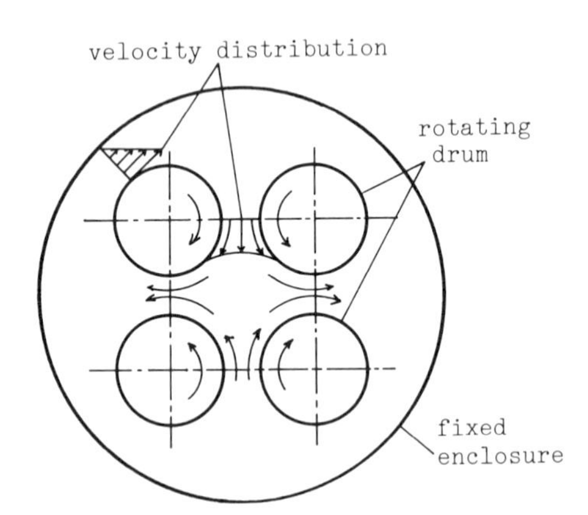

FIGURE 2. Outline of four rotating drums with same direction and velocity each other in a fixed enclosure.

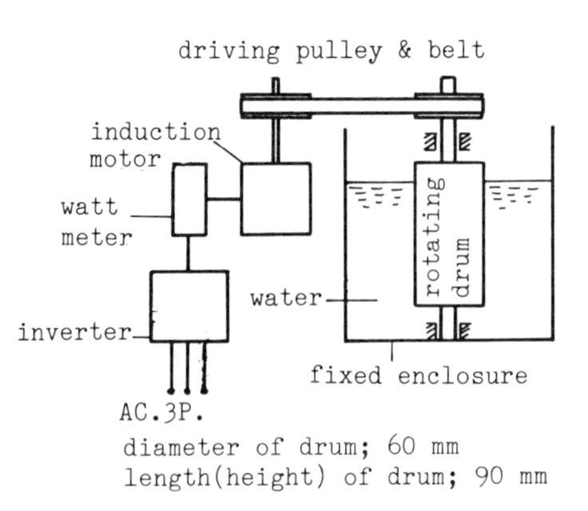

FIGURE 3. Experimental apparatus for single rotating drum in one fixed enclosure.

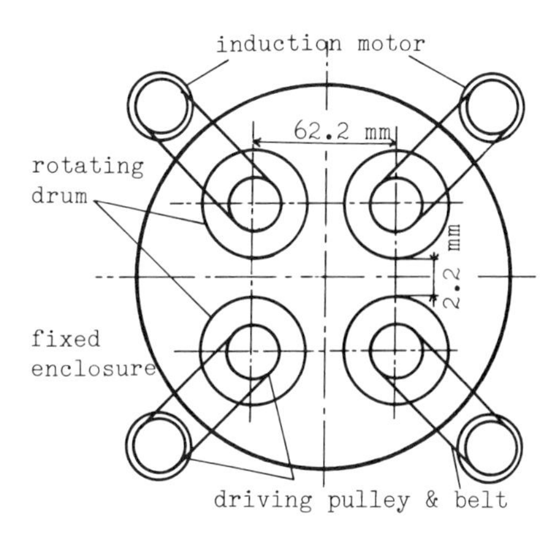

FIGURE 4. Experimental apparatus for four rotating drums.

FIGURE 3. The cylindrical drum is assembled in a large water basin which is a cylindrical type. The drum is driven by an induction motor through a driving belt and the revolution is controlled by means of an electric current frequency by an inverter. The load for rotation of the drum is measured by a Watt meter (electric power meter) under the experimental conditions, i.e. water level in the basin and the revolution of the drum, etc.

The experimental apparatus concerning multi rotating cylindrical drums is shown in FIGURE 4. Four rotating cylindrical drums are employed in this experiment. The drums are placed symmetrically to the center of the cylindrical water basin and the four couples composed with the same drum and drive-motor as the single drum system in FIGURE 3 are assembled in the enclosure. The revolution of the respective drums is controlled at same time by the same method of the single rotating drum system.

4. THE EXPERIMENT

4.1 The experiment concerning a single rotating drum.

A skill method is necessary for the measurement of resistance of the outside surface on the rotating cylindrical drum. Then the experiment was proceeded by the following method ; the load for rotation of the drum was measured for the revolution per unit time at the top level and bottom level of water on the drum surface (FIGURE 5). The measured load includes load for rotation of the bottom surface of the drum, many moving mechanical parts and electrical parts. The difference of the measured load under the both conditions is considered to be approximately the load for rotation of the outside surface of the drum except the bottom surface.

Data of the load, i.e. power P_1 with respect to the revolution of the single rotating drum are shown in FIGURE 6. And the friction factor f_1 are calculated from Eq.(7) and the result of FIGURE 6, and are shown in FIGURE 7 for Reynolds

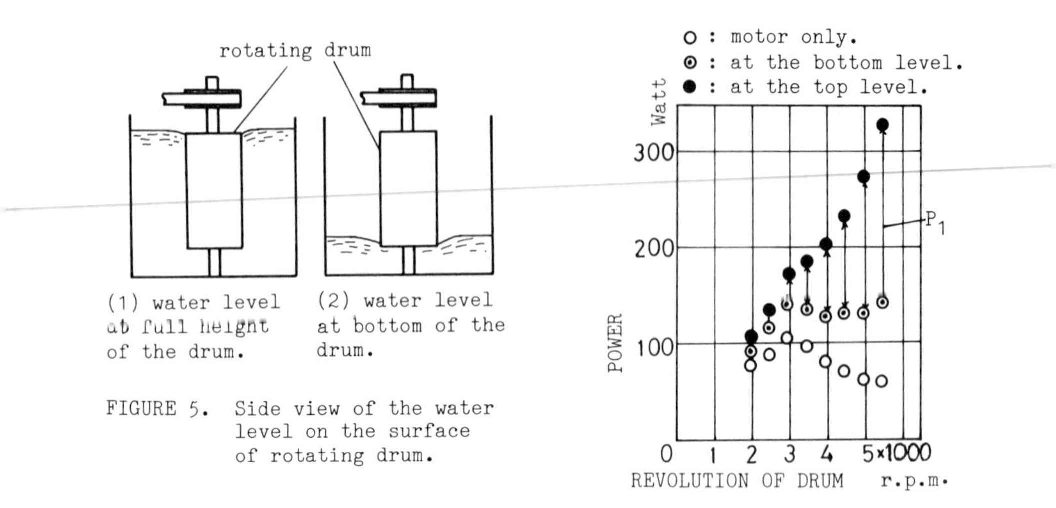

FIGURE 5. Side view of the water level on the surface of rotating drum.

FIGURE 6. Power for rotation of a drum under the conditions of water level.

numbers Re_1($=vr/\nu$), Re($=v\delta/\nu$) and Re*($=[(1 + \delta^*/2)/(1 + \delta^*)^2]Re$). It is understood that the friction factor f_1 has nearly the same values as the data by Kanagawa's experiment [7] from FIGURE 7.

4.2 The experiment concerning four rotating drums.

The power for resistance on the outside surface of a single rotating cylindrical drum in four rotating drums in one fixed enclosure was measured by the same method as the above mentioned experiment in the system of a single rotating drum. Load for rotation of one of the four drums was measured at the top level and the bottom level of water on the drum surface, where, all cylindrical drums were driven with the same velocity and the surfaces of the drums were moved with the same direction each other. The difference of the measured load under the both conditions can be considered to be approximately the load for rotation of only the outside surface on the drum.

Data of the load, i.e. power P_4 with respect to the revolution of the drum are shown in FIGURE 8 for the system of four rotating drums. And the friction factor f_1 for velocity of the outside surfaces on the drums are calculated from Eq.(8) and the result of FIGURE 8, and the values for Reynolds number Re_4 ($= vr_0/\nu$) are shown in FIGURE 7, where r_0 is a radius of envelope of the outside surfaces of the four drums. It is assumed that the fluid flow in the envelope of the outside surfaces on the all rotating drums is roughly same velocity. Therefore the envelope is assumed to be one rotating drum, and δ_4 is hypothetica gap between the envelope and the fixed enclosure.

To investigate the behavior of the fluid flow in the circumstances between the rotating cylindrical drums and one fixed enclosure, a particle tracing metho were employed in the experiment. Many micro bubbles of hydrogen were generated by electrolysis of water from cathode wires stretched between the drum surfaces or the surface of the drum and the inside wall of the enclosure, and the trace of the bubbles was photographed from upside to downward. A typical behavior of the fluid flow based on the photograph is shown in FIGURE 10.

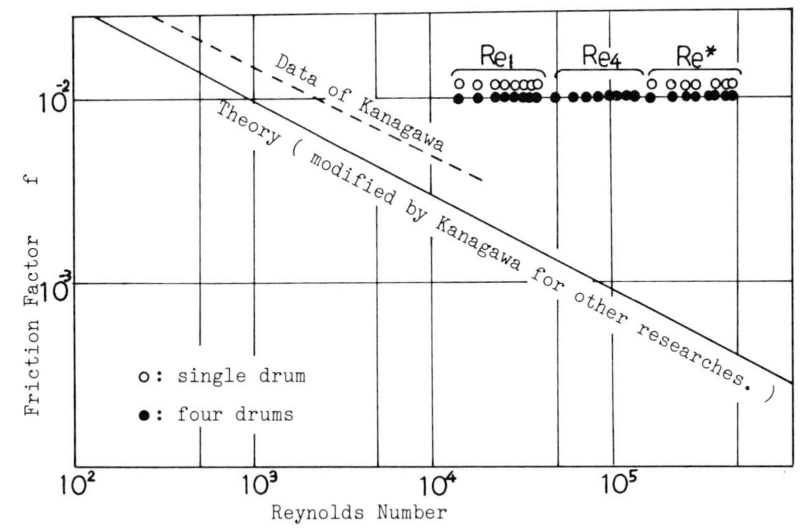

FIGURE 7. Friction factor for Reynolds number.

5. CONSIDERATION OF THEORY AND EXPERIMENT

Now it is understood that fluid flow resistance of the outside surface on a rotating cylindrical drum in a system of four rotating cylindrical drums in a fixed enclosure with the same velocity and direction as the outside surface of the drums is a little smaller than fluid flow resistance of outside surface on one rotating cylindrical drum in one fixed enclosure at the same dimension and velocity as the outside surface on the drum from the experiment, as shown in FIGURE 9. And the ratio $F_4 (= P_4/P_1)$ of fluid flow resistance of the surface on rotating drum in the system of four rotating drums to the system of one rotating drum is approximately 0.83 with the same dimension of drums and revolution per unit time from FIGURE 9. The ratio is not smaller than expected.

From photographic observation, the behavior of fluid flow of the system of four rotating drums in one fixed enclosure shows nonuniform velocity distribution at the entrance of the narrow gaps and at the most narrow gaps between two rotating drums. There are some stagnation points in parts of the crossing flow from each narrow gaps where the four streams concentrate to the center of the fixed enclosure and exit through the narrow gaps toward the inside wall of the enclosure. And a little turbulence is created in the divergent channel where the flow exit from the narrow gap.

It is understood that the parts of stagnation of flow and the divergence of the fluid flow from the narrow gaps induce considerably fluid flow resistance. The conditions that the fluid flow resistance is induced between the rotating cylindrical drums and the inside surface of one fixed enclosure are considered the same as the system of one rotating cylindrical drum in one fixed enclosure. The difference of flow velocity between at the surface of the drum and at the center of the gap is rather much as shown in FIGURE 10. And it is considered that the fluid flow velocity at the exit from the gap is decreased by inside surface on the fixed enclosure and the flow is accelerated again at the entrance of the gap by the rotation of both surfaces of the drums.

The fluid flow resistance is classified into three groups; the first group is a

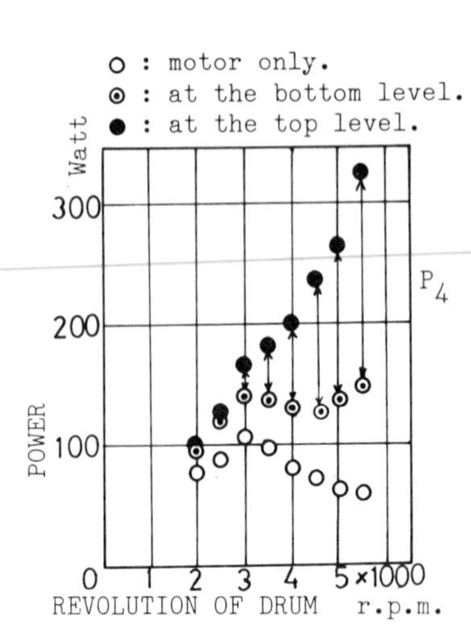

FIGURE 8. Power for rotation of a drum in four rotating drums system.

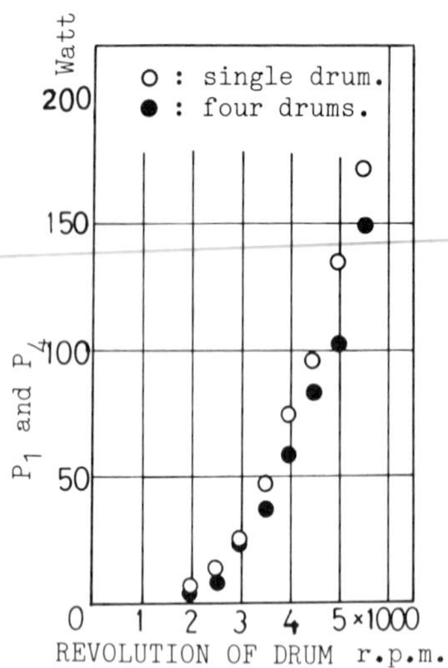

FIGURE 9. Power P_1 and P_4 for fluid flow resistance of drum surface.

low resistance group in the area of the entrance of the narrow gaps and the gaps and is concerned with a third region of the surfaces on rotating drums as shown in FIGURE 11 because the fluid flow in the region is a convergent type and has no relative velocity for the surfaces on the rotating drums. The second is an usual resistance group to which the region between the outside surfaces on the drums and the inside surface on the fixed enclosure belongs, and the area is concerned with a quarter section of the inside surface of the fixed enclosure or a third region of the surfaces on rotating drums because the inside surface on the fixed enclosure is not moved, and there are some convergent and divergent channels in this region and relative fluid motion. The third is a little higher resistance group to which the exit region of the flow in the narrow gaps belongs, and the group is concerned with a third region of the surfaces on rotating drums because the fluid flow in this region is a divergent type and has low relative velocity for the surfaces on rotating drums.

As the results, if many sets of four drums systems with the same velocity and direction of surfaces on drums are arranged in one fixed enclosure, as shown in FIGURE 12, (number of drums = 4n, n; integer. the number of drums = 64, in this example), it may be expected to decrease the fluid flow resistance of the surfaces on the rotating drums in the region of center group, i.e. 16 drums, because the region is surrounded by the rotating drums with the same surface velocity and direction.

Not strictly speaking, the ratio F_{4n} of fluid flow resistance of the surafce on the rotating drum in those systems, i.e. 4n drums, to it in the system of one rotating drum would be smaller than F_4, for example, F_{4n} is roughly estimated to $0.83[\ 25 - (\ 5 - 2\ )^2\ ]/25 = 0.53$ in $n = 5^2 = 25$, i.e. 100 drums because the fluid flow resistance would be mostly depend on the outside subassemblies among the total assemblies with four rotating drums as a rule. Of course, the above mentioned articles are not strict and F_{4n} has to be obtained by experiment in the future.

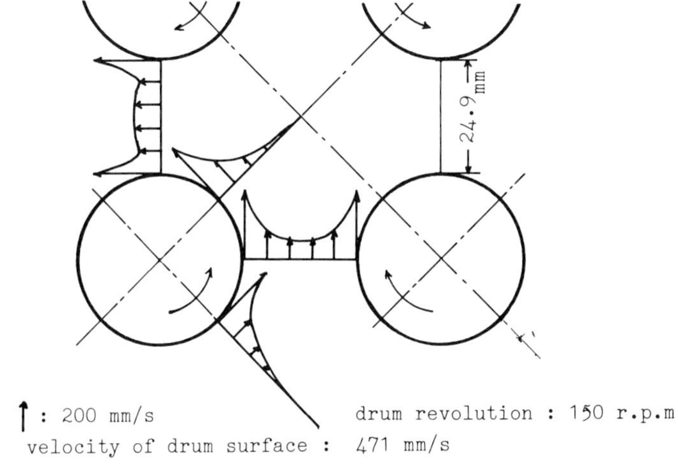

FIGURE 10. Fluid flow pattern and velocity distribution of flow in system of four rotating drums.

6. CONCLUSION

One method of reducing fluid flow resistance in system of rotating cylindrical drums in a fixed enclosure with the same velocity and direction of surfaces of rotating drums has been presented and investigated by the authors.

Fluid flow resistance in this system has been investigated experimentally, and no considerable decrease of the resistance has been found because the fluid flow is accelerated through the gap between the rotating drums, in convergent channels, and the velocity decreases through the exit of the gap to the inside wall of the enclosure, in divergent channels, that is, the change of fluid flow velocity is considerable large. So, it is necessary for the improvement to smooth the fluid flow.

However, one idea or mechanism is proposed for reducing fluid flow resistance in the system of rotating cylindrical drums in one fixed enclosure, based on the result of the experiment. Assemblies of many numbers of rotating drums with the same velocity and direction of surfaces of the drums in one fixed enclosure, for example 100 drums assembled, are considered as an example. The fluid flow resistance of the rotating surfaces in this system is estimated to be half of the resistance in a system of one rotating drum.

7. ACKNOWLEDGEMENT

The help and assistance of Mr. Inayoshi in performing the experiments is greatly appreciated. Thanks are also due to Iwata Tosoki Co. Ltd. for their invaluable help and advice.

div.;divergent channel region.
con.;convergent channel region.

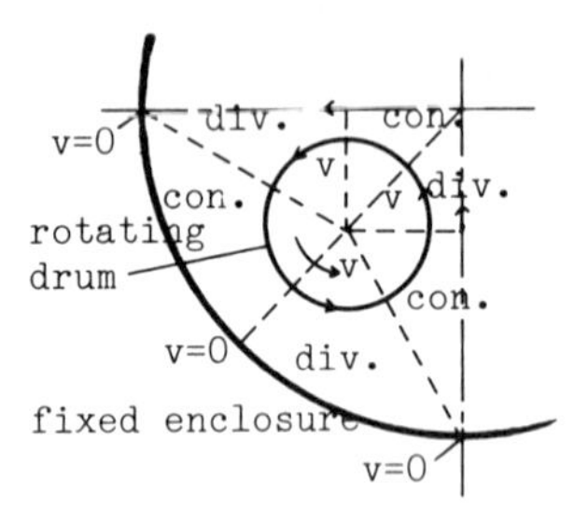

FIGURE 11. Schematic region of fluid flow in system of four rotating drums.

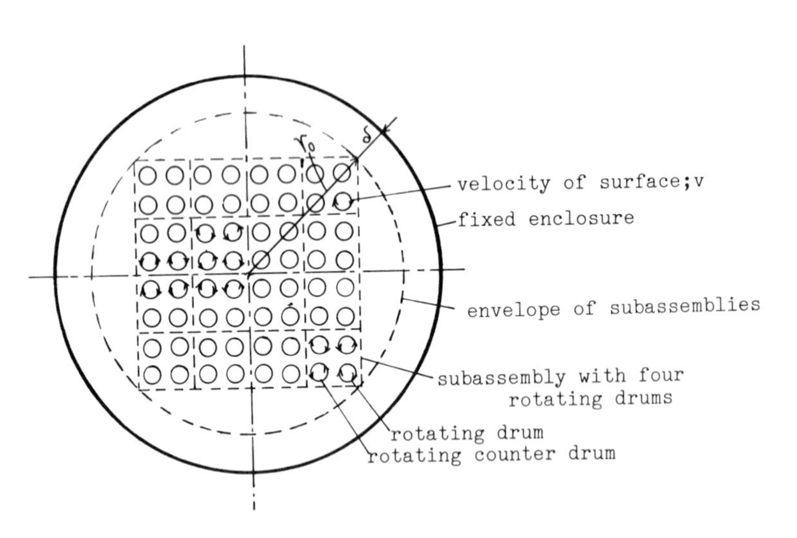

FIGURE 12. System of multi subassemblies with four rotating drums in a fixed enclosure. (ex. $n = 4^2 = 16$, number of drums $= 4n = 64$)

8. NOMENCLATURE

F_4 ; ratio of fluid flow resistance of surface on a rotating drum in system of four rotating drums to it in system of one rotating drum.
F_{4n} ; ratio of fluid flow resistance of surface on a rotating drum in system of four times of n rotating drums to it in system of one rotating drum.
f ; fluid flow friction factor in surface on a rotating drum shown in Eq.(1).
f_1 ; fluid flow friction factor in surface on a rotating drum shown in Eq.(2).
n ; number of subassemblies with four rotating drums.
P_1 ; power for resistance in unit height (length) of surface on a rotating drum for system of one rotating drum. (W/m)
P_4 ; power for resistance in unit height (length) of surface on a rotating drum for system of four rotating drums. (W/m)
Δp ; hypothetical pressure drop in unit length of fluid flow channel.
Re ; Reynolds number for fluid flow between a rotating drum and a fixed enclosure. ($v\delta/\nu$)
Re_1 ; Reynolds number for fluid flow between a rotating drum and a fixed enclosure. (vr/ν)
Re^* ; Reynolds number for fluid flow between a rotating drum and a fixed enclosure. ($Re(1 + \delta^*/2)/(1 + \delta^*)^2$)
Re_4 ; Reynolds number for fluid flow between envelope of total subassemblies of four rotating drums and a fixed enclosure.
r ; radius of a drum. (m)
r_0 ; radius of envelope of total subassemblies of four rotating drums. (m)
v ; velocity of fluid flow or velocity of surface on rotating drums. (m/s)
δ ; gap between a rotating drum and a fixed enclosure or gap between envelope of subassemblies of four rotating drums and a fixed enclosure. (m)
δ^* ; δ/r.
ρ ; density of fluid. (kg/m^3)
τ ; frictional shearing stress on a surface of a rotating dum in unit area.
ν ; kinematic viscosity of fluid.

9. REFERENCES

1. Stuart, J. T., *Journal of Fluid Mechanics*, 4, pp. 1, 1958.

2. Davey, A., *Journal of Fluid Mechanics*, 14, pp. 336, 1962.

3. Batchelor, G. K., *Journal of Fluid Mechanics*, 7, pp. 416, 1960.

4. Taylor, G. I., *Proc. Roy. Soc.*, A 157, pp. 546, 1936.

5. Wendt, F., *Ing. Arch.*, 4, pp. 557, 1933.

6. Donnelly, R. J., *Proc. Roy. Soc.*, A 246, pp. 312, 1957.

7. Kanagawa, A. etal., *J. Atom. Ener. Soc. Japan*, 3, pp. 118, 1968.

Local Nonsimilar Solution for Forced Convection from a Flat Plate in a Rotating Flow Field

G. J. HWANG and C. M. YAU
Department of Power Mechanical Engineering
National Tsing Hua University
Hsinchu, Taiwan 300, Republic of China

ABSTRACT

This paper presents a local non-similar solution for forced convection from a flat plate in a rotating flow field. The rotating flow field can be visualized near curved surfaces in a rotating fluid machinery or observed inside a rotating cylindrical fluid container. The length of the flat plate, 2L is much smaller than the radius of rotation of the fluid, R_0, therefore the potential flow field of the rotating fluid is not seriously disturbed. The free stream velocity and the pressure distribution along the plate are obtained by using the potential flow theory. The plate is heated and maintained at a temperature T_w which may be higher or lower than the free stream temperature T_0. The momentum and energy equations in the boundary layer are transformed into non-similar equations and solved by using a differential-difference approach. This approach is also verified by applying this method to the Howarth's retarded flow and uniform surface mass transfer problems. Comparions are made with the local similarity, the two-equation and the three-equation models. Excellent agreements between the present solution and the results from the three-equation model are found. This paper shows velocity and temperature distributions for various positions along the plate. The friction factors and the Nusselt numbers are also presented for Pr = 0.7, L/R_0 = 0.001, 0.01, 0.05, 0.1 and 0.25.

INTRODUCTION

The applications of boundary layer flow theory and similarity solution have been widely utilized in the twenty century since the flow near a flat plate in a uniform flow field was originally investigated by Blasius in 1908. In 1931 Falkner and Skan generalized the similarity solution to the analysis of the wedge flow problem. There are many theories and discussion regarding the boundary layer flow reported in books written by Schlichting [1], Rosenhead [2], and Kays and Crawford [3].

To fulfill the requirements in similarity transformation, the boundary conditions in both velocities and temperature must be restricted in some special forms. Unfortunately, most of the problems do not satisfy these requirements. Therefore, the methods for non-similar solutions were developed. Hartnett and Irvine [4], and Gebhart [5] discussed the non-similar methods extensively. In general the non-similar solutions fall into two categories i.e., the local similar solution and the local non-similar solution.

With some assumptions ordinary differential equations can be derived and the local similar solution is obtained independently at any streamwise location. The computer time required for solving the ordinary differential equation is much less than the time required for solving the original partial differential equation. Koh and Price [6] utilized the local similar solution to analyze the boundary layer heat transfer of a rotating cone in a flow. Dhir and Lienhard [7] also applied the same method for film condensation with variable gravity and body shape. Cheng [8] obtained the similarity solutions for mixed convection from horizontal impermeable surfaces in saturated porous media and discover the governing parameter of the problem is $Ra/(PrRe)^{1/3}$. Schneider [9] examined the combined forced and free convection flow over a horizontal plate and found that the similar solution exists with the condition $T_W \sim 1/\sqrt{x}$.

In certain problems the terms neglected in the differential equations in the local similar solution play an important role in the solution. Thus the local non-similar solution considers the terms omitted in the local similar solution and improves the accuracy of the solution. Sparrow et al. [10] developed a two-equation model and a three-equation model to approximate the local non-similar solution. The Howarth's retarded flow and surface mass transfer in boundary layer near a flat plate were examined and the solutions agreed excellently with other results. Sparrow and Yu [11] also applied the same method to analyze the thermal boundary layer with a high degree of accuracy. Minkowycz and Sparrow [12] studied the local non-similar solutions for natural convection on a vertical cylinder and compared the results with the integral solution of Hama et al. [13] The other studies discussing the local non-similar solution can be found in literature [14-17].

This paper investigates a local non-similar solution for forced solution from a flat plate in a rotating flow field with a large radius of rotation. The momentum and energy equations are transformed into non-similar equations and solved by using a differential-difference approach. A fourth-order Runge-Kutta method is used to solve the differential equation at each position along the plate. Smith and Clutter [18] used similar idea but employed a predictor-corrector method to solve the Howarth's retarded flow problem.

THEORETICAL ANALYSIS

Consider a steady two-dimensional incompressible rotational flow field, the tangential velocity is U_0 at a radius $r = R_0$. The relationships between the velocities referring the Cartesian coordinates and the polar coordinates shown in Fig. 1 are

$$U' = -(U_0/R_0)r\sin\phi = -(U_0/R_0)Y' \text{ and } V' = (U_0/R_0)r\cos\phi = (U_0/R_0)X' \qquad (1)$$

From the potential flow theory, the pressure distribution can be obtained as

$$P = \rho(U_0/R_0)^2(X'^2+Y'^2)/2 \qquad (2)$$

where ρ is the density of the fluid. A flat plate of length 2L which is much smaller than the radius of rotation is located at $r = R_0$ in the flow field. After the coordinate transformation and considering equation (2) and the equation $dP/dX = -\rho U_x dU_x/dX$ from the potential flow theory, the dimensionless velocity distribution at the edge of boundary layer along the plate is

$$u_x = U_x/U_0 = [1 - \gamma^2 (1 - \xi)^2]^{\frac{1}{2}} \qquad (3)$$

where $\gamma = L/R_0$ is the ratio of the half length and the radius of rotation, and $\xi = X/L$ is the dimensionless coordinate along the X-direction. It is noted that for a small value of γ or $R_0 >> L$ the rotating flow field will not be seriously disturbed by the existence of the flat plate, and the rotating potential flow field may be regarded as the potential flow field outside the boundary layer in the vicinity of the flat plate.

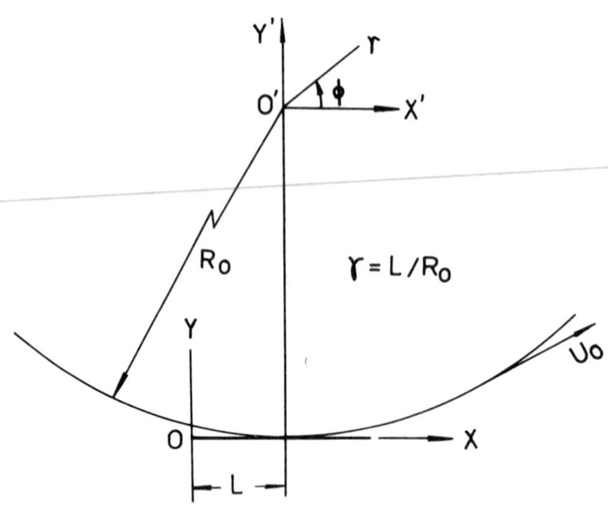

FIGURE 1. The coordinate systems

Due to viscosity and thermal diffusivity, momentum and thermal boundary layers grow gradually from the leading edge. By introducing the dimensionless transformations

$$\xi = X/L,\ y = YRe_L/L,\ u = U/U_0,\ v = VRe_L/U_0 \qquad (4)$$

$$P = P/\rho U^2 \text{ and } \theta = (T - T_0)/(T_w - T_0)$$

where $Re_L = U_0L/\nu$ is the Reynolds number, and defining $u = \partial\psi/\partial y$ and $v = -\partial\psi/\partial\xi$, the boundary equations can be derived as,

$$\frac{\partial\psi}{\partial y}\frac{\partial^2\psi}{\partial\xi\partial y} - \frac{\partial\psi}{\partial\xi}\frac{\partial^2\psi}{\partial y^2} = u_X\frac{du_X}{d\xi} + \frac{\partial^3\psi}{\partial y^3} \qquad (5)$$

$$\frac{\partial\psi}{\partial y}\frac{\partial\theta}{\partial\xi} - \frac{\partial\psi}{\partial\xi}\frac{\partial\theta}{\partial y} = \frac{1}{Pr}\frac{\partial^2\theta}{\partial y^2} \qquad (6)$$

where $Pr = \nu/\alpha$ is the Prandtl number. It is noted that the continuity eqiation is satisfied automatically, and the compression work and the viscous dissipation are not considered in deriving energy equation (6). The term $u_X\ du_X/d\xi$ in the R.H.S of momentum equation (5) is the pressure derivative along the X-direction and can be derived from equation (3).

For the solution of the boundary layer equation, it is useful to introduce the similarity variables,

$$\eta = y/\sqrt{\xi/u_X},\quad f(\xi,\eta) = \psi/\sqrt{u_X\xi} \qquad (7)$$

Then equations (5) and (6) become,

$$f''' + A(\xi)ff'' + B(\xi)(1-f'^2) = \xi(f'\partial f'/\partial\xi - f''\partial f/\partial\xi) \qquad (8)$$

$$\theta'' + PrA(\xi)f\theta' = Pr\xi(f'\partial\theta/\partial\xi - \theta'\partial f/\partial\xi) \qquad (9)$$

where $A(\xi) = (1+\xi u_X'/u_X)/2$, $B(\xi) = \xi u_X'/u_X$ and ' indicates the derivative in the η-direction. The first term in the L.H.S. of equation (8) represents the viscous force, the second term is the inertia term, and the third term is derived from the pressure and inertia terms. The terms in the R.H.S. of equatio (8) stand for the ξ-direction variation of the inertia force. The first and second terms in the L.H.S. of energy equation (9) represent respectively the conduction and advection terms. The term in the R.H.S. of equation (9) is the ξ-direction variation of the advection term.

There are a third-order derivative in the η-direction and a first order derivative in the ξ-direction for function f in equation (8). Thus equation (8) requires three boundary conditions in the η-direction and one boundary condition in the ξ-direction. Similarly equation (9) needs two boundary conditions in the η-direction and one condition in the ξ-direction. The corresponding boundary condition in η-direction are

$$\eta = 0, \; f = f' = \theta - 1 = 0 \qquad (10)$$
$$\eta \to \infty \quad f' - 1 = \theta = 0$$

For the boundary conditons in ξ-direction, one may consider the conditons at $\xi = 0$. By setting $\xi = 0$ in equations (8) and (9), one obtains the equations

$$f''' + ff''/2 = 0 \qquad (11)$$

$$\theta'' + Prf\theta'/2 = 0 \qquad (12)$$

It is noted that the solutions f and θ of equations (11) and (12) satisfying boundary conditions (10) are the boundary conditions in the ξ-direction for equations (8) and (9).

METHODS OF SOLUTION

There is a basic assumption in the local similarity model that the terms in the R.H.S. of equations (8) and (9) can be neglected by considering $\xi \to 0$ at a position near the leading edge or at a down-stream position all the terms $\partial f'/\partial\xi$, $\partial f/\partial\xi$ and $\partial\theta/\partial\xi$ approach zero. Therefore only the terms in the L.H.S. of equation (8) and (9) remain. Noting that the coefficients $A(\xi)$ and $B(\xi)$ are functions of ξ only, at each ξ position, the functions f and θ can be solved along the η-direction as one does in the similarity solution.

There are two methods in the local non-similarity model to solve the complete non-similar equations (8) and (9). Sparrow et al. [10] developed two-equation and three-equation models or a model involving even higher number of equations.

The implicit difference forms are written along a line between line i and line i+1 as

$$(f'''_i + f'''_{i+1})/2 + A(\xi_i + \Delta\xi/2)(f_i + f_{i+1})(f''_i + f''_{i+1})/4 + B(\xi_i + \Delta\xi/2)[1 - (f'_i + f'_{i+1})^2/4]$$
$$= (\xi_i + \Delta\xi/2)[(f'_i + f'_{i+1})(f'_{i+1} - f'_i) - (f''_i + f''_{i+1})(f_{i+1} - f_i)]/(2\Delta\xi) \qquad (13)$$

$$(\theta''_i + \theta''_{i+1})/(2Pr) + A(\xi_i + \Delta\xi/2)(f_i + f_{i+1})(\theta'_i + \theta'_{i+1})/4$$
$$= (\xi_i + \Delta\xi/2)\,[(f'_i + f'_{i+1})(\theta_{i+1} - \theta_i) - (\theta'_{i+1} + \theta'_i)(f_{i+1} - f_i)]/(2\Delta\xi) \qquad (14)$$

where $\Delta\xi$ is the step size in the ξ-direction between line i and i+1.

Since all the values and their derivatives of f and θ at $\xi = 0$ are known by solving equations (11) and (12) and the boundary conditions (10), the values and their derivatives of f and θ can be obtained at next step. The process is repeated until the solutions for all the positions ξ and η are obtained.

The solution at each ξ position is solved by using a fourth-order Runge-Kutta method. The boundary conditions $f'-1 = \theta = 0$ at $\eta \to \infty$ are changed to initial conditons f'' and θ' at $\eta = 0$. The initial guess at $\eta = 0$ is then corrected by using the value at $\eta \to \infty$ and the Newton-Raphson shooting method. In order to obtain a reliable and convergent solution, the range of η_{max} for the condition of $\eta \to \infty$ and the step sizes $\Delta\eta$ and $\Delta\xi$ are tested. Table 1 shows the results.

TABLE 1. Test for η_{max}, $\Delta\eta$ and $\Delta\xi$

γ	ξ	$\Delta\xi$	η_{max}	$\Delta\eta$	f"(0)	-θ'(0)
			7.0		0.3355	0.2933
0.1	0.3	0.1	8.0	0.02	0.3355	0.2931
			9.0		0.3355	0.2931
				0.05	0.3345	0.2922
0.1	0.2	0.1	8.0	0.02	0.3346	0.2930
				0.01	0.3346	0.2931
		0.2			0.3349	0.2931
0.1	0.2	0.1	8.0	0.02	0.3346	0.2929
		0.52			0.3346	0.2928

It is seen from this table that a combination of $\eta_{max} = 8$, $\Delta\eta = 0.02$ and $\Delta\xi = 0.1$ is selected to ensure an accuracy of result up to at least three significant figures. It is also noted that in solving the solution along the η-direction at each ξ position, the conditions for f' and θ at η_{max} are set as

$$|f'-1|_{\eta_{max}} < 1 \times 10^{-5} \sim 5 \times 10^{-4}$$

and (15)

$$|\theta|_{\eta_{max}} < 1 \times 10^{-5}$$

respectively. At a large value of ξ, the conditions f' = 1 and θ = 0 at η_{max} are very sensitive, and the criterion is set at 1×10^{-3}.

In the present study, it takes 480 sec for the non-similar solution on a CDC CYBER-172 computer system for a value of γ. But it takes about 3000 sec in applying the two-equation model for the same case. On the contrary it consumes only 100 sec on the CDC system for the local-similar solution. It may be concluded that the present investigation gives a simple and time saving computation technique in obtaining the non-similar boundary layer solution.

RESULTS AND DISCUSSION

To ensure the accuracy of the present method, the solutions for Howarth's retarded flow [10] and flat plate heat transfer with a uniform surface mass transfer [11] are illustrated in Fig. 2 for comparison. Fig. 2(a) shows the

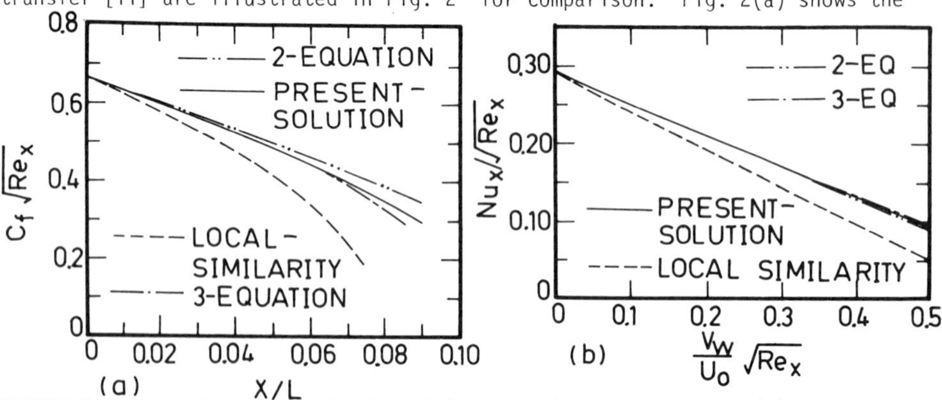

FIGURE 2. Comparisons of solutions (a) Howarth's retarded flow (b) flat plate heat transfer with a uniform surface mass transfer

results of wall shear stress versus dimensionless coordinates X/L. It is seen that the local similarity solution is only valid at the vincinity of $X/L = 0$ which is a singular point. The solution from the two-equation model deviates from the one by using the present method at $X/L \approx 0.3$. The solution from the three-equation model agrees well with the present solution up to $X/L = 0.65$. Fig. 2(b) depicts the relation between $Nu_x/\sqrt{Re_x}$ and $\sqrt{Re_x}\ V_w/u_0$. The value V_w/u_0 is the ratio of the wall mass transfer velocity and the main flow velocity. A trend similar to the one shown in Fig. 2(a) is observed. It may be concluded that both the present solution and the solution from the three-equation model present flow and heat transfer results with a high degree of accuracy and the present method can be employed for a further study.

Fig. 3 shows the velocity distribution at various positions along the ξ-direction for the case of $\gamma = 0.001$, 0.01, 0.05, 0.1 and 0.25. One can observe that the velocity profile for the case of $\gamma = 0.001$ coincides with the one from the Blasius solution because of its extremely large radius of rotation. For the case of $\gamma = 0.01$ and 0.05, a slight deviation from the Blasius solution is observed at $\eta > 4.0$ near the external potential flow field. For the case of $\gamma = 0.1$, the acceleration and deceleration of the potential flow along the flat plate affect the velocity profile at $\eta \geq 1.0$ a position near the plate. For the case of $\gamma = 0.25$ the influence of the potential flow on the velocity profile is clearly observed. A separation of the velocity profile is seen at $\xi = 1.6$ and 2.0.

The growth of momentum boundary layer along the flat plate in the ξ-direction is shown in Fig. 4. The momentum boundary layer thickness is defined as the thickness at which the velocity equals to 0.99 of its free stream velocity. The

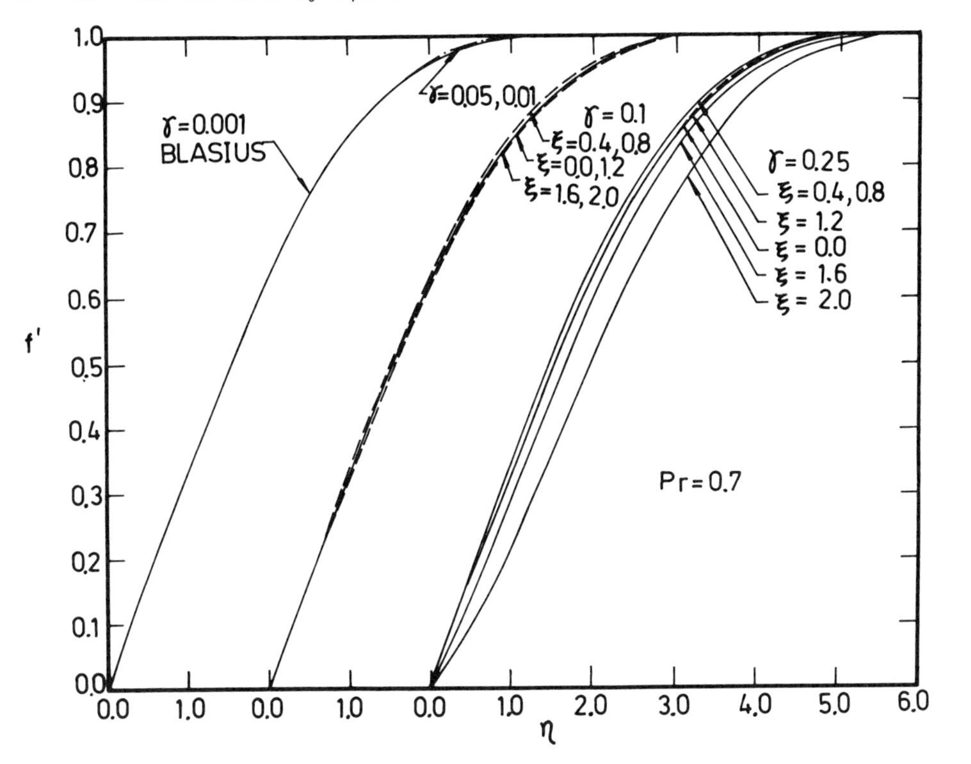

FIGURE 3. The velocity distribution

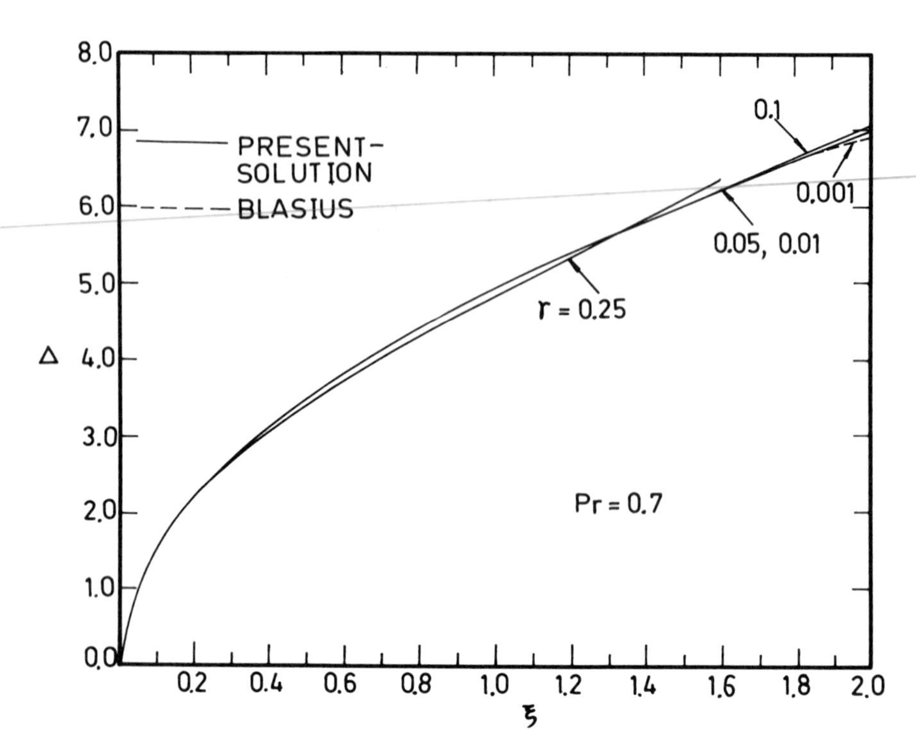

FIGURE 4. The growth of momentum boundary layer

expression is

$$\Delta = \eta_{f'=0.99} \cdot \sqrt{\xi/u_x} \tag{16}$$

It is seen from this figure that the momentum boundary layer thickness increases monotonically with the ξ-coordinate. All the curves for $\gamma = 0.001 \sim 0.25$ lie together with the curve from the Blasisus solution with a small difference of 2.5 percent only. This observation indicates that the existence of the accelerating and decelerating potential flow field would not affect much on the momentum boundary layer thickness.

Fig. 5 depicts the relationship between the friction factor $C_f\sqrt{Re_x}/2$ and ξ for the cases of $\gamma = 0.25 \sim 0.001$. The friction coefficient is defined as $C_f = \tau_0/(\frac{1}{2}\rho U_x^2)$ and the value $C_f\sqrt{Re_x}/2$ is equal to $f''(0) = 0.332$ in the Blasius solution for all ξ. For the case $\gamma > 0$, the friction factor increases at a small ξ position but decreases at a large ξ position. This trend is more pronounced as the ratio γ increases. The maximum value of friction factor appears at $\xi \approx 0.5$ and the minimum value appears at the trail edge of the plate $\xi = 2.0$. For the case of $\gamma = 0.25$, it presents a 8.4% increase in the maximum value of friction factor at the position $\xi \approx 0.5$.

The dimensionless temperature distribution at various positions along the ξ-direction for the case of $\gamma = 0.001$, 0.01, 0.05, 0.1 and 0.25 is shown in Fig. 6. One can see that the dimensionless temperature profile for the case of $\gamma = 0.001$ coincides with the one from the Blasius solution. For the case of $\gamma = 0.01$ and 0.05, a slight difference is shown at $\eta > 4.5$ near the edge of thermal boundary layer. For the case of $\gamma = 0.1$, the acceleration and deceleration of the potential flow along the flat plate affect the dimensionless temperature at a position near the plate. For the case of $\gamma = 0.25$, the influence of the external potential flow on the dimensionless temperature profile is clearly observed.

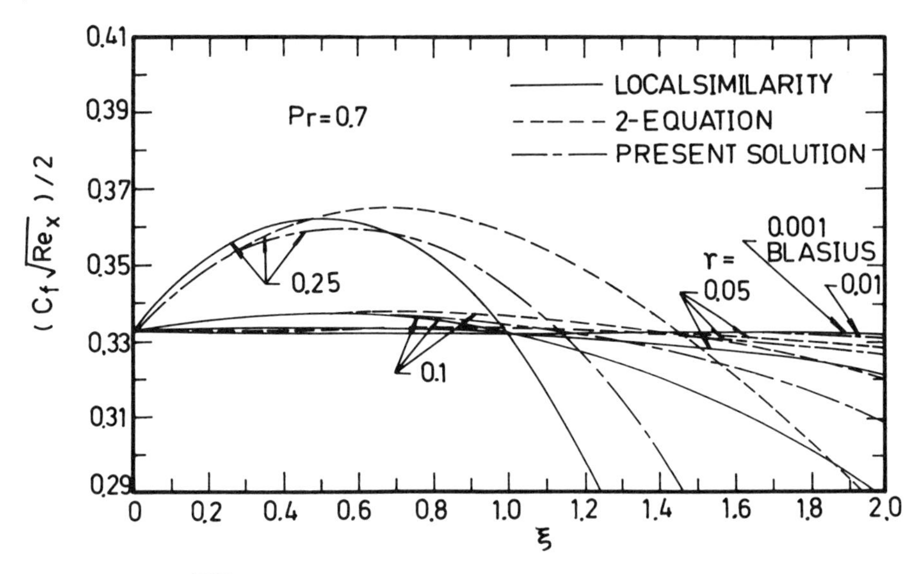

FIGURE 5. $C_f \sqrt{Re_x}/2$ vs. ξ

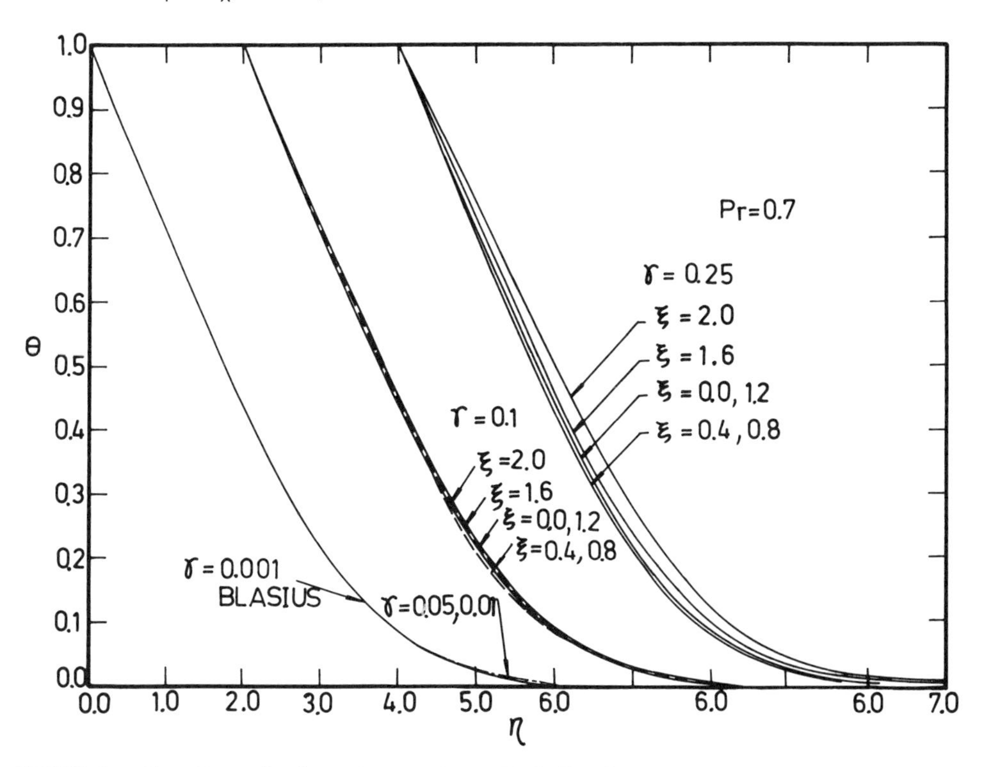

FIGURE 6. The dimensionless temperature distribution

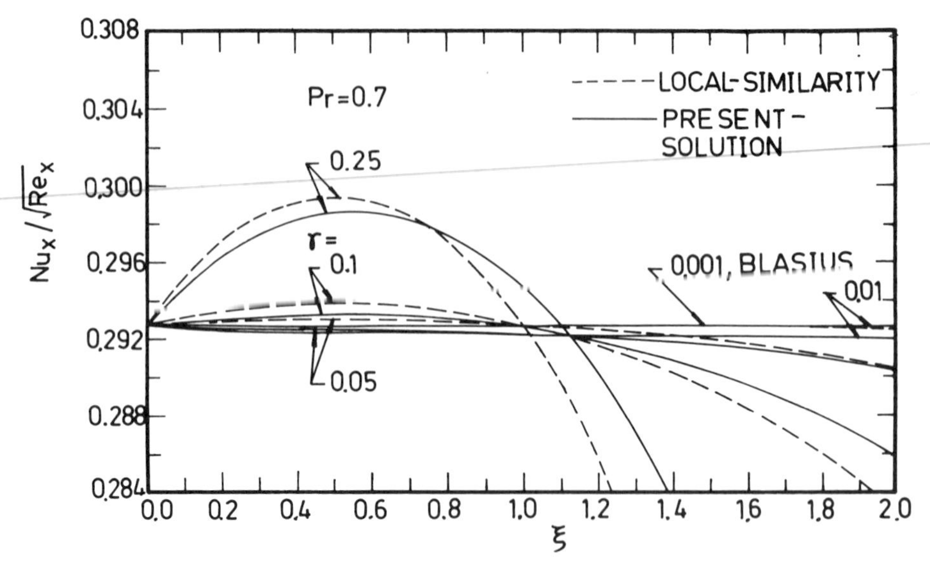

FIGURE 7. $N_{u_x}/\sqrt{Re_x}$ vs. ξ

Fig. 7 presents the relationship between $-\theta'(0) = N_{u_x}/\sqrt{Re_x}$ and ξ for the cases of $\gamma = 0.25 \sim 0.001$. The Nusselt number is defined as $N_{u_x} = hX/K$ and $N_{u_x}/\sqrt{Re_x} = 0.293$ is observed for the Blasius solution for all values of ξ. For the case of $\gamma > 0$, the Nusselt number increases at a small ξ position but decreases at a large ξ position. This trend is more pronounced as the ratio γ increases. The maximum value of Nusselt number appears at $\xi \approx 0.5$ and the minimum value appears at the end of the flat plate $\xi = 2.0$. For the case of $\gamma = 0.25$, it presents a 2% increase only in the maximum value of Nusselt number at $\xi \approx 0.5$.

CONCLUDING REMARKS

1. This paper presents a solution for forced convection from a flat plate with length 2L which is much smaller than the radius of rotation of the external flow field. The present results indicate that for the case of $\gamma = 0.1$ it shows a 6.6% difference in the friction factor and a 2.4% difference in the heat transfer result from the Blasius solution.

2. For the case of $\gamma > 0$, a maximum friction factor $C_f\sqrt{Re_x}/2$ and a maximum heat transfer rate $Nu_x/\sqrt{Re_x}$ are observed at $\xi \approx 0.5$. Minimum values are seen at the end of the flat plate $\xi = 2.0$.

3. The present implicit finite-difference method requires only an initial value in applying the Runge-Kutta method at each position in the ξ-direction but it requires two initial values in the two-equation model. It may be concluded that the present investigation gives a simple and time saving computation technique in obtaining the non-similar boundary layer solution.

REFERENCES

1. Schlichting, H., Boundary-Layer Theory, 7th edition, McGraw-Hill, New York, 1979.

2. Rosenhead, L., Laminar Boundary Layers, Oxford University, London Press, 1963.

3. Kays, M., and Crawford, E., Convective Heat and Mass Transfer, 2nd edition, McGraw-Hill, New York, 1980.

4. Hartnett, J.P., and Irvine, Jr, T.F., Advances in Heat Transfer, vol. 4, pp. 317-446, Academic Press, New York, 1967.

5. Gebhart, B., Heat Transfer, McGraw-Hill, New York, 1971.

6. Koh, J.C.Y., and Price, J.F., Nonsimilar Boundary-Layer Heat Transfer of a Rotating Cone in Forced Flow, Jounral of Heat Transfer, Trans. ASME, Series C, vol. 89, pp. 139-145, 1967.

7. Dhir, V.K., and Lienhard, J.H., Similar Solution for Film Condensation With Variable Gravity or Body Shape, Journal of Heat Transfer, Trans. ASME, Series C, vol. 95, pp. 483-486, 1973.

8. Cheng, P., Similarity Solutions for Mixed Convection from Horizontal Impermeable Surfaces in Saturated Porous Media, Int. J. of Heat and Mass Transfer, vol. 20, pp. 893-898, 1977.

9. Schneider, W., A Similarity Solution for Combined Forced and Free Convection Flow over a Horizontal Plate, Int. J. Heat Mass Transfer, vol. 22, pp. 1401-1406, 1979.

10. Sparrow, E.M., Quack, H., and Boerner, C.J., Local Non-Similarity Boundary-Layer Solutions, AIAA Journal, vol. 8, pp. 1936-1942, 1970.

11. Sparrow, E.M., and Yu, H.S., Local Nonsimilarity Thermal Boundary Layer Solutions, Journal of Heat Transfer, Trans. ASME, Series C, vol. 93, pp. 328-334, 1971.

12. Minkowycz, W.J., and Sparrow, E.M., Local Nonsimilar Solutions for Natural Convection on a Vertical Cylinder, Journal of Heat Transfer, Trans. ASME, Series C, vol. 96, pp. 178-183, 1974.

13. Hama, F.R., Recesso, J.V., and Christianens, J., The Axisymmetric Free Convection Temperature Field Along a Vertical Thin Cylinder, Journal of the Aero/Space Sciences, vol. 93, pp. 191-194, 1970.

14. Chen, T.S., Sparrow, E.M., and Mucoglu, A., Mixed Convection in Boundary Layer Flow on a Horizontal Plate, Journal of Heat Transfer, Trans. ASME, Series C, vol. 99, pp. 66-71, 1977.

15. Hasan, M.M., and Eichhorn, R., Local Nonsimilarity Solution of Free Convection Flow and Heat Transfer from an Inclined Isothermal Plate, Journal of Heat Transfer, Trans. ASME, vol. 101, pp. 642-647, 1979.

16. Chen, T.S., and Strobel, F.A., Combined Heat and Mass Transfer in Mixed Convection over a Horizontal Flat Plate, Journal of Heat Transfer, Trans. ASME, vol. 102, pp. 538-543, 1980.

17. Minkowycz, W.J., and Cheng, P., Local Non-Similar Solutions for Free Convection Flow with Uniform Lateral Mass Flux in a Porous Medium, Letters in Heat and Mass Transfer, vol. 9, pp. 159-168, 1982.

18. Smith, A.M.O., and Clutter, D.W., Solutions of the Incompressible Laminar Boundary Layer Equations, AIAA Journal, vol. 1, no. 9, pp. 2062-2071, 1963.

Theoretical and Experimental Studies on Transport Phenomena in Co-Rotating Disks

WEN-JEI YANG
Department of Mechanical Engineering & Applied Mechanics
University of Michigan
Ann Arbor, Michigan 48109

S. MOCHIZUKI
Department of Mechanical Engineering
Tokyo University of Agriculture & Technology
Koganei, Tokyo, Japan

Y. S. SIM
Division of Thermal Hydraulics, KAERI
Chungnam, Korea

ABSTRACT

Theoretical and experimental studies are conducted to determine the heat transfer and friction loss performance in laminar and turbulen flow through co-rotating parallel disks. Theoretical results are obtained by a finite difference numerical method with a SIMPLER algorithm. A modified single-blow technique is employed to obtain experimental results with the aid of an automated data reduction system. Theory is compared with experiments. The effects of rotation on flow, pressure, and thermal behavior in the radial flow are determined.

INTRODUCTION

Many types of rotating machinery are associated with heat and mass transfer. Gas and steam turbines are typical examples of the rotating devices in which heat transfer takes place. Various mixing devices are employed to promote the mass transfer process. Often, these devices are characterized by a geometrical shape so complicated that the fabricating and testing of their prototype becomes the only means for determining their performance characteristics. Various simple models have been used to explore the basic principle and mechanisms of transport phenomena in rotating units such as the free disk, rotor-stator systems, and rotating cavities (1). An extensive review of the literature pertinent to the subject is available in reference (1) and thus will not be repeated here.

The present study is concerned with heat transfer and pressure drop characteristics in laminar and turbulent flow through rotating parallel circular disks. A pair of the disks is employed in the theoretical model (2-4), while experimental study utilizes multiple disk assemblies (5, 6), Theoretical results are obtained by a finite-difference method with the SIMPLER algorithm, whereas a modified single-blow technique with a computer-assisted data-reduct system is utilized to get experimental results. Theoretical prediction compares well with experiments. Rotational effects on flow pressure and thermal behavior in the radial flow are determined.

THEORY

Theoretical Model and Numerical Integration

The physical system to be studied consists of a pair of parallel circular disks with an opening of $2r_1$ at the center, as shown in Fig. 1. They rotate with respect to the center at an angular velocity Ω. A fluid enters normally through the openings and then flows radially out through the spacing s. Cylindrical coordinates (r, θ, z) are employed with the origin fixed at the center O of the lower disk. Let r_1 and r_2 denote the inner and outer radii of the disks, respectively. The flow is assumed to be steady, incompressible, axisymmetrical, and constant in all physical properties. Both laminar and turbulent flow are considered. A two-equation $\kappa-\varepsilon$ - turbulence model is used in turbulent flow.

With the use of the velocity components (u, v, w) and pressure P, the governing transport equations for turbulent, as well as laminar flow can be expressed in a unified form as

$$\frac{1}{r}\frac{\partial}{\partial r}(r\,u\,\phi) + \frac{\partial}{\partial z}(v\,\phi) = \frac{1}{r}\frac{\partial}{\partial r}\left(r\,\lambda\,\frac{\partial\phi}{\partial r}\right) + \frac{\partial}{\partial z}\left(\lambda\,\frac{\partial\phi}{\partial z}\right) + B \tag{1}$$

The source term B and diffusion coefficient λ for each general variable ϕ are defined in Table 1. The continuity equation reads

$$\frac{1}{r}\frac{\partial}{\partial r}(r\,u) + \frac{\partial v}{\partial z} = 0 \tag{2}$$

In addition to uniform disk surface temperature and no slip on disk walls, the velocity and temperature profiles at the inlet are assumed uniform. Pressure is determined from a velocity field.

All governing equations were reduced to a set of algebraic finite-difference equations using the discretization method called the hybrid scheme, which is a combination of the central and upwind schemes. These discretized governing equations were then solved by the SIMPLER (Semi-Implicit Pressure-Linked Equations Revised) algorithm with certain modifications. Details of numerical integration are available in references 2 - 4 amd thus will not be repeated here.

Theoretical Results and Discussion

The following dimensionless parameters are employed in the theoretical study:

$$Nu = \frac{hs}{k};\ \overline{Nu} = \frac{\overline{h}s}{k};\ Re = \frac{\dot{m}}{s\mu};\ Ta = \frac{s^2\,\Omega}{\nu};\ Re_1 = \frac{u_1 s}{\nu}$$
$$G_x = \frac{(R^2 - R_1{}^2)}{Re_Q\ Pr};\ R = \frac{2r}{s};\ R_t = \frac{r_1\,\Omega}{u_1} \tag{3}$$

(i) Laminar flow case

The unique feature in the stationary disk system is that an increase in the flow area causes (i) a decrease in the radial velocity and (ii) a possible increase in pressure. The latter depends on the counter-balance between two opposite trends: pressure drop due to the viscous effect and pressure rise due to the radius (i.e. area) effect. Both effects diminish along the radial passage. Figure 2 (a) shows the local pressure distribution which varies with the

Table 1: Definition of variables and coefficients in transfer equations

Φ	λ	B
u	ν	$-\frac{1}{\rho}\frac{\partial p}{\partial r} + \frac{1}{r}\frac{\partial}{\partial r}(r\nu\frac{\partial u}{\partial r}) + \frac{\partial}{\partial z}(\nu\frac{\partial v}{\partial r}) - 2\nu\frac{u}{r^2} + \frac{w^2}{r}$
w	ν	$\frac{\nu}{r}\frac{\partial w}{\partial r} - \frac{1}{r^2}\frac{\partial}{\partial r}(\nu r w) - \frac{uw}{r}$
v	ν	$-\frac{\partial p}{\partial z} + \frac{1}{r}\frac{\partial}{\partial r}(\nu r\frac{\partial u}{\partial z}) + \frac{\partial}{\partial z}(\nu\frac{\partial v}{\partial z})$
κ	λ_κ	$G - \varepsilon$
ε	λ_ε	$\varepsilon(C_1 G - C_2\varepsilon)/\kappa$
T	α	0

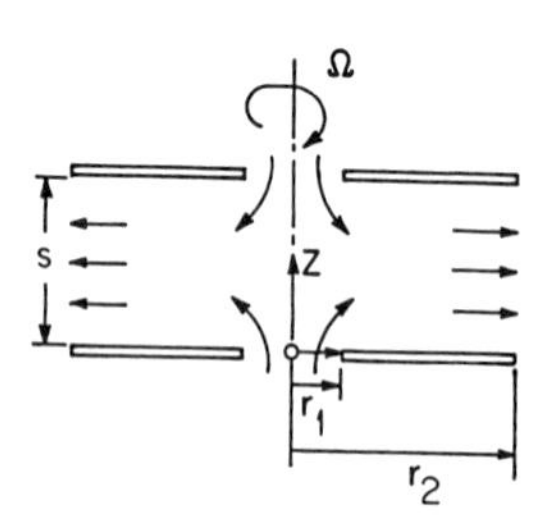

Fig. 1 A schematic of test core

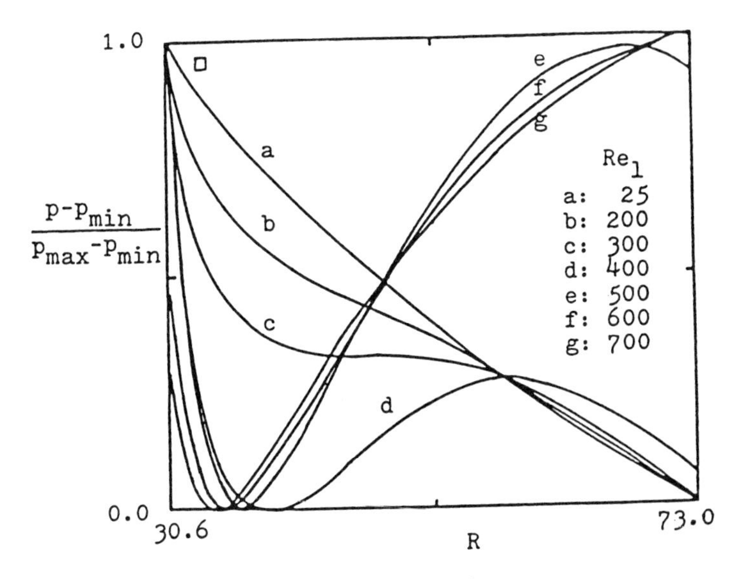

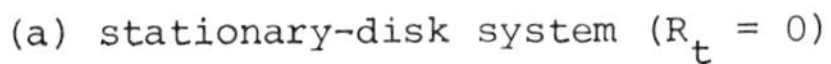

(a) stationary-disk system (R_t = 0)

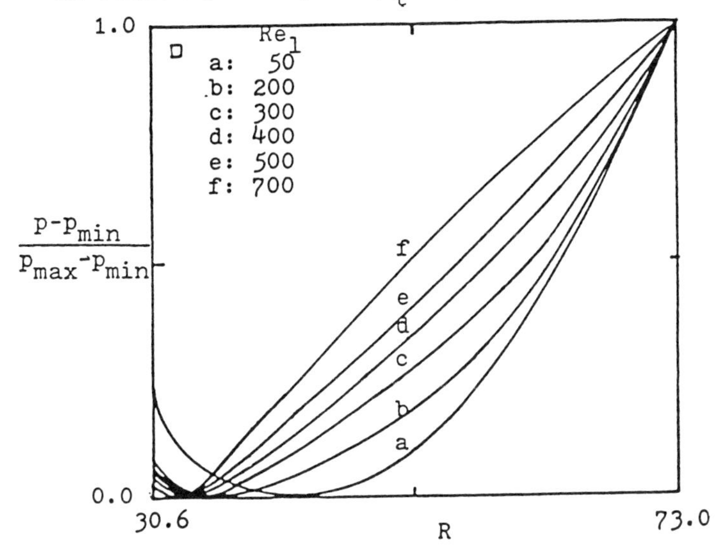

(b) rotating-disk system (R_t = 2.0)

Fig. 2 Reynolds number effect on pressure distribution in laminar flow for R_1 = 30 and R_2 = 75.5

flow rate. In high Re flow, the pressure build-up becomes so rapid that the flow separates, i.e. generation of separation bubbles. The dimensionless pressure drop within the disk is negative for Re_1 less than 500, and becomes positive for Re_1 exceeding 500, as seen in Fig. 3 (a). It increases with Re_1.

The rotation of the disks produces the centrifugal force $\rho w^2/r$ and the Coriolis force $u\,w/r$ in the flow, whose effect grows with the radial distance. The radial velocity diminishes, while the tangential velocity grows with increasing radius. In the entrance region, the viscous effect is dominant with only minor influence from centrifugal force. Therefore, like in the stationary disk system or in the ordinary boundary layer, the fluid near the walls is decelerated while the accelerated core flows toward the channel center, Fig. 4 (a). It is followed by a parallel movement of the fluid. Further downstream, the viscous force subsides and the centrifugal effect becomes dominant. The unique features of the centrifugal effect include (i) shifting the location of the accelerated flow core (i.e. peak radial velocity) from the channel center toward the wall region and (ii) directing the entire fluid to move toward the walls. The net effect is a larger volume of flow from the channel center onto the walls, rather than a reduced volume of flow in the boundary layer parallel to the walls as in the stationary system. Consequently, fluid pressure downstream from the entrance region increases, as shown in Fig. 2 (b) for the R_t = 2.0 case. The diffusion of angular momentum from the disk to the fluid is illustrated in Fig. 4 (b). Figure 3 (b) shows that disk rotation causes a reduction in the dimensionless pressure drop within the disk as Re_1 is increased.

Figure 5 shows that the local Nusselt number experiences a steep fall in the entrance region due to the development of thermal boundary layers on the walls. In the stationary disk system, the thermal boundary layer becomes fully developed within a short distance irrespective of Re and the value of Nu approaches 3.7. When the system is in rotation, Nu increases continuously along the flow passage after reaching a minimum value. The heat transfer enhancement is induced by the fluid flow directed from the channel center to the wall region with an increased flow rate in the wall region and a higher fluid-wall temperature difference. In the absence of rotation, the thermal boundary layer would have lower flow volume and less fluid-wall temperature difference. A higher flow volume (Re), a higher disk rotation (n), or a small disk opening (R_1) yiel higher heat transfer performance.

(ii) Turbulent flow case

Turbulent flow in a radial disk channel shares the same basic featu with laminar flow. The increase in flow area contributes to the decrease of radial flow momentum, resulting in the pressure buildup Rotation supplies angular momentum to the flow through viscous diffusion, developing a centrifugal force field to raise the pressure. As in laminar flow, rotation reshapes both the distribution and direction of the through flow. However, small-scale fluctuations of the velocity and pressure in turbulent flow enhance both momentu

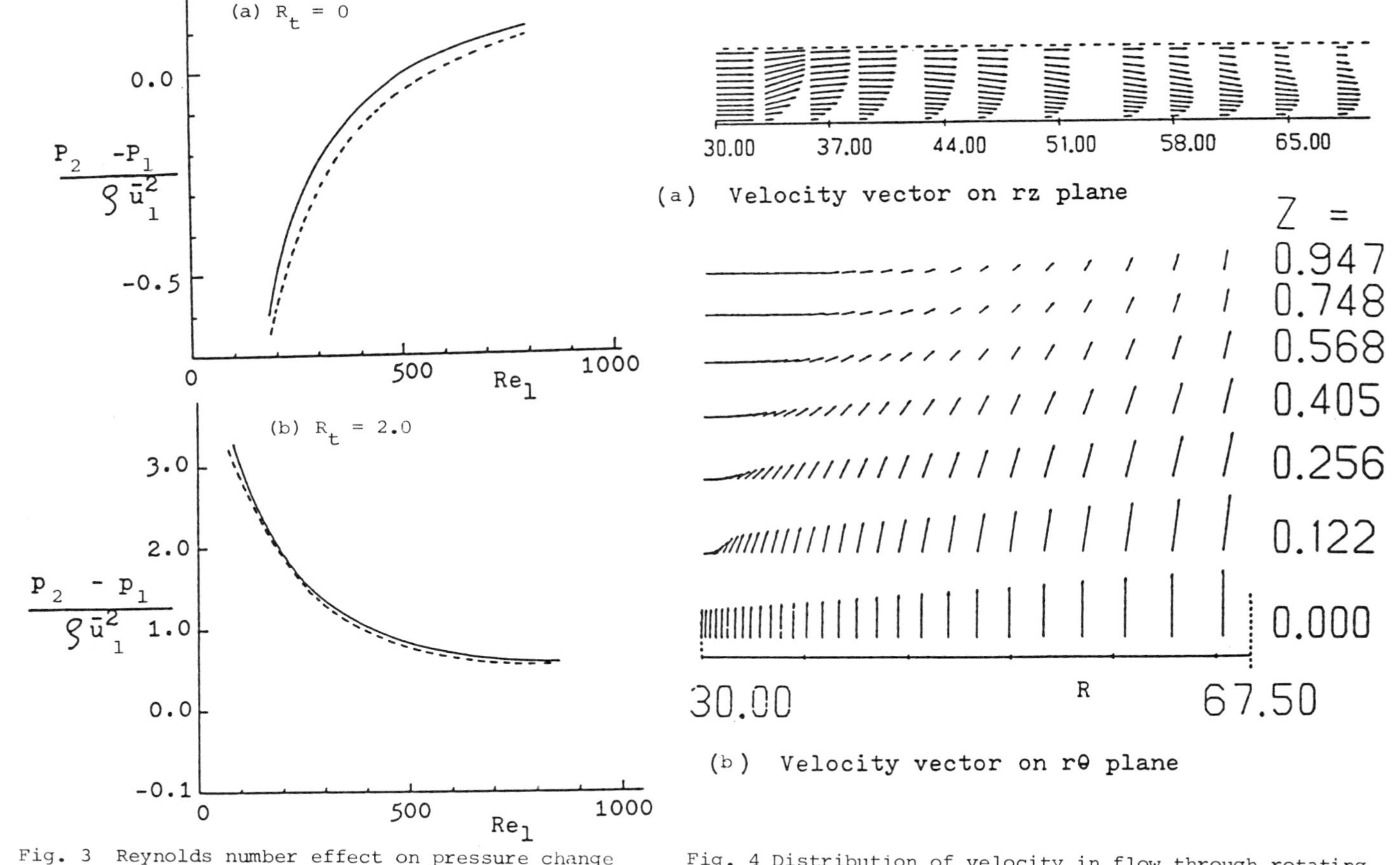

Fig. 3 Reynolds number effect on pressure change through disk channel with $R_1 = 30$ and $R_2 = 75.5$ ——— at center line ; - - - adjacent to disk walls

Fig. 4 Distribution of velocity in flow through rotating rotating-disk system with $R_1 = 30$, $R_2 = 67.5$, $Re_1 = 267$ and $R_t = 2.0$

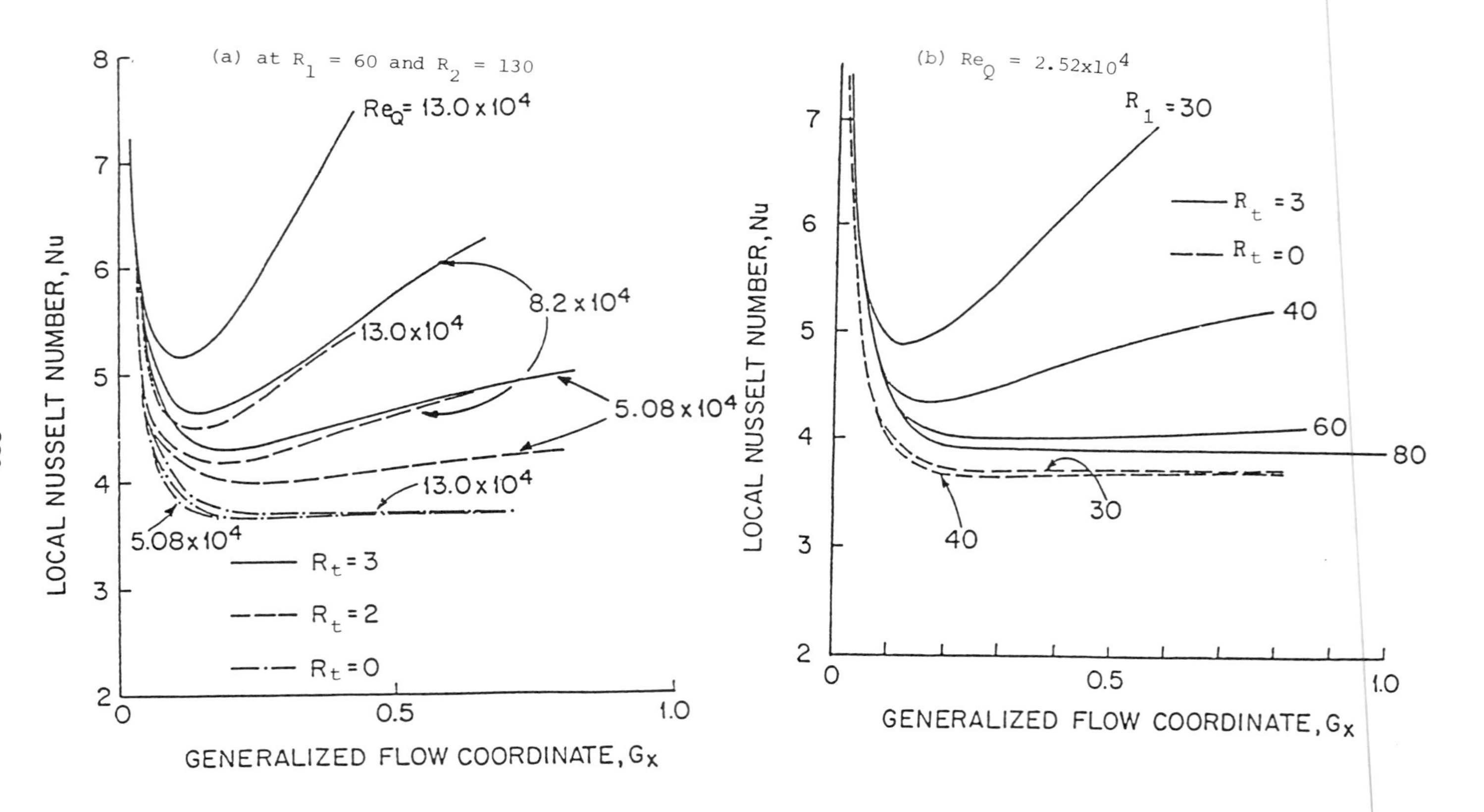

Fig. 5 Effects of disk geometry, Re_Q and R_t on local Nusselt number for laminar flow

and heat transport, which are not uniform in the flow field.

In the stationary disk system, the velocity profile near the exit takes the shape of a fully-developed turbulent boundary layer flow, namely a steep change in the wall region and a gradual contour in the central region. The fluid pressure is nearly uniform across the flow cross-section but increases continuously through the radial passage as seen in Fig. 6 (a), for Re_1 ranging from 10,000 to 100,000. In the low Re_1 region, ΔP increases with an increase in Re_1, but rises gradually with Re_1 thereafter, Fig. 7 (a). Figure 8 shows that the local Nusselt number Nu falls sharply in the entrance region but continues to diminish downstream. It implies that turbulent flow through a stationary disk system requires much more distance for thermal development than in the laminar flow case.

Rotation causes a substantial change in the shape of local pressure distribution as seen in Fig. 6 (b). The centrifugal force contributes to a steep pressure rise in the exit region of disk core. The pressure distribution varies with the flow rate. Figure 7 (b) demonstrates that the dimensionless pressure drop decreases with Re_1 and levels off at high values of Re_1. Rotation causes a steeper fall of Nu in the entrance region followed by a recovery in Nu, which increases with the radius. The extent of sharp Nu drop in the entrance region and the subsequent Nu recovery in the downstream enhances with an increase in the rotational speed R_t or the flow rate Re_1.

For both laminar and turbulent flow cases, the average Nusselt number is defined as

$$\overline{Nu} = \int_A Nu \, dA/A \qquad (4)$$

The average value will be used for comparison with experiments.

EXPERIMENTS

Test Apparatus and Procedure

Figure 9 illustrates a schematic of the parallel-disk assembly. It consisted of multiple disks with openings in the center, which rotated as a unit. Each disk was made of aluminum 0.5 mm in thickness with 160-mm i.d. (d_1) and variable o.d. (d_2). The corner piece was shaped to produce a uniform flow profile at the inlet to the disk assembly. Three values of the disk spacing (s) of 1.5, 2.5, and 3.5 mm were combined with four variations in d_2, namely 240, 280, 310, and 345 mm to fabricate twelve test cores. The number of disks (N) in each test core was altered depending upon s.

Figure 10 is a schematic illustration of the entire test apparatus. With the flow rate adjusted by the damper, air was supplied to the test core from the turbofan through a calming chamber. The air flow rate was monitored by the Pitot tube and calculated from the reading of the manometer. After passing through the diffuser, screen, and honeycomb in sequence, the air was heated by the nichrome-wire heating

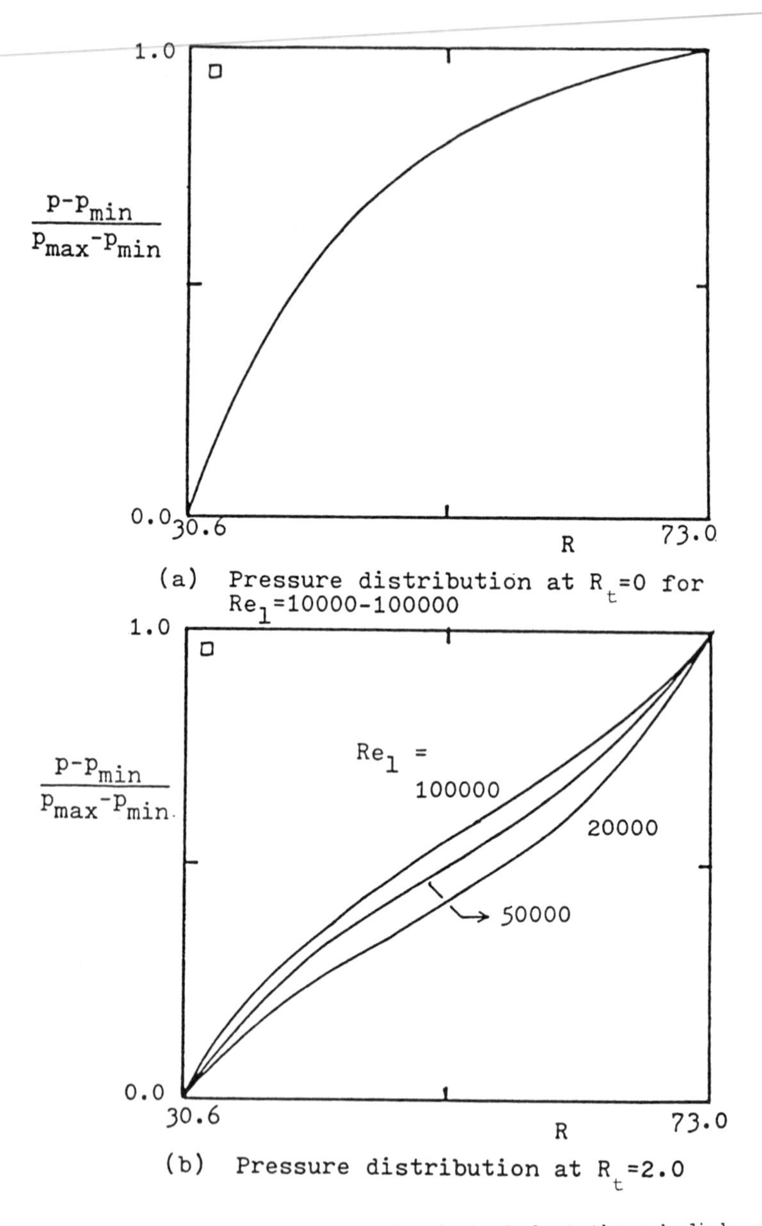

Fig. 6 Pressure distribution in turbulent through disks with R_1 = 30 and R_2 = 75.5

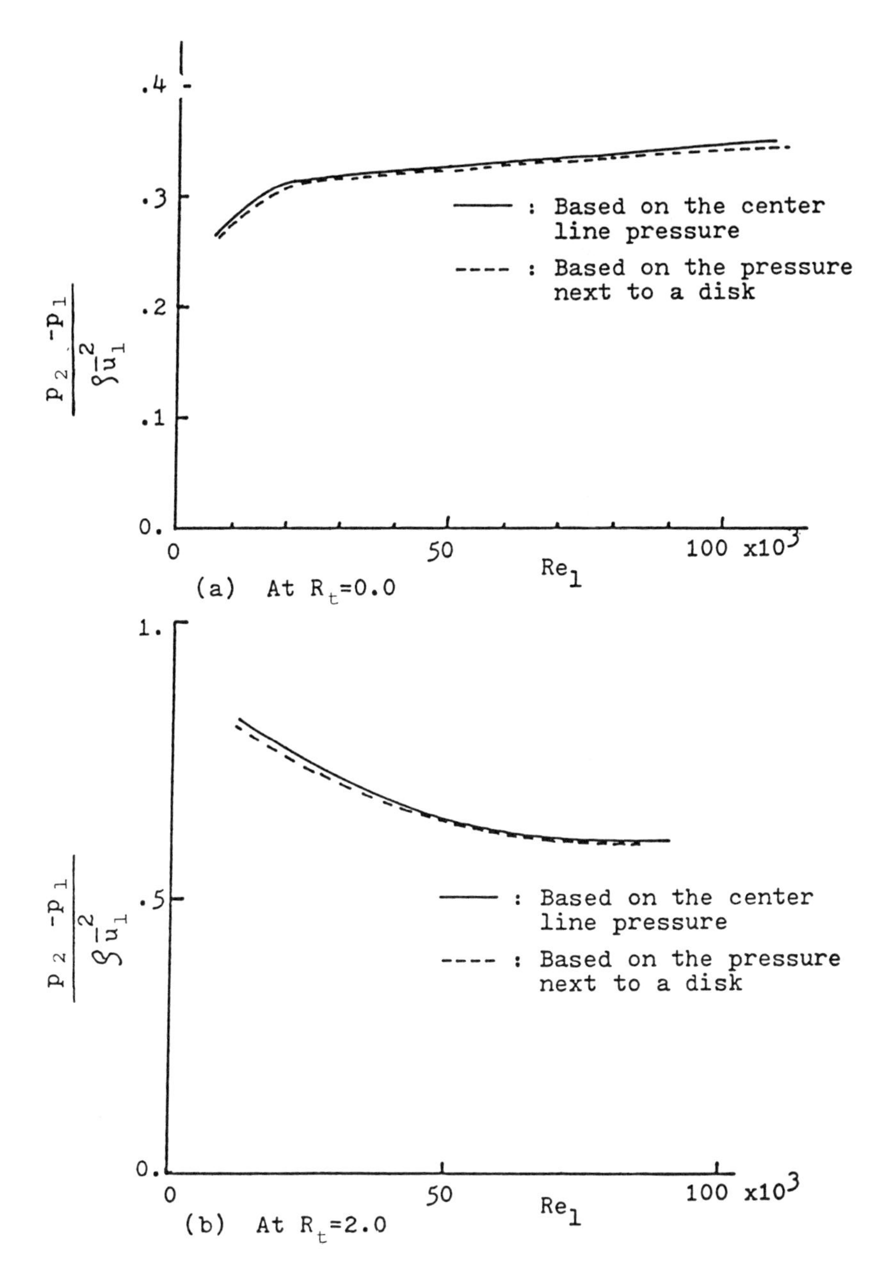

Fig. 7 Pressure change in turbulent flow through disks with R_1 = 30 and R_2 = 75.5

(a) $Re_1 = 100,000$

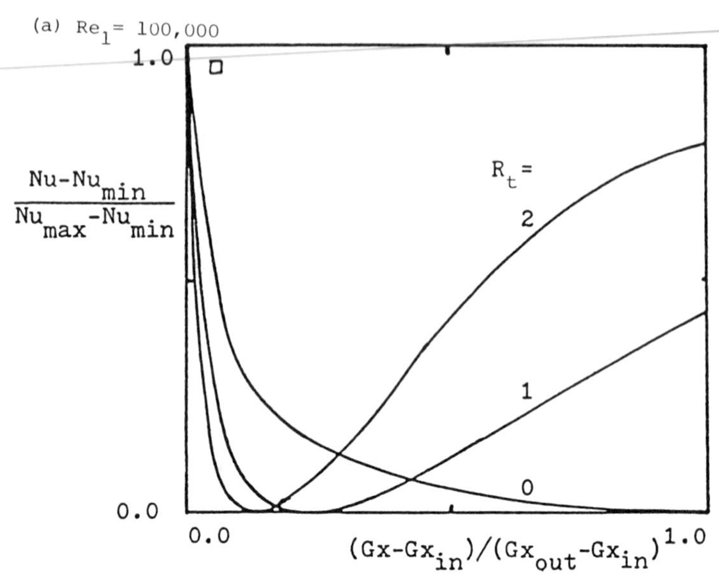

(b) $Re_1 = 10,000$ and $80,000$

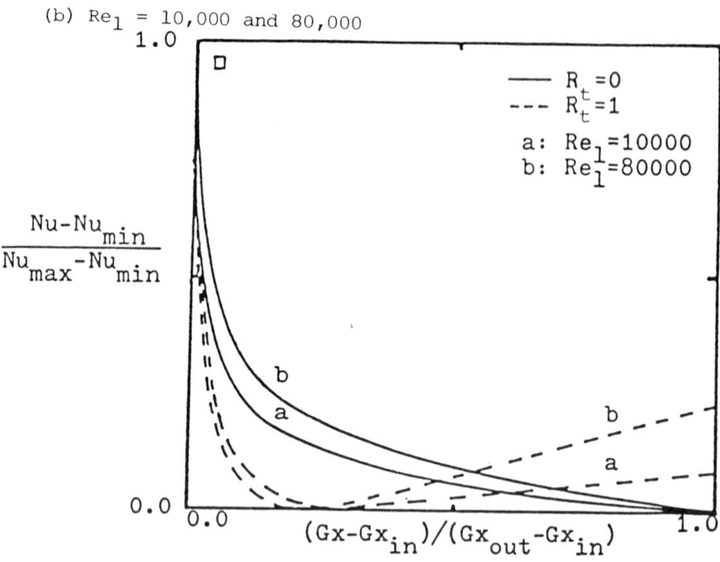

Fig. 8 Effects of Re_1 and R_t on distribution of local Nusselt number for turbulent flow in disks with $R_1 = 60$ and $R_2 = 130$

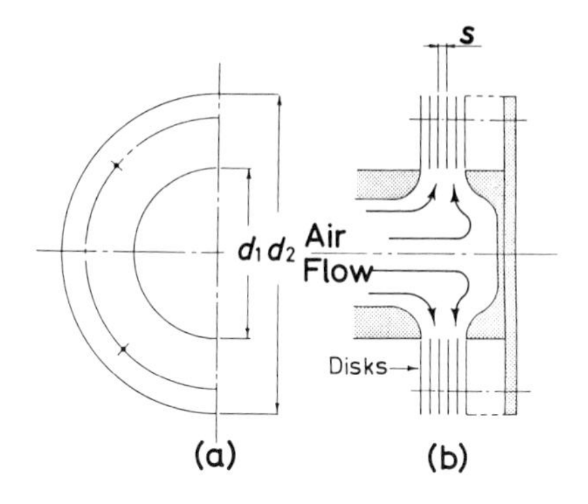

Fig. 9 A schematic of test core

Fig. 10 A schematic of test apparatus

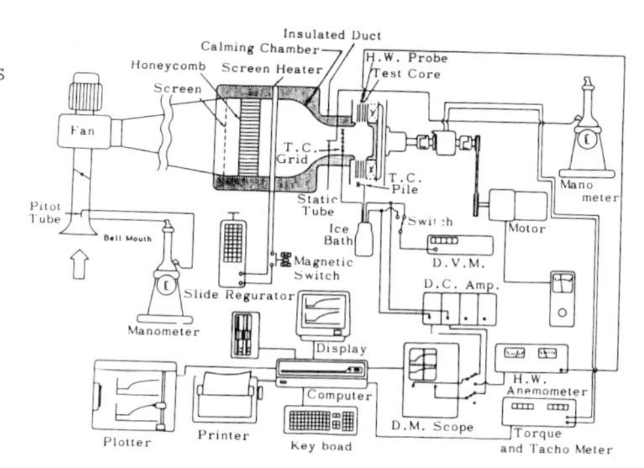

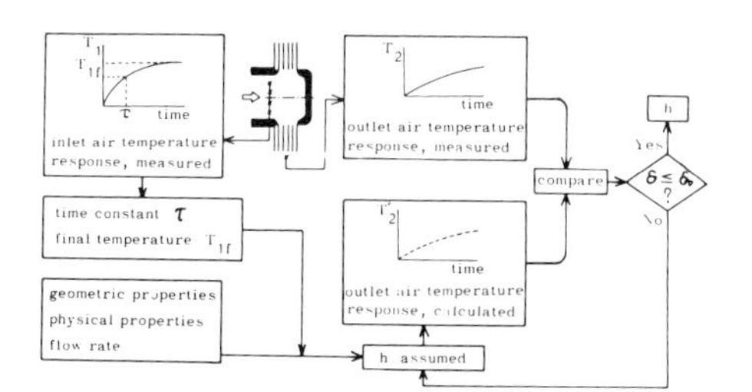

Fig. 11 Flow diagram for Automatic evaluation of h

screen installed in the calming chamber. A step change in power input to the heating screen was produced by the closing of the magnetic switch. The heated air entered the test core through the insulated duct. A thermocouple grid placed in the duct was used to measure the inlet air temperature, while a thermocouple pile monitored the mean outlet air temperature around the test core. Steady state temperatures were measured by the digital voltmeter, while transient temperature measurements were amplified before being fed into the digital memory scope. The static pressure drop across the test core was measured by the Pitot tube in the duct through the manometer. For rotating tests, the rotating speed of the test core was regulated by the variable-speed motor and measured using the tachometer.

A computer-assisted data reduction system was used to automatically evaluate the mean heat transfer coefficient $\bar{h}$ from the measured air temperature responses at the core inlet and exit. Figure 11 is the flow diagram showing the procedure. The inlet air temperature T_1 exhibited the first-order characteristics in response to a step power input to the heating screen. Its time constant τ and final steady-state air temperature T_{1f} were determined from the response curve. With τ, T_{1f}, the geometrical dimension of the test core, physical properties of the air and disk material, and the flow rate as the input data, the theoretical outlet air temperature response $T_2'(t)$ was calculated for an assumed value of the mean heat transfer coefficient (7). The calculated $T_2'(t)$ was compared with the recorded exit air temperature curve $T_2(t)$ at several specified time instants. The deviation between the calculated and measured exit air temperature is defined as

$$\sigma = \left| T_2'(t) - T_2(t) \right| / T_2'(t) \qquad (5)$$

If the deviations at all specified time instants were equal to or less than a desired value σ_o, the assumed $\bar{h}$ would be the value that is desired. Otherwise, a new $\bar{h}$ was assumed and the above procedure would be repeated until the criterion $\sigma \leq \sigma_o$ is fulfilled.

Experimental Results and Discussion

Twelve disk assemblies were fabricated and tested for the rotation speed n ranging from 0 to 25 cycles per second and the flow rates up to $u_m D_H/\nu = 13{,}000$. u_m denotes the radial velocity at the midpoint $(d_1 + d_2)/4$. Only representative results are presented here to determine the effects of n, Re, and disk geometry on the transport performance.

Figures 12-(a) and (b) show the effects of n and u_m on the average heat transfer coefficient $\bar{h}$. The disk assembly has s = 1.5 mm, d_1 = 160 mm and d_2 = 345 mm. At lower velocities, $\bar{h}$ is linearly proportional to n. At a high value of u_m, $\bar{h}$ is enhanced more rapidly with its rotational speed at low values of n and practically level

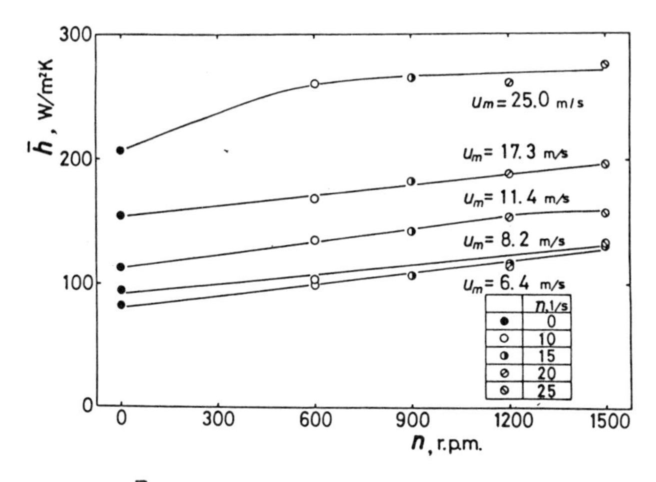

(a) $\bar{h}$ versus n

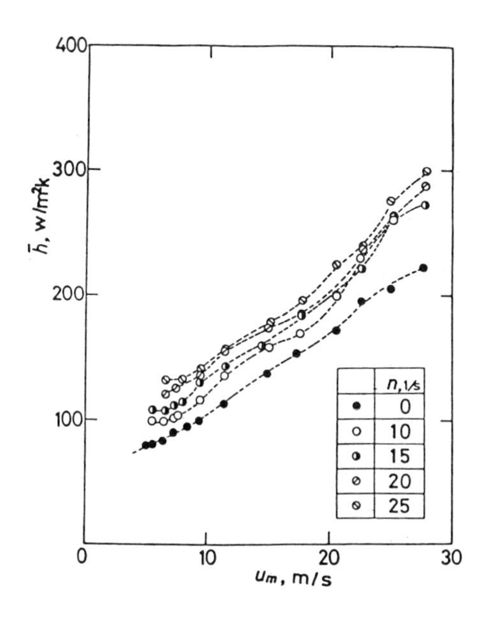

(b) $\bar{h}$ versus u_m

Fig. 12 Effects of n and u_m on average heat transfer coefficient $\bar{h}$

off at higher rotation. Figure 12 (b) indicates that $\overline{h}$ is progressively augmented with an increase in u_m at every rotational speed. It is not shown here but for the same disk size and operati conditions, an enlargement of the disk spacing (equivalent to a reduction in the flow length-hydraulic diameter ratio) results in a retardation of heat transfer performance.

In dimensionless correlation, Fig. 13 reveals that rotation causes a substantial reduction in f and a moderate enhancement in j. Here the Fanning friction factor f and the heat transfer factor j are defined as

$$f = \frac{PD_H}{2\ \rho u_m^2 L}; \ j = \frac{\overline{Nu}}{RePr^{1/3}} \tag{6}$$

L denotes the flow length $(d_2 - d_1)/2$. The effects of the rotation speed on f and j diminish with an increase in Re. When the flow reaches the transition-turbulent range, disk rotation plays a very minor role on transport performance.

The method described in reference (8) was used to evaluate the unce tainities in the test data. The physical properties, geometric mea surements of the test core, temperature and pressure measurements were taken into consideration. The uncertainities in f and j were estimated to be approximately 3 percent and 6 percent, respectively

COMPARISON BETWEEN THEORY AND EXPERIMENTS

Figure 14 presents a comparison of laminar heat transfer performanc between theory and experiments. It is observed that theory agrees well with test results at a low Re_1 range for all disk speeds. As Re_1 increases, theory underpredicts the heat transfer performance. The discrepancy grows with Re_1, although theory and experiments hol on to their own consistent trend. The discrepancy can be attribute to some differences between the theoretical and experimental models such as (i) the core geometry, namely s and d_2; (ii) the boundary conditions at the core inlet and exit as well as the wall thermal condition; and (iii) the flow patterns. Theory treated the dual-stream flow influx, while the single-stream flow influx was employe in experiments.

In theory, a laminar flow analysis is imposed throughout the entire radial passage. In reality, however, a self-sustained vortex-induc oscillating flow and the subsequent transition from laminar to turbulent flow (9) may have occurred in the stationary disk assembly. The situation is prone to take place as Re_1 increases. In rotating disk assemblies, certain unknown processes may have triggered the transition to initiate turbulence in the flow. Also superimposed i Fig. 14 are theoretical results for slug flow and Poiseuille flow, assuming they prevail through the stationary disk channels. The present theoretical curve for the no-rotation case falls betwen two ideal cases. The experimental line for the no-rotation case seems to approach the Poiseuille flow as Re_1 is reduced.

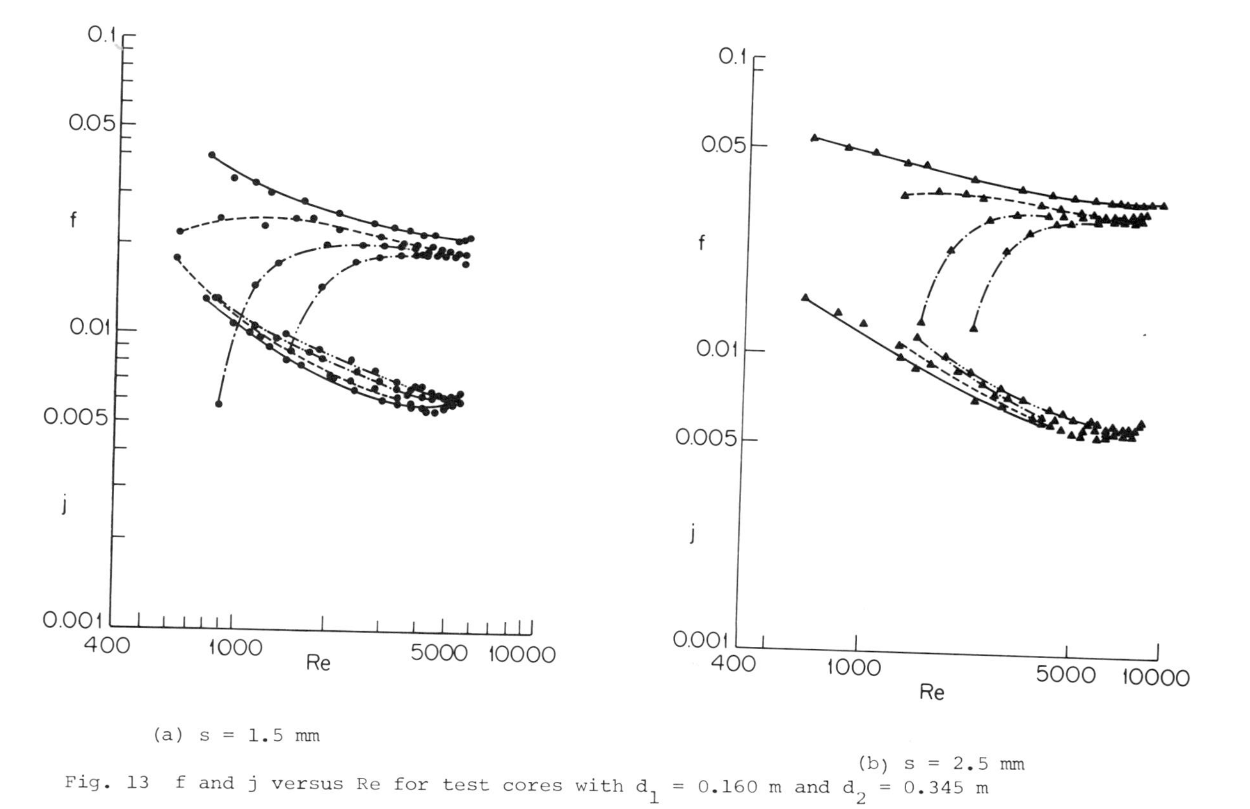

(a) s = 1.5 mm

(b) s = 2.5 mm

Fig. 13 f and j versus Re for test cores with d_1 = 0.160 m and d_2 = 0.345 m

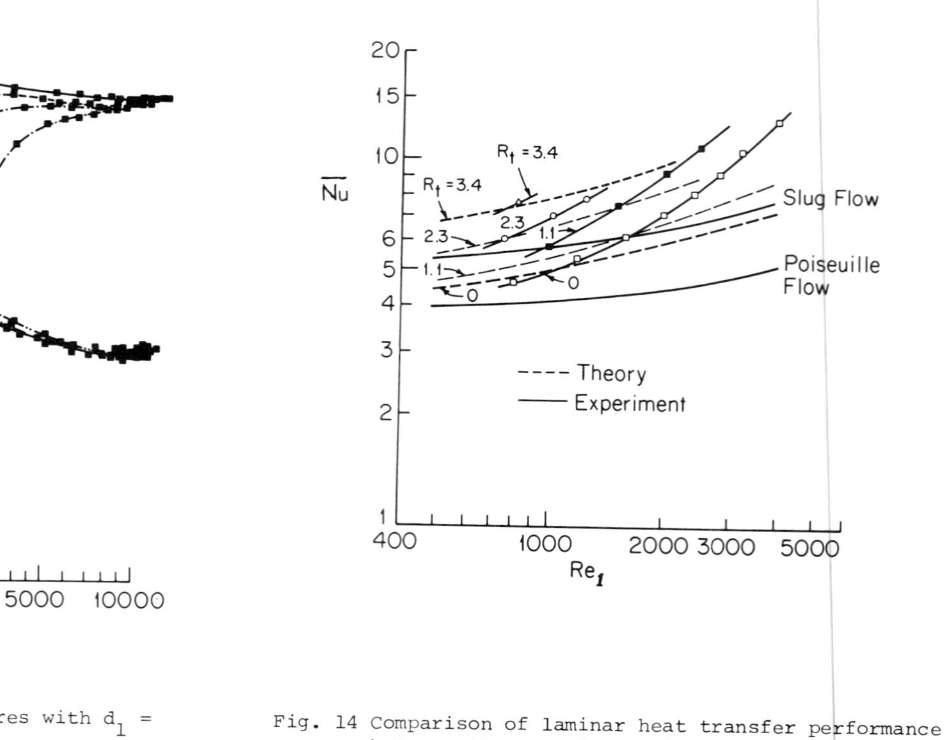

Fig. 13 f and j versus Re for test cores with d_1 = 0.160 m and d_2 = 0.345 m

Fig. 14 Comparison of laminar heat transfer performance between theory and experiments

CONCLUSIONS

Both theoretical and experimental studies have been performed to investigate the transport phenomena in radial flow through co-rotating parallel disks. A finite-difference method has been employed to obtain numerical results for both laminar and turbulent flow. A modified single-blow technique has been utilized to determine the heat transfer and friction loss performance with the aid of a computer-assisted data reduction system. It is concluded from the study that

(i) The automated data reduction device results in the improvement of test accuracy, saving labor and time.

(ii) Theory predicts that rotation causes both a shift of the accelerated core flow from the channel center to the wall regions and a directional change of the entire fluid to flow toward the disk walls.

(iii) The rotational effect, which grows with the radial distance, causes an increase in the fluid pressure. Only in laminar flow does the viscous effect become important near the entrance region, resulting in a v-shaped pressure distribution in the disk core.

(iv) The re-shaping of flow by the centrifugal force induces an increased convective flow toward the wall and a high fluid-wall temperature difference, resulting in an increase in the local heat transfer performance. A sharp decline in the local heat transfer coefficient in the entrance region, followed by a surge in the heat transfer rate downstream, results in a v-shaped performance curve.

(v) In the laminar flow regime, an increase in the disk rotation results in a moderate increase in heat transfer performance and a substantial decrease in friction loss. When the flow enters the transition-turbulent regime, the rotational effect becomes unimportant.

(vi) At low flow, theoretical prediction of heat transfer performance agrees well with experiments at all tested rotational speeds. The discrepancy between theory and experiments grows with the flow rate. The reasons for such a discrepancy are explained.

ACKNOWLEDGEMENT

The study was partially supported by the Japanese Ministry of Education under Grant C 1981 and by the U.S. National Science Foundation under Grant Number ME 80-18031.

NOMENCLATURE

A	disk surface area, m^2
B	source term as defined in Table 1
C_1, C_2	empirical constants
D_H	hydraulic diameter = 2s, m
d	disk diameter, m; d_1, inner; d_2, outer; d_m, mean, $(d_1 + d_2)/m$

f — Fanning friction factor as defined by equation (6)

G — function in κ - ε-turbulence model [3], m^2/s^3

G_x — generalized flow coordinate = $\pi(r^2 - r_1^2)/Re_Q Pr$, m^2

h — local heat transfer coefficient, $W/(m^2 \cdot K)$

$\bar{h}$ — average heat transfer coefficient, $W/(m^2 \cdot K)$

j — heat transfer factor

k — fluid thermal conductivity, $W/(m \cdot K)$

L — flow length = $(d_2 - d_1)/2$, m

N — number of disks in test core

n — number of disk rotation, Hz

Nu — local Nusselt number = hs/k

$\overline{Nu}$ — average Nusselt number = $\bar{h}s/k$ in experiments, = $\int_A Nu \, dA/A$ in theory; Nu_{max}, maximum; Nu_{min}, minimum

P — pressure, Pa; P_{max}, maximum; P_{min}, minimum

P — pressure drop within test core, Pa

Pr — Prandtl number

R — $2r/s$; $R_1 = 2r_1/s$, inner; $R_2 = 2r_2/s$, outer

Re — Reynolds number = $u_m D_H/\nu$; $Re_1' = u_1 s/\nu$, inlet; $Re_Q = \dot{V}/(\nu s)$, through-flow

R_t — rotation number = $r_1 \Omega/u_1$

r — radial distance, m; r_1, inner; r_2, outer

s — disk spacing, m

T — fluid temperature, C; T_1, inlet; T_{1f}, final steady state at inlet; $T_2(t)$, measured value at exit; $T_2'(t)$, calculate value at exit

Ta — Taylor number = $s^2 \Omega/\nu$

u — radial velocity, m/s; u_1, inlet; u_m, mean at midpoint $(d_1 + d_2)/4$

$\dot{V}$ — volume flow rate, m^3/s

v — normal (axial) velocity, m/s

w — tangential (angular) velocity, m/s

Z — $2z/s$

z — axial coordinate, m

Greek Letters

α — thermal diffusivity, m^2/s (effective value in turbulent flow case)

ε — turbulence energy dissipation rate, m^2/s^3

θ — angular coordinate

κ — turbulence kinetic energy

λ — diffusion coefficient as defined in Table 1, m^2/s; λ_κ and λ_ε , physical properties in κ - ε-turbulence model [3]

μ — dynamic viscosity, Pa·s

ν — kinematic viscosity, m^2/s (effective value in turbulent flow case)

ρ	fluid density, kg/m^3
σ	$\lvert T_2'(t) - T_2(t)\rvert / T_2'(t)$; σ_o, a desired value
ϕ	general variables as defined in Table 1
Ω	rotational velocity, rad/s

REFERENCES

1. Owens, J.M., "Fluid Flow and Heat Transfer in Rotating Disc Systems," in Heat and Mass Transfer in Rotating Machinery (edited by D.E. Metzger and N.H. Afgan), Hemisphere, Washington, D.C., pp. 81-103, 1984.

2. Sim, Y.S. and Yang, Wen-Jei, "Numerical Study on Heat Transfer in Laminar Flow Through Co-Rotating Parallel Disks," International Journal of Heat and Mass Transfer, in print, 1985.

3. Sim, Y.S. and Yang, Wen-Jei, "Turbulent Heat Transfer in Co-Rotating Annular Disks," Numerical Heat Transfer Journal, in print, 1985.

4. Sim, Y.S., "A Numerical and Experimental Study on the Flow and Heat Transfer Characteristics in Co-Rotating Disk Systems," Ph.D. Thesis, Department of Mechanical Engineering and Applied Mechanics, University of Michigan, Ann Arbor, Michigan, 1983.

5. Mochizuki, S. and Yang, Wen-Jei, "Heat Transfer and Friction Loss in Laminar Radial Flows Through Rotating Annular Disks," Journal of Heat Transfer, Vol. 103, pp. 212-217, 1981.

6. Mochizuki, S., Yang, Wen-Jei, Yagi, Y., and Ueno, M., "Heat Transfer Mechanisms and Performance in Multiple Parallel Disk Assemblies," Journal of Heat Transfer, Vol. 105, pp. 598-604, 1983.

7. Liang, C.Y. and Yang, Wen-Jei, "Modified Single-Blow Technique for Performance Evaluation on Heat Transfer Surfaces," Journal of Heat Transfer, Vol. 97, pp. 16-21, 1975.

8. Kline, S.J. and McClintock, F.A., "Describing Uncertainty in Single Sample Experiments," Mechanical Engineering, Vol. 75, pp. 3-8, 1953.

9. Mochizuki, S. and Yang, Wen-Jei, "Self-Sustained Radial Oscillating Flows Between Parallel Disks," Journal of Fluid Mechanics, in print, 1985.

Heat Transfer for Flow through Simulated Labyrinth Seals

DARRYL E. METZGER and RONALD S. BUNKER
Mechanical and Aerospace Engineering Department
Arizona State University
Tempe, Arizona 85287

ABSTRACT

An experimental study has been conducted to measure local convectiv heat transfer for flow through a narrow two-dimensional gap with on of the bounding walls containing multiple two-dimensional transvers cavities. The situation models flow through the elements of labyri seals commonly used to reduce undesirable flow around rotating components in turbomachinery. In high temperature turbomachinery applications, knowledge of the heat transfer characteristics of fl leaking through the seals is needed in order to accurately predict seal dimensions and performance as affected by thermal expansion. There is currently little information available to aid the designe either in predicting the heat transfer magnitudes or in understand the basic convection heat transfer mechanisms present. This is particularly true for the cavity side of the seal; almost all available information is in the form of overall average heat trans on the opposite side of the seal. The object of the present study to provide information on the character of local variations in hea transfer on the cavity side and the influence of changes in cleara and cavity size on these variations.

NOMENCLATURE

A	Heat transfer surface area
c_p	Fluid specific heat
C	Seal clearance
D	Cavity or groove depth
D_h	Channel hydraulic diameter at gap
h	Convection heat transfer coefficient, $= q / A(t_p - t)$
k	Test surface thermal conductivity
k_f	Fluid thermal conductivity
Nu	Local Nusselt number, $= hC / k_f$
Nu_m	Average Nusselt number over the cavity floor
q	Surface heat transfer rate
Re	Reynolds number, $= \rho VC / \mu$
t	Local test surface temperature
t_p	Plenum temperature
U	Fundamental solution, see Eq. (3)
V	Mean velocity at gap
W	Cavity Width in streamwise direction
x	Cavity floor coordinate

y	Cavity side wall coordinate
α	Test surface thermal diffusivity
θ	Time
μ	Fluid dynamic viscosity
ρ	Fluid density
τ	Time step

INTRODUCTION

Labyrinth seals are commonly used as non-contact seals to reduce undesirable flows between adjacent stationary and rotating surfaces in many types of rotating machinery. Figure 1 shows three representative seal configurations out of numerous variations that are found in practice. In all cases the overall idea is to reduce the leakage flow for a given pressure differential across the seal by creating multiple constrictions in series along the flow path.

Labyrinth seals in high temperature turbomachinery applications pose special problems for the designer [1,2] since the seal geometry cannot be specified independent of the leakage flow through the seal. The seal clearances are determined in part by thermal expansions of the seal elements which in turn are dictated in part by the convection heat transfer rates between the leakage flow and the seal components. Unfortunately, there is very little information available on convection in even the most simple labyrinth seal geometry. It is generally accepted that the relative motion between adjacent seal surfaces usually has a negligible influence on both the pressure loss and heat transfer characteristics [3,4], although supporting experimental evidence, particularly for heat transfer, is meager [5].

In axial flow turbomachinery stages, the phenomenon of leakage flow over the tips of compressor and turbine blades from pressure to suction sides of the blades has considerable commonality with the labyrinth seal situation. Flow is through a narrow gap between two surfaces in relative motion. A recent study [6] has demonstrated that the relative motion of the opposite surface has a negligible influence on convection heat transfer to the blade tips. This lends additional credence to the idea of investigating labyrinth seal behavior with stationary modeling, and this approach was taken in the present investigation. Several geometrical variations of the most simple seal configuration (Figure 1a) were investigated experimentally with both the grooved and smooth sides of the model held stationary.

In the seal configuration tested, each of the grooves is a simple rectangular cavity, and it is important to note that flow and heat transfer in such cavities have been the subjects of continuous investigation over many years, eg [7-13]. In all cases the cavities studied have been installed in wind tunnel walls where the cavity is open to a well developed, zero pressure gradient flow over an otherwise smooth surface. The cavities in a labyrinth seal differ from those previously studied by virtue of the confined nature of the geometry. They are, in effect, completely enclosed rectangular volumes both supplied and relieved from narrow short slots at the top side corners. However, a recent study by the present authors [14], modeling the heat transfer on a turbine blade tip with a single groove, shows that the heat transfer characteristics of single confined cavities are qualitatively quite similar to those of the

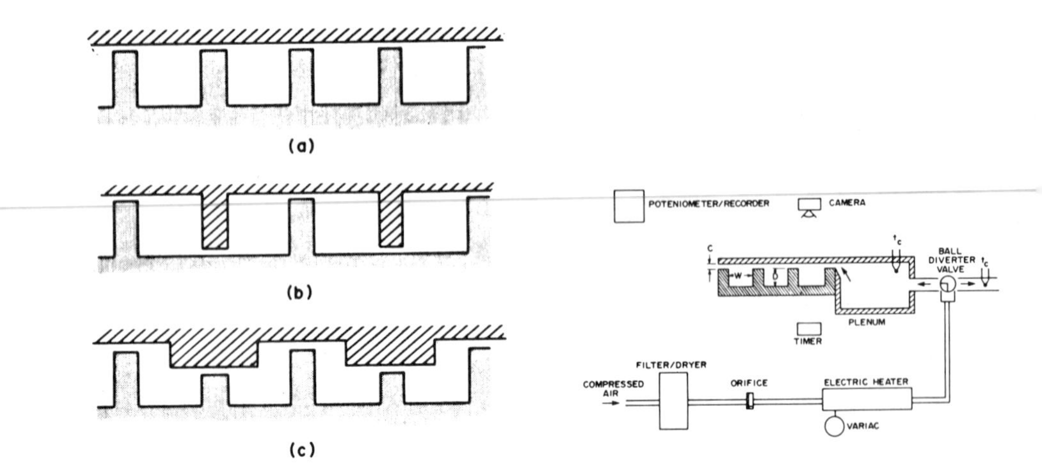

FIGURE 1. Typical Seal Configurations

FIGURE 2. Apparatus Schematic

unenclosed cavities previously studied. It is anticipated that the heat transfer in the multiple cavities of labyrinth seals will also exhibit similar behavior.

The objective of the present study was to acquire local heat transf information on the grooved or cavity side of seals in sufficient detail to provide at least a preliminary understanding of the convection mechanism. The present report covers only the initial results with a simple seal configuration. However, the experimenta technique employed is well suited for the study of other, more complex, seal geometries.

EXPERIMENTAL APPARATUS AND PROCEDURES

The experiments make use of melting patterns on thin replaceable coatings on the test surfaces in the presence of a heated air strea to determine the local surface heat transfer rates. Details of the method and the procedures used have been previously given [15,16], only a brief description will be repeated here.

Figure 2 shows a schematic of the test apparatus. Laboratory compressed air, filtered and dried, is metered through an ASME standard orifice and supplied to a heating section and following diverter ball valve. In operation the heated air is first diverte away from the test section and at the same time the test section is shielded from the heated flow so that it remains uniformily at the laboratory ambient temperature. The ball valve remains in the diverted position until a steady state temperature (above the melt point of the coating applied to the test surfaces) has been achiev in the diversion channel. At that time the valve is used to sudde route the heated air flow through the test section.

The test section construction is entirely transparent acrylic plastic and consists of a plenum chamber and interchangable seal models held between two end wall plates and sealed with O-rings. Figure 3 shows a cross section view of one of the seal models indicating measuring locations and coordinate orientation. Each seal model has three identical cavities in series, separated by identical separating baffles between the cavities. In all cases the cavity width in the mean flow direction is 0.75 in. and the baffle thickness in the mean flow direction is 0.375 in. Two separate models with cavity depths, D, of 0.40 and 0.75 in. were used in the tests, each in conjunction with two values, 0.1 and 0.2 in., of the clearance C. A flat surface, D=0, was also used with clearances of 0.1, 0.2 and 0.3 in. The span of the cavities in the direction normal to the cross section shown in Figure 3 is 3.25 in.

A thin layer of the coating material (approximately 0.002 in.) is sprayed evenly on the bottom and sides of the cavities before each test. Coating was also applied to the tops of the baffles (locations 1 and 2, 6 and 7, 11 and 12, 16 and 17). The coating is a commercially available product (Tempil Industries, S. Plainfield, N.J.) with a nominal melting point of 109F. Calibration tests were performed with the coating batch used in the tests with a resulting measured mean melting point temperature of 109.5F for fifteen samples. The lowest and highest melting temperatures measured were 109.1 and 109.6F. In both the calibration tests and in subsequent test program use, the melting is determined visually from the property of the coating to change from opaque white to transparent upon melting. A black background facilitates the melting determination.

The test is started by diversion of the heated flow to the test section, and the resulting conduction of heat into the test section walls has been numerically simulated [13] on a finite element code using specified heat transfer coefficients of the magnitude and spatial variation expected for the experiments. For these conditions together with physical properties for acrylic plastic, the simulations show that the depth of heating into the wall over the expected test duration is less than the wall thickness. In addition, lateral conduction in the wall has a negligible effect on the local surface temperature response. At any surface point, the wall temperature can thus be represented by the classical one-dimensional response of a semi-infinite medium to the sudden step application of a convecting fluid at temperature t_p:

$$(t - t_i)/(t_p - t_i) = 1 - \exp(h^2\alpha\Theta/k^2)\,\mathrm{erfc}(h\sqrt{\alpha\Theta}/k) \qquad (1)$$

If each surface point of interest were subjected to a true step increase in t_p, then measurement of the required times to reach the known phase change temperature allows solution of Eq. (1) for the heat transfer coefficients. This is the essence of the method, with the thin coating of phase change material providing a means of acquiring an array of temperature-time pairs over the surface.

However, in actual internal flow experiments, the wall surfaces will not experience a pure step change in air temperature because of the transient heating of the upstream plenum chamber and duct walls. Nevertheless, Eq. (1) is a fundamental solution that can be used to represent the response to a superposed set of elemental steps in t_p

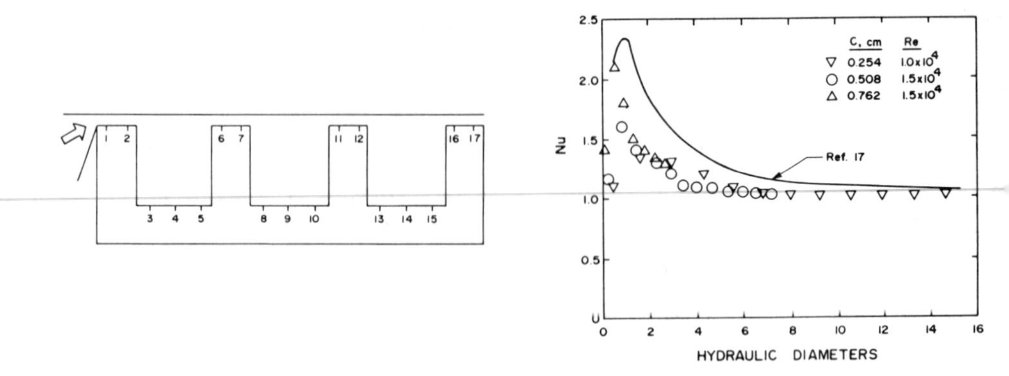

FIGURE 3. Model Cross Section and Measuring Locations

FIGURE 4. Normalized Results, D = 0

arranged to represent the actual air temperature rise:

$$t - t_i = \sum_{i=1}^{N} U(\Theta - \tau_i)\, \Delta t_p \tag{2}$$

where:

$$U(\Theta - \tau_i) = 1 - \exp\left[\frac{h^2 \alpha(\Theta - \tau_i)}{k^2}\right] \operatorname{erfc}\left[\frac{h\sqrt{\alpha(\Theta - \tau_i)}}{k}\right] \tag{3}$$

In the present experiments, air temperature is determined from t plenum thermocouple measurement. The t_p variation with time is recorded and approximated by steps, and the resulting superposed solution, Eqs. (2) and (3), is solved for the local surface heat transfer coefficients, using observed local melting times. Meltin patterns were recorded photographically at discrete times with a motor-driven camera during the test transient. The time required melting at a given location was determined from visual comparisons between pairs of adjacent photographs in the sequence.

RESULTS AND DISCUSSION

Figure 4 presents the present results obtained with the flat test surface (D=0) in the absence of cavities. The Nusselt numbers hav been normalized with expected fully developed channel values, and presented as a function of the distance, in hydraulic diameters, downstream of the clearance gap entrance. For comparison, the sol line shows results from [17], as presented in [18] for an abrupt contraction entrance. The present results display a similar maxim heat transfer location downstream of the entrance, but they are in general not as high above fully developed values. Such behavior s reasonable since the present geometry has an abrupt entrance on on

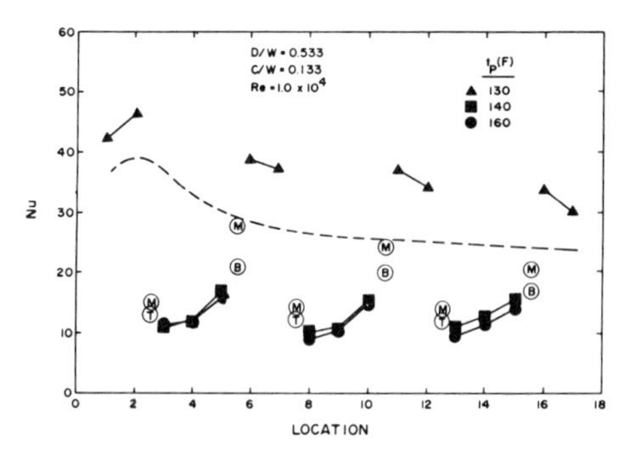

FIGURE 5. Results, D/W = 0.533, C/W = 0.133, Re = 1.00 x 10^4

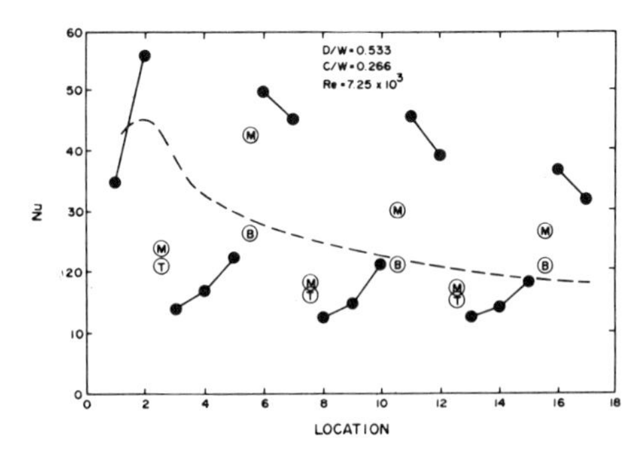

FIGURE 6. Results, D/W = 0.533, C/W = 0.266, Re = 7.25 x 10^3

one side of the channel. The good agreement with expected values in the downstream region provides confidence in the experimental techniques employed. A detailed treatment of the uncertainties in the technique is given in [15], based on the methods of [19]. For the present tests, the uncertainty in local Nusselt number is estimated to be ±10%.

Repeatability in the test procedure is considerably better than ±10%, as demonstrated in [14]. This is illustrated in Figure 5 which presents results for the shallow cavity, D/W = 0.533, and small clearance, C/W = 0.133. In the cavity bottom, the test was rerun a second time with a new application of coating and a different plenum temperature. In general, the experimental uncertainty at a given location can be minimized by choosing a temperature potential that is appropriate for the level of heat transfer rate at that locality. Temperature potentials that are too large result in short melting times and larger uncertainties. Those that are too small result in melting times larger than the time required for the thermal transient to penetrate the test wall thickness, compromising the validity of the assumed semi-infinite wall response. During the test program, if too short or too long melting times are found on a surface region of interest, the test is re-run with a different plenum temperature.

The distribution of heat transfer on the test surfaces shown in Figure 5 is typical of all of the present results. For reference the dashed line shows the Nusselt number values that would be expected if the grooved surface were replaced with a smooth surface. Relatively high heat transfer rates are present on the top surface of the upstream baffle, and these have the same general magnitude and character as those measured with the flat test surface. From location 2 on the top surface to location 3 at the upstream end of the groove floor, heat transfer decreases by a factor of four. It then increases toward the downstream end of the floor. This general increase in Nusselt number across the floor of the groove is consistent with behavior observed in unenclosed cavities.

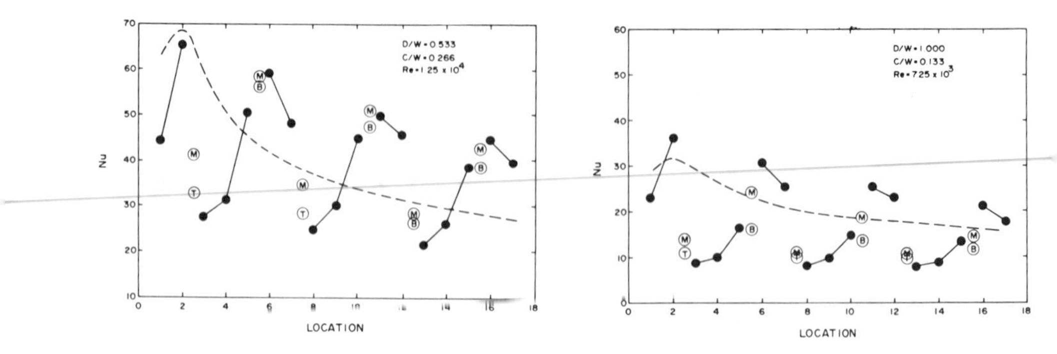

FIGURE 7. Results, D/W = 0.533, C/W = 0.266, Re = 1.25 X 10^4

FIGURE 8. Results, D/W = 1.000, C/W = 0.133, Re = 7.25 x 10^3

Also shown on Figure 5 are two Nusselt number values for each baffl side wall, representing the values observed at mid-height (M) and a either the top (T) or the bottom (B). In general these indicate he transfer that increases from bottom to top on the baffle upstream facing walls and from top to bottom on the downstream facing walls. The variation over the downstream facing walls is smaller than that the upstream facing walls. Again, this behavior is consistent with that reported for unenclosed cavities.

Heat transfer on the top of the second baffle is back up to nearly level on the top of the first baffle. In this case, however, and a all succeeding baffles, the highest value of Nu is on the upstream end, rather than at the downstream end as is the case for the top o the first baffle. This behavior supports a notion that the source the downstream clearance gap flows is largely fluid moving downstre adjacent to the smooth surface, partially impinging on and flowing down the downstream cavity walls toward the cavity floors. Thus th would be little or no flow separation at the downstream gap entranc as is the case in the first gap.

An almost identical pattern is repeated for the second and third cavities, although moving in the mean flow direction there is in general a small decrease in Nusselt number between corresponding points on the groove and baffle surfaces. This decrease results fr the fact that all heat transfer coefficients in the present study a based on the plenum temperature. Depending on the geometry and Reynold's number, there is a decrease in driving potential of from 10% between adjacent cavities. If heat transfer coefficients were based on a regional mixed mean temperature rather than the upstream plenum value, then the Nusselt number values for the first, second, and third cavities are generally identical within experimental uncertainty.

Figures 6 through 10 show the balance of the results acquired in th present study. Although the parameter ranges covered in the tests limited, some general observations can be made that should help designers in estimating the amount of heat entering or leaving a component through the labyrinth seal area.

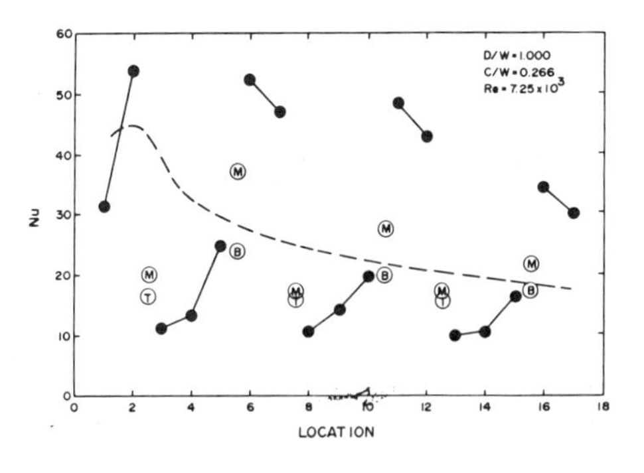

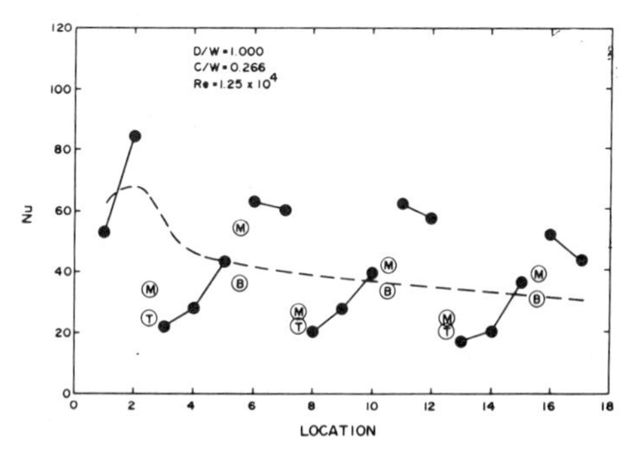

FIGURE 9. Results, D/W = 1.000, C/W = 0.266, Re = 6.25 x 10^3

FIGURE 10. Results, D/W = 1.000, C/W = 0.266, Re = 1.25 x 10^4

The total amount of heat transfer consists of the sum of three parts: the groove floors, the baffle sides, and the baffle tops. Heat transfer at the baffle tops is high, characteristic of duct entry flows, and should be reasonably accurately predictable as such.

Heat transfer on the groove floor consistently shows an increase from front to rear, but the average level varies significantly from one geometry to another. This variation appears to be mainly tied to the relative clearance, C/W, with larger clearance values leading to higher average heat transfer on the groove floor. This is shown in Figure 11 where a fairly good correlation of the average floor heat transfer is achieved by assuming a 0.8 power Reynolds number dependence. The results from the present study are consistent with the trend shown by the single cavity results of [14], also shown on Figure 11. There is little effect of D/W apparent in this range (0.5 ≤ D/W ≤ 1.0), and this again is consistent with behavior in unenclosed cavities.

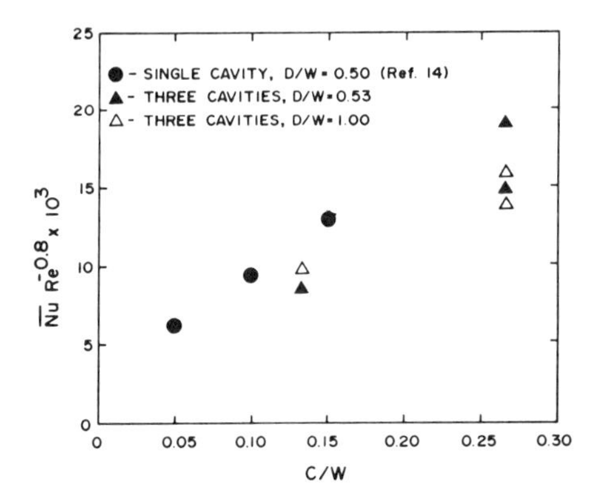

FIGURE 11. Average Groove Floor Heat Transfer

Heat transfer rates over the downstream facing baffle wall are associated closely with the rates at adjacent locations on the floor and exhibit relatively small variations over the wall. Rates on the upstream facing baffle wall show a much larger variation. They appe to vary monotonically from near the baffle top surface value at the top of the wall to near the adjacent floor value at the bottom of th wall.

SUMMARY

Convection heat transfer in a series of rectangular cavities has bee measured as a stationary model of a labyrinth seal. Very little information is presently available on heat transfer characteristics for the baffle side of the seal, and the present study is an attempt to provide some understanding of the phenomena and guidance for design. The study covers only a limited range of parameter variati for the most simple seal geometry. Within these limits, the follow are the principal findings.

1. Local Nusselt number distributions on the groove floors and baff sides have characteristics similar to those of unenclosed cavities.

2. Average heat transfer on the tops of the baffles can be predicte as an entry flow to a narrow gap.

3. Average heat transfer over the groove floors correlates as $Re^{0.8}$ and shows a strong C/W effect and little D/W effect over the ranges $0.05 \leq C/W \leq 0.27$, $0.5 \leq D/W \leq 1.0$.

4. Heat transfer rates on the baffle sides can be predicted once th baffle top and groove floor values are established.

5. The narrow clearance gap between cavities effectively uncouples flow in one cavity from a dependence on the details of flow in adjacent cavities. As a result, Nusselt number distributions are t same for all cavities, if local mixed mean temperature is used as t driving potential.

REFERENCES

1. Henneke, D.K., "Heat Transfer Problems in Aero-Engines," Heat Mass Transfer in Rotating Machinery, D.E. Metzger and N.H. Afg Eds. Hemisphere, Washington, D.C., 1984, pp. 353-379.

2. "Seal Technology in Gas Turbine Engines," AGARD-CP-237, 1979.

3. "Duct Systems-Labyrinth Seals," Fluid Flow Data Book, General Electric Company, 1974.

4. "Forced Convection-Labyrinth Seals," Heat Transfer Data Book, General Electric Company, 1976.

5. Shvets, I.T., Dyban, E.P. and Khavin, V.Y., "Heat Exchange in Labyrinth Seals of Turbine Rotors," Energomashinostroenie, 12, 1963, pp. 8-11.

6. Mayle, R.E. and Metzger, D.E., "Heat Transfer at the Tip of an Unshrouded Turbine Blade," Proceedings of 7th International Heat Transfer Conference, 3, 1982, pp. 87-92.

7. Seban, R.A., "Heat Transfer and Flow in a Shallow Rectangular Cavity with Subsonic Turbulent Air Flow," International Journal of Heat and Mass Transfer, 8, 1965, pp. 1353-1368.

8. Roshko, A., "Some Measurements of Flow in a Rectangular Cutout," NACA TN No. 3488, 1955.

9. Chapman, D.R., "A Theoretical Analysis of Heat Transfer in Regions of Separated Flow," NACA TN 3792, 1956.

10. Charwat, A.F., Dewey, C.F., Roos, J.N. and Hitz, J.A., "An Investigation of Separated Flows - Part II: Flow in the Cavity and Heat Transfer," Journal of the Aerospace Sciences, 28, 1961, pp. 513-527.

11. Fox, J., "Heat Transfer and Air Flow in a Transverse Rectangular Notch, "International Journal of Heat and Mass Transfer, 8, 1965, pp. 269-279.

12. Yamamoto, H., Seki, N. and Fukusako, S., "Forced Convection Heat Transfer on Heated Bottom Surface of a Cavity," Journal of Heat Transfer, 101, 1979, pp. 475-479.

13. Aung, W., "An Interferometric Investigation of Separated Forced Convection in Laminar Flow Past Cavities," Journal of Heat Transfer, 105, 1983, pp. 505-512.

14. Metzger, D.E. and Bunker, R.S., "Cavity Heat Transfer on a Transverse Grooved Wall in a Narrow Flow Channel," ASME/AIChE National Heat Transfer Conference, 1985.

15. Larson, D.E., "Transient Local Heat Transfer Measurements in 90° Bends Using Surface Coatings Having Prescribed Melting Points," MS Thesis, Arizona State University, 1983.

16. Metzger, D.E. and Larson, D.E., "Use of Melting Point Surface Coatings for Local Convection Heat Transfer Measurements in Rectangular Channel Flows with 90 Deg Turns," to be published in Journal of Heat Transfer, Trans. ASME.

17. Boelter, L.M.K., Young, G., and Iverson, H.W., "An Investigation of Aircraft Heaters XXVII - Distribution of Heat Transfer Rate in the Entrance Region of a Tube," NACA TN 1451, 1948.

18. Kays, W.M. and Crawford, M.C., Convective Heat and Mass Transfer, 2nd Ed., McGraw-Hill, New York, 1980.

19. Kline, S.J. and McKlintock, F.A., "Describing Uncertainties in Single Sample Experiments," Mechanical Engineering, Vol. 75, January, 1953.

GENERAL TOPICS

Cooling of Superconducting Electric Generators by Liquid Helium

W. NAKAYAMA and H. OGATA
Mechanical Engineering Research Laboratory, Hitachi, Ltd.
502 Kandatsu, Tsuchiura, Ibaraki, Japan

INTRODUCTION

Superconducting generators have a great potential in future electric supply systems in increasing the efficiency of generators and in enhancing the stability of power network systems (1). Recognition of possible advantages over gas-cooled and water-cooled generators has led research institutes and manufacturers in several countries to wage substantial research and development efforts. Fig.1 shows the electric power capacities of the test generators already built, under construction, or in the planning stage. Since earlier attempts at MIT (2), steady improvements in the design of generators have been made, and experience of generator operation has been accumulated.

Among many technical problems, crucial to the ultimate success in the development of generators of low construction and operation costs is the accurate prediction of heat transfer in the rotor where the field windings have to be maintained at a specified cryogenic temperature. The present paper reviews the recent advances in heat transfer study on liquid helium cooling of the rotors in the authors' laboratory. Also included are the references to the related works reported by other investigators.

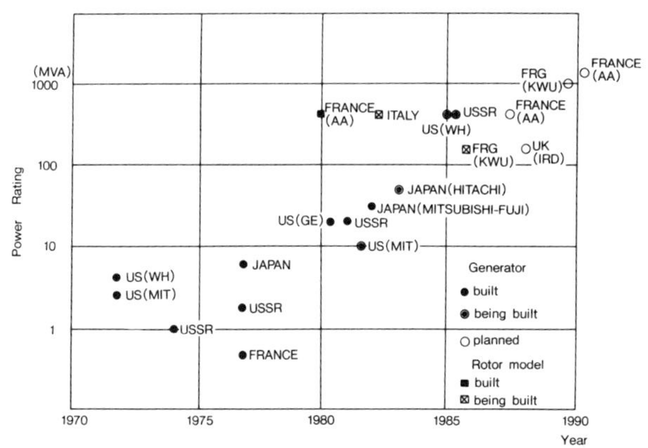

FIGURE 1. Superconducting generators built or being planned in the world

Fig.2 shows the longitudinal cross sections of the superconducting generators employing two different schemes of driving helium flow; the one under externally applied pressure head (Fig.2(a)), and the other utilizing the body force generated by density variation of helium in a strong centrifugal acceleration field (Fig.2(b)). The field windings (1) are mounted on the torque tube (4) which provides a mechanical support against tortional and vibrational forces during the generator operation. There are two shields constituting the outer shells of the rotating dewar; the outer shield (3) designed to protect the superconducting field windings from the disturbance of magnetic field incurred by external sources, and the internal shield (2) serving as thermal radiation shield. The functions of the shields (2) and (3) just described are of course only the primary ones; the shield (2) also serves as absorber of magnetic field fluctuations, and the shield (3) is forming the boundary between the external environment of room temperature and vacuum environment inside. The space (13) inside the rotor, except for those of helium passage, has to be maintained at a vacuum pressure of around 10^{-6} Torr ($\sim 10^{-4}$ Pa) in order to secure thermal insulation. The seal (10), located between the helium supply coupling (6) and the rotating shaft, has to be robust enough to maintain the required vacuum for a long period of generator operation. The magnetic-fluid seal has been employed in most of the test generators.

Before embarking on the explanation about the flow of helium, the sources of heat load to be carried by helium will be noted. During the steady operation, the largest heat load comes from heat leaks through the ends of the torque tube. There is a large temperature gradient along the torque tube, the one end at room temperature and the other close to the field windings at temperature of liquid helium. There are other routes of incoming heat, through radiation and conduction across the shields, and also conduction through minor structural members. Among the sources of transient heat generation of heat, the most serious one would be the AC loss under a great surge of magnetic field intensity caused by three-phase short circuit in the external power grid. The design of a cooling

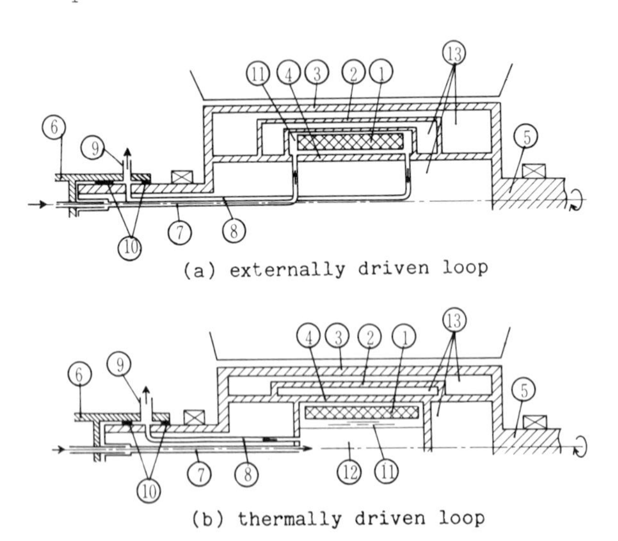

FIGURE 2. Structure of a superconducting generator

system should meet the requirements of removing all those anticipated heat leaks and heat generation.

Fig. 2 (a) and (b) show only the major routes of helium. The actual route is branched out to several subroutes in the rotor, including those passages for cooling the torque tube. In Fig. 2 (a), helium is introduced through the coupling (6) at the end of the rotor shaft, led through the coolant passages in the field windings (1) under a pressure head imposed from outside. A calculation of the centrifugally produced pressure build-up, and a resultant temperature rise of helium, at the location of the field winding shows that, in this scheme, the inlet temperature of helium to the rotor has to be well below the critical temperature for maintenance of superconductivity. In the scheme of Fig. 2 (b), the field windings are immersed in a reservoir of liquid helium (11) which is held against the inner wall of the torque tube by centrifugal force. Heat removed by convection currents of liquid in the passages of the field windings is brought to the free surface of the liquid reservoir, and carried away in the phase of vapor. The vaporized helium is used to cool the torque tube, and if the vapor passage is properly designed, as proposed by Bejan in his thesis (3), the pressure in the vapor core (12) in the rotor can be maintained at a reduced level. This obviates the needs to supply subcooled helium at the inlet owing to the equilibrium relationship between reduced pressure and low temperature. The driving potential for creation of a reduced pressure is generated by the density difference between warm and cold helium in radial passages provided at different axial locations on the torque tube. Those radial passages are not shown in Fig. 2 (b), however, the principle could be inferred from what are reported in a later section. The driving potential can be made large enough to drive helium convection all the way from the inlet (7) to the outlet (9) at the extreme end of the rotating shaft (5).

In common to those different schemes and their variants, there are needs for study on heat transfer to helium in a rotating system in order to make the design of a machine on a solid basis of the data of flow and heat transfer. Table 1 summarizes the problems pertaining to the helium flow system and the two major components. In the following sections, we proceed from the reporting of the basic data of heat transfer to the discussion of the problems in the flow system.

TABLE 1. Thermohydraulic problems in the rotors of superconducting generators

System and components	Problems
Helium convection loop	• How to drive the helium flow • Prediction of the rate of helium flow in a convection loop
Field windings	• Distribution of helium flow in the matrix of passages revolving around an axis • Heat transfer to flows of supercritical helium in a centrifugal acceleration field of several thousand Gs
Torque tubes	• Counter flow heat exchange between the incoming heat leak and the ongoing helium flow

Finally in this section, notes will be made of distinctive natures of heat transfer to helium in a rapidly spinning system. Liquid helium has a large coefficient of volumetric expansion $\sim 0.1\ K^{-1}$ at ~ 4 K, and this coupled with a high centrifugal acceleration gives rise to a very high Rayleigh number. For example, in a coolant passage of 10 mm in diameter revolving in a circle of 500 mm in radius at the rate of 377 radian/sec (3600 rpm), the imposition of only 0.1 K of ΔT results in a Rayleigh number of $10^{11} \sim 10^{12}$. This implies a very strong effects of the body force on flow and heat transfer which can hardly be expected with other coolants. Pressurized hydrogen, a coolant commonly used in generators today, the Rayleigh number becomes $\sim 10^{4}$ under the same condition, and for water, $\sim 10^{7}$. The theories and the experimental data available in the literature, recently summarized in (4) and (5), are mostly the results of the works aimed at the cases of gas or water cooling. Whether one can extend them to the present case has not yet been fully investigated.

Another important problem involved in the helium cooling is a wide range of pressure experienced by helium as it flows radially in the rotor. The critical point of helium is at 5.2 K and 2.26 x 10^{5} Pa, so that, in a radial distance of 500 mm in the rotor spinning at a rate of 377 rad/sec, the equilibrium phase of helium changes from vapor at the center to supercritical near the periphery where the pressure ratio P/Pc could be as high as 30. The behavior of flow in such an operating condition has not yet been fully understood.

What are summarized in this review paper do not provide comprehensive answers to the above problems, however, they are intended to serve as useful guides to the designers on a basis of the current state of the knowledge. The difficulties experienced in the laboratory works, and posing a serious barrier to further advances of the investigation, reside in the measurements of heat flow in a rotating test rig which has to be thermally insulated by a high vacuum. The estimate of the accuracy of measurements itself is a challenging task. It should be borne in mind that the data of heat transfer coefficient reported herein have an estimated accuracy range of 50 per cent at low heat flux ends to 20 per cent at high heat flux ends, and those of flow rate 5 - 20 per cent.

THERMALLY DRIVEN CONVECTION IN CENTRIFUGAL FIELDS

Considered in this section is the convection of liquid helium in a confined space where a heat dissipating surface is set perpendicular to the direction of centrifugal acceleration, and heat is transferred to the surrounding cold surfaces. This is a possible situation in the coolant passages of the field windings especially where the scheme of thermally driven convection (Fig. 2 (b)) is employed. Fig. 3 shows the test cell where heat is transferred by convection of helium in a closed cell (6). The heat transfer surface is the end face of a copper cylinder of 6 mm in diameter which is cemented by epoxy-resin to the inner wall of the cell. An enamelled constantan wire of 0.2 mm in diameter is wound around the cylinder to serve as electrical heater. The temperature sensors made of carbon resistor (Allen-Bradley, 100 Ω, 1/8 W) are embedded in the cylinder and the cell wall. The cell is immersed in a pool of liquid helium where the temperature is maintained at 4.2 K, and revolves around an axis. The radial distance between the heat transfer surface and the rotation axis is set at 113 mm. In this construction of the cell, helium in the cell is sealed from the exterior reservoir of helium which serves as heat sink of a large heat capacity. The pressure of helium in the cell during a test run was determined by adding the pressure rises caused by rotation and heating to the initial sealing pressure. Different cell pressures in a range

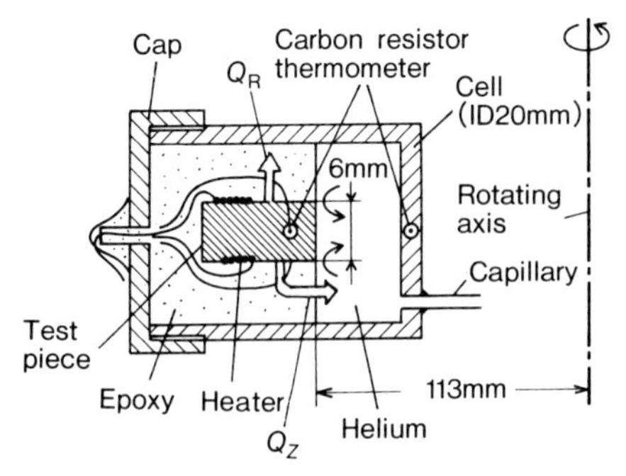

FIGURE 3. **Heat transfer cell attached to the end of a rotating arm**

of 0.15 - 0.7 MPa were produced by changing the initial sealing pressure.

The heat tranfer data were reduced into the non-dimensional forms; the Nusselt number based on the diameter of the cylinder D,

$$Nu = \frac{q\,D}{k\,\Delta T} \quad , \tag{1}$$

and the Rayleigh number based on the centrifugal acceleration $a = r\,\omega^2$,

$$Ra = \frac{D^3\,\rho^2\,g\,\beta\,\Delta T}{\mu^2} \cdot \frac{C_p\,\mu}{k} \cdot \frac{a}{g} \quad , \tag{2}$$

Refer to the Nomenclature at the end of the paper for the meaning of the symbols. The temperature difference ΔT is that between the heat transfer surface and the bulk of helium in the cell. The temperature at the heat transfer surface, T_w, was determined by performing the finite element analysis of a temperature field in the cylinder and the resin. The bulk temperature of helium in the cell, T_b, was estimated by the analysis of heat transfer to the exterior reservoir, and it was found to be very close to 4.2 K. The estimated ΔT occupies 95 per cent of a temperature potential between the test surface and the exterior reservoir. The physical properties k, ρ, μ and C_p are estimated at the bulk temperature, and the coefficient of volumetric expansion β (7) at $(0.25\,T_w + 0.75\,T_b)$. The least-square of the data produced the following correlation for supercritical helium,

$$Nu = 0.028\,Ra^{0.404} \tag{3}$$

which correlates the data within a band of ± 12 per cent.

It is interesting to see how the data obtained by the present experiment compare with those of other rotating test-rig configurations or those in the natural gravity field. In Fig. 4, the data reported by various authors are compared. Refer to the present authors' paper (6) for the specific sources of the data. There included are the data obtained with discs, cylinders and spheres, and the characteristic length in the definition of Nu and Ra is diameter for a horizontal cylinder and a sphere, and height for a vertical cylinder. The data obtained in the rotating systems are those of Scurlock and Thornton (8) and Haseler et al. (9). In the experiments of those authors, helium is sealed in a radial pipe with the heat dissipating element located at the end, therefore, helium is given a larger freedom to circulate than in the

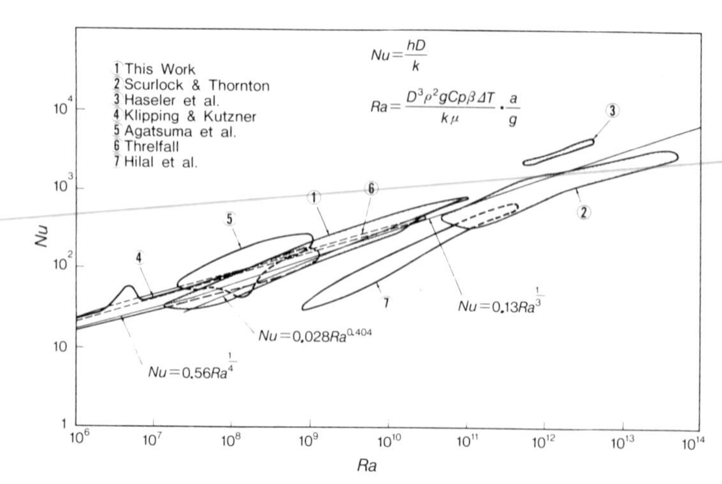

FIGURE 4. **Heat transfer by buoyancy driven convection**

present system. The attempts in the conversion of the data reported in the original literature to the form comparable each other in Fig. 4 met with the difficulty in finding the measured data or the definition of ΔT in some of the original sources. Despite the uncertainty arising from this difficulty, the data of various authors are seen to cluster around the conventional formulas for natural convection in the gravity field (10)

$$Nu = \begin{cases} 0.56 Ra^{1/4} & \text{for } Ra \leq 10^9 \\ 0.13 Ra^{1/3} & \text{for } Ra > 10^9 \end{cases} \tag{4}$$

where the acceleration term in the definition of Ra is the centrifugal acceleration a for the data of rotating systems. The Nusselt number computed from equation (3) agrees with the one from equation (4) within the possible experimental accuracy.

On the basis expounded by Fig. 4, we recommend the use of equation (4) for the estimation of heat transfer coefficients on the parts of the rotor where heat transfer surfaces have different geometries and the spaces of various sizes for helium convection are present. The length spanning the heat transfer surface is to be taken as characteristic length in Nu and Ra, and ΔT is the difference between the surface temperature and the mean temperature of helium in a space near the surface.

FORCED CONVECTION IN CENTRIFUGAL FIELDS

Forced convection in a channel revolving around a parallel axis is found not only in an externally driven loop (Fig. 2 (a)) but also in a thermally driven loop (Fig. 2 (b)) where a large driving potential is generated elsewhere. Prediction of heat transfer in such a channel is important to the design of coolant passages in the field windings. There have been reported no experimental data obtained with cryogenic helium in a revolving channel except those by the present authors' group (11).

The test channel is a copper circular pipe of 5 mm in inside diameter, 2 mm in wall thickness and 500 mm in length. Fig. 5 shows the construction of the test section, where the test tube is mounted on a drum by the flanges made of fiber reinforced plastic. The drum, not shown in a full view in Fig. 5, has shafts on both sides, and it contains all the necessary components such as a counter-weight on the opposite side of the test tube, a pipe line for helium supply, the leads of the temperature sensors and heaters, and on the rotating shafts, the seals to maintain vacuum in the space between the stationary container and the drum, and the slip-rings to take out temperature signals or feed the electric power to the heater. The radial distance between the center of the tube and that of the drum is 250 mm.

A nichrome wire in an insulation sheath of 1 mm in diameter is wound around the tube over the length of 300 mm to serve as electric heater. The mixed mean temperature of helium is measured at the inlet and the outlet of the test tube, where the temperature sensors of carbon resistor are inserted in the mixing cups. The measurement of wall temperatures requires an elaborate arrangement. As shown in a sketch in Fig. 5, a copper holder carries four carbon resistors imbedded in the tabs on the periphery, and it is soldered to the outside of the test tube. There are four such holders attached to the heated section of the tube at an axial interval of 100 mm.

Thermal insulation is provided by wrapping the test tube by a foil of polyester having deposition of thin aluminum film, and covering the assembly of the tube and the drum by a copper shell of 1 mm thick. This shell is cooled by the flow of cold helium to 20 K. This is further shielded by the outer stationary shell which is cooled to 80 K by liquid nitrogen.

The pressure in the test tube was estimated from the pressure measured upstream of the supply coupling at the end of the rotating shaft and the pressure rise due to centrifugal acceleration. The flow rate of helium was measured by the rotameters at a point downstream of the outlet of the rotor, where helium was already warmed up to room temperature and its pressure approached the atmospheric.. The heat input to helium flow in the test section is the sum of the power input to the heater and the heat leak from outside; the latter was estimated from the readings of the thermocouples imbedded in the key structural members.

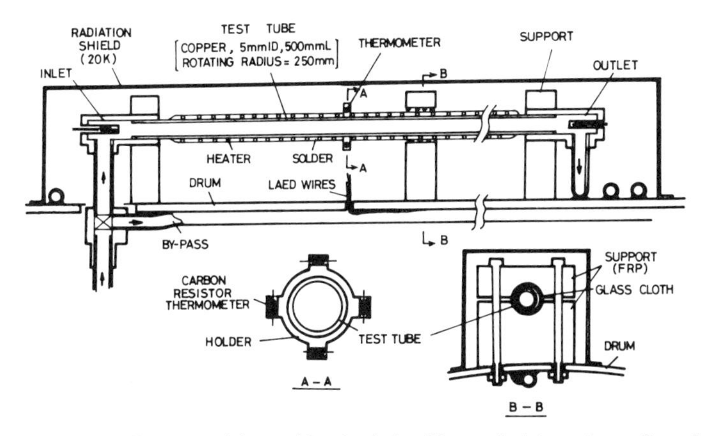

FIGURE 5. Test section attached to the rotating drum for forced convection experiments

Great care was taken to supply single-phase (gaseous, liquid) or supercritical helium to the test tube, and also to conduct the experiment in a stabilized thermal environment. This was accomplished by the combined effects of cold helium diverted to cool the rotating drum and the auxiliary heater on the supply line. The ranges of the experimental parameters are as follows : the pressure in the test section, 0.1 - 0.26 MPa; the temperature of helium, 4.2 - 12 K; the heat flux 77 - 4,000 W/m^2 ; the flow rate of helium in the test tube, 0.035 - 0.47 g/s (Re = 2,700 - 78,000) ; the rate of revolution, 0 - 25.2 Hz (a/g = 1 - 640).

The heat transfer coefficient was obtained by dividing the heat flux q by the difference (ΔT) between the measured wall temperature (T_w) and the bulk temperature of helium (T_b) both at the distance of 200 mm from the inlet. The bulk temperature T_b was estimated from the inlet temperature and the heat input to the helium flow up to the specified location. In terms of the length (ℓ) to diameter (d) ratio, the chosen location is at ℓ/d = 40, hehce, the heat transfer coefficient thus obtained was assumed to be that for a fully developed flow.

The data will now be shown in terms of the non-dimensional parameters; the Nusselt number

$$Nu = \frac{q\,d}{\kappa\,\Delta T}$$

and the parameters derived to account for the effect of thermally driven secondary flows. The parameter for turbulent flows (12) is Re/Γ^2, where Re is the Reynolds number $Re = \dot{m}d/\mu$, $\dot{m}$ the mass velocity, and

$$\Gamma = Re^{18/11}\,(Gr\,Pr^{2/3})^{-10/11}$$

Gr = ($\rho^2 g \beta \tau d^4 / \mu^2$) (a/g), τ temperature gradient along the tube axis, Pr Prandll number. The parameter for laminar flows (13) is $Ra\,Re$ ($= Gr\,Pr\,Re$). All the physical properties are evaluated at the bulk temperature T_b .

Fig. 6 (a) shows the experimental data for turbulent flows. The data are plotted in the ratio Nu/Nu_o vs. Re/Γ^2, where Nu_o is the Nusselt number for a stationary pipe (14),

$$Nu_o = 0.023\,Re^{0.8}\;Pr^{0.4}\;\left(\frac{T_w}{T_b}\right)^{b} \tag{6}$$

$$b = -\left\{\frac{1}{4}\log_{10}\left(\frac{T_w}{T_b}\right) + 0.3\right\}$$

Also shown are the data obtained in the non-rotating state (a/g = 1) as well as those obtained with another test tube having a larger diameter of 7.5 mm. The curves are the predictions computed from the formula developed by Nakayama (12),

$$\frac{Nu}{Nu_o} = \frac{1.132\,Pr^{2/3}}{Pr^{2/3} - 0.05}\left(\frac{Re}{\Gamma^2}\right)^{1/20}\left\{1 + \frac{0.061}{(Re/\Gamma^2)^{1/5}}\right\} \tag{7}$$

The experimental difficulties caused scatter of the data, however, the data are seen to cluster around the predictions. Close inspection of the data revealed that some obtained at high Gr and low Re depart systematically from the predictions for turbulent flow. Taking cognizance of the effect of body forces on the transition from laminar to turbulent flow reported in the literature (15, 16), laminarization of the flow is suspected, and those data at high Gr and low Re are compared in Fig. 6 (b) with the predictions for laminar flow (13):

$$\left.\begin{aligned} \frac{Nu}{Nu_0} &= \frac{0.180}{E}\left(E-1+\frac{1}{3E}\right)^{1/5}(RaRe)^{1/5}\frac{1}{1+\dfrac{1}{10\,EPr}} \\ E &\equiv \frac{1}{5}\left(2+\sqrt{\frac{10}{Pr^2}-1}\right) \quad \text{for} \quad Pr \le 1 \quad ; \quad Nu_0 = \frac{48}{11} \end{aligned}\right\} \quad (8)$$

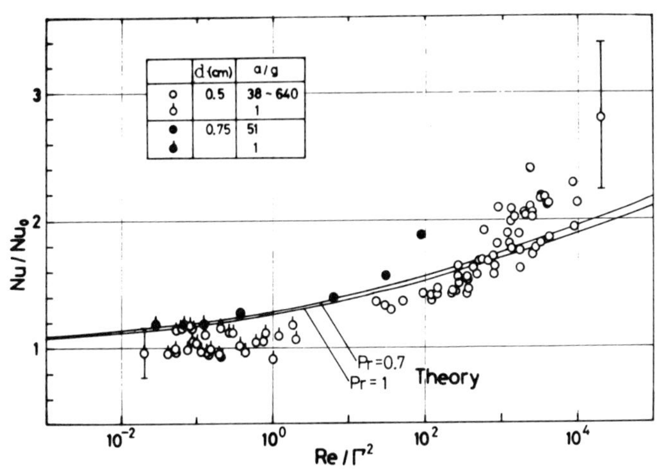

(a) comparison with the predictions for turbulent flow regime,

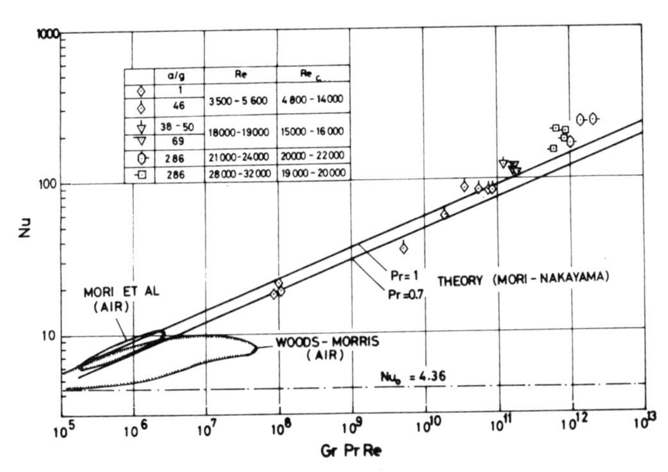

(b) comparison with the predictions for laminar flow regime

FIGURE 6. Comparison of the heat transfer data with the analytical predictions

Also shown are the data obtained with air (17, 18). With a due consideration on the experimental difficulties, one may conclude that the agreement between the experimental data and the predictions is fair. Fig. 6 (b) demonstrates that very high values of $Ra\,Re$ can be reached by cryogenic helium in a revolving channel. Heat transfer coefficient increased by the effect of secondary flows can be more than 100 times the one in a stationary system. Such intense enhancement of heat transfer has not been observed experimentally with other conventional coolants.

Equations (7) and (8) are now used to demarcate the condition of transition between turbulent and laminar flows. The Nusselt numbers computed from the both equations are plotted in a graph of $Nu - Re$ with Gr as parameter, the intersections of the curves for turbulent and laminar flows are located, and a set of the intersection points are correlated by the following equation,

$$Re_c = 790\ Ra^{0.185} \tag{9}$$

Concluding this section, we recommend the use of equation (7) when $Ra \geq Re_c$, and equation (8) when $Re < Rec$. The channel considered above has a circular cross section, while actual coolant passages could be rectangular. In a rectangular channel, especially those having a large aspect ratio, the flow is likely to be different from the one in a circular channel (19, 20). The heat transfer to cryogenic helium flow in such a channel is open to future investigation.

Channels running radially are also present in the rotor. They are, however, short in terms of l/d, and the flow at the inlet is expected to have a strong swirling component after having passed through a sharp bend. In such situations the rotation seems hard to yield a large effect on heat transfer. The data of Schnapper and Hofmann (21) obtained with liquid helium in a radial pipe show little effect of rotation.

CONVECTION LOOP IN THE ROTOR

In this section, first, the two elementary problems bearing importance in the design of the cooling system will be addressed; the flow distribution among the coolant passages of the field winding, and the pressure drop in a revolving bend passage. Then, the experiment performed with a thermally driven convection loop on a rotating rig will be reported.

Flow Distribution in the Coolant Passages of the Field Winding

As mentioned in the previous section, the cooling system must absorb transient heat generation at a time of three-phase short circuit accident. The scheme employed in many of the rotor designs to meet this requirement is to maintain steady convection of helium in the coolant passages at a flow rate sufficiently high to absorb anticipated heat generation during an accident. For this it is important to make sure that the coolant flow is uniformly distributed among multiple coolant passages. Fig. 7 illustrates the apparatus for flow visualization experiments to study the distribution of flows among the passages. This is a model of the rotor cut in half at the middle of its length, set vertical with the cut section at the bottom. The coolant channels are formed by attaching the spacers in a specified pattern to the inner cylinder and covering them by a transparent acrylic cylinder. In a real generator, the field windings molded in a saddle shape are to be mounted over the coolant passages. Two continuous channels are provided along the top and bottom peripheries of the drum, where several radial portholes are distributed to connect the coolant passages to the coolant reservoir inside the drum. The primary objective of this experiment was to see the effect of the Coriolis force on the distribution of flows. To do this, the test fluid needs not be helium; the only requirement is an approximate correspondence of the Rossby number, $V/(d\omega)$, where V is the fluid velocity and $d\omega$ represents the magnitude of the Coriolis force, between the model and the real machine. In this experiment, aqueous solution of thymol blue (0.1 per cent) is used as model fluid. The space inside the drum as well as the coolant passages are filled with this solution, and the whole assembly is rotated about the vertical axis. The driving potential of convection is generated by heat applied from the heater which is imbedded at the top rim of the drum. In a centrifugal acceleration field, the warm fluid flows radially inward to the reservoir in the drum through the raidal portholes, and with this concur the upward flows in the multiple passages and the flows through the portholes at the bottom from the reservoir to the ring space.

The solution of thymol blue changes its color by the addition of a small amount of hydrochloric acid (to yellow) or sodium hydroxide (to blue) with little change in density. A capillary tube inserted in the drum is designed to release a trace of hydrochloric acid or sodium hydroxide at the bottom of the drum, so that one can observe the front of changed color in the coolant passages. Fig. 8 shows the photographs taken with a strobo light at the time t. The rotational speed was 29 rad/sec, and the equivalent diameter of the coolant passages was 3.7 mm. The velocity at the inlet portholes, estimated from the advancing speed of the color front was 2.2 mm/sec, hence, the Rossby number at the portholes in this example was 0.02 . Such a small Rossby number means that the inertial momentum of fluid is negligible compared to the Coriolis force, and the fluid has enough time to adjust its angular momentum to the value of solid body rotation as it flows radially, that is, the angular speed of fluid particles is almost equal to the speed of the drum. Once in the ring space upstream of the parallel passages, the fluid behaves as if it flows in a stationary system and shows little preference in the entry to the divided passages. A note is made of the fact that for the effect of Coriolis force to appear in a flow system, there has to be a radial movement of fluid of an appreciable scale, which is absent in this example of radially shallow channels. The above argument explains the observed uniform

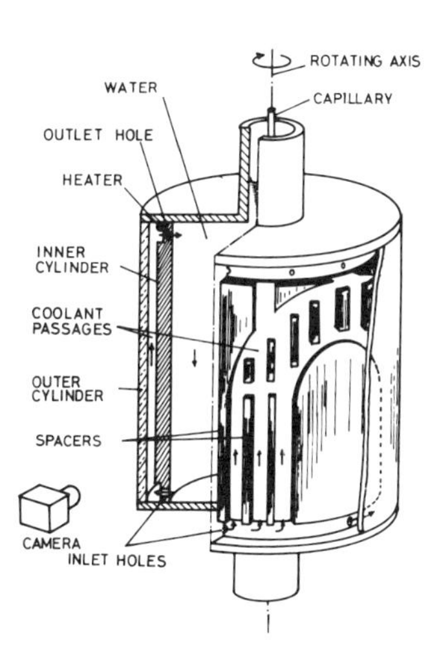

FIGURE 7. Model of coolant passages in the field winding for flow visualization experiments

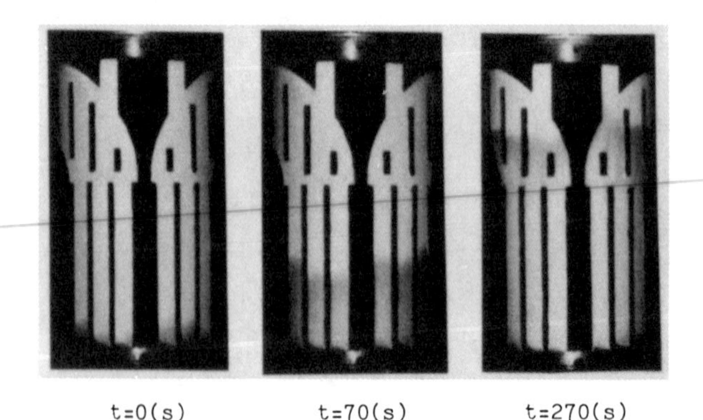

FIGURE 8. Photographs of advancing flows in the model coolant passages revolving at 277 rpm (heater power = 141 W)

distribution of flows in the passages in Fig. 8, and offers a guarantee for uniform coolant distributions in real generators where the Rossby numbers are much smaller than the one in this model experiment.

Resistance to Flow Through Revolving Bends

The estimate of the resistance to coolant flows is critically important to guarantee reliable operation of generators. Among the various sections of the coolant paths, those posing a substantial percentage in the total resistance are bends or elbows. The presence of a number of bends in the rotor is unavoidable because of the basic structural requirements on the field windings and the support components.

The flow resistance in bends depends on several factors; geometry of a bend (radius of curvature, cross section), rotational speed, flow velocity, and configuration of a fluid passage upstream of a bend. Data on this problem are too scarce to present a summary on the effects of those factors on the pressure drop across the bend. Nakayama et al. (22) proposed a correlation for the ratio of the resistance coefficient in a revolving bend (ζ) to that in a stationary bend (ζ_0),

$$\frac{\zeta}{\zeta_0} = 2.3\, Ro^{-0.22} \quad \sim \; (\omega/V)^{-0.22} \tag{10}$$

where Ro is the Rossby number based on the mean velocity of fluid V. Murakami and Kikuyama (23) correlated their data in a form

$$\frac{\zeta}{\zeta_0} \propto \omega^{b} \tag{11}$$

where b is a function of the radius of revolution, and ω is measured in radian per second. The value of b in a typical generator design becomes 1.4, so that equations (10) and (11) give widely differed predictions at high ω. In the authors' work on superconducting generators, it was decided to rely on the data of Murakami and Kikuyama, because their data contain those of high ω. We propose a new correlation which yields a reasonably lower prediction for ζ than the one from equation (11).

$$\frac{\zeta}{\zeta_0} = 5 \left(\frac{De\ \omega}{2V}\right) \quad (12)$$

This is of course a tentative recommendation. The resistance coefficient could vary greatly by the location of a bend as well as a modification of the bend geometry, as reported by Csillag (24).

Experiment of Thermally Driven Convection in a Rotating Loop

The purpose of the experiment reported in this section is to verify the formulas developed to predict the flow rate in a rotating loop. Fig. 9 (a) shows the schematic drawing of the apparatus, and Fig. 9 (b) shows the picture of the rotor rig. In the middle of the apparatus is a drum ⑯ where liquid helium is held against its inner wall by centrifugal acceleration. The radial depth of helium, monitored by the liquid level sensor (a 0.25 mm dia. NbTi wire) extending normal to the inner wall, is maintained at 10 mm by controlling the rate of helium supply through the port ①. Mounted on the rotor rig are two loops, I (⑥) and II (⑭), formed by copper tubes. The inside diameter of the tube of the loop I is 3 mm, and that of the loop II is 7.5 mm. The loop I has two openings to the liquid helium reservoir, while the loop II has an open end at the center of rotation in the drum and another end leading to the exhaust port ②. The tube in the loop I running parallel to the rotation axis, about 400 mm long on the revolving beams and 250 mm apart from the rotation axis, is equipped with heaters and temperature sensors in a manner similar to the one shown in Fig. 5. The heaters are also wound around the radial sections of the loop I. Those heaters are independently powered, so that, by controlling the power input to the heaters, thremally driven convection is produced in a determined direction.

Liquid helium first enters the drum, then flows into the loop I. Evaporation takes place at the free surface of the reservoir in the drum as heated helium from the loop I reaches there. Gaseous helium then enters the loop II, and finally exits to the atmosphere. In what follows, the data obtained with the loop I will be reported.

The rotational speed was set at 12.5 Hz (750 rpm), and the pressure in the helium resrvoir at 0.123 MPa (at the center). Heat was added to the parallel section of the loop I in a range of up to 0.2 W, with power to the other heaters shut off. Measurements were made of the heat flux q, and the difference between the wall temperature at about the middle of the parallel section and the temperature of helium at the inlet to that section, ΔT. From measured q and ΔT, the flow rate in the tube was computed by use of equations (6) and (7). Namely, the parallel section was used as a velocity sensor on the basis of those equations.

Experiments were also performed using liquid nitrogen as working fluid. The experiment with nitrogen is much easier than the one with helium, and with the significant reduction of heat leaks from the outside, the accuracy in the estimation of heat input to the flow is expected to improve. The pressure in the reservoir drum was maintained at 0.1 MPa (at the center), and the radial depth of liquid nitrogen in the drum was set at 10 - 20 mm during rotation at the rates of 7.2 (430), 14.2 (850), and 16.7 Hz (1000 rpm). Several combinations of power input to the separate heaters were employed; power input to the heater on the parallel section was increased up to 12 W, and the heaters on the radial

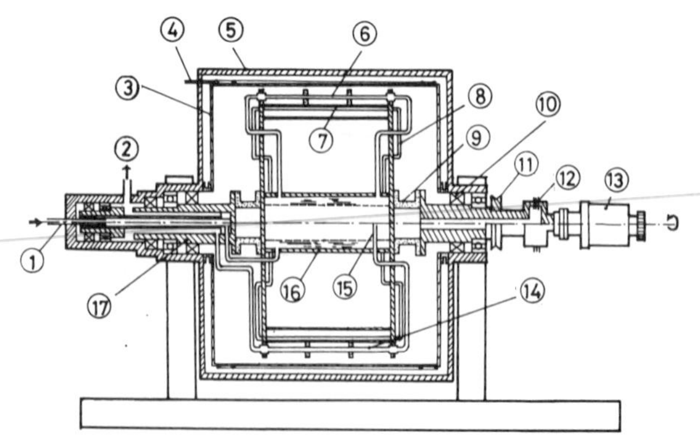

(a) construction

(b) photograph of the test section

FIGURE 9. Experimental apparatus for measurements of the flow rate in a rotating convection loop

sections were powered in a range of up to 7 W. The convection velocity was estimated from the total power input and the temperature rise of nitrogen between the inlet and the outlet of the heated section.

The data are shown in Fig. 10, where the convection velocity in the tube (V) is plotted against the total heat load (Q). Groups of helium data are shown by the rectangles enveloping the scattered data. Among the nitrogen data, soild symbols are the data obtained at 1000 rpm, open circles at 850 rpm, and a triangle at 430 rpm. The solid curves are the predictions obtained as follows.

The driving potential for convection is generated by the density difference between fluids in the two radial sections of the loop.

$$\Delta P = \left[\int \rho r \omega^2 \, dr\right]_{in} - \left[\int \rho r \omega^2 \, dr\right]_{out} , \tag{13}$$

where the first integral is performed over the cold radial leg, and the second over the warm leg. The decrease of fluid density in the warm leg is readily estimated from the heat input to the flow. This pressure head is balanced by the resistance to the flow,

$$\Delta P = \zeta \cdot \frac{\rho V^2}{2} \quad , \tag{14}$$

where ζ is the resistance coefficient. The largest components of ζ in this experiment are the resistance coefficients at the bends.
A bend in the loop I has a radius of curvature of 10 mm, its ratio to the tube's inner radius is 2. There are six such bends on the loop. Equation (12) was used to estimate the increase in the resistance coefficient due to rotation, with ζ_0 determined in a copy of the rotating loop which was built separately for stationary tests. Also taken into account are the resistance coefficients in the straight sections of the loop.

Most of the experimental data of convection velocity fall above the predictions as seen in Fig. 10. Especially in the case of helium, the driving potential is likely to be enhanced by nucleation of bubbles in the warm leg. In spite of this uncertainty, we recommend the use of the developed formulas, the primary one in this experimental situation being equation (12), in the design of a cooling system, since the method gives predictions which are on the conservative side from the designer's point of view. In the design of a cooling system for a real machine, the pressure drop increase in parallel paths due to rotation has to be given a full attention, since it occupies a significant percentage in the total resistance against flow. We recommend the use of the formulas of resistance coefficient developed in (12) for turbulet flow and in (13) for laminar flow.

Convection loop experiments at much higher rotational speeds were performed at the University of Southampton (25 - 28), however, with a much simplified configuration. The loop consists of one or two radial arms, with a small heating element at or near the periphery of a rotation plane. A correlation was proposed in terms of the Nusselt number, the Rayleigh number, and the stratification factor $S = (dT/dr) \cdot (L/\Delta T)$, where dT/dr is the temperature gradient due to the centrifugal compression, and L the charcteristic length of the heater (28).

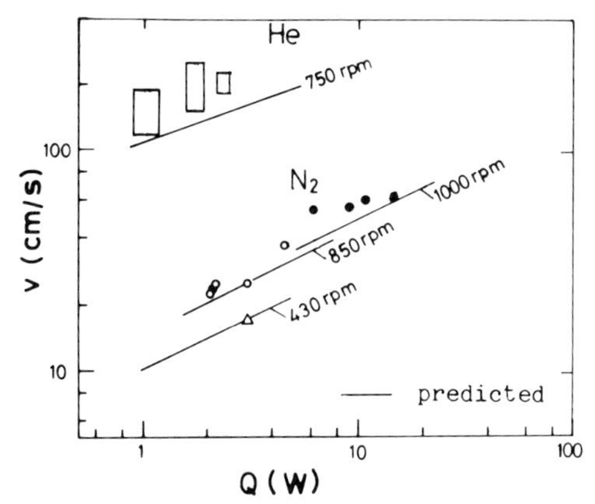

FIGURE 10. Measured and predicted convection velocities vs. heat load

$$Nu = 0.17\ Ra^{0.3}\ (1 + S^{1/3}) \tag{15}$$

The stratification factor was introduced to correlate the data better than by a conventional $Nu - Ra$ correlation (27). Equation (15) may be applicable for a heat dissipating component located directly downstream of a radial coolant channel.

Concern about the flow instability in a rotating loop has been expressed (3, 28). The instability was actually observed with a simple loop of two consecutive radial arms (25). With a proto-type machine of 50 MVA capacity which has been built in the authors' company, no experience about the fluctuating behavior of coolant flow has been reported until now.

CONCLUSIONS

A series of laboratory experiments produced the data of heat transfer which are applicable to the design analysis of a cooling system on the rotor of a superconducting generator. The following recommendations are made on the basis of the present work and the works reported in the literature.

(1) For the heat dissipating components cooled by thermally driven convection of helium in a confined space, heat transfer coefficients can be estimated from equation (4). Where the heat transfer surface is exposed to helium at the end of a long radial channel, equation (15) may be used as another reference source.
(2) The coolant channels for the field windings have the sections extending in parallel to the axis of rotation having large length-to-diameter ratios. The heat transfer coefficients can be estimated from equations (6) - (8).
(3) For successful operation of the scheme of thermally driven convection (self-pumping), care should be exercised to fully account for the increase in the resistance against coolant flows due to rotation. Equation (12) is proposed to estimate the increase in resistance coefficients at the bends which are present along the coolant path. For the resistance coefficients in the channels running parallel to the axis of rotation, refer to (12, 13).

More works are needed to make the recommendations on a more solid basis, especially with a rotor rig capable of rotating at high speeds. Such an apparatus requires the complexity of a real machine, therefore, would become highly expensive. A note is also made here of the need to investigate the flow and heat transfer in channels having cross sections other than circular ones.

Acknowledgement

This paper presents a summary of the works conducted by the heat transfer research group at the Mechanical Engineering Research Laboratory, HITACHI, LTD. A full account is to be found in the latter author's (H. Ogata) Ph. D. dissertation submitted to Nagoya University. The authors thank our co-workers, particularly Messrs K. Fujioka and A. Yasukawa, for their cooperation.

Nomenclature

a = $r\omega^2$, centrifugal acceleration
Cp = specific heat at constant pressure
D = diameter of the heating surface
d = inside diameter of a coolant channel
g = natural gravity
Gr = Ra/Pr, rotational Grashof number
k = thermal conductivity
l = length of a coolant channel
$\dot{m}$ = mass velocity
Nu = Nusselt number, equation (1)
Pr = Prandle number
ΔP = pressure drop
q = heat flux
r = radius of rotation
Ra = Rayleigh number based on a, equation (2)
Re = Reynolds number
ΔT = temperature difference
T = temperature
V = mean velocity of coolant
β = coefficient of volumetric expansion
Γ = parameter, equation (5)
ω = angular speed of rotation
μ = viscosity
ρ = density
τ = temperature gradient along the tube axis
ζ = $\Delta P/(\rho V^2/2)$, resistance coefficient

REFERENCES

1. Woodson, H. H., Smith, Jr., J. L., Thullen, P., and Kirtley, J. L., The Application of Superconductors in the Field Windings of Large Synchronous Machines, IEEE Trans., Vol. PAS-90, no. 2, pp. 620-627, 1971.
2. Thullen, P., Dudley, J.C., Greene, D. L., Smith, Jr., J. L., and Woodson, H. H., An Experimental Alternator with a Superconducting Rotor Field Winding, IEEE Trans., Vol. PAS-90, no. 2, pp. 611-619, 1971.
3. Bejan, A., Refrigeration for Rotating Superconducting Windings of Large A-C Electric Machines, Cryogenics, Vol. 16, no. 3, pp. 153-159, 1976.
4. Mori, Y., and Nakayama, W., Secondary Flows and Enhanced Heat Transfer in Rotating Pipes and Ducts, Heat Transfer in Rotating Machinery, ed. D. E. Metzger and N. H. Afgan, pp. 3-24, Hemisphere, Washington, D. C., 1984.
5. Morris, W. D., Heat Transfer and Fluid Flow in Rotating Coolant Channels, John Wiley & Sons, Chichester, England. (Research Studies Press), 1982.
6. Ogata, H., and Nakayama, W., Heat Transfer to Subcritical and Supercritical Helium in Centrifugal Acceleration Fields, 1. Free Convection Regime and Boiling Regime, Cryogenics, Vol. 17, no. 4, pp. 461-470, 1977.
7. McCarty, R. D., Thermophysical Properties of Helium -4 from 2 to 1500 K with Pressures to 1000 Atmosheres, NBS TN-631, National Bureau of Standards, 1972.
8. Scurlock, R. G., and Thornton, G. K., Pool Heat Transfer to Liquid Helium in High Centrifugal Acceleration Fields, Int. J. Heat Mass Transfer, Vol. 20, no. 1, pp. 31-40, 1977.
9. Haseler, L. E., Ball, A., and Scurlock, R. G., Experiments on Helium Refrigerant at High Speed of Rotation, Proc. ICEC 6, pp. 414-417, IPC Science

and Technology Press, 1976.
10. Fujii, T., and Imura, H., Natural Convection Heat Transfer from a Plate with Arbitrary Inclination, Int. J. Heat Mass Transfer, Vol. 15 no. 4 pp. 755-767, 1972.
11. Ogata, H., Fujioka, K., Nakayama, W., and Sato, S., Thermal Design of a Crogenic Rotor for a 50 MVA Superconducting Generator, Proc. ICEC 7, pp. 276-281, IPC Science and Technology Press, 1978.
12. Nakayama, W., Forced Convective Heat Transfer in a Straight Pipe Rotating Around a Parallel Axis (2nd Report, Turbulent Region), Int. J. Heat Mass Transfer, Vol. 11, no. 7, pp. 1185-1201, 1968.
13. Mori, Y., and Nakayama, W., Forced Convective Heat Transfer in a Straight Pipe Rotating Around a Parallel Axis (1st Report, Laminar Region), Int. J. Heat Mass Transfer, Vol. 10, no. 9, pp. 1179-1194, 1967.
14. Sleicher, C. A., and Rouse, M. W., A Convenient Correlation for Heat Transfer to Constant and Variable Property Fluids in Turbulent Pipe Flow, Int. J. Heat Mass Transfer, Vol. 18, no. 5, pp. 677-683, 1975.
15. Trefethen, L., Flow in Rotating Radial Ducts, GE Report No. 55 GL 350-A, General Electric Company, 1957.
16. Ito, H., and Nambu, K., Flow in Rotating Straight Pipes of Circular Cross Section, Trans. ASME, Ser. D, J. Basic Engineering, Vol. 93, no. 3, pp. 383-394, 1971.
17. Mori, Y., Fukuda, T., and Yanatori, M., Forced Convective Heat Transfer in Centrifugal Fields (2nd Report), Trans. JSME, Series 2, Vol. 39, no. 324, pp. 2484-2485, 1973.
18. Woods, J. L., and Morris, W. D., An Investigation of Laminar Flow in the Rotor Windings of Directly-Cooled Electric Machines, J. Mech. Eng. Sci., Vol. 16, no. 6, pp. 408-417, 1974.
19. Mori, Y., and Uchida, Y., Study on Forced Convective Heat Transfer in Horizontal Flat Plate Channels, Trans. JSME, Vol. 31, no. 230, pp. 1511-1520, 1965.
20. Levy, E., Brown, G., Neti, S., and Kadambi, V., Laminar Pressure Drop in a Heated Rectangular Duct Rotating About a Parallel Axis, ASME Paper no. 84-WA/HT-59, 1984.
21. Schnapper, C., and Hofmann, A., Heat Transfer to Helium in Rotating Channels with Natural and Forced Convection, International Institute of Refrigeration Conference, Commission A1-2, Zurich, 1978.
22. Nakayama, W., Fujioka, K., and Watanabe, S., Flow and Heat Transfer in the Water-Cooled Rotor Winding of a Turbine Generator, IEEE Trans. Vol. PAS-97, no. 1, pp. 225-231, 1978.
23. Murakami, M., and K. Kikuyama, I., Flow in Pipes Under the Influence of Coriolis Forces, Trans. JSME, Series 2, Vol. 37, no. 302, pp. 1959-1971. 1971.
24. Csillag, I. K., Experimental Study of Water Flow in Full-Size Test Rig for Water-Cooled Turbo-Generator Rotor, Proc. Institute Mechanical Engineers, Vol. 181, pt. 1, no. 3, pp. 53-73, 1966-67.
25. Scurlock, R. G., Stevens, F. A., and Utton, D. B., The Zero Mechanical Work Thermal Pump in the High Speed Rotating Frame and Its Associated Instabilities, Proc. ICEC 7, pp. 373-377, IPC Business Press, 1978.
26. Khan, W. I., and Scurlock, R. G., Heat Transfer to Rotating Helium at Tip Speeds of 120 m/s, Proc. ICEC 8, pp. 623-627, IPC Science and Technology Press, 1980.
27. Scurlock, R. G., and Utton, D. B., A Modified Correlation for Heat Transfer by Natural Convection to Liquid, Proc. ICEC 7, pp. 399-402, IPC Business Press, 1978.
28. Khan, W. I., Heat Transfer and Flow Instability in a Superconducting Machine Rotating at 3000 RPM, Cryogenics, Vol. 24, no. 1, pp. 11-14, 1984.

Recovery Effects on Heat Transfer Characteristics within an Array of Impinging Jets

L. W. FLORSCHUETZ and C. C. SU
Department of Mechanical and Aerospace Engineering
Arizona State University, Tempe, Arizona

INTRODUCTION

When impinging jets are utilized for internal cooling of gas turbine components the overall cooling scheme configuration may be such that the jets are subject to a crossflow. Even if the cooling air is supplied to the component at a single temperature, the crossflow air approaching a jet may be at a higher temperature than the jet air because of upstream heat addition to the air comprising the crossflow. In addition to the effect of the crossflow on the flow field of the impinging jet, which may in turn affect the heat rate at the impingement surface even if the crossflow temperature is identical to the jet temperature, there will also be the effect of the crossflow temperature relative to the jet temperature on the impingement surface heat rate.

Most prior studies of heat transfer to single impinging jets or single spanwise rows of impinging jets subject to a crossflow were performed with the crossflow temperature essentially identical to the jet temperature [1,2,3]. Bouchez and Goldstein [4], however, did study the effect on impingement heat transfer of crossflow temperature relative to jet temperature for a single circular jet.

Two-dimensional arrays of circular jets impinging on a heat transfer surface opposite the jet orifice plate produce conditions in which individual jets or rows of jets in the array are subject to a crossflow the source of which is other jets within the array itself. In gas turbine applications the flow from the jets is often constrained to exit essentially in a single direction along the channel formed by the jet orifice plate and the impingement surface. Experimental studies of impingement surface heat transfer for such configurations were reported, for example, by Kercher and Tabakoff [5], Florschuetz et al. [6], and Saad et al. [7]. In these studies the effect of the temperature of the crossflow approaching a spanwise jet row within the array relative to the jet temperature was not explicitly determined. In fact, such a determination cannot be made from this type of test when the jet air source is from a single plenum and the form of the thermal boundary condition at the impingement surface (e.g., uniform temperature or uniform flux) is fixed. Under such conditions the crossflow temperature approaching a given spanwise row within the array cannot be independently varied.

Saad et al. [7] reported spanwise average, streamwise resolved Nusselt numbers for one array geometry at a single jet flow rate for three different initial crossflow rates approaching the array from upstream extension of the channel formed by the jet orifice plate a the impingement surface. The magnitude of the initial crossflow temperature relative to the jet temperature was not indicated. Presumably the temperature of the air in the initial crossflow plen was the same as that for the air in the main jet array plenum.

Florschuetz, Metzger, and Su [8] reported experimental results for two-dimensional arrays of circular jets with an initial crossflow approaching the array (Fig. 1). The initial crossflow originated from a separate plenum so that its flow rate and temperature could independently controlled relative to the jet flow rate and temperature. Spanwise average, streamwise resolved (regional average) Nusselt numbers and values of a parameter, η, representing the influence of initial crossflow temperature relative to jet temperature were determined as a function of overall array flow parameters for a range of geometric parameter values.

Subsequently, the data was further analyzed in an attempt to determine regional average Nusselt numbers and η values defined solely in terms of parameters associated with the individual spanwi row opposite the given region. The objective was to determine if t application of the results in this form could be generalized to app to individual rows of a larger class of arrays or sub-arrays having similar geometry but an arbitrary number of spanwise rows.

It had been concluded that for the test conditions utilized recove effects were not significant in influencing the Nusselt numbers an the η values defined in terms of overall array parameters. Based o this prior conclusion, the early analysis of the data in terms of individual row parameters also did not separately consider recovery effects. Apparently anomalous behavior of the reduced data (particularly the η parameter) for several geometric parameter sets led to the realization that although the influence of the recovery effect was normally small when considered relative to overall arra parameters, the same was not always true when considered in terms o individual row parameters.

A reformulation of the data reduction scheme so as to account for

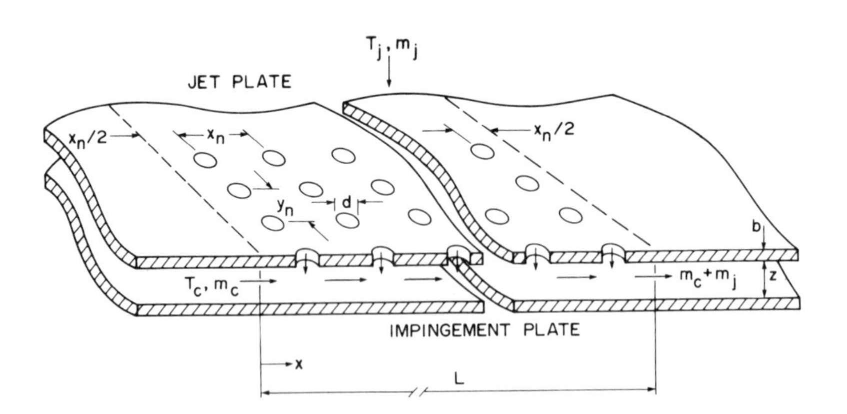

FIGURE 1. Basic test model geometry.

recovery effects (combined with the use of some additional test results) eliminated the anomalous behavior, and permitted the evaluation of recovery factors. This paper presents illustrative results for regional average recovery factors associated with individual spanwise rows within a two-dimensional array as a function of flow and geometric parameters. The significance of recovery effects as they affect regional average heat transfer characteristics at the impingement surface within a two-dimensional array are also examined.

FORMULATION OF THE PROBLEM

The basic test model geometry and nomenclature are shown schematically in Fig. 1. Most of the jet arrays tested in the presence of an initial crossflow had uniform inline hole patterns as illustrated in Fig. 1. However, two jet arrays were also tested for corresponding staggered patterns in which alternate spanwise rows were offset by one-half a spanwise hole spacing.

For steady-state conditions, heat rates over regional areas covering the entire span of the impingement surface, but having a streamwise length of one streamwise hole spacing, and centered opposite spanwise hole rows could be measured. Thus, the regional average heat flux, $\bar{q}$, associated with any given spanwise row of the array (Figs. 2 and 3) could be determined. It was desired to obtain the basic set of heat transfer characteristics for the case of constant fluid properties. Hence, the tests were conducted at relatively small temperature differences; e.g., the maximum surface-to-jet-temperature difference utilized was about 35 K.

Overall Array Domain

First consider the regional heat fluxes as a function of parameters associated with the overall array (Fig. 2). The total jet flow rate (m_j) and the initial crossflow rate (m_c) are specified. The mixed-mean total temperature of the jet flow at the jet exit plane (T_j) and the total temperature of the initial crossflow (T_0) at the entrance to the array (x=0) are also specified since these mixed-mean total temperatures will normally be available based on energy balances carried out upstream of these boundaries. Since in the present case the objective is to cool the surface by designing impinging jets into the system, the jet flow (always present) is considered to be the primary flow, while the initial crossflow (which, in general, may or may not be present) is considered the secondary flow. Thus, it is convenient to consider (T_s - T_j) as the

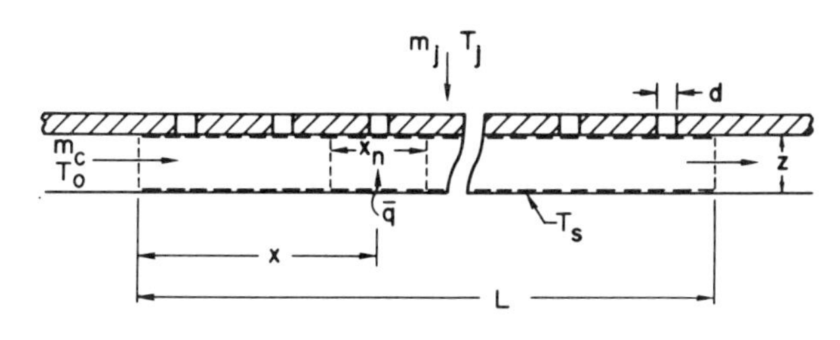

FIGURE 2. Definition of overall array domain and associated parameters.

primary temperature potential and consider the condition T_0 differe
from T_j as a secondary effect. Working from the differential energ
equation and boundary conditions written in terms of total
temperature for the overall array domain indicated in Fig. 2,
retaining the dissipation term, but assuming constant fluid
properties and a uniform specified impingement surface temperature
(T_s), it may be shown [9] using linear superposition arguments that
the regional average heat flux can be expressed in the form

$$\overline{q} = (k/d)Nu[(T_s - T_j) - \eta(T_0 - T_j)] + \epsilon \quad (1)$$

The constant ϵ is the heat flux which would occur if all three
temperatures were equal. $Nu = hd/k$ may be regarded as the Nusselt
number and h as the heat transfer coefficient for the special case
when $T_0 = T_j$ and recovery effects are absent; and η may be regarded
as a fluid temperature influence factor reflecting the strength of
the influence on the heat flux when T_0 differs from T_j. The jet ho
diameter is d and the fluid thermal conductivity is k.

It is customary in the heat transfer literature when considering
recovery effects to define a recovery temperature and a normalized
form, the recovery factor. The recovery temperature is normally
defined as the steady-state surface temperature corresponding to a
zero surface heat flux. However, in the present problem, the zero
heat flux surface temperature for a given jet temperature will not
only be influenced by recovery effects but also by the level of the
crossflow temperature relative to the jet temperature. It is also
noted that we are considering a uniform impingement surface
temperature boundary condition and are concerned with regional
average heat fluxes. Therefore, the recovery temperature, T_{rec}, is
defined as the surface temperature for a zero mean heat flux, $\overline{q}$ =
0, under the condition that T_0 is the same as T_j. In the present
problem the recovery factor could be defined in terms of either
characteristic crossflow or characteristic jet flow conditions. It
is here defined in terms of jet flow conditions because the jet flo
is considered the primary flow:

$$r = (T_{rec} - t_j)/[(G_j/\rho)^2/2c_p] \quad (2)$$

where t_j is the static temperature and ρ is the density at the jet
exit plane. Considering Eq. (1) with the above definitions of T_{rec}
and r, r may be expressed in terms of ϵ as

$$r = 1 - (\epsilon/h)/[(G_j/\rho)^2/2c_p] \quad (3)$$

By dimensional analysis based on the governing differential equatio
and boundary conditions for the velocity and temperature fields it
may be shown [9] that the three dimensionless parameters, Nu, η, an
r, for computing regional average heat fluxes based on Eq. (1) may
considered to depend at least on the following parameters associate
with the overall array:

Geometric parameters (x/L, x_n/d, y_n/d, z/d, L/x_n)

Flow and fluid parameters ($\overline{Re}_j$, m_c/m_j, Pr)

Here $\overline{Re}_j = \overline{G}_j d/\mu$ is the array mean jet Reynolds number, where $\overline{G}_j$ is
the mean jet mass flux over the array, and Pr is Prandtl number. I
addition there is a dependence on the normalized velocity and

temperature profiles of the initial crossflow at the entrance to the array, but this effect turns out to be insignificant except in some cases when the initial crossflow is dominant. In general, there may also be a dependence on hole pattern; i.e. inline vs. staggered.

Individual Spanwise Row Domain

Now consider $\bar{q}$ as a function of parameters associated with the domain of individual spanwise row *n* as specified in Fig. 3. The mass flux at the jet exit plane (G_j) and the mean mass flux for the crossflow approaching row *n* (G_c) are specified. As in the case of the overall array domain the mixed-mean total temperature at the jet exit plane (T_j) is specified as the characteristic jet temperature. The characteristic crossflow temperature is specified as the mixed-mean total temperature ($T_{m,n}$) at the channel cross-section located at the upstream edge of the impingement suface region immediately opposite row *n*; i.e., one-half a streamwise hole spacing upstream of row *n*. In terms of these parameters, the regional average heat flux opposite row *n* may be expressed as [9]

$$\bar{q} = (k/d)Nu_r[(T_s - T_j) - \eta_r(T_{m,n} - T_j)] + \epsilon_r \qquad (4)$$

The subscript *r* is used on the dimensionless parameters $Nu_r = h_r d/k$, η_r, and r_r to distinguish them from the corresponding quantities previously considered as a function of overall array parameters. A recovery temperature $T_{rec,r}$ and recovery factor r_r are defined in analogous fashion to T_{rec} and r. The relationship between r_r, $T_{rec,r}$, ϵ_r, and h_r is the same as given by Eqs. (2) and (3) for the unsubscripted quantities. Nu_r, η_r, and r_r may be considered as functions of the following individual row parameters [9]:

Geometric parameters (x_n/d, y_n/d, z/d)

Flow and Fluid parameters (Re_j, G_c/G_j, Pr)

where $Re_j = G_j d/\mu$. Here also there will, in general, be a dependence on the normalized velocity and temperature profiles of the crossflow at the entrance to the individual row domain.

EXPERIMENTAL APPROACH AND DATA REDUCTION

The details of the experimental facility were reported previously [8]. The basic experimental approach as it relates to the problem formulation summarized above will be outlined here. For a given test run, a basic jet orifice plate geometry and channel height was selected. These are specified in the following form: $B(x_n/d, y_n/d, z/d)I$ where B designates the particular jet plate length

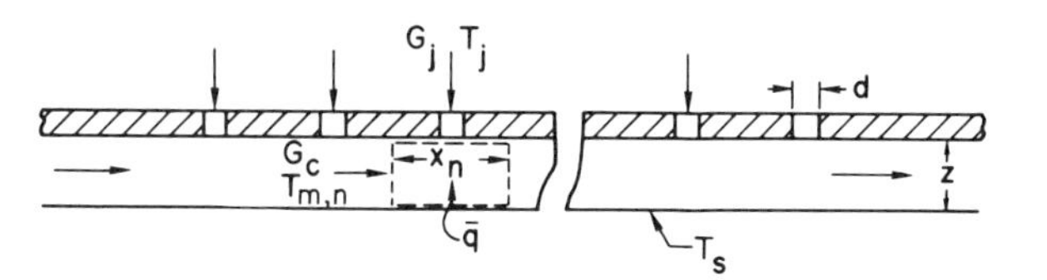

FIGURE 3. Definition of individual spanwise row domain and associated parameters.

(L = 12.7 cm) and I designates the inline hole pattern for which mo of the initial crossflow tests were conducted (S is used to designa a staggered hole pattern). Typically the jet plates had ten spanwi rows of holes (L/x_n = 10). Jet hole diameters were either 0.127 or 0.254 cm. Once the geometry was fixed, setting the total jet flow rate and the initial crossflow rate resulted in a set of fixed valu for all of the independent dimensionless parameter sets summarized the preceding section. The measured distributions of the jet and crossflow mass fluxes (G_j and G_c) over the spanwise rows, needed fo representing heat transfer characteristics in terms of individual r parameters, were reported in [10]. The exhaust pressure at the exi of the jet array channel was one atmosphere.

Referring to Eq. (1) it is clear that for the fixed conditions described in the preceding paragraph, measurement of three independent data sets ($\overline{q}, T_s - T_j, T_0 - T_j$) would permit determination of Nu, η, and ϵ. Originally these data sets were obtained as follows. The value of T_j was fixed nominally at ambient temperature level. Then a uniform maximum T_s was set such that $T_s - T_j \simeq 35$K by individually adjusting $\overline{q}$ at each copper segment of the heat transfer test plate, including those in the initial crossflow channel. The segment surface areas of length x_n in the streamwise direction were the areas over which the regional average heat fluxe could be controlled and measured. The initial crossflow plenum temperature was fixed roughly midway between T_j and the maximum T_s. This condition gave one data set. Keeping T_j and the initial crossflow plenum temperature fixed, a second set was obtained by adjusting each $\overline{q}$ to roughly half the prior values, and a third set by cutting $\overline{q}$ essentially to zero. T_0 was determined for each data set by an energy balance over the initial crossflow channel utilizi the measured heat inputs and initial crossflow plenum temperature. In the original data reduction an equation in the form of (1) was utilized, but with ϵ neglected. Thus, one of three data sets was redundant and was used to check the calulated results for Nu and η, with typically good consistency. This was taken as an indication that there was no need to account for recovery effects when evaluating Nu and η.

When this data was reanalyzed to obtain heat transfer characteristi in terms of individual row parameters based on Eq. (4) with $T_{m,n}$ evaluated via an energy balance, ϵ_r was neglected just as ϵ had bee Results for Nu_r and η_r again typically showed good consistency when the redundant data set was utilized. However, anomalous behavior o these results for some cases (detailed later in the paper), observe when they were plotted against the flow parameters for the individu rows, led to the realization that something was awry in spite of th apparent consistency of the redundant data sets. At this point the data was reanalyzed using Eq. (4) with ϵ_r included. Three data set permitted the calculation of Nu_r, η_r, and ϵ_r, but the results in sc cases then showed highly random scatter with some completely unrealistic magnitudes. It was concluded that the three originally obtained data sets were ill-conditioned in these cases. Hence, one of the three data sets for each case had to be replaced. The thir set obtained with $\overline{q}$ set essentially at zero was rerun, but this time the initial crossflow plenum temperature was maintained at a value approximately the same as the jet temperature. The first an second sets were retained for use in conjunction with the newly obtained third set. This combination was not ill-conditioned, and led to well behaved results for Nu_r and η_r, as well as values for ϵ

The revised combination of data sets was then also reduced using Eq. (1) to obtain values of Nu, η, and ϵ in terms of overall array parameters. There was only a small influence on the results for Nu and η compared to the values from the prior combination of data sets.

With ϵ_r (and ϵ) determined, r_r (and r) were calculated using values of ρ based on one-dimensional adiabatic flow through the jet holes. Values of r and r_r were essentially identical. Note that the determination of Nusselt numbers and η values is independent of ρ.

RESULTS AND DISCUSSION

Illustrative results for Nu and η as a function of streamwise location measured from the entrance to the array are shown in Fig. 4a for a (10,8,1)I jet array with ten spanwise rows of holes. Because

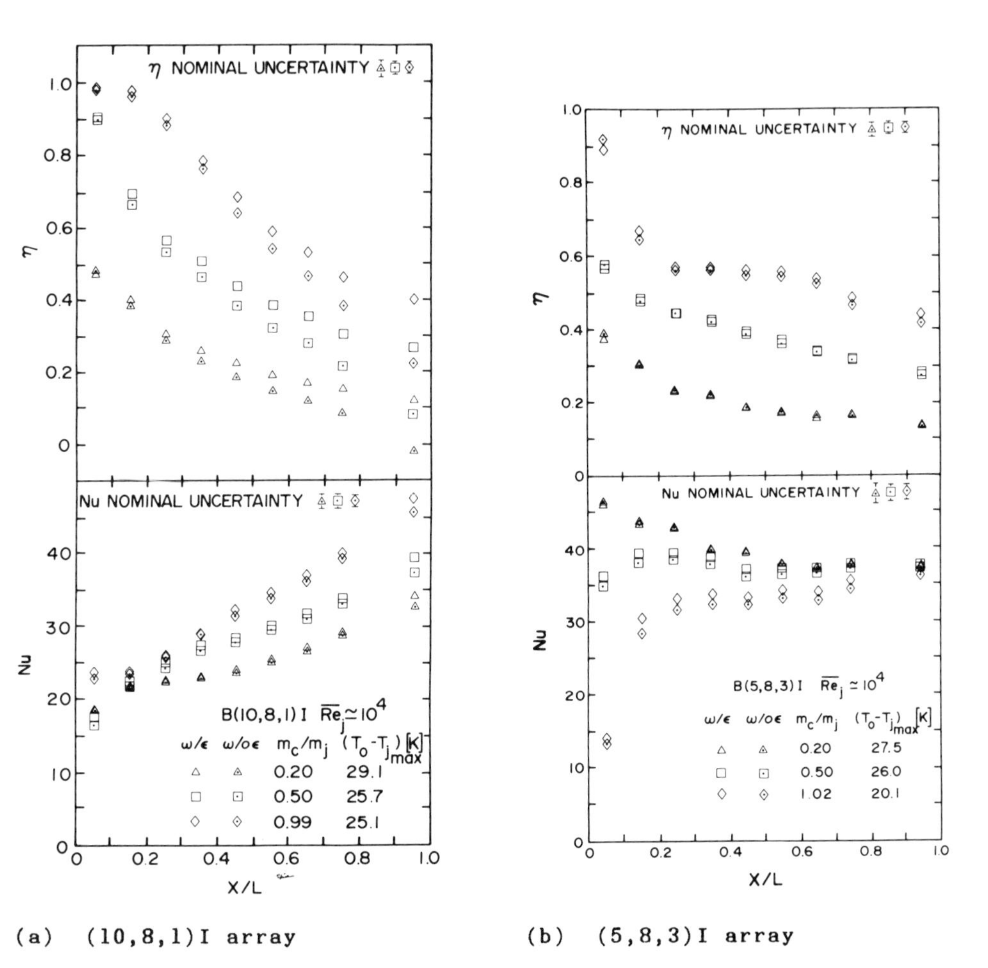

(a) (10,8,1)I array (b) (5,8,3)I array

FIGURE 4. Comparisons of Nu and η defined in terms of overall array parameters with and without retaining ϵ in data reduction.

of the narrow channel height of one hole diameter for this geometry the jet flow distribution is highly nonuniform [10]. Even for the smallest m_c/m_j of 0.2 the jet mass flux at the last (downstream) ro is about twice the value at the first (upstream) row, while for th largest initial crossflow tested, m_c/m_j of unity, the same factor i about seven. Even for this case with large downstream jet and crossflow velocities, the effect on Nu of including ϵ (open data symbols, denoted w/ϵ in Fig. 4a) versus neglecting ϵ (symbols with center point, denoted $w/o\epsilon$) in the data reduction via Eq. (1) is within experimental uncertainty. The effect on η becomes increasingly noticeable downstream, the maximum difference being about 0.18. For a specified heat flux, jet temperature, and initia crossflow temperature the corresponding change in magnitude of the calculated surface temperature would be 18% of the initial crossflo to-jet temperature difference. This may or may not be significant depending on the particular design application.

The same type of comparison is shown in Fig. 4b at the same mean je Reynolds number for a jet array geometry, (5,8,3)I, for which the j flow distribution remains approximatily uniform, even at the larges initial crossflow rate. In this case the effect on both Nu and η o retaining versus neglecting ϵ is not significant. This was typical for most of the initial crossflow test cases.

Now consider the heat transfer parameters Nu_r and η_r defined for th domain of an individual spanwise row. These parameters are shown plotted as a function of G_c/G_j in Figs. 5 and 6 for the same two je array geometries discussed above. Since the raw data was drawn fro tests conducted at a fixed mean jet Reynolds number for the array, the jet Reynolds numbers associated with individual spanwise rows varied depending on the jet flow distribution for the given geometr and m_c/m_j. It had been found from prior noninitial crossflow resul [6] and verified from initial crossflow tests that the Nusselt numb dependence on Reynolds number could be accounted for according to $Re_j^{0.73}$. This interpolation formula was used to adjust the Nu_r values from a given test to the same individual row Reynolds number $Re_j = 10^4$, for examining the dependence on G_c/G_j as in Figs. 5 and In the case of η_r, it was found to be essentially insensitive to Re to within experimental uncertainty, so that no adjustment to these values was made for plotting against G_c/G_j.

The comparisons shown in Fig. 5 for the (10,8,1)I jet array show th the effect of neglecting ϵ_r in data reduction based on Eq. (4) has little effect on Nu_r, but the effect on η_r for this extreme case is quite significant. The η_r values reduced neglecting ϵ_r (those with center point shown) indicate a behavior as a function of G_c/G_j that appears quite irregular. However, when ϵ_r is retained the η_r value (the open points) form a pattern through which a monotonic curve could reasonably be drawn which could be extrapolated to zero as G_c/G_j goes to zero and asymptotically approaches one as G_c/G_j goes infinity. An exception is the point from the $m_c/m_j = 0.50$ case (th square points) at the first row of the array. This may be attribut to the effect of the normalized velocity and temperature profiles a the entrance to the domain associated with each individual row. These profiles would be spanwise uniform approaching the first row the array (the initial crossflow), but not approaching rows within the array. The effect is not noticeable for the $m_c/m_j = 0.20$ and cases because in the first case G_c/G_j is small at the first row and the influence of the crossflow is small, while in the second case

G_c/G_j is very large over the first three rows and the crossflow completely dominates resulting in $\eta_r = 1$. This dominance of the crossflow is verified by the Nu_r results for these rows which agree quite closely with prior results for a fully developed turbulent flow in a parallel plate channel with one side heated [11]. For the (5,8,3)I jet array the results in terms of individual row parameters shown in Fig. 6 show little effect of the neglect of ϵ_r.

Attention is now directed to typical results for recovery factors. Values of r_r resulting from tests with the (5,8,3)I geometry are plotted in Fig. 7 as a function of Re_j for several fixed values of

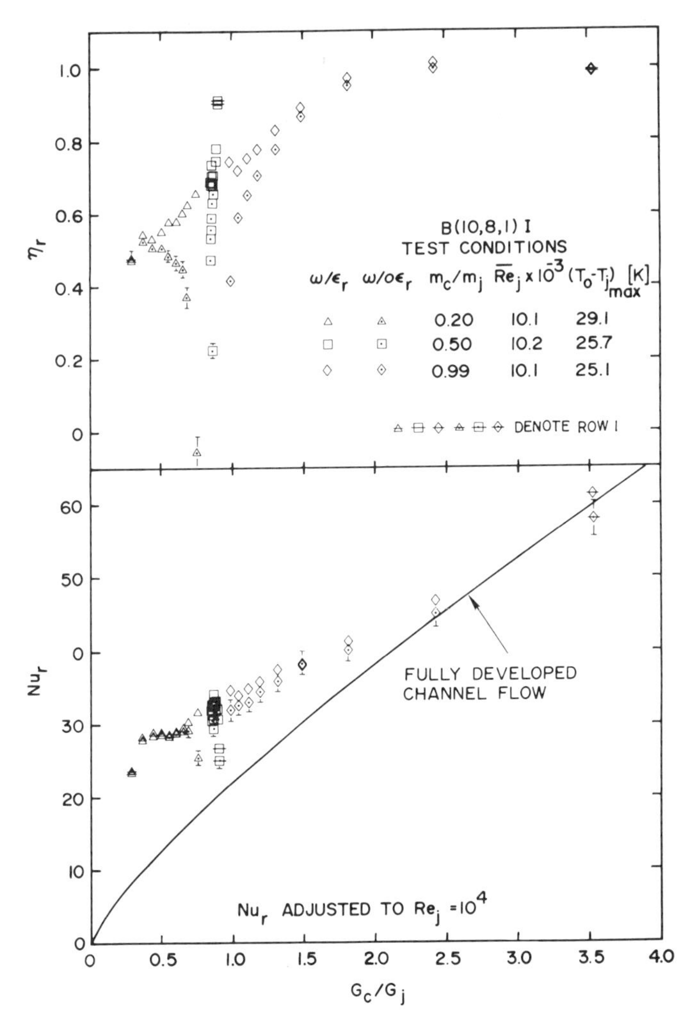

FIGURE 5. Comparisons of Nu_r and η_r defined in terms of individual spanwise row parameters with and without retaining ϵ_r in data reduction — (10,8,1)I array.

G_c/G_j. Except for the first row of the array the recovery factors appear independent of Re_j. Results for other geometries also showe the recovery factor essentially independent of Reynolds number, except in some cases for the first row values.

Based on the above conclusion the effect of G_c/G_j on r_r could be examined without regard to Re_j. The effect of G_c/G_j is illustrated in Fig. 8 for the (5,8,3)I geometry and in Fig. 9 for the (10,8,1)I geometry. For (5,8,3)I the value of r_r considering experimental uncertainties appears to fall essentially at unity, with the exception of row 1. For this geometry the value of G_c/G_j does not exceed 0.6. A regional average value of unity seems quite reasonab for a row of jets within an array subject to a relatively small crossflow velocity. At row 1 values larger than unity are obtained but it appears that a curve drawn through these points would extrapolate to unity as G_c/G_j goes to zero. Since the recovery factor is defined in terms of a recovery temperature for equal mixe mean total temperatures of the two fluid streams, the static temperature of the crossflow will be larger than that of the jet fl for G_c/G_j less than unity. With these two fluid streams mixing as they interact with the surface it is possible that the recovery temperature could acheive a value greater than the total temperatur of the jet flow. Local recovery factors greater than unity for a single jet in a crossflow were reported by Sparrow et al [2] who

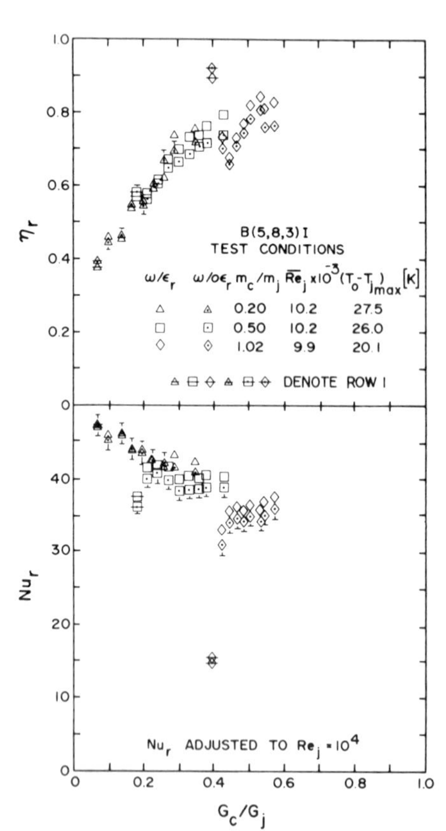

FIGURE 6. Comparisons of Nu_r an η_r defined in terms of individual spanwise row parameters with and without retaining ϵ_r in data reduction — (5,8,3) array.

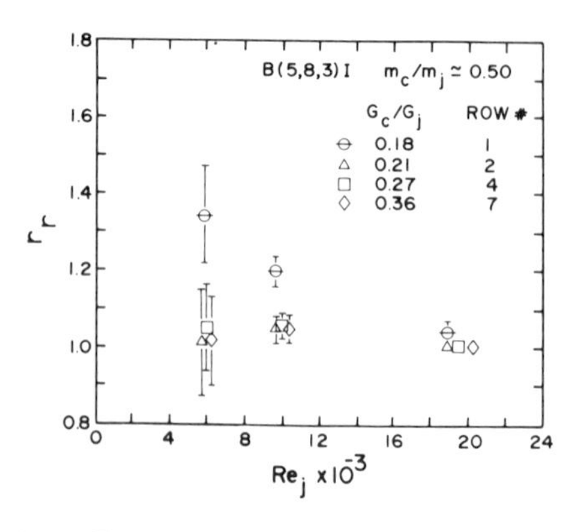

FIGURE 7. Effect of jet Reynolds number on recovery factor.

suggested a similar explanation. A single jet in a crossflow or jets in the first row of an array with an initial crossflow both have spanwise uniform crossflow streams approaching. Rows within an array, i.e., rows downstream of the first row, even when subject to a crossflow at the same mixed-mean total temperature as the jet flow, tend to mix most directly with the crossflow originating from the immediately upstream jet rows. This may explain why the recovery factors for downstream rows within an array sometimes differ from the values at the first row. That is, as noted in the problem formulation section, the recovery factor values will, in general, depend on the normalized velocity and temperature profiles of the approaching crossflow. As already noted, these profiles for downstream rows within the array will certainly be different from the spanwise uniform profiles approaching the first row.

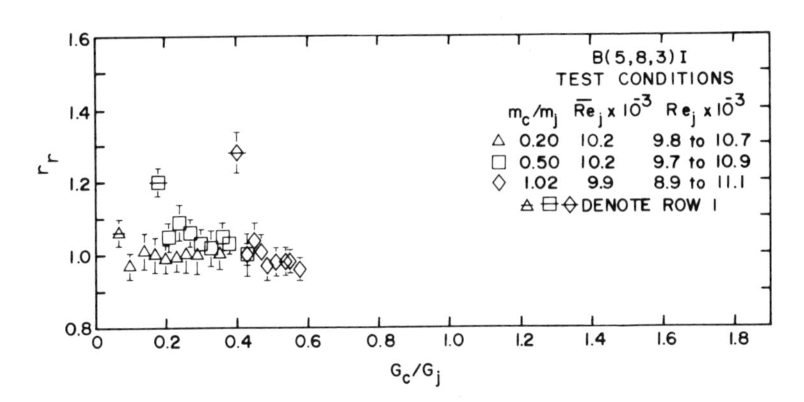

FIGURE 8. Effect of crossflow-to-jet mass flux ratio on recovery factor – (5,8,3)I array.

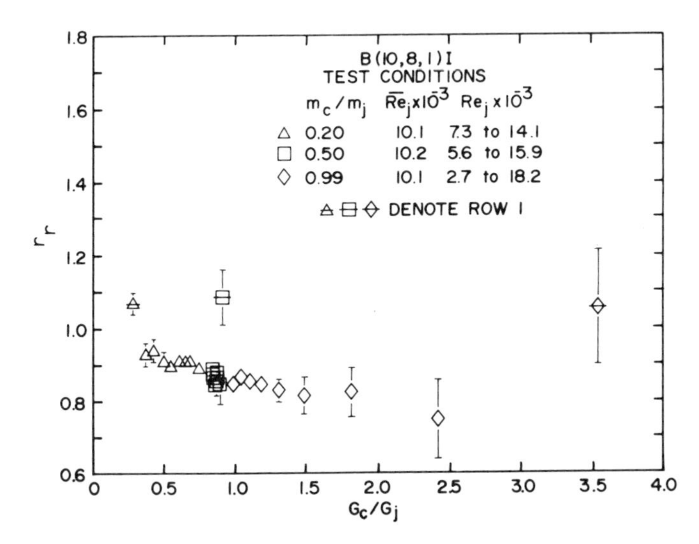

FIGURE 9. Effect of crossflow-to-jet mass flux ratio on recovery factor – (10,8,1)I array.

Recovery factors for the (10,8,1)I geometry (Fig. 9) are slightly below unity (row 1 excepted) for the smaller values of G_c/G_j and decrease slowly with increasing G_c/G_j. This decreasing r_r as the crossflow becomes more dominant is quite reasonable. The value of of about 0.85 in the vicinity of Gc/Gj=1 is reasonably consistent with results for boundary flows with free streams parallel to the surface and also with results for channel flows such as the circula duct results reported by McAdams et al [12]. Note that a curve dra through these points would again extrapolate to unity as G_c/G_j goes to zero and the jet flow dominates. The row 1 values are again greater than unity with the explanation for the values at G_c/G_j les than unity possibly being similar to that noted in the previous paragraph. However, it should be recognized that as G_c/G_j increase the definition of r_r in terms of jet flow parameters becomes inappropriate since the crossflow becomes dominant and approaches t characteristics of a pure channel flow.

The recovery factor results presented above are typical of those found for other geometries tested including two corresponding staggered hole patterns for which the values were essentially the same as for their inline counterparts [9].

CONCLUDING REMARKS

Illustrative results for the three heat transfer parameters Nusselt number (Nu_r), temperature difference influence factor (η_r), and recovery factor (r_r) associated with individual spanwise rows of a two-dimensional array of impinging jets have been presented. Becau of the large temperature differences involved recovery effects will normally not be of significance in the gas turbine application of such results. The possible importance of considering recovery effects in determining heat transfer coefficients for jet impingeme based on data from tests conducted at nominally ambient pressure an temperature levels with relatively small temperature differences is generally well recognized (usually accounted for by using a recover temperature in defining the heat transfer coefficient). However, i cases where the jet is subject to a crossflow having a different temperature from the jet flow, the possible significance of recover effects is perhaps not as fully recognized. That is, when one is attempting to isolate the influence of crossflow temperature relati to jet temperature, the importance of recovery effects may take on added significance. The separation of recovery effects and crossfl temperature effects for a single impinging jet in a crossflow was addressed in the work of Bouchez and Goldstein [4]. The present results obtained for two-dimensional jet arrays emphasize rather dramatically that under some conditions where recovery effects may not be significant in evaluating Nusselt numbers, these effects may be quite significant in evaluating crossflow temperature effects (a represented here by the η_r parameter). It is also evident that values of the individual row parameters Nu_r, η_r, and r_r evaluated f a given set of the independent individual row parameters (Re_j, G_c/G_j x_n/d, y_n/d, z/d) may differ significantly for the first upstream row an array as compared with those for downstream rows within the arra This is attributed to differences in the normalized velocity and temperature profiles approaching the respective jet rows. After th first row this effect appears to diminish very rapidly.

REFERENCES

1. Metzger, D.E., and Korstad, R.J., Effects of Crossflow on Impingement Heat Transfer, *Journal of Engineering for Power*, vol. 94, pp. 35-41, 1972.

2. Sparrow, E.M., Goldstein, R.J., and Rouf, M.A., Effect of Nozzle-Surface Separation Distance on Impingement Heat Transfer for a Jet in a Crossflow, *Journal of Heat Transfer*, vol. 97, pp. 528-533, 1975.

3. Goldstein, R.J. and Behbahani, A.I., Impingement of a Circualar Jet with and without Cross Flow, *Int. J. Heat Mass Transfer*, vol. 25, pp. 1377-1382, 1982.

4. Bouchez, J.P. and Goldstein, R.J., Impingement Cooling from a Circular Jet in a Crossflow, *Int. J. Heat Mass Transfer*, vol. 18, pp. 719-730, 1975.

5. Kercher, D.M., and Tabakoff, W., Heat transfer by a Square Array of Round Air Jets Impinging Perpendicular to a Flat Surface Including the Effect of Spent Air, *Journal of Engineering for Power*, vol. 92, pp. 73-82, 1970.

6. Florschuetz, L.W., Truman, C.R., and Metzger, D.E., Streamwise Flow and Heat Transfer Distributions for Jet Array Impingement with Crossflow, *Journal of Heat Transfer*, vol. 103, pp. 337-342, 1981.

7. Saad, N.R., Mujumdar, A.S., Abdel Messeh, W. and Douglas, W.J.M., Local Heat Transfer Characteristics for Staggered Arrays of Circular Impinging Jets with Crossflow of Spent Air, ASME/AIChE National Heat Transfer Conference, Orlando, ASME Paper 80-HT-23, 1980.

8. Florschuetz, L.W., Metzger, D.E., and Su, C.C., Heat Transfer Characteristics for Jet Array Impingement with Initial Crossflow, *Journal of Heat Transfer*, vol. 106, pp. 34-41, 1984.

9. Su, Chiung-Chieh, *Heat Transfer Characteristics for Two-Dimensional Arrays of Impinging Jets with Initial Crossflow*, Ph.D. dissertation, Arizona State University, Tempe, December 1984.

10. Florschuetz, L.W. and Isoda, Y., Flow Distributions and Discharge Coefficients for Jet Array Impingemnet with Initial Crossflow, *Journal of Engineering for Power*, vol. 105, pp. 296-304, 1983.

11. Kays, W.M. and Crawford, M.E., *Convective Heat and Mass Transfer*, 2nd ed.,Chapter 13, McGraw-Hill, New York, 1980.

12. McAdams, W.H., Nicolai, A.L., and Keenan, J.H., Measurements of Recovery Factors and Coefficients of Heat Transfer in a Tube for Subsonic Flow of Air, *Transactions AIChE*, vol. 42, pp.907-925, 1946.

Developing Turbulent Flow in a Curved Square Duct

T.-M. LIOU and C.-H. LIU
Department of Power Mechanical Engineering
National Tsing Hua University
Hsinchu, Taiwan, Republic of China

ABSTRACT

Experimental measurements of mean velocity, turbulence intensity and pressure in various cross-stream planes have been made in a square-sectioned, 90-degree bend of 1.5 radius ratio for the developing turbulent flow with the Reynolds number of 4.7×10^4 (Dean number of 2.7×10^4). The technique used was laser-Doppler velocimetry which is capable of accurately measuring the secondary flow.

The results show that the axial mean velocity is characterized by the free-vortex profile in the first 60 degrees of the bend and by the doubly peaked velocity profile around the bend exit when the inlet flow to the short upstream tangent is almost axially uniform. The secondary flow is found to be characterized by four-vortex and six-vortex structures around the bend exit. Large turbulence intensity is found in the regions where Dean vortices appear and where large radial velocity associated with the normal vortices of the secondary motion flows. The increase of the wall pressure from the inner wall to the outer wall is found to be approximately linear in the bend. The adverse axial pressure gradient near the inner wall does not lead to flow separation in our curved duct.

1. INTRODUCTION

Fluid flow in cruved ducts occurs in many engineering applications such as inlet ducts of side dump combustors, blade passages of turbomachinery and ducts of air conditioning systems. The existence of a curved duct changes the flow structures in the straight-duct sections upstream and downstream of it. The appearance of the secondary flow in curved ducts results in the redistribution of streamwise main flow, a pressure loss and enhanced heat transfer at the duct wall. Investigation of the magnitude of these effects is therefore important in engineering design.

The low frequency response of the Pitot tube, the calibration problems and the contradiction between probe size and direction resolving capability of the hot-wire or hot-film anemometer make the laser-Doppler velocimetry (LDV) a popular technique in recent years. Using LDV, Agrawal, et al. [1] measured the laminar water flow in a 180° curved circular pipe and both Humphrey, et al. [2,3] and Taylor, et al [4] measured the laminar and turbulent water flows in 90° curved square ducts. Their results show the strong effects of the inlet velocity distribution on the flow development and the strength of the secondary flow. For air flow only Holt, et al. [5], to authors' knowledge, studied the turbulent flow in a 90° curved square duct using LDV. However, as they mentioned, there are some spurious results in their measurements.

In this study, measurements were made in the air flow of a 90° curved square duct with a short upstream tangent. A practical value of radius ratio 1.5 [6,7] was chosen. The development of turbulent flow was measured since the flow in a finite bend is almost always a developing flow. Another purpose of this work is to provide a data base for the evaluation of both numerical techniques and turbulence models for the complicated three dimensional spiral flow.

2. EXPERIMENTAL EQUIPMENT AND CONDITIONS

2.1 Experimental System

The curved square duct and LDV experimental set-up is shown in schematic form in Figure 1. The velocity and pressure measurements were made in the LDV-Combustion Laboratory at National Tsing Hua University. Air was drawn into the upstream bend tangent through a flow straightener and four screens in the settling chamber and a bell-mouth entry (10 to 1 contraction) by a blower at the downstream end. The inlet flow established in this way was an essentially inviscid, uniform velocity profile. The fluid then flowed into the curved duct, a downstream bend tangent, a flow straightener, a flowmeter (rotameter), a bellows, a diffuser, and exhausted by the blower (3300 rpm/3 phase/1kW).

The LDV optics were set up in a dual beam forward scattering configuration. A linearly polarized 15-mW helium-neon laser (wavelength 6328 Å) provided the coherent light source. This beam was split into two parallel beams of equal intensity by a beamsplitter. A Bragg cell was used to cause a 40 MH_Z frequency shift on one of the beams. The frequency shift is used to eliminate the directional ambiguity which is essential if there is flow reversal as is expected in a curved duct. The resulting pair of beams was then passed through a 120 mm focal-length lens at a beam separation distance of 50 mm. The focused beams entered the curved duct through the transparent plexiglass wall, intersected inside the duct giving a probe volume with dimensions of 0.52 mm by 0.097 mm, and then passed through another side wall into the beam traps. The light scattered from the seeding particles was collected by a receiving optical package consisted of a 250 mm focal-length lens, a convex lens, a concave lens,

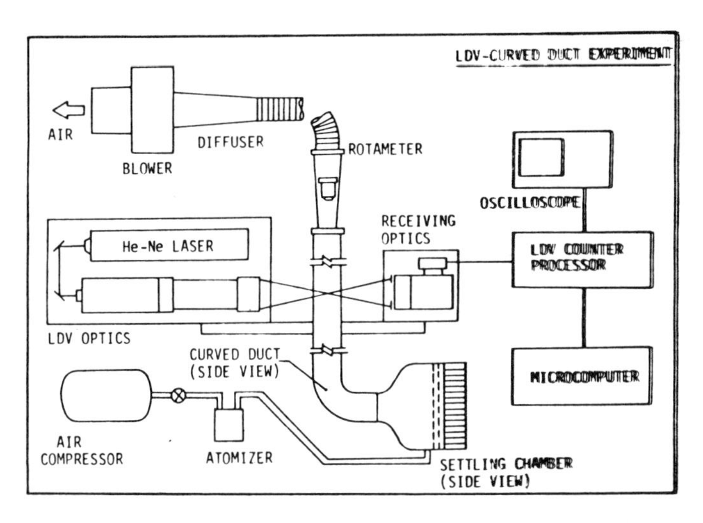

FIGURE 1. Schematic drawing of overall LDV-curved duct experimental system.

and a 45-degree mirror to reflect the collected light into a photomultiplier. The detected signal was electrically downmixed to the appropriate frequency shift (2 MH_Z in the present work). Then a counter processor with 2 nsec resolution was used to process the Doppler signal. All of the LDV optics and electronics are manufactured by TSI. The Doppler signal was monitored on an oscilloscope and the digital output of the counter processor was fed directly to a micro-computer for storage and analysis. The seed particles were introduced into the air stream by four atomizers symmetrically located on the four walls of the settling chamber. The atomizers were operated by filtered compressed air and water and generated water droplets of 2 μm diameter or smaller.

A micromanometer (model MM3, Flow Corporation, $\pm$ 0.0004 g/cm^2) was used for static pressure measurements along the walls of the curved duct. To measure pressure, the micromanometer was connected to each pressure tap (1 mm in diameter) on the model and the liquid level was read.

2.2 Experimental Conditions

The configuration of the curved duct model, coordinate systems, and dimensions are sketched in Figure 2. The duct was made of a 5-mm plexiglass and consisted of a 90° bend of mean radius 60 mm and of radius ratio 1.5. The upstream and downstream tangent lengths were 100 mm and 1100 mm respectively. The dimensions of the cross-section was 40 mm x 40 mm.

The velocity measurements were made in 14 planes (R* - Z* planes) normal to the curved duct walls. In each of these data were collected only in the symmetrical half. The symmetry was verified from the previous work [8]. These data planes were located at x_H = -1.0 and -0.5, respectively in the upstream tangent, at every 15° within the bend, and at x_H = 0.5, 1.0, 2.0, 10.0 and 22.0, respectively in the downstream tangent. In each data plane the velocity measurements were made at 95 points for the axial component and at 60 points for the radial component. Within a given data plane, the probe volume was brought as near as 1 mm to 4 mm, depending on the existence of the dead zone, from the wall for

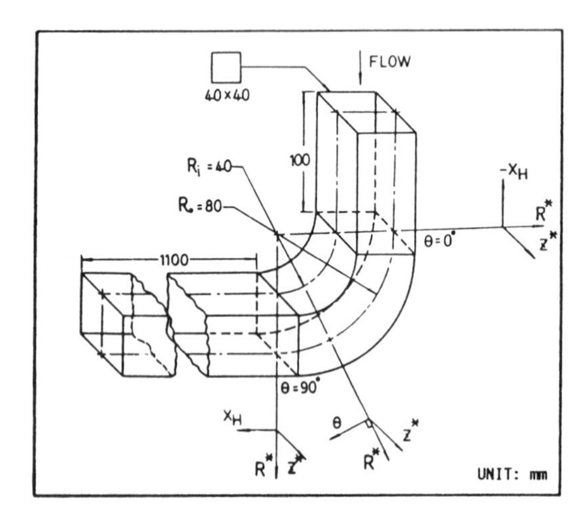

FIGURE 2. Sketch of co-ordinate systems and dimensions of 90° square duct with tangents.

measurements. The measuring locations were chosen to give a clear pattern of the secondary flow. The mean velocity 17.28 m/s measured at upstream plane x_H = -1.5 was used as a reference to normalize the experimental results. This velocity corresponds to a Reynolds number of 4.7 x 10^4, indicating the flow to be turbulent. The corresponding Dean number for the bend was 2.7 x 10^4.

The pressure measurements were taken at every 0.5 hydraulic diameter starting from plane x_H = -1.5 in the upstream tangent, at every 15° within the bend and at every 0.5 hydraulic diameter up to plane x_H = 2.0 in the downstream tangent. In each of these static pressures were measured at 11 points along the inner wall, side wall and outer wall. All the measured pressure values were relative to a reference value taken at location (R^* = 1.0, θ = 15°, Z^* = 0) since the pressure gage is a differential type micromanometer.

3. RESULTS AND DISCUSSION

The mean velocity and turbulence intensity were calculated from the probability distribution function of the measurements. There were typically 2000 to 4000 measurements at each measuring location. The corresponding statistical error was between 0.1% to 3.5% in the mean velocity and between 2.2% to 3.1% in the turbulence intensity for 95% confidence level.

3.1 Mean Axial Velocity and Turbulence Intensity

The profiles of the mean axial velocity and the corresponding turbulence intensity in the four axial planes, located at Z^* = 0.0, 0.4, 0.7 and 0.9, respectively, are shown in Figure 3. As expected the fine-mesh screens in the settling chamber and the bell-mouth entry provide an essentially inviscid, flat mean axial velocity profile and a low turbulence level at upstream station x_H = -1.0. An exception occurs at Z^* = 0.9 (2 mm from the side wall) where wall and corner effects are dominant. At the bend entrance (θ = 0°) the mean axial velocity profile indicates the effect of curvature becomes evident. Since the upstream tangent of the bend is short and the strong favorable pressure gradient generated by the bell-mouth entry thins the boundary layer the flow within the curved duct, at least in the first 60°, is nearly potential. The balance between the centrifugal force and the radial pressure gradient then causes a free-vortex type mean axial velocity profile. In the downstream bend tangent the effect of curvature decreases gradually. The axial mean velocity profile tends to gradually approach the fully developed turbulent flow.

There are several observations worthy of our interest. First, a closer look of the axial flow behavior near the inner and the outer wall respectively is given in Figure 4 where the corresponding wall static pressure is also included for the purpose of comparison. These are typical results taken in the axial symmetry plane. As can be seen near the inner wall the axial mean velocity increases with the axial distance before θ = 30°, decreases after θ = 30° up to x_H = 0.5, and then increases after x_H = 0.5 with a maximum approximately at θ = 30° and a minimum at x_H = 0.5. On the other hand, near the outer wall the axial mean velocity decreases before θ = 30° and then increases after θ = 30° with a minimum at θ = 30°. These behaviors can be explained from the associated pressure data. In curved ducts centrifugal force causes a radial pressure gradient (will be shown later) which results in a pressure increase near the outer wall of the curved duct, starting at x_H = -0.5 and rising to a maximum at θ = 45°. Therefore, in region from x_H = -0.5 to θ = 45° the fluid is opposed by an adverse pressure gradient. In contrast, the radial pressure gradient results in a pressure decrease near the inner wall of the curved duct. For this reason at the inner wall the pressure decreases to θ = 45° and then rises again. That is, an adverse pressure gradient also exists from θ = 45° to x_H = 1.5. The adverse

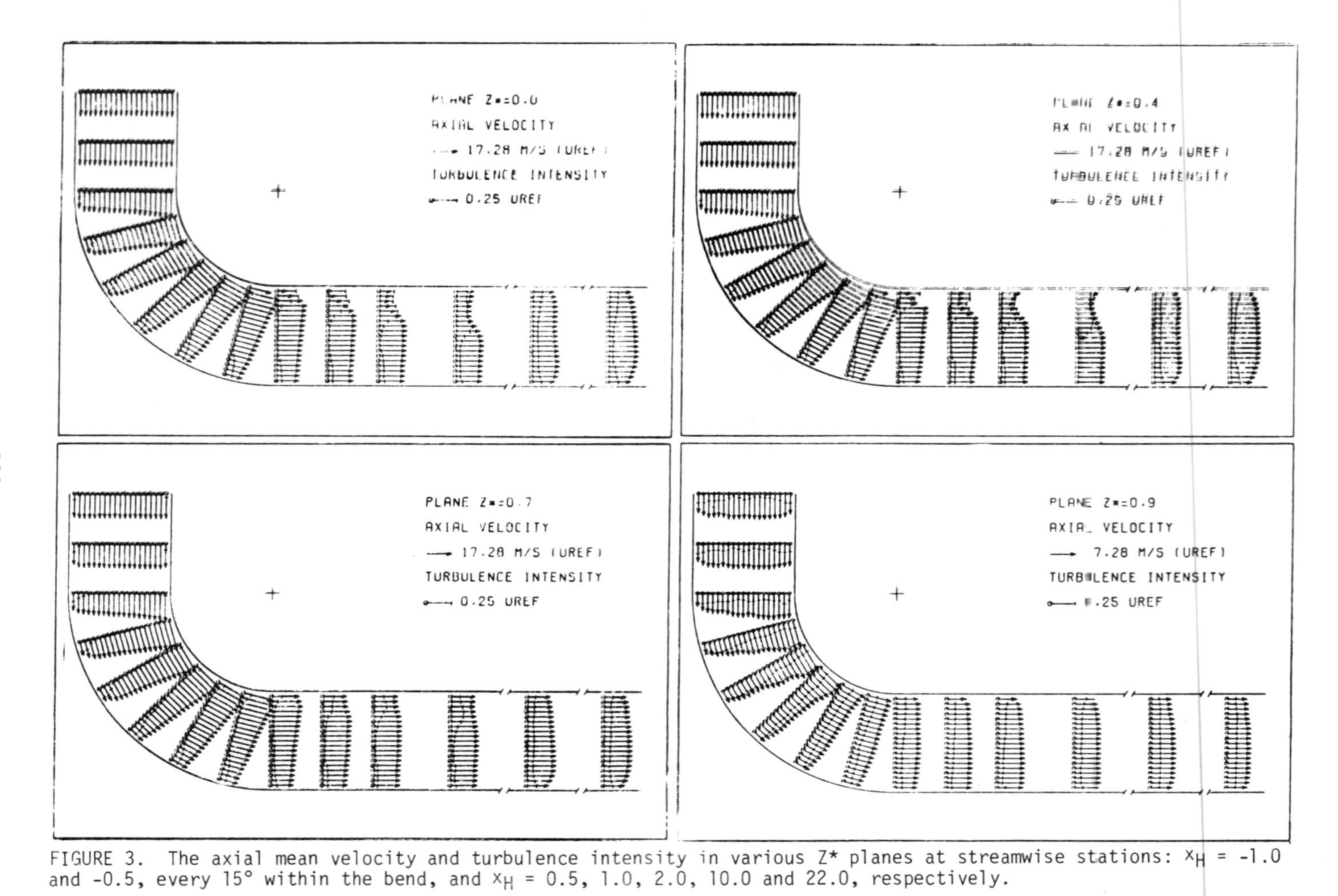

FIGURE 3. The axial mean velocity and turbulence intensity in various Z* planes at streamwise stations: x_H = -1.0 and -0.5, every 15° within the bend, and x_H = 0.5, 1.0, 2.0, 10.0 and 22.0, respectively.

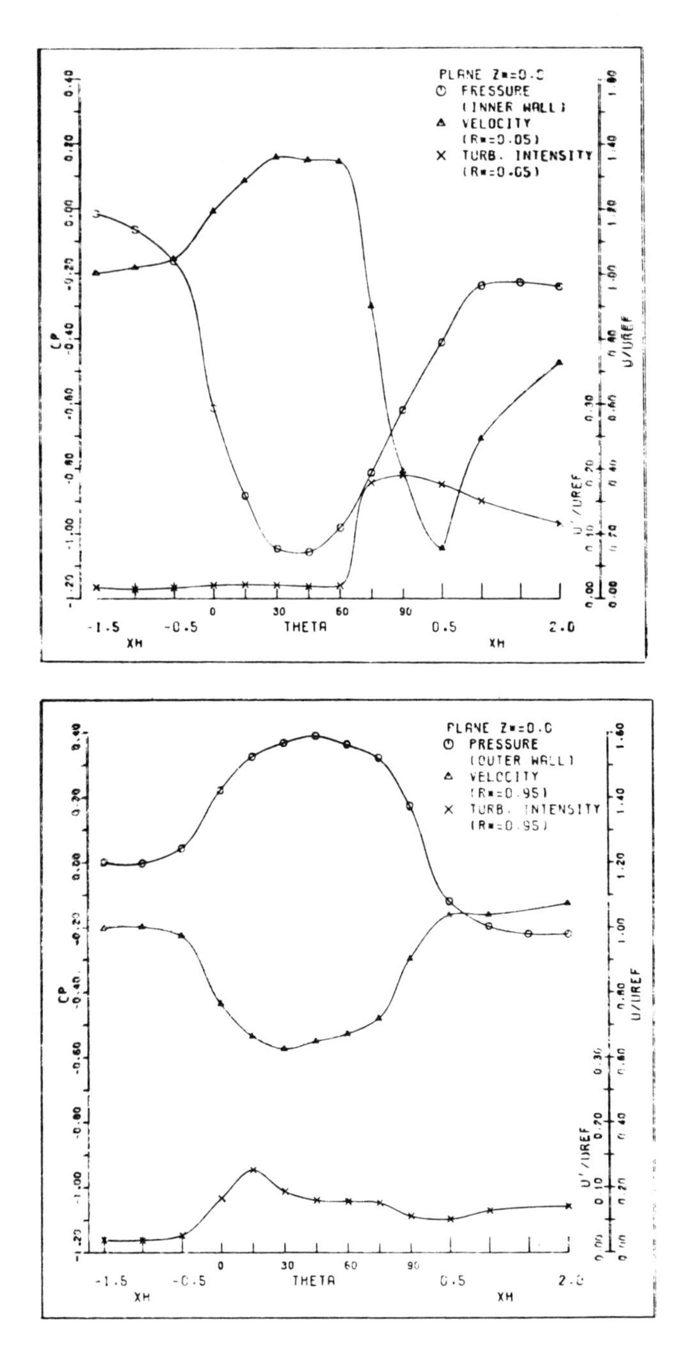

FIGURE 4. The axial mean velocity, turbulence intensity and wall pressure versus x_H or θ in the axial symmetry plane ($Z^* = 0$) at $R^* = 0.05$ and 0.95, respectively.

pressure gradient causes the decrease of the axial mean velocity and in turn the high turbulence intensity as shown in Figure 3 and 4.

Second, normal to the streamwise direction near the inner wall in the region of adverse pressure gradient the pressure is lower near the axial symmetry plane (as will be shown later). In other words, near the axial symmetry plane the fluid is opposed by a steeper adverse pressure gradient and therefore the axial mean velocity is lower near the axial symmetry plane close to the inner wall. This, together with the steep velocity gradient near the side wall, results in a doubly peaked mean axial-velocity profile in the R* - Z* plane as shown in Figure 5. Doubly peaked mean axial-velocity profile has also been observed by others in the curved square ducts both laminar flow [2,4] and turbulent flow [3,4]. In this work the lowest mean axial velocity between the double peaks is located at (x_H = 0.5, Z* = 0, R* = 0.05) and has a value only .15 U_{REF}. The highest turbulence intensity is located at (X_H=0.5,Z*=0,R*=0.15) and has a value .21 U_{REF}.

Third, Figure 3 shows the influence of the bend on the fluid flow exists only about one hydraulic diameter upstream from the bend due to the acceleration of the bell-mouth entry and a large distance downstream which is out of our measuring range x_H = 22.0.

3.2 Radial Mean Velocity and Turbulence Intensity

Figure 6 shows the distributions of the radial mean velocity and turbulence intensity in the 12 planes normal to the axial stream. As one can see at bend entrance, $\theta = 0°$, the radial mean velocity flows towards the inner wall to maintain the mass conservation since the axial mean velocity profile is an irrotational-vortex type distribution. At $\theta = 15°$ a weak clockwise vortex starts to appear around the right corner. The strength and size of this vortex grow continuously and the radial mean velocity near the side wall penetrates towards the inner wall as the flow proceeds downstream. That the corresponding turbulence intensity is higher near the outer wall indicating the turbulent energy transfer towards the wall is enhanced. Higher turbulence intensity is also found near the side wall where flow reversal occurs. The penetration of the large radial velocity illustrates the faster recovery of the mean axial velocity downstream from the bend exit along the inner wall in the planes of Z* = 0.0 and 0.4 (see Figure 3). Near the side wall the radial mean velocity is larger, which reaches a largest value 0.35 U_{REF} at $\theta = 75°$, since the axial mean velocity is lower there and therefore acted by a larger centrifugal force-radial pressure gradient imbalance. Because the developing flow gives a thin boundary

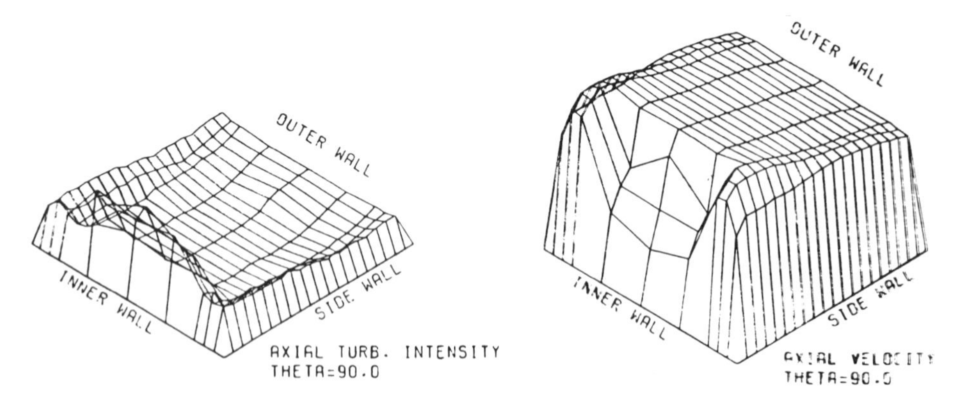

FIGURE 5. Doubly peaked mean axial-velocity and corresponding turbulence intensity profiles at $\theta = 90°$.

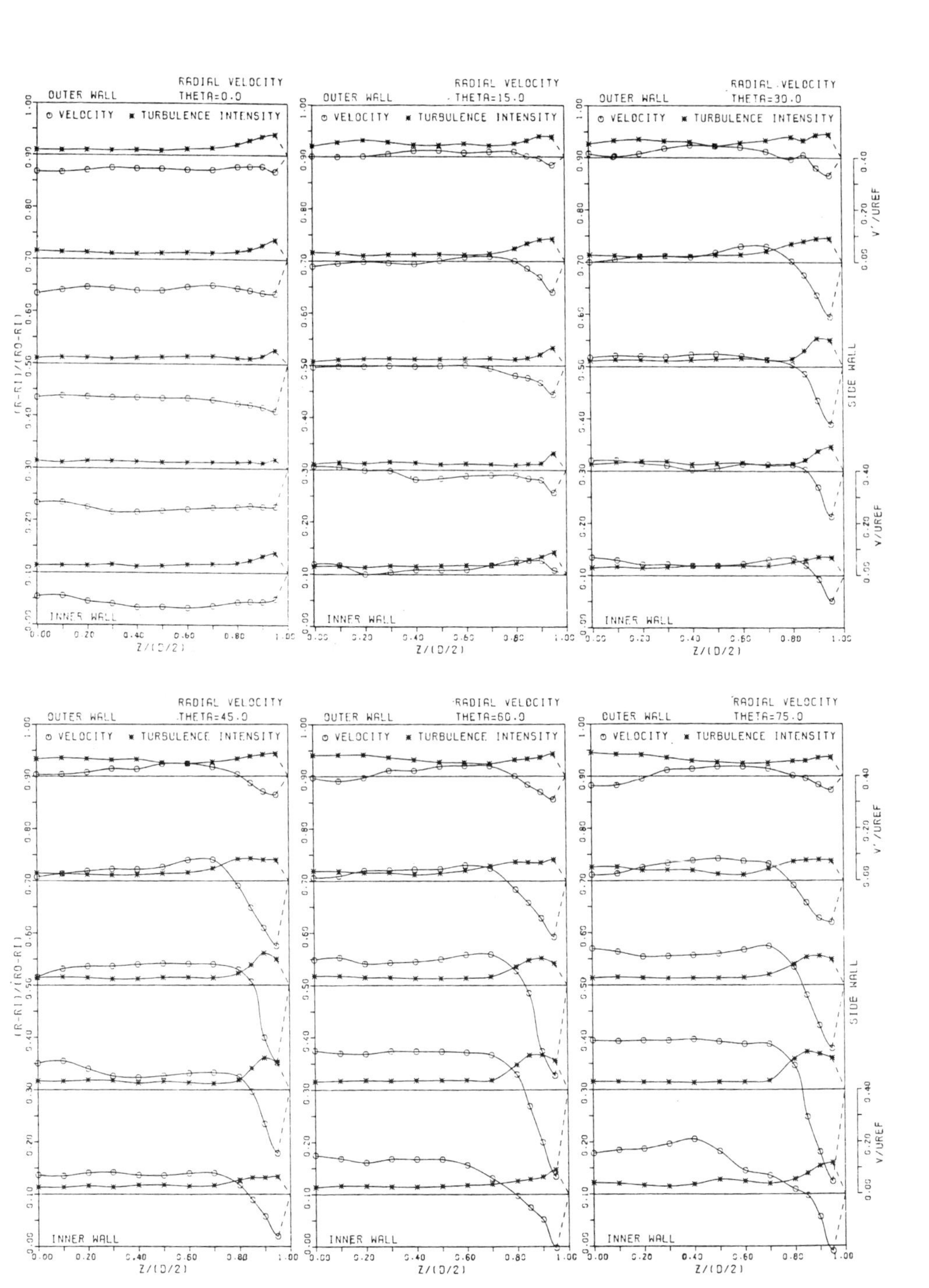

FIGURE 6. Radial mean velocity and turbulence intensity profiles at various streamwise stations.

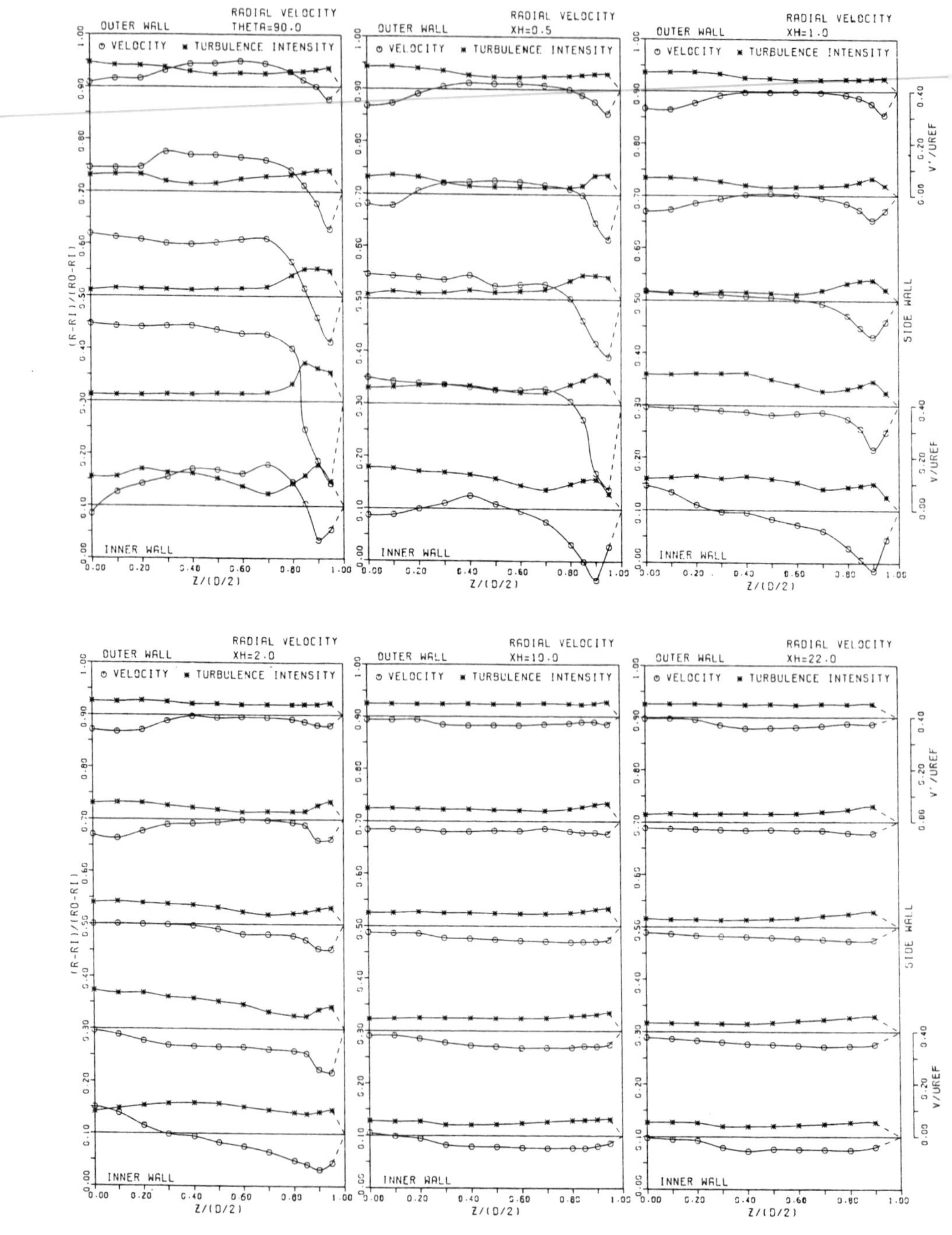

FIGURE 6. Radial mean velocity and turbulence intensity profiles at various streamwise stations. (continued)

layer the large radial velocity is confined to a narrow region near the side wall. This results in a vortex center closer to the side wall.

In addition to the described large clockwise vortex there is a growing counter-clockwise small vortex appearing near the outer wall close to the axial symmetry plane at $\theta = 60°$ and $75°$. An interesting behavior is it disappears at bend exit ($\theta = 90°$) and then reappears at $x_H = 0.5$. After $x_H = 0.5$ it grows continuously and then decays eventually at large distance downstream. The secondary flow pattern characterized by four-vortex structure (i.e. an additional pair of Dean vortices) in the whole cross-section has also been predicted theoretically and observed experimentally using flow visualization by Cheng, et al. [9,10] for the case of laminar flow in curved square ducts and inferred, from both yaw measurements and the shape of the pressure contours by Rowe [11] for the case of turbulent flow in an 180° circular pipe.

The most interesting feature of the measured secondary motion is the presence of the third additional vortex, which is smaller and counterclockwise, near the inner wall close to the axial symmetry plane at $\theta = 90°$ and $x_H = 0.5$ respectively. It increases in strength at $x_H = 0.5$ where the radial turbulence intensity reaches a maximum of $0.16\ U_{REF}$ and then disappears after $x_H = 0.5$. It is worth noting that this third additional vortex is trapped in the cone-shape low velocity zone of the doubly peaked axial-velocity profile (see Figure 5). This observation may suggest that there is some correlation between the existence of the third additional vortex and the doubly peaked axial-velocity profile. There are now six vortices, the normal twin counter-rotating vortices are distorted to accommodate the two additional pairs of Dean vortices, in the entire cross-section at $\theta = 90°$ and $x_H = 0.5$ respectively. The largest strength ratio of these three pairs of vortices is approximately 10 (normal vortices) to 2 (Dean vortices near the outer wall) to 1 (Dean vortices near the inner wall).

3.3 Pressure Distribution and Loss

The wall pressure is plotted versus axial distance for radial positions $R^*=0.025$, 0.250, 0.500, 0.750 and 0.975 in Figure 7. As one can see the existence of the bend causes a pressure loss. The bend pressure-drop coefficient recommended by Ward-Smith [12] is used to express this loss. It has a value of 0.146 based on the measurements taken from this study. Figure 7 also shows that a radial pressure gradient exists and the largest pressure difference between the outer wall and the inner wall is at $\theta = 45°$ plane. Through this radial pressure gradient the secondary flow is driven and the bend influences the flow in the bend tangents.

Another observation is that initially a favorable pressure gradient develops on the suction surface, and an adverse gradient on the pressure surface, causing the observed acceleration and deceleration of the fluid near the respective surfaces. These trends are reversed after $\theta \simeq 45°$. The adverse pressure gradient near the inner wall may lead to separation. It was reported [12] that separation and reversed flow occur in 90°-bend having a radius ratio of about 1.5 or less for ducts of both square and circular sections. Figure 3 measured from this study with a radius ratio of 1.5 does not show any axial-flow separation near the inner wall. Another way to look the increase of pressure from the suction surface to the pressure surface is to plot the wall pressure versus circumferential distance, from inner wall through side wall to the outer wall, at various axial stations. This is presented in Figure 8. As can be seen the cross-stream increase of the pressure is approximately linear within the bend. Also the larger values of radial pressure gradient are associated with the larger radial velocities at $\theta = 60°$ and $\theta = 75°$ as one would expect.

4. CONCLUSIONS

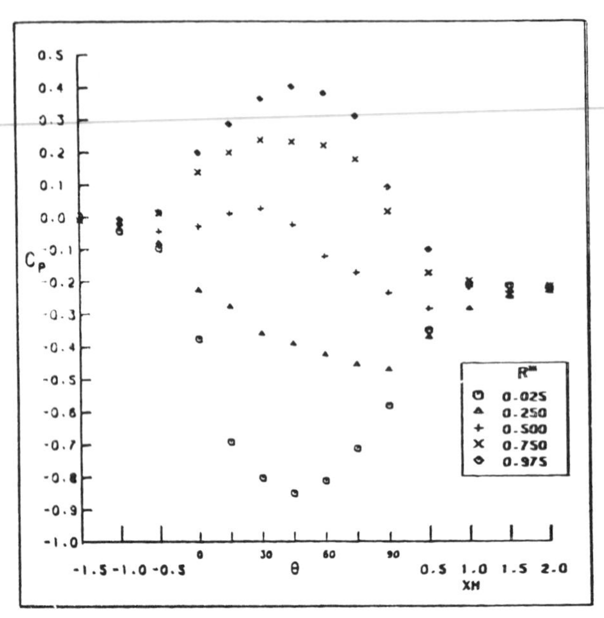

FIGURE 7. Wall static pressure versus axial distance at various radial positions

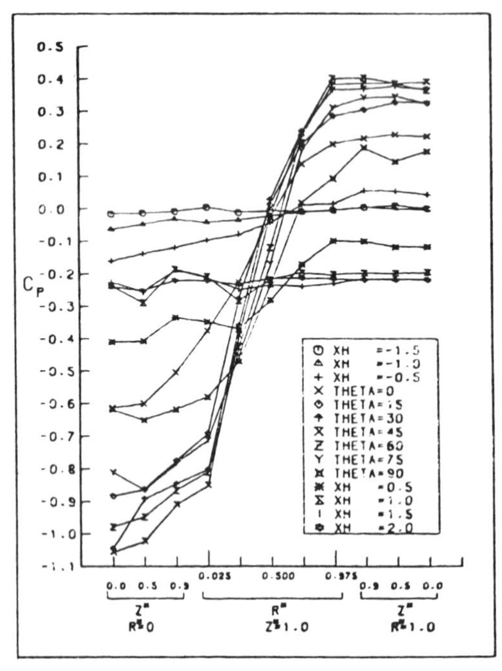

FIGURE 8. Wall static pressure versus circumferential distance at various axial positions.

1. The effect of a 90° square bend with a short upstream tangent on the turbulent flow passing through it has been reported in detail in this study in terms of mean velocity, turbulence intensity and pressure characteristics.

2. The mean axial velocity is found to form a profile of free-vortex type in the first 60° of the curved duct when the short upstream tangent has an inlet condition of almost uniform axial-velocity distribution. Around the bend exit the axial flow is characterized by the doubly peaked velocity profile which could result in an axial velocity as low as 0.15 U_{REF} in the cone-shape low velocity zone.

3. The large radial mean velocity of the secondary motion is found to be confined to a narrow region near the side walls and reaches a maximum of 0.35 U_{REF}. The most interesting observation is the existence of 4-vortex and 6-vortex structures after $\theta = 60°$. The largest strength ratio of the three pairs of vortices is approximately 10 to 2 to 1.

4. It is found that turbulence is generated in the regions of adverse pressure gradient where the flow is decelerated. Relatively large turbulence intensity is also found in the regions where Dean vortices appear, reaching a maximum of 0.16 U_{REF}, and where large radial velocities associated with the normal vortices of the secondary motion flow.

5. The increase of the wall pressure from the inner wall to the outer wall is found to be approximately linear in the bend. The adverse axial pressure gradient near the inner wall does not lead to flow separation in our curved duct.

ACKNOWLEDGEMENTS

Support for this work was partly provided by the National Science Council (Contract NSC74-0401-E007-01).

NOMENCLATURE

C_P	=	pressure coefficient $C_P \equiv (P - P_{REF})/\frac{1}{2}\rho U_{REF}^2$
D	=	hydraulic diameter
P	=	wall static pressure
P_{REF}	=	reference value of P (x_H = -1.5, R* = 1.0, Z* = 0)
R	=	radial coordinate direction
R_i	=	radius of curvature of inner wall
R_O	=	radius of curvature of outer wall
R*	=	normalized radial coordinate $R^* \equiv (R - R_i)/(R_O - R_i)$
U	=	axial mean velocity
U_{REF}	=	reference value of U
V	=	radial mean velocity
x_H	=	axial distance along straight duct, expressed in hydraulic diameters
Z	=	spanwise coordinate direction
Z*	=	normalized spanwise coordinate $Z^* \equiv Z/\frac{1}{2}D$
θ	=	axial coordinate direction
ρ	=	air density

REFERENCES

1. Agrawal, Y., Talbot, L. and Gong, K., Laser Anemometer Study of Flow Development in Cuvred Circular Pipes, J. Fluid Mech., Vol. 85, part 3, pp. 497-518, 1978.

2. Humphrey, J.A.C., Taylor, A.M.K.P. and Whitelaw, J.H., Laminar Flow in a Square Duct of Strong Curvature, J. Fluid Mech., Vol. 83, pp. 509-527, 1977.

3. Humphrey, J.A.C., Whitelaw, J.H. and Yee, G., Turbulent Flow in a Square Duct with Strong Curvature, J. Fluid Mech., Vol. 103, pp. 443-463, 1981.

4. Taylor, A.M.K.P., Whitelaw, J.H. and Yianneskis, M., Curved Ducts with Strong Secondary Motion: Velocity Measurements of Developing Laminar and Turbulent Flow, J. Fluid Eng., Vol. 104, pp. 350-359, 1982.

5. Holt, M., Flores, J. and Turi, P.J., Measurements of Air Flow in a Curved Pipe Using Laser-Doppler Velocimetry, 1st Int'l Symp. on Appl. of LDA to Fluid Mech., Lisbon, Protugal, 1982.

6. Air Conditioning Refrigerating Data Book, The American Society of Refrigerating Engineers, 10th edition, 1957.

7. Jones, W.P., Air Conditioning Engineering, 2nd ed., Edward Arnold Ltd., 1973.

8. Liu, C.-H. and Liou, T.-M., A Study of Velocity and Pressure Distributions in a Curved Square Duct, Proceedings of the Eighth National Conference on Theoretical and Applied Mechanics, Tainan, Taiwan, R.O.C., December 1984.

9. Cheng, K.C., Lin, R.-C. and Ou, J.-W., Fully Developed Laminar Flow in Curved Rectangular Channels, J. Fluid Eng. Trans., ASME, Vol. 95, pp. 41-48, 1976.

10. Cheng, K.C., Nakayama, J. and Akiyama, M., Effect of Finite and Infinite Aspect Ratios on Flow Patterns in Curved Rectangular Channels, in Flow Visualization (Tokyo, Japan, 1977), Hemisphere Pub. Corp., pp. 181-186, 1979.

11. Rowe, M., Measurements and Computations on Flow in Pipe Bends J. Fluid Mech., Vol. 43, pp. 771-783, 1970.

12. Ward-Smith, A.J., Internal Fluid Flow, Clarendon Press, Oxford, 1980.

Contact Thermal Resistance between Two Rotating Horizontal Cylinders

SHINICHIRO YAMAZAKI and AKIHIRO SHIMIZU
Department of Mechanical Engineering
Tokyo National College of Technology, Tokyo Japan

AKIRA TSUCHIDA
Department of mechanical Engineering, Faculty of Engineering
Seikei University, Tokyo, Japan

INTRODUCTION

In general, we have many manufacturing processes continuously to temper running material on hot rollers, in steel mills, synthetic fiber msnufacturing plants and so on. The heat transfer characteristics, through the contact points between rotating rollers and running material, have not always been investigated satisfactorily, in spite of its significance to realize a certain quality of products. Many investigations have been performed about the thermal resistances through the contact points among stationary bodies, but scaresely those between moving bodies touched to each other. Estimating experimentally the thermal resistance as above mentioned, we chose a system of two horizontal rotating cylinders contacted parallel to each other, as one of the most simplified and fundamental configuration, and then presented the resistance as a function of the rotating speed and the contact load. Three kinds of the material were applied to these cylinder, aluminium, brass and mild steel. Furtheremore, we try to discuss the convection phenomena observed around the contact zone which happen to effect on the resistance.

DEFINITIONS TO ESTIMATE THE THERMAL RESISTANCE

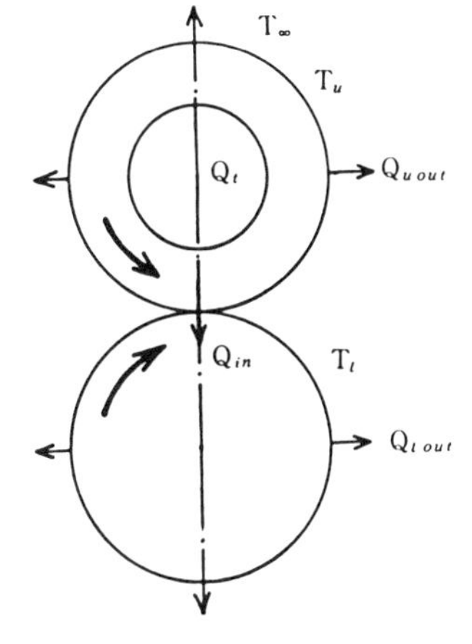

FIGURE 1. Physical model of two rotating cylinders contacted to each other.

Fig.1 shows the physical model of two rotating cylinders contacted to each other. Over-all heat generation at the upper cylinder, the heat loss from the surface and that of the lower one are represented as Q_t [W], Q_{uout} [W] and Q_{lout} [W], respectively. The heat transfered from the upper to the lower through the contact line is denoted as Q_{in} [W], then we obtain under a steady state as follows:

$$Q_t = Q_{uout} + Q_{lout} \tag{1}$$

Now, the average heat transfer rate from the cylinder to the ambient, the perimetric area, the temperature of the surface and the ambient temperature are represented as α [$W/m^2 K$] , A [m^2] , T [K] and T_∞ [K] , respectively. Then Q_{uout} and Q_{lout} are shown as follows:

$$Q_{uout} = \alpha_u A (T_u - T_\infty) \tag{2}$$

$$Q_{1out} = \alpha_1 A (T_1 - T_\infty) = Q_{in} \tag{3}$$

where suffixes u and 1 denote the upper and the lower, respectively. Put $\alpha = \alpha_u = \alpha_1$, and then Q_t and α result in the following equations.

$$Q_t = \alpha A (T_u + T_1 - 2 T_\infty) \tag{4}$$

$$\alpha = Q_t / A (T_u + T_1 - 2 T_\infty) \tag{5}$$

On the other hand, from the equation (3), we obtain

$$Q_{in} = \alpha A (T_1 - T_\infty) \tag{6}$$

Then, the thermal resistance r_c [m K / W] is represented as follows:

$$r_c = (T_u - T_1) / q_{in} \tag{7}$$

where $q_{in} = Q_{in} / L$ and L shows the length.
Now, the thermal conductance h_c is

$$h_c = \frac{1}{r_c} \tag{8}$$

then, we define Nusselt number N_{uh} as

$$N_{uh} = h_c / \lambda \tag{9}$$

where λ [W / m K] denotes the thermal conductivity of the ambient air at the temperature $(T_1 + T_\infty) / 2$.

EXPERIMENTAL APPARATUS AND MEASUREMENT

The Thermal Resistance

Fig.2 and 3 show the general view of the apparatus and the heating device in the cylinder in detail, respectively.

The pair of cylinders has 90 mm O.D. and 150 mm length, made of aluminium, brass and mild steel (S25C) and is arranged to be heated from the upper cylinder to the lower. The upper heating cylinder is mounted in its inner wall (thickness: 15 mm) with the electrical heat source, composed of Nichrome wire (0.18 mm D, 46 Ω / m) wound around the Bakelite cylinder (60 mm O.D.) along the groove (1.1 mm depth) engraved on its surface. To keep the Bakelite cylinder as isothermal state, two kinds of device are applied; first, another Nichrome heat source with the same capacity as the heater mounted at the outerside ; second, mica-insulated heaters sandwitched with Bakelite disks (20 mm thick) are installed at the ends. By adjusting the voltages loading to these heaters,we try to keep heat losses as little as possible.

For the temperature measurement, copper-constantan thermocouples (0.2 mm D.) are installed at six points marked as x in Fig.3 . Also two couples are mounted on the lower cylinder and another is for the ambient temperature measurement.

The upper cylinder is followed by the rotation of the lower, driven by the motor. The surface of the cylinder is finished as fine as 0.8S in its roughness, and always kept clean to avoid error of the measurement.

The experiment were carried out under the conditions that the rotating speed and the contact load range from null to 600 (rpm) and from 608 to 1471 [N / m], respectively.

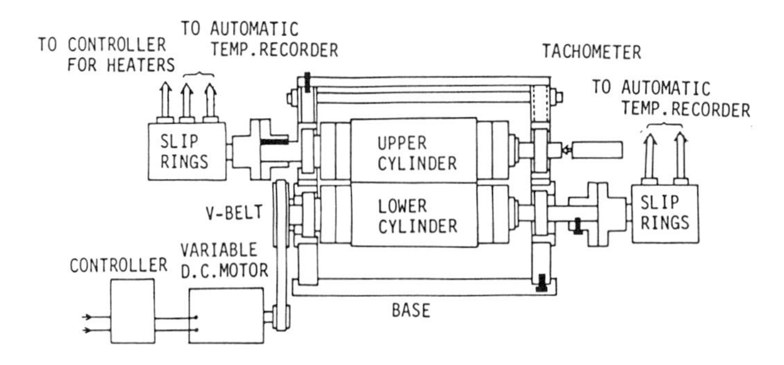

FIGURE 2. General view of the experimental apparatus

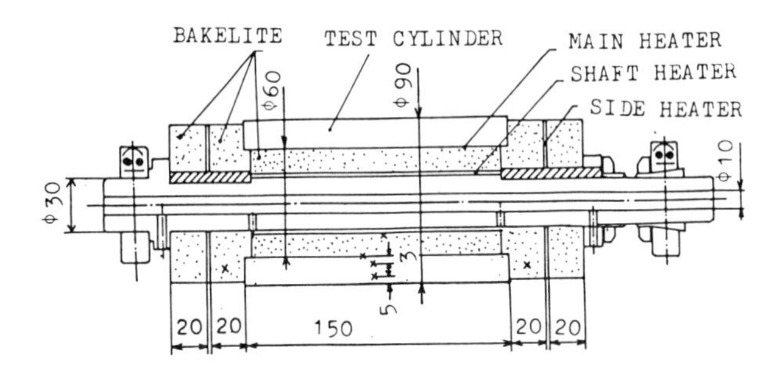

FIGURE 3. Detail of the upper cylinder

The Convection Phenomena around the Contact Line

The convection phenomena were observed by means of the smoke test, Schlieren's method and the temperature distribution. For the convenience, a pair of smaller sized aluminium cylinders (68 mm O.D.) is applied.

Smoke test. The smoke stream, gradually risen by natural convection, is introduced under the lower cylinder and then flows up along the surfaces of two cylinders. The stream trajectory is always kept bright by means of the narrow slitted light, projected parallel to the stream, and then recorded as higher sensitive photograph (ASA6400).

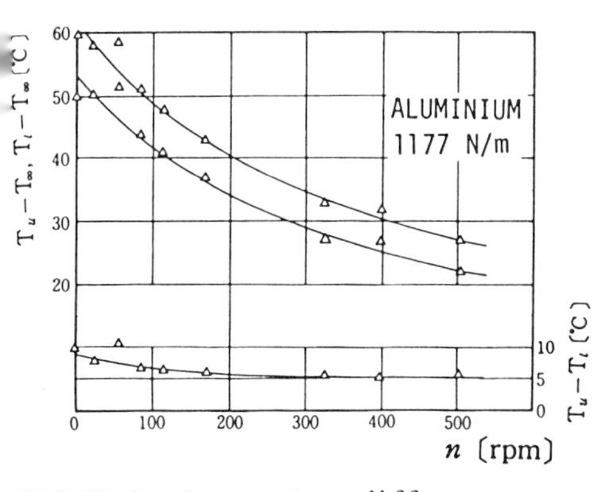

FIGURE 4. Temperature differences and rotating speed

Temperature mesurement. The measurement device is made of a copper-constantan thermocouple (0.2 mm D.),whose junction is welded to the point of the needle fixed on the three dimensional cathetometer.

RESULTS AND DISCUSSION

Some experimental results are shown in Fig.4-6 as the relation between the

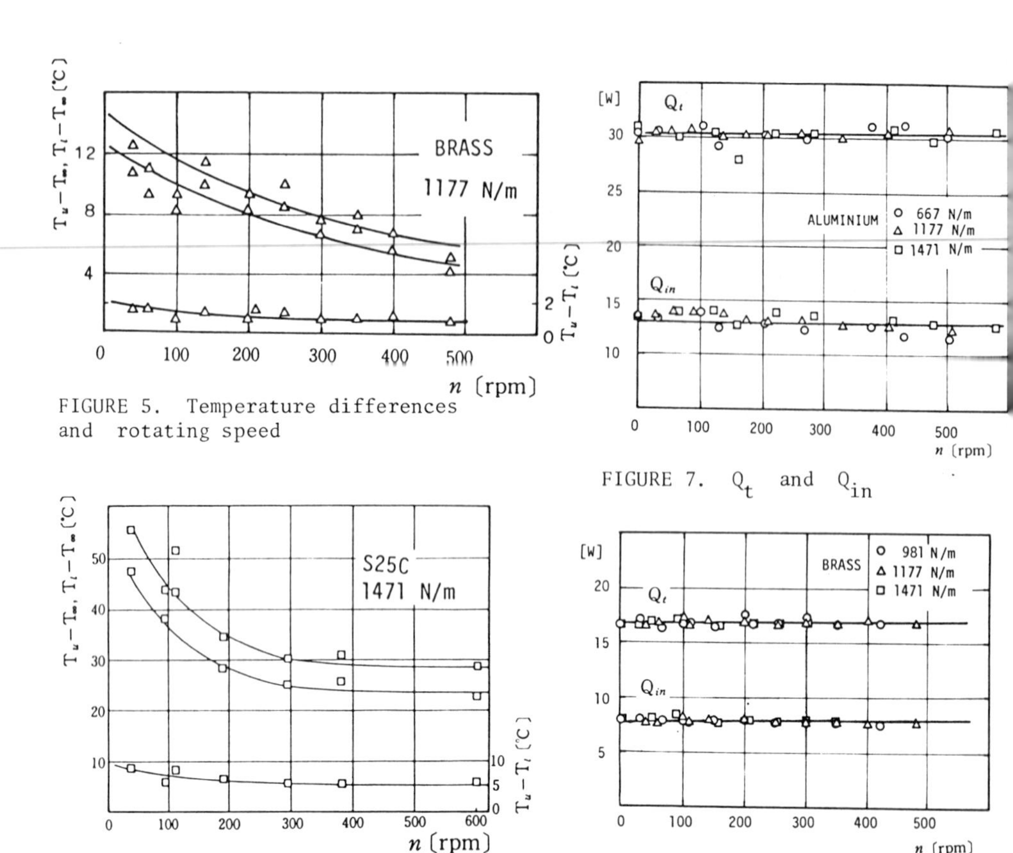

FIGURE 5. Temperature differences and rotating speed

FIGURE 7. Q_t and Q_{in}

FIGURE 6. Temperature differences and rotating speed

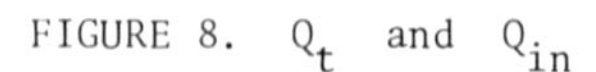

FIGURE 8. Q_t and Q_{in}

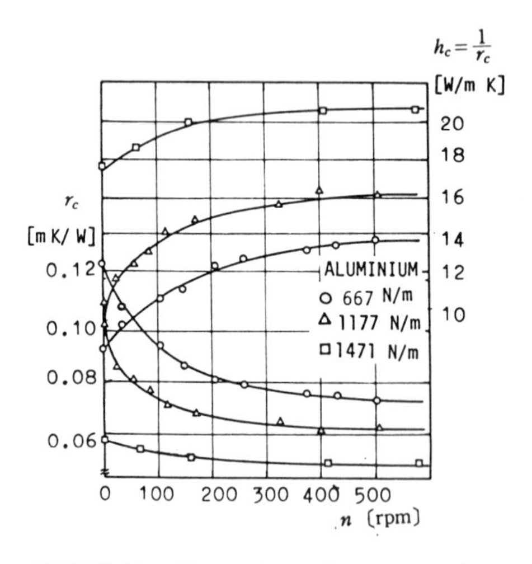

FIGURE 9. Thermal resistance and coductance against rotating speed

rotating speed 'n' and the temperature differences 'T_u-T_∞','T_1-T_∞' and 'T_u-T_1'as a parameter of the cylinder material unde the constant contact load. These results show that the temperature difference '$\Delta T=T_u-T_1$' decreases gradually to a certain value as the rotating speed 'n' increases.

Fig.7 and 8 show the relation between Q_t and Q_{in}. Q_t and Q_{in} keep constant in spite of changing the load 'P' and the rotating speed 'n'.

Fig.9-10 show the relation between the the value 'n' and the thermal resistance 'r_c', Especially about the material, aluminium and brass,the relations between the value 'n' and the conductance 'h_c' are added. IN these cases, the results show that the value 'r_c' converges to a constant value as the value'n' comes up to 300 rpm, and also that the value'r_c' decreases as the load 'P' increases. Concerning mil

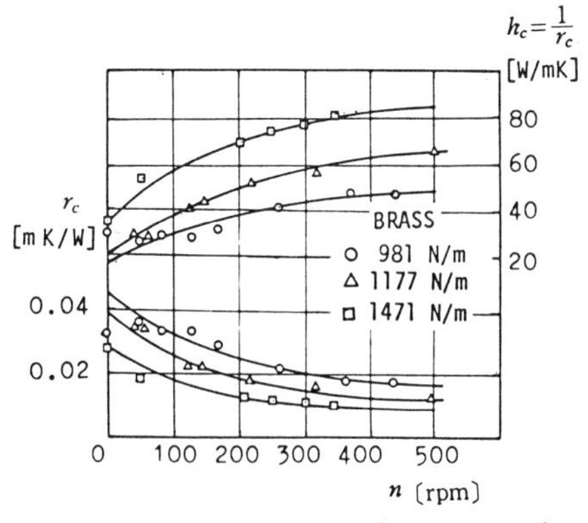

FIGURE 10. Thermal resistance and conductance against rotating speed

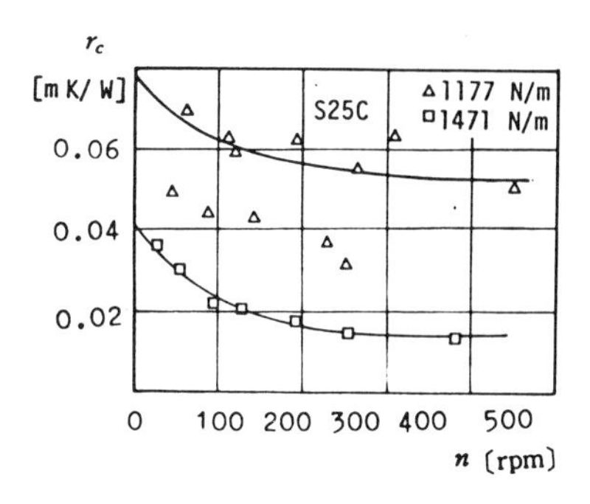

FIGURE 11. Thermal resistance against rotating speed

steel the plots of the value 'r_c' scatter on the broad area, caused of surface oxidization under the higher temperature of the cylinder. It is a well-known fact that the thermal resistance changes its value extremely according to the oxidization and the contact load among the stationary bodies. And also it is well-known that the resistance decreases as the material hardness decreases, that is, the value 'r_c' decreases in order of the hardness of material; mild steel, aluminium and brass (Brinell's hardness 144, 117 and 99).

Now, we try to discuss the reason why the value 'r_c' decreases gradually to a certain constant value as the value 'n' increases. As shown in Fig.12, the heat rate Q_{in} transfered from the upper to the lower depends on the heat rate Q_{in_1} through the turely contacted area by the heat conduction and the heat rate Q_{in_2} through the stagnation zone near the contact line by the heat conduction by the heat convection.

$$Q_{in_1} = \lambda\, a_1 \frac{T_u - T_l}{l} \tag{10}$$

$$Q_{in_2} = \alpha\, a_2 (T_u - T_l) \tag{11}$$

l = apparent distance between two cylinders.
a_1 = true contact area
λ_1 = thermal conductivity of the cylinder
α = apparent heat transfer coefficient
a_2 = apparent area corresponding to the heat transfer

Therefore,

$$Q_{in} = Q_{in_1} + Q_{in_2} = \lambda\, a_1 \frac{T_u - T_l}{l} + \alpha\, a_2 (T_u - T_l) = Q_{lout} = \alpha\, A\, (T_l - T_\infty)$$

$$= \Delta T \left(\lambda \frac{a_1}{l} + \alpha\, a_2 \right) = \alpha\, A\, \Delta T_1 \tag{12}$$

where $\Delta T = T_u - T_l$, $\Delta T_1 = T_l - T_\infty$

Hence,

$$h_c = \frac{1}{r_c} = \frac{q_{in}}{\Delta T} = \frac{1}{L} \left(\lambda \frac{a_1}{l} + \alpha\, a_2 \right)$$

$$= \frac{1}{L}\, \alpha\, A\, \Delta T_1 \tag{13}$$

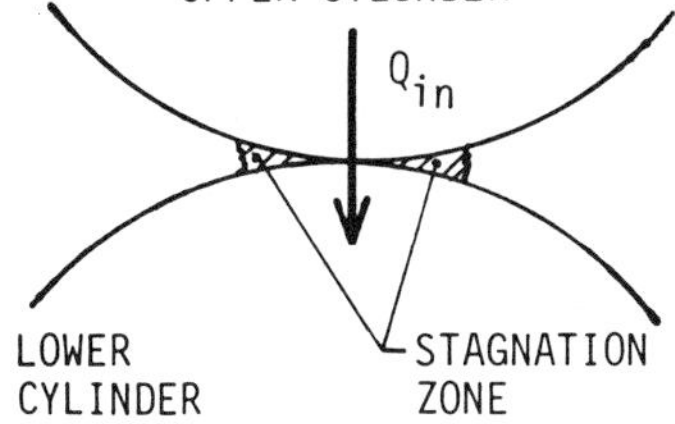

FIGURE 12. Heat flow through the contact surface

From equation (13), under the constant load 'P', the term ' $\lambda a_1/l$ ' can be regarded as constant because of smaller order of a_1, l, and λ, even though the value 'n' increases.

On the other hand, the flow rate through the clearance between two cylinders can be estimated too little to effect the heat transfer rate 'α' in spite of increasing 'n'. However, the area corresponding to the heat transfer by the convection comes up larger, resulting in that 'αa_2' increases to a certain value as the value 'n' increases. That is, the value 'h_c' increases to a certain constant value as the value 'n' increases. Now, the reason why the value 'r_c' comes to smaller value as the load 'P' increases, could be understood as the value 'h_c' comes larger as the term $\lambda a_1/l$ in equation (13) increases because of increasing true contact area 'a_1' and, at the same time, the term αa_2 happens not to decrease.

Flow visualization and temperature distribution measurement near the contact surfaces between both cylinders were tried. Fig.13 (a-h) show the stream pattern by smoke test. Under rotating speed such low as 30 rpm, the stream near the contact surfaces is dragged at the rolling-in side and blows out through the clearance to the rolling-out side. At the rolling-out side, the stream coming up along the cylinder surface is encountered the stream passed through the contacts clearance, resulting in a small eddy. This phenomena fades out as the rotating speed increases, as shown in Fig. 13 (c),(d). Meanwhile, the stream at the rolling-in side along the upper cylinder begins to separate away almost in horizontal direction at 200 rpm, and at the rolling-out side, according to Coanda's effect, the stream separates from the surface of the lower cylinder and stick to the upper surface. And then, at 300 rpm, we can observe the stream comes to turbulence in horizontal direction at the rolling-in side and a clear Coanda's effect happens at the rolling -out side.

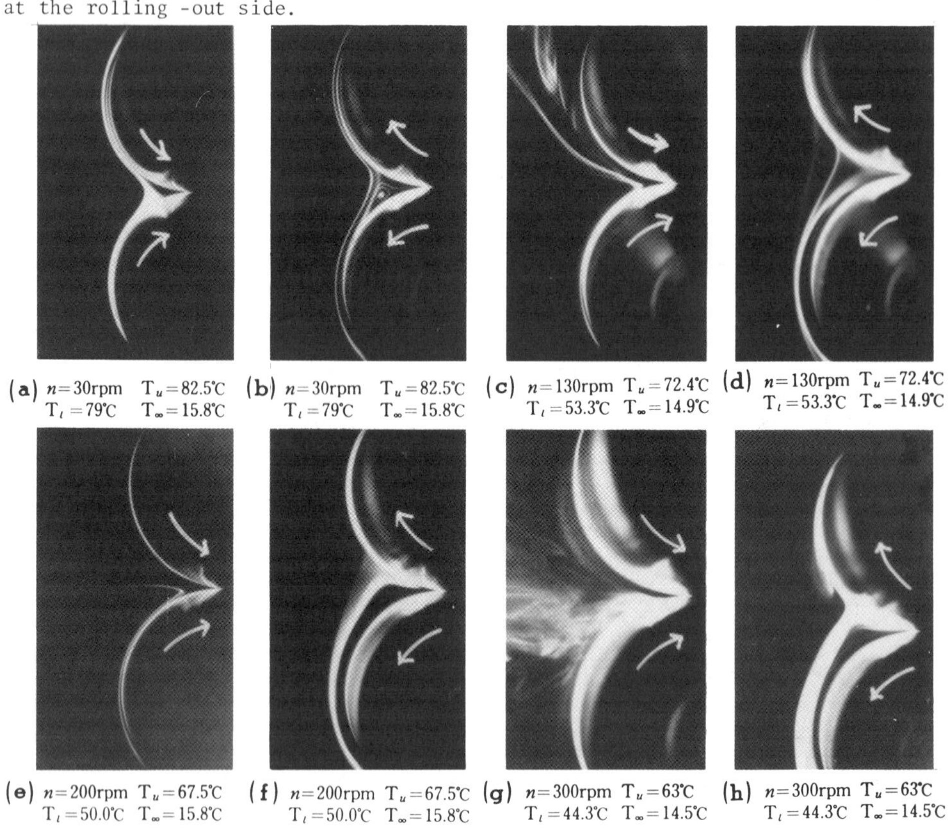

(a) n=30rpm T_u=82.5°C T_l=79°C T_∞=15.8°C
(b) n=30rpm T_u=82.5°C T_l=79°C T_∞=15.8°C
(c) n=130rpm T_u=72.4°C T_l=53.3°C T_∞=14.9°C
(d) n=130rpm T_u=72.4°C T_l=53.3°C T_∞=14.9°C
(e) n=200rpm T_u=67.5°C T_l=50.0°C T_∞=15.8°C
(f) n=200rpm T_u=67.5°C T_l=50.0°C T_∞=15.8°C
(g) n=300rpm T_u=63°C T_l=44.3°C T_∞=14.5°C
(h) n=300rpm T_u=63°C T_l=44.3°C T_∞=14.5°C

FIGURE 13. Stream pattern by smoke test

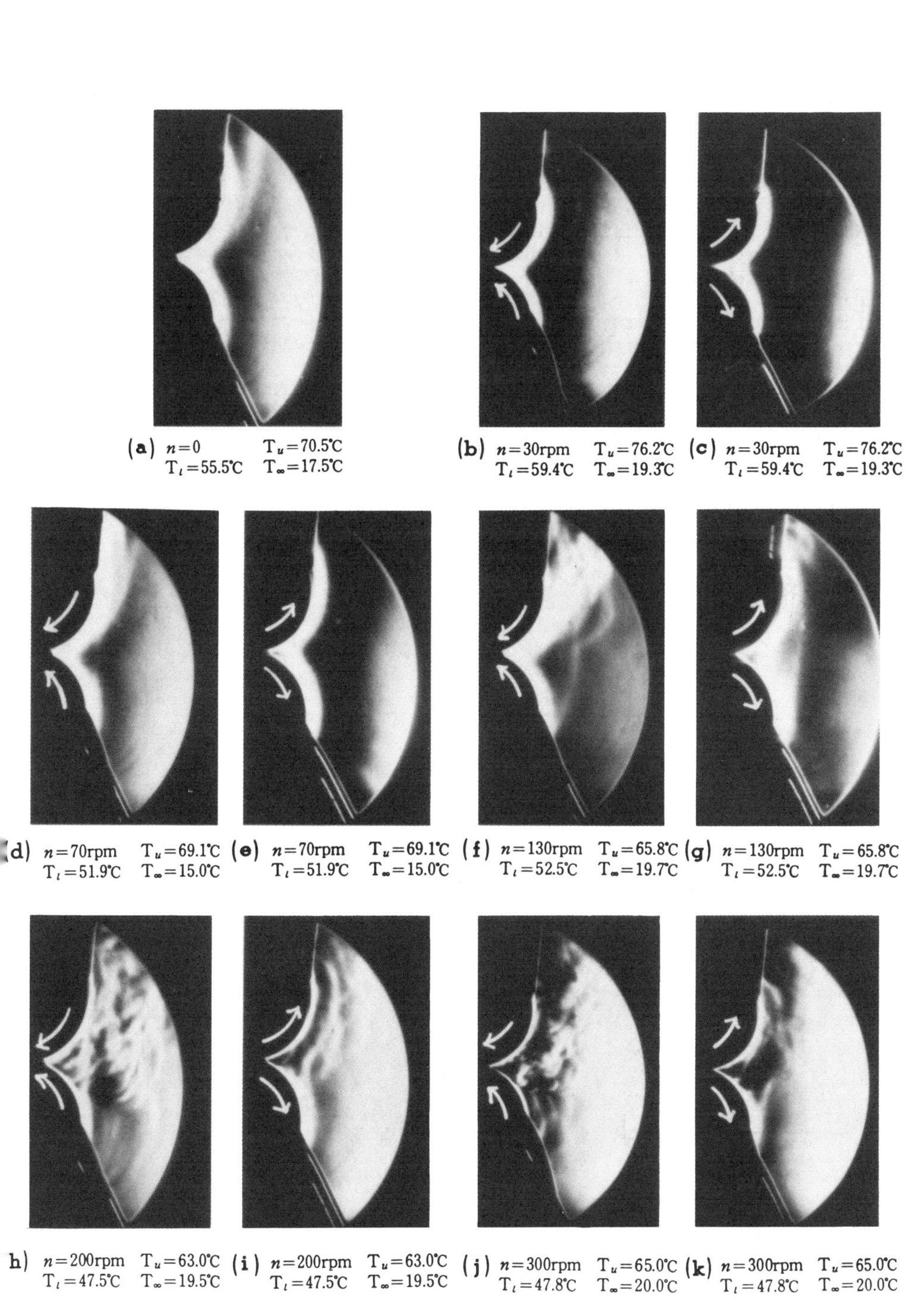

(a) $n=0$ $T_u=70.5°C$ $T_l=55.5°C$ $T_\infty=17.5°C$

(b) $n=30$rpm $T_u=76.2°C$ $T_l=59.4°C$ $T_\infty=19.3°C$

(c) $n=30$rpm $T_u=76.2°C$ $T_l=59.4°C$ $T_\infty=19.3°C$

(d) $n=70$rpm $T_u=69.1°C$ $T_l=51.9°C$ $T_\infty=15.0°C$

(e) $n=70$rpm $T_u=69.1°C$ $T_l=51.9°C$ $T_\infty=15.0°C$

(f) $n=130$rpm $T_u=65.8°C$ $T_l=52.5°C$ $T_\infty=19.7°C$

(g) $n=130$rpm $T_u=65.8°C$ $T_l=52.5°C$ $T_\infty=19.7°C$

(h) $n=200$rpm $T_u=63.0°C$ $T_l=47.5°C$ $T_\infty=19.5°C$

(i) $n=200$rpm $T_u=63.0°C$ $T_l=47.5°C$ $T_\infty=19.5°C$

(j) $n=300$rpm $T_u=65.0°C$ $T_l=47.8°C$ $T_\infty=20.0°C$

(k) $n=300$rpm $T_u=65.0°C$ $T_l=47.8°C$ $T_\infty=20.0°C$

FIGURE 14. Schlieren's photographs

Fig.14 (a-k) shows approximate temperature distribution by the photo-contrast observed by means of Schlieren's method. At the stationary state (a), the thermal boundary layer of the upper seems to be formed as similar to the lower. But concerning the rolling-in side the layer at the upper is observed to increase its thickness in comparison with that at the lower as the rotating speed increases and then clearly burst into turbulence beyond 200 rpm. Concerning the rolling-out side, the layer at the lower keeps thicker than that at the upper and this trend seems to be independent of the value 'n'.

Now, the temperature distributions around the cylinders are shown in Fig.15(a-f). At the rolling-in side, the isothermal line comes to separate from the upper and is observed turbulent as the value 'n' increases. This means that the upper cylinder is cooled by the induced air effected by the lower. Concerning the rolling-out side, the temperature near the cylinder surface decreases as the value 'n' increases. To observe the thermal behavior near the contact zone, putting the origin of the co-ordinates at the contact point and the co-ordinate x in horizontal direction, the temperature at x=-10 mm (at the rolling-in side) shows higher than that of the lower, but the temperature at x=+10 mm (at the rolling-out side) shows lower than that of the lower.

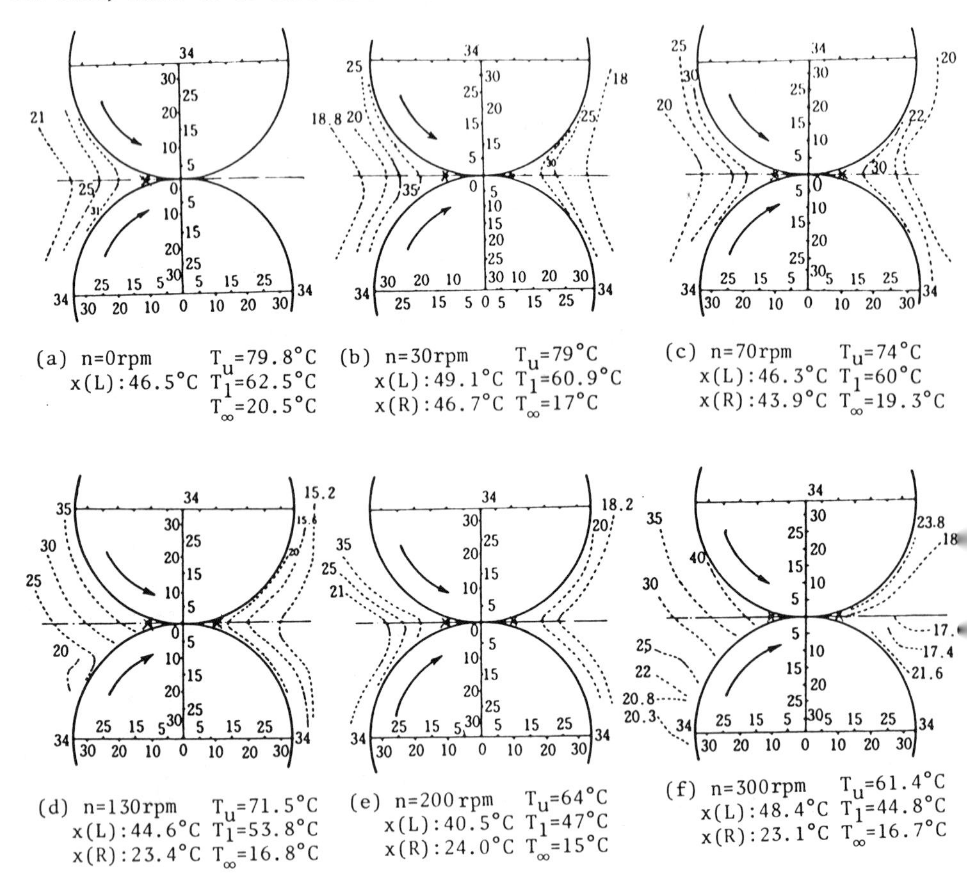

(a) n=0rpm T_u=79.8°C x(L):46.5°C T_1=62.5°C T_∞=20.5°C

(b) n=30rpm T_u=79°C x(L):49.1°C T_1=60.9°C x(R):46.7°C T_∞=17°C

(c) n=70rpm T_u=74°C x(L):46.3°C T_1=60°C x(R):43.9°C T_∞=19.3°C

(d) n=130rpm T_u=71.5°C x(L):44.6°C T_1=53.8°C x(R):23.4°C T_∞=16.8°C

(e) n=200rpm T_u=64°C x(L):40.5°C T_1=47°C x(R):24.0°C T_∞=15°C

(f) n=300rpm T_u=61.4°C x(L):48.4°C T_1=44.8°C x(R):23.1°C T_∞=16.7°C

x(L):left point marked x , x(R):right point marked x

FIGURE 15. Temperature distributuons around the cylinders

From the results of observation above mentioned, it can be understood that the thermal phenomena seems to depend on the contactive heat transfer through the contact area, not on the convective heat transfer.

It is known that under the stationary state the contact thermal resistance becomes smaller than the value measured at the beginning when the measurement is repeated under the former low contact pressure after high contact pressure was loaded. Fig.16 is a result measured for examining such a point about two rotating cylinder. As a consequence, the result shows similar trend to the cace of the stationary state.

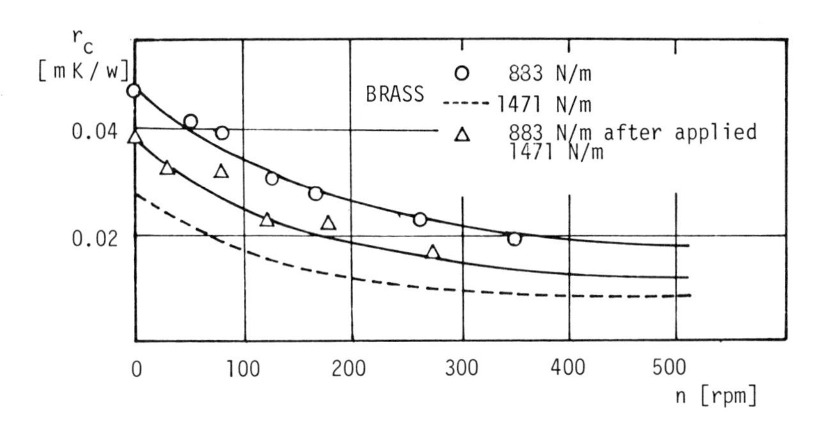

FIGURE 16. Contact thermal resistance after applied high contact pressure

The effect of frictional heat generated on the contact sufaces between two rotating cylinders were also measured, but the heat generated on the contact surfaces was to the surrounding air by the convection and there was no influence to both cylinder.

Fig.17 and 18 show the relation between N_{uh} (Nusselt number based on the thermal conductance defined by the equation (9)) and R_e (rotating Reynolds number).

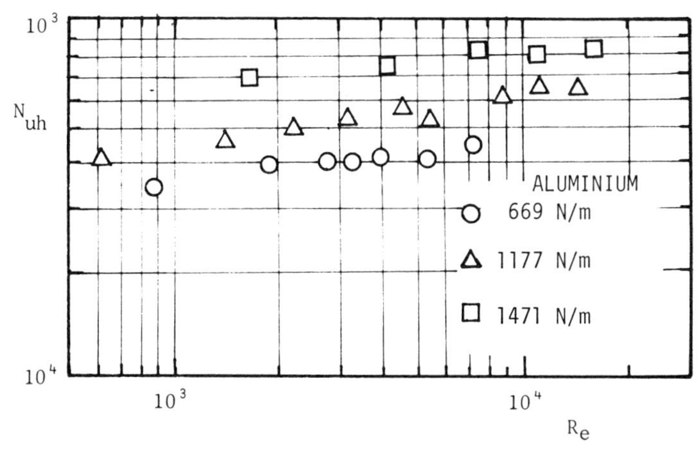

FIGURE 17. The relation between Nusselt number and rotating Reynolds number

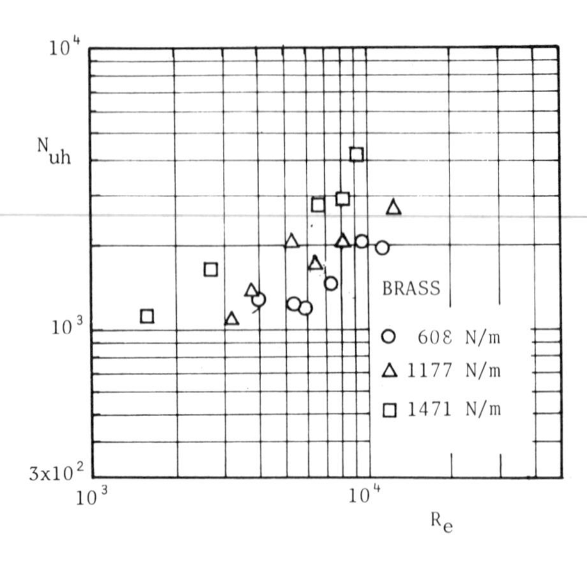

FIGURE 18. The relation between Nusselt number and rotating Reynolds number

As the result, that is described as the following equation (14),

$$N_{uh} = C\ R_e^{\ n} \tag{14}$$

where C is a constant.

CONCLUSION

From the experimental investigation about the thermal resistance between two horizontal rotating cylinders, we come to the following conclusions:

1. The thermal resistance decreases as the rotating speed of the cylinders increases.
2. The thermal resistance decreases,as the contact load increases and after the contact pressure is loaded high.
3. The thermal resistance shows the value proportional to the hardness of the material.
4. The relation berween Nusselt number based on the heat conductance and rotating Reynolds number based on the rotating speed is represented as equation (14).

Impeller Shroud to Casing Leakage Flow Simulations in the Space Shuttle Main Engine High Pressure Fuel Pump

MUNIR M. SINDIR
Rockwell International
Rocketdyne Division
Canoga Park, California 91304

ABSTRACT

Quasi 3-D Navier-Stokes calculations were carried out for the Space Shuttle Main Engine high pressure fuel pump to simulate the impeller shroud to casing leakage flow. This flow geometry was modeled as an axisymmetric cavity flow with a stationary surface representing the casing, and a rotating surface denoting the impeller. A 63 by 81 node mesh provided sufficient resolution in the regions of greatest flow variations and reduced the effects of numerical diffusion. The turbulence field was closed with the high Reynolds number form of the k-ε model supplemented with wall functions in the vicinity of the walls. Finally, a parametric study quantified the effects of through mass flow changes on this leakage flow.

INTRODUCTION

Accurate prediction of internal turbomachinery flow dynamics is essential in identifying possible failure modes and establishing the coupling between cyclic loading and structural dynamics. Quasi 3-D Navier-Stokes calculations were performed in this context for the Space Shuttle Main Engine (SSME) high pressure fuel pump (HPFP) to simulate the impeller shroud to casing leakage flow.

This leakage flow follows the path indicated in Figure 1, entering from the tip of the impeller and exiting near the hub. This geometry was modeled as an axisymmetric cavity flow with a stationary surface representing the casing and a rotating surface denoting the impeller. The flow parameters, including the cavity gap L, cavity height R, angular velocity Ω of the impeller, and the inlet flow conditions were determined from the design case.

The Cavity Flow Code (CFC) developed for this analysis is a derivative of the author's STEP family of computer programs [1]. These general purpose Navier-Stokes codes use the basic structure and discretization technique of the TEACH family of computer programs [2] assembled at Imperial College, London. The STEP codes solve 2-D and quasi 3-D steady and time-dependent elliptic partial differential equations through an iterative under-relaxation procedure based on an integral control volume analysis with hybrid upwind finite differencing and staggered grids. Both cartesian and axisymmetric grids with arbitrary spacing can be specified. These codes solve for the primitive mean flow variables (U, V, W, P) and the turbulence parameters (k and ε) from differential transport equations, and for the Reynolds stresses ($\overline{u_i u_j}$) from algebraic

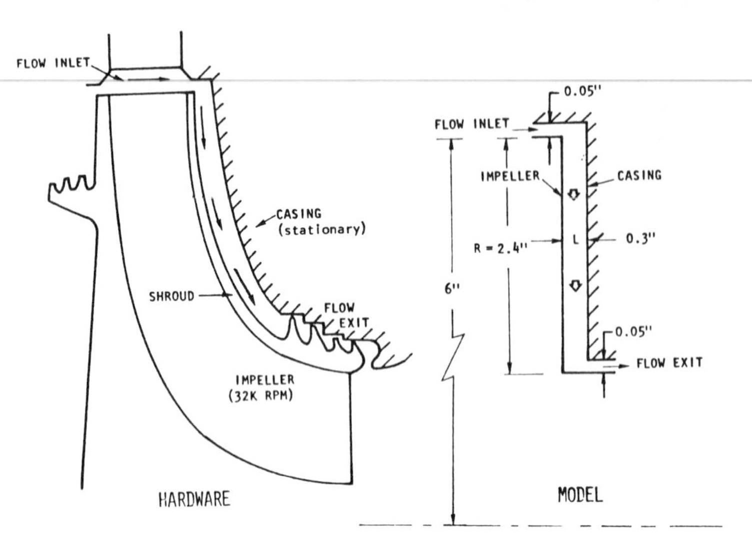

Figure 1. Flow geometry.

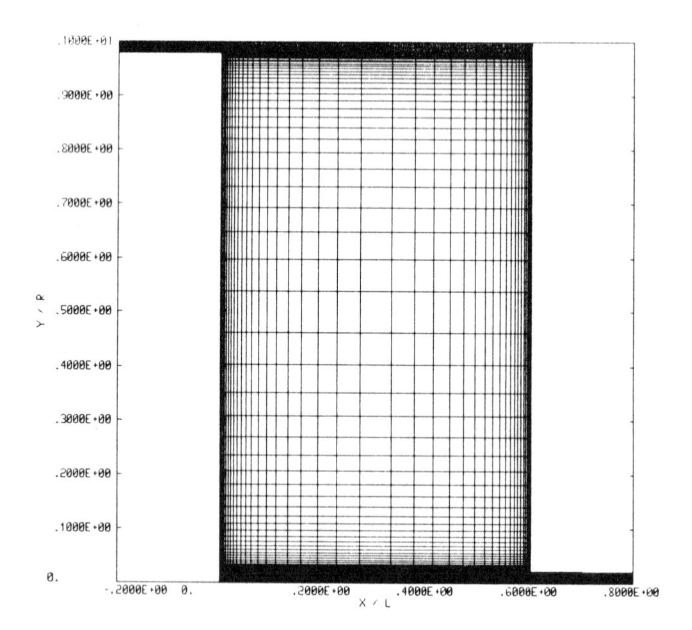

Figure 2. Mesh.

relationships. The turbulence field is closed with the k-ε [3] or the more advanced algebraic Reynolds stress models [1,4]. These codes were tested extensively on complex turbulent flows including recirculating [1,5,6,7,8] and swirling [9,10] flows, and wall jets in adverse pressure gradients [6,10]. The STEP codes are described in detail in [1].

The cavity flow geometry shown in Figure 1 was built into the CFC code for the HPFP impeller shroud to casing flow predictions. The rotational boundary condition for the impeller was modeled as a no-slip velocity condition for the computational cells adjacent to the rotating surface. A 63 by 81 node nonuniform mesh was set up with at least 10 grid points in each of the wall boundary layers. This provided sufficient resolution in the regions of greatest flow variations and enabled placement of the first grid point at a distance X/L of 0.002 (0.00015 m) from the wall. This assured y^+ values in the range of 100-300, a requirement for the successful use of wall functions. This mesh is shown in Figure 2. For these calculations, the turbulence field was closed with the high Reynolds number form of the k-ε model supplemented with wall functions in the vicinity of the walls. Convergence required approximately 2000 iterations for a total computation time (CYBER 875 CPU) of 85 minutes for the design case. The off-design low and high mass flow cases (for the through mass flow study) required 20 percent higher and 10 percent lower computation times, respectively.

GOVERNING EQUATIONS AND CLOSURE MODELS

The governing equations of motion for turbulent axisymmetric flows are:

Conservation of Mass

$$\frac{\partial \rho}{\partial t} + \rho \left\{ \frac{\partial u}{\partial x} + \frac{1}{r}\frac{\partial}{\partial r}(rV) \right\} = 0 \quad (1)$$

Axial Momentum

$$\rho \frac{\partial U}{\partial t} + \rho U \frac{\partial U}{\partial x} + \rho V \frac{\partial U}{\partial r} = -\frac{\partial P}{\partial x} + \frac{\partial}{\partial x}\left[2\mu \frac{\partial U}{\partial x} - 2/3\mu\, \emptyset - \rho \overline{u^2} \right]$$

$$+ \frac{1}{r}\frac{\partial}{\partial r}\left[r \left\{ \mu \left(\frac{\partial U}{\partial r} + \frac{\partial V}{\partial x} \right) - \rho \overline{uv} \right\} \right] \quad (2a)$$

Radial Momentum

$$\rho \frac{\partial V}{\partial t} + \rho U \frac{\partial V}{\partial x} + \rho V \frac{\partial V}{\partial r} - \rho \frac{W^2}{r} = -\frac{\partial P}{\partial r} + \frac{\partial}{\partial x}\left[\mu \left(\frac{\partial u}{\partial r} + \frac{\partial v}{\partial x} \right) - \rho \overline{uv} \right]$$

$$+ \frac{1}{r}\frac{\partial}{\partial r}\left[r \left\{ 2\mu \frac{\partial V}{\partial r} - 2/3\mu\emptyset - \rho \overline{v^2} \right\} \right]$$

$$- \frac{1}{r}\left[2\mu \frac{V}{r} - 2/3\mu\emptyset - \rho \overline{w^2} \right\} \Big] \quad (2b)$$

Tangential Momentum

$$\rho \frac{\partial W}{\partial t} + \rho U \frac{\partial W}{\partial x} + \rho V \frac{\partial W}{\partial r} + \rho \frac{VW}{r} = \frac{\partial}{\partial x}\left[\mu \frac{\partial W}{\partial x} - \rho \overline{uw} \right]$$

$$+\frac{1}{r^2}\frac{\partial}{\partial r}\left[r^2\left\{\mu r\frac{\partial}{\partial r}\left(\frac{W}{r}\right)-\rho\overline{vw}\right\}\right] \quad (2c)$$

$$\emptyset \equiv \frac{\partial U}{\partial x}+\frac{1}{r}\frac{\partial}{\partial r}(rV)$$

where U, V, W are the axial, radial, and tangential components of the mean velocity, P is the pressure, ρ and μ are the fluid density and dynamic viscosity, respectively. $\overline{u^2}$, $\overline{v^2}$, $\overline{w^2}$, $\overline{uv}$, $\overline{uw}$, $\overline{vw}$ are the Reynolds stresses which need to be modeled to close the turbulence field.

k-ε Model

The k-ε model achieves closure by relating the Reynolds stresses to the mean strain rate through the Boussinesq approximation:

$$-\rho\,\overline{u_i u_j}=\mu_t\left(\frac{\partial U_i}{\partial x_j}+\frac{\partial U_j}{\partial x_i}\right)-2/3\,\delta_{ij}\,\rho k \quad (3)$$

The effective viscosity appearing above, μ_t, is defined in terms of a characteristic length and velocity. If this length is taken as the turbulence length scale, $k^{3/2}/\varepsilon$, and the velocity as $k^{1/2}$, μ_t can be expressed as

$$\mu_t \equiv c_\mu \rho k^2/\varepsilon \quad (4)$$

where k is the turbulence kinetic energy, ε is the dissipation rate, and c_μ is a constant of proportionality. The Reynolds stresses are then defined as

$$-\rho\,\overline{u}^2=2\mu_t\frac{\partial U}{\partial x}-2/3\,\rho k \quad (5)$$

$$-\rho\,\overline{v}^2=2\mu_t\frac{\partial V}{\partial r}-2/3\,\rho k \quad (6)$$

$$-\rho\,\overline{w}^2=2\mu_t\frac{W}{r}-2/3\,\rho k \quad (7)$$

$$-\overline{\rho uv}=\mu_t\left(\frac{\partial U}{\partial r}+\frac{\partial V}{\partial x}\right) \quad (8)$$

$$-\overline{\rho uw}=\mu_t\frac{\partial W}{\partial x} \quad (9)$$

$$-\overline{\rho vw}=\mu_t\, r\frac{\partial}{\partial r}\left(\frac{W}{r}\right) \quad (10)$$

k and ε Transport Equations

The k-ε model, as discussed, requires evaluation of the turbulent kinetic energy and its dissipation rate to define the turbulent time and length scales.

The high Reynolds number forms of the k and ε transport equations used in this study are

$$\rho \frac{\partial k}{\partial t} + \rho U \frac{\partial k}{\partial x} + \rho V \frac{\partial k}{\partial r} = \rho G - \rho \varepsilon + \frac{\partial}{\partial x} D_{k_x} + \frac{1}{r}\frac{\partial}{\partial r}(r D_{k_r}) \tag{11}$$

$$\rho \frac{\partial \varepsilon}{\partial t} + \rho U \frac{\partial \varepsilon}{\partial x} + \rho V \frac{\partial \varepsilon}{\partial r} = \rho\, \varepsilon/k\, (c_{\varepsilon_1} G - c_{\varepsilon_2}\, \varepsilon) + \frac{\partial}{\partial x} D_{\varepsilon_x} + \frac{1}{r}\frac{\partial}{\partial r}(r D_{\varepsilon_r}) \tag{12}$$

where

$$D_{k_x} = \left(\frac{\mu_t}{\sigma_k} + \mu\right)\frac{\partial k}{\partial x} \tag{13}$$

$$D_{k_r} = \left(\frac{\mu_t}{\sigma_k} + \mu\right)\frac{\partial k}{\partial r} \tag{14}$$

$$D_{\varepsilon_x} = \left(\frac{\mu_t}{\sigma_\varepsilon} + \mu\right)\frac{\partial \varepsilon}{\partial x} \tag{15}$$

$$D_{\varepsilon_r} = \left(\frac{\mu_t}{\sigma_\varepsilon} + \mu\right)\frac{\partial \varepsilon}{\partial r} \tag{16}$$

$$G = \frac{\mu_t}{\rho}\left[2\left\{\left(\frac{\partial U}{\partial x}\right)^2 + \left(\frac{V}{r}^2\right) + \left(\frac{\partial V}{\partial r}^2\right)\right\} + \left(\frac{\partial U}{\partial r} + \frac{\partial V}{\partial x}\right)^2 + \left(r\frac{\partial}{\partial r}\left(\frac{W}{r}\right)\right)^2 + \left(\frac{\partial W}{\partial x}\right)^2\right] - 2/3\, k \left[\frac{\partial U}{\partial x} + \frac{V}{r} + \frac{\partial V}{\partial r}\right] \tag{17}$$

σ_k and σ_ε are turbulent Prandtl numbers for k and ε, respectively. These are defined in Table 1 with model constants c_{ε_1} and c_{ε_2}.

TABLE 1. Turbulence model constants

c_μ	=	0.09
c_{ε_1}	=	1.44
c_{ε_2}	=	1.92
σ_k	=	1.00
σ_ε	=	1.22

Wall Function Treatment

Most turbulence models including the present version of the k-ε model are devices for high Reynolds number flows. However, in the vicinity of solid boundaries where the velocities are small, the low Reynolds number effects previously neglected become significant and should be accounted for. This can be accomplished either by solving the low Reynolds number form of the transport equations or by developing wall functions that introduce these effects into the existing high Reynolds number models. Chieng and Launder [12] found that the first option required vast amounts of computer time due to the slow convergence characteristics of the low Reynolds number models. On the other hand, a new wall function treatment proposed by the same authors was shown to incorporate these effects with practically no increase in computing time. An expanded version of this treatment was used in the present study. Details of this approach are given in [1].

RESULTS

Three variations of the HPFP impeller shroud to casing flow corresponding to different inlet conditions (Table 2) were calculated. Case 1 is the design flow and Cases 2 and 3 represent, respectively, lower (LMF) and higher (HMF) than design through flow rates. These three cases were selected to bound the HPFP operation range and parameterize performance in terms of through mass flow. The low mass flow case also served as a code verification test case since it is a good approximation to rotating disks in housings for which experimental correlations exist [12]. No detailed experimental data are as yet available for the design case.

TABLE 2. Inlet flow conditions - fuel H_2

	$M_{in.}$ (kg/s)	$U_{in.}$ (m/s)	$W^1_{in.}$ (m/s)
Case 1			
Design Flow	0.583	7.96	362
Case 2			
Low Mass Flow - LMF	0.0583	0.796	362
Case 3			
High Mass Flow - HMF	2.915	39.8	362

[1]Taken as 70 percent of the tip speed.

Calculations for the design case indicate a strong secondary flow in the cavity with boundary layers on both the impeller and the casing. These boundary layers bracket a large stagnant region with no appreciable axial or radial velocity components as shown in Figure 3. This stagnant region rotates at 60 to 78 percent of the local shaft speed (Figure 4) depending on the radial location, Y/R, inside the cavity. Axial velocities do not exceed 10 percent of the inlet velocity (U_{ref}) for the bulk of the flow (Figure 5), and vary very little axially, except at the high and low Y/R regions where the inflow

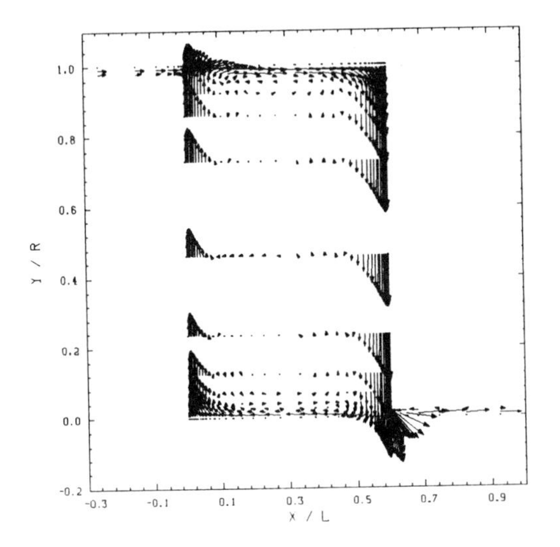

Figure 3. Velocity vectors - design case.

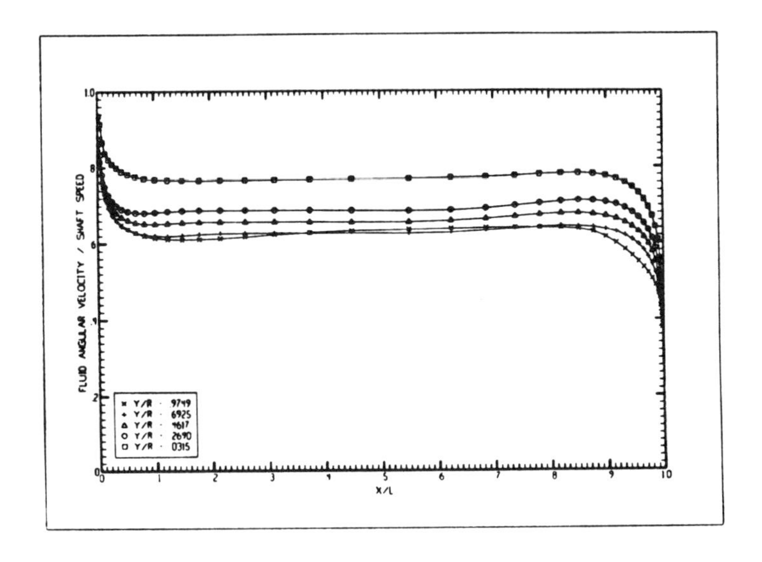

Figure 4. Tangential velocities.

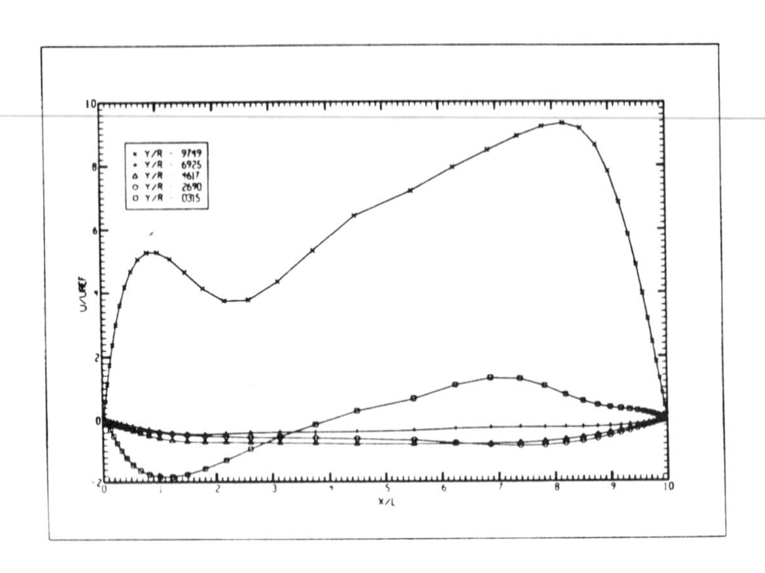

Figure 5. Axial velocities.

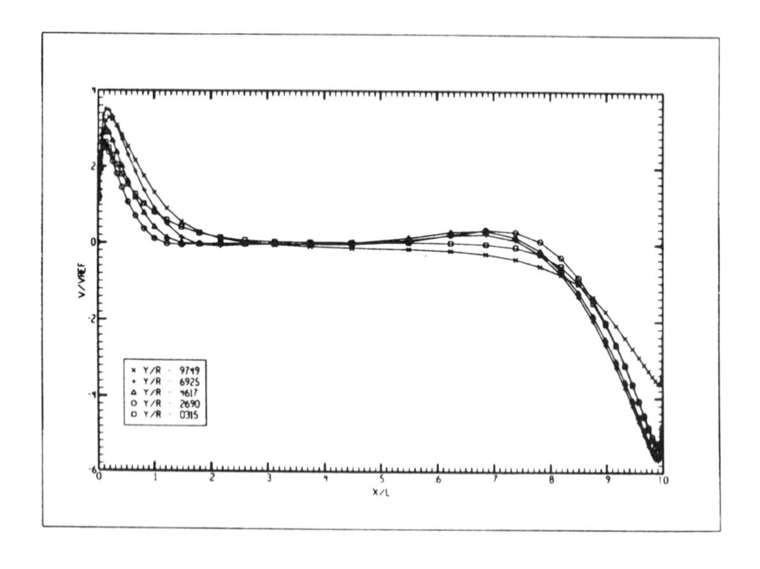

Figure 6. Radial velocities.

and outflow conditions affect the flow behavior. Radial velocities are large in the boundary layers (Figure 6) where the fluid near the impeller is forced outwards by the centrifugal acceleration and a compensating flow is set up inwards in the boundary layer on the stationary casing. This creates the strong secondary flow in the cavity. As shown in Figure 7, there is essentially no axial pressure gradient at any location. However, this flow maintains a significant pressure drop radially.

The gross effects of the through mass flow changes are shown graphically in Figures 3, 8, and 9 for the design, low mass flow (LMF), and high mass flow (HMF) cases, respectively. High mass flow rates appear to diminish the boundary layer thickness and the radial velocities near the impeller, and produce a more Couette like flow. For this case, the incoming flow jets into the cavity and follows the contour of the casing as a wall jet before exiting at relatively high speeds near the hub. This reduces the secondary flow in the cavity. W/W_{shaft}, however, increases as depicted in Figure 10, primarily due to the larger influx of tangential momentum corresponding to the higher through flow rate. HMF also produces the largest radial pressure drop as shown in Figure 11. The LMF case, on the other hand, behaves more like a true rotating disk in a housing with a strong secondary flow and well established boundary layers of roughly equal thickness on both the impeller and the casing. The stagnant region between the boundary layers rotates at 56 percent of the shaft speed as compared to 73 percent for the HMF case. The design case, as expected, lies between the LMF and HMF results.

CONCLUDING REMARKS

This work demonstrates that complex multi-dimensional turbomachinery problems can now be calculated realistically with full Navier-Stokes formulations, reasonable turbulence models, and fine meshes, for modest computation times. Future advances in CFD and computer technology will make numerical simulation even a more powerful tool in turbomachinery design and analysis. The following conclusions can be drawn from this study on the HPFP leakage flow:

1. The flowfield is completely dominated by the centrifugal forces resulting from the rotational motion of the impeller. This sets up a strong secondary flow in the cavity with boundary layers on both the impeller and the casing. The stagnant region between the boundary layers rotates at 60 to 78 percent of the local shaft speed depending on the radial location. There is only a radial pressure drop in the cavity with no pressure changes in the axial direction.
2. Higher through flow rates reduce the boundary layer on the impeller and form a wall jet that follows the contour of the casing before exiting. Lower through flow rates, on the other hand, behave more like a true rotating disk in a housing with a strong secondary flow and well established boundary layers on both the impeller and the casing. Higher through flow rates also increase both the rotational speed of the stagnant region and the radial pressure drop in the cavity. The design case lies between these LMF and HMF results.
3. Complex flowfields with recirculation zones, such as the present case, are especially susceptible to numerical diffusion which degrades the accuracy of the calculations. In this work, refining the mesh from a 42 by 71 to a 63 by 81 node grid changed the results by up to 20 percent.

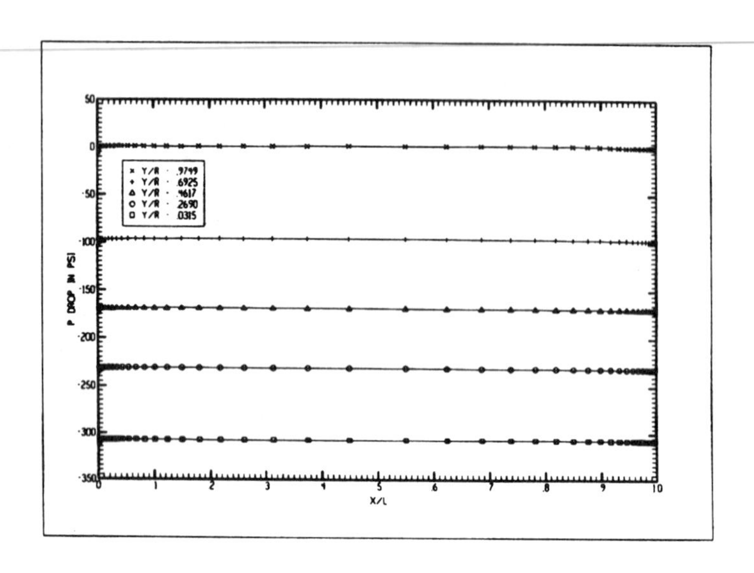

Figure 7. Pressure drop profiles

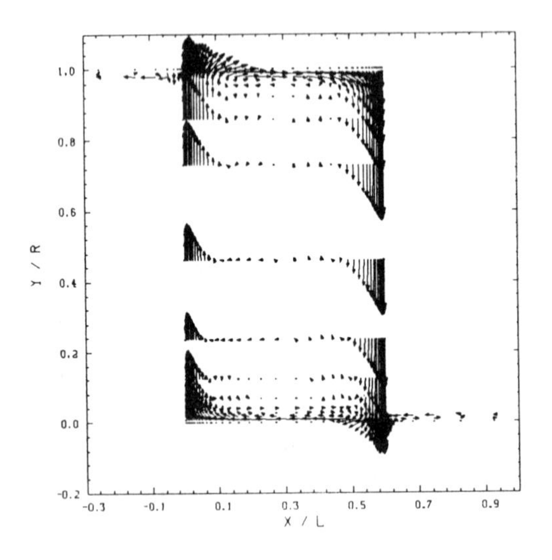

Figure 8. Velocity vectors - LMF case.

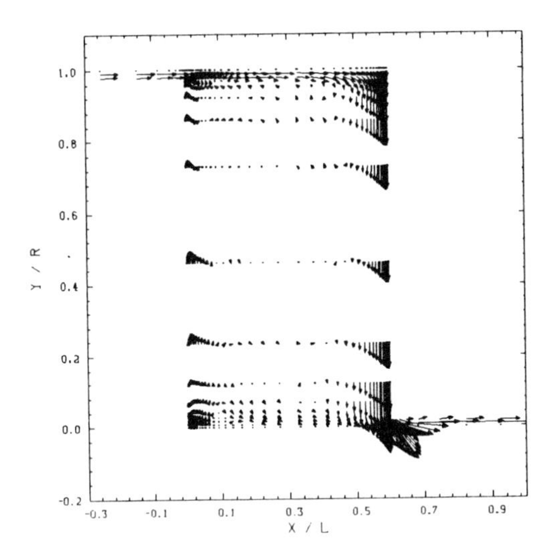

Figure 9. Velocity vectors - HMF case.

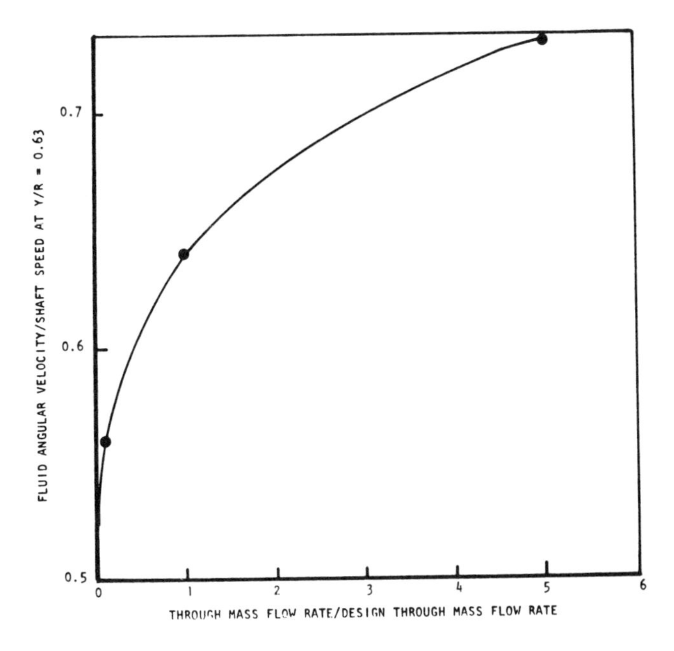

Figure 10. Effects of through mass flow changes on tangential velocity.

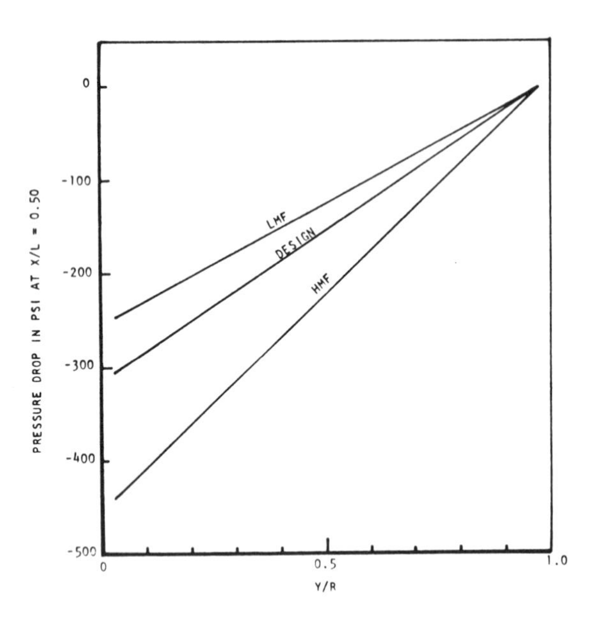

Figure 11. Effects of through mass flow changes on pressure drop.

REFERENCES

1. Sindir, M. M., "Numerical Study of Turbulent Flow in Backward-Facing Step Geometries - Comparison of Four Models of Turbulence," Ph.D, Thesis, University of California - Davis, June 1982.

2. Gosman, A. D. and Ideriah, F. J. K., "TEACH-2E: A General Computer Program for Two-Dimensional, Turbulent Recirculating Flows," Department of Mechanical Engineering Internal Report, Imperial College, University of London, 1978.

3. Jones, W. P., and Launder, B. E., "The Prediction of Laminarization with a Two-Equation Model of Turbulence," Int. J. Heat and Mass Transfer, Vol. 15, 1972, pp. 301-314, 1972

4. Rodi, W., "The Prediction of Free Turbulent Boundary Layers by Use of a Two-Equation Model of Turbulence," Ph.D. Thesis, University of London, 1972.

5. Sindir, M. M., "Calculation of Deflected-Walled Backward-Facing Step Flows: Effects of Angle of Deflection on the Performance of Four Models of Turbulence," ASME Paper 83-FE-16, 1983.

6. Sindir, M. M., "Effects of Expansion Ratio on the Calculation of Parallel-Walled Backward-Facing Step Flows: Comparison of Four Models of Turbulence," ASME Paper 83-FE-10, 1983.

7. Sindir, M. M., and Harsha, P. T., "Assessment of Turbulence Models for Scramjet Flowfields," NASA Contractor Report 3643, November 1982.

8. Launder, B. E., Leschziner, M. A., and Sindir, M. M., "The UMIST/UCD Computations of Complex Turbulent Flows," Proc. Stanford Conference on Complex Turbulent Flows, Stanford University, California, 1982.

9. Sindir, M. M. and Harsha, P. T., "Turbulent Transport Models for Scramjet Flowfields," NASA Contractor Report on Contract NAS1-15988, November 1983.

10. Harsha, P. T., Wang, T-S, Sindir, M. M., and Edelman, R. B., "Analytical Models for Industrial Gas Burner Design," GRI Topical Report on Contract No. 5011-345-0100, January 1983.

11. Chieng, C. C., and Launder, B. E., "On the Calculation of Turbulent Heat Transport Downstream of an Abrupt Pipe Expansion," Numerical Heat Transfer, Vol 3, pp. 189-207, 1980.

12. Schlichting, H., Boundary Layer Theory, McGraw-Hill, New York, 1968.

ACKNOWLEDGEMENT

This work was performed at the Rocketdyne Division of Rockwell International for the Rotating Machinery Analysis unit. The work was supported by NASA Space Shuttle Contract NAS8-27980 from the George. C. Marshall Space Flight Center.

Finite Difference Solution for a Generalized Reynolds Equation with Homogeneous Two-Phase Flow

M. J. BRAUN
Department of Mechanical Engineering
University of Akron
Akron, Ohio 44325

R. L. WHEELER III
Department of Mechanical Engineering
University of Akron
Akron, Ohio 44325

R. C. HENDRICKS
NASA Lewis Research Center
MS-23-2
Cleveland, Ohio 44135

R. L. MULLEN
Department of Civil Engineering
Case Western Reserve
Cleveland, Ohio 44106

1. INTRODUCTION

Cavitation plays a most important role in hydraulic turbomachinery used today. The phenomenon, is closely associated with the thermodynamic state of the working fluid and has been regarded, as far as journal bearings are concerned, to be the result of the fracture of the thin fluid film located between the bearing and the rotating journal. Braun and Hendricks [1,2] explained through the experiments they performed both the fluid fracture and incipient cavitation, during the startup. They also observed the propagation of the cavity to its final equilibrium steady-state form. For the case of hydrodynamic bearings, there is a wealth of information regarding the film rupture and formation of cavity. Hendricks, Mullen and Braun [3] proposed an analytical approach to explain the rupture of the film and its subsequent propagation. Sommerfeld [4], Swift [5], Stieber [6], Coyne and Elrod [7] and Floberg [8] analyzed the limiting conditions of the film rupture.

Face seals and squeeze film dampers represent another category of hydrodynamic devices which exhibit cavitation as a form of a two-phase system. However, most of the studies have treated the region where the fluid lubricant ruptures (cavitates) as being at zero pressure. The assumption of vanishing fluid pressure in the cavitation zone leads to erroneous results [9-11]. In actuality the pressure in the cavitation zone while falling below saturation, never reaches zero, and furthermore varies throughout the cavity. It has been shown by Braun [12], Kauzlarich [10] and Parkins [9] that under certain cavitation conditions, the fluid throughout will be in a two-phase homogeneous state and provide a significant pressure contribution to the system as a whole.

During flashing, pressures fall beneath the saturation pressure and vapor is produced throughout the entire mass of the depressurized fluid. Zuber and Dougherty [13] have derived an equation for a dispersed two-phase lubrication film. They have introduced a sink/source term in their continuity equation which depends on the strength of condensation/evaporation induced by the temperature and pressure conditions.

We shall expand here on their work by applying the concept to the generalized Reynolds equation as it was presented by Dowson [14] and Fowles [15] and couple it to the kinetic theory of nucleation, to yield a model for the sink/source term. The Reynolds equation with variable properties is coupled with a set of energy equations representing heat balances for the rotating shaft (3-D), fluid wedge (2-D) and the stationary outer bearing (3-D).

2. ANALYSIS

2.1 Geometry and Assumptions

The geometry used is described in Fig. 1. The shaft's arbitrary position in the bearing is defined by the position of O_j with respect to O_b in Fig. 1b and by the rotations β_x around the O'X' and β_y around O'Y' respectively, Fig. 1a. The following assumptions were made:

i) The radius of curvature of the journal and bearing are much larger than the thickness of the fluid film. This allows us to neglect the curvature effects, and permits the oil film to be unwrapped, with one of the surfaces represented by the plane y=0. The system of coordinates will be cartesian in nature with x and z representing the circumferential (unwrapped) and axial directions respectively. The clearance h will be

$$h = h(x, z, t) \tag{1}$$

ii) The lubricant is considered a Newtonian fluid and the inertia and body forces are neglected.

iii) There is no slip between the fluid and a moving solid boundary.

iv) The magnitude of the velocities derivative $\partial u/\partial x$ and $\partial v/\partial z$ is large when dimensionally compared to other velocity gradients.

v) The working fluid properties, the density ρ and the dynamic viscosity η are

$$\begin{aligned} \rho &= \rho(x,y,z) \\ \eta &= \eta(x,y,z) \end{aligned} \tag{2}$$

where ρ and η represent the properties of the mixture.

vi) The effect of the bubbling gas/vapor out of the liquid is to change the effective density of the liquid-gas (vapor) mixture which is considered homogeneous.

2.2 Mixture Quality Equation

In view of the assumptions made at (v) and (vi), the continuity equation for the homogeneous mixture is (using Einstein index notation):

$$\frac{\partial \rho}{\partial t} + \frac{\partial}{\partial x_i}(\rho u_i) = 0 \qquad i = x,z \tag{3}$$

A separate equation is written for the vapor phase which ensures continuity and accounts for the vapor source term. Denoting the quality of the mixture by α the vapor phase equation is

$$\frac{\partial (\rho\alpha)}{\partial t} + \frac{\partial}{\partial x_i}(\rho\alpha u_i) = \dot{m}_{g+} \qquad (i = x,z) \tag{4}$$

In Eq. 4, the right hand side, $\dot{m}_{g+}$, represents the vapor mass flux generation term and the group $(\rho\alpha)$ can be expressed as

$$\rho\alpha = (\rho_f \rho_g \alpha)/[\rho_g(1-\alpha) + \rho_f\,\alpha] \tag{5a}$$

and

$$\eta = \eta_g \, \eta_f / [\eta_f \alpha + \eta_g (1-\alpha)] \tag{5b}$$

Expanding Eq. 4 and accounting for Eq. 3 one can rewrite Eq. 4 as

$$\rho \frac{\partial \alpha}{\partial t} + u_i \frac{\partial \alpha}{\partial x_i} = \dot{m}_{g+} / \rho \tag{6}$$

where ρ_f and ρ_g are the densities of the liquid (f) and gas/vapor (g or v) phase respectively.

One final operation sees Eq. 5 being introduced into Eq. 6. The result is an equation for the evaluation of the change in quality with respect to time and space, and as a function of a forcing function represented by the vapor generation term $\dot{m}_{g+}$.

$$[\rho \frac{\partial \alpha}{\partial t}] + u_i \frac{\partial \alpha}{\partial x_i} - \alpha \, \dot{m}_{g+} \left(\frac{1}{\rho_g} - \frac{1}{\rho_f}\right) = \frac{\dot{m}_{g+}}{\rho_f} \tag{7}$$

If the quality α is solely an implicit function of the spatial coordinates $\alpha = \alpha[p(x,z),T(x,z)]$ then the first term on the left hand side of the Eq. 7 vanishes. The source term $\dot{m}_{g+}$ will be determined using the kinetic theory [16] as it was applied to the evaporation and condensation phenomena by Schrage [17] and Collier [18]. It has been generally accepted that the state of equilibrium between a liquid and its vapor phase is not static, but results rather from a process of condensation and evaporation which occur simultaneously and at equal rates, at the interface of the two phases, Fig. 2. However, just a slight shift in pressure (or temperature) from saturation, will result in net mass fluxes of condensate or evaporative flux. During flash conditions when the pressure drops precipituously throughout the fluid, the temperature of the fluid remains much above the saturation temperature corresponding to the new pressure. Thus, the equilibrium moves towards the right on the curves of Fig. 2 and a net evaporation mass flux results.

Using the model initiated by Schrage [17] the flashed mass of liquid can be calculated using

$$\dot{m}_{g+} = \int_{-\infty}^{+\infty} \int_{-\infty}^{+\infty} \int_{0}^{+\infty} v_x m \tilde{\rho} \; dv_x dv_y dv_z \tag{8}$$

where m is the mass of a molecule, v_x is the velocity of the molecules as it crosses an interface, and where $\tilde{\rho}$ is the Maxwell velocity distribution function.

$$\tilde{\rho} = N \, (\beta^3 / \Pi^{1.5}) \exp\,[-\beta^2 (v_x^2 + v_y^2 + v_z^2)] \tag{9}$$

where

$$\beta^2 = m/(2\, RT/N_o) \tag{10}$$

R = universal gas constant

N_o = Avogadro number

N = number density of molecules

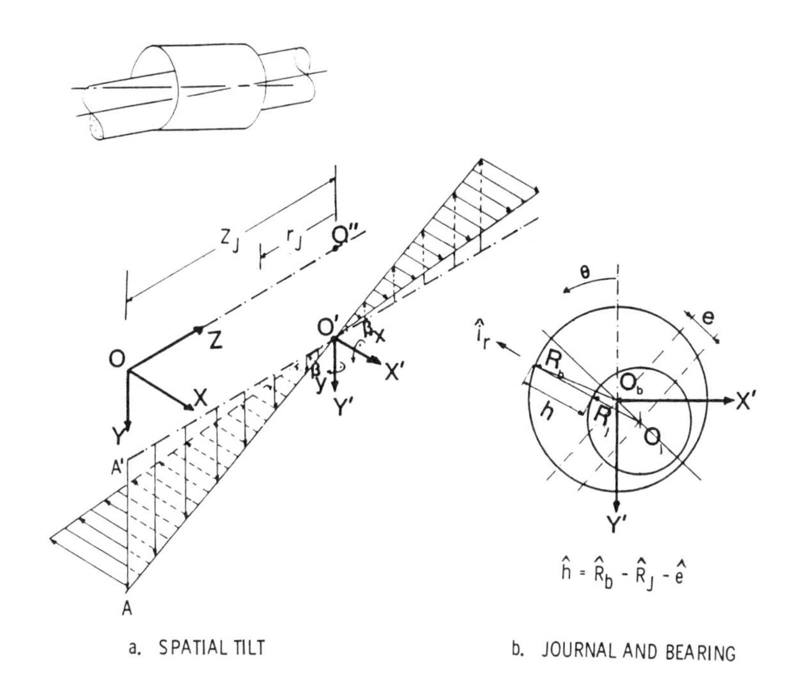

FIGURE 1 THREE-DIMENSIONAL GEOMETRY OF THE JOURNAL BEARING

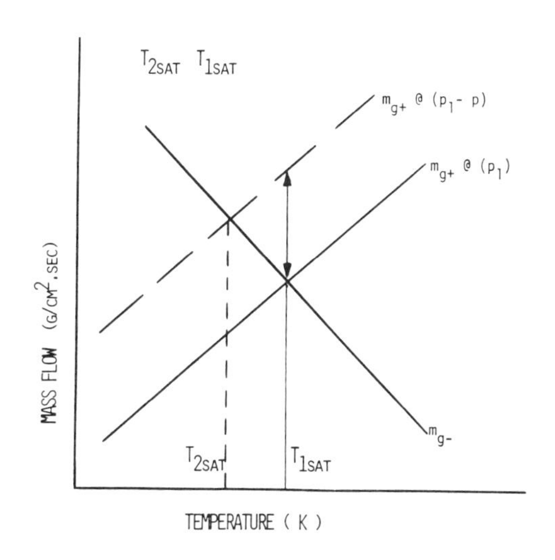

FIGURE 2 EVAPORATION-CONDENSATION KINETIC THEORY DIAGRAM

Assuming that the flashed vapor respects the perfect gas law, upon integration of Eq. 8, one obtains

$$\dot{m}_{g+} = P_g(M/2\Pi RT_g)^{.5} \tag{11}$$

which explicitly relates the vapor mass generation to the thermodynamic state of the substance. Should one consider the evaporation-condensation exchange as it was described earlier and shown in Fig. 2, one has to modify Eq. 11 such that a NET mass evaporated can be calculated. As shown by both Schrage [17] and Collier [18]

$$\dot{m}_{net} = (\frac{M}{2\Pi R})^{.5} [(P_f/T_f^{.5}) \sigma_e - (P_g/T_g^{.5})\sigma_c \Gamma] \tag{12}$$

where σ_e, σ_c and Γ are correction factors for evaporation and condensation, M is the molecular weight and P and T are pressure and temperatures for the gas/vapor (g) and liquid (f) phases. For this latter case, Eq. 12, the right hand side of Eq. 7, will be replaced by $\dot{m}_{net}/\rho_f$, in order to account for mass traffic to and from the liquid.

2.3 The Generalized Reynolds Equation

In the light of assumption (i) through (iv) and using the development of Dowson [14] and Fowles [15], Braun, Mullen and Majumdar [19] derived a generalized Reynolds equation with variable properties

$$\frac{\partial}{\partial x_i}(\frac{\partial p}{\partial x_i}) + \frac{1}{\hat{m}_2}\frac{\partial \hat{m}_2}{x_i}(\frac{\partial p}{\partial x_i}) + \frac{3}{h}\frac{\partial p}{\partial x_i}(\frac{\partial h}{\partial x_i})$$

$$= \frac{U}{\hat{m}_2 h^3}[\frac{\hat{m}_1}{F_o}\frac{\partial h}{\partial x_i} + h\frac{\partial}{\partial x_i}(\frac{\hat{m}_1}{F_o})] + \rho\frac{\partial h}{\partial t}, \quad i=x,z \tag{13}$$

where U is the shaft velocity and

$$\hat{m}_2 = \{\frac{F_1}{F_o}\hat{m}_1 - \int_o^{h(x,y)} \rho [\int_o^Y ydy/\eta g_c]dY\} \frac{1}{h^3} \tag{14a}$$

$$\hat{m}_1 = \{\int_o^{h(x,z)} \rho [\int_o^Y ydy/\eta g_c]dy\} \frac{1}{h} \tag{14b}$$

$$F_1 = \int_o^{h(x,z)} ydy/\eta g_c \tag{14c}$$

$$F_0 = \int_o^{h(x,z)} dy/\eta g_c \tag{14d}$$

The expressions in the accolade parenthesis of Eq. 14a and 14b are presented by Fowles [15] as quantities m_2 and m_1. We replaced m_1 and m_2 by the modified quantities

$$\hat{m}_2 = m_2/h^3$$
$$\hat{m}_1 = m_1/h \tag{15}$$

It becomes evident that in Eq. 13, $\hat{m}_2$ and $(\hat{m}_1/F_o)$ preserve a physical significance, i.e., the effective weighted kinematic viscosity and effective weighted density respectively. Once the pressure field yielded by Eq. 13 is obtained, one can proceed and calculate the flow velocity profiles.

In the circumferential direction

$$u(x,Y,z) = A(x,z) \int_0^{Y(x,z)} dy/\eta g_c + \frac{\partial p}{\partial x} \int_0^{Y(x,z)} y dy/\eta g_c \tag{16}$$

and in the axial direction

$$v(x,Y,z) = B(x,z) \int_0^{Y(x,z)} dy/\eta g_c + \frac{\partial p}{\partial z} \int_0^{Y(x,z)} y dy/\eta g_c \tag{17}$$

where

$$A(x,z) = \frac{U - (\partial p/\partial x) \int_0^{h(x,z)} y dy/\eta g_c}{\int_0^{h(x,z)} dy/\eta g_c} \tag{18a}$$

$$B(x,z) = - \frac{(\partial p/\partial z) \int_0^{h(x,z)} y dy/\eta g_c}{\int_0^{h(x,z)} dy/\eta g_c} \tag{18b}$$

Separate evaluation of the fluid velocity profiles is necessary for the evaluation of the energy convective terms in the lubricant energy equation.

2.4 The Energy Equation

As mentioned earlier there are three separate domains which have to be considered: (a) the rotating shaft, (b) the lubricant, and (c) the stationary outer bearing. A general equation can describe the energy balance for all three domains at once:

$$\frac{\partial}{\partial t} (\rho e) + u_i \frac{\partial}{\partial x_i} (\rho e) = - p \frac{\partial}{\partial x_k} u_k + \frac{\partial}{\partial x_i} (k \frac{\partial T}{\partial x_i}) + \Phi \tag{19}$$

where e is internal energy, k is the conductivity and Φ represents the energy dissipation term.

<u>The shaft and the bearing.</u> Due to the fact that $h \ll R$ (assumption (1)) we were able to neglect the radius of curvature for the fluid wedge. However, this simplification is not possible either in the case of the shaft or the stationary outer bearing. In both cases cylindrical coordinates have been used, Fig. 3.

For the rotating shaft we assume: (I) steady state, (II) incompressibility, (III) no energy generation or dissipation, and (IV) no convective energy transport in the axial (z) and radial (r) directions; then Eq. 19 becomes

$$\rho c\left[\frac{v_\theta}{r}\frac{\partial T}{\partial \theta}\right] = k\left[\frac{1}{r}\frac{\partial}{\partial r}\left(r\frac{\partial T}{\partial r}\right) + \frac{1}{r^2}\frac{\partial^2 T}{\partial \theta^2} + \frac{\partial^2 T}{\partial z^2}\right] \tag{20}$$

In Eq. 20 the right hand side contains the terms for the energy flow due to conduction with r, θ and z representing the radial, circumferential and axial directions respectively. The left hand side is the energy convected due to the shaft rotation with c being the specific heat and v_θ the circumferential velocity of the shaft. For the stationary bearing (v_θ=0) Eq. 19 simplifies to the Laplace equation for conduction in cylindrical coordinates, since all the assumptions mentioned above for the shaft continue to apply.

$$\frac{1}{r}\frac{\partial}{\partial r}\left(r\frac{\partial T}{\partial r}\right) + \frac{1}{r^2}\frac{\partial^2 T}{\partial \theta^2} + \frac{\partial^2 T}{\partial z^2} = 0 \tag{21}$$

For the fluid film contained between the rotating shaft and the bearing, in accordance with the assumptions which led to the unwrapping of the film, a cartesian system of coordinates has been used. Assuming (I) steady state, (II) compressible effects, and (III) neglecting the convective transport effects in the radial direction Eq. 19 becomes

$$u\frac{\partial}{\partial x}(\rho c T) + v\frac{\partial}{\partial z}(\rho c T) = -p\left(\frac{\partial u}{\partial x} + \frac{\partial v}{\partial z}\right) + \frac{\partial}{\partial x}\left(k\frac{\partial T}{\partial x}\right) + \frac{\partial}{\partial z}\left(k\frac{\partial T}{\partial z}\right) + \Phi \tag{22}$$

where p represents the pressure and Φ the energy dissipation term.

3. METHOD OF SOLUTION

The numerical algorithm involves an iterative procedure which connects the field equations presented in the previous section with equations of state for evaluation of the working fluid transport and physical properties. Figure 4 presents a simplified flowchart of the calculations. The Reynolds equation is solved first, yielding field pressures, p, and the axial and circumferential velocities, u and v, of the fluid film. The temperature of the shaft, fluid and bearing are then calculated. The velocities u and v calculated from the Reynolds equation are used in the calculations of the convective energy terms of Eq. 22. With the pressures and temperatures known at every point of the film grid, the properties can be re-evaluated and then reintroduced during the next iteration in the Reynolds and energy equations. When the quality has to be calculated Eqs. 7 and 12 are evaluated for the determination of the source term $\dot{m}_{g+}$ and quality α. Once quality has been calculated, properties of the homogeneous two phase can be obtained.

4. RESULTS

The results presented here are for a hydrodynamic bearing with a shaft radius r = 38.1 mm and a length ℓ = 76.2 mm. The circumferential clearance c = .0762 mm. The fluid used was an oil of dynamic viscosity η = .448 dyn.sec/cm^2.

The cases presented herein are particular cases of Eqs. 13 and 22. More specific the properties are constant. Therefore the compressibility term is eliminated as is the third dimension effect in the conductivity, k $\partial T/\partial y$, in Eq. 22. The effects of Eq. 7 and vapor generation term are not considered

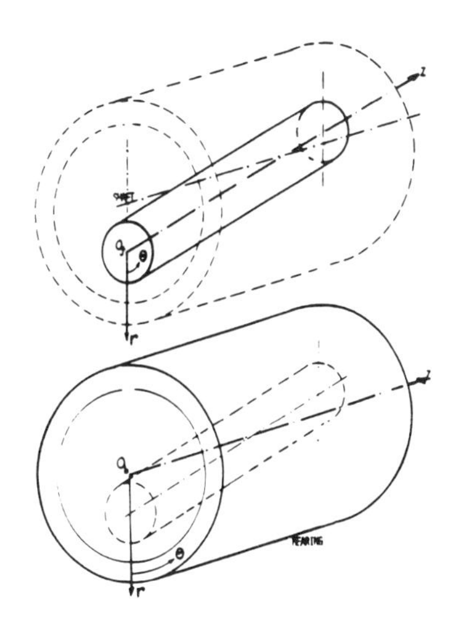

FIGURE 3 CYLINDRICAL SYSTEM OF COORDINATES FOR THE SHAFT AND THE BEARING

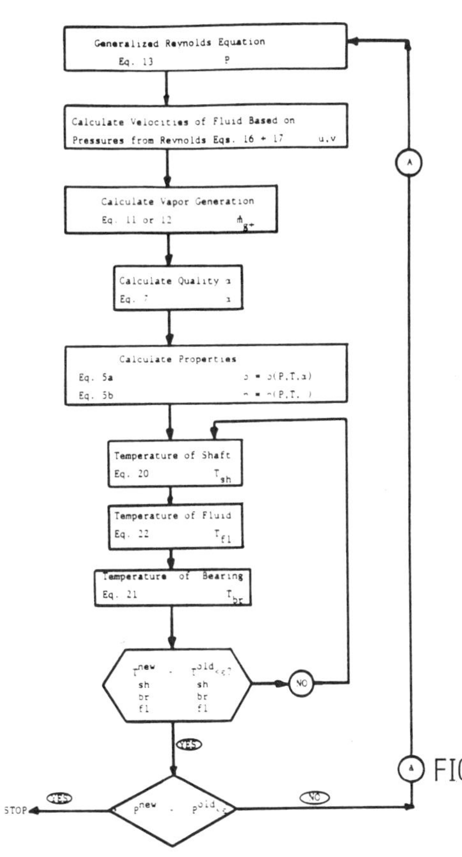

FIGURE 4 FLOWCHART FOR COMPUTATION OF THE REYNOLDS AND ENERGY EQUATION

for the numerical results presented here. However, the formation of cavity zones and effects of inverted flow are detailed and discussed.

Figures 5 and 6 present the pressure and temperature maps for a bearing rotating at 1000, 2000 and 6000 rpm, with eccentricity (ε_x=.5, ε_y=.5) and no tilt. In Fig. 5 one can see the development of the pressure profiles in the axial and circumferential directions as the angular velocity is increased. In all three cases the prescribed inlet pressure is 689×10^3 N/m^2 and the exit pressure is 620×10^3 N/m^2. In the divergent part of the fluid film one can clearly notice the delineation of the cavitation zone, as the pressure has been set there equal to the atmospheric pressure. Through this zone the only flow permitted is the one carried by the Couette effect. The Poiseuille induced flow component is nonexistent since $\partial p/\partial x_i=0$. The maximum pressures obtained in the region of the minimum clearance increases considerably with the angular velocity starting at 728×10^4 N/m^2 at 1000 rpm and peaking at 409×10^5 N/m^2 at 6000 rpm. Preliminary results (not shown here) indicate a noticeable decrease in the peak pressures when variable properties are considered. That occurs due to a decrease in viscosity as the fluid temperature rises with increased rotation and heat dissipation.

Figure 6 which presents the temperature map shows rather large increases in temperature with rotation. The fluid enters at 27° C. The boundary conditions (temperatures) are prescribed on the portions where the fluid is entering the gap and are allowed to float freely where the fluid exits. Such treatment is justified by the large cell Peclet numbers which were calculated. In effect, they ensure a parabolic behavior of the energy equation at the expense of the elliptic conduction terms which are overridden due to large velocities of the fluid in the axial and circumferential directions. Again, one has to mention here that the peak temperatures 58° C at 1000 rpm and 1718° C at 6000 rpm decrease considerably in variable properties computations, due to a marked decrease in the viscosity with increased temperatures.

Figures 7 and 8 are presenting the same type of data as Figs. 5 and 6 but the shaft has an angular tilt (β_x=.4, β_y=.4) wich is responsible for the shifting forward of the peak pressures and the much higher temperatures obtained.

Finally, Fig. 9 presents a composite of the development of temperatures at the shaft surface for a given temperature distribution in the fluid and a large heat transfer coefficient at the fluid shaft interface 3.4 W/cm^2°C. The two carpet maps at the top of the picture present the temperatures at one grid point inside the shaft (R=30.4 mm) and at the shaft surface respectively (R=38.0 mm). It is apparent from a number of numerical experiments we have performed, that the heat transfer coefficients at shaft-fluid interface and bearing-fluid interface are instrumental in shaping the temperature profile of the fluid. That can be easily understood if one compares the thermal capacities of the shaft and bearing with that of the thin fluid film. In effect, one can say that if the heat transfer coefficients are large, the bearing will function isothermally, with the practical consequence of rather constant properties and large load lift capabilities.

5. CONCLUSION

The work presented herein endeavored to relate elements of two-phase flow and kinetic theory to the modified generalized Reynolds equation and to the energy equation. The purpose was to create a unified model which could simulate the pressure and flows in a journal bearing, hydrostatic journal bearing, or squeeze film damper when a two-phase situation occurs due to sudden fluid

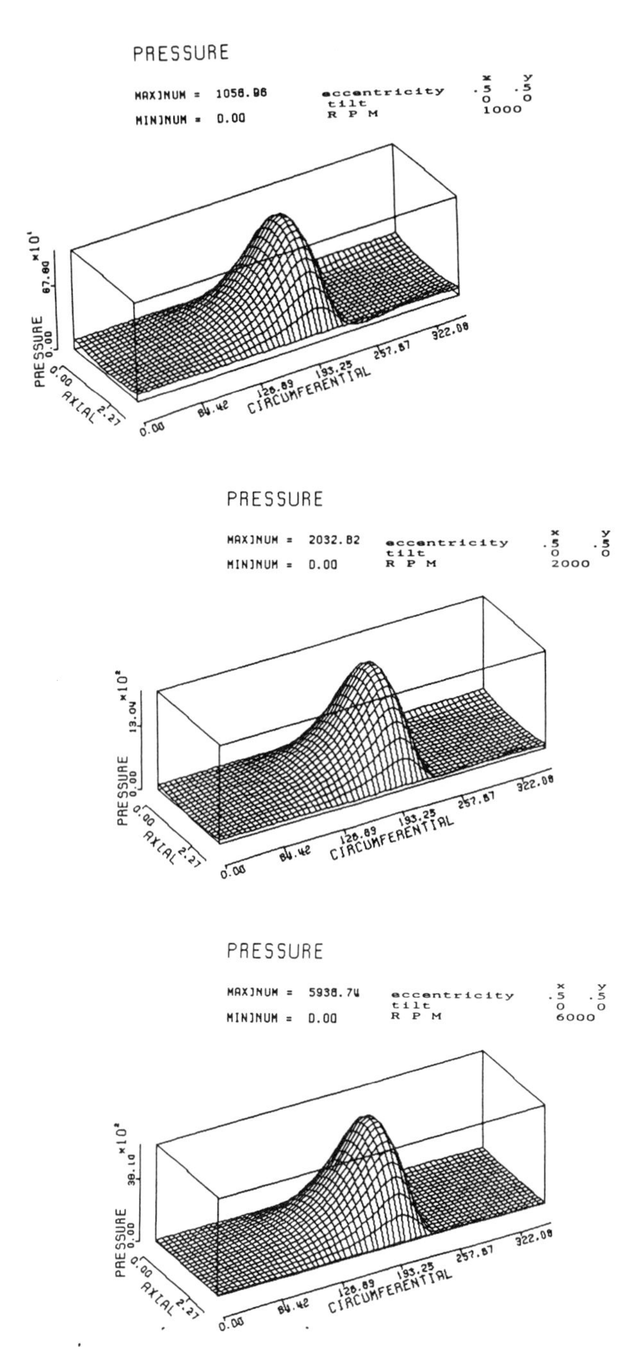

FIGURE 5 DIMENSIONLESS PRESSURE DISTRIBUTION IN THE BEARING (NO TILT)

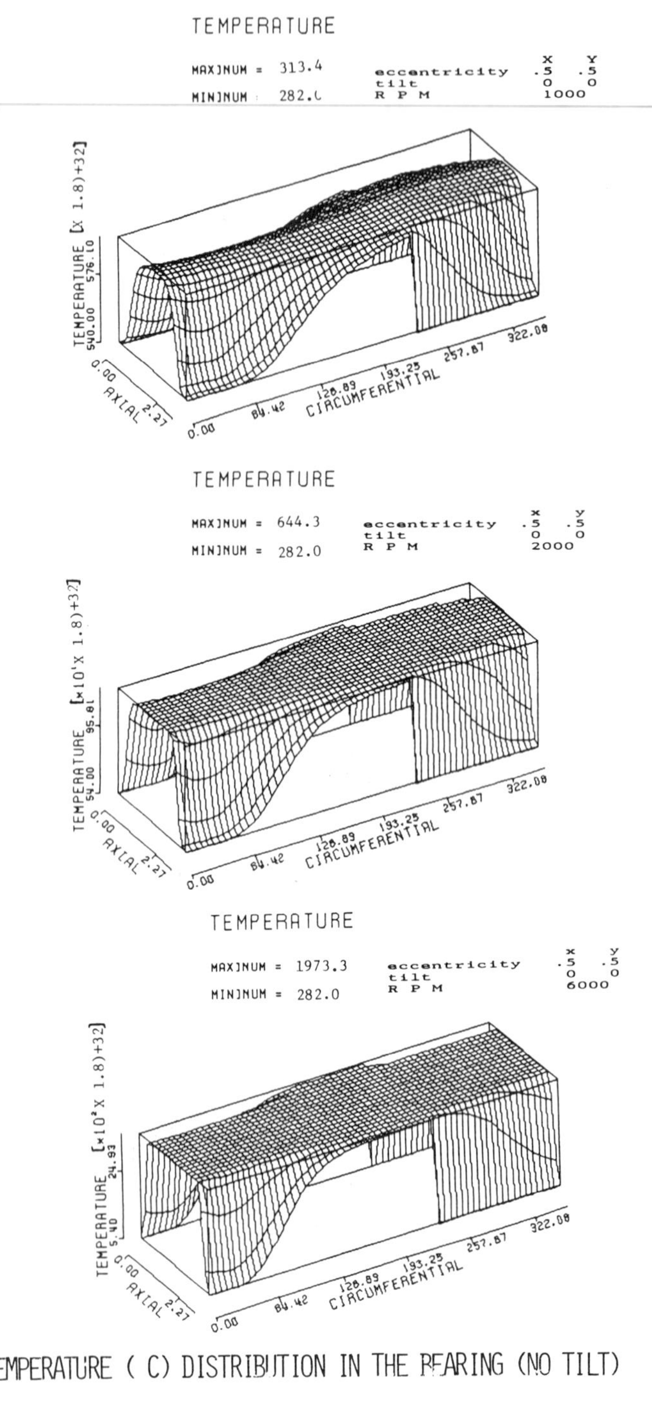

FIGURE 6 TEMPERATURE (C) DISTRIBUTION IN THE BEARING (NO TILT)

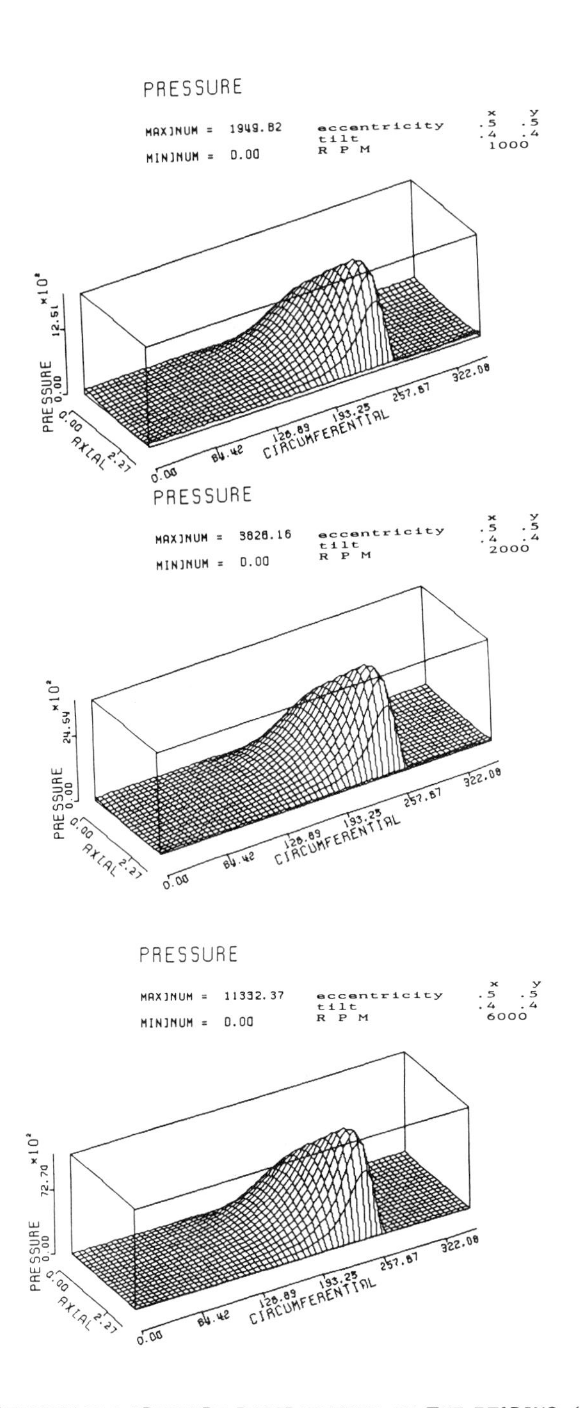

FIGURE 7 DIMENSIONLESS PRESSURE DISTRIBUTION IN THE BEARING (WITH TILT)

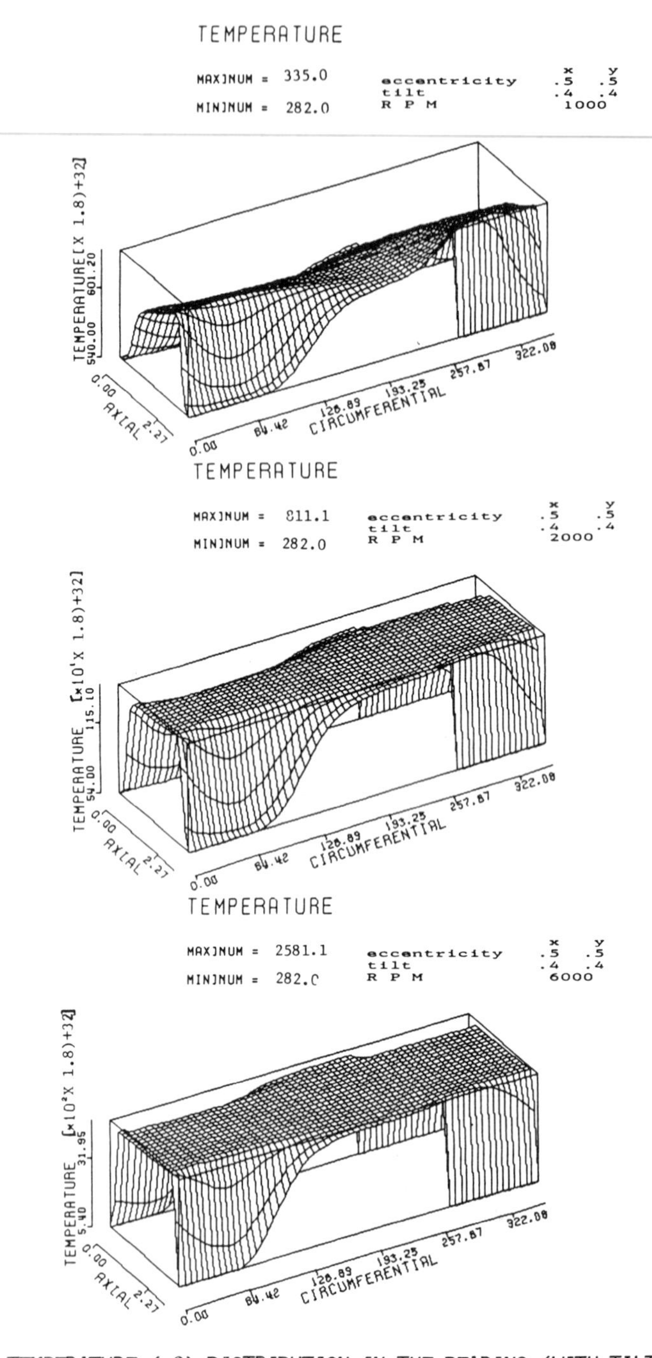

FIGURE 8 TEMPERATURE (C) DISTRIBUTION IN THE BEARING (WITH TILT)

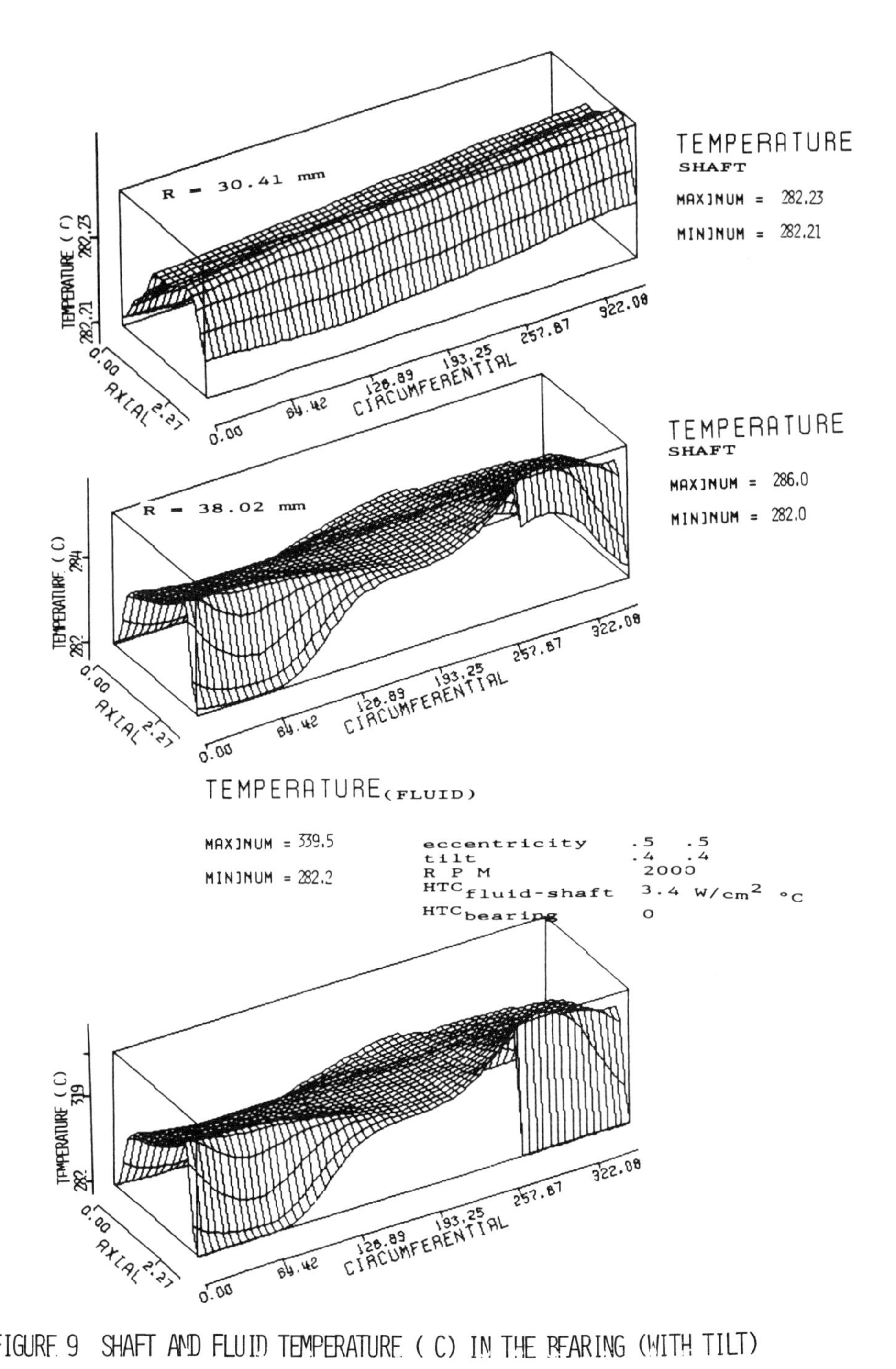

FIGURE 9 SHAFT AND FLUID TEMPERATURE (C) IN THE BEARING (WITH TILT)

depressurization and heat generation. The numerical examples presented are particular cases of the more general theory we have developed. They provide a good checkup of the algorithm for constant properties and at the same time an insight into the effect of the shaft fluid heat transfer coefficient on the temperature profiles. It is evident that the different level of pressures which can be attained for the same angular velocity depends on whether the bearing is isothermal or nonisothermal.

The upwind differencing which was used in the solution of the fluid temperatures proved to be indispensable towards obtaining a realistic profile, as well as numerical convergence of the algorithm.

6. REFERENCES

1. Braun, M.J., Hendricks, R.C., "An Experimental Investigation and Some Analytical Considerations Concerning the Vaporous/Gaseous Cavity Characteristics of an Eccentric Shaft Seal/Bearing", ICHMT Symposium Proceedings, Dubvovnik, Yugoslavia, 1982.

2. Braun, M.J., Hendricks, R.D., "An Experimental Investigation of the Vaporous/Gaseous Cavity Characteristics in an Eccentric Journal Bearing", A.S.L.E. Transactions, Volume 27, pp 1-14, 1984.

3. Hendricks, R.C., Mullen, R.L., Braun, M.J., "Analogy Between Cavitation and Fracture Mechanics", ASME/JSME Thermal Engineering Conference Proceedings, Volume 1, pp 35-43, Honolulu, March 1983.

4. Sommerfeld, A., "Zur Hydrodynamische Theorie der Schmiermittelreibung", Z. Math. Phys., 50, 1904.

5. Swift, H.W., "Stability of Lubricating Films in Journal Bearings", Proc. Inst. Civil Engrs. (London) 233, 1932.

6. Stieber, W., Das Schwimmlager, VDI, 1933, Berlin.

7. Coyne, J.C. and Elrod, H.G., "Conditions for the Rupture of a Lubricating Film, Part I: Theoretical Model", J. Lubr. Tech., 92, pp 451-456.

8. Floberg, L., "On Hydrodynamic Lubrication with Special Reference to Sub-Cavity Pressures and Number of Streamers in Cavitation Regions", Acta Polytechnica Scandinavica, ME, Series 19, pp 1-35, 1965.

9. Parkins, D.W., May-Miller, R., "Cavitation in an Oscillatory Oil Squeeze Film", Journal of Lubrication Technology, Vol. 106, pp 360-366, July 1984.

10. Kauzlarich, J.J., "Hydraulic Squeeze Bearing", Trans. ASLE, Vol. 15, 1, pp 37-44.

11. Haber, S., Etsion, I., "Analysis of a Squeeze Film Damper with Bubbly Lubricant", Technion, Israel Institute of Technology, EEC-104, August 1983

12. Braun, M.J., Mullen, R.L., Hendricks, R.C. and Wheeler, R.L., "Development and Discussion of a Two-Phase Homogeneous Model for a Generalized Reynolds Equation", Cavitation and Two-Phase Forum, Albuquerque, June 1985

13. Zuber, N., Dougherty, D.E., "The Field Equations for Two-Phase Reynolds Film Flow with Change of Phase", ASLE Trans., Vol. 25, Nr1, pp 108-116.

14. Dowson, D., "A Generalized Reynolds Equation for Fluid Film Lubrication", Int. J. Mech. Sci., Vol. 4, pp 159-170, 1962.

15. Fowles, P.E., "A Simpler Form of the General Reynolds Equation", Journal of Lubrication Technology, October 1970, pp 661-662.

16. Lee, J.W., Sears, F.W., Turcotte, I.L., "Statistical Thermodynamics", Addison-Wesley Publishing Co., 1963.

17. Schrage, R.W., "A Theoretical Study of Interphase Mass Transfer", Ph.D. Thesis, Columbia University Press, 1953.

18. Collier, J.G., "Convective Boiling and Condensation", pp 303-310, McGraw Hill, 1972.

19. Braun, M.J., Mullen, R.L., Majumdar, A.K., "A Three Dimensional Numerical Method with Application to Tribological Problems", 10th Leeds-Lyon Symposium on Tribology, 1985, and "Developments in Numerical Experimental Methods Applied to Tribology", Butterworth Publishers.

Comparison of Generalized Reynolds and Navier Stokes Equations for Flow of a Power Law Fluid

R. L. MULLEN
Department of Civil Engineering
Case Western Reserve University
Cleveland, Ohio 44106

A. PREKWAS
CHAM of North America
Huntsville, Alabama 35805

M. J. BRAUN
Department of Mechanical Engineering
University of Akron
Akron, Ohio 44325

R. C. HENDRICKS
NASA Lewis Research Center
Cleveland, Ohio 44135

ABSTRACT

This paper compares a finite element solution of a modified Reynolds equation with a finite difference solution of the Navier Stokes equation for a power law fluid. Both the finite element and finite difference formulation are reviewed. Solutions to spiral flow in parallel and conical geometries are compared. Comparison with experimental results are also given. The effects of the assumptions used in the Reynolds equation are discussed.

1. SYMBOLS

A(x,y)	see Equation (10)
B(x,y)	see Equation (11)
B^{ij}	see Equation (31)
B^u	see Equation (41)
E^u	see Equation (41)
C	power law viscosity coefficient
C1,C2,C3,C4	see Equation (34)
e_i ; e^i	basis vectors
g	determinant of g_{ij}
g_c	gravitation constant
h	height of fluid
H	Hilbert space
J	Jacobian

k	coefficient of cubic term in non-newtonian model
K_{IJ}	stiffness matrix
m_1	generalized density
$1/m_2$	generalized viscosity
n	unit outward normal
P	pressure
Rb	radius of ceterbody
S_ψ	source term
U	relative surface velocity
u	circumferential component
V	axial component of fluid velocity
x,y,z	coordinates
X,Y	coordinates, see Equations (29) and (34)
α	exponent in constitutive relation
γ	test function
Γ	general diffusion term
ρ	density
μ	viscosity constant in power law model
η	equivalent newtonian viscosity
ζ ; ξ	general coordinates
ψ	general cartesian coordinate
Φ	shape function
τ	shear stress

2. INTRODUCTION

The accepted governing equations for the prediction of fluid velocities and pressures in thin films such as occur in tribological problems is the Reynolds equation (Ref. 1). This equation is a simplified form of the Navier-Stokes equation applicable when the film thickness is much less than the other dimensions. Terms such as inertia and body forces are assumed small. The pressure distribution through the thickness is therefore uniform. Generally, the Reynolds equation can accurately model the behavior of lubricating films where moderate pressures and velocities are encountered. Most engineers believe they have good intuition on the range of problems that can be solved by Reynolds equation when using the Newtonian fluid assumption. However, this judgement is lost when solving for problems with non-newtonian fluid behavior. In this paper, a comparison between the flow predicted by a finite element solution of a general Reynolds solution and a finite difference solution to the Navier Stokes equation is presented.

A review of non-newtonian constitutive relations has been presented by Bird (Ref. 2). In this paper the power-law-model will be assumed to characterize non-newtonian behavior of thin films.

Several researchers have investigated non-newtonian behavior in lubricating films. Finite difference solutions for the two dimensional problem with a constitutive relation in the form

$$\tau + k\tau^3 = \frac{\partial u}{\partial x} \tag{1}$$

was presented by Swamy, Prabhu, and Rao (Ref. 3). They calculated stiffness and damping coefficients by a finite difference method and showed a decrease in the linearized stiffness and damping coefficients for shear thinning fluids. Gorla (Ref. 4) calculated thermal and pressure fields in a thrust bearing using a grade two fluid by a fourth order Runge-Kutta method. Integral expressions for the one dimensional problem with a power law fluid was presented by Singh and Singh (Ref. 5). Inertia effects in power-law fluids were studied by Elkouh (Ref. 6) for squeeze films and Khaderian Vachon (Ref. 7) for radial outward flow. Christenson and Sabiel (Ref. 8) presented an analysis of a slider bearing in non-newtonian flow. A survey of finite element analysis of fluid film lubrication was compiled by Huebner (Ref. 9).

In the next section, the governing equations for Reynolds model are developed. In section 4, the discrete formulation used for Navier Stokes solutions is given. In section 5, applications of these equations are presented along with comparison made. Conclusions regarding this work are then presented in section 6.

3. REYNOLDS EQUATION SOLUTION

3.1 Reynolds Equation

The geometry under consideration is shown in Figure 1. The height of the fluid is identical to the clearance height h. The analysis assumes A) steady-state conditions, B) constant temperature thermophysical properties, C) a gap height $h << R_b$ 4) laminar flow with no entrance, inertia, or body force effects, and D) no shaft motion in the direction perpendicular to the plane of rotation. Taking into consideration these assumptions the generalized Reynolds equation may be written, (Ref. 10),

$$\frac{\partial}{\partial x}\left(m_2 \frac{\partial P}{\partial x}\right) + \frac{\partial}{\partial y}\left(m_2 \frac{\partial P}{\partial y}\right) = U \frac{\partial}{\partial x}(h\, m_1) \tag{2}$$

where

$$m_2 = m_1\, h \int_o^h \frac{z}{\eta}\, dz - \int_o^h \frac{\rho}{g_c} \int_o^z \frac{z'}{\eta}\, dz'\, dz \tag{3}$$

$$m_1 = \frac{\int_o^h \frac{\rho}{g_c} \int_o^z \frac{1}{\eta}\, dz'\, dz}{h \int_o^h \frac{1}{\eta}\, dz} \tag{4}$$

with P(x,y) given on Γ_1 (Dirichlet surface) and $\frac{\partial P}{\partial n}$ on Γ_2 (Neumann surface).

If η and ρ are constant, the values of m_1 and m_2 are

$$m_1 = \frac{\rho}{2g_c} \tag{5}$$

$$m_2 = \frac{h^3\, \rho}{12 g_c\, \eta} \tag{6}$$

and equation (2) becomes the standard Reynold equation.

$$\frac{\partial}{\partial x}\left(h^3 \frac{\partial P}{\partial x}\right) + \frac{\partial}{\partial y}\left(h^3 \frac{\partial P}{\partial y}\right) = 6\eta U \frac{\partial h}{\partial x} \tag{7}$$

Once the pressure field is determined from equation (2), the velocity profile can be obtained from the following relations

$$u(x,y,z) = A(x,y) \int_o^z \frac{1}{\eta}\, dz' + \frac{\partial P}{\partial x} \int_o^z \frac{z'}{\eta}\, dz' \tag{8}$$

$$v(x,y,z) = B(x,y) \int_o^z \frac{1}{\eta}\, dz' + \frac{\partial P}{\partial y} \int_o^z \frac{z'}{\eta}\, dz' \tag{9}$$

where

$$A(x,y) = \frac{U - \frac{\partial P}{\partial x} \int_o^h \frac{z}{\eta}\, dz}{\int_o^h \frac{1}{\eta}\, dz} \tag{10}$$

$$B(x,y) = \frac{- \frac{\partial P}{\partial y} \int_o^h \frac{z}{\eta}\, dz}{\int_o^h \frac{1}{\eta}\, dz} \tag{11}$$

Since the above expressions are valid for a viscosity and density that are sufficiently smooth, the power law constitutive relation can be introduced by defining an equivalent newtonian viscosity η. The strain rate, shearing force relation in two dimensions for a power law fluid is

$$\tau_{xz} = \mu \left[\left(\frac{\partial u}{\partial z}\right)^2 + \left(\frac{\partial v}{\partial z}\right)^2 \right]^{\frac{\alpha-1}{2}} \frac{\partial u}{\partial z} \tag{12}$$

$$\eta = \mu \left[\left(\frac{\partial u}{\partial z}\right)^2 + \left(\frac{\partial v}{\partial z}\right)^2 \right]^{\frac{\alpha-1}{2}} \tag{13}$$

By introducing

$$\tau_{yz} = \mu \left[\left(\frac{\partial u}{\partial z}\right)^2 + \left(\frac{\partial v}{\partial z}\right)^2 \right]^{\frac{\alpha-1}{2}} \frac{\partial v}{\partial z} \tag{14}$$

the strain rate, shearing force relationship can be written in a pseudo newtonian form.

$$\tau_{xz} = \eta \frac{\partial u}{\partial z} \tag{15}$$

$$\tau_{yz} = \eta \frac{\partial v}{\partial z} \tag{16}$$

By combining the definition of η (Eq. 14) with the velocity fields in equations (8,9) the following recursive relationship for η (x,y,z) can be written

$$\eta(x,y,z) = \mu \left\{ \frac{1}{\eta^2} \left[\frac{U - \frac{\partial P}{\partial x} \int_0^h \frac{z}{\eta} dz}{\int_0^h \frac{1}{\eta} dz} + \frac{\partial P}{\partial x} z \right]^2 + \frac{1}{\eta^2} \left[z \frac{\partial P}{\partial y} - \frac{\partial P}{\partial y} \frac{\int_0^h \frac{z}{\eta} dz}{\int_0^h \frac{1}{\eta} dz} \right]^2 \right\}^{\frac{\alpha-1}{2}} \tag{17}$$

The resulting boundary value problem for the pressure field is then given by equations (17), (2), (3), (4).

3.2 Discretization

The finite element discretizaion will be based on a weak form of equation (2). Find $P \varepsilon H'_g$ for all $\gamma \varepsilon H'_o$ such that

$$\int_\Omega \left[\frac{\partial \gamma}{\partial x} m_2 \frac{\partial P}{\partial x} + \frac{\partial \gamma}{\partial y} m_2 \frac{\partial P}{\partial y} - U h m_1 \frac{\partial \gamma}{\partial x} \right] d\Omega = 0 \tag{18}$$

where H'_g is the subset of the Hilbert space of function H^1 which satisfies the boundary condition on Γ.

The finite element discretization is then obtained by approximating the function γ and P over small regions by a polynominal forms

$$\gamma(x,y) = \gamma_I \Phi_I(x,y) \tag{19}$$

$$P(x,y) = P_I \Phi_I(x,y) \tag{20}$$

where Φ are the element shape functions and subscripts I and J denote nodal values. Repeated subscripts imply summation. Although it is not required, for simplicity the value of the film thickness will also be expressed using the element shape functions

$$h(x,y) = h_I \Phi_I(x,y) \tag{21}$$

Assuming the polynomial forms in (19-21), equation (18) can be rewritten,

$$K_{IJ} P_J = U_I \tag{22}$$

where

$$\int_{\Omega} \left(\frac{\partial \Phi_I}{\partial x} m_2 \frac{\partial \Phi_J}{\partial x} + \frac{\partial \Phi_I}{\partial y} m_2 \frac{\partial \Phi_J}{\partial y} \right) d\Omega = K_{IJ} \tag{23}$$

$$\int_{\Omega} \left(\frac{\partial \Phi_I}{\partial x} U h_J \Phi_J m_1 \right) d\Omega = U_I \tag{24}$$

Although equation (22) is written as a system of linear equations, it represents only part of the solution procedure since the coefficient m_1 and m_2 which enter in the calculation of K_{IJ} are nonlinearly dependent on the pressure field. Successive approximations to the solution are obtained from updating the values at each Gauss point in the numerical integration of equations (23) and (24) employing the following recursive expression

$$\eta^{i+1} = \mu [\eta^i]^{(1-\alpha)} \left(\left[\frac{U - \frac{\partial P^i}{\partial x} \int_o^h \frac{z}{\eta^i} dz}{\int_o^h \frac{1}{\eta^i} dz} + \frac{\partial P^i}{\partial x} z \right]^2 + \left(\frac{\partial P^i}{\partial y} \right)^2 \left[z - \frac{\int_o^h \frac{z}{\eta^i} dz}{\int_o^h \frac{1}{\eta^i} dz} \right]^2 \right)^{\frac{\alpha-1}{2}} \tag{25}$$

where the superscript i indicates the iteration number.

The new value of η, is then used in equations (3) and (4) to obtain new values of the coefficients m_1 and m_2. The new values of m_1 and m_2 are then used in equation (22) to update the pressure field. This procedure is repeated until the change in the values of η is less than a small number

$$\eta^{i+1} - \eta^i \leq \varepsilon \eta^{i+1} \tag{26}$$

4. NAVIER STOKES EQUATION

4.1 Navier Stokes Equation

The steady state, viscous fluid flow equations in the divergence form are

$$\text{div} (\rho \underset{\sim}{v}) = 0 \tag{27}$$

$$\text{div} (\rho \underset{\sim}{v} \underset{\sim}{v} + \underset{\approx}{\tau}) = 0 \tag{28}$$

where ρ is the fluid density, $\underset{\sim}{V}$ is the velocity vector and $\underset{\approx}{\tau}$ is the shear tensor (includes pressure P otherwise include (-grad P)).

To simulate the flow in complex passages equations (27) and (28) should be expressed in general curvilinear coordinates. Let X^i be fixed cartesian coordinates and ψ^i be curvilinear coordinates with nonzero Jacobian of the transformation, $J= |dX^i/d\psi^i|$. Symbols e_i and e^i are used to describe the covariant and contravariant base vectors in the new coordinate system. The metric tensors g_{ij} are defined such that the differential element of arc length ds is obtained from the relationship

$$(ds)^2 = g_{ij} \, d\psi^i \, d\psi^j \tag{29}$$

The continuity and momentum equations in the transferred coordinates are expressed in the following form (see Ref. 11-16 for more details):

$$d(\rho v^i \sqrt{g}\,)/d\psi^i = 0 \tag{30}$$

and

$$\left[\frac{dB^{ij}}{d\psi^i} + B^{i\ell} \, \Gamma^j_{i\ell}\right] e_j = 0 \tag{31}$$

where

$$B^{ij} \equiv \left[\rho \, v^i \, v^j + \tau^{ij}\right] \sqrt{g}$$

and

$$\tau \equiv \tau^{ij} \, \bar{e}_i \, \bar{e}_j$$

and where Γ^l_{ij} is the Christoffel symbol of the second kind.

The above equations are commonly used as a basis in several applications (Refs. 14,15). It is seldom noticed, however, that while the continuity equation is in conservation law form, the momentum equation misses this form due to the source like term ($B^{ij} \, \Gamma^j_{il}$). In the present computations, a strongly conservative form (Refs. 11,12,16) have been used with the momentum equation expressed as follows:

$$\frac{d}{d\psi^j}\left[B^{ij}\left(\frac{dx^m}{d\psi^i}\right)\right] = 0 \tag{32}$$

X^m is the position vector in 3-space.

4.2 Discretization

The continuity and momentum equations were first approximated by the system of the first order upwind finite difference equations derived on the control volume method.

For example, the discrete conservation equation for dependent variable Ψ will be developed. An upwind differencing is used to approximate the derivatives of the convection terms and a central differencing approximation for the diffusion terms.

The governing hydrodynamic equation for the nonorthogonal grid in strong conservative form is

$$d(\rho U\Psi)/d\zeta + d(\rho V\Psi)/d\xi$$

$$-d\,[\Gamma\,(C_1\,\Psi_\zeta - C_2\,\Psi_\xi)]/(J d\zeta) \tag{33}$$

$$-d\,[\Gamma\,(C_3\,\Psi_\xi - C_4\,\Psi_\zeta)]/(J d\xi) = JS_\Psi$$

where U,V are contravariant velocity components, $\Psi = \{u,v\}$ is the dependent variable, ζ and ξ are the transformed coordinates, $\Psi_\zeta = \partial\Psi/\partial\zeta$ and where

$$C_1 = X_\xi^{\,2} + Y_\xi^{\,2}$$

$$C_2 = C_4 = X_\zeta\,X_\xi + Y_\zeta\,Y_\xi$$

$$C_3 = X_\zeta^{\,2} + Y_\zeta^{\,2} \tag{34}$$

$$J = X_\zeta\,Y_\xi - X_\xi\,Y_\zeta$$

The cartesian velocity components are solved for from equation (33) while contravariant components U and V are calculated as secondary variables from

$$U = u\,Y_\xi \quad -v\,X_\zeta$$

$$V = v\,X_\zeta \quad -u\,Y_\xi \tag{35}$$

A staggered grid system, shown in figure (2), is used to obtain the finite difference form of equation (33).

An upwind differencing technique for convective terms and central differencing technique for diffusive terms yields the finite difference formula:

$$a_p\,\Psi_p = \sum a_j\,\Psi_j + S_\Psi \qquad j = N,S,E,W \tag{36}$$

where

$$a_p = \sum a_j \tag{37}$$

the 'link' coefficients are calculated for each neighbor, e.g.,

$$a_E = [\langle 0_1 - \rho U_E\rangle + \Gamma\,C_1\,/\,(J\,\Delta\,\zeta)]\;\Delta\xi \tag{38}$$

.... etc.

The 'viscous cross-derivative' terms of equation (33) are transferred to the source term.

The pressure correction equation is obtained from the continuity equation,

$$(\rho U \Delta\xi)_e - (\rho U \Delta\xi)_\omega +$$

$$(\rho V \Delta\zeta)_n - (\rho V \Delta\zeta)_s = 0 \qquad (39)$$

by expanding the contravariant velocity components viz:

$$U = U^* + (B^u Y_\xi - B^v X_\xi) P'_\xi +$$

$$(E^u Y_\xi - E^v X_\xi) P'_\xi \qquad (40)$$

where

$$B^u = -Y_\xi / a_p^u$$

$$E^u = Y_\zeta / a_p^u \qquad (41)$$

.... etc.

Once the continuity equation is solved the U and V velocity components are corrected (from eq. 8) and u,v are obtained (from eq. 35). The iterative process follows all steps of the SIMPLE* procedure of Patankar and Spalding (Ref. 17).

In the present case the metrics are calculated from simple algebraic formulations based on the following transformation formulation

$$X = X(\zeta) \quad ; \quad Y = Y_S + (Y_N - Y_S)$$

$$X_\zeta = dX/d\xi \quad ; \quad X_\xi = 0$$

$$Y_\xi = (Y_N - Y_S) \quad ; \quad Y_\zeta = \xi dY_N/dX + (1 - \xi) dY_S/d\xi$$

Here Y_S, Y_N are 'south' and 'north' wall distances measured from the axis of symmetry. The equations are solved in the physical domain with control cell volume and cell face areas calculated for cylindrical polar coordinates.

5. RESULTS AND DISCUSSION

5.1 Description of Test Cases

The finite difference calculations have been performed to predict the spiral flow in an annular channel shown in figure 3. A steady state problem of viscous non-newtonian flow in the annular channel with the inner or outer wall rotating has been considered. The dimensions of the computational domain, shown in figures 1 and 3, are: R,b = 4.425 cm (1.742 in), h = 1.28 cm (.504 in), L = 6 cm (2.36 in) and four inner wall inclination angles, β = 0.1, 1.0, 5., and 15 degrees.

For the non-newtonian flow calculations the viscosity has been calculated based on the following power law:

$$\eta = C_{\mu} \left[\left(\frac{du}{dz} \right)^2 + \left(\frac{dv}{dz} \right)^2 \right]^{\frac{\alpha-1}{2}} \qquad (42)$$

where u and v are the velocity components in the axial and circumferential directions respectively x and y are the axial and circumferential coordinates, and z is the radial coordinate. The fluid is a solution of 3-weight percent polyisobutane dissolved in a practical grade cis-trans mixture of Decalin; the material constants are given in table 1. Herein the fluid viscosity constants, C_{μ} and α were assumed as 9.4 and 0.77 respectively (Refs. 18, 19).

The shear stress dependence on the strain is expressed in pseudo-newtonian form,

$$\tau_{xz} = \eta\, du/dz$$

$$\tau_{yz} = \eta\, dv/dz \qquad (43)$$

As the basic test case, a fully developed flow with prescribed pressure gradient $dP/dx = 981$ dynes/cm^3 has been simulated in the cylindrical annular channel, (with zero wall inclination angle $\beta = 0$). In the second step, a developing flow in a conical annulus with prescribed inlet and exit pressures has been considered.

5.2 Discussion of Results

In the finite difference (FD) calculations the full momentum equations were solved, while in the reported finite element (FEM) studies, Reynolds approximation equations were employed. The purpose of this study was to compare two approaches and to gain insight into the importance of the inertia (convective) terms on the non-newtonian rotating flows in turbomachines (bearings, seals, etc.).

5.2.1. Newtonian comparison. Calculations (FD and FEM) of the rotating flow with constant viscosity $\mu = 1$ g/cm/s and density $\rho = 1$ g/cm^3 'exactly' reproduced the Couette flow analytical solution (see Ref. 20, pages 94-95) and will not be presented herein.

5.2.2 Nonnewtonian spiral flow. With the power law exponent (1/3) an analyticl solution to the viscous flow between two rotating cylinders can be determined. The flow becomes strongly dependent on the gap thickness and the flow rate becomes proportional to (1/h**3), i.e., the smaller the gap the larger the flow.

Assuming the fluid to be separan with power law constants ($\alpha = 0.333$ and $\mu = 250$ dyne-sec^2/cm^2, and the geometry that of figure 3, figure 4 illustrates the agreement between the FEM, FD, and anaytic1 treatments. The FEM and analyticl solution agree to less than 0.1 percent and the difference between the FEM and FD solutions may be due to the wall function treatment for surface roughness which is not a part of the FEM or analytical treatment.

In a separate treatment, the gap height was decreased by a factor of 10 which means that the flow rtes should have increased by 1000; however from the FD analysis the viscosity only changes by a factor of 4 and the flow actually decreases. In contrast, the FEM solution is in reasonable agreement with theory. The discrepency is being investigated.

5.2.3 Coaxial cylinders (Schowalter case). Since, in our previous publication, ref. 21, the verification studies of the FEM indicated good agreement between the experimental and numerical predicted results (see figures 5 and 6), the FD results will be directly compared with the FEM predictions. The geometric configuration is that illustrated in figure 3.

Figure 7 presents a comparison of the velocity magnitudes as calculated from the Reynolds solution (FEM) and the Navier-Stokes solution (FD). The parameters are the same as for figures 5 and 6, i.e., the outer wall rotational speed was 100 rpm and the axial pressure gradient dp/dx = 819 dyne/cm^3. The comparison indicates relatively good agreement between FD and FEM calculations. The discrepancy is more visible between the duct center and the outer wall.

In the FEM calculations the maximum of the axial velocity, u_{max}, is located at the center of the duct. The results of the FD calculations show that, u_{max} is located closer to the inner wall.

The minor difference in the axial velocity profile shift resulted in slightly more pronounced differences in the circumferential velocity profile v.

5.2.4 Conical centerbody-cylindrical housing. The second computational test included spiral flow in the conical section (figure 1) with prescribed pressures as the inlet and out boundary conditions, (P_1 = 819 dynes/ cm^2 and P_e = 0.). It is assumed that the housing is cylindrical and either the housing or the centerbody is rotating. The centerbody is conical with a fixed clearance at L/2, figure 1. The cone angles are varied between 0 and 15 degrees with corresponding rotational speed of 100 rpm and the same fluid was used as in the first case (5.2.3), see table 1. The flow was developing in the conical section due to the shear stress, wall friction, and rotation. A uniform computational grid with 10 nodes in the axial direction and 50 nodes in circumferential direction has been used in both FEM and FD calculations. The calculated mass flow rates are given in Tables 2 and 4.

Table 2 provides the comparison of the flow rates in g/s for various centerbody tapers as calculated by the FEM or Reynolds solution method for shear thinning ($\alpha < 1$), newtonian ($\alpha = 1$), and dilitant ($\alpha > 1$), fluids. Hendricks, Ref. 22, presents experimental data for newtonian flow of gases through nonrotating-convergent-taper seal passages with taper angles less than 4 degrees. The flow rate data of Ref. 22, increase with increasing angle, but $\bar{A}/A_o$ also increased with taper angle as A_o was held constant. While a direct comparison is difficult to make the flow rates for tapers less than 15 degrees are related to $\bar{A}/A_o$.

Table 3 provides a comparison of the flow rates in g/s between the Reynolds (FEM) and the Navier-Stokes (FD) approaches for different centerbody wall inclination angles. The fully developed solution flow rate is less than the developing flow solution. The ratio of the flow rates indicates a maximum near 3 degrees and increasingly significant departures to 15 degrees when compared to the newtonian case.

The results (table 3) are for a rotating housing with a stationary centerbody (following Schowalter, Refs, 18, 19); however another case of interest is rotating the centerbody with the housing stationary. The Reynolds solution does not change (as the mean velocity would not change). The Navier-Stokes solution on the other hand indicates some radial flows and in general show flows (table 4) which are lower than those of table 3. As the cone angle increases to 18-20 degrees there appears to be a crossover, and the flows with rotating centerbody become greater.

5.2.5 Flow details (Navier-Stokes (FD) solutions). Figure 8 represents the axial (u) and circumferential (v) velocity profiles for selected centerbody inclination angles, and provide more detail for the data of tables 3 and 4 for housing and centerbody rotation respectively. The fluid properties are those of table 1 and the profiles of Fig. 8 represent flow details at $(x/L) = 0.95$.

It is apparent that increasing the axial velocity first increases with inclination angle and then decreases for both figure 8 (a) and 8(b) (centerbody/housing rotation respectively). The newtonian axial velocity profile is given for comparison.

While the axial velocity profiles are similar, the circumferential velocity profiles are significantly different, e.g., (inner concave, outer convex). As the flow becomes developed there is a larger variation in the circumferential velocity. At large cone angles, the circumferential velocity is linear since the effective viscosity is nearly constant. The effective viscosity in the fully developed flow exhibits a large increase at the center of the annulus. The flow appears to be destabilized by inner wall rotation.

5.2.6 Dilitant flows. In contrast to shear thinning flows, the viscosity of a dilatant flow increases with shear rate and suggests an analogy with laminar to turbulent flow transition and with turbulent flows in general.

For the geometry of figure 3, with an assumed hypothetical working fluid, $\alpha = 1.05$, $\mu = 9.4$ dyne-sec/cm^2, FD solutions for $\beta = 0$ and $\beta = 15$ degrees are illustrated in figure 9. The velocity jump at the rotating wall perhaps represents a problem with the assumed form of the wall function; this problem is being investigated.

6. SUMMARY

Variations in passage geometry affect flow rates in nonconventional manner, even for newtonian flows, and must be taken into consideration both in the design of turbomachines and their inlet/exit passages.

The flow rates for shear thinning fluids in conically converging passages with the outer wall rotating are greater than with inner wall rotating at small angles of convergence; at large angles of convergence the opposite is true.

As the flow becomes developed, decreasing beta ($\beta \to 0$) there is a larger variation in the circumferential velocity. At large cone angles, the circumferentail velocity is linear since the effective viscosity is nearly constant. The effective viscosity in the fully developed flow exhibits a large increase at the center. The velocity profiles reinforce these conclusions.

The calculated mass flow rates indicate that the development length of the flow correlates with the power law exponent

The peak axial velocity increases as,

a) the area at the exit is reduced (i.e., increasing cone angle).

b) the flow develops (i.e., decreasing cone angle).

(b) becomes more critical than (a) in the geometry considered for this study at a 3 degree cone angle. Further, both the mass flow rate and the velocity profiles show a peak at a 3 degree cone angle.

The shear thinning fluids are better modeled by creeping flow assumptions than their newtonian counterparts.

At large cone angles the length available for the flow to develop is short, therefore the fully developed flow assumptions (Reynolds equation) are poor.

Can dilitant fluids be modeled by a creeping flow formulation? This will require experimental and analytical verification.

The major disagreements between the FEM and FD solutions may be related to the assumed form of the wall function and surface roughness representation; the discrepancy is being investigated.

7. REFERENCES

1. Reynolds, O. "On the Theory of Lubrication and Its Application to Mr. Beauchamp Tower's Experiments, Including an Experimental Determination of Viscosity of Olive Oil", Philosophical Transactions of the Royal Society, London, Vol. A1777, 1816 pp 157-234.

2. Bird, R.B., Armstrong, R.C., and Hassageur, O.,"Dynamics of Polymeric Liquids", John Wiley and Sons, New York, 1977.

3. Swamy,T.N., Prabhu,B.S. and Rao, B.V.A., "Stiffness and Damping Characteristics of Finite Width Journal Bearings With a Non-Newtonian Film and Their Application to Instability Predictions", WEAR, Vol. 32(1975) pp. 379-390.

4. Gorla, R.S.R. "Thermal Characteristics of a Non-Newtonian Fluid in a Porous Thrust Bearing" WEAR, Vol. 67, 1981, pp 351-359.

5. Singh, P. and Singh, C. "Non-Newtonian Squeeze Films in Spherical Bearings" WEAR, Vol. 68, 1981, pp 133-140.

6. Elkouh, A.F. "Fluid Inertia Effects in Non-Newtonian Squeeze Films", ASME Journal of Lubrication Technology 1976, pp 409-411.

7. Khader, M.S. and Vachon, R.I. "Inertia Effects in Laminar Radial Flow of Power Law Fluids", Int. J. Mech. Sci., Vol. 15, 1973, pp 221-225.

8. Christensen, R.M., and Saibel, E.N. "Normal Stress Effects in Visco-elastic Fluid Lubrication", J. Non-Newtonian Fluid Mechanics, Vol. 7, 1980, pp 63-70.

9. Huebner, K.H. "Finite Element Analysis of Fluid Film Lubrication - A Survey", Finite Elements in Fluids, Volume 2, Wiley Interscience, New York, 1975,pp. 225-254.

10. Braun, M.J., and Mullen, R.L., "A Three-dimensional Numerical Method with Applications to Tribological Problems", Proceeding of the 10th Leeds-Lyon Symposium on Triblogy, 1983.

11. Aris, R. "Vectors, Tensors, and the Basic Equations of Fluid Mechanics", Prentice Hall, 1972.

12. Eiserman, P.R., "A Coordinate System for a Viscous Cascade Analysis," J. Comp. Phys. V.26, 307-338, 1978.

13. Vinkour, M., "Conservation Equations of Gasdynamics in Curvilinear Coordinate Systems," J. Comp. Phys., V.14, 105-125, 1974.

14. Rhie, C.M. and Chow, W.L., "Numerical Study of the Turbulent Flow Past an Airfoil with Trailing Edge Separation", AIAA Jourl., V21, 1983.

15. Demirdzic, I., Grosman, A.D., Issa, R.I., "A Finite Volume Method for the Prediction of Turbulent Flow in Arbitrary Geometries," Proc. of 7-th International Conf. on Numerical Methods in Fluid Dynamics, Stanford Univ. Press, June 1980.

16. Anderson, J., Preiser, S. and Rubin, E., "Conservation Law Form of the Equations of Hydrodynamics in Curvilinear Coordinate Systems", J. Computational Physics, V2, 1968, p 279.

17. Patankar, S.V., "Numerical Heat Transfer and Fluid Flow", Hemisphere Press, 1980.

18. Dierckes, A.C. Jr. and Schowalter, W.R., "Helical Flow of a Non-Newtonian Polyisobutylene Solution", I and EC Fundamentals, Vol. 5, May 1966, pp 263-271.

19. Rea, D.R., and Schowalter, W.R, "Velocity Profiles of a Non-Newtonian Fluid in Helical Flow", Trans. Soc. of Rheology, Vol. 11, 1967, pp 125-143.

20. Bird, R.B., Steward, W.E. and Lightfoot, E.N., "Transport Phenomena", John Wiley, 1960.

21. Mullen, R.L., Braun, M.J., and Hendricks, R.C., "FEM for a Reynolds Equation Using a Power Law Fluid. To be presented, ASLE/ASME Joint Lubrication Conference, Atlanta GA, Oct. 1985.

22. Hendricks, R.C.,"Some Flow Characteristics of Conventional and Tapered High Pressure Drop Simulated Seals, ASLE/ASME Joint Lubrication Conference, Dayton OH, Oct. 1979.

TABLE 1. MATERIAL PROPERTIES OF 3-WEIGHT PERCENT POLYISOBUTANE DISSOLVED IN A PRACTICAL GRADE CIS-TRAN MIXTURE OF DECALIN.

Reference		$dyne\text{-}sec^{\alpha}/cm^2$	α
Dierckes [18]		9.4	.77
Rea [19]	mean shear rate	6.93 $\pm$ .08	.825 $\pm$ .003
	high shear rate	10.97 $\pm$ .95	.725 $\pm$ .018
	low shear rate	5.84	.872

TABLE 2. REYNOLDS EQUATION SOLUTION FOR MASS FLOW RATES AS A FUNCTION OF CONE ANGLE FOR VARIOUS POWER LAW FLUIDS (g/s). NONROTATING-TAPERED CENTERBODY WITH CONSTANT DIAMETER HOUSING ROTATING AT 100 RPM.

Cone Angle	Power Law Exponent α				Area Ratio
	0.77	0.95	1.0	1.05	A,bar/A,o
0.0	1295.8	515.88	418.12	342.49	1.0
0.1	1295.7	515.86	418.11	342.49	0.9988
0.5	1294.7	515.46	417.78	342.20	0.9940
1.0	1291.7	514.26	416.74	341.3	0.9879
2.	1279.5	509.2	412.61	337.8	
3.	1259.3	500.91	405.76		0.9624
5.	1195.5	474.73	384.15	314.0	0.9352
10.	914.07	359.75	289.56		0.8605
15.	517.95	199.67	158.89	127.4	0.7768

A,bar/A,o = (A,in + 2A,out)/3, limited to small convergent angles.

TABLE 3. A COMPARISON OF NAVIER-STOKES (FD) AND REYNOLDS (FEM) EQUATIONS SOLUTIONS FOR MASS FLOW RATES AS A FUNCTION OF CONE ANGLE FOR A POWER LAW FLUID WITH α = 0.77. (g/s). NONROTATING-TAPERED CENTERBODY WITH CONSTANT DIAMETER HOUSING ROTATING AT 100 RPM.

Test Case	Inclination	FEM	FD	FEM/FD
Fully developed flow	0	1295.8	1292	
Developing flow	0	------	1322	.980
Conical developing flow	0.1	1295.7	1321	.981
	1.	1291.7	1316	.982
	3.	1259.3	1285	.980
	5.	1195.5	1228	.974
	10.	914.1	973	.939
	15.	517.95	568	.91
newtonian (α = 1.0)	15.	159.	225	.71

TABLE 4. MASS FLOW RATES AS A FUNCTION OF CONE ANGLE FOR A POWER LAW FLUID WITH α = 0.77. (g/s). NONROTATING CONSTANT DIAMETER HOUSING ROTATING WITH TAPERED CENTERBODY ROTATING AT 100 RPM. A,bar/A,o the same as table 2.

Test Case	Inclination	FEM	FD
Fully developed flow	0	1295.8	1281
Developing flow	0	-----	1308
Conical developing flow	0.1	1295.7	1302
	1.	1291.7	1298
	3.		1269
	5.	1195.5	1214
	10.		963
	15.	517.95	570

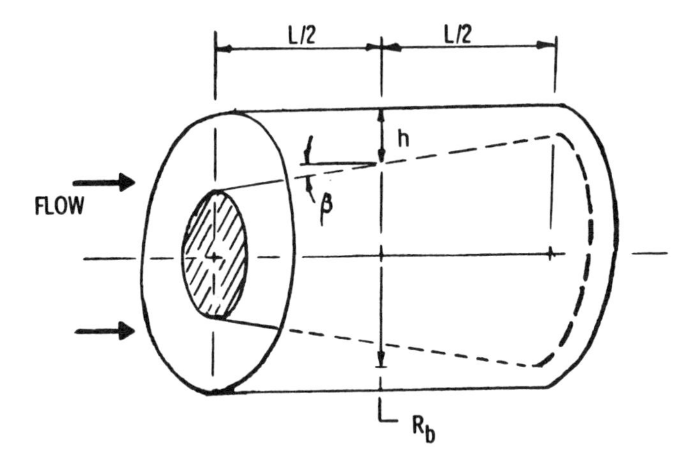

Figure 1. Schematic of the Cylindrical Housing with Tapered Centerbody.

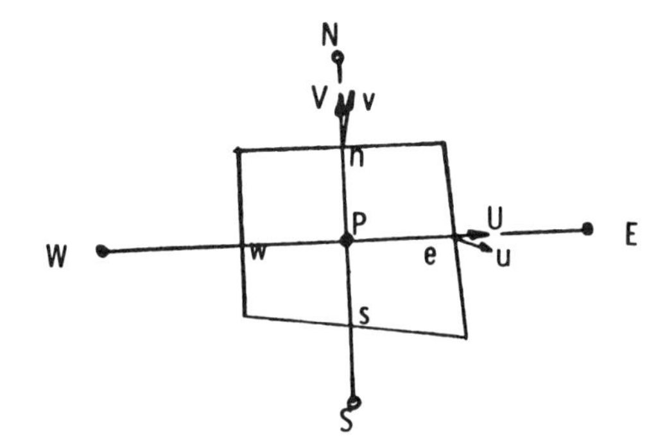

Figure 2. Staggered Grid Discretization for Navier-Stokes Solution

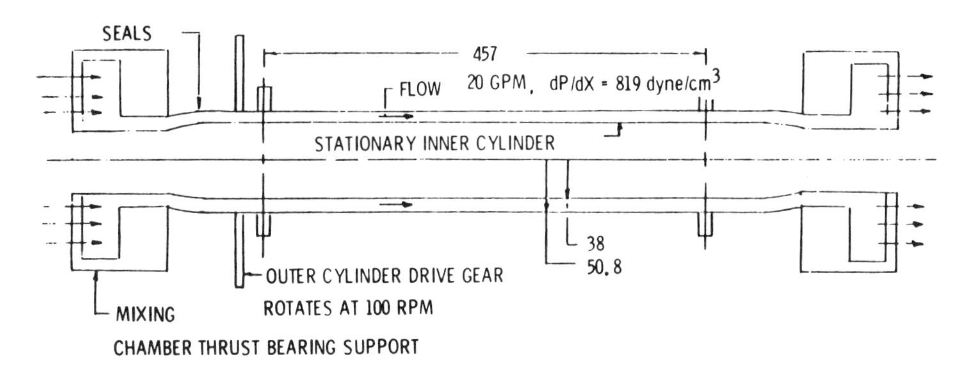

Figure 3. Schematic of Helical Flow Apparatus, Refs, 18,19. Dimensions are in mm.

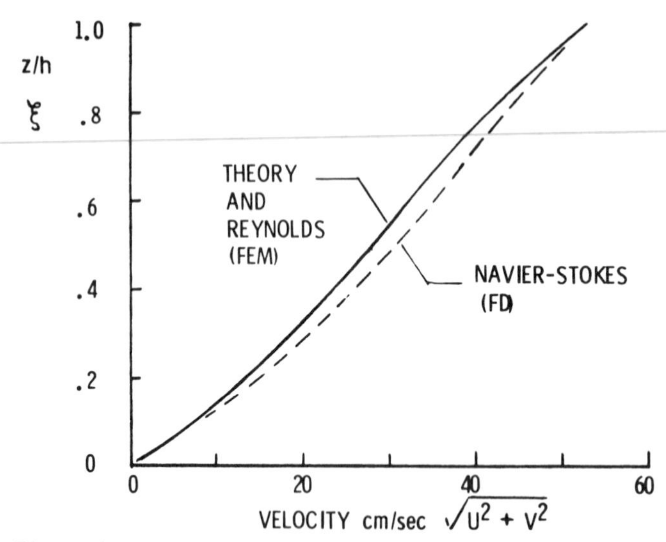

Figure 4. Velocity magnitude as a function of radial position; geometry of Fig. 3. The fluid is separan, $\alpha = 1/3$, $\mu = 250$ dyne-sec^2/cm^2.

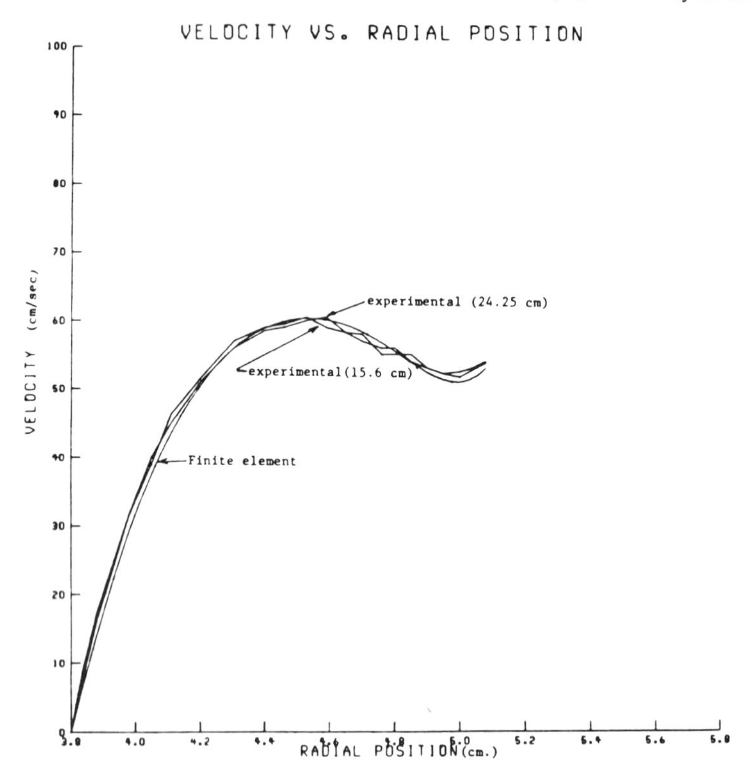

Figure 5. Velocity magnitude as a function of radial position; data of Ref. 19.

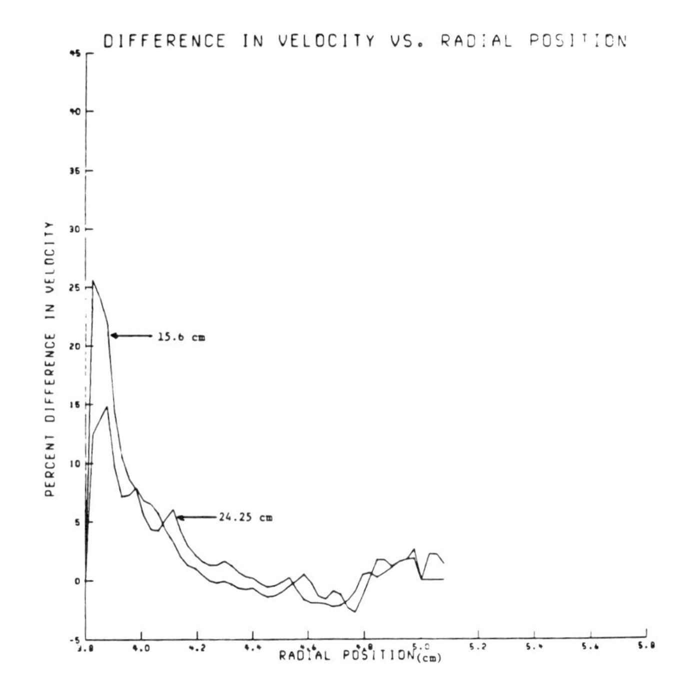

Figure 6. Difference between calculated (FEM) and experimental velocity magnitude; data of Ref. 19.

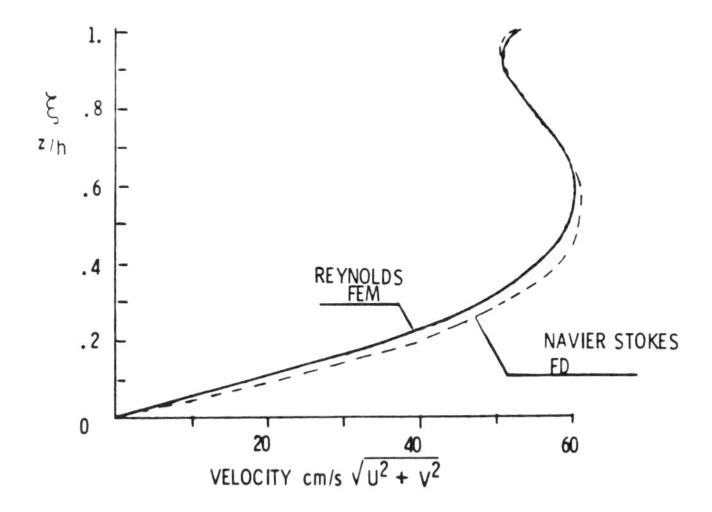

Figure 7. Comparison of velocity magnitudes predicted by FD and FEM solution procedures.

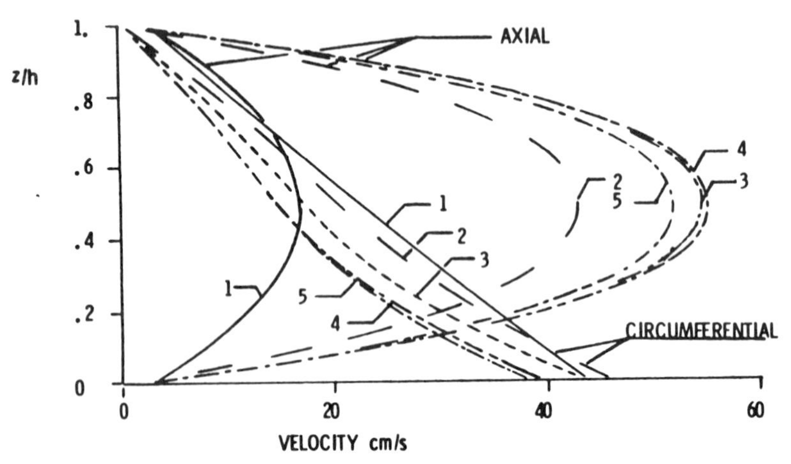

Figure 8. Navier-Stokes Solution for Axial and Circumferential Velocity Profiles for selectd parameters.

a. Inner wall rotating

b. Outer wall rotating

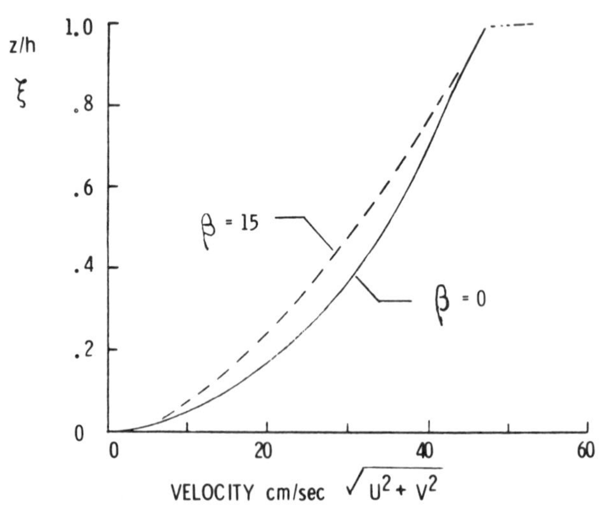

Figure 9. Velocity magnitude as a function of radial position or developing flow for a hypothetical dilitant fluid, $\alpha = 1.05$, $\mu = 9.4$ dyne-sec /cm^2. Geometry is that of Fig. 3.

TURBINES AND COMPRESSORS

Unsteady, Three-Dimensional Flow through a Cascade of Fluctuating Lifting-Lines

SAEED FAROKHI
Aerospace Engineering Department
The University of Kansas
Lawrence, Kansas 66044

ABSTRACT

The problem of unsteady, three-dimensional flow through a cascade of fluctuating lifting lines is analytically formulated for a perfect fluid. A general two-dimensional approach is due to Hawthrone, who proposed a linearized perturbation analysis, for the solution of the 2-D problem, utilizing the method of generalized functions. The present theory is an extension of Hawthorne's approach, which includes the effect of spanwise variation of circulation along the lifting lines on the induced flow field at the cascade. The results in the three-dimensional approach show a significant departure from the 2-D limit. For comparison, the three-dimensional steady flow through a cascade of lifting lines has been separately calculated and is shown that the unsteady solution, in the limit of $\omega \rightarrow 0$ (i.e., angular velocity of lift fluctuation approaching zero), reduces to that of 3-D steady case. Furthermore, it is shown that with increasing reduced frequency, the 3-D fluctuating induced velocity at the cascade increases, the minimum being at the 2-D steady point. Significant effect of three-dimensionality on the unsteady flow through cascades is shown to be a decrease in the phase angle between the fluctuating lift and the induced flow at the blades.

INTRODUCTION

Theoretical investigation of unsteady, three-dimensional fluid flow through an axial turbomachinery blade row is a complex task. A major simplifying assumption is the omittance of viscous stresses from the equations of motion, which limits the scope of analysis to inviscid fluids. The compressibility effect, though not as complicated as viscosity, is often in the initial stages of the theoretical development of such kinds neglected. A further idealization introduced originally by Prandtl, the so-called lifting-line theory, is now a widely used concept in the analysis of two and three-dimensional flows in turbomachines [1,2]. Hawthrone utilized the above assumptions and developed a two-dimensional theory which accounts for the effects of lift-fluctuations in a turbomachinery cascade [3]. The present attempt is an extension of Hawthrone's theory to three-dimensions.

The unperturbed upstream flow is assumed to be steady, uniform and one-dimensional,

$$\underline{V}_{-\infty} = \hat{i}\, V_o \tag{1}$$

Thus, the "base" flow contains no vorticity. The blades are modeled as an infinite set of periodically-spaced lifting-lines of finite height, h, encased between two parallel planes (shown in Figure 1). To evaluate the effects of unsteady, spanwise loading variations upon the induced flow at the cascade, the lifting-lines are assumed to be of variable strength along their span and harmonically fluctuate in time. The lift variation is chosen to be

$$L(z,t) = L_o \cdot \cos(kz) \cdot \exp(i\omega t) \quad , \; k \equiv \pi/h \tag{2}$$

with a half-period cosine function describing the spanwise distribution of lift. This is dictated by the solid wall boundary condition, which requires that $\partial\Gamma/\partial z$ (or equivalently, $\partial L/\partial z$) to vanish at the solid walls. Higher harmonics in z, i.e. a Fourier cosine series representation of the spanwise lift distribution could describe a general behavior in the z-direction, however for the sake of simplicity only the first term in the series is taken to represent the spanwise lift variation. The source of lift fluctuations could be blade vibrations or mutual interference effects between neighboring blade rows. With the above description of lift, the wakes would contain two components of vorticity, namely the spanwise component due to the time variation of circulation and streamwise, as a result of non-uniform loading along the span. The shed wakes, in the absence of viscosity, remain in thin sheets and are assumed to be convected downstream by the steady, unperturbed far-upstream flow. This is in accordance with the classical aerodynamic theory (e.g. thin airfoils) and is permissable to apply to cascade flows if

a) the steady, spanwise-mean circulation, $\bar{\Gamma}$, is small and
b) the unsteady amplitude of circulation, Γ_o, is small.

These conditions guarantee no mean-turning of the flow through the cascade. In the present analysis, the steady component of the spanwise-mean circulation, $\bar{\Gamma}$, is set identically equal to zero and Γ_o is assumed to be small (i.e. $\Gamma_o/sV_o \ll 1$). Thus, a linearized perturbation approach may be applied to the up- and downstream flow regions of the lifting-lines with the solutions matched at the plane of the cascade.

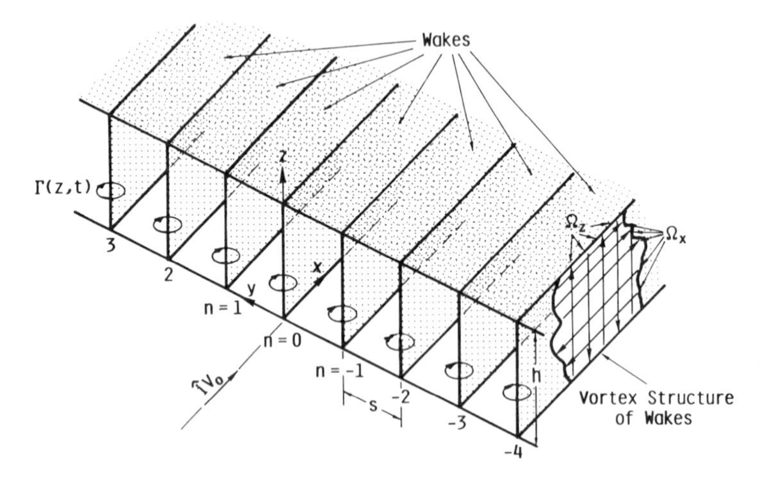

Figure 1. Cascade of lifting-lines and the trailing wakes.

THEORY

Formulation of the flow is based on the kinematical as well as dynamical equations of motion and the associated boundary conditions.

Kinematics of Flow

A general description of the velocity field, in a small perturbation analysis, can be represented as

$$\underline{V} = \underline{\bar{V}} + \underline{\tilde{V}} + \text{higher order terms} \tag{3}$$

where

$$|\underline{\tilde{V}}|/|\underline{\bar{V}}| = O(\varepsilon) \quad , \varepsilon \ll 1$$

The perturbation velocity field, $\underline{\tilde{V}}$, may be represented as the sum of a potential and a rotational field, i.e.

$$\underline{\tilde{V}} = \nabla\phi + \underline{A} \tag{4}$$

where ϕ is the scalar potential and $\underline{A}$ the vector potential, describing respectively, the irrotational and rotational velocity components of the perturbation velocity field. Thus, equation (3) may be rewritten as

$$\underline{V} = \underline{\bar{V}} + \nabla\phi + \underline{A} \tag{5}$$

In the present analysis, the time-mean or base flow is assumed to be uniform and one-dimensional, i.e.

$$\underline{\bar{V}} = \hat{i}\, V_o \tag{6}$$

By taking the curl of equation (5), the vorticity field $\underline{\Omega}$ is written in terms of vector potential $\underline{A}$, as

$$\underline{\Omega} \equiv \nabla \times \underline{V} = \nabla \times \underline{A} \tag{7}$$

Now, the continuity equation for incompressible fluids demands that

$$\nabla \cdot \underline{V} = 0 \tag{8}$$

which applied to (5) yields

$$\nabla^2\phi = -\nabla \cdot \underline{A} \tag{9}$$

The above equation is the Poisson equation for the perturbation potential, ϕ, with the rotational term $-\nabla \cdot \underline{A}$ as the "driving" source. The equation (9) is solely derived through kinematical considerations, however its solution depends on the dynamical description of the flow field.

Dynamics of Flow

In the absence of viscosity, the Euler momentum equation governs the fluid motion,

$$\rho\, D\underline{V}/Dt = -\nabla p + \underline{F} \tag{10}$$

where $\underline{F}$ represents the body force acting on the fluid.

Now, by substituting equations (5) and (6) in (10) and some manipulation, we get

$$\left(V_o\, \partial/\partial x + \partial/\partial t\right) \underline{A} = \underline{F}/\rho \tag{11}$$

(for more details see Hawthorne [3]).

The body force $\underline{F}$, acting on the fluid, is everywhere zero except at the position of lifting-lines, and is thus presented by a periodic (Dirac) delta function,

$$\underline{F} = -\hat{j}\, L_o \cos(kz) e^{i\omega t}\, \delta(x) \sum_{n=-\infty}^{+\infty} \delta(y - ns) \tag{12}$$

where $\hat{j}\, L_o \cos(kz)e^{i\omega t}$ is the fluctuating force acting on the lifting-lines. At this point, a note on the dimensional consistency of equation (12) is in order and the reader is reminded that the Dirac's delta function has dimension of $[L^{-1}]$ and thus the body force $\underline{F}$ has dimensions of $[ML^{-2}\, T^{-2}]$, i.e. force per unit volume, as required by equation (10).

By substituting equation (12) in (11), the following expression for the vector potential $\underline{A}$ is derived by inspection,

$$\underline{A} = -\hat{j}\, \frac{L_o}{\rho V_o} \cdot \cos(kz) \cdot \exp(-i\lambda\xi)\, H(x) \sum_{n=-\infty}^{+\infty} \delta(y - ns) \tag{13}$$

where $\xi \equiv x - V_o t$ and $\lambda \equiv \omega/V_o$.

The ratio $-L_o/\rho V_o$ is recognized as the amplitude of the circulation, Γ_o. The vorticity field expressed by equation (7), can be written as

$$\underline{\Omega} = \hat{i}\, \Omega_x + \hat{j}\, \Omega_y + \hat{k}\, \Omega_z \tag{14a}$$

$$\text{where } \Omega_x = -\Gamma_o\, k \sin(kz) \cdot \exp(-i\lambda\xi)\, H(x) \sum_{n=-\infty}^{+\infty} \delta(y - ns) \tag{14b}$$

$$\Omega_y = 0 \tag{14c}$$

$$\Omega_z = -\Gamma_o \cos(kz) \cdot \left[-i\lambda \exp(-i\lambda\xi)\, H(x) + \exp(i\omega t)\, \delta(x)\right] \sum_{n=-\infty}^{+\infty} \delta(y - ns) \tag{14d}$$

As expected, the vorticity field is concentrated in "thin" sheets (the wakes of zero thickness) emanating from and downstream of the lifting-lines. The streamwise component of vorticity, Ω_x, is clearly due to spanwise variation (i.e. $\partial/\partial z$) of lift and Ω_z, the unsteady component of vorticity results from the temporal behavior (i.e. $\partial/\partial t$) of circulation. It is interesting to note that the first term in the bracket in equation (14d) is the convected spanwise vorticity in the wake, the so-called free-vorticity, where the second term in (14d) describes the "bound" vortices, or the lifting-lines themselves.

Perturbation Flow Fields

The governing kinematical equation of motion throughout the flow field is derived to be the Poisson equation

$$\nabla^2\phi = -\nabla \cdot \underline{A} = -\Gamma_o \cos(kz) \cdot \exp(-i\lambda\xi) \cdot H(x) \sum_{n=-\infty}^{+\infty} \delta'(y - ns) \tag{15}$$

which in the upstream region reduces to the Laplace equation (potential flow)

$$\nabla^2\phi^u = 0 \tag{16}$$

Since, $H(x) \equiv 0$ at $x < 0$, and downstream of the lifting-lines ($x > 0$ and $H(x) \equiv 1$) the Poisson equation is

$$\nabla^2\phi^d = -\Gamma_o \cdot \cos(kz) \cdot \exp(-i\lambda\xi) \sum_{n=-\infty}^{+\infty} \delta'(y - ns) \tag{17}$$

Upstream Solution ($x < 0$). The solution of the Laplace equation governing the perturbation potential in the upstream region may be written in terms of Fourier series as

$$\phi^u = \cos(kz) \cdot \exp(i\omega t) \sum_{n=-\infty}^{+\infty} C_m^u \exp(\lambda_m x) \exp(i2m\pi y/s) \tag{18}$$

where $C_m^u \equiv$ Fourier coefficients and $\lambda_m \equiv \sqrt{(2m\pi/s)^2 + k^2}$, decay parameter. Solution (18) satisfies:

a) the solid end-wall boundary condition ($\partial\phi^u/\partial z = 0$ at $z = 0$ and h)
b) periodicity of the flow along the cascade (i.e., y-direction)
c) exponential decay to zero as $x \to -\infty$ and
d) harmonic fluctuation in time.

The Fourier coefficients, C_m^u, must be determined through matching conditions at the plane of the cascade ($x = 0$, or yz - plane).

Downstream Solution ($x > 0$). The solution to the Poisson equation (17) has two components, namely

$$\phi^d = \phi_h^d + \phi_w^d \tag{19}$$

where the homogeneous ϕ^d, or ϕ_h^d satisfies the Laplace equation

$$\nabla^2\phi_h^d = 0 \tag{20}$$

and the "wake" contribution to the ϕ^d, or ϕ_w^d, is the "particular" solution of

$$\nabla^2\phi_w^d = -\Gamma_o \cos(kz) \cdot \exp(-i\lambda\xi) \sum_{n=-\infty}^{+\infty} \delta'(y - ns) \tag{21}$$

The homogeneous solution is similar in mathematical form to the upstream solution

$$\phi_h^d = \cos(kz) \cdot \exp(i\omega t) \sum_{n=-\infty}^{+\infty} C_m^d \exp(-\lambda_m x) \exp(i2m\pi y/s) \tag{22}$$

To establish the wake-potential, ϕ_w^d, the periodic δ' - function is Fourier analyzed according to M. J. Lighthill [4]

$$\sum_{n=-\infty}^{+\infty} \delta'(y - ns) = \frac{d}{dy}\left[\frac{1}{s} \sum_{n=-\infty}^{+\infty} \exp(i2n\pi y/s)\right] = \frac{i}{s} \sum_{n=-\infty}^{+\infty} (2n\pi/s) \exp(i2n\pi y/s) \qquad (23)$$

and by inspection, the solution of (21) is

$$\phi_w^d = \cos(kz) \cdot \exp(-i\lambda\xi) \sum_{n=-\infty}^{+\infty} A_n \exp(i2n\pi y/s) \qquad (24)$$

$$\text{where } A_n = + \frac{i\Gamma_o}{s} \cdot \frac{2n\pi/s}{(2n\pi/s)^2 + \alpha^2} \quad \text{and} \quad \alpha^2 \equiv k^2 + \lambda^2 \qquad (25)$$

The Fourier coefficients A_n are determined by substituting equation (24) in (21)

Matching Conditions (at x = 0). To obtain a unified solution throughout the flow field, the individual solutions derived for up- and downstream of the cascade need to be matched at the plane $x = 0$. The conditions for consistency of the perturbations are of kinematical and dynamical nature, i.e., the mass flux through $x = 0$ plane must be identical for both up- and downstream solutions (the kinematical condition) and in the absence of spanwise force at $x = 0$ plane, the spanwise component of the induced velocities is continuous across the cascade (the dynamical condition). These conditions require

$$\phi_x^u(x = 0^-) = \phi_x^d(x = 0^+) \quad (26), \text{ and} \quad \phi_z^u(x = 0^-) = \phi_z^d(x = 0^+) \qquad (27)$$

Application of (26) and (27) to the perturbation potentials ϕ^u and ϕ^d yields the unknown Fourier coefficients C_m^u and C_m^d. These are

$$C_m^u = \frac{A_m}{2}(1 - i\lambda/\lambda_m) \quad (28), \text{ and} \quad C_m^d = -\frac{A_m}{2}(1 + i\lambda/\lambda_m) \qquad (29)$$

Now, with the knowledge of these Fourier coefficients, the perturbation velocities throughout the flow field are determined.

Induced Flow at the Cascade (x = 0 plane). To evaluate the change of the flow angle, at the position of lifting-lines, induced by the trailing vortex sheets, we need to eliminate the self-induced effects of "bound" and free-vortices at $x = 0$ plane. In Prandtl's lifting-line theory, this is achieved by taking the principal value of the downwash integral. In the unsteady lifting-line theory presented here, the self-induced effect of the "bound" vorticity is eliminated by taking the mean-value of the induced velocities up- and downstream of the cascade, i.e.

$$\tilde{V}_y(x = 0) = \frac{1}{2}\left[\tilde{V}_y^u(x = 0^-) + \tilde{V}_y^d(x = 0^+)\right] \qquad (30)$$

Now, by substitution for $\tilde{V}_y^u$ amd $\tilde{V}_y^d$ in equation (30) and extensive manipulations (shown in Appendix A), we arrive at

$$\frac{\tilde{V}_y(x = 0, y = ns)}{\cos(kz) \cdot \exp(i\omega t)} = + \frac{\Gamma_o}{2s} \cdot \frac{\alpha s}{2} \cdot \coth\left(\frac{\alpha s}{2}\right) + \frac{i\lambda\Gamma_o}{s} \sum_{1}^{\infty} \frac{1}{\lambda_m} \cdot \frac{(\alpha s)^2}{(2m\pi)^2 + (\alpha s)^2}$$
$$- \frac{i\lambda\Gamma_o}{s} \sum_{1}^{\infty} \frac{1}{\lambda_m} \qquad (31)$$

The first term in the above expression is the quasi-steady downwash (shown in Appendix B to be the steady, three-dimensional downwash, when α is replaced by π/h). The second term, is the contribution of unsteady vortices in the wakes to the induced velocity at the position of lifting-lines (note that this series converges as $\sim 1/m^3$) and finally the last term in (31) is the self-induced velocity, diverging as $\sim 1/m$, caused by the free-vortices in the spanwise direction at the plane x = 0. Examination of equation (14d) reveals that the strength of the free-spanwise vorticity at the position of lifting-lines is proportional to $i\lambda\Gamma_o/s$ (since the Heaviside step-function, $H(x) \equiv 1/2$ at $x = 0$). Thus, the wake-induced velocity at the cascade is

$$\left[\left(\frac{\Gamma_o}{2s}\right)\left(\frac{\alpha s}{2}\right)\coth\left(\frac{\alpha s}{2}\right) + \frac{i\lambda\Gamma_o}{s}\sum_{1}^{\infty}\frac{1}{\lambda_m}\frac{(\alpha s)^2}{(2m\pi)^2 + (\alpha s)^2}\right]\cos(kz)\cdot\exp(i\omega t) \tag{32}$$

It is interesting to note that the groupings in the above expression have counterparts in the cascade flows and unsteady wing theory, namely $(\Gamma_o/2s)$ is the familiar downwash downstream of a cascade of lifting lines, with spacing s and strength Γ_o and $\alpha s/2$ is the "modified" reduced frequency, which accounts for the effect of blade span as well the unsteadiness on the 3-D induced flow.

RESULTS AND DISCUSSIONS

The flow through a three-dimensionally fluctuating cascade of lifting-lines has been mathematically formulated. The induced flow at the cascade, caused by the trailing wakes, is shown to be a function of two parameters, namely $\pi s/2h$ and $\omega s/2V_o$. The real component of the non-dimensional induced velocity at the lifting-lines has been derived in closed form, i.e., $(\alpha s/2)\coth(\alpha s/2)$, and is plotted in Figure 2. By varying the three-dimensionality parameter, $\pi s/2h$, the effect of blade height on the induced velocities is seen to be a notable growth in the downwash, with the minimum at the 2-D limit. This effect may be traced back to the existance of streamwise trailing vortices in the wakes, due to spanwise variation of circulation. It is interesting to note that in the 2-D, steady flow limit, i.e. $\pi s/2h = \omega s/2V_o = 0$, the induced velocity at the cascade is $\Gamma_o/2s$, as predicted by the circulation theorem. With the increasing degree of unsteadiness, i.e. as $\omega s/2V_o$ increases, the real component of the induced velocities approach a 45° asymptote as shown in Figure 2.

The imaginary component of the non-dimensional induced velocity at the cascade is expressed in terms of a converging series ($\sim 1/m^3$) and the summation of the first five terms in the series is plotted in Figure 3. A growing trend in the magnitude of the imaginary component is observed with respect to increasing reduced frequency parameter, $\omega s/2V_o$. The effect of three-dimensionality is seen to be a reduction of the imaginary component of the induced velocity, which expressed in terms of the phase angle, θ, with respect to the fluctuating lift, leads to a substantial reduction of the phase angle, over a wide range of reduced frequencies (see Figure 4).

The above results, even though derived for a perfect fluid, point to the inadequecy of two-dimensional analysis to predict the behavior of an inherently three-dimensional flow phenomena. A further refinement of the theory calls for the inclusion of inlet flow shear, to model the wakes of upstream blade rows. Extension of the present theory to include the viscous effects is the most difficult task and would require a computational approach.

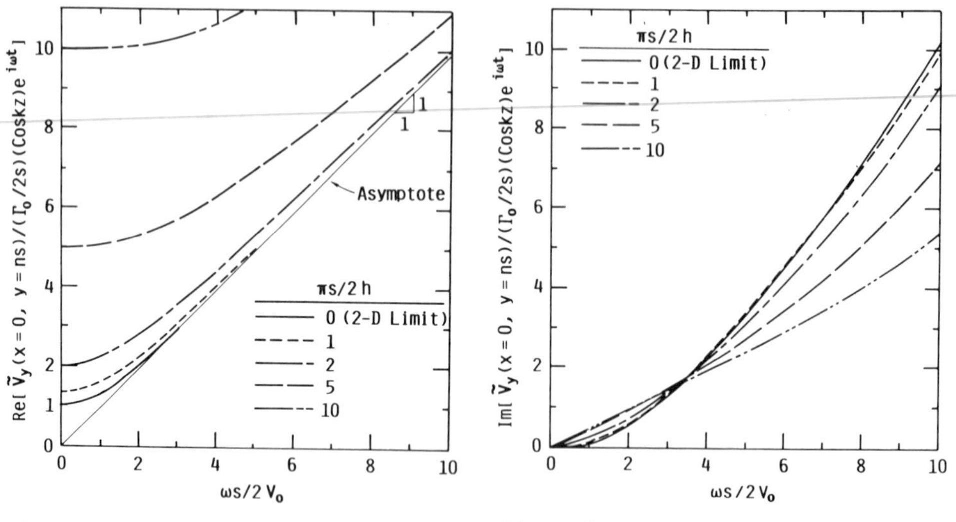

Figure 2.

Figure 3.

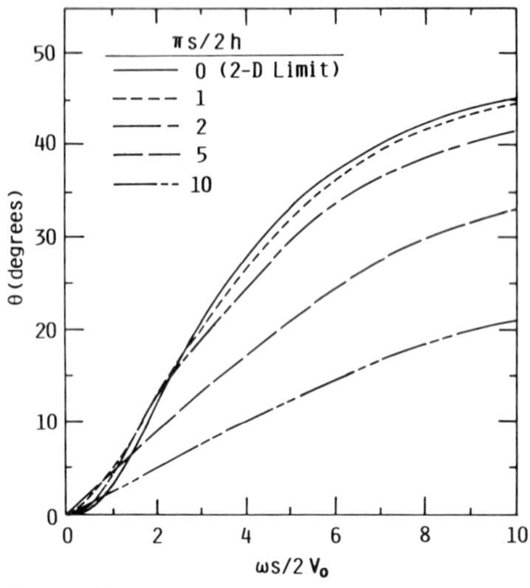

Figure 4.

Figure 2. Real component of the non-dimensional induced velocity at the cascade.

Figure 3. Imaginary component of the non-dimensional induced velocity at the cascade.

Figure 4. Phase angle between the induced velocity and fluctuating lift.

NOMENCLATURE

Latin

$\underline{A}$	Vector potential
A_m, C_m	Fourier coefficients
D/Dt	$\partial/\partial t + V\partial/\partial x$,
$\underline{F}$	Body force
h	Blade height
H(x)	Heaviside step function
i	$\sqrt{-1}$
$\hat{i}$, $\hat{j}$	unit vectors in x and y
Im	Imaginery part
k	π/h
L	Lift per unit span
L_o	Amplitude of lift
p	Static pressure
Re	Real part
s	Blade spacing
t	Time
$\underline{V}$	Velocity vector
x,y,z	Cartesian coordinates

Greek

α	$\sqrt{k^2 + \lambda^2}$

Γ	Circulation
Γ_o	Amplitude of circulation
θ	Phase angle between lift and induced velocity
ϕ	Scalar potential
λ	ω/V_o
λ_m	$[(2m\pi/s)^2 + k^2]^{1/2}$
$\delta(x)$	Dirac's delta function
ε	Small quantity, $\ll 1$
ω	Angular velocity of lift fluctuations
ρ	Fluid density
∇	Gradient operator
$\nabla\cdot$	Divergence operator
$\nabla\times$	Curl operator
∇^2	$\partial^2/\partial x^2 + \partial^2/\partial y^2 + \partial^2/\partial z^2$
ξ	$x - V_o t$, wake coordinate
$\underline{\Omega}$	Vorticity vector

Subscripts

$-\infty$	Far-upstream
$-$	Vector
x,y,z	Component of a vector, derivative of scalar function
h	Homogeneous solution
w	"Wake" or particular solution
o	Amplitude or initial-value

Superscripts

'	Derivative
$-$	Time or spanwise average
$\sim$	Perturbation quantity
u	Upstream of cascade
d	Downstream of cascade

REFERENCES

1. McCune, J. E., Hawthrone, W. R., The Effect of Trailing Vorticity on the Flow Through Highly-Loaded Cascades, *Journal of Fluid Mechanics*, vol. 74, pp. 721-742, 1976.
2. McCune, J. E., Dharwadkar, S. P., Lifting-Line Theory for Subsonic Axial Compressor Rotors, MIT Gas Turbine Lab. Report No. 110, July 1972.
3. Hawthrone, W. R., Flow Through Moving Cascades of Lifting-Lines with Fluctuating Lift, *Journal of Mechanical Engineering Sciences*, vol. 15, no. 1, 1973.
4. Lighthill, M. J., *Introduction to Fourier Analysis and Generalized Functions*, Cambridge University Press, 1958.
5. Meschkowski, H., *Unendliche Reihen*, p. 146, Hochschultaschenbuecher-Verlag, Bibliographisches Institut AG, Mannheim, 1962.

APPENDIX A: DERIVATION OF THE INDUCED VELOCITY AT THE LIFTING-LINES

The upstream perturbation flow is described by $\underline{\tilde{V}}^u = \nabla\phi^u$ therefore

$$\tilde{V}^u_y = \phi^u_y = \frac{\partial}{\partial y}\left[\sum_{m=-\infty}^{+\infty} C^u_m \exp(+\lambda_m x) \cdot \exp(i2m\pi y/s)\cdot \exp(i\omega t)\cdot \cos(kz)\right] \tag{A1}$$

The y-component of the downstream perturbation velocity evaluated at $x = 0^+$ is

$$\tilde{V}^d_y\,(x = 0^+) = \phi^d_y(x = 0^+) + \Gamma_o\cos(kz)\cdot\exp(i\omega t)\sum_{n=-\infty}^{+\infty}\delta(y - ns) \tag{A2}$$

and may be substituted in equation (30), and evaluated at the lifting lines, $y = ns$, to get

$$\tilde{V}_y(x = 0,\ y = ns) = \frac{1}{2}\left[\sum_{-\infty}^{+\infty}\left(C^u_m + C^d_m + A_m\right)(i2m\pi/s) + \frac{\Gamma_o}{s}\right]\cos(kz)\cdot\exp(i\omega t) \tag{A3}$$

Now, substituting for C^u_m, C^d_m, and A_m from equations (28), (29), and (25), respectively, in (A3) and some simplifications, we get

$$\tilde{V}_y(x = 0,\ y = ns) = \left[\frac{\Gamma_o}{2s} + \frac{\Gamma_o}{s}\sum_1^\infty \frac{(\alpha s)^2}{(2m\pi)^2 + (\alpha s)^2} + \frac{\Gamma_o}{s}\sum_1^\infty \frac{i\lambda}{\lambda_m} + \frac{\Gamma_o}{s}\sum_1^\infty \frac{i\lambda}{\lambda_m}\cdot\frac{(\alpha s)^2}{(2m\pi)^2 + (\alpha s)^2}\right]\cos(kz)\cdot\exp(i\omega t) \tag{A4}$$

The following summation identity (reference [5])

$$\sum_{k=1}^\infty \frac{x}{k^2 + x^2} = \pi\ \coth(\pi x) \tag{A5}$$

simplifies the first series in (A4) and results in equation (32).

APPENDIX B: DERIVATION OF STEADY, THREE-DIMENSIONAL INDUCED VELOCITY

In a similar analysis to the unsteady flow, we assume that

$$\underline{V}_{-\infty} = \hat{i}\ V_o \quad \text{(B1)}, \qquad \text{and} \qquad L(z) = L_o\ \cos(kz) \tag{B2}$$

By representing the velocity field as

$$\underline{V} = \hat{i}\ V_o + \nabla\phi + \underline{A} \tag{B3}$$

and identical manipulations, as in the unsteady analysis, we get

$$\nabla^2\phi^u = 0 \quad \text{(B4)}, \qquad \text{and} \qquad \nabla^2\phi^d = -\nabla\cdot\underline{A}^d = -\Gamma_o\ \cos(kz)\cdot\sum_{-\infty}^{+\infty}\delta'(y - ms) \tag{B5}$$

The solutions to (B4) and (B5) are, respectively,

$$\phi^u = \sum_{m=-\infty}^{+\infty} C_m^u\ \exp(\lambda_m x)\ \cos(kz)\cdot\exp(i2m\pi y/s) \tag{B6}$$

and

$$\phi^d = \sum_{m=-\infty}^{+\infty}\left[C_m^d\ \exp(-\lambda_m x) + \frac{i2m\pi\Gamma_o}{(2m\pi)^2 + (ks)^2}\right]\cos(kz)\cdot\exp(i2m\pi y/s) \tag{B7}$$

The matching conditions (26) and (27) applied to (B6) and (B7) will determine the unknown Fourier coefficients C_m^u and C_m^d, i.e.

$$C_m^u = \frac{+\ i\ m\pi\ \Gamma_o}{(2m\pi)^2 + (ks)^2} \quad \text{(B8)}, \qquad \text{and} \qquad C_m^d = \frac{-\ im\pi\ \Gamma_o}{(2m\pi)^2 + (ks)^2} \tag{B9}$$

The induced velocity is

$$\tilde{V}_y(x = 0) = \frac{1}{2}\left[\tilde{V}_y^u(x = 0^-) + \tilde{V}_y^d(x = 0^+)\right] = \frac{1}{2}\left[\phi_y^u(x = 0^-) + \phi_y^d(x = 0^+) + A^d(x = 0)^+\right] \tag{B10}$$

where upon substitution from (B6) and (B7) and similar manipulations as described in Appendix A we get

$$\tilde{V}_y(x = 0,\ y = ns) = \frac{\Gamma_o}{2s}\cdot\left(\frac{ks}{2}\right)\cdot\coth\left(\frac{ks}{2}\right)\cdot\cos(kz) \tag{B11}$$

Temperature and Surface Behavior of Water Film Driven by Co-Current Steam Flow

BONG-HWA SUN and GENE E. SMITH
Department of Mechanical Engineering and Applied Mechanics
The University of Michigan
Ann Arbor, Michigan 48109

ABSTRACT

Co-current steam-liquid film flow was studied in a low-pressure steam tunnel. Water flow rate and inlet water temperature were varied to study their effects on film temperature on turbine blades. Instantaneous film thickness data were collected by a computer-based data acquisition system with electric conductivity thickness sensors. Transitions of film flow patterns were determined by photographic study.

The experimental data have been compared with available models and relevant experimental data. It is concluded that:

1. Water film temperature on blade surface is equal to the local saturated steam temperature and is independent of water flow rate and inlet water temperature.
2. Mikielewicz's model can be used to model transition between rivulet and continuous film flow.
3. Ishii's model can be used to model transition between continuous film and atomization film flow.
4. Universal velocity profile model can be used to model shear driven film flow.
5. Normal distribution function cannot be used to describe film thickness variation.

The results of this study provides better understanding of water film behavior on blade surfaces in the final expansion stage of large steam turbine. In addition, these studies also provide useful information for shear-driven film flow which has been encountered in many heat and mass transport devices such as emergency cooling systems for reactor cores, cooling film for reentrant vehicle nose cones, process towers in chemical industry, etc.

INTRODUCTION

The liquid film flows over a solid surface have been found in many engineering devices in the fields of petroleum, chemical, nuclear and power industries. These film flows can be driven by (1) body force, such as centrifugal flore [1-3], gravity force [4-6], electric and magnetic force [7-8], (2) surface force, such as surface tension force [9] and interfacial shear force [10-12], and (3) combination of body and surface force [13-14].

Though steam driven film flows occur and have direct application in the steam turbine, only a few papers have studied the physical phenomena [15]. Thus, the necessity of studying water films driven by cocurrent high speed steam flow is apparent.

Since the surface behavior and temperature of liquid films affect heat, mass and momentum transport rates significantly, the present paper reports an experimental study on water film flows driven by a low pressure high-speed steam stream. All test data are collected using an advanced apparatus at the University of Michigan. In addition, the experimental results are compared with some pertinent theoretical models.

EXPERIMENTAL FACILITY AND CONDITIONS

Wet steam (near saturation) was drawn from a building heating boiler at about 1.6 bar abs., and expanded into a rectangular, transparent (Lexan) test section of constant flow area (Fig. 1). A thin flat-plate tapered-end blade was inserted parallel to the flow, and along the center-plane (Fig. 2).

Test section steam pressure was maintained at about 0.15 bar (abs.) to simulate the steam conditions at the low pressure stage of a steam turbine. The steam was at a nearly saturated state [16]. Water was injected via a transverse slot (Fig. 2) to form a thin liquid film along the blade. It traversed a heating and mixing tank for temperature control before entering the test section.

Film thickness was measured by electrical conductivity gages. The output from an a.C. Wheatstone bridge was passed thru a signal conditioner to remove the 10 KHz carrier wave (Fig. 3). Signals then entered a real-time microcomputer. Thickness data were sampled by a real-time computer-based data acquisition system at a selected frequency (2 KHz) and were recorded on floppy disks. Analysis and graphical work were processed on an Amdahl 5860 computer (Michigan Computing Center). High-speed still and motion picture photography were used to study the film behavior.

RESULTS AND DISCUSSIONS

1. Water Film and Turbine Blade Temperature

The film flow rate (Q) and the inlet temperature effects on film temperature were studied. It is found that water film temperature, which is equal to the saturated steam temperature ($\sim$1 C deviation), (Table 1), is independent of inlet water temperature for both $T_{inlet} > T_{sat}$ and $T_{inlet} < T_{sat}$ (Table 2). It is also concluded that Q (flow rate) has no significant effects on the film temperature (Table 2). This is very useful information for improving the steam turbine design because it provides the quantitative information on liquid films which can be used to estimate how much heat is necessary to get rid of liquid film on stators. Another conclusion is that the blade temperature without liquid film (dry) was at least 15°C above the saturated steam temperature (Table 3), which confirms the existence of a heated sublayer on the nozzle wall of a steam turbine [17]. Thus, direct condensation of liquid on the blade is unlikely.

2. Surface Behavior of Liquid Films

Three different flow regimes, i.e., rivulet film flow, continuous film flow and atomized (shed droplet) film flow, may be seen in the present horizontal shear-driven film.

In case of the rivulet flow, the film is interrupted by discrete rivulets. Non-uniform heat and mass transfer rates due to these "dry spots" have attracted some interest in determining a transition between the rivulet and continuous regimes. Both the photos (global qualitative information) and the probability distribution density functions (local quantitative information) of the instantaneous thickness have been used to determine the transition point between the film flow regimes. A typical thickness trace for the rivulet flow is shown on Fig. 4.a.

In the "continuous" regime, the liquid film covers the entire wall surface. A typical thickness trace for the continuous film flow is shown in Fig. 4.b. The photo for the continuous film flow (water flow rate = 200 cc/min, steam velocity = 45 m/s) is shown on Fig. 5.a. Wavelets exist on the film surface. In general, wavelets may increase the mass, momentum, and heat transfer rates significantly.

In the "atomized" regime, droplets are entrained by the driving steam (or gas) flow. Such an atomization is especially important to mass and momentum transfer. Photos were used here to determine its onset. A typical photo for the atomized flow is shown in Fig. 5.b, in which a white blur appear on the liquid film surface.

Both the still and the movie pictures as well as film thickness traces are used to study surface behavior. Flow conditions and the corresponding film behavior are summarized in Table 4. The general trends are:

a) Transitions from the rivulet to continuous film flows occur at $Q < 100$ cc/min.
b) Wave lengths decrease as Q increases at a fixed steam velocity, V, and decrease as V increases at a fixed Q. The maximum wave lengths are < 30 mm.

3. Rivulet Transition Criteria

Three different methods, i.e., small perturbation [18-19], force balance [20-21], and minimum energy model [22], have been used for the prediction of the transition between the continuous and rivulet film flows.

Hartley and Murgatroyd [20], H-M model, have proposed a model based on the balance of surface tension and pressure forces at the liquid stagnation point between the dry and wetted regions and find the critical film thickness and the flow rate as:

$$h_c = 1.82(\sigma(1-\cos\theta)/\rho_f)^{1/3}(\mu_f/\tau_i)^{2/3}$$

$$Q_c = 3.3\ (\sigma(1-\cos\theta))^{2/3}(\rho_f\mu_f/\tau_i)^{1/3}$$

where h_c is the critical film thickness; σ, surface tension; Q_c, the corresponding critical flow rate; μ_f, liquid viscosity; τ_i, interfacial shear; ρ_f liquid density; and θ, contact angle. Mikielewicz and Moszynski [22], M-M model, proposed the minimum power model (transition occurs when the sum of film surface and kinetic energy is minimum) to obtain the relationship for circular segment rivulet cross sections with linear liquid velocity profile assumed. They found that the critical thickness is:

$$h^+ = 3.2^{-2/3}\left(\frac{\theta}{\sin\theta-\cos\theta}\right)^{2/3} h^{2/3} g(\theta)^{2/3}\ \frac{\sin\theta}{\phi(\theta)}$$

where

$$h^+ = \frac{\rho_f \tau_i^2 \bar{h}^3}{\mu_f^2 \rho}$$

$$g(\theta) = \frac{1}{4}\cos^3\theta - \frac{8}{13}\cos\theta + \frac{15}{8}\theta\ \sin\theta - \frac{3}{2}\theta\sin\theta\ ,$$

$$\phi(\theta) = \sin\theta - \frac{1}{3}\sin^3\theta - \theta\cos\theta$$

With the critical thickness measurement and these theoretical models, the liquid contact can be calculated (Table 5). The substantial discrepancy between the prediction (H-M model) and the measurements of the contact angle is probably due to an over-simplification of the physical conditions, e.q., neglecting the possible rivulet waviness and the variation of the contact angles which depend on variable wetting conditions. However, the predicted contact angle by the M-M model agrees fairly well with the actual contact indicating that the M-M model is good for predicting the rivulet transition.

4. Onset of Atomized Flow

The unbalance of the surface forces (pressure, shear and surface tension) and the body forces (gravity, centrifugal and magnetic) may provoke an atomization process.

The models based on (1) shearing of the wave crests by a driving gas and (2) undercutting of waves by driving the flow were developed by Ishii and Grolmes [23]. they found that the critical gas velocity for the onset of atomization is proportional to $Re_f^{-1/3}$ for $160 < Re_f < 1635$, and to $Re_f^{-1/2}$ for $2 < Re_f < 160$. It becomes independent of Re_f for $Re_f > 1635$.

The present data are compared with several existing empirical relations and the Ishii's prediction in Fig. 6. The discrepancy among the data may be due to the difference in the definitions of the inception criteria, entrainment detection methods, and geometry.

It is found that the Ishii's analytical model is in good agreement with the present data.

5. Average Film Thickness

The average values of the film thickness, $\bar{h}$, are computed from thickness data and are given in Table 5. Two important trends are noted:

a) $\bar{h}$ increases with Q for a fixed V. For V = 325 m/s; $\bar{h}$ increases from 16 μm to 56 μm as Q increases by 6 X (100 to 600 cc/min).
b) Film thickness, $\bar{h}$, increases as V is decreased at constant Q. For example, at Q = 200 cc/min, $\bar{h}$ increases from 21.6 μm to 251 μm (12 X) as V decreases from 325 to 105 m/s (3 X).

The smallest measured $\bar{h}$ is 16 μm for V = 325 m/s and Q = 100 cc/min. The largest $\bar{h}$ is 360 μm for V = 105 m/s and Q = 600 cc/min. Obviously, Q and V affect $\bar{h}$ significantly.

6. Velocity Profile in the Liquid Film

Von Karman proposed a universal velocity distribution equation for turbulent pipe flow.

$u^+ = y^+$ for $0 < y^+ < 5$

$u^+ = -3.05 + 5.0 \, \mathrm{Ln} y^+$ for $5 < y^+ < 30$

$u^+ = 5.5 + 2.5 \, \mathrm{Ln} y^+$ for $30 < y^+ < \infty$

where, $u^+ = u/u^*$, u, local velocity;

u^*, shear velocity = $\sqrt{\dfrac{\tau_0}{\rho_f}}$;

$$y^+ = \frac{u^* \rho_f y}{\mu_f} \; ; \qquad y = \text{distance from the wall.}$$

If these relations hold in the film flow, then, the average film thickness can be estimated from the above expressions and the liquid flow rate. Both the estimated average film thickness and the measured average film thickness are plotted on Fig. 7 in terms of liquid Reynolds Number Re_f and $\bar{\bar{h}}$,

$$\frac{\rho_f \sqrt{\tau_0/\rho_f}\,\bar{h}}{\mu_f} .$$

The solid line represents the results obtained by assuming "universal velocity profile" in the film. It appears that such a correlation can be applied with reasonable accuracy for the Newtonian fluids such as water. Thus, the Von Karman's universal velocity profile can be used to model the velocity distribution in thin liquid film flow.

7. Statistical Distribution Function

The chi-square test is used to determine how well theoretical distribution compared with empirical distribution obtained from sample data.

The statistic quantity χ^2 (chi-square), which is a measure of discrepancy between observed and expected frequencies, is defined as:

$$\chi^2 = \sum_{j=1}^{k} \frac{(O_j - E_j)^2}{E_j}$$

where O_j, is the observed frequency and E_j is the expected frequency.

If $\chi^2 = 0$, the observed and theoretical frequencies coincide. However, there is discrepancy when χ^2, the greater is the discrepancy between observed and expected frequencies.

Table 6 shows "goodness of fit" results for statistical tests. It is demonstraed that the normal distribution models cannot be used for correlating the high shear force-driven film thickness in the present data. However, the normal distribution function has been reported to model the gravity-driven film successfully.

CONCLUSIONS

The following important conclusions have been drawn:

1) The water film temperature is equal to the local saturated steam temperature (0.5 ∿ 1°C deviation), and is independent of the inlet water temperature and flow rate. In addition, experiments confirm that a superheat sublayer (16°C) existed on the dry blade surface.
2) A comparison between the existing theoretical analyses and the present data for critical conditions of the rivulet flow show that the M-M model is acceptable. However, a better model is desirable, taken into account the dynamic contact angle and rivulet waviness.

3) The Ishii's model can be used to model the transition between the atomized and continuous films.
4) The Von Karman's universal velocity profile for single phase boundary layer flow can be used for modelling the velocity profile in the thin film flow driven by a shear force.
5) It is found that the normal distribution function cannot be used for correlating film thickness data.

ACKNOWLEDGEMENTS

The authors would like to thank the Department of Mechanical Engineering and Applied Mechanics, The University of Michigan, for financial support. Thanks are also due to Ms. Ruth Howard for her assistance in preparing the paper.

REFERENCES

1. El-Masri, M. A., etc., "On the Design of High Temperature Gas Turbine Blade Cooling Channels", ASME Paper, 78-GT-29, 1978.

2. Inujuka, M., "Flow Characteristics of a Liquid Film on an Inclined Rotating Surface", Int. Chem. Engr., vol. 21, no. 2, 1981.

3. Dakin, J. T., "The Dynamics of Thin Liquid Films in Rotating Tubes: Approximate Analysis", ASME, 78-WA/FE-9, 1978.

4. Duckler, A. E., etc., "Characteristics of Flow in Falling Liquid Films", Chem. Engr. Prog., 48, pp. 557-563, 1952.

5. Hewitt, G. F., "The Breakdown of the Liquid Film in Annular Two-Phase Flow", Int. J. Heat Mass Transfer, vol. 8, pp. 781-791.

6. Chu, K. J., "Statistical Characteristics of Thin, Wavy Films: Part II Studies of the Substrate and Its Wave Structure", AIChE J., vol. 20, no. 4, pp. 695-706, July 1974.

7. Choi, H. Y., "Electrodynamic Condensation Heat Transfer", J. Heat Transfer, v. 96, pp. 455, 1981.

8. Hsu, Yu Kao, "Laminar Film Condensation Over a Vertical Circular Cylinder with Effect of Electrical Field", ASME 78-WA/HT-49, 1978.

9. Hinkebein, T. E., "Surface Tension Effects in Heat Transfer Through Thin Liquid Films", Int. J. Heat Mass Transfer, vol. 21, pp. 1241-1249, 1978.

10. Hanratty, T. J., "Interaction Between a Turbulent Air Stream and a Moving Water Surface", AIChe J., vol. 3, no. 3, pp. 298-304.

11. Henstock, W. H., "The Interfacial Drag and the Height of the Wall Layer in Annular Flows", AIChE J., vol. 22, no. 6, pp. 990-1000, Nov. 1976.

12. Fukano, T., "Liquid Films Flowing Concurrently with Air in Horizontal Duct (2nd report, Characteristics of Two-Dimensional Wave and Disturbance Wave Regions)", B. of JSME, vol. 24, no. 191, pp. 78-789, May 1981.

13. Blass, E., "Gas/Film Flow in Tubes", Int. Chem. Eng. vol. 19, no. 2, pp. 183-195, Apr. 1979.

14. Suzuki, S., "Behavior of Liquid Films and Flooding in Counter-Current Two-Phase Flow, Part 1. Flow in Circular Tubes", Int. J. Multiphase Flow, vol. 3, pp. 517-532, 1977.

15. Kim, W., etc., "Investigation of Behavior of Thin Liquid Films with Co-Current Steam Flow", Proc. Two-Phase Flow and Heat Transfer Symposium, Oct. 18-20, 1976, U. of Miami.

16. Kim, W., "Steam Turbine Erosion Related Problems", Ph.D. thesis, 1978.

17. Traupel, W., "Zur Theories dert Nassdamfturbine", Scheweiz, Bauzq, vol. 77, pp. 324, 1957.

18. Cohen, L. S., "Generation of Waves in the Concurrent Flow of Air and a Liquid", AIChE J., pp. 138-144, Jan. 1965.

19. Yih, Chia-Shun, "Stability of Liquid Flow Down an Inclined Plane", The Physics of Fluids, vol. 6, no. 3, pp. 321-334, March 1963.

20. Hartley, D. E., "Criteria for the Break-up of Thin Liquid Layers Flowing Isothermally over Solid Surfaces", Int. J. Heat Mass Transfer, vol. 7, pp. 1003-1015, 1964.

21. Bankoff, G. S., etc., "Dryout of a Thin Heated Liquid Film", Proc. Int. Heat Mass Transfer Center Seminar, Dubronik 1978, Hemisphere Publishing Co., 1978.

22. Mikielewicz, J., "Minimum Thickness of a Liquid Film Flowing Vertically Down a Solid Surface", Int. J. Heat Mass Transfer, vol. 19, no. 7, pp. 771-776, 1976.

23. Ishil, M., "Inception Criteria for Droplet Entrainment in Two-Phase Concurrent Film Flow", AIChE J., vol. 21, pp. 308-317, March 1975.

TABLE 1. Comparison of Saturated Steam Temperature and Water Film Temperature

Steam Velocity (m/s)	Saturated Steam Temperature °C	Water Film Temperature °C
325	57.8	58.9
285	56.1	56.7
210	56.7	57.8
150	58.3	57.8
105	56.7	56.8

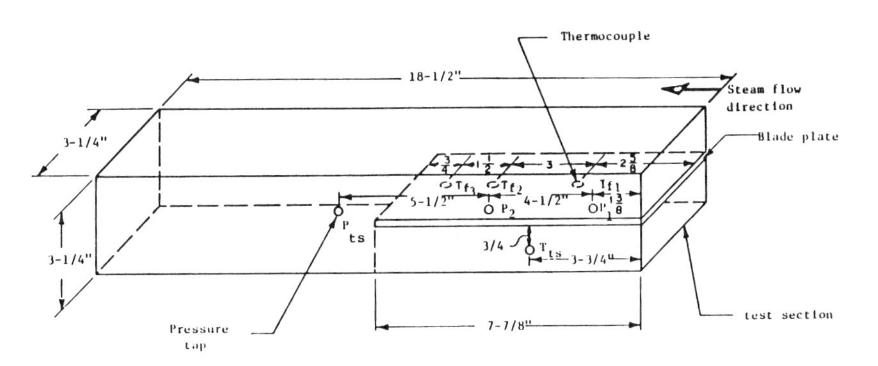

FIGURE 1. Drawing of test section

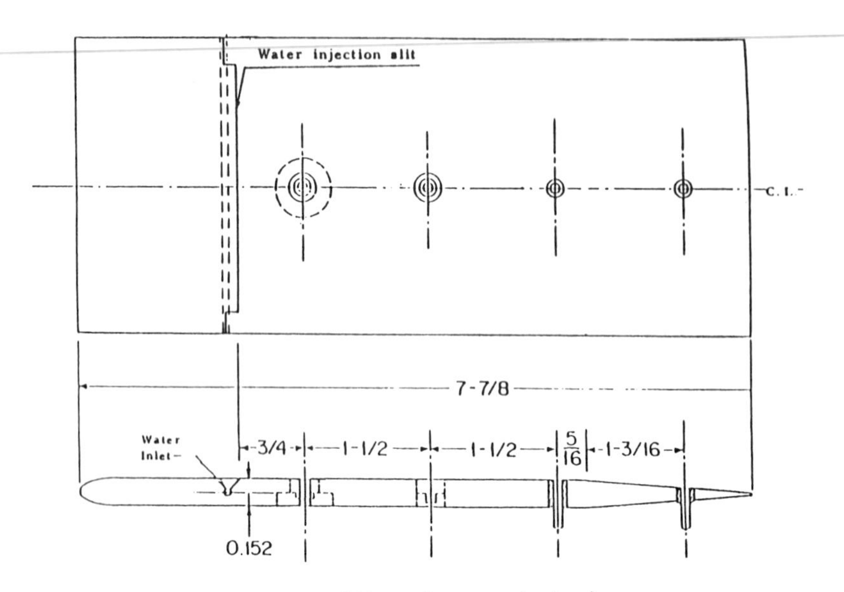

FIGURE 2. Test section blade (dimension are inches)

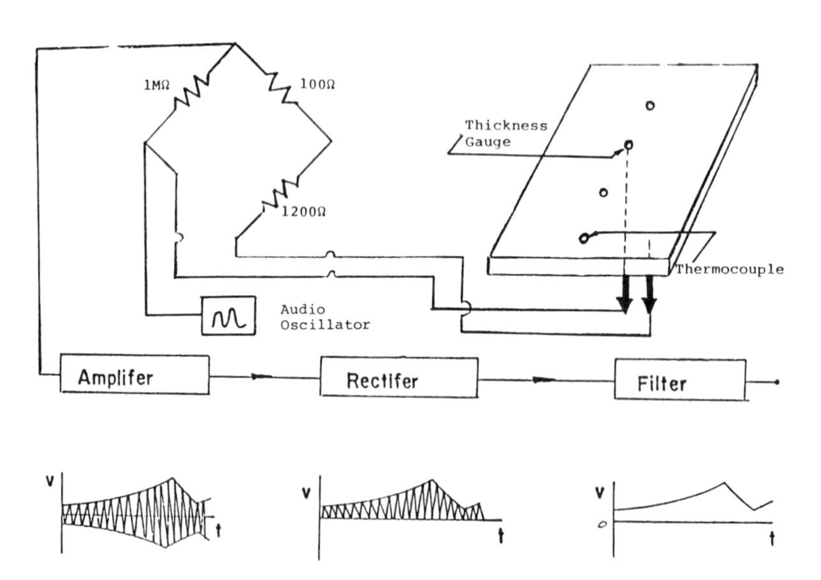

FIGURE 3. Schematic of thickness detection system

TABLE 2. Water Film Temperature Under Various Conditions

Steam Velocity (m/s)	Water Film Flow Rate (cc/min)	water Inlet Temperature °C	water Film Temperature °C
370	100	21.0	50.0
370	100	50.0	49.4
370	200	21.0	48.9
370	200	52.2	49.4
370	400	21.0	50.0
370	400	47.8	49.4
370	600	21.0	51.5
370	600	52.2	50.0
340	100	20.0	43.9
340	100	38.8	44.4
340	200	20.5	43.9
340	200	59.4	44.4
340	400	21.0	45.0
340	400	56.7	45.6
340	600	20.5	45.6
370	600	55.6	45.6
270	100	15.6	37.8
270	100	58.9	36.7
270	200	15.6	37.8
270	200	55.0	37.8
270	400	15.6	37.8
270	400	56.7	37.8
270	600	15.6	37.8
270	600	53.8	38.3

TABLE 3. Superheated Blade Temperature

Steam Velocity (m/s)	Superheated Temperature ΔT, °C
325	25
285	15
210	15
150	30
105	35

ΔT= Blade Temperature- Saturated Steam Temperature

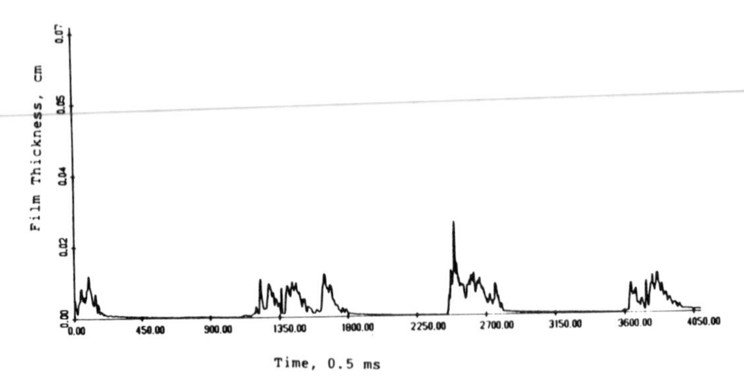

FIGURE 4a. Film thickness trace for V_s = 150 m/s, Q = 100 cc/min (water film)

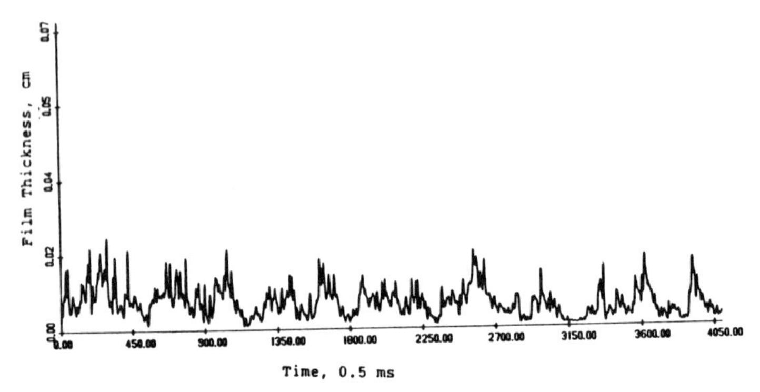

FIGURE 4b. Film thickness trace for V_s = 150 m/s, Q = 300 cc/min (water film)

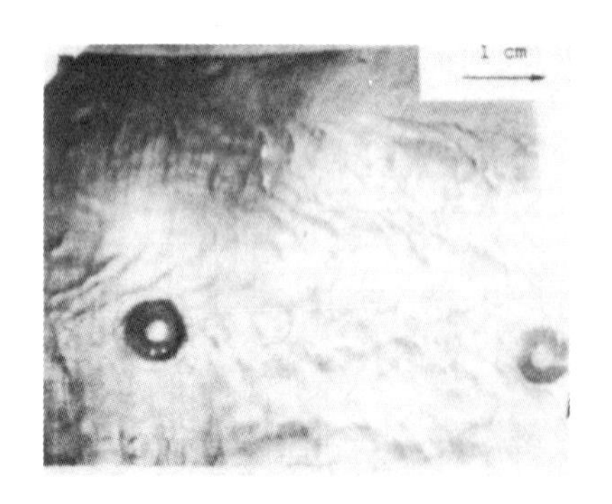

V = 45 m/s, Q= 200 cc/min.

(a)

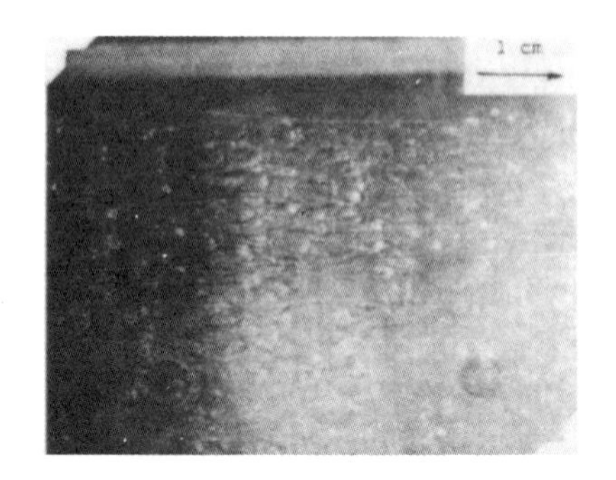

V = 210 m/s, Q = 400 cc/min.

(b)

FIGURE 5. Photos for continuous film and atomized film flow

TABLE 4. Flow Conditions for Water Film

Plate Number	Steam Velocity (m/s)	Liquid Flow Rate cc/min	Flow Feature
A1.1	25	50	R
A1.2	25	100	W
A1.3	25	200	W
A1.4	25	300	W
A1.5	25	400	W
A1.6	25	600	W
A2.1	45	80	R
A2.2	45	100	W
A2.3	45	200	W
A2.4	45	300	W
A2.5	45	400	W
A3.1	105	50	R
A3.2	105	100	W
A3.3	105	200	W
A3.4	105	300	W
A3.5	105	400	W
A3.6	105	600	A
A4.1	150	30	R
A4.2	150	100	W
A4.3	150	200	W
A4.4	150	300	W
A4.5	150	400	W
A4.6	150	600	A
A5.1	210	50	R
A5.2	210	100	W
A5.3	210	200	W
A5.4	210	300	W
A5.5	210	400	A
A5.6	210	600	A
A6.1	285	30	R
A6.2	285	50	R
A6.3	285	80	R
A6.4	285	100	W
A6.5	285	300	W
A6.6	285	600	A
A7.1	325	50	R
A7.2	325	100	W
A7.3	325	200	W
A7.4	325	300	W
A7.5	325	400	A
A7.6	325	600	A

R:Rivulet Flow

W:Continuous Wavy Film

A:Atomization

TABLE 5. Criteria of Rivulet Transition

Water Film				
U_s	105	210	285	325
Q	100	100	100	100
τ_*	28	97	168	213
h^+	0.187	0.078	0.039	0.027
θ(M-M) deg	~36	~28	~25	~20
θ(H-M) deg	~14.3	~9.2	~6.5	~5.4

U_s: Steam velocity (m/s); Q: Liquid flow rate (cc/min)

τ: Shear stress ($dyne/cm^2$); h^+: $\rho_f \tau^2 h^2/\mu^2 \sigma$

θ(M-M): Contact angle calculated from Mikielewicz-Moszynski model

θ(H-M): Contact angle calculated from Hartley-Murgatroyd model.

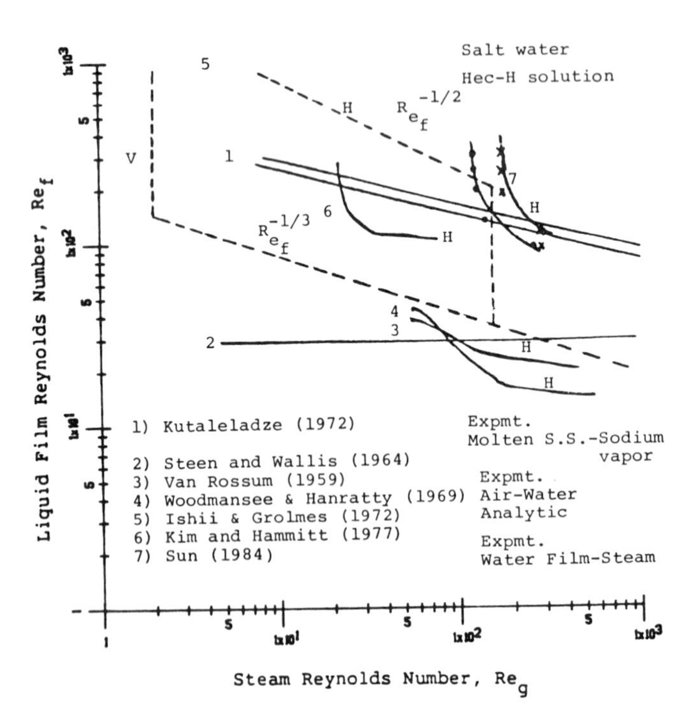

FIGURE 6. Comparison of various atomization inception.

TABLE 6. Chi-square test results for film thickness data

Solution	Steam Velocity m/s	Liquid Flow Rate cc/min	χ, Chi-Square value
Tap Water	325	600	9658
Tap Water	285	600	8073
Tap Water	210	600	2594
Tap Water	150	200	3104
Tap Water	150	300	1156
Tap Water	150	600	962
Tap Water	105	600	556
AP-30-Water	325	600	7328
AP-30-Water	285	600	9787
AP-30-Water	210	600	2193
AP-30-Water	150	200	11342
AP-30-Water	150	300	8027
AP-30-Water	150	600	936
AP-30-Water	105	600	832
Hec-H-Water	325	600	2062
Hec-H-Water	285	600	600
Hec-H-Water	210	600	360
Hec-H-Water	150	200	1874
Hec-H-Water	150	300	747
Hec-H-Water	150	600	380
Hec-H-Water	105	600	349

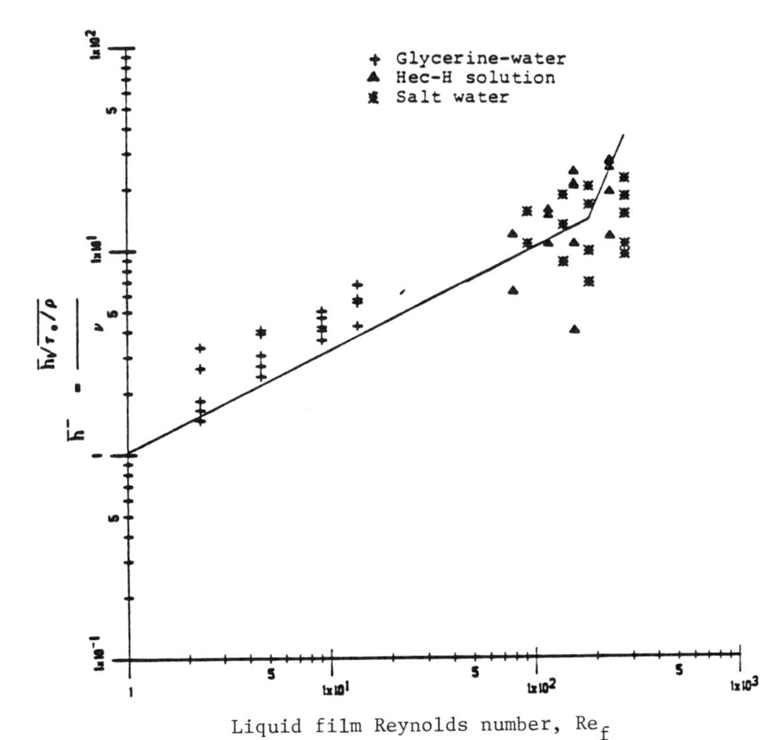

Liquid film Reynolds number, Re_f

FIGURE 7. Nondimensional film thickness, $\bar{h}^-$, vs Re_f

Generalized Transient Rotor Thermal Stress

WILLIAM G. STELTZ
Consulting Engineer
1206 Forge Road
Cherry Hill, New Jersey 08034

INTRODUCTION

The development of turbine rotor thermal stresses during transient operation depends on many variables: rotor geometry, material thermal and mechanical properties, initial temperatures, and rates of change of the ambient fluid and rotor surface temperature are among the the most important. The determination of the resultant rotor thermal stress can be achieved by a transient calculation procedure done on a case by case basis the results of which are governed by these variables.

Significant analysis capability and time saving has been achieved as a result of the development of a generalized transient rotor thermal stress analysis. This generalized approach combines the variables into three dimensionless quantities which are easily displayed on a single plot. Particular values for a given problem can be converted to dimensionless quantities and the dependent variable, e.g. rotor thermal stress, can then be determined directly from the appropriate dimensionless quantity.

APPLICATION AND BOUNDARY CONDITIONS

Figure 1 is a sketch of a typical steam turbine rotor showing critical rotor elements subject to transient thermal stresses. The stress analysis assumes that rotor dimensions can change according to material expansion characteristics but that the boundaries remain plane surfaces and the material retains its elastic behavior.

The thermal analysis impresses a ramp change of ambient steam temperature having an initial value known with respect to the rotor body initial temperature. The surface convective film coefficient is sufficiently high so that the rotor surface temperature follows the steam temperature within a few degrees. The initial steam temperature can be greater or less than the initial rotor body temperature and the impressed fluid temperature ramp may either increase or decrease with time. Figure 2 displays variations of boundary conditions included by this generalized analysis. It can be shown that the case of a step change in steam temperature - i.e., when the ramp rate is zero - is also represented by this analysis.

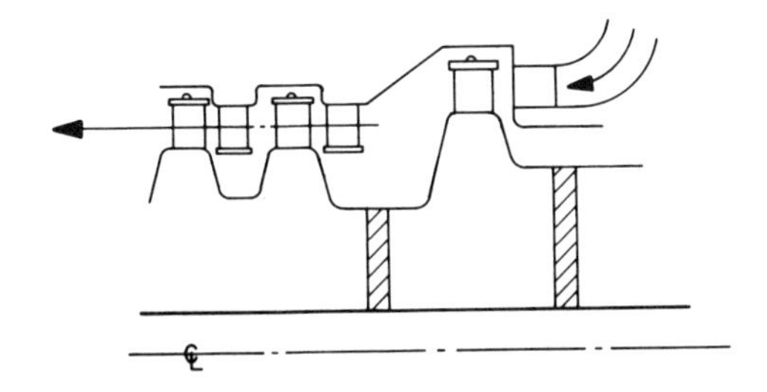

FIGURE 1. Typical steam turbine rotor configuration.

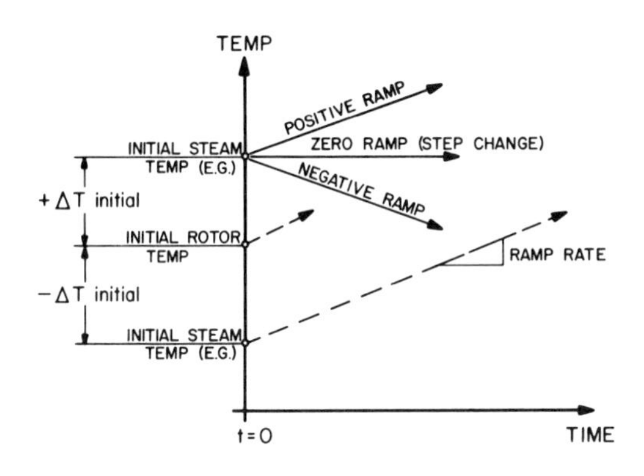

FIGURE 2. Steam temperature boundary condition variations.

THERMAL AND MECHANICAL ANALYSES

The heat transfer analysis satisfies the one-dimensional transient governing equation

$$\frac{1}{r}\frac{\partial}{\partial r}\left(r\frac{\partial T}{\partial r}\right) = \frac{\rho c}{k}\left(\frac{\partial T}{\partial t}\right) \qquad (1)$$

and describes the thermal characteristics of the rotor body. Energy transferred between the rotor body surface and the steam is dependent on the Newton rate equation

$$q = h\,A\,\Delta T \qquad (2)$$

where $\Delta T = T_{steam} - T_{surface}$

The solution technique is based on a finite element synthesis of the rotor body with computations done at time increments as required by the particular problem in order to assure numerical stability.

Fundamental dimensionless quantities of pertinence to transient rotor thermal stress analysis are the Nusselt and Fourier numbers which relate parameters in convective and conductive heat transfer processes. The numeric ranges of these parameters are as follow:

Fourier: $\alpha_{th}\, t/(d/2)^2$ 0 to 1.5

Nusselt: hd/k 90 to 270

When the rotor body temperature distribution has been determined at any value of time, the rotor surface thermal stress is computed according to

$$\sigma = \frac{\phi E \alpha_{ex}}{(1-\nu)} (T\ avg - T\ surface) \tag{3}$$

Where T avg is the rotor body mass average temperature and ϕ is a stress concentration factor. Similarly, the bore thermal stress can be computed using the bore temperature in place of the surface temperature.

GENERALIZED ANALYSIS

By means of dimensional analysis concepts a generalized transient rotor thermal stress analysis has been developed relating the parameters shown in Table 1. These parameters include individual variables as well as groupings of variables combined in familiar and logical sets. In particular, the thermal diffusivity is one of the most familiar in heat transfer analysis and the mechanical properties parameter represents the operator used to determine rotor thermal stress as a function of rotor temperature distribution.

The three resultant dimensionless quantities include a modified Fourier number which is in terms of diameter rather than radius. The other two quantities combine the two driving forces of this transient process - the initial temperature difference and the ambient fluid temperature ramp - with the thermal, mechanical, and size parameters. These dimensionless quantities are:

$$\pi_1 = \alpha_{th}\, t/d^2 \tag{4}$$

$$\pi_2 = \alpha_{th} \frac{\Delta T}{Ramp} /d^2 \tag{5}$$

$$\pi_3 = \frac{\alpha_{th}}{\frac{\phi E \alpha_{ex}}{(1-\nu)}} \frac{\sigma}{Ramp} /d^2 \tag{6}$$

The convective heat transfer quantity Nusselt number (or Biot number) which itself is dimensionless does not appear. It is of significance to the transient process but as a fourth variable results in an unnecessarily more complex system less amenable to visualization and analysis. The effects of Nusselt number have been determined and are treated in a separate section.

The definition of these dimensionless quantities is consistent with the nomenclature and description of boundary conditions described in Figure 2. A conceptual plot relating the three dimensionless quantities is shown schematically in Figure 3.

As an aid to the maintenance of parameters of engineering interest, the working plot of Figure 4 was developed. Material and geometric parameters have been ratioed by reference values and the dimensionless quantities now assume particular units: the abscissa is in minutes, the ordinate in $(lb/in^2)/(F/hr)$ and the parameter in F/(F/hr). Figure 4 is based on constant values of thermal diffusivity, modulus of elasticity, coefficient of thermal expansion, Poisson's ratio, and stress concentration factor. Figure 4 also assumes the existence of a rotor bore which for all cases is a fixed proportion of the outer diameter in order to maintain geometric similitude. The full extent of this working plot is shown on a condensed scale in Figure 5. All parameters asymptotically approach the same ordinate value.

For comparison with Figure 4, Figure 6 presents a similar working plot for a rotor without a bore, the full extent of this plot is displayed in Figure 7.

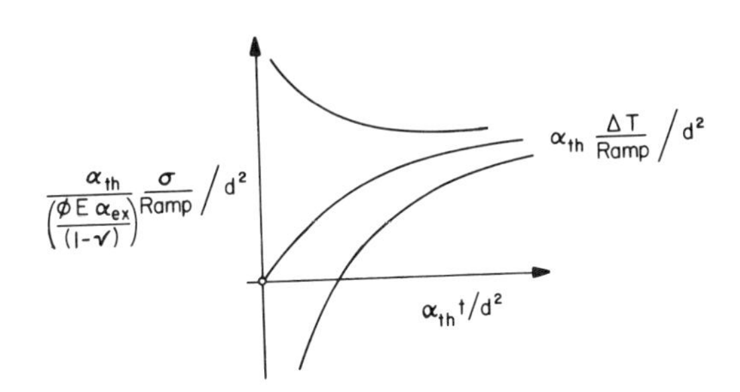

FIGURE 3. Generalized transient rotor thermal stress plot.

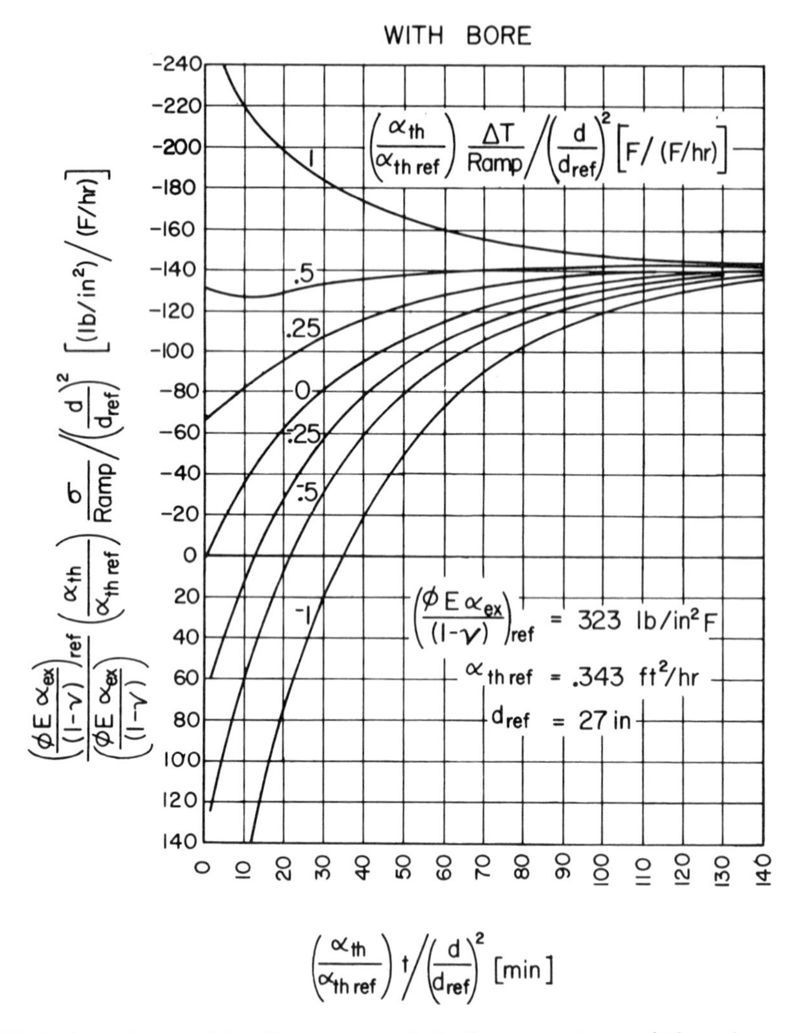

FIGURE 4. Generalized stress plot for a rotor with a bore.

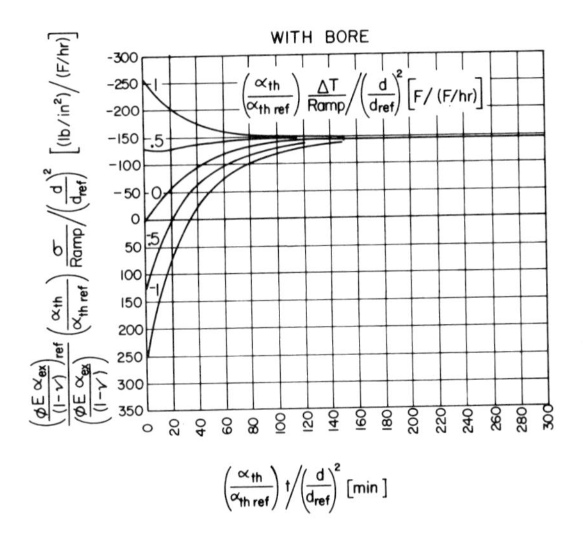

FIGURE 5. Generalized stress plot for a rotor with a bore.

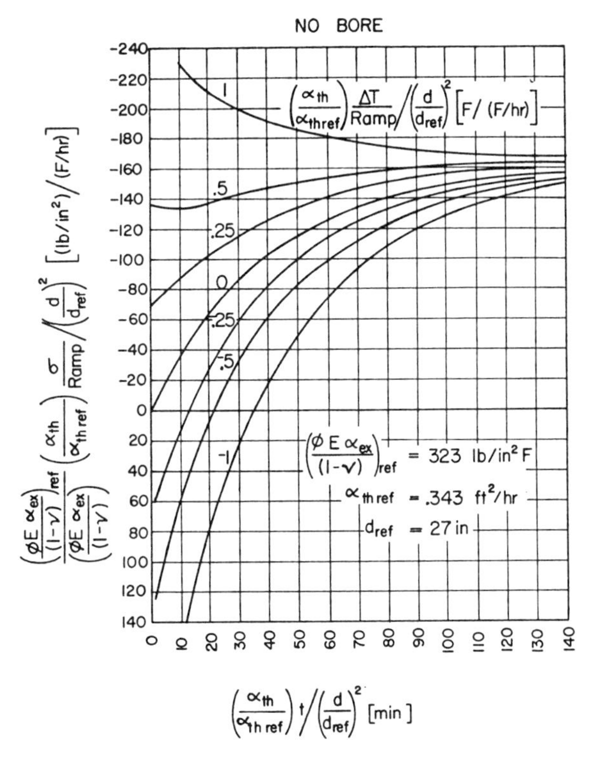

FIGURE 6. Generalized stress plot for a rotor with no bore.

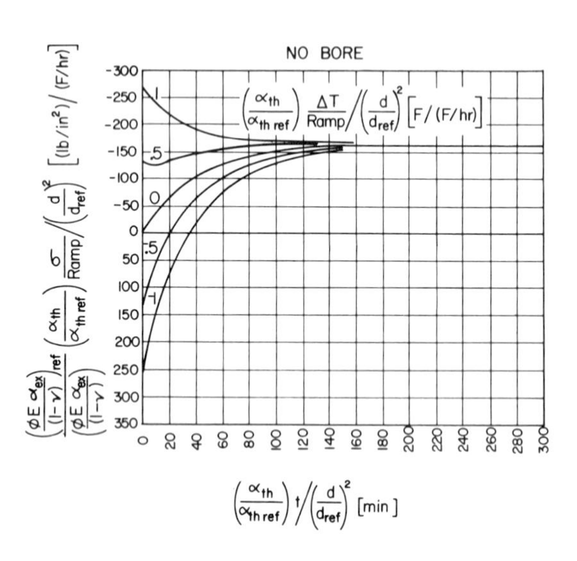

FIGURE 7. Generalized stress plot for a rotor with no bore.

STEP CHANGE ANALYSIS

The case of a step change in steam temperature is also represented by the generalized transient rotor thermal stress plots of Figures 4, 5, 6 & 7. In the nomenclature of these figures, ΔT is the value of the step change while the ramp rate is physically zero. It can be shown that compatibility with the generalized plots requires that the ramp rate assume a value of 100 for use in both the ordinate and parameter evaluation. The resultant surface thermal stress can then be determined as the difference in ordinate values between the appropriate parametric value and the zero value parameter. This relationship is shown schematically in Figure 8.

APPLICATION TO CYCLIC LIFE CURVES

The generalized transient rotor thermal stress plots can be used to develop (or verify) cyclic life curves such as those generally supplied with steam turbine operating information. These life curves can be plotted in terms of steam temperature ramp rate vs rotor surface temperature change with parameters of cyclic life - Figure 9 is a schematic version of these plots. The cyclic life values represent rotor material life expenditure per cycle which is equivalent to a value of stress (or cyclic stress range) experienced by the rotor surface in response to impressed steam temperature variations.

The value of maximum surface stress occurs at the time steam temperature stops changing - shown also in Figure 9. This maximum value of stress can be determined directly from the generalized transient rotor thermal stress plots of Figures 4 or 6. Use of the time at the end of the ramp (when the stress maximizes) the temperature ramp rate, the rotor geometry, and knowledge of the material properties allows the direct calculation of rotor surface thermal stress using the zero parameter line of Figure 4 (or Figure 6 for the no bore case).

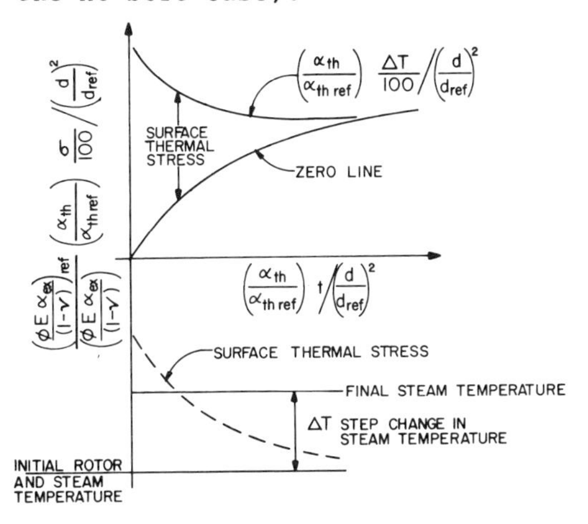

FIGURE 8. Step change analysis.

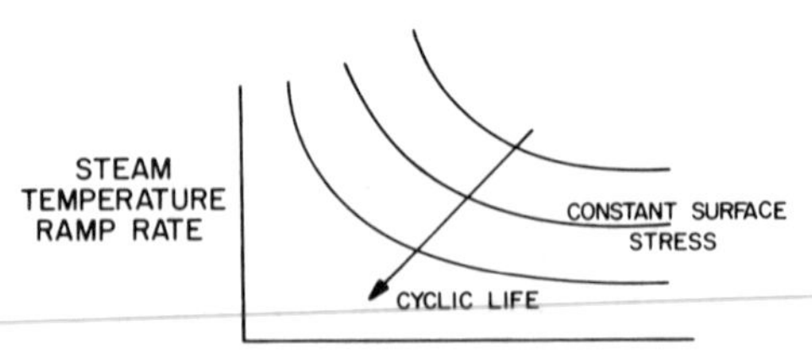

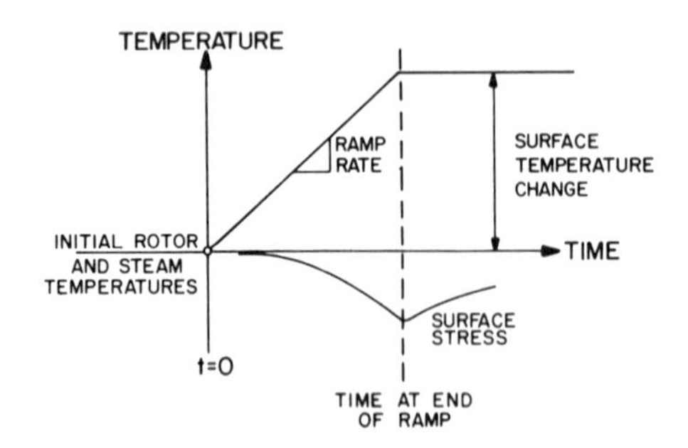

FIGURE 9. Cyclic life curves.

For example, if a stress level of 25 ksi corresponds to a life of 10,000 cycles and the material properties and rotor geometry are the same as the reference values of Figure 4, then the following tabulation defines the relationship between steam temperature ramp rate and the rotor surface temperature change required to develop a 25 ksi rotor surface thermal stress.

Ramp F/hr	Stress/Ramp $(25000\ lb/in^2)/(F/hr)$	Time minutes (Figure 4)	Temperature Change F
180	139	128	384
190	132	91	288
200	125	74	247
225	111	53	199
250	100	42	175
275	91	35.5	163
300	83	30.5	153
325	77	27	146

Similarly, the next tabulation is for the case without a bore making use of Figure 6.

Ramp F/hr	Stress/Ramp $(25000\ lb/in^2)/(F/hr)$	Time minutes (Figure 6)	Temperature Change F
160	156	125	333
170	147	93	264
180	139	76	228
200	125	58	193
225	111	45.5	171
250	100	38	158
275	91	32	147
300	83	28	140
325	77	25	135

These results are plotted in Figure 10 which corresponds in form to the schematic plot of Figure 9.

EFFECT OF NUSSELT NUMBER

Intuitively, and as a result of dimensional analysis reasoning, the heat transfer film coefficient at the rotor surface is a factor that must be considered in the analysis. The corresponding dimensionless quantity is the Nusselt number.

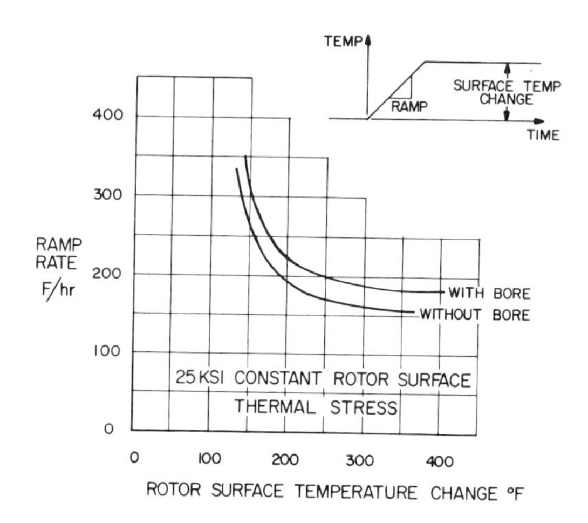

FIGURE 10. Cyclic life curves for rotor with and without a bore at a constant surface stress of 25 ksi.

Figures 4 to 7 are applicable to a range of Nusselt number from 90 to 270. For the numeric values actually used in the computations the convective heat transfer coefficient is about 1000 Btu/hr-ft^2-F. This high a value is typical of high pressure steam turbine operation around design conditions. During part load operation and startup and shutdown procedures, the film coefficient will reduce to a lesser value. The result is a rotor surface temperature that is further removed from the steam temperature. A separate study was made to determine the effect of film coefficient (Nusselt number) on the generalized transient rotor thermal stress plots. These results are shown for the zero and -.5 parametric values in Figure 11 which is drawn to the same scale as Figures 4 and 6. The range of Nusselt number was determined according to the following tabulation:

Film Coefficient h (Btu/hr-ft^2-F)	Range of Nusselt Number hd/k
250	20-35
1000	90-135
2000	180-270

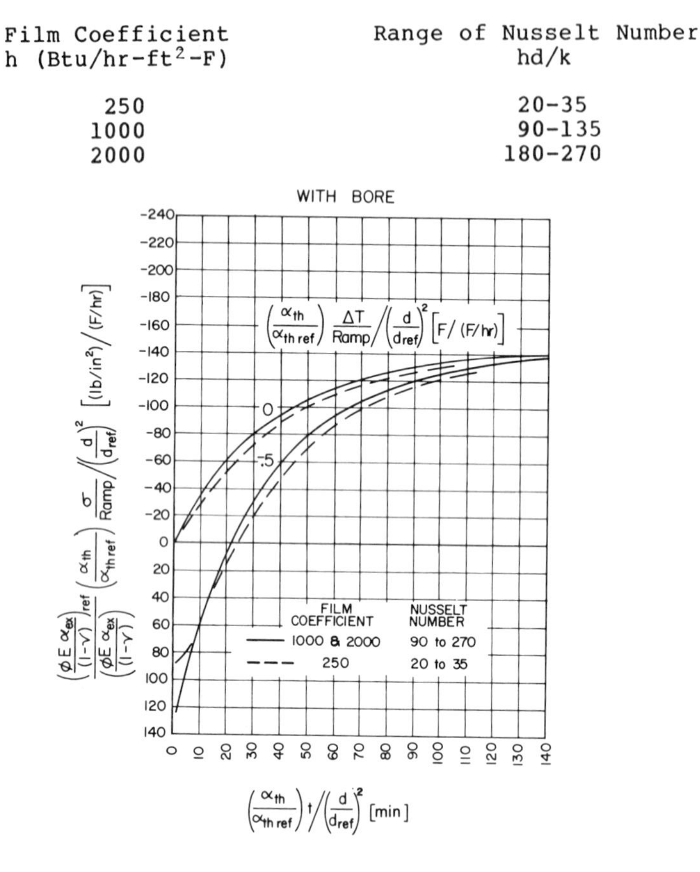

FIGURE 11. The effect of the surface convective heat transfer coefficient on the generalized stress quantity.

A direct comparison of ordinate values (stress) in Figure 11 shows a general reduction of surface stress due to the lag of rotor surface metal temperature resulting from the low surface film coefficient of 250 Btu/hr-ft^2-F. The following tabulation describes the difference in stress (ordinate value) as a percentage of the high film coefficient results.

Percent Difference in Ordinate Value

Time (minutes)	Zero Parameter
5	22.5
10	15.3
20	11.2
40	6.7
60	4.3
90	2.2

During low load operation in a steam turbine the rotor surface film coefficient may reduce as the .8 power of the mass flow rate. At 25% load the film coefficient would be about one-third of its value at 100% load.

Film Coefficient h at 100% Load (Btu/hr-ft^2-F)	Film Coefficient h at 25% Load (Btu/hr-ft^2-F)
1000	330
2000	660

This estimating process can be used to approximate the rotor surface film coefficient and Figure 11 can be used as a guide for the determination of a more exact stress value at lower values of film coefficient.

ACCURACY

The accuracy of the working plots, Figures 4 to 7, is directly dependent on the numerical computations used to determine the rotor body temperature distribution. These finite element procedures are in turn a function of the number of radial increments used in the calculations. An infinite number would match an exact solution of the governing differential equation. The effect of the number of radial increments on computed stress level is shown in Figure 12 which is a plot of the asymptotic value of the stress level vs the number of radial increments for the case of the rotor with a bore.

The asymptotic value approaches its own limit as the number of radial increments increases. This limiting value is approximately 135 from Figure 12. The plots of Figures 4 to 7 are based on 20 radial increments resulting in an asymptotic value difference of about 5%, (142-135)/135. The use of 20 increments is based on this level of asymptotic value difference and the practical fact that computational time increases approximately as the square of the ratio of number of radial increments.

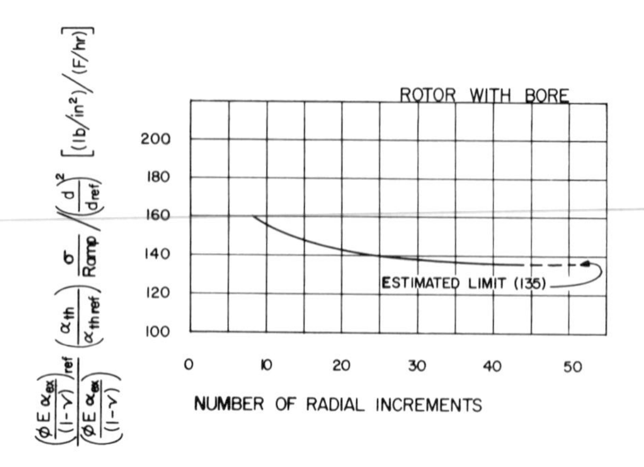

FIGURE 12. Asymptotic value of the stress quantity as a function of the number of radial increments used in computation.

At lesser values of time, say between 10 and 50 minutes, the difference in stress values is less than 5% and decreases to zero as time approaches zero. In effect, the use of 20 radial increments yields results (stresses) which are within 5% of the exact values. This built-in engineering inaccuracy must be considered with respect to knowledge of values of the individual variables, e.g., the heat transfer film coefficient, thermal diffusivity, mechanical properties, and especially the stress concentration factor.

CONCLUSIONS

A generalized transient rotor thermal stress analysis has been developed. Based on dimensional analysis concepts the generalized approach allows the determination of transient rotor thermal stress for arbitrary combinations of ambient temperature ramp rate, initial temperature difference between rotor and ambient fluid, material mechanical and thermal properties, and rotor size. The solution to this problem has been achieved by the use of three dimensionless quantities which can be displayed on a single plot.

Working plots are developed for rotors with and without a bore and their application to the determination of rotor cyclic life is described. Reference material properties and rotor size have been chosen for application to steam turbine rotor analysis although the plots are general and applicable to any rotor.

The concepts and dimensionless quantities developed are of a general nature and can be utilized in the analysis of other situations such as thick walled piping (i.e., a rotor with a large bore), pressure vessels used in steam generators (boiler drums and headers), casings used in pumps, compressors, turbines, etc. Application to geometries other than cylindrical can also be conceived, for example, spheres and hemispheres, flat plates, structural members and so on.

ACKNOWLEDGMENTS

The author would like to acknowledge Power Dynamics, Inc., Broomall, PA for permission to publish this material, and the Electric Power Research Institute, Palo Alto, CA who funded this work under EPRI RP-1184-3. The assistance of Transamerica Delaval Inc., Trenton, NJ in the preparation of this manuscript is also gratefully acknowledged.

NOMENCLATURE

A	area
c	material specific heat capacity
d	diameter
E	material modulus of elasticity
h	convective heat transfer coefficient
k	material thermal conductivity
q	heat transfer rate
r	radius
t	time
T	temperature
ΔT	temperature difference
α_{ex}	material coefficient of thermal expansion
α_{th}	material thermal diffusivity
ϕ	stress concentration factor
π	dimensionless quantity
ρ	material density
σ	rotor surface thermal stress
ν	Poisson's ratio

TABLE 1 - Dimensional Analysis Parameters

Variable	Symbol	Dimensions
Rotor Thermal Stress	σ	$\frac{M}{LT^2}$
Mechanical Properties	$\frac{\phi E \alpha_{ex}}{(1-\nu)}$	$\frac{M}{LT^2 F}$
Thermal Diffusivity	α_{th}	$\frac{L^2}{T}$
Diameter	d	L
Ramp Rate	Ramp	$\frac{F}{T}$
Initial Temperature Difference	ΔT	F
Time	t	T

Resultant Dimensionless Quantities

$$\pi 1 = \alpha_{th}\ t/d^2$$

$$\pi 2 = \alpha_{th}\ \frac{\Delta T}{Ramp}\ /d^2$$

$$\pi 3 = \frac{\alpha_{th}}{\frac{\phi E \alpha_{ex}}{(1-\nu)}}\ \frac{\sigma}{Ramp}\ /d^2$$

Measurements of the Wilson Point and the Steam Condensate Droplet Size and Number Density at Low Expansion Rates

P. V. HEBERLING
Corporate Research and Development
General Electric Company
Schenectady, New York 12301

T. H. McCLOSKEY
Electric Power Research Institute
Palo Alto, California 94303

R. A. KANTOLA
Power Generation Group
General Electric Company
Lynn, Massachusetts 01910

1. INTRODUCTION

As steam expands through a turbine, its temperature drops and eventually crosses the saturation line, thus becoming subcooled. Moisture subsequently forms in the turbine, initiated by the phenomenon of homogeneous nucleation. The onset of homogeneous nucleation in steam is commonly called the Wilson point. The expansion of subcooled or wet steam degrades the efficiency of a turbine by a number of thermodynamic and mechanical aspects of wet steam/blading interactions, which are discussed in detail by Gyarmathy [1]. A recovery of even a small fraction of the moisture-related efficiency losses in steam turbines would result in a substantial savings to the utilities over the lifetime of a turbine.

Spurred by the tremendous cost reductions in computing power, the field of computational fluid mechanics has recently progessed to where three-dimensional flows in complicated geometries can be modeled in detail. Application of the models is resulting in improved turbine blade and endwall contour designs in the dry stages of turbines. Extension of these models to the wet steam region of turbines is limited by our ability to reliably model the condensation process. Specific quantities characterizing the condensation process that are important to steam turbine design engineers are the thermodynamic conditions at the Wilson point, the number density of droplets that results from the nucleation process, and the droplet growth history. The dependence of the above quantities on the rate of change of the thermodynamic conditions is also of considerable importance to turbine design engineers. The earliest attempts to quantify the spontaneous condensation of steam were the experiments performed by Hirn, Cazin, Bauman, and Stodola [2]. These pioneering studies have been followed by continuing research motivated principally by the need to accurately assess and reduce the efficiency loss that occurs when subcooled or wet steam is expanded in a turbine. Prior steam condensation data [3-8], for the most part, have been obtained in rather short convergent-divergent (C-D) nozzles. The resultant expansion rates were generally very high and the condensation often occurred in the supersonic portion of the nozzle, where interactions between the condensation front and the shock patterns complicate the interpretation and the usefulness of the data. The use of C-D nozzles meant that each expansion rate required a different nozzle, creating experimental difficulties. Low expansion rates require a larger C-D nozzle and therefore higher steam flow rates, especially at high pressures. Consequently, there are very few prior studies of steam condensation at low expansion rates typical of large steam turbines, and these were done at low pressures.

With the nucleation tube, the apparatus used in this study, many of the problems of the C-D nozzles have been overcome. The flow is subsonic and low expansion rates at high or low pressures are easily obtained. Experimentally, the expansion rate is easily changed. The goal of

* This research was supported by the Electric Power Research Institute under Contract RP735.

the work reported herein is to determine the thermodynamic conditions at the onset of nucleation and measure the subsequent droplet growth history and number density at the low expansion rates and the high pressures typical of large nuclear powered steam turbines. These measurements characterize the physics of spontaneous steam condensation without the effects of turbulence, sheared flows, temperature gradients, shock waves, and contaminants. Attenuation measurements at two wavelengths of laser light and transient pressure measurements are used to determine the Wilson point, the number density of droplets, and the droplet growth rate. It should be emphasized that the present study, with its use of two-color attenuation, does not require the return to equilibrium assumption to close the data reduction calculations.

2. STEAM CONDENSATION MODEL

The classical result for the nucleation rate derived by Volmer in 1939, Zeldovich in 1942, and Frenkel in 1946, is used for the theory-data comparisons in this study. An equation for the nucleation rate by itself does not allow for the termination of the droplet formation process in expanding steam. Additional models are required, including a description of the droplet growth rate. The following is a general description of the algorithm used in this study.

The entropy of the steam is determined from the initial pressure and temperature of the superheated steam. The entropy is assumed to be constant throughout the condensation process. Steam properties are calculated using the equations recommended by the International Formulation Committee of the Sixth Conference on the Properties of Steam [9]. Extrapolations into the supersaturated region are included in these equations. Hedbaeck [10] has demonstrated the validity of these extrapolations. The measured pressure versus time data is smoothed before the thermodynamic conditions at times intermediate to the measured pressures are interpolated at 0.5 μs intervals.

The program marches forward in time down the pressure decay curve. Once the steam becomes supersaturated, the nucleation rate is calculated at every time step. The droplets formed at each time step are treated as individual groups. Each group has a constant number of droplets and a diameter that is adjusted at each subsequent time step using the equation derived by Gyarmathy for the droplet growth rate [11]. Up to one thousand groups of drop sizes are used in the program. The total moisture is calculated by summing up the numbers and sizes of droplets in each class. The temperature of the steam is calculated and the result compared to its value at the previous time step to determine whether the subcooling has reached its maximum, the criterion for the Wilson point. If the Wilson point has not been reached, the program moves on to another time step. Once the Wilson point is reached, the droplet groups are combined into a single group with the same total number density and the corresponding mass average diameter for the subsequent droplet growth calculations. Y, the condensate mass, is calculated from n, the droplet number density, and d, the droplet diameter, at every time step. Y_e, the available moisture, is calculated from the pressure and the entropy at each time step. The assumption of a monodisperse cloud for the calculations subsequent to the Wilson point has very little effect on the results. No new droplets are produced and the droplet diameters tend to approach each other in diameter because the smaller droplets grow faster than the larger droplets.

3. EXPERIMENTAL SYSTEM

The apparatus used in this study employs an unsteady expansion wave to obtain a controlled subsonic expansion of initially superheated steam. The resultant isentropic adiabatic cooling produces supersaturated steam in which condensation can readily be studied. The gas dynamic process corresponds to the driver section of a shock tube or shock tunnel. In order to relate the concept to this study, the apparatus has been designated as the nucleation tube. The nucleation tube is closed at one end and sealed with a rupture diaphragm at the other. The tube used in this study has a square bore, 1 inch on each side. Immediately ahead of the diaphragm is a smaller diameter orifice plate. A closed dump tank is connected to the downstream side of the diaphragm. This basic design is represented in the schematic of Figure 3-1. When the diaphragm is ruptured, the expansion wave centered at the orifice/diaphragm station propagates through the steam. The measurement section contains windows that are flush with the inside walls. The pressure transducer is also flush with the inside wall and in the same plane, perpendicular to the tube axis, as the windows. A protrusion or recess in the tube would generate reflections of the transient expansion waves. The entire apparatus is modular in construction. The controlling parameters of a particular wet steam turbine stage can be duplicated by varying the length of the tube, the position of the measurement section along the length of the tube, the orifice area, and the initial temperature and pressure of the steam in the tube. The nucleation tube is surrounded by an insulated furnace consisting of multizoned heaters, control equipment, and viewing ports. The furnace viewing ports double as lenses for the optical instrumentation system.

The steam generator is a vertical electric boiler that can provide steam up to 1000 psia and 560 °F. Distilled water is fed to the boiler. To further remove impurities, the boiler water is recirculated through a demineralizer (Sybron/Barnstead Nanopure). The purity of the boiler water is monitored with an Orion Research sodium monitor and a Barnstead resistivity meter. Results from the latter are reported as conductivities. As the boiler feed water leaves the demineralizer, its conductivity is always less than 2×10^{-7} ohm^{-1} and usually 5.5×10^{-8} ohm^{-1} cm^{-1}, or that of absolutely pure water. These conductivities are within the typical recommended range for boiler feed water in high-pressure, high-temperature steam power plants. To quantify the steam purity, steam is allowed to flow through the heated nucleation tube and then is condensed in a heat exchanger. On-line analyzers measured both conductivity and the sodium ion concentration of the condensate with typical values ranging from 5×10^{-7} to 5×10^{-6} ohm^{-1} cm^{-1} and 0.3 to 0.6 ppb Na, respectively.

Laser light attenuation at two wavelengths is used to determine the properties of the condensate droplet cloud. The attenuation technique measures the transmitted light intensity, I, that is not scattered by the droplets as the incident beam of initial intensity, I_o, traverses the test

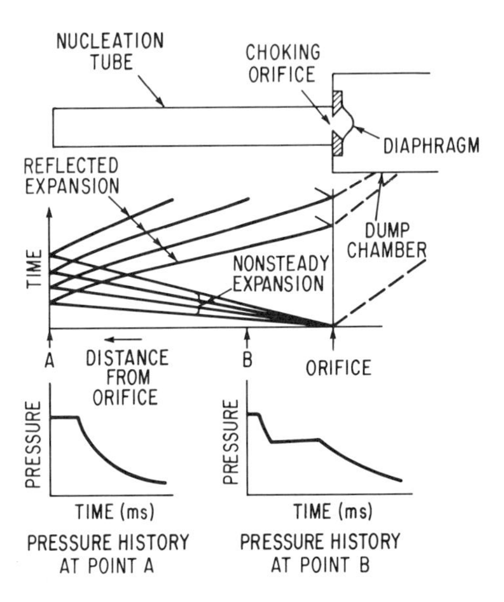

FIGURE 3-1. Nucleation tube concept for steam condensation studies.

section along a path of length l. The integration of the light removed by individual scattering events yields the result

$$\frac{I}{I_o} = \exp\left[-n \frac{\pi d^2}{4} l K\right] \tag{3-1}$$

where n is the droplet number density, d is the droplet diameter, and K is the scattering efficiency. This equation is variously referred to as the Beer-Lambert law or Bouguer's law. In this form, it assumes a monodisperse droplet cloud. Equation 3-1 does not apply to results obtained with an optical system that does not discriminate between scattered and unscattered light or when significant multiple scatterings occur. Multiple scattering occurs at high attenuations and causes single scattered light to be rescattered back to, and to be received by, even a well-designed detector. The attenuation measurements are made at the wavelengths 457.9 nm in the blue and 632.8 nm in the red. Taking the ratio of Equation 3-1 applied to the red and blue wavelengths yields

$$\frac{\ln\left[\frac{I}{I_o}\right]_r}{\ln\left[\frac{I}{I_o}\right]_b} = \frac{K_r}{K_b} \tag{3-2}$$

The left side is determined experimentally. The right side, a single-valued function of diameter up to 1.5 μm is derived from Mie light scattering theory [12]. The result is shown in Figure 3-2.

A detailed description of the optical system can be found in a report on an earlier phase of this study [13]. The measurement of I_o, the intensity of the incident red and blue wavelengths, was at first accomplished by simply averaging I (after the test section) over a short period of time prior to any condensation. This method failed to account for the variations in the laser output over the time scale of the test event. To alleviate the effects of laser noise and some sources of electronic noise, a second photomultiplier system to monitor the incident light intensities (before the test section) during the transient was added. The modified attenuation system is illustrated in Figure 3-3. The voltage signals corresponding to I and I_o for both wavelengths are divided in real time by a pair of Analog Devices analog dividers, Model 429B, and the quotient is recorded. This system extends the useful range of the attenuation measurements to values of $I/I_o \approx 0.99$. Extending the range of the attenuation measurements was necessary because of the constant path length used for the measurements at pressures or densities that varied over several orders of magnitude.

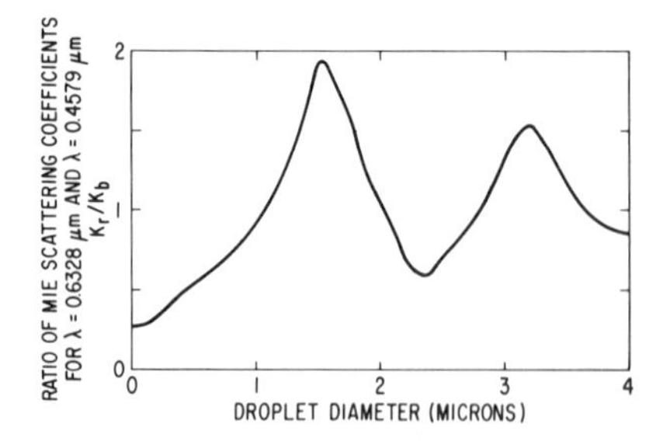

FIGURE 3-2. Attenuation ratio versus droplet diameter.

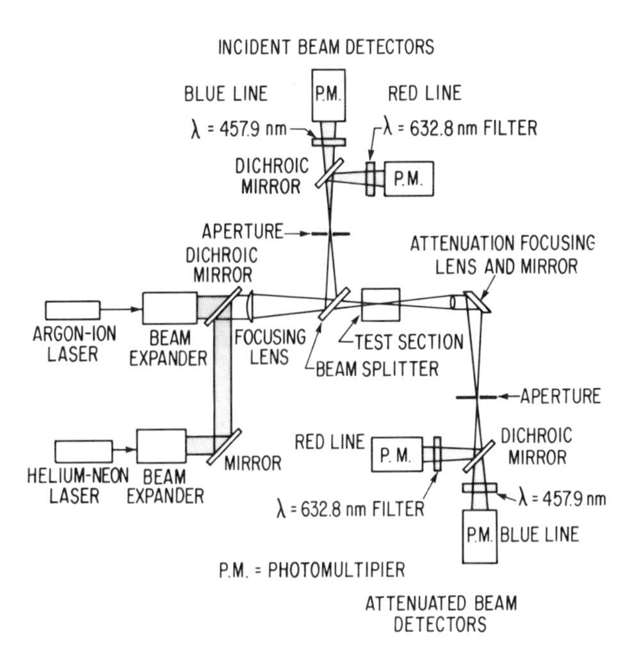

FIGURE 3-3. Optical arrangement including incident light intensity monitoring system.

Before heating the nucleation tube, the optical system is aligned. Alignment is checked after heating the tube due to the proximity of the optics to the tube. Saturated steam is admitted into the nucleation tube, where it is heated to, and maintained at, 20 to 100 °F of superheat. Some steam is purged through the tube to remove impurities. High-temperature piezoelectric pressure transducers are used to measure the pressure transient. The initial steam pressure is measured with a bourdon tube pressure gage calibrated with a dead weight tester. The initial steam temperature is measured with thermocouples embedded in the nucleation tube wall.

The test is initiated by bursting the diaphragm, which causes an expansion wave to travel toward the measurement site. The measurement site for the work reported here is located 1.9 inches from the closed end. Because of the steam velocity, droplets drift into and out of the measurement site. Therefore, the droplets seen at the start of nucleation are not those observed near the end of the observed interval of droplet growth. However, the pressure history is very nearly equal for all neighboring points; therefore, the droplet growth histories are also nearly equal, and this convection effect can be neglected. Equally important is the absence of any steep density gradients that would tend to bend the laser beam and yield an erroneous attenuation signal. The pressure and attenuation signals are recorded over the interval of interest using a waveform digitizer, which is triggered on the pressure signal. The data are fed to a computer after the experiment. The steam temperature and other thermodynamic properties are calculated from the measured initial conditions and pressure decay using the model discussed in the last section. The droplet diameter, number density, and condensate mass histories are calculated from the attenuation measurements, using Equations 3-1 and 3-2 and the relationship in Figure 3-1. Photomultiplier tube nonlinearity is taken into account in the calculation of the number density. These results are compared to the predictions of the model. The results are plotted by the computer.

4. RESULTS

Results obtained from the attenuation method for a typical high-pressure run with a moderate expansion rate are shown in Figures 4-1 and 4-2. The steam was expanded from the initial pressure and temperature of 516 psia and 495 °F along the 1.4825 Btu/lbm°R isentrope at the rate of 1200 s^{-1}. The resulting conditions at onset are typical of the last stages of the high-pressure section of a nuclear powered steam turbine. The droplet diameter, the logarithm to the base ten of the number density and the pressure are plotted in Figure 4-1. The available moisture, Y_e, and the ratio of the measured moisture to the available moisture, Y/Y_e, are plotted in Figure 4-2.

A characteristic of the runs of this study is that there is a small amount of condensate detected before the onset of homogeneous nucleation. In this run, it is readily apparent in the form of about 10^6 droplets/cm^3. This condensate appears to result from heterogeneous nucleation on impurities in the steam. The heterogeneous droplet diameters are scattered because the attenuation levels are very low and sensitive to noise, the droplets may not be monodisperse, and they may be larger than 1.5 μm. From the tabulated results, the pressure is 225 psia and the available moisture is 4.4% at onset. The expansion rate of 1200 s^{-1} is also read from tabulated results calculated from pressure data that has been smoothed. In this high-pressure run, multiple scattering appears to begin affecting the data at a Y/Y_e of about 0.06. In summary, several results are obtained from a single experimental run. Heterogeneous nucleation effects are seen, and to some extent, quantified. The droplet growth history is observed, the number density is determined, and the Wilson point is determined.

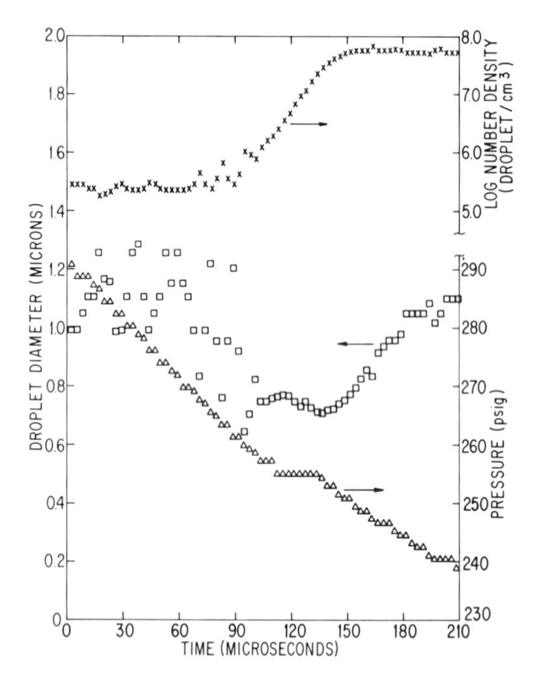

FIGURE 4-1. Pressure, droplet diameter and number density versus time. Initial pressure = 516 psia. Initial temperature = 495 °F. Expansion isentrope = 1.4825 Btu/lbm°R. Expansion rate = 1200 s^{-1}. Pressure at onset = 255 psia. Available moisture at onset = 4.4%.

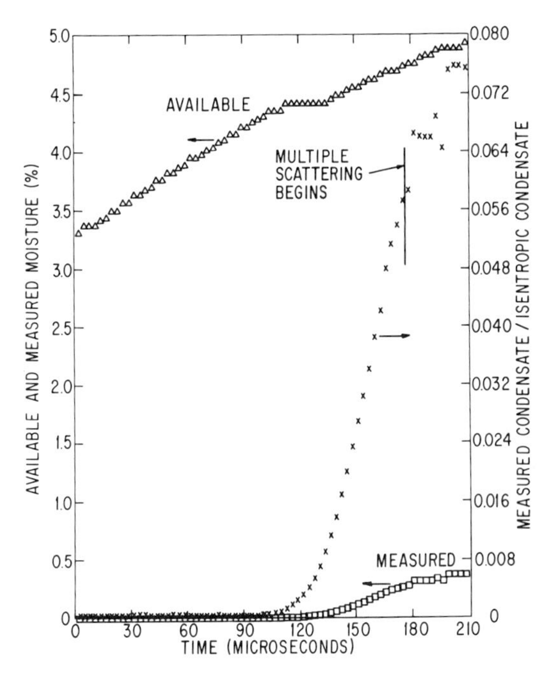

FIGURE 4-2. Available moisture, measured moisture and their ratio versus time for the conditions of Figure 4-1.

Despite some early scatter in these data, they are extremely good. The droplet diameter exhibits the expected qualitative dependence on time. The number density is essentially constant after onset as expected from a physical basis. If any of the numerous assumptions underlying the attenuation method were not satisfied or the measurements themselves were in error in some way, one would not, in general, expect to obtain a constant number density. The same reasoning applies to the possibility of determining a Y/Y_e greater than unity. Significantly, Y/Y_e never exceeds unity in this run or in any other run of this study. The physical consistency of this ratio is a considerable accomplishment when one considers that no independent measurement of the condensate is used to normalize the data. Y and Y_e are obtained by totally independent methods. Y is determined by light attenuation measurements with no calibration, but relying solely on Mie scattering theory predictions for a single droplet applied to what is assumed to be a noninteractive cloud of monodisperse droplets. Y_e is determined from pressure measurements assuming an isentropic expansion. Their values are compared and found to be physically consistent, in that the ratio Y/Y_e increases to unity as equilibrium is approached. To the authors' knowledge, such a comparison between measured and available moisture was first published in the preceding phase [13] of this study.

In prior nucleation studies, experimentalists generally have used the first indication from a pressure or optical sensor as the criterion for the onset of homogeneous nucleation. Using the first indication of a sensor as the Wilson point is valid only if the sensor is insensitive to early condensation due to heterogeneous nucleation. The attenuation method of this study, using optical sensors, has been developed to the level where it is sensitive to condensate formed when there is less than 1% available moisture and, therefore, long before the true Wilson point. A new criterion for the Wilson point is required.

The number density of the heterogeneous and homogeneous droplets differs by an average of two orders of magnitude. This results in a corresponding change in slope in the droplets of Y and Y/Y_e at the onset of homogeneous nucleation, which allows one to easily discriminate between the two types of condensation. The condensate mass is more closely related to the raw attenuation measurement and has less scatter than n. The criterion adopted for the Wilson point in this study is to extrapolate the straight portion of the Y/Y_e plot after the sudden change in slope back to zero condensate. The result is taken as the time of the Wilson point. The thermodynamic conditions at this time are read from tabulated computer output.

Figure 4-3 shows an example of the application of the new Wilson point criterion. In this run, the steam was expanded from the initial pressure and temperature of 515 psia and 495 °F along the 1.4829 Btu/lbm°R isentrope at an expansion rate of 600 s^{-1}. The extrapolation yielded a time of onset of 460 μs on the arbitrary time scale. At this time, the tabulated computer output listed the pressure and available moisture as 247 psia and 4.6%, respectively. Although in some runs, the extrapolation was slightly arbitrary, the uncertainty in the pressure and moisture at onset was generally less than ±1% of the pressure and ±0.1% moisture, respectively.

5. THEORY AND DATA COMPARISON

Two Wilson point measurements have been presented: 4.6% available moisture at an expansion rate of 600 s^{-1}, and 4.4% at 1200 s^{-1}, both approximately along the 1.48 Btu/lbm°R isentrope. These results are typical of the measurements of this study made at high pressure over that range of expansion rates. At these conditions, the model based on classical nucleation theory predicts Wilson points in the range of 2.7% to 3.0%.

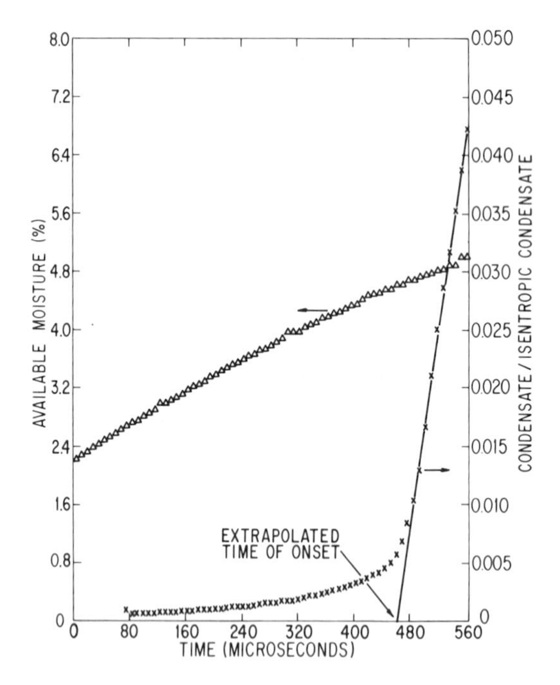

FIGURE 4-3. Method of determining Wilson point by extrapolation of the ratio of measured moisture to available moisture curve. Initial pressure = 515 psia. Initial temperature = 495 °F. Expansion isentrope = 1.4829 Btu/lbm°R. Expansion rate = 600 s^{-1}. Pressure at onset = 247 psia. Available moisture at onset = 4.6%.

In a part of this study not presented here, measurements also were made at low pressures. These Wilson point results are in good agreement with the few measurements of others at these expansion rates. Gyarmathy and Meyer [14] determined the Wilson point in steam expanded at a rate of 1000 s^{-1} in a nozzle. Petr [6] determined the Wilson point in similar experiments at an expansion rate of 2000 s^{-1}.

6. CONCLUSIONS

The most significant new result is the observation at high pressures and low expansion rate of a large discrepancy between the measured Wilson point and that predicted by a model based on classical nucleation theory. This combination of high pressure and low expansion rate is typical of nuclear powered large steam turbines. The deviation is significant to the engineer, inasmuch as the steam is able to attain a much higher subcooling than had previously been assessed, resulting in a correspondingly higher efficiency loss. The detailed Wilson point, droplet diameter, and number density results of this study, including comparisons to two models for steam condensation, are to be published in an Electric Power Research Institute report [15]. These results greatly extend the range of pressure and expansion rate over which steam condensation measurements are available. The unexpected dependencies of the data on these parameters provide the theoretician with new tests for models of nucleation and droplet growth. With the present lack of a model that adequately predicts these measurements, the engineer must treat the results empirically and simply interpolate the data. There is an immediate need to verify the results of this study with an independent experiment. Depending on the resources of the experimentalist, a nucleation tube experiment may be the only practical means of obtaining the data. However, a new tube could be optimized for high-pressure experiments with a shorter optical path length. Better yet would be a rectangular cross section allowing two possible path lengths.

7. ACKNOWLEDGEMENTS

The technical support of several individuals was vital to the success of this study. Dr. C. Murray Penney continued his consulting from the earlier phase of this study on the droplet sizing system. The experimental measurements were performed by Mr. Frank Bowden. Ms. Elaine Birbilis wrote the computer code used for the data reduction. Messrs. Bernard Loyd and Richard Wells wrote the computer code used for the theoretical model predictions. The research in this paper was supported by the Electric Power Research Institute under Contract RP735.

8. REFERENCES

1. G. Gyarmathy, *Grundlagen einer Theorie der Nassdampfturbine,* Dissertation ETH Zurich, Juris-Verlag (1962), English translation USAF-FTD, Dayton, Ohio, Report TT-63-785, revised 1966.

2. A. Stodola, *Steam and Gas Turbines,* (translated from 6th edition), McGraw-Hill Book Company, New York, 1927.

3. T. Krol, Results of Optical Measurements of Diameters of Drops Formed Due to Condensation of Steam in a Laval Nozzle, (in Polish), *Trans. Inst. Fluid Flow Mech.* (Poland), no. 57, p. 19, 1971.

4. G.A. Saltanov, L.I. Seleznev, and G.V. Tsiklauri, Generation and Growth of Condensed Phase in High-Velocity Flows, *Int. J. Heat Mass Transfer 16*, p. 1577, 1973.

5. G. Gyarmathy and F. Lesch, Fog Droplet Observations in Laval Nozzles and in an Experimental Turbine, *Proc. of the Institution of Mechanical Engineers 184*, London, p. 29, 1970.

6. V. Petr, Measurement of an Average Size and Number of Droplets During Spontaneous Condensation of Supersaturated Steam, *Proc. of the Institution of Mechanical Engineers 184*, London, p. 22, 1970.

7. G. Gyarmathy, H.P. Burkhard, F. Lesch, and A. Siegenthaler, Spontaneous Condensation of Steam at High Pressure: First Experimental Results, *Proc. of the Institution of Mechanical Engineers 187*, London, p. 192, 1973.

8. A. Smith, Wilson Point Experiments in Steam Turbines, Third Conference on Steam Turbines of Great Output, Gdansk, Poland, September 1974.

9. IFC 1967 Formulation for Industrial Use, Issued by the International Formulation Committee of the Sixth International Conference on the Properties of Steam, 1967.

10. A.J.W. Hedbaeck, The Behavior of the State Equation IFC - 1967 for Supersaturated States of Steam, VDI Bericht No. 361, p. 193, 1980.

11. G. Gyarmathy, Condensation in Flowing Steam, in *Two-Phase Steam Flow in Turbines and Separators*, ed. M.J. Moore and C.H. Sieverding, ed., p. 165, Hemisphere Publishing Company, London, 1976.

12. R. Penndorf, New Tables of Mie Scattering Functions for Spherical Particles, Geophysical Research Paper No. 45, Part 6, Air Force Cambridge Research Laboratory, Bedford, Massachusetts, 1965.

13. R.A. Kantola, Condensation in Steam Turbines, Electric Power Research Institute Report CS-2528, Palo Alto, California, 1982.

14. G. Gyarmathy and H. Meyer, Spontaneous Condensation Phenomena, Part 2: Experimental Investigations on the Influence of the Expansion Rate on the Fog Formation in Supersaturated Steam, *V.D.I. Forschungsheft 508,* 1965.

15. P.V. Heberling, Condensation in Steam Turbines, Electric Power Research Institute Final Report, Project RP735-3, Palo Alto, California (to be published).

An Experimental Study of Waves in Rotating Liquid Films

A. T. KIRKPATRICK
Mechanical Engineering Department
Colorado State University
Fort Collins, Colorado 80523

M. A. EL-MASRI
Mechanical Engineering Department
Massachusetts Institute of Technology
Cambridge, Massachusetts 02139

J. F. LOUIS
Aeronautics and Astronautics Department
Massachusetts Institute of Technology
Cambridge, Massachusetts 02139

INTRODUCTION

Interest in liquid films in channels rotating about an orthogonal axis has increased recently due to applications in turbine rotor blade cooling. In this application, a liquid film is spun radially outward along the interior cooling tubes of a rotating turbine blade [1]. The liquid film in the rotating channel is pulled radially outward by a centrifugal body force, $g = \omega^2 R$. It also experiences a coriolis body force $g_c = 2\ \omega u$ which acts perpendicular to the flow direction, and pushes the film to one side of the channel, as shown in Figure 1.

The effects of rotation on liquid film behavior are not well known, and thus, knowledge about the film behavior in this geometry is limited. Analytical studies of the rotating film stability [2] and boiling heat transfer [3] have been reported. The hydraulic stability studies predict that instabilities resulting in roll wave formation can occur at rotational Froude numbers greater than $5^{1/2}$. Experimental data on the actual film behavior are lacking.

The experimental data that are of interest are the instantaneous film thickness, average, maximum, and minimum film thickness, and the wave frequency and speed. In addition, the spectral density and the amplitude probability distribution add to the quantitative description of the wave motion. By nondimensionalizing the liquid film continuity and momentum equations, it can be shown that these parameters are functions of the film Reynolds number, $4\ uh/\nu$, the ratio of inertial to viscous forces; and the film Froude number, $(u/2\ \omega h)^{1/2}$, the ratio of inertial to rotational forces.

Experiments in the related case of inclined plane flow are plentiful, see for example References [4,5]. These experiments indicate that the film flow becomes wavy at Reynolds numbers above about 4, and turbulent at Reynolds numbers above about 1000. Above Reynolds numbers of 1000, the flow has been characterized as a series of turbulent slugs or waves, i.e., a somewhat stochastic as opposed to a periodic phenomena with a single frequency component. The wave inception distance increases from about 400 film thicknesses at Re = 1000 to about 650 film thicknesses at Re = 10,000. A turbulent steady film flow analysis has been performed by Dukler [6]. His results can be parameterized as

$$h/(\nu^2/g)^{1/3}=0.205Re^{0.548} \tag{1}$$

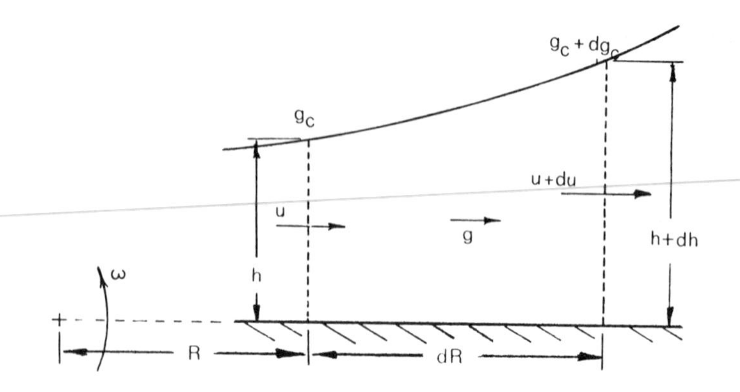

Figure 1. Schematic of Film Flow

where $(\nu^2/g)^{1/3}$ is a length scale based on a balance between the viscous and streamwise gravitational body force.

The present experiment was designed to measure in detail the effects of flow rate and rotation on waves in liquid films. Time series records of the film thickness were obtained at two different positions in a rotating channel. The spectral density, amplitude probability distribution, and cross-correlation were computed. From this information the average, maximum, minimum and average film thickness; and wave frequency and speed were deduced and correlated with the Reynolds number.

EXPERIMENTAL APPARATUS AND PROCEDURE

A rotating channel was built and instrumented with conductance probes. A schematic of the channel is shown in Figure 2. The channel rotated in the horizontal plane, and was enclosed in a plastic shroud. The channel was powered by a 2 hp variable speed motor and was supported by two bearing assemblies. The electrical connections from the lab passed through a set of slip rings into a hollow section in the channel support shaft and out onto the channel. The channel was machined from a single piece of aluminum and was 0.35 meters in radius and 3.175 cm wide. A piece of plexiglass flat stock was fitted into slots in the aluminum to form the vertical flat surface upon which the water film ran. To shield the water film from the air flow, a plastic cover was used on the open side of the channel. For balancing purposes, a dummy channel was placed radially symmetric (180°) with the instrumented channel. The channel assembly was also dynamically balanced. A tapered trailing edge was added to reduce aerodynamic drag.

Filtered water was pumped to the channel through flowmeters, a rotating seal above the channel, and a stainless steel diffusion nozzle. The nozzle was designed to release the water 20 cm from the center of rotation, so the rotational forces would be much larger than the local gravity field. The nozzle opening was 1 mm high, and parallel with the channel, so the liquid would extend across the width of the channel. Numerical calculations using the slope equation derived in [3] indicated that the flow will attain a normal thickness in about 1 cm, i.e., 100 normal film thicknesses. The water was pulled by the rotational forces along the channel and exited at the channel tip onto a shroud. The shroud collected the water for drainage to a reservoir and pump. Rotational speeds greater than 500 rpm were required to make the local gravity field less

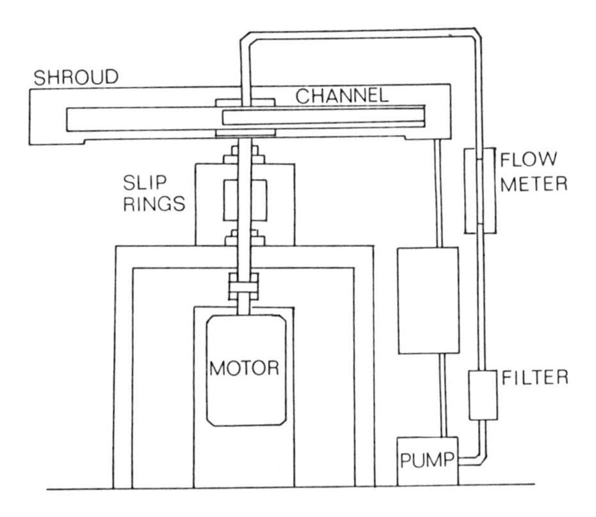

Figure 2. Schematic of Channel

than 2% of the centrifugal force at the nozzle exit.

The experimental variables were the rotational speed of the channel, ω, and the water film flow rate, $\Gamma = uh$. The rotational speed was measured by a magnetic pick up and a digital counter. The uncertainty in the rotational speed measurement was 5-10 rpm. The water flow rate was measured by a series of flowmeters, with an uncertainty of $\pm 2\%$. A test sequence began with calibration of the probes, powering the channel to the desired rotational speed, pumping the water to the channel, sampling the film thickness repeatedly, shutting off the water supply to the channel, stopping the channel rotation, and finally, performing a post sample probe calibration.

The liquid films in this rotational geometry were very thin, with average thicknesses of about 10^{-4} m, so a noninvasive means of thickness measurement was required. In this experiment, the thickness of the liquid film was measured with conductance probes. Conductance probes have been used successfully in previous experiments [7] for liquid film thickness measurements.

The conductance probes were set flush with the flat plexiglass bottom surface, so the water film was not disturbed by the probe. The water film was made electrically conducting by addition of an electrolyte, so its conductance was proportional to thickness. When a voltage was applied across the probes, there was a current flow in the liquid film from one probe to another. The resulting current was converted to a voltage signal by including a small resistance in series with the probes. The effective resistance of the water was about $10^5\,\Omega$ for the probe geometry used, and for the amount of electrolyte used, which was 0.1% by volume. The frequency of the applied voltage was 10,000 Hz, resulting in a capacitive reactance between the probes of $10^8\,\Omega$, much larger than the water film resistance, and thus a negligible path for current flow. The aluminum channel was anodized to render it nonconducting.

One problem with conductance probes was uncertainty as to the active area of measurement. In this experiment, guard electrodes were introduced on either side of the thickness probes to confine the induced current to a well defined area. By symmetry, the guard electrodes confined the field of the thickness probes to a region equal to half the distance between the guard and thickness

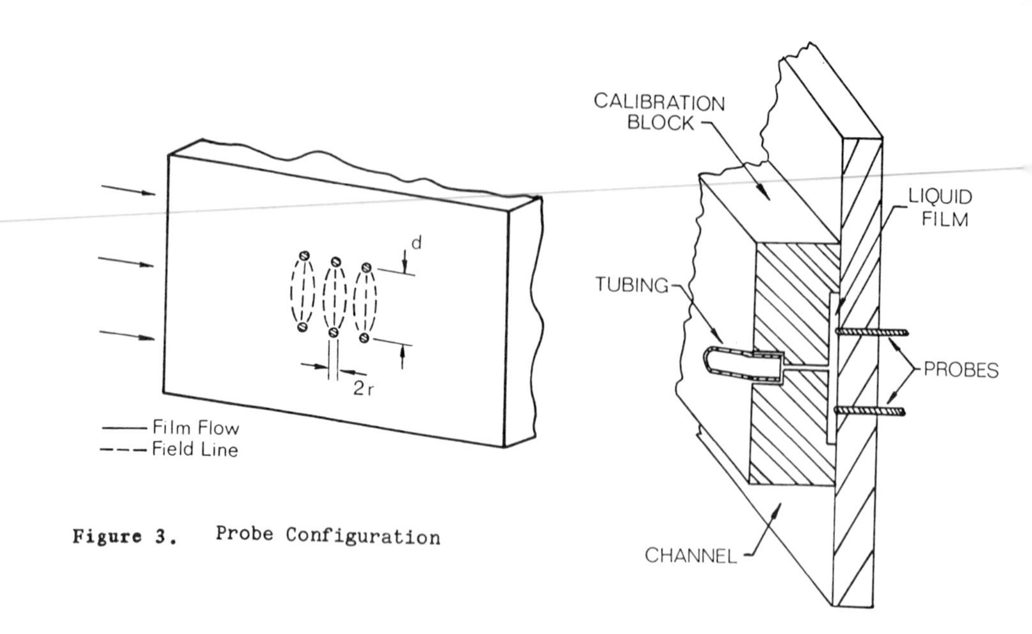

Figure 3. Probe Configuration

Figure 4. Probe Calibration

probe. The probe configuration is shown in Figure 3. The electrodes in each probe assembly were located 7 mm apart in the spanwise direction and 1.75 mm apart in the streamwise direction. Each electrode was made of 0.75 mm diameter stainless steel. Probe assemblies were located at radial distances 0.241 m and 0.300 m, respectively, equivalent to about 400 and 1000 film thicknesses from the nozzle.

The calibration of the probes was accomplished by the use of plexiglass blocks machined to a given thickness as shown in Figure 4. The electrolyte solution was introduced into the calibration block by means of a hole drilled in the back which was fitted with surgical tubing. The calibration block was placed above the thickness probes, filled with electrolyte solution, and the resulting signal from the circuitry recorded. The smallest calibration block was 0.25 mm. For the Reynolds numbers of interest, it was possible for the minimum thickness to be below this value if the rotational speed was greater than 1500 rpm, so the channel rotation speed was always below 1500 rpm, and greater than 500 rpm. The initial and final calibrations before and after a test were averaged to produce the calibration values. The difference between the initial and final values was typically 2-3%.

The analog thickness instrumentation was connected to a CAMAC A/D converter, which in turn was connected to a VAX 11/750 computer. The digital sampling rate was 4000 Hz, and the sample period was about one half second, resulting in a frequency resolution of 1/2 Hz, and a total number of 2048 samples for each data point. The digitizing equipment used consisted of a CAMAC 2264 H eight big waveform digitizer, CAMAC 32 k memory module, and Kinetic Systems Computer Interface.

The digitized film thickness traces stored on the computer disk were converted from voltage to thickness using linear interpolation between calibration points. The 2048 samples were then analyzed for their waveform content. An amplitude probability density distribution, defined as the relative fraction that the liquid thickness was in a given thickness band, was computed by dividing the thickness from zero to maximum thickness into thirty sections and tabulating the time the flow spent in each section. The time averaged thickness was computed from the statistical average of the 2048 samples. The maximum and minimum thicknesses were computed by locating the many extremum values to obtain an average maximum and average minimum.

The spectral density was computed using a Fast Fourier Transform (FFT) program. The FFT program calculated the frequency spectrum of a digitized time series record by transforming 2048 points in the time domain into the frequency domain, and plotted the frequency spectrum on an electrostatic plotter. The frequency at which the wave had a maximum on the power spectrum was chosen as the average frequency, f.

The wave speeds were calculated by use of the cross correlation routine in the FFT subroutine. The cross correlation routine calculated the time lag between records which maximized the similarity between them. Thus for two probes separated a distance x apart, the wave speed was given by $c = \frac{x}{\tau}$, where τ is the cross correlation time lag. The software packages were validated by digitizing various sine waves which were produced from a signal generator.

RESULTS

Experiments were conducted in the Reynolds number range from 900 to 20,000, and rotational speeds from 500 to 1500 rpm (8.3 to 25 hz). The value of the rotational Froude number was determined only after the thickness measurements were performed, so the results are presented primarily in terms of Reynolds number and the ratio of the centrifugal to coriolis force, g/g_c. In addition, the Froude number of the film did not turn out to be a strong function of the flow rate and rotational speed, since it was proportional to $\omega^{1/6}$ and $Re^{0.04}$. At Re = 5000, and ω = 1000 rpm, the Froude number based on the average thickness was about 5.

Stroboscopic photographs were taken by darkening the laboratory, and using a magnetic pickup to trigger a strobe lamp once, and recording the image through an open shutter of a Graftex camera. The photographs generally showed a pebbly-like surface at all Reynolds numbers and rotational speeds tested.

A time series record is shown in Figure 5. The thickness is not constant, but varies in time. A range of maximum wave heights and minimum thicknesses are evident. The waves are not symmetric as the front portion of the wave is steeper than the back portion.

A spectral density distribution is shown in Figure 6. A range of frequencies can be seen on the plot. When the experiment was being debugged, a wave frequency equal to the rotational frequency was present. This was subsequently eliminated by changing the inlet nozzle design. The point of maximum spectral density was taken as the dominant frequency. For the flow rates of this experiment, the dominant frequency was usually in the range of 150 to 200 hz.

A cross-correlation between the signals from two probes is shown in Figure 7. The point of maximum correlation was chosen as the characteristic wave transit time between probes. Since the cross-correlation is not a sharp spike, the disturbances can not be characterized as single periodic wave motion.

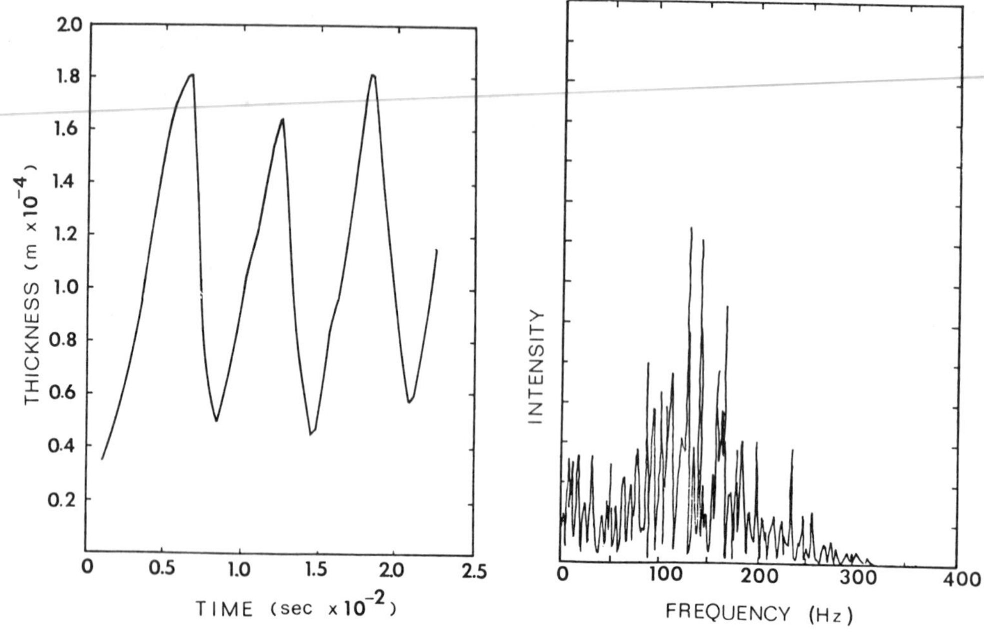

Figure 5. Time Series Record (Re = 8368, ω = 1000 rpm)

Figure 6. Spectral Density Profile (Re = 5579, ω = 750 rpm)

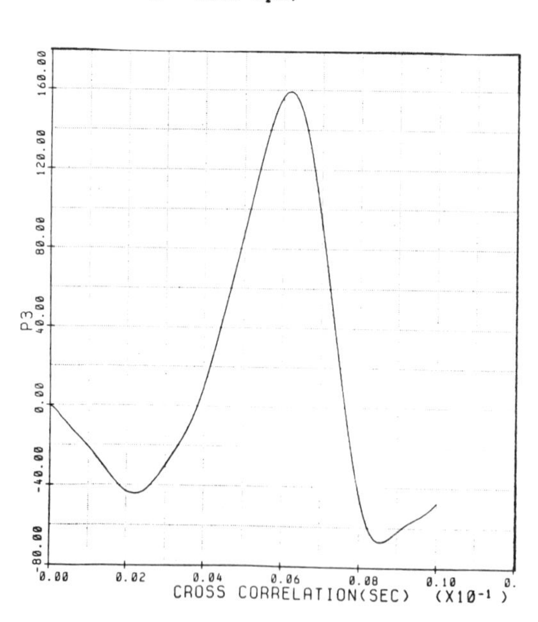

Figure 7. Cross Correlation Profile (Re = 6974, ω = 527 rpm)

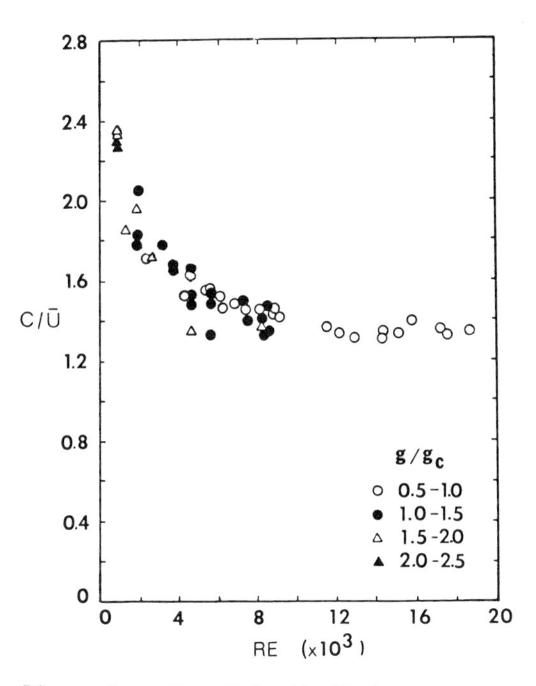

Figure 8. Wave Velocity Plot

Using the cross-correlation results to determine characteristic wave speeds, Figure 8 was prepared. It is a graph of dimensionless wave speed versus Reynolds number. To be consistent with previous work the parameter used to nondimensionalize the wave speed was the theoretical film velocity for undisturbed film flow, obtained from equation (1). The dimensionless wave speed decreases with increasing Reynolds number, consistent with previous results for inclined film flow [4]. If the experimental average velocity is used to nondimensionalize the wave speed, the dimensionless wave speed is of the order 1, varying from .95 to 1.05.

The amplitude probability diagram is shown in Figure 9. The diagram is not symmetric, indicating that the time mean thickness is not the average of the maximum and minimum thicknesses. The maximum thicknesses are of the order of twice the average thickness, and the minimum thicknesses are of the order of one half the average thickness.

The average thickness is given in Figure 10 as a function of Reynolds number and rotation rate. Also plotted is the turbulent steady flow equation, equation (1). A least squares fit through the data is

$$h/(\nu^2/g)^{1/3} = 0.168\ Re^{0.54} \qquad (2)$$

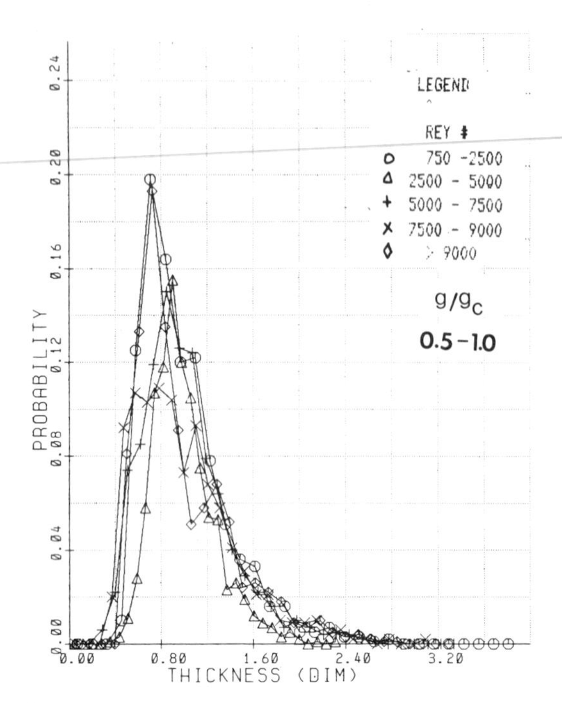

Figure 9. Amplitude Probability Plot

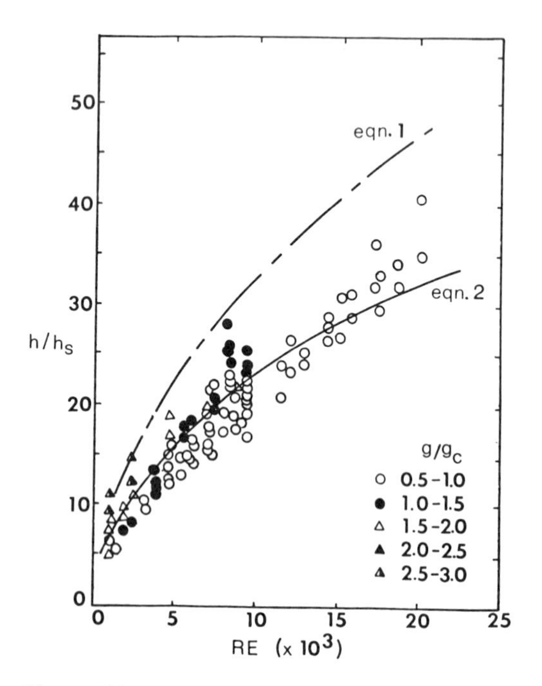

Figure 10. Average Thickness Plot

The average thicknesses for the unsteady flow are about 30% less than the analytical steady flow equation. This can be qualitatively explained by the non-linear flowrate versus thickness relationship: $\Gamma \sim h^2$, so $d\Gamma/dh \sim 2h$. Positive excursions from a mean thickness have a higher differential flowrate than negative excursions. Therefore a wavy flow of mean thickness h will have a higher flowrate than a steady flow with the same thickness.

SUMMARY AND CONCLUSIONS

The basic result of the experimental work is to show that finite size wave disturbances can exist in liquid films in a channel rotating about an orthogonal axis. For film Reynolds numbers about 1000, the flow can be characterized as unsteady, with a definite band of spectral components. The time series records of the instantaneous film thickness indicate that the waves have a maximum height of about twice the average film thickness, and a minimum thickness of about half the film thickness. The dimensionless wave speeds which have a value of about 1.5, decrease slightly with increasing Reynolds number. The average flow rate for the wavy flow was higher than that for an equivalent steady flow with the same mean thickness.

ACKNOWLEDGEMENTS

The research was performed in the M. I. T. Energy Laboratory. The able assistance of Mr. Gerrald Power is gratefully acknowledged.

References

1) Grondahl, C. M., Dudley, J. C., and Miller, H. E., "Gas Turbine Bucket Design with Water Cooling," ASME paper 83-HT-12.

2) Kirkpatrick, A., El-Masri, M., and Louis, J. F., "Wave Motion in Liquid Films on Rotating Plates," to appear in March 1985 ASME J. of Applied Mechanics.

3) El-Masri, M. A., and Louis, J. F., "On the Design of High Temperature Gas Turbine Blade Cooling Channels," ASME J. Engineering for Power, Vol. 100, No. 4, Oct. 1978, pp. 586-591.

4) Brauner, N., and Maron, D. M., "Characteristics of Inclined Thin Films, Waviness and the Associated Mass Transfer," Int. J. Heat Mass Transfer, Vol. 25, No. 1, Jan. 1982, pp. 99-110.

5) Fulford, G. C., "The Flow of Liquids in Thin Films," Adv. in Chem. Eng., Vol. 5, 1964, p. 151.

6) Dukler, A. E., "Fluid Mechanics and Heat Transfer in Vertical Falling Film Systems," Chemical Engineering Progress Symposium Series, Vol. 56, No. 30, 1960.

7) Hewitt, G. F. and Hall Taylor, N. S., Annular Two Phase Flow, Pergamon Press, Oxford, 1970.

Measurement of Unstable Flows under Stalled Conditions in the Vaneless Diffuser of a Centrifugal Compressor

SHIMPEI MIZUKI and C. W. PARK
Hosei University
Koganei, Tokyo, Japan 184

B. E. L. DECKKER
University of Saskatchewan
Saskatoon, Canada S7N OWO

INTRODUCTION

The existence of complex flows within the impeller passages of centrifugal compressors has, for several years, led to attempts to quantify them. They arise from the effects of the Coriolis force, streamline curvature, adverse pressure gradients in wall boundary layers leading to flow separation, tip leakage flow, rotating stall, and so on. Many of these effects have not yet been clarified but advances in their understanding, theoretically and experimentally, have contributed usefully not only to the design of these machines but also have led to a better appreciation of the nature of these flows.

To measure flow patterns in the flow passages of a rotating impeller, two methods are available [1]. One method is to measure the patterns by suitable sensors installed in a passage itself, that is, in a framework of rotating relative coordinates; the other is to measure them from a stationary position, or in absolute coordinates.

The use of hot-wire sensors and pressure transducers mounted in the rotating impeller passage offers a well tried and convenient method of measuring flow patterns. However, their use is not without disadvantages as the mechanical traverse mechanism in the impeller passage can be complicated while, from inertial considerations, the rotational speed of the impeller has to be kept reasonably low to accommodate the mountings for the sensors and traverse mechanism. Such a system could be expensive to install and, moreover, measurements could be laborious if comprehensive flow patterns are to be obtained. Wherever feasible, the use of hot-wire anemometry to measure flow patterns from stationary coordinates is considered by the authors to be the most convenient and economical method for that purpose.

Flow patterns in rotating impeller passages operating near design point conditions have been measured by one of the authors [2,3] as well as by other investigators [4], and their details, to some extent, are well understood. On the other hand, few measurements have been made under conditions of stalling. For that reason, experiments were undertaken to measure the flow patterns in the vaneless diffuser of a compressor at the onset of rotating stall and at fully stalled conditions in order to gain some insight into the flow in the impeller passages. Since the flow-patterns derived from stationary coordinates downstream of the impeller exit were found to be influenced by rotating stall cells, their effect on time-averaged flow patterns in rotating relative coordinates has been examined and discussed.

EXPERIMENTAL CONSIDERATIONS

Figure 1 is a sectional view of the vaneless diffuser of the experimental compressor in which are shown the spanwise locations in the flow passage at which measurements with a hot-wire probe and initially with a pitot tube were made. The compressor was designed for low mass rates of flow and for low rotational speeds to ensure that the measurements made were of acceptable accuracy. The vaneless diffuser was relatively large, its overall diameter being 700 mm, and discharged directly into the atmosphere. The impeller had 26 radial vanes with splitters and was 294.5 mm in diameter.

The axial symmetry of the time-averaged pressure and velocities in the impeller and in the vaneless diffuser in stationary coordinates was confirmed by measurement at static pressure taps along three meridional planes 120 deg. apart from entry to the impeller to exit from it and in the vaneless diffuser. The locations of the measuring stations in the vaneless diffuser are shown in Figure 1.

A pitot-tube and I-shaped hot-wire probe were used to measure in stationary coordinates the time-averaged velocity distributions in the vaneless diffuser. The pitot-head was measured by means of an inclined manometer (5:1). The time-averaged patterns obtained by using the two probes were similar except near the casing and hub where reversed flows due to stall-cells were present. From an examination of these patterns, it was clear that a pitot-tube could not be used for the measurement of fluctuating flows with reversed velocities in the stall-cells. Hence, in the discussion that follows, reference will be made only to the time-averaged flow patterns obtained by means of the hot-wire probe. The linearizer (Kanomax, Model 1013) of the hot-wire anemometer (Kanomax, Model 1009, 1011, and 1021) was considered to be adequate for the present application.

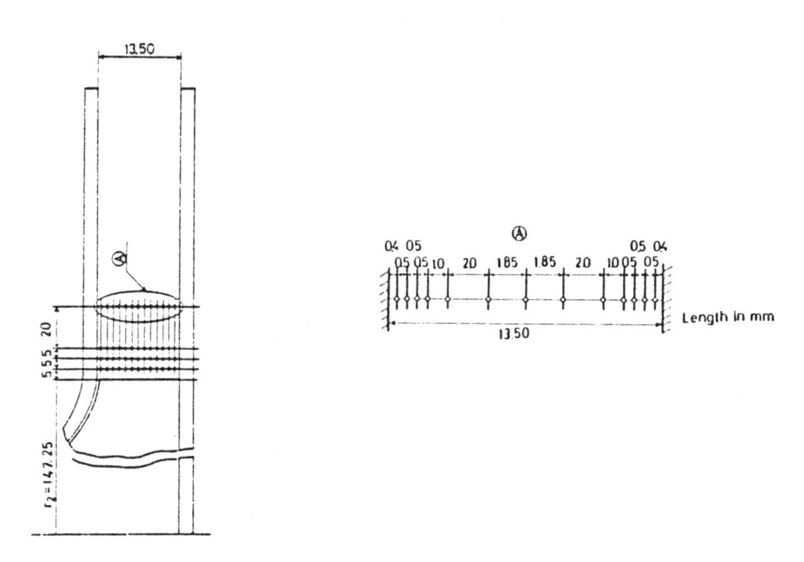

FIGURE 1. Vaneless diffuser-half-section showing locations of static pressure taps in meridional plane and spanwise probe positions.

The hot-wire probe was rotated about its axis by a stepping-motor mounted on the mechanical traverse unit. The stepping motor was driven either by manual input signals from a preset controller (Oriental Motor, OSD006) or by signals from the mini-computer (Kanomax, REALEX 16) through Direct Input/Output interfaces (Kanomax, Model 1333). The position of the probe in the diffuser passage (Figure 1) was monitored digitally (Sony, LF 20) with an accuracy of 0.01 mm. The output signals from all the sensors were monitored by synchroscopes.

A photo-tachometer (Sagawara, DT-102A) was located near the exit of the vaneless diffuser and focussed on the tips of the impeller blades. Thus, a trigger pulse for each passing blade was easily obtained. By using this pulse as an external trigger for the real-time digital-correlator (Kanomax, SAI 42A), the hot-wire signals of one blade pitch were divided into 100 points and digitized at each point in the 256 levels from the minimum to the maximum voltage. The trigger pulses from the passing blades enabled a synchronized summation to be made 1024 times (39.4 revolutions of the impeller) and the results were then displayed by means of an X-Y plotter. Signals from the hot-wire probe were also transmitted to the Fast Fourier-Transform Analyzer (Ono Sokki, CF400) and the frequency spectra of the fluctuations derived.

The hot-wire probe was rotated from 0 deg. through $\pm$ 5 deg. and $\pm$ 10 deg. about the time-averaged mean flow direction. For each setting angle synchronized summations were obtained as described above and the results stored in the memory of the mini-computer. From the 100 digitized points obtained from one blade pitch 25 were selected, and the relationship between the corresponding output voltages and setting-angles of the hot-wire probe was obtained for each of these selected points by spline-fit curves. In this way, maximum output voltages and the corresponding flow angles were computed on blade-to-blade

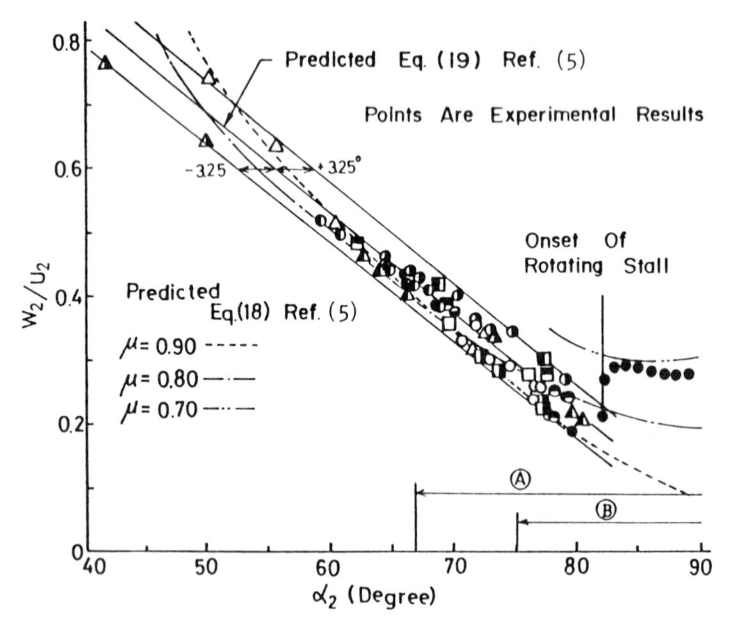

FIGURE 2(a). Variation of circumferential relative velocity with flow exit angle. Experimental results compared with predictions, Ref. [5], A - some points are surging, B - all points for surging.

surfaces. Based on these results, the radial and circumferential components of the relative velocity at a position within the passage section were obtained directly or by interpolation.

Since the data acquisition system included A/D converters (Kanomax, Model 1337) and the Direct Input/Output interfaces it would have been possible to carry out the synchronized summations and Fast Fourier analysis using the mini-computer. However, it was more convenient to use the real-time digital correlator and Fast Fourier-Transform analyzer for this purpose.

With the existing instrumentation, it would have been possible, also, to use slanted-wire probes for measurement of the Reynolds stresses under normal circumstances [8], but, since the experiments were carried out at low mass flow rates and rotating stall cells were present, this was thought not to be feasible. In general, when stall-cell rotation is different from impeller rotation, the velocity distribution in the impeller passages varies in the circumferential direction and, if a slanted-probe were employed, it would have been necessary to isolate the effects of stall-cell rotation from that of the turbulence if precise time-dependent data were to be obtained. Velocity correlations between adjacent blade passages would have required a real-time data processing system with the sensors mounted in rotating blade passages.

The natural frequency of the instruments employed in this investigation was 20

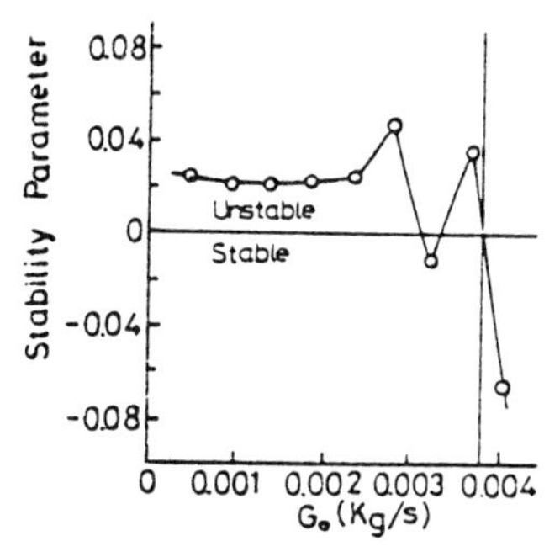

FIGURE 2(b). Onset of rotating stall using Stability Parameter.

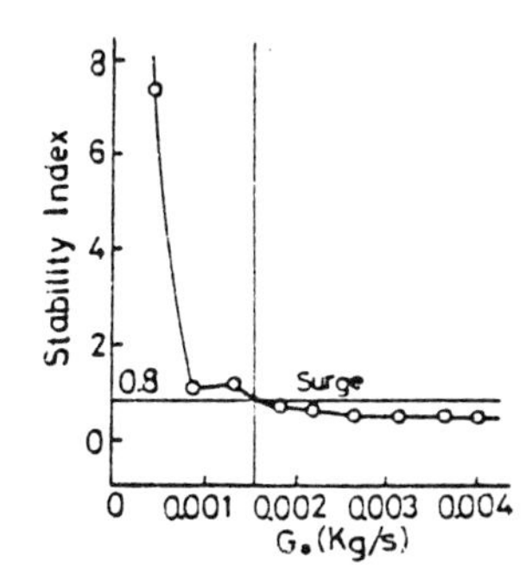

FIGURE 2(c). Onset of surging using Stability Index.

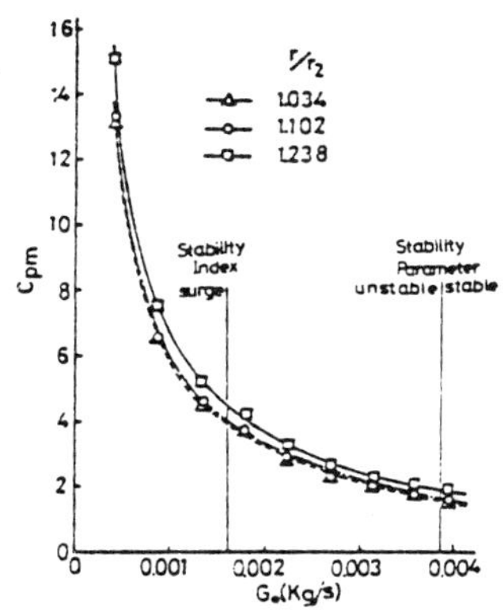

FIGURE 2(d). Variation of static pressure recovery coefficient with mass flow rate.

kHz (τ = 50 μs). The rotational speed of the impeller in these experiments was 1600 rpm, and the corresponding blade-passing frequency of the rotor was 693.3 Hz (τ = 1.4 ms).

OPERATING CONDITIONS OF EXPERIMENTAL COMPRESSOR

Experiments were performed for a range of low flow rates corresponding to the onset of rotating stall through to surging. Because the compressor was designed with a small system-volume, phenomena generally associated with surging were absent even at extremely small flow rates.

The onset of rotating stall was identified by the same relationship between the non-dimensional relative velocity and impeller exit flow angle as given in Ref. [5]. The pertinent figure in that reference has been reproduced in Figure 2(a) and the results of the present experiments have been superimposed on it. The sudden increase in the non-dimensional relative velocity with increase in the flow angle at exit is clearly seen and signals the onset of rotating stall. Because the rotational speed in the present experiments is relatively low, the onset of rotating stall occurs at a larger value of the flow exit angle. In conventional compressors, exit angles of that magnitude would normally be associated with surging as demarcated in the figure. The slip factor also decreases as the non-dimensional relative velocity is increased.

The onset of rotating stall occurred at the corrected mass flow rate G_o = 0.00359 kg/s, and this was also confirmed by the time-averaged flow patterns obtained by measurement in stationary coordinates at exit from the impeller. The mass flow rate at which the onset of unstable flow is predicted by the Stability Parameter [6] is G_o = 0.00379 kg/s as seen in Figure 2(b). These values of the mass flow rates, however, are near enough to one another that for practical purposes, the mass flow rate G_o = 0.00359 kg/s may be considered as signalling the onset of rotating stall.

The mass flow rate at which surging occurs is predicted by the Stability Index [7] to be G_o = 0.00153 kg/s, as shown in Figure 2(c). In the present experiments, however, the usual manifestations of surging were not evident even at the extremely small mass flow rate G_o = 0.00045 kg/s when, in fact, the measured flow patterns tended towards equilibrium. The absence of large fluctuations in velocity can only be associated with the small system-volume of the experimental compressor and reinforces the view that surging is a resonance phenomenon.

Increases in the static pressure (expressed as the pressure recovery coefficient, C_{pm}) radially in the vaneless diffuser as the mass flow rate is decreased is shown in Figure 2(d). The trends in these curves are reflected in variations in the flow patterns over the range of mass flow rates from the onset of rotating stall to near-equilibrium at G_o = 0.00045 kg/s.

EXPERIMENTAL RESULTS AND DISCUSSION

The spanwise distributions of the relative radial velocities on blade-to-blade surfaces and the corresponding circumferential velocities are shown in Figure 3(a) and Figure 3(b), respectively. These have been derived in the manner described in the previous section from time-dependent measurements made in stationary coordinates at a distance of 10 mm downstream of the impeller exit (r/r_2 = 1.068) and at the mass flow rate corresponding to the onset of rotating stall, G_o = 0.00359 kg/s. Similarly, the spanwise distributions of the relative radial and circumferential velocities are shown in Figure 4(a) and 4(b) for the smaller mass flow rate of G_o= 0.00135 kg/s, when rotating stall

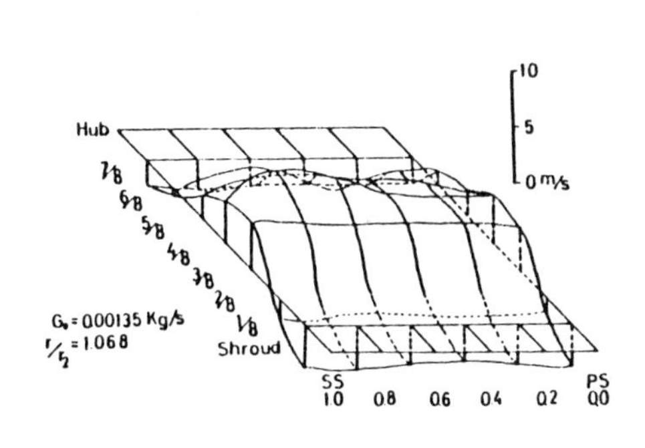

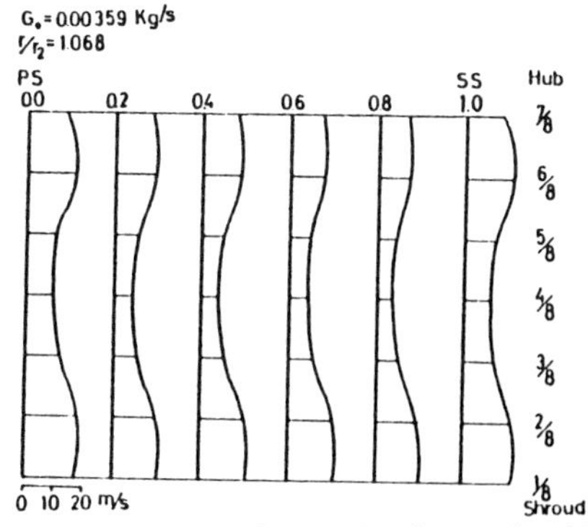

(a) Radial relative velocity. (b) Circumferential relative velocity.
FIGURE 3. Spanwise radial and circumferential velocity distributions at onset of rotating stall, 10 mm from impeller tip.

is well developed, at the same position in the vaneless diffuser. The existence of large stall-cells are seen in Figure 4(a) near the shroud (casing) and hub. In contrast, in Figure 3(a) at the onset of rotating stall, only the region near the hub is affected. Variations of the relative radial velocity components on meridional planes appear to be larger than variations in the corresponding circumferential velocity components. Actually, however, the absolute values of the variations in the circumferential components are larger as may be seen in Figure 3(b) and Figure 4(b). In Figure 4(a), it is evident that the through-flow occupies the central portion of the passage between shroud and hub. This is due to the blockage effects that arise from the existence of the large stall-cells. Because the relative circumferential velocity components are relatively large near the casing and hub, where the corresponding radial components are reversed, it was possible to measure the time-averaged velocity distributions of the stall-cells. If the stall-cells near the suction-side and near the casing were composed, like wake flow, of low energy fluid the slip velocity would be small in spite of the low through-flow velocity [8], or the relative radial velocity components. However, flow within the stall-cells exhibits high slip velocities which is quite different to that expected in a wake flow. On the other hand, the slip velocity is decreased in

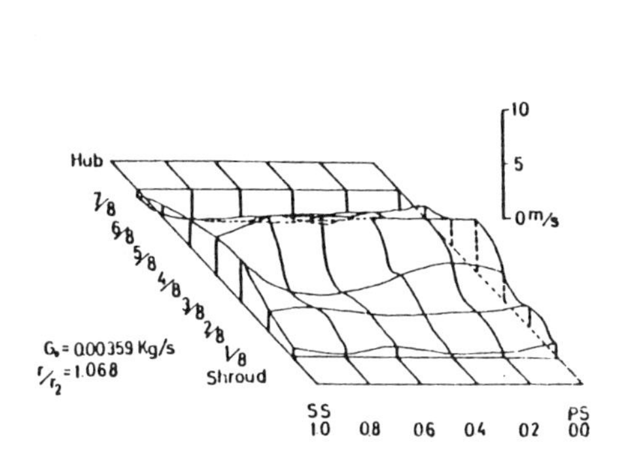

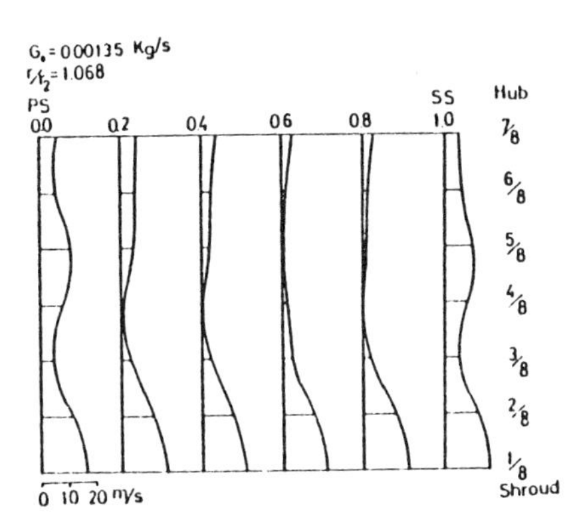

(a) Radial relative velocity. (b) Circumferential relative velocity.
FIGURE 4. Spanwise radial and circumferential velocity distributions at fully developed stall, 10 mm from impeller tip.

the main through-flow which, again, is different to that associated with jet flows. Thus, flows in which rotating stall-cells are present exhibit characteristics that are altogether different to the jet and wake flows found under conditions near the design point.

Figure 5 summarizes the directions and magnitudes of the relative radial and circumferential velocity components on spanwise stream surfaces between the shroud (casing) and hub. They have been derived from time-averaged measurements made in stationary coordinates at the positions shown, that is, from $r/r_2 = 1.034$ to $r/r_2 = 1.238$, or at distances varying from 5 mm to 35 mm from the impeller tip, and at two mass flow rates, $G_o = 0.00359$ kg/s at the onset of rotating stall and $G_o = 0.00135$ kg/s for fully developed stall. In the actual instantaneous flow patterns, fluctuations due to rotating stall-cells would be superimposed on the time-averaged flow patterns so that the velocities would fluctuate more strongly. Stall-cell rotation is 0.6071 times that of impeller rotation as may be seen in Figure 8(a) or Figure 8(b). Variations in flow direction near the shroud (1/8-surface) and hub (7/8 surface) are due to boundary layer development at increasing distances from the impeller tip. Increases in the circumferential component near the shroud and hub at the low mass flow rate are due to development of the rotating stall-cells downstream of the impeller. The associated flow patterns are skewed and strongly three-dimensional. This explains why the pitot-tube was found to be unsatisfactory for time-averaged measurements as compared with a hot-wire probe.

Periodicity associated with impeller rotation in time-averaged flow patterns in relative rotating coordinates has been investigated. In Figure 6, the weakest hot-wire signals at 5 mm downstream of the impeller tip for the mass flow rate $G_o = 0.00045$ kg/s, has been superimposed on the trigger-pulse extending over one blade pitch. The periodicity is clearly related to impeller rotation. Using the absolute maximum amplitude of fluctuation of the strongest hot-wire signals at the flow rate $G_o = 0.00359$ kg/s, at the onset of rotating stall, which occurs at midspan in the blade passage as a reference quantity, other maxima corresponding to different mass flow rates and to three stream surfaces between shroud (casing) and hub have been non-dimensionalized and plotted against radial position in the vaneless diffuser as in Figure 7. Although in

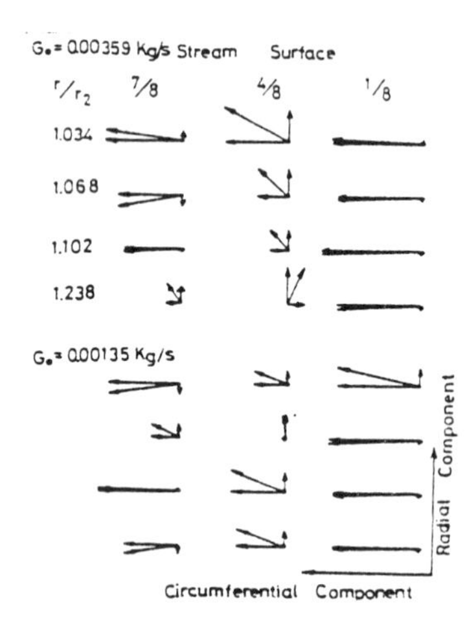

FIGURE 5. Variations of relative velocity components with distance from impeller tip.

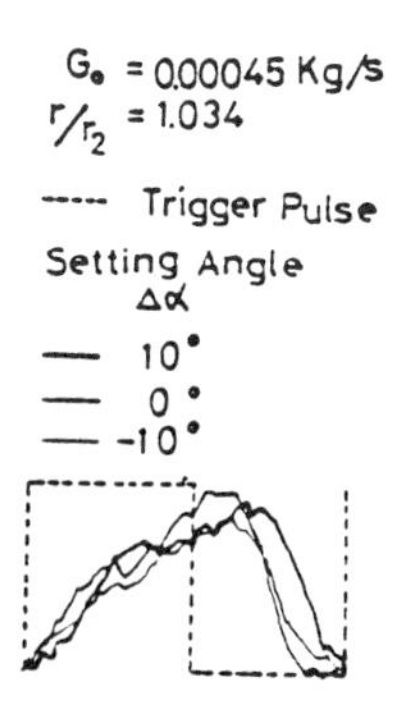

FIGURE 6. Weakest signal (synchronized summation) 5 mm from impeller tip for G_O = 0.00045 kg/s.

G₀ Kg/s	0.00045	0.00135	0.00359
Shroud Side	▲	■	●
4/8 Surface	◭	◧	◑
Hub Side	△	□	○

FIGURE 7. Attenuation of periodicity and amplitude ratios of fluctuations with distance from impeller tip.

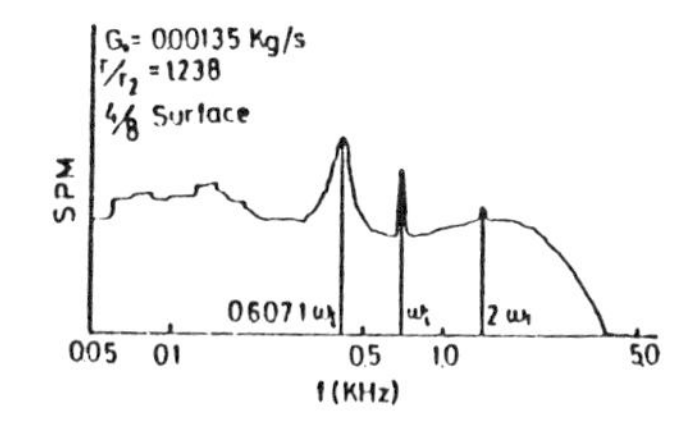

(a) Frequency spectra on 4/8-surface, G_O = 0.00135 kg/s.

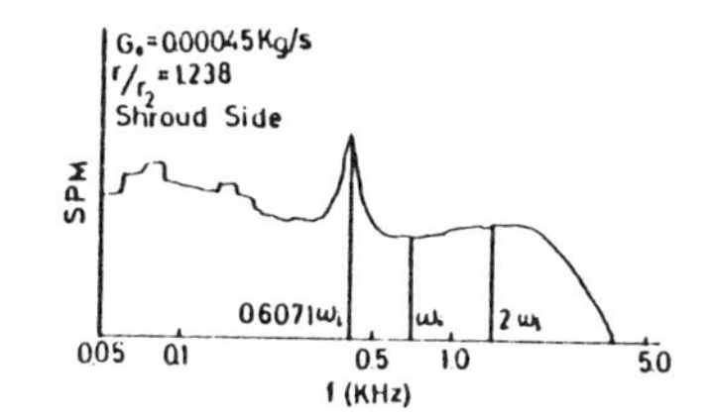

(b) Frequency spectra on 1/8-surface, G_O = 0.00045 kg/s.

FIGURE 8. Frequency spectra on two surfaces at decreasing mass flow rates.

the figure a 'critical' value of H/H_{max} = 0.20 has, to some extent, been arbitrarily chosen, nevertheless, it is clear that the amplitude ratios not only diminish at farther distances from the impeller tip but the periodicity, noted in Figure 6, is also lost. The implication here is that growth of the rotating stall-cells due to decrease in the mass flow rate and in proximity to the shroud (casing) and hub attenuates the effect of impeller rotation. Results of frequency spectra analysis of the hot-wire signals on the stream-surface at midspan (4/8) at the mass flow rate G_o = 0.00135 kg/s in the fully stalled condition are shown in Figure 8(a). The influence of impeller rotation is clearly seen in this figure. On the other hand, in Figure 8(b) at the smaller mass flow rate G_o = 0.00045 kg/s on a stream-surface near the shroud (casing) there is no effect of impeller rotation. The results in these figures may be compared with those in Figure 7 at the radial position ℓ/ℓ_T = 1.238, or 35 mm from the impeller tip.

CONCLUSIONS

The synchronized summation method has been employed in this investigation to derive time-averaged flow patterns in rotating relative coordinates from time-averaged measurements in stationary coordinates. It cannot be used to obtain instantaneous fluctuations due to rotating stall-cells in relative coordinates as it would be necessary to use trigger-pulses to indicate rotation of the stall-cells. This restriction would apply also to the use of laser anemometry and other hot-wire anemometer techniques [9]. On the other hand, if the measurements made are synchronous with impeller rotation, that is, if one signal is obtained from the sensor per revolution, and a similar technique as that used in this investigation is employed to indicate the angular position of the measurement in the blade passage, the data acquisition system will become very elaborate. Alternatively, a computer with a large memory may be employed, but the data would have to be carefully examined to isolate the fluctuations due to the stall cells. This would present severe problems in signal processing, but it is intended to attempt such a solution in the near future.

With regard to the flow patterns that have been measured, quantities, such as, the non-dimensional relative velocity and flow exit angle derived from time-averaged values have effectively indicated the onset of rotating stall and surging when used in calculating the Stability Index and Stability Parameter, respectively.

A periodicity associated with impeller rotation is present in the time-averaged flow patterns for the larger mass flow rates used in the investigation, particularly, at midspan. Rotating stall-cells are present at the impeller tip and their subsequent growth downstream in the vaneless diffuser attenuates the influence of impeller rotation. At very small mass flow rates and near the shroud (casing) and hub, the periodicity is suppressed. This behaviour has been confirmed by frequency spectra analysis.

NOMENCLATURE

C_{pm} = static pressure recovery coefficient $(= (P_S - P_2)/(P_2 - P_1) \times 100)$

G = mass flow rate

G_0 = corrected mass flow rate $(= G(P_{a0}/P_a)\sqrt{T_{a0}/T_a})$

H = amplitude of hot-wire signal wave

H_{max} = maximum value of H

ℓ = length along casing from impeller inlet

ℓ_T = value of ℓ from impeller inlet to tip

r = radius

P_S = static pressure

PS = pressure surface

SS = suction surface

SPM = frequency spectra

W = relative velocity

U = impeller peripheral speed

α = angle between absolute velocity and radial direction

μ = slip factor

ω = angular speed

Subscripts

a = ambient condition

0 = standard conditions (293.16^{O}K, 102.050 kPa)

1 = at impeller inlet

2 = at impeller exit

REFERENCES

1. "Measurement Method in Rotating Components of Turbomachinery", Edited by B. Lakshiminarayana, ASME Symposium Volume, 1980.

2. Mizuki, S., Ariga, I., and Watanabe, I., "Investigation Concerning the Blade Loading of Centrifugal Impellers", ASME Paper, 74-GT-143, 1974.

3. Mizuki, S., Ariga, I., and Watanabe, I., "A Study on the Flow Mechanism within Centrifugal Impeller Channels", ASME Paper, 75-GT-14, 1975.

4. Eckardt, D., "Instantaneous Measurements in the Jet-Wake Discharge Flow of a Centrifugal Compressor Impeller", Trans. ASME, Series A., Vol. 97-3, pp. 337-346, 1975.

5. Mizuki, S., and Imai, H., "A Study Concerning Performance Characteristics of Centrifugal Compressors", ASME Paper 85-GT-97, 1985.

6. Dean, Jr., R.C., "The Fluid Dynamic Design of Advanced Centrifugal Compressors", Creare TN-185, 1974, pp. 68, 1974.

7. Rodgers, C., "Impeller Stalling as Influenced by Diffusion Limitations", ASME Symposium Volume, "Centrifugal Compressor and Pump Stability, Surge and Stall", pp. 37-67, 1976.

8. Dean, Jr. R.C.,and Senoo, Y., "Rotating Wakes in Vaneless Diffusers", Trans. ASME, Series D, p. 563, 1960.

9. Kuroumaru, M., and Ikui, T., "Measurements of Three-Dimensional Flow behind a Rotor by Periodic Sampling Method", (in Japanese), Trans. ASME, Vol. 48-427, pp. 408-417, 1983.

Flow Field Determination at Axial Pump Impeller Tip Section

K. K. WONG and J. P. GOSTELOW
School of Mechanical Engineering
The New South Wales Institute of Technology
Sydney, Australia

INTRODUCTION

In most applications the principal limitation on the performance of an axial-flow pump is its cavitation-free operating range, characterized by the nett positive suction head (NPSH). The adverse effects of cavitation are not restricted to impaired performance; noise and vibration levels tend to increase and mechanical integrity of components can be jeopardised, sometimes severely. Cavitation may occur in the inlet region or, in some instances, in the stator blades; however the most usual source of cavitation occurrence is the impeller blading, specifically the tip section.

There would, in principle, appear to be scope for an approach to the re-design of impeller blading having the objective of delaying the onset of cavitation or at least localizing and controlling its effects. Cavitation occurs when the local pressure falls below the vapor pressure and the approach would be to design the blading in such a way as to ensure that severe suction peaks were avoided and that local static pressures would remain as high as possible when operating in conditions susceptible to cavitation.

If three-dimensional effects are ignored the problem is analogous to that of wing and blade design in compressible flow where it may be desired to achieve the highest possible inlet Mach Number without provoking the inception of shock waves. It might therefore be considered that an extension of modern purpose-designed airfoil theory to the hydrodynamic environment should result in significant improvements. This approach has, of course, been used in applications as diverse as hydrofoils and propellors and has achieved success in the delay of cavitation inception. Despite these successes it is not clear that this approach will necessarily succeed in the complex flow environment of the axial flow pump. The flow over the blading of an axial pump is neither steady nor two-dimensional and certainly not inviscid. It might be ineffective to design a pump impeller having a blade section purposely-designed to delay cavitation onset if the mechanism of cavitation were other than inception at the suction peak of a two-dimensional blade.

There was therefore seen to be scope for a research investigation aimed at clarifying the mechanisms of cavitation in the impeller of an axial-flow pump in order to provide a rational basis for re-design of the blading. A program of work was initiated using the axial-flow pump facility at the New South Wales Institute of Technology. It was considered necessary to be able not only to make measurements of the flow field over the blading but also to have the capability to optically view the flow field. The approach adopted was for the existing pump rig to be provided with a transparent casing bowl for observation and photography. Mounted on this bowl were to be one or more blocks permitting a tube-mounted pressure transducer to sequentially communicate with casing pressure taps.

THE PUMP RIG

The rig is composed of 400 mm flanged pipework in a horizontal rectangular closed loop (Fig. 1). A flow straightener situated upstream of the pump and cascade vanes at the corners were used for reducing head loss and turbulence. Variation of system NPSH was achieved using a vent valve to control system pressure. Discharge butterfly valves were used for varying the system resistance.

FIGURE 1. The axial flow pump rig.

An axial-flow pump stage of 300 mm nominal diameter was mounted in the pump bowl section. The pump was driven by a 75 kW electric motor at about 1470 rpm. A transparent bowl was centrifugally cast from Epirez-135 epoxy to replace the existing cast-iron bowl. The five-bladed impeller was a standard industrial unit and dimensional tolerances were therefore not of research quality. The geometry of the impeller is as shown in Table 1 and Figure 2.

TABLE 1 Geometry of Pump Impeller

	Tip	Hub
Chord Length, mm	155	140
Pitch, mm	192	107
Pitch-Chord Ratio	1.239	0.764
Stagger Angle, degrees	72	45
Blade Thickness, mm	10	10
Diameter, mm	304	168
Nominal Tip Clearance, mm	0.76	

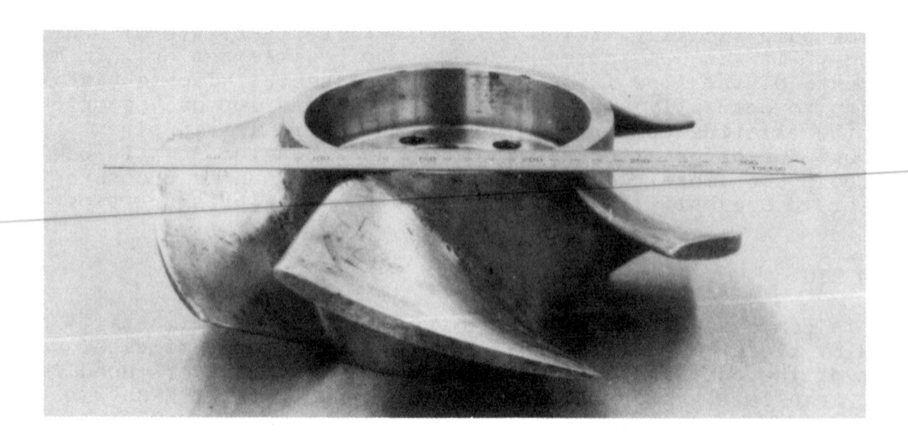

FIGURE 2. The test impeller.

Two perspex blocks (Fig. 3) were machined and attached with epoxy to the outside surface of the bowl. Pressure taps of 1 mm diameter were drilled through the block and the transparent bowl. A total of twelve taps were spaced between the leading and trailing edges.

FIGURE 3. Mounting block for pressure transducer tube.

CASING PRESSURE DISTRIBUTIONS

In order to obtain information on the blade loading and flow field at the rotor tip the technique of using an indexable tube-mounted pressure transducer was used. The technique will firstly be described and then some preliminary measurements of rotor tip pressure distribution.

The pressure transducer used was a strain gage type made by Gaeltec. The probe (Fig. 4) consisted of a stainless-steel cylindrical housing with a bleed tube allowing excess water to escape. Two 'O'-rings were situated on either side of the sensing-hole for isolating pressure signals from nearby pressure taps. Similar pressure transducers had previously been calibrated dynamically. The response had been observed to be uniform over the frequency range of interest. Static calibration only was used for these investigations using a dead-weight tester.

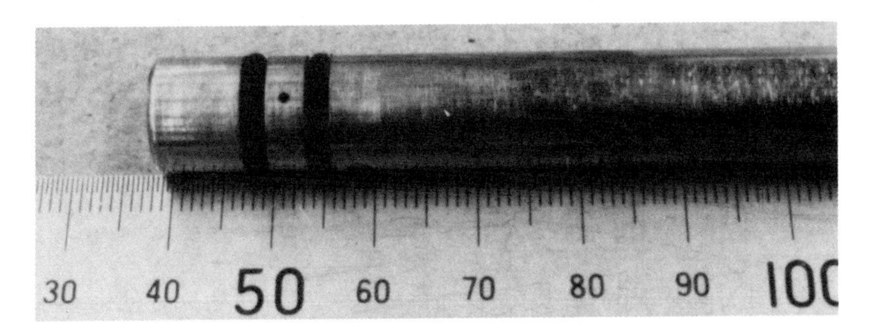

FIGURE 4. Pressure transducer probe tube.

Using the tube-mounted transducer it was possible for the transducer to communicate with any one of twelve pressure taps over the rotor tip. In any axial location the signal thus produced constituted the circumferential variation of static pressure. A typical trace is presented in Figure 5 and reveals the strongest suction pressure on the blade suction surface, followed by a rise in static pressure across the passage; the highest pressure levels may, or may not, coincide with the pressure surface but there is then a sudden drop in pressure corresponding to the passage of the blade. Once a complete set of traces had been obtained for each condition these were then intercepted at the deduced blade surface locations and a pressure distribution around the blade was generated.

It was not found necessary to apply filtering techniques to the signals. In particular although the technique of digital phase lock averaging [1] was available these signals were found to be relatively repeatable, even in the trailing edge region; the technique was therefore not applied.

Data on casing pressure distributions have, thus far, only been obtained under conditions having a positive suction head, well-removed from incipient cavitation. This is because of the high pressure levels generated when cavitation bubbles collapse, which could instantaneously damage the transducer diaphragm. The results do not therefore provide information on blade behavior in cavitation but rather indicate loadings, pressure gradients and regions of maximum suction pressure during normal operation.

A pressure distribution obtained at moderate loading is presented in Figure 6. The distribution presents no problems of interpretation and shows a fairly strong suction peak in the leading edge region.

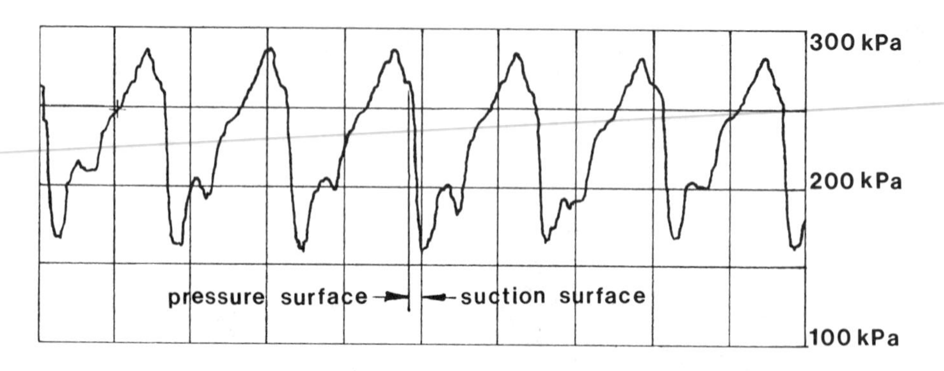

FIGURE 5. Typical trace for pressure variation across passage.

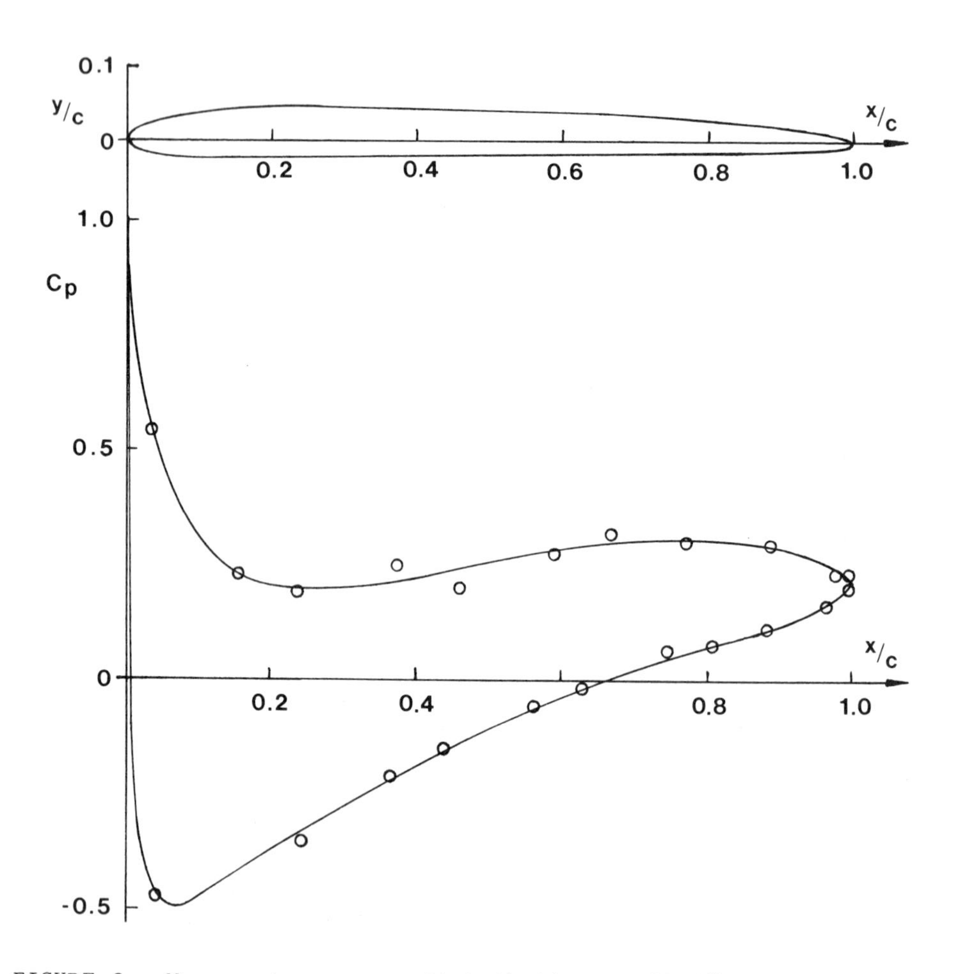

FIGURE 6. Measured pressure distribution at 93.8% flow.

CAVITATION OBSERVATIONS

The principal objective of the work was observation of cavitation and this was achieved stroboscopically by eye and by camera. Still photography was used with stroboscopic lighting. The camera was a Nikon F3 with a 55mm micro-lens and a minimum focusing distance of 250 mm. A Nikon MD-4 six frame/sec motor-drive was synchronized with the stroboscopic light.

For synchronization, magnets were attached to the shaft coupling. One magnet activated the motor-drive and the other a strobotac. Positioning of the "frozen" blade image through the transparent bowl was adjusted by varying the location of the magnet on the shaft coupling. The time required for the impeller to make one full revolution was just over 1/30th of a second. Hence, the shutter speed of the camera was set at that interval allowing the maximum time delay before the next triggering. The magnet triggering the motor-drive was located at 180 degrees opposite to that triggering the strobotac. This arrangement allowed more than 1/60th of a second for the triggering of the camera shutter prior to the triggering of the strobotac.

Photographs of the rotor tip cavitation behavior were obtained under a wide range of flow conditions. The two independent variables were blade loading and absolute pressure level, obtained by discharge throttling and by vent valve control of system pressure respectively. Results presented were taken firstly, with atmospheric inlet pressure and a full range of discharge throttling to cover the characteristic between maximum flow and stall conditions, and secondly, with a wide open discharge throttle for a range of values of inlet pressure.

Cavitation photographs are firstly presented for the experiments with atmospheric inlet pressure (constant NPSH of 10.2 m) but a wide range of discharge throttle settings. Starting with the discharge throttle wide open four photographs are presented in Figure 7 for loadings up to stall. The rotor tip cavitation bubbles show up clearly against the dark background and in a sense, at atmospheric inlet pressure, the cavitation bubbles act mainly as a means of flow visualization of the complex three dimensional flow conditions at the rotor tip. However on closer inspection it is clear that genuine local cavitation effects are being observed. An implication of the presence of such marked regions of cavitating flow under these conditions is that the levels of static pressure in the region of the rotor tip suction surface are significantly lower than those elsewhere on the blade. The principal area of cavitation appears to be a region of the suction surface mainly in the forward portion of the blade and a subsequent downstream region which tends to become detached from the blade surface. This is the main vortex associated with the tip clearance region (the scraping vortex) and is the principal region of very low pressures in the blade row.

The main change in the configuration of the tip leakage vortex as the pump is throttled up the characteristic is that the region of bubbles appears to separate earlier from the surface and at a steeper angle. At the higher loading conditions the region is

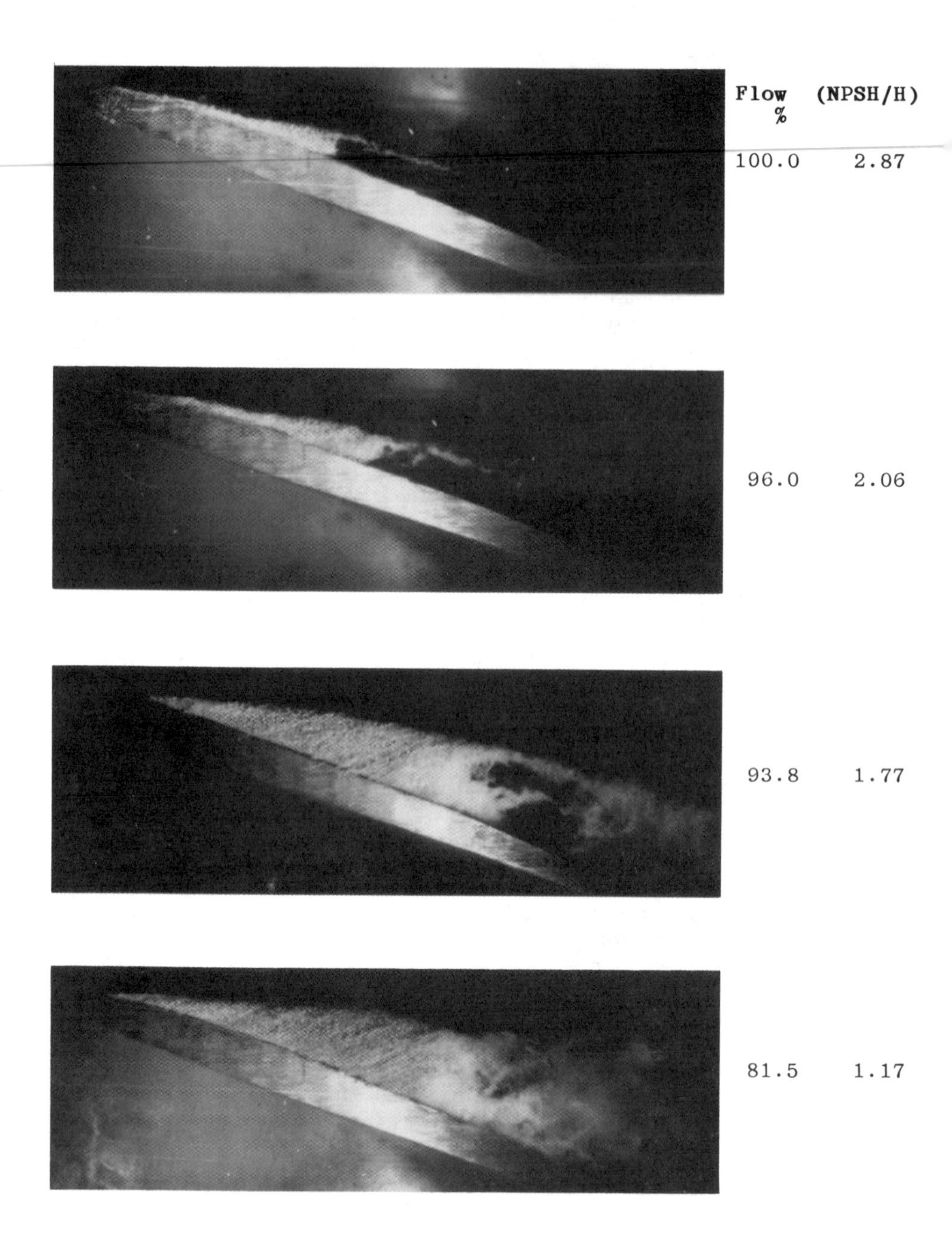

FIGURE 7. Effect of throttle variation, from wide open to maximum loading condition.

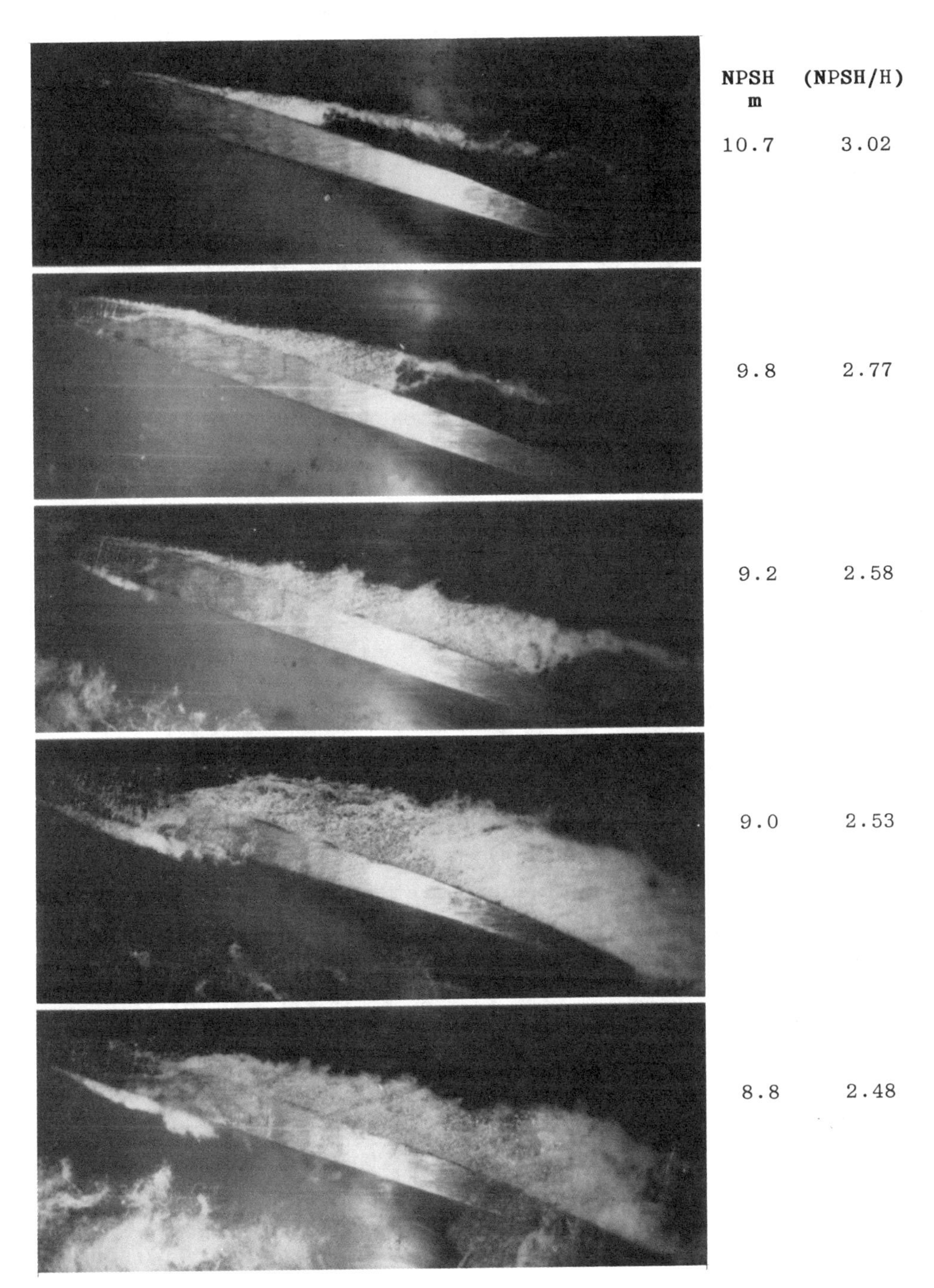

FIGURE 8. Variation of flow conditions with inlet throttling.

terminated by the frothy white zone of bubble collapse. Visual observations and other photographs have shown this zone of bubble collapse to be very abrupt; further inboard its termination of ribbons of cavitation has the appearance of a shock wave. This also appears to be the extent of the attached cavitation zone and downstream of this region there only remains a detached vortex core. Another feature deserving of attention is the helical nature of the bubble paths in the cavitation zone. This is also associated with the leakage vortex and bubbles have been observed streaming from the actual blade tip into the helix.

Figure 8 presents the variation of flow conditions when the discharge throttle remains wide open (100% flow condition) and the inlet pressure is varied. As inlet pressure is reduced the cavitation region extends in the downstream direction. The edge of the zone, which was well defined in the sheet cavitation of Figure 7, becomes ragged and unsteady. Although the photographs give an appearance of stability of location in practice the zone oscillated [2] with slugs of cavitating fluid moving rapidly downstream.

The increasing tendency for cavitation to occur over the blade tip at the leading edge is also noteworthy. For low inlet pressures cavitation also occurs on the pressure surface and at the lowest inlet pressure it becomes clear that the clearance vortex is established over the tip, from pressure surface to suction surface, in the leading edge region. Whereas evidence of this leading edge region tip vortex was present under most conditions tested it was particularly acute for low inlet pressures.

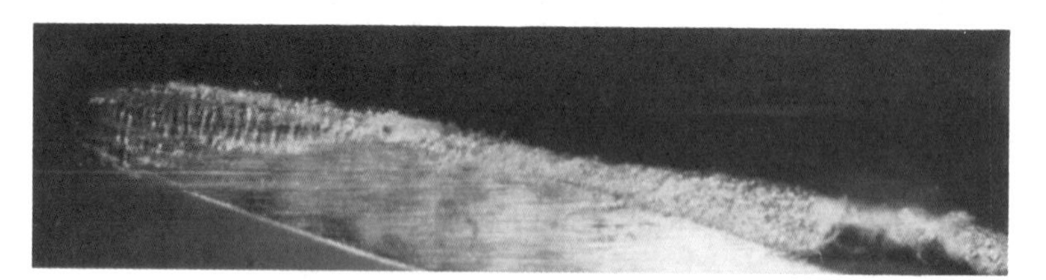

FIGURE 9. Periodicity in flow over leading edge.

An interesting feature is the apparent periodicity in flow through the tip clearance region, as evidenced in Figure 9. This persistent behavior appeared to be more than simply the streaming of bubbles from nucleation sites and some observations suggest that the waves are connected with the helical structure of the cavitation sheet downstream. Related behavior is present in the work of Rains [3] where it was demonstrated that the sharp pressure surface blade edge was the cause of cavitation in the clearance region. The observations also relate to the identification by Majka [4] of "rope-like" separation vortices in the clearance region.

Figure 10 presents photographs taken in stalled operation of the pump at a high inlet head. It is clear that under these conditions the suction surface leading edge region experiences an intense suction spike, sufficient to cause local cavitation. This spike also appears to cause separation from the leading edge, probably of the short bubble type, and the cavitation trace gives an indication of the severe separation present, at least at the blade tip. The first photograph shows the bursting process and the second the ensuing backflow.

FIGURE 10. Leading edge separation under stalled conditions.

CONCLUSIONS

Studies of the flow field in the rotor tip region of an axial flow pump have been conducted. Measurements of rotor tip pressure distribution have revealed a conventional pressure distribution with the peak suction pressure occurring at the leading edge.

Visual and photographic observations of cavitation tend to confirm previous observations, especially with regard to the strength of the tip vortex and its predominant role in the cavitation behavior of the blading. Bubble collapse tended to occur in a thin and clearly defined shock-like region. Inception of the tip vortex appeared to occur predominantly over the rotor tip and there was evidence of strong cavitation tendencies, periodicity and vortex formation over the tip at the leading edge. Observations in stalled operation revealed leading edge separation behavior with a tendency to backflow around the leading edge.

The assistance of Warman International Ltd. in donating the pump and test rig is gratefully acknowledged.

REFERENCES

1. Gostelow, J.P. A New Approach to the Experimental Study of Turbomachinery Flow Phenomena. Trans ASME, Journ. Eng. for Power, Jan. 1977

2. Wade, R.B. and Acosta, A.J. Investigation of Cavitating Cascades. Trans ASME, Journ. Basic Eng., Dec. 1967

3. Rains, D.A. Tip Clearance Flows in Axial Flow Compressors and Pumps. Caltech Hydro. and Mech. Eng. Labs. Rept. No. 5, June 1954

4. Majka, K. An Experimental Study of End-Wall Flow in a Compressor Cascade. Ph.D. thesis, Indian Inst. of Sci., June 1982

Experimental Research on Effects of Deflectors on Power and Static Torque Performances of Savonius Rotor

H. MURAI
Institute of High Speed Mechanics
Tōhoku University
2-1-1 Katahira, Sendai, 980 Japan

H. WATANABE and S. ONUMA
Institute of High Speed Mechanics
Tōhoku University
2-1-1 Katahira, Sendai, 980 Japan

M. KATAOKA
General Research and Development Center
Tōhoku Electric Power Company, Inc.
7-2-1 Nakayama, Sendai, 980 Japan

INTRODUCTION

A Savonius rotor wind turbine, though often used for a small- or medium-scale power source because of its advantages over a horizontal axis wind turbine, such as simple construction, large torque, lack of necessity for overspeed control, acceptance of wind from any direction without orientation etc., has the disadvantage of low efficiency or power coefficient compared with wind turbines of other types. And, even in recent years, experimental studies [1], [2] were conducted for optimizing the shape of the rotor vane and the overlap distance of the rotor vane pair to obtain the conclusions different from each other. On the other hand, the attempt for improving the efficiency by equipping simple deflectors in the immediate vicinity of the Savonius rotor has been reported [3] to have failed to obtain any increase in power output.

This report is going to present the experimental results which showed the considerable increase in the power coefficient by equipping deflectors around the Savonius rotor and the influences of the shape of the deflectors, the shape of the rotor vanes and the wind direction on the performance of the turbine, and to show the working mechanisms of the deflectors and the rotors.

EXPERIMENTAL APPARATUS AND METHODS

The circulatory water tunnel [4] was used for convenience of visualizing flow around the deflectors and the rotor. The tunnel has the measuring section of breadth of 200 mm and height of 1,200 mm, and the size of the wind turbine model was decided as shown in Fig. 1 in order to minimize the blockage effect of the tunnel walls, especially breadthwise. Two shapes of model were adopted; one recommended by Littler [1] (named by rotor No. 2), and the other by Kahn (No. 3). The model was so installed at the measuring section as the rotating shaft was situated at the center height, kept breadthwise horizontally, and connected with the measuring and controlling system of the output torque and the number of revolution at one end penetrating the tunnel wall (Fig. 2).

The deflectors consisting of four or two sheets of plates distributed around or arranged upstream the rotor apart from each other by $\pi/2$ radian (Fig. 1), and having two kinds of profile forms of the element plate, one being circular arc and the other flat plate, were used in order to examine the number and shape of deflectors. The symbols, for examples, No. 2-4C and No. 3-2F express the cases of the rotor No. 2 with four circular arc deflectors and the rotor

No. 3 with two flat plate deflectors.

The output torque were measured by using the set of the strain gage-type pickup ①, and the dynamic strain meter. The number of revolutions was measured by using the optical set of the multi-holed disc ②, the He-Ne laser beam, the photo-transistor and the electronic counter, and was controlled by controlling the exciting current. The optical set was also used for detecting the rotation angle of rotor.

Flow patterns around the deflectors and the rotor were visualized by injecting air bubbles through the branched probe composed of the seven nickel pipes of 0.3 mm inner diameter apart by 15 mm with each other laid upstream by 330 mm from the center of the model shaft. The injecting velocity of air bubbles was controlled by controlling the air pressure to be the same as the velocity of ambient water. The obtained diameter of air bubble was ranged between 0.5 mm and 1.5 mm. The lift of a streakline between the streamwise positions corresponding to upstream and downstream edges of the four deflectors was 10 mm.

The free stream velocity was kept constant at 3 m/s throughout the experiment, considering the rigidity of the model. The angle of the chord of the frontest deflector against the free stream velocity (wind direction, β) was fixed at every 15° (partly 7.5°) between 0° and 60°. The output power, C_P varying with the rotation angle of the rotor, α was calculated from the measured dynamic torque recorded on a magnet tape and its mean value, $\bar{C}_P$ was obtained by averaging C_P in many rotations and plotted at every 0.1 of the tip speed ratio between 0.3 and 1.5. The static torque, C_T was also measured fixing the rotor at every 6° of rotation angle. The flow pattern was photographed by 1/500 sec exposure at every about 10° of rotation angle of the rotating rotor and at every 6° of rotation angle of the fixed rotor.

The photographed flow patterns were used not only for observing and understanding the flow aspects around and on the surfaces of the deflectors and the rotor vanes but also for analysing the velocity and torque distributions along the rotor vanes by using the following relations; the angle γ of the absolute velocity against the radius vector of a point on the vane surface is measured by the streakline at the point. The relative velocity of water to the vane surface at the point is the resultant of the above absolute velocity and the relative velocity of water to the vane surface at the point due to rotation of the vane surface, and is directed to the tangent at the point if the flow has not been separated which fact can be recognized by the flow pattern. And the magnitude of the relative velocity v is calculated as

$$v = \omega x \cos\gamma/\sin(\gamma-\tau),$$

where ω, x and τ are angular velocity of rotor rotation, magnitude of radius vector and angle between tangent and radius vector at the point. The pressure difference between the upper and lower surfaces of rotor vane at the point gives the elementary torque on the rotor shaft

$$\Delta T = (\rho/2)(v_u^2-v_l^2)x \cos\tau \, \Delta s,$$

where suffixes u and l correspond values on the upper and lower surfaces, and Δs denotes elementary length along the vane surface.

EXPERIMENTAL RESULTS AND DISCUSSIONS

Output Power Performance

Fig. 3 (a) and (b), and 4 (a) and (b) show the variations of the mean power coefficient, $\bar{C}_P$ versus the tip speed ratio, λ for the two rotors No. 2 and No. 3, and for the two cases when four deflectors and two deflectors were equipped, respectively, at given inlet angles of wind, β indicated in the Figures. The original performances of rotors when no deflectors is equipped are shown for comparisons. Considerable changes of the performance, increases of $\bar{C}_P$ and decreases of the effective working range of λ by equipping the deflectors, and their changes with β can be observed.

The variations of the maximum $\bar{C}_P$, $\bar{C}_{Pmax}$ and the λ where $\bar{C}_P$ becomes maximum, λ_{max} with β in the cases of No. 2-2C, No. 2-4C, No. 3-2C and No. 3-4C and the cases of No. 2-2F, No. 2-4F, No. 3-2F and No. 3-4F are shown in Figs. 5 (a) and (b) compared with the original maximum $\bar{C}_P$'s of the two rotors, being different little with each other. The differences of influences on the performance between the two rotors and the profile forms of deflectors can clearly be seen. The increases of $\bar{C}_{Pmax}$ are 0.115, 0.090, 0.060 and 0.035 in the cases of No. 2-2C, No. 3-2C, No. 2-4C and No. 3-4C, but the increases reduce to 0.055, 0.060, 0.020 and 0.020 in the cases of No. 2-2F, No. 3-2F, No. 2-4F and No. 3-4F, respectively. β where $\bar{C}_{Pmax}$ becomes maximum is 15° or 37.5°. Although the variation of $\bar{C}_{Pmax}$ with β loses one of the merits of a vertical axis wind turbine, acceptability of wind from any direction, the increases of $\bar{C}_{Pmax}$ by 30 % in the range of 30° of β in the case of No. 2-2C, by 15 % in 30° in the case of No. 2-4C, by 25 % in 40° in the case of No. 3-2C, and by 10 % in 30° in the case of No. 3-4C can be considered to be a large merit of equipping deflectors.

Working Mechanisms of Rotor and Deflectors

In order to present the materials for establishing a mathematical model of the flow pattern around the rotor with or without the deflectors, and for optimising the shape of the rotor vane and/or the separation distance between the vanes, the shape and arrangement of the deflectors, and the suitability between the rotor and the deflectors, the working mechanisms of the rotor and the deflectors were investigated by using the visualized flow patterns and the C_P-α characteristics.

The rotor No. 2 without the deflectors. Figs. 6 (a), (b), (c) and (d) show the flow patterns around the rotor No. 2 with no deflector at λ_{max}. Fig. 6 (a) corresponds to α where the instantaneous output power coefficient C_P becomes maximum versus α, α_{max}, and Fig. 6 (c) corresponds to α of minimum C_P, α_{min}. The C_P-α characteristics of the original rotor No. 2 at λ_{max} can be seen in Fig. 7 from which α's where C_P becomes maximum or minimum have been found. At α_{max} ($\cong$30°, Fig. 6 (a)), the incomming flow is separated at the outer edge of the advancing vane (concave to the free stream). Water in the separated space, surrounded by the separated stream line starting at the outer edge of the advancing vane and the vane surface facing the free stream through the central half of the receding vane (convex to the free stream), rotates with the rotor shedding a vortex with probably some volume of water corresponding to the volume reduction of the separated space at this α. The pressure in the separated space can be estimated to be not low by using the flow direction along the separated stream line.

A part of the free stream, divided downwards after bamping to the separated space, flow along the convex surface of the advancing vane with fairly high speed and is hardly separated from the vane surface until it bamps to the concave surface of the receding vane. The resulted negative pressure distribution on the convex surface of the advancing vane and the momentum change of stream by the concave surface of the receding vane produce a large torque around the rotating axis.

Fig. 6 (b) shows the flow separation from the convex surface of the advancing vane, which reduces the magnitude of negative pressure on the surface and the momentum of the stream bamping to the concave surface of the receding vane to reduce the torque. Furthermore, it also shows the flow stagnation on the convex surface of the receding vane, which increases the negative torque.

At α of minimum C_P (Fig. 6 (c)), a small separations of the incomming flow at the outer edge and the expansion of the separated region on the convex surface of the advancing vane, and the resulted reduction of the momentum of the separated flow can be observed. The increase of the momentum of the free stream on the convex surface of the receding vane can also be observed.

In Fig. 6 (d) it can clearly be seen that the momentum of the free stream on the convex surface of the front (new advancing) vane and the moment arm are reduced, and the acceleration of the incomming flow on the convex surface of the advancing vane and the followed momentum change on the concave surface of the receding vane have begun to be enlarge to cause the increase of the output torque.

The rotor No. 3 without the deflectors. Fig. 8 (a), (b), (c) and (d) show the visualized flow patterns at α_{max}, α_{min} and intermediate α's around the rotor No. 3. The flow patterns at the corresponding α's are very similar with each other, respectively, except the existence of circulatory flows in the split space between the pair of vanes of the No. 3 rotor and the flow patterns near them.

The C_P-α characteristics, which can be seen in Fig. 9, differs from that of No. 2 in that C_{Pmax} and the absolute value of C_{Pmin} of the rotor No. 3 are smaller than those of the rotor No. 2, α_{max} and α_{min} of the former are smaller and larger than those of the latter, respectively, and the C_P-α curve of No. 3 is concave upwards during the decrease of C_P versus α while that of No. 2 is convex upwards during the increase of C_P versus α.

That α_{max} and λ_{max} of the rotor No. 3 are smaller than those of the rotor No. 2 can be understood through Fig. 6 (a) and Fig. 8 (a) to be due to that the curvature at the outer edge and the following shape of the convex surface of the former are fit for smaller α than the latter. That C_{Pmax} of the former is smaller than that of the latter is due to that the deflecting angle and the mass of the flow leaving the advancing vane, and bamping to and leaving the receding vane are smaller than those of the latter, other than that λ_{max} of the former is smaller than that of the latter.

Fig. 8 (b) shows, in comparison with Fig. 6 (b), that the flow separation from the convex surface of the advancing vane progresses more gradually for the rotor No. 3 than for the rotor No. 2, and a moment arm of the negative torque caused by the stagnation pressure of the free stream and the rotating velocity, and the stagnation pressure itself on the convex surface of the receding vane are smaller for the former than for the latter, which causes the more gradual decrease of C_P with increasing α and the smaller absolute value of C_{Pmin} at larger α_{min} (Fig. 8 (c)) for the rotor No. 3.

The comparison of Fig. 8 (d) with Fig. 6 (d), shows that with increasing α from α_{min} to the following α_{max} the moment arm of the stagnation pressure on the fore part of the convex surface of the front (new advancing) vane becomes larger for the rotor No. 3 than for the rotor No. 2, and the negative pressure on the following part of the surface grows later for the former than for the latter.

Working mechanisms of deflectors. Fig. 10 (a)~(d) show the flow patterns around the deflectors of different numbers and shapes at two wind incidence angles, β's, respectively without any rotor. They explain the roles of the deflectors being in concentrating the incomming flow within a limited space, increasing and deflecting the flow velocity, and forming a flow channel between a vane surface, each of which plays a part, in giving the output power or the static torque, varying with the rotation angle and depending on the shapes of the rotor vane and the deflectors.

Increasing the flow velocity in the working space of the two vanes and diminishing the flow velocity in the expending space of the receding vane (Fig. 10 (a)) is useful for increasing the output power. The remarkable increases of C_{Pmax}'s in the cases of No. 2-2C and No. 3-2C at $\beta=15°$ is thus obtained Fig. 11 (a) and Fig. 12 which have ruled the large increases of $\bar{C}_{Pmax}$'s. That $\bar{C}_{Pmax}$ of No. 3-2C is smaller than $\bar{C}_{Pmax}$ of No. 2-2C is due to that C_{Pmax} of No. 3-2C is smaller than C_{Pmax} of No. 2-2C which is caused by the difference between the momentum change on each concave surface of the receding vane, and that α of the second minimum of C_P in either case is restricted to be about the same with each other by the position of the outer edge of the advancing vane relative to the lower boundary of the concentrated flow (Fig. 11 (b)).

As β becomes larger, the upper deflector becomes to work more effectively, and the incidence angle of the incomming flow to the advancing vane becomes larger (Fig. 10 (b)) to give the larger torque on the advancing vane in the smaller range of α till the flow along the convex surface of the advancing vane begins to separate from the surface. And the above effect is realized by the rotor No. 3 which works more effectively at a negative incidence to the relative velocity and a larger incidence without the flow separation (Fig. 13).

As for the four deflectors, the downstream ones obstruct the course of the flow concentrated by the upstream ones to give the smaller velocity to the flow, although the lower downstream deflector plays a part in keeping the flow along the convex surface of the advancing vane unseparated at an appropreate α.

In the case of the flat plate deflectors, the lower boundary of the concentrated flow is more unstable than one in the case of the circular arc deflectors, which makes C_{Pmax} around λ_{max} be smaller to result $\bar{C}_{Pmax}$ around λ_{max} be smaller.

Static Torque Performance

Fig. 14 shows the variation of the static torque coefficient, C_T versus the incidence angle of the free stream to the rotor, α for every cases written in. The group of C_T's for the rotor No. 2 lies almost always under that for the rotor No. 3, which is due to the difference of the mechanisms of generating the torque between the two rotors. In the cases of the rotor No. 2 the flow always separates from the upper and lower edge(s) or surface(s) at any α (Fig. 15 (a) and (b)). And the static torque is merely the difference between the positive and negative torques generated by the downward and upward flows, respectively. In the cases of the rotor No. 3, however, the third torque, which is always positive, generated by the flow through the separated split contributes to the static torque (Fig. 16). C_T-α performances changes with the changes of the shape and arrangement of the deflectors and β. The existence of α's where C_P's vary shapely in the cases of four deflectors is noticed. Fig. 17 (a) and (b) show variations of C_{Tmax}'s and α_{Tmax}'s (α where C_T becomes maximum) with β.

CONCLUSION

(1) The power coefficient of Savonius rotor could be increased remarkably by equipping deflectors around it. The S-type rotor recommended by Littler [1] was more effective than the separated vanes-type rotor recommended by Kahn [2] for equipping deflectors in this experiment, although the latter was superior to the former as originals.
(2) The circular arc was superior to the flat plate for the profile form of deflectors. And two was superior to four for the number of deflectors.
(3) The maximum power coefficient varied with the wind direction with respect to the deflectors. The amounts of improvement varied with the profile form and the number of deflectors for each rotor.
(4) The static torque coefficient could also be increased by equipping deflectors.
(5) The extent of improvement and the order of superiority of the number of deflectors varied with wind direction for each rotor.
(6) The working mechanisms of the rotor and the deflectors were investigated, which would be expected to be useful for establishing the flow models around the rotor with or without the deflectors and optimising the shape of the rotor, and the profile form, number and arrangement of the deflectors.

NOMENCLATURE

C_P: power coefficient ($=2T\omega/\rho V^3 DH$) C_T: torque coefficient ($=4T/\rho V^2 D^2 H$)
D: rotor swept diameter H: height of rotor T: torque
V: freestream velocity
α: angle of incidence between diametric line connecting tipes of rotor vanes; positive sense is taken in the direction of rotor rotation.
β: inclination of chord line of frontest deflector to freestream; positive sense is taken in the direction of rotor rotation.
ρ: density of fluid ω: angular velocity of rotation
λ: tip speed ratio ($=D\omega/2V$)

REFERENCES

1. Littler, R. D., Further Theoretical and Experimental Investigation of the Savonius Rotor, Thesis (B, E) Univ. of Queensland, 1975

2. Khan, M. H., Model and Prototype Performance Characteristics of Savonius Rotor Windmill, Wind Engineering, 2-2, 1978, pp. 75-85

3. Baird, J. P. and Pender, S. F., Optimization of a Vertical Axis Wind Turbine for Small Scale Applications, 7th Australasian Hydraulics and Fluid Mechanics Conference, Brisbane, 1980, pp. 431-434

4. Murai, H., Shimizu, S. and Onuma, S., Experimental Research on Hydro-elastic Characteristics of Flexible Supercavitating Hydrofoils, 2nd Intern. Conf. on Cavitation, London, 1983, pp. 205-210

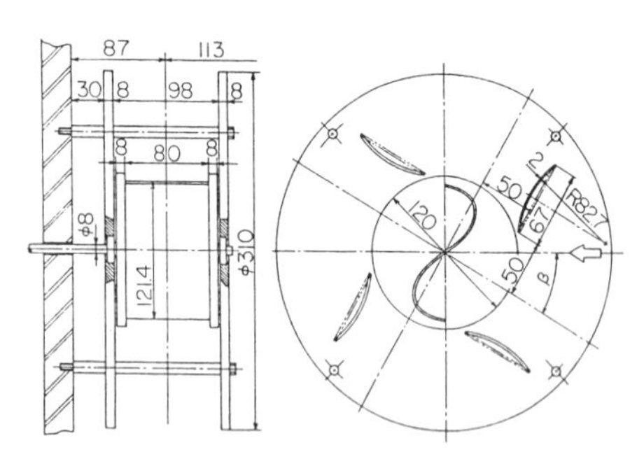

FIGURE 1. Wind turbine model equipped with deflectors

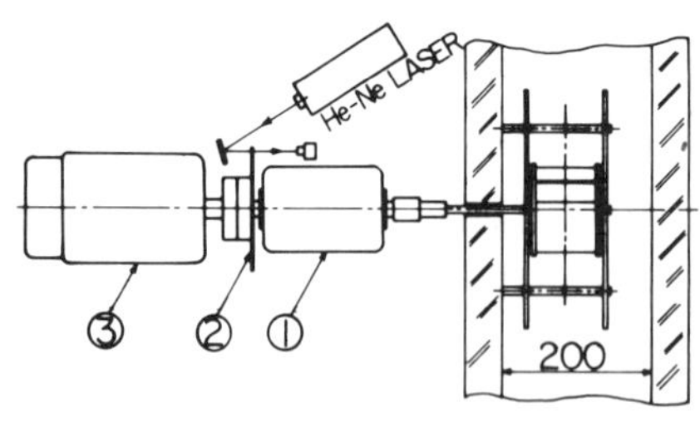

FIGURE 2. Measuring and controlling system of torque and number of revolution

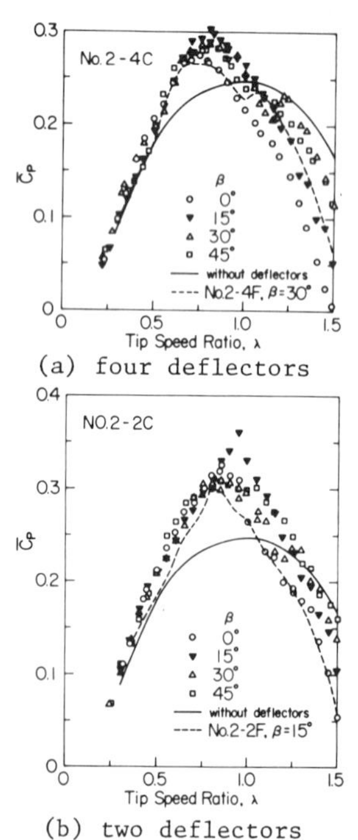

FIGURE 3. Power-tip speed ratio performance, No. 2

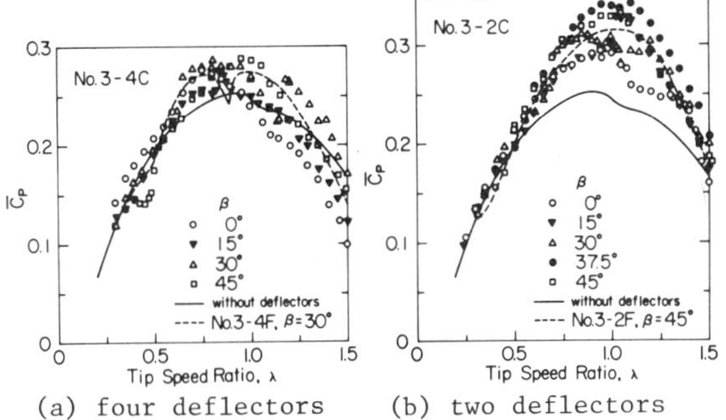

(a) four deflectors (b) two deflectors

FIGURE 4. Power-tip speed ratio performance, No. 3

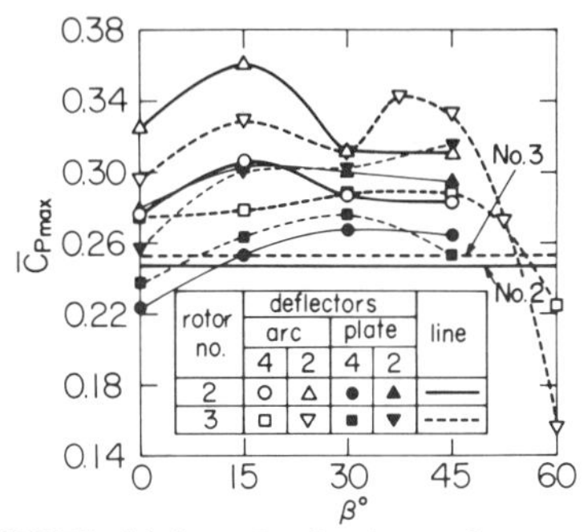

FIGURE 5(a). Variation of maximum power coefficient with wind direction

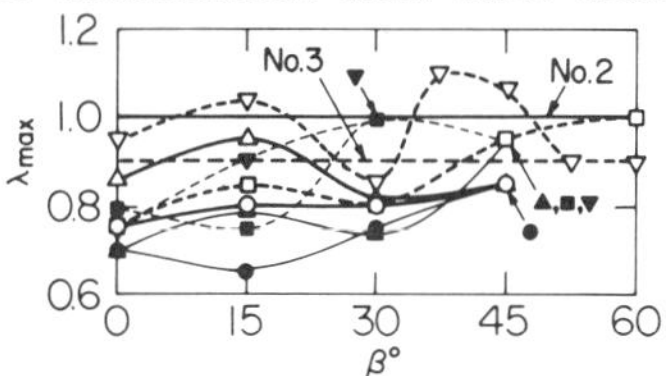

FIGURE 5(b). Variation of tip speed ratio at $\bar{C}_{Pmax}$ with wind direction

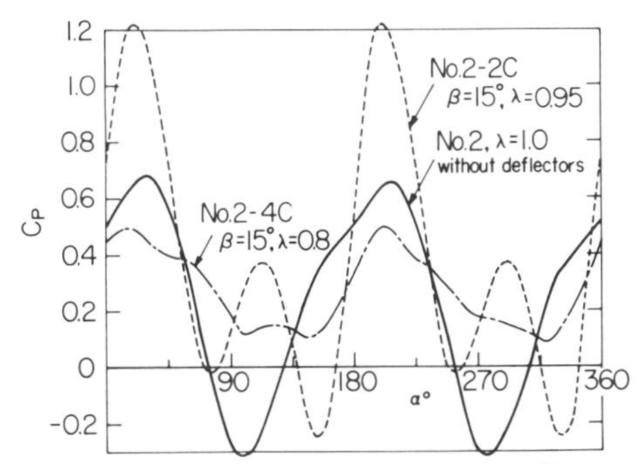

FIGURE 7. Variation of power coefficient with incidence angle to rotor, No. 2

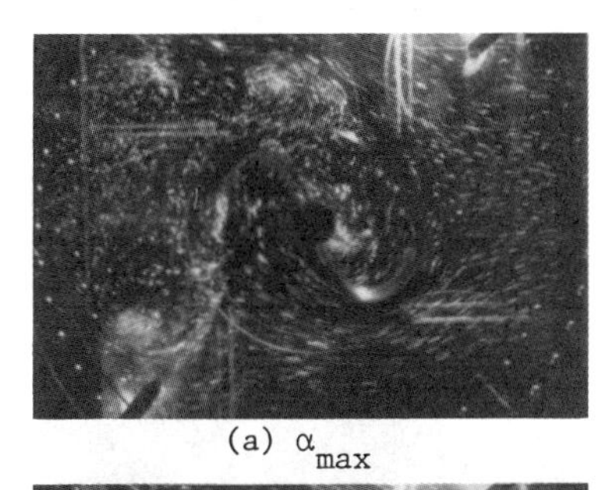

(a) α_{max}

(b) $\alpha_{max} < \alpha < \alpha_{min}$

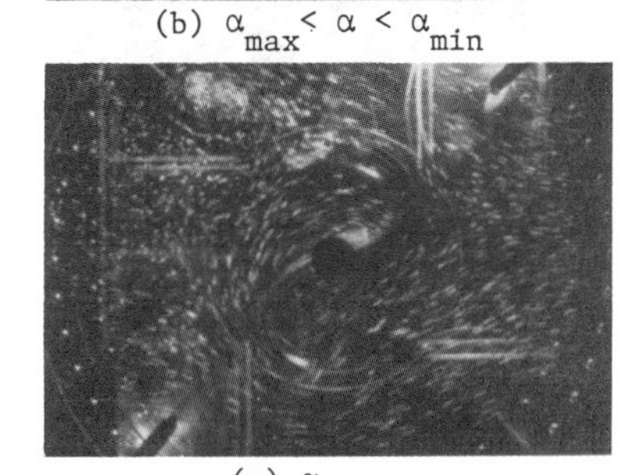

(c) α_{min}

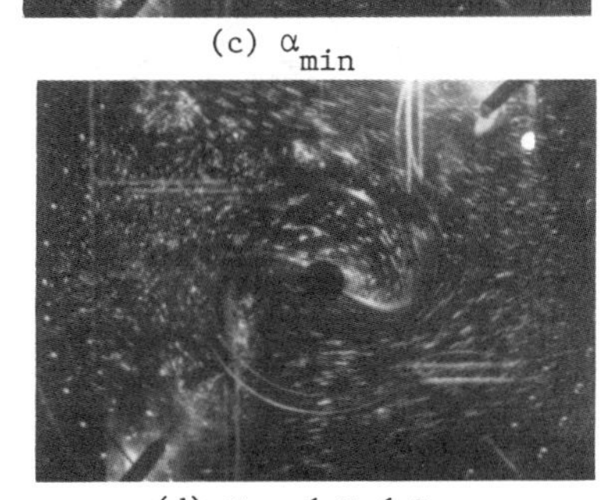

(d) $\alpha_{min} < \alpha < \alpha_{max}$

FIGURE 6. Visualized flow patterns around rotor No. 2

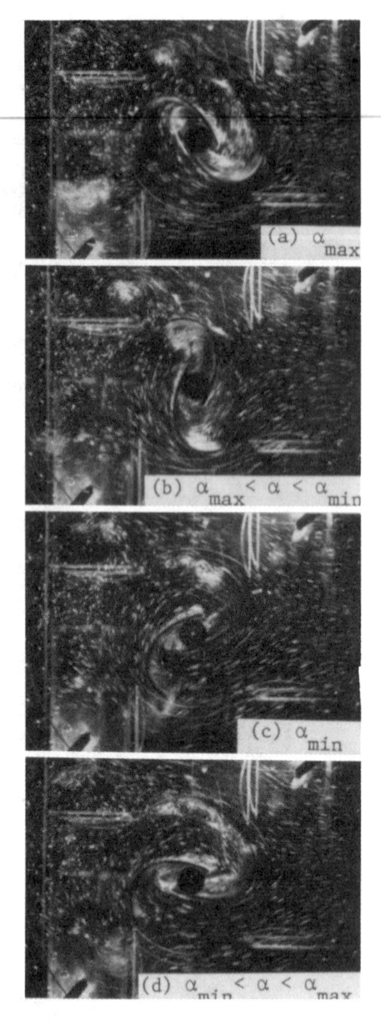

FIGURE 8. Visualized flow patterns around rotor No. 3

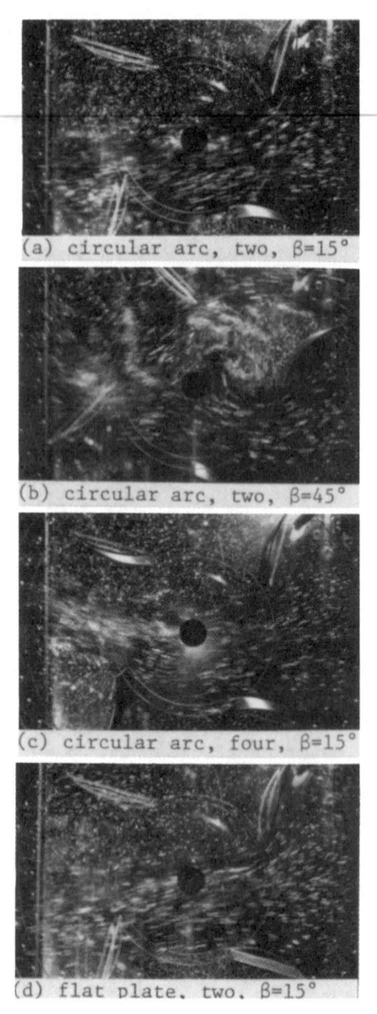

FIGURE 10. Visualized flow patterns around deflectors without rotor

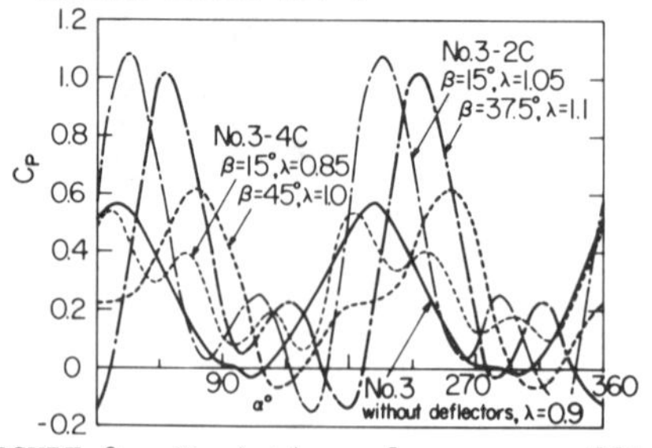

FIGURE 9. Variation of power coefficient with incidence angle to rotor, No. 3

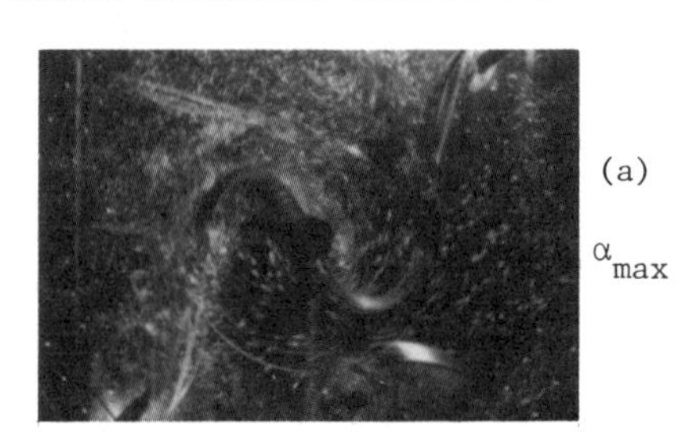

FIGURE 11. Visualized flow patterns, No. 2-2C

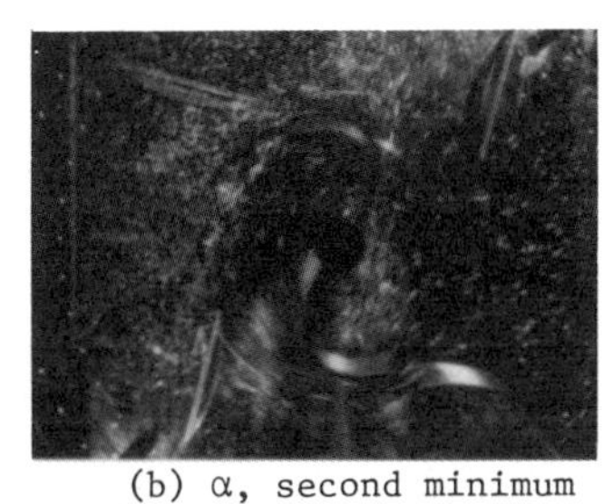

(b) α, second minimum

FIGURE 11. Visualized flow patterns, No. 2-2C

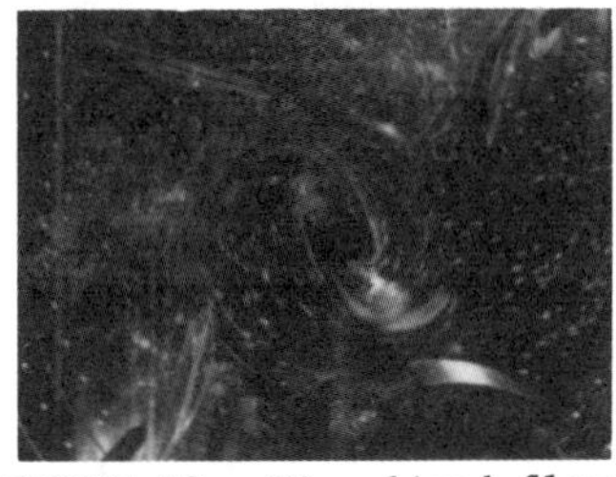

FIGURE 12. Visualized flow pattern, No. 3-2C, β=15°, α_{max}

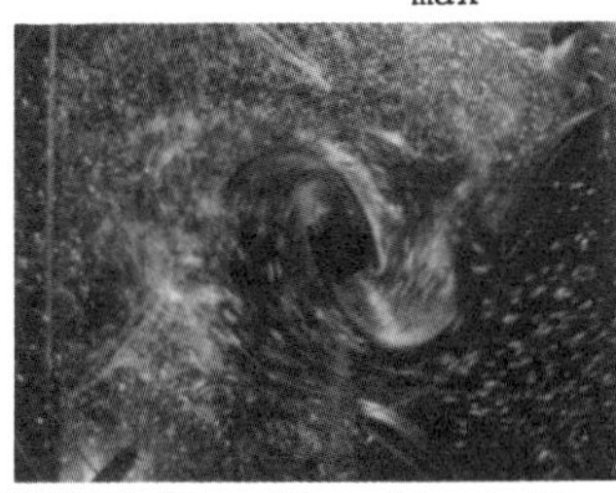

FIGURE 13. Visualized flow pattern, No. 3-2C, β=37.5°, α_{max}

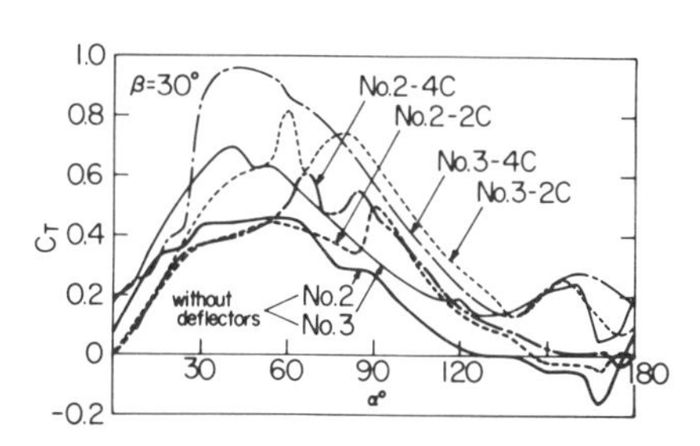

FIGURE 14. Variation of torque coefficent with incidence angle to rotor

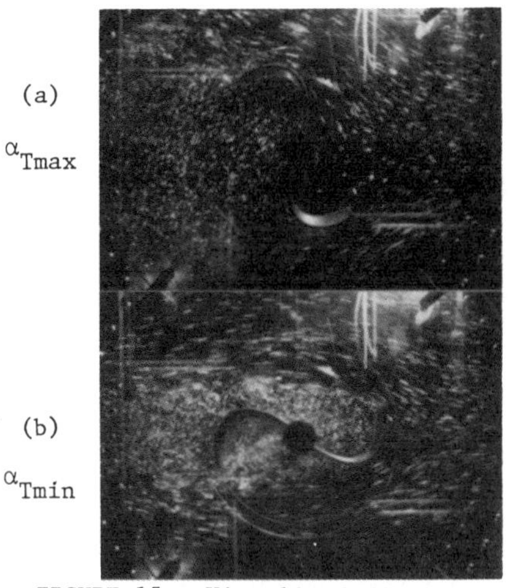

FIGURE 15. Visualized flow patterns, static No. 2

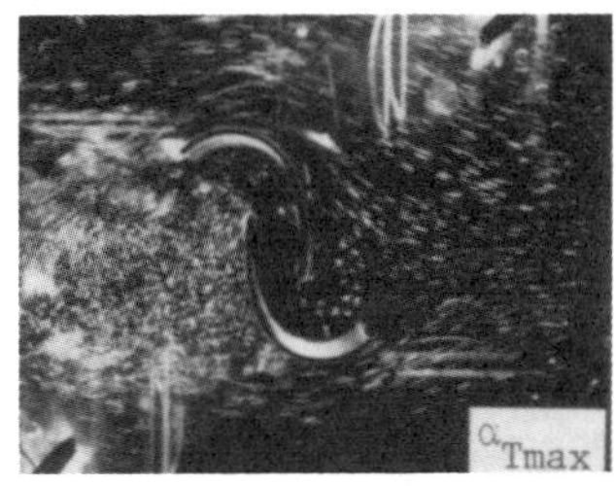

FIGURE 16. Visualized flow pattern, static, No. 3,

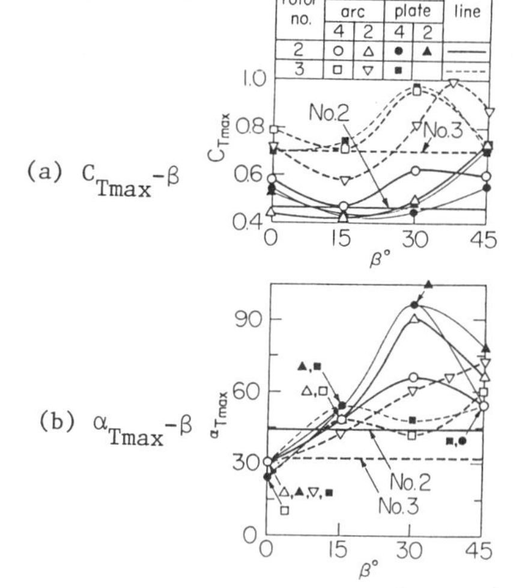

FIGURE 17. Variations of C_{Tmax}'s and α_{Tmax}'s with β

Author Index

Subject Index